Contents

Foreword

The Second International Conference on Environmental Future, held by invitation of the Government of Iceland in their capital city of Reykjavik through the week of 5–11 June 1977, was a natural outcome of the first such Conference, which had been held in 1971 by invitation of the Government of Finland. As recounted in the published proceedings*, that pioneering event was convened in the Finnish capital, Helsinki, but took place mainly in the stimulating atmosphere of the 'central' town of Jyväskylä.

With remarkable percipience H.E. Ahti Karjalainen, who as Prime Minister of Finland had been Chairman of the Finnish Organizing Committee supporting that first International Conference on Environmental Future, in his Foreword to the published proceedings, laid the foundations for this its successor when he wrote: 'It was encouraging to note that this first conference of its kind voted unanimously at its conclusion to continue with further such attempts [to look into the future of the world's environments] and under the same leadership at intervals of a few years henceforth.' This 'would seem most important in view of the need that politicians and the world's governments are feeling increasingly for guidance in environmental matters. I deeply hope that we shall rapidly achieve concrete results and all together be able to do more for rational development *and* the environment—for positive change of human existence and the quality of life on Earth. . . . I also hope that these proceedings will be a useful contribution to the coming United Nations Conference on the Human Environment and for enlightened reading throughout the world.'

These hopes seem to have been fulfilled to some degree—at least to the extent of our being persuaded that our efforts and the much more massive ones of the UN 'Stockholm' Conference, together with others having similar objectives, have helped to alert decision-makers very widely to the dangers that beset our world. These stem basically from ever-increasing human population pressures, with their concomitant exhaustion of resources and pollution and other environmental ills. The progress towards saving mankind's world from itself has been limited, but at least we can point to the fact that seemingly every government worth its salt now has a ministry or department of environment—whatever it may be called and actually do. Also, the general public has been widely alerted to environ-

**The Environmental Future: Proceedings of the first International Conference on Environmental Future, held in Finland from 27 June to 3 July 1971*, edited by Nicholas Polunin. Macmillan, London & Basingstoke, and Barnes & Noble, New York: xiv + 660 pp., illustr., 1972.

mental problems and threats through impending shortages and inculcated consciousness of what is apt nowadays to be termed 'ecology'. Meanwhile some of us have preserved from those earlier conferences and much additional deliberation the conviction that considerable latitude still remains for further economic growth and development in an ecologically sound setting. Hence the theme of our Second Conference, namely, 'Growth Without Ecodisasters', although we have felt it prudent to retain a query at the end of the title of these its proceedings.

Thus we embarked on this 2nd ICEF (as it has been widely referred to for short) with the feeling that, despite the parlous state of Planet Earth environmentally and otherwise, the right decisions and vigorous actions could save much of this still beautiful world in a state which we would like future generations to enjoy *ad infinitum*. But some of these decisions and actions are urgent, and eternal vigilance will be needed in respect of others.

N.P.

Preface

The origins of these International Conferences on Environmental Future have already been described in the Preface to the proceedings of the first Conference, as cited in the footnote to the Foreword to the present volume. Each Conference is guided by an International Steering Committee having a written constitution which requires that it 'shall consist of not fewer than 8 members who shall be resident in 6 or more different countries representing at least four continents' (Clause 1), and 'at their first meeting they shall elect a Chairman from among their number' (Clause 2). Matters of quorum, procedure and invitations are also laid down, and the final clause (No. 9) stipulates that 'The Committee for each Conference shall be disbanded at its end but without prejudice as to possible re-election of individuals in the future'.

The International Steering Committee of the 2nd ICEF was established soon after the first such Conference and met at least once yearly until early in 1977. One of the leading lights of the first ICEF, and indeed of the biological world, Professor Jean-Georges Baer, of Switzerland, was its Chairman until his untimely death early in 1975, following which Dr Thomas Adeoye Lambo, of Nigeria and Switzerland, Deputy Director-General of the World Health Organization, was elected to fill the vacancy. The other members, in alphabetical order, were Mrs Elisabeth Mann Borgese (USA and Malta), Professor Jean Dorst (France), Dr Erik Jensen (Malaysia and Switzerland), Professor Mohammed Kassas (Egypt), Professor Viktor Abramovich Kovda (USSR), H.E. Ambassador Haraldur Kröyer (Iceland),* Dr Nedumangathu Kesava Panikkar (India), the undersigned as Secretary-General and Vice-Chairman, Dr Sidney Dillon Ripley (USA), Mr Peter Bennet Stone (UK and Switzerland), and the Hon. Beatrice E. Willard (USA).

Early in 1976 the Government of Iceland, as confirmed in the Prime Minister's letter of 9 April addressed to the Secretary-General, decided to invite the Conference to take place in their capital city of Reykjavik in September of that year. This proved too short a time for adequate financing and organization, and so the Conference was postponed until 5–11 June, 1977, when it was finally held. Meanwhile the Prime Minister had generously undertaken to be Patron of the Conference and a supporting Icelandic National Committee had been set up under Professor Gunnar G. Schram and Dr Sturla Fridriksson. The Scientific Patron of the Conference from

*Formerly represented by the Icelandic Chargé d'Affaires, Mr Kornelius J. Sigmundsson.

early on was H.E. Professor Friedrich T. Wahlen, sometime President of the Swiss Confederation, and the President of the Conference was Professor Linus Pauling, Nobel Laureate for Chemistry and for Peace. Sponsors of the Conference were the Government of Iceland and the (Swiss-based but international) Foundation for Environmental Conservation, and co-sponsors were the US Energy Research and Development Administration of Washington, DC, and the Holcomb Research Institute, of Indianapolis, Indiana.

Overseas participation in the Conference was strictly by invitation—either of individual environmental scientists or others who were usually outstanding leaders in their fields, or of certain *international* organizations that were chosen by the Steering Committee and invited to send their chief or a 'qualified representative'. Apart from a limited number of leading participants whom we managed to help financially to attend, and some others who were sent by the United Nations Environment Programme from several of the less-developed countries, people came at their personal expense or with backing from their own or other sponsoring institutions. They were adjured to speak quite freely in their personal capacities, and in general seemed to do so—often with a display of lively enthusiasm. Indeed it was stated jocularly that the Conference seemed to 'vibrate quite chronically'.

As long-planned, the Conference lasted for almost a week. It took place mainly in the small and cosy theatre of the leading Hotel Loftleidir, on the outskirts of Reykjavik, starting on the Sunday evening of 5 June with a get-together drinks party (also to celebrate Environment Day), and terminating with a State Luncheon, given by the Prime Minister in the Icelandic President's absence abroad, on the following Saturday. Relieved by a memorable mid-Conference excursion to Thingvellir, 'the World's Oldest Parliament', contributed by the Icelandic Minister of Culture, there were nine half-days of concentrated work, each divided into two sessions, plus a final plenary session on the Saturday morning. Each of the 18 sessions normally consisted of the Chairman's introduction (at most 5 minutes), résumé for about 15 minutes of the keynote paper (usually circulated earlier), and 3 or 4 panellists commenting for up to 5 minutes each from different standpoints. The panellists, especially, were charged with emphasizing future prospects and desirable action. This strategy with good chairmanship commonly left about an hour for free discussion to end each session—which, to judge from the paucity of adjournments, was usually sufficient.

In addition to the regular and usually very full sessions, there were various dinners and receptions and also two special evening events. The first of these was a 'workshop' on the controversial theme of 'Is Iceland Polluted?' which commenced with two papers from each side of the controversy, followed by an open debate which extended far into the night; at the end the independent (Canadian) chairman was roundly applauded for declining to take a vote and the packed theatre seemed happy that it was a 'drawn match'. (The event being a predominantly Icelandic one on a touchy subject of mainly local interest, it has been decided not to publish details in these Proceedings.) The other special evening event was the first Baer–Huxley Memorial Lecture, established in memory of Jean Baer and Sir Julian Huxley, whose deaths two years earlier had deprived the environmental movement of two of its greatest figures. Being in urgent demand, Mr Maurice Strong's splendid lecture on 'The International Community and the Environment' was published quickly in *Environmental Conservation* (Vol. 4, No. 3, pp. 165–72, Autumn 1977) and is reprinted in updated form on pages 611–25 of this book,* followed by a poem by our Foundation's Poet Laureate and then the Chairman's introductory remarks.

A privately organized international conference on this scale (nearly 150 participants) can only be carried through in the present state of the world with the help and support of numerous friends, and in this we were extremely fortunate. The biggest item by far in financing was the sale to the Government of Canada of our long-unused and practically unwanted library of arctic books, etc., and the next biggest at the time was a generous cheque from the Icelandic Aluminium Corporation (ISAL) which was presented quite unexpectedly following the 'drawn match' referred to above. Later, a contract with the United Nations Environment Programme helped substantially towards publication.

Other generous and especially valued donations came from the Government of Iceland, from the Conference's Scientific Patron, H.E. Professor Friedrich T. Wahlen, from the Charlpeg Foundation (through Foresta Institute for Ocean and Mountain Studies), from a participating colleague Mr Erico Nicola, from AMAX Inc., from Mr Brewster Hanson, and from Mr Yves Oltramare, while substantial and sometimes crucial help in other ways came from the United Nations Environment Programme, Professor F. Kenneth Hare, Dr Thomas F. Malone, Dr Gordon C. Butler, Sir Harold P. Mitchell, Bt, Professor Donald J. Kuenen, Dr Bent Juel-Jensen, Mr Maurice F. Strong, Professor Reid A. Bryson, Miss Angela Gibbs, Mrs Helen E. Polunin, and Mr Nicholas V. C. Polunin. Others whose helpful generosity was most welcome included the Conference's Patron (the Prime Minister of Iceland), the Icelandic Minister of Culture, the Lord Mayor of Reykjavik, British Petroleum Limited, H.E. Ambassador Jamil M. Baroody, Banque Populaire Suisse (Swiss Volksbank), Mr and Mrs Peter B. Stone, the Hon. Beatrice E. Willard, Mrs Patricia Prater, Mr Marcel Disch, Mr and Mrs J. Haley Booth, Mr Stanley Dempsey, Dr Erik Jensen, Captain Robert I. Maxwell, H.E. Ambassador Kenneth East, Miss Joanna Pinches, Manager Emil Gudmundsson, and the Nature Conservancy Council (Great Britain).

Mme Gail Staehli has been steadfast with her skilful secretarial help through the years and Drs W. Ted Hinds and Randall G. Gossen earned our special gratitude for their exhibits during the Conference and their stalwart support in other respects. The prompt work of the Imprimerie du Journal de Genève was especially appreciated, for which the Assistant Chief du

*In view of the great length of these proceedings, and as a considerable proportion of the points are made in the discussions of the 18 Conference Sessions which are published in this book, it has been decided to omit the discussion of the Memorial Lecture.

Travail, Monsieur Daniel Charlet, is to be particularly thanked, while also warmly appreciated was the dedication of our travel agent Monsieur Jean-Pierre Baer. Many others contributed in various ways and degrees to make the Conference memorable and, it seemed, not unsuccessful: their participation will usually be indicated in the general text and discussions. Finally on the scene came my able and dedicated Personal Assistant, Mrs Rona Tiller, without whose skilful application to detail this book, on top of my other duties, could not possibly have been delivered to the primary publishers on time.

As for the proceedings of the actual Conference, these follow the next item, entitled 'Opening of the Conference', and comprise 18 chapters corresponding to the normal sessions. Each chapter or session consists of a keynote paper (sometimes in two parts, contributed by as many authors), followed by the 'cream' of the ensuing discussion. This latter starts with the Chairman's introduction (actually given earlier) and continues with his rejoinder (if any) and then normally come the comments made from different viewpoints by the three or four panellists sitting on the platform and usually selected for their special interests bearing on different aspects of the main topic. Thereafter comes an open discussion, mainly from the floor but interspersed with remarks from the platform, and finally, often, some response from the keynoter and/or panellists and a summing-up by the Chairman.

With so much said and recorded throughout an extremely active near-week, and only limited space allowed by or negotiated with the publishers, much of the general discussion, particularly, has had to be discarded (*see* below); but it is hoped that most of what was really worth while has been retained. In contrast to the situation only six years earlier, when all but one of the papers were received well before the Conference and the publishers had imposed no limitation in length, on this occasion only half the keynoters submitted their papers in time for proper processing. The other half brought their drafts with them or sent them to Iceland and so, doing our best towards remedying this situation as the sessions rolled on, we personally missed most of the Conference, although we had the impression (and were assured by people who had been at both), that this one was of a generally higher standard than its predecessor. Consequently it is all the more deplorable that, basically owing to the general strike to which the local populace was being subjected at the time, much of the recording of the first session and bits of two others were lost or defective, although Mrs Patricia Prater did a remarkably fine job of transcribing the rest after attempts in Iceland had largely failed. We apologize to speakers whose efforts may thus have been spoiled or even wasted, except in debate. Otherwise the things which went wrong during the actual Conference were minor, and generally the occasion seems to have been widely enjoyed in the cool and stimulating northern atmosphere of almost perpetual daylight.

Despite the above limitations and many other problems—including the death of an eminent keynoter—encountered in laying on the Conference and the shock of a close associate fainting alarmingly on the platform during it, so much material was accumulated that, for these proceedings, there has had to be drastic compression or deletion of much of what was said and even of some of what was written.* This has especially been the case where material was extraneous or repetitious, while in other cases items were combined. As on the previous occasion, any really substantial change from what was actually said (apart from total omission) is placed in square brackets in the published version of the discussions, and this seems to have become accepted practice. Those who are accustomed for example to the formal procedures of United Nations and other intergovernmental conferences must not assume that everything recorded in the papers was, if not challenged in the discussion, tacitly accepted by everyone present; on the contrary, it is one of the valuable attractions of such free exchanges that controversial issues can be raised in them with impunity.

NICHOLAS POLUNIN, Secretary-General & Editor,
International Conferences on Environmental Future,
15 Chemin François Lehmann,
1218 Grand-Saconnex,
Geneva, Switzerland.

10 April 1978

*As the papers in these proceedings range from scientific surveys to philosophical essays, and space is at a premium, it has been decided to dispense with the usual lists of literature references in certain cases in which they do not seem necessary for background reading, justification of statements, or future perspectives. In some other cases, references are kept to a bare minimum. Finally, something should be noted which one hopes is less obvious: of the nearly 80 participants who actually spoke publicly during the Conference, about 12 failed to return their transcripts after due checking and correction—probably owing to loss in the mails—and this has led to some guessing or even discarding of interventions to avoid perpetration of possible errors. As a further safety-measure, the primary publishers very kindly undertook to send to all speakers (as well as, of course, to authors of papers) proofs of their contributions, so we feel that we really did everything possible, within the constraints of limited time, in the interests of accuracy.

Opening of the Conference

INTRODUCTION by the Secretary-General

Your Excellencies, Distinguished Participants, and Friends:—The Icelandic National Committee of our Conference have asked me, as Secretary-General, to introduce to you our distinguished Patron, Prime Minister Geir Hallgrímsson. This I consider it a signal honour to do—to introduce an [outstanding leader] to an eminent gathering of world authorities on a vital theme *in his own country*.

His Excellency Geir Hallgrímsson is a lawyer by profession and was Lord Mayor of Reykjavik for several years before becoming Prime Minister of Iceland in 1974. He is leader of the Independence Party, the largest political party in Iceland, and has led his gallant country in their struggles over fishing rights in which, to the joy of their friends and admirers throughout the world, they have strongly prevailed. It gives me great pleasure to call on you, Prime Minister, to give us some encouragement—if, in your wisdom, you can—to persist in our own struggle to help the world to save itself from itself. The Environmental Movement must win through.

WELCOMING ADDRESS by the Prime Minister of Iceland, H.E. Geir Hallgrímsson

Ladies and gentlemen, Mr Secretary-General:—I take great pleasure to welcome you all to this Second International Conference on Environmental Future, which now today convenes here in Iceland. Six years ago the first Conference was held in Finland and yielded fruitful results. Last year the organizers decided to select Iceland for their second gathering and to accept our invitation to meet here. I can only applaud their predilection for the Nordic countries as a meeting place for so many of the world's renowned environmentalists, who are gathering here today.

To an Icelander the theme of the Conference sounds singularly apt and timely: 'Growth Without Ecodisasters'. Iceland is a country which has been going through a cycle of rapid economic growth in the last three decades. This development has indeed brought us considerable material benefits. We are but few in a country of great expanses, as you may have gathered on your arrival in Iceland; so we have been spared many of the environmental disasters which are usually linked with rapid economic growth—which we need to seek in order to attain a living standard comparable with that of our neighbours, and to secure our political and economic independence. This, however, does not mean that we are unmindful of the dangers which are prevalent in this area of Man's relations with Nature, and of the need to discover ways that will allow us to proceed without harming or possibly destroying those values of life which are found in an unspoiled environment.

This Conference will help to keep us alert in these matters, serve as an added impetus in our campaign for a cleaner and healthier environment in this country, and remind us of the prime importance of enhancing our growth as a nation in concord with Nature but not against her. We have, indeed, much in this country to preserve and to take care of. This is a semi-arctic environment, which I believe Sir Julian Huxley said was on the borderline of the inhabitable world. We have no quarrel with that evaluation so long as we are on the habitable side of that borderline, and therefore we have long since decided to make this environment as hospitable as is humanly possible.

A part of that is a resolve to preserve and cultivate to the greatest possible extent our sensitive vegetation and the numerous bird-sanctuaries throughout the island, as well as to keep the ocean resources, which are the basis of our livelihood, free from local and foreign industrial contamination and overexploitation. Conserving the ocean resources around Iceland is indeed our foremost environmental preoccupation, and therefore Iceland was among the first nations who raised the issue of ocean overexploitation and pollution in the United Nations. Our endeavours in environmental matters are expressed in numerous ways by laws and regulations in different fields, and recently my Government has prepared a comprehensive code of environmental law which we hope our Parliament will study thoroughly and enact subsequently.

The themes you will discuss and debate at the Conference in the next few days are varied and wide-ranging. But they have a common denominator in so far as they all are timely and have a direct bearing upon Man's relationship with his environment. I hope I am neither too impatient nor too optimistic when I interpret the theme of your Conference, 'Growth Without Ecodisasters', to presuppose growth and to emphasize that ecodisasters can be averted. I want to express the wish that your deliberations will confirm such a conclusion. Only by reasoning together can we come to a common conclusion on how to secure a world which is both materially bountiful and also a healthy and happy place to live in. It is my sincere hope that this Conference will advance us towards this common goal.

I now declare the Second International Conference on Environmental Future open, and extend a specially cordial welcome to all the distinguished participants coming from abroad, while wishing them a pleasant stay in our country. Thank you, Mr Chairman, thank you Mr Secretary-General.

Secretary-General

I thank you, Prime Minister, for your gracious and encouraging words and, in the name of the International Steering Committee of our Conference and of the Conference itself as its Secretary-General, express the hope that you will remain with us awhile and enter into our discussion of vital themes. We also hope that you will find time at least to visit us and add lustre to our proceedings—especially on our final morning following which we most gratefully accept your kind invitation to the State Luncheon. [We are delighted to note that you accept to be listed as a full participant of the Conference.]*

Now before I hand over to our Conference President, Professor Linus Pauling, to start the first session, I would like to express warm appreciation for the quite numerous cables and other messages of goodwill and good wishes that have come to us from usually quite unexpected sources in various parts of the world. A recurrent theme in them is one of encouragement and hope that we shall come out with some strong recommendations as to what should be done to save the world from environmental disasters. With Professor Pauling's help and that of our wives, we have chosen four of these messages—from the world's international summit and three different continents—for presentation to you on this auspicious occasion. Allow me, please, to read them in the order of increasing brevity:

(1) *From His Excellency Professor Friedrich T. Wahlen* [sometime President of the Swiss Confederation who by his plan had practically fed Switzerland during the war, formerly Deputy Director-General of the United Nations Food and Agriculture Organization, and our Conference Scientific Patron, but who, like his predecessor Sir Julian Huxley on our first such occasion, is unfortunately unable to attend for reasons of health],* Berne:

> It is highly creditable that the Government of Iceland and the Foundation for Environmental Conservation have made it possible to hold a Second International Conference on Environmental Future. The conservation and now ever-more-urgent restoration of a healthy environment protecting plants, animals, and Man, is the greatest task of our time, demanding very diverse efforts on many levels. Scientists, whose task it is to expose the threats to the biosphere and the means to save it, bear a primordial responsibility. Through the presentation of the results of their research at events like the Reykjavik Conference, they succeed in bringing thinking people to acute awareness of existing dangers and the means of preventing them.
>
> Yet this small but fortunately growing group of people who are not only thinking but ready for action, is not in a position to carry through the necessary measures. The entire human population of the world must be made aware of these dangers which might take on unexpectedly destructive proportions in the next few decades. Only a movement from the ranks, carried by the large majority of the citizens, can bring the responsible members of the governments, politicians, and professional circles that are chiefly preoccupied with economic success, to

*Throughout this work, square brackets are used to indicate that the words within them were not necessarily spoken on the occasion involved.

take the necessary action even if vested interests are thus inexorably touched upon.

The degree of the progressive pollution of the biosphere grows largely parallel with the extent of economic expansion. Attempts must therefore be made to establish an upper limit for the exponential economic growth. However, in striving to preserve the international faculty of competition, no state will be prepared to go it alone. To ensure fair competition at the international level, a limit must be set in the framework of an international agreement. In my opinion, about 1% of economic growth yearly would be an acceptable solution.

So the conservation of a life-upholding biosphere is a task of universal character, towards the solution of which the individual citizen, the economic circles concerned, politicians, governments, and international organizations, must work together. The foundations of this movement, however, are to be furnished by science, which bears a marked political responsibility in this field. I hope and expect that the Second International Conference on Environmental Future will bring a fruitful contribution to this high goal.

(2) *From Dr Kurt Waldheim* (Secretary-General of the United Nations), New York:

Your Excellencies, Distinguished Participants:

It gives me great pleasure to extend my warm greetings and best wishes for a successful conference on the environmental future. To the Government of Iceland and the Foundation for Environmental Conservation I extend my sincere appreciation for their sponsorship.

It is most important that the impetus towards preservation of the environment, begun five years ago to this day at the United Nations Conference in Stockholm, be maintained. To this end, your contributions are essential. I am also happy to note that the United Nations Environment Programme was able to help . . . your Conference.

As we tackle the environmental problems of today we need to be alerted to those that are likely to develop in the future and your meeting serves a very useful purpose in this regard.

Again, I extend to you my best wishes for a productive and successful meeting.

(3) *Cable from Mrs Indira Gandhi* (former Prime Minister of India), New Delhi:

Nicholas Polunin, Conference on Environmental Future, Reykjavik—Deeply regret unable attend Conference good wishes for success.

It is wrongly propagated that there is a conflict between progress and protection of the environment. Progress can not be equated with immediate profit but with developmental activities which raise standard of living, relieve drudgery, bring beauty into daily life.

Progress and healthy survival are dependent on our ability to conserve the environment and preserve the balance of nature. Urbanization must be controlled and better planned. This implies better coordination with and improvement of rural conditions.—Indira Gandhi.

(4) *From H.E. William H. Barton* (Ambassador and Permanent Representative of Canada to the United Nations—and President of the UN Security Council), New York:

On the occasion of the Reykjavik Meeting of the International Conferences on Environmental Future, I would like to extend greetings and good wishes for success in your important work. The environmental future of Man is one of the most vital issues requiring the continuing attention of policy-makers in the world today, and the work you are doing will be of the greatest importance to the future of mankind. I wish you well.—W. H. Barton.

Secretary-General

These and many other messages have warmed our hearts and surely propelled us in the right direction for doing whatever we can to further the cause of biospheral survival. And lest anybody remarks that many of us here were in the environmental movement long before Stockholm, it should be explained that the impetus to which the UN Secretary-General alluded was *political*—also that the United Nations Environment Programme has helped us very substantially by sending to our Conference some truly excellent people from developing countries. To the extent that I know them and they actually turn up—some are with us already—I shall be surprised if they do not prove to be among our liveliest and soundest interveners.

The single page of notes for participants, printed only this morning because of the general strike, having now been circulated, I hope we may take it as read, and so proceed to consider our slate for the Resolutions Committee. This has been discussed informally among some of our 'inner circle' and, subject to your amendation and formal approval, is as follows:

Professor Linus Pauling, the President of our Conference, as Chairman;
Professor Donald J. Kuenen, President of IUCN and Rector Magnificus of the University of Leiden;
Professor Gunnar G. Schram, Chairman of the Icelandic Environment Protection Committee and of the Icelandic National Committee of the 2nd ICEF, also National Adviser to the UN Conference on the Law of the Sea;
Dr Thomas F. Malone, Director of our co-sponsor the Holcomb Research Institute;
Dr E. Barton Worthington, Scientific Director of the International Biological Programme throughout its life and now of its publications; and
Dr Letitia E. Obeng, Senior Programme Officer of the United Nations Environment Programme.

These being mere—though I hope enlightened—suggestions at this stage, if anybody has any others of any kind, please speak out or forever hold your peace.

There being no dissensions and no further nominations for this responsible and onerous task, may we accept this as an *election* that is approved unanimously? Yes! I thank you for your confirmation and wish them all good fortune and satisfaction in coming out with what many of us—including the past and present Executive Directors of the United Nations

Environment Programme—are waiting for in the way of a strong statement with guidelines for action.

Finally, it is my satisfaction and honour to step down and hand over to the Chairman of our first session, who is our most eminent Conference President and surely not needful of introduction to anybody: Professor Linus Pauling.

Session 1

INTRODUCTION: ECONOMIC GROWTH AND THE BIOSPHERE

Keynote Paper Part I

An Economist's View and Questions

by

JAN TINBERGEN

Professor Emeritus of Development Planning, University of Rotterdam;
sometime *Director of the Central Planning Bureau, Government of the Netherlands;*
recipient of the first *Nobel Memorial Prize for Economic Science:*
Haviklaan 31, Den Haag, Netherlands.

ECONOMIC GROWTH: PAST AND PROSPECTS

For centuries, up until about 1940, economic growth had been slow to moderate. It had been slow until the industrial revolution, which started around Adam Smith's birth or the independence of the United States or the French Revolution. From then on it had been moderate. During the nineteenth century, the average annual growth of the products of the industrializing countries reached 2–3%. At the same time their population growth was of the order of 1% p.a., leaving *per caput* growth of real income at an average of 1–2%. From figures given by Zimmerman (1964), we find that between 1860 and 1913 the rate of population growth for the rest of the world amounted to 0.6% p.a. About 1940, for a typically developing country such as India, that rate had become about 1%.

The picture changed considerably between 1950 and today. The rate of growth of population in the industrialized countries between 1960 and 1970 remained in the neighbourhood of 1%, but in the developing countries it had become more than 2.5%. In contrast, the growth-rate of product *per caput* attained almost 4% annually in the developed countries, as against 2% in the developing countries.

So, because of both the high rate of growth of *per caput* income in the developed countries and of the high rate of growth of population in the developing ones, world production rose by some 4.5% p.a. in the 1960s. This means a doubling of the growth-rate of world production as compared with the figure for the nineteenth century. Correspondingly, the pressure on the biosphere exerted by the human population increased more rapidly in the period 1950–70 than ever before, and it is this problem which has brought great concern to many ecologists.

In order to discuss desirable policies, we must make an attempt to estimate the developments to be expected in the future.Forecasting is a difficult job at any time, but it has become even more difficult in the past few years because of the unexpected recession. Assuming that the recession will be overcome, the main question to ask is what the continuation of the 1960–70 rate of growth would mean. It would mean a doubling of production every 16 years, and everybody agrees that this would sooner or later become intolerable. But the inhuman conditions in which a considerable part of the population of the underdeveloped world lives, calls for an acceleration of their production. Hence the necessity for the developed world to reduce its growth-rate in comparison with the 1960s. There is all the more reason to emphasize this necessity since, of the total world production in 1970, in United States dollars, of some 3.6 billion in the European sense (trillion in the American sense), more than 80% in nominal terms, or some 60% in real terms, was consumed by the developed world, whose population constituted, in 1970, less than one-third of world population at that time.

If we analyse the consequences of the two targets just mentioned—an accelerated growth of the production in developing countries and a reduction in growth of the developed countries—the resulting forecasts depend, as a matter of course, on the quantitative specification given to both targets. The lowest alternative worth mentioning seems to be one where the poor world only continues to grow at the observed rate for 1960–70, that is at 5.5% p.a., whereas the rich world returns to the rate of 2.5% p.a. as before 1914. For world production in nominal terms this means a growth-rate of 3%—in real terms of 3.7%—involving still a doubling in 23 years or 19 years, respectively. From the point of view of the poor nations this is unacceptable, however, and the illustrative figures mentioned in the RIO Report to the Club of Rome (1976) constitute a growth of world production per annum of 4.3%—again a doubling in 16 years.

Thus, however uncertain future production may be, we must envisage the likelihood of a rate of growth in real terms of around 4% p.a. The only important reduction in this figure that is conceivable for the coming decades can be attained by either (*a*) a drastic fall in population increase, or (*b*) a drastic fall in consumption *per caput* in the rich countries. In the present paper we will stick to a figure of some 4% annual growth.

Impact on the Biosphere: Qualitative Aspects

Already in the past, accelerated demographic and economic development has contributed to local deterioration of the biosphere, the natural environment in which we live. Pollutions of air and water, affecting both human and other animal life, have become increasingly frequent and are spreading. Major lakes have been deoxygenated, killing important fish populations. Over-fishing in the North Sea has required drastic measures already. In the Southern Hemisphere, whales have been caught in much larger numbers than is desirable if viable populations are to be maintained. Erosion and desertification have not only hit important areas on three continents but are threatening an increasing number of countries. One of the recent discussions is about the prospects for Java, the most populated island of Indonesia.

With the economic prospects described in our opening section, let us discuss somewhat more systematically the various impacts which a continued demographic and economic growth is likely to exert on the biosphere.

Population growth, even without growth in well-being, is bound to increase the pressure on natural resources proportionately. Growth in well-being will affect the pressure in two different ways. First, an increase will take place more or less proportionately to the rate of increase in income. Secondly, there will be a shift towards consumption of a more luxurious character. Beforehand, it is not certain whether this shift will induce a higher or a lower pressure on the biosphere. But we do know, from recent experience, of a number of activities which have an unfavourable impact.

One general feature is the rapid increase in the use of energy and the polluting effects that go with it. The three main uses of more and more energy are for motor traffic, for heating of buildings, and for manufacturing industry. Among the last, some of the most intensive energy-consuming industries are the heavy industries (metallurgy and basic chemicals) and paper production. Also, motor traffic in the restricted sense entails the transportation of oil in tankers of increasing size, with a rapidly increasing number of accidents, implying a special type of ocean and beach pollution. Yet another bottleneck is the consumption of fresh water by both households and a wide range of industries.

A number of quantitative studies have been made of the increased demand for energy and water. Some of these show reassuring counterforces. Thus, in the last century, the use of energy per unit of gross national product in the USA had shown a steep growth between 1880 and 1920, when it had reached more than twice the level of 1880, followed by a fall to 1.5 times the 1880 figures after World War II (Schurr, 1976).

Also, anti-pollution measures have been taken, by the setting of maximum standards for the content of poisonous components in air and water, resulting in the installation of anti-pollution equipment in factories or purifying installations in the water system. Estimates of the costs needed to keep pollution under control in future decades, made in the USA, Japan and elsewhere, leave one with the impression that some 5% of gross national product may suffice to keep the environment clean. It is also likely that human inventiveness will assist us in restructuring production and consumption so that we become less dangerous to ourselves and to the biosphere.

Even so, the probability of an ongoing process of economic growth requires continual caution, research, and directives, in order to keep the ecological situation under control and to bring it under control where it has run out of hand.

Need for Quantitative Data at the Macro-level

A major problem which ecologists are facing is that of measurement at the macro-level. Whereas technicians and economists have developed at least crude figures about alternative future developments in their fields, and crude ideas about the impact of conceivable policies in their realms, this is not so for ecologists. They are able to point out qualitatively some of the tremendous dangers mentioned above, and many others, but they are not yet able to estimate the extent of the disasters, or the extent to which some remedies must be used. Their subject-matter is so much more complicated than the subject of non-living matter or the subject of economics, that the type of aggregation into macro-figures which is applied in technological and economic research has still to be developed.

Answers must be sought to questions of the following type. What surface of the world must be kept in its natural condition in order to constitute a sufficiently large 'gene reserve' to keep the biosphere in a state of equilibrium? What quantities of man-made cleaning processes are necessary?

Economists and politicians do understand that measures to protect the biosphere are called for. They do not understand, however, that the relative frequency of the innumerable species of plants and animals should be brought back to what it was before. In fact the equilibrium in the biosphere, they feel, has been subject to continuous change, and the point at stake is to arrive at an equilibrated further development rather than to return to previous states. In order that the future development be equilibrated, economists do understand that there may be certain ceilings or bottoms in the flows between the biosphere—without human artefacts—and human society; but we should know the order of magnitude of such ceilings or bottoms.

An example of the sort of questions I have in mind is supplied by a study by Buringh *et al.* (1975), which estimates the 'absolute maximum food production of the world'. The figure they arrive at, on the basis of quality of soil, climate and availability of water, is about 30 times the food production of 1970. The question to be asked of ecologists seems to me to be: how much too high do you think this figure is, and because of what ecological factors not included in this inquiry?

It seems to me that research programmes directed at answering part of this question should have a high priority.

References

Buringh, P., Heemst, H. D. J. van & Staring, G. J. (1975). *Computation of the Absolute Maximum Food Production of the World.* Agricultural University, Department of Tropical Soil Science, Wageningen, Netherlands: iii + 59 pp., illustr.

Reshaping the International Order [cited as RIO] (1976). *A Report to the Club of Rome.* E. P. Dutton, New York, N.Y.: vi + 325 pp., illustr.

Schurr, S. H. (1976). *The Energy Connection.* Resources for the Future, Washington, D.C.: No. 53 [not available for checking].

Zimmerman, L. J. (1964). *Arme en Rijke Landen* [Poor and Rich Countries—in Dutch]. Albani, The Hague, Netherlands: 163 pp.

Keynote Paper Part II

A Biologist's View and Warnings

by

DONALD J. KUENEN

President of the International Union for Conservation of Nature and Natural Resources (IUCN); Rector Magnificus and Professor of Zoology, University of Leiden, Stationsveg 46, Leiden, Netherlands.

INTRODUCTION

In 4,500 million years the Earth with its biosphere has evolved to a degree which daily astonishes those who take the time, or whose profession it is, to have a closer look at it. There are at least 200,000 species of higher plants and many more lower ones, and probably at least 2 million species of animals, currently living on the Earth. Some flourish in immense multitudes, while others barely survive on the brink of extinction. In addition, we find the remains of innumerable extinct species of animals and plants which populated the Earth in former times.

The process of evolution began in the inorganic surface-layer of the primeval Earth and could be kept going because plants developed that extraordinary process known as photosynthesis, in which the energy of the sun is used to build up complex organic molecules. These molecules store the radiated energy in a structural form which makes it possible for animals to use this energy to build up their own bodies, each with its specific proteins.

During the process of evolution the diversity of species has increased, their complexity has grown, and while a great many forms gradually became extinct, they also were the stock from which ever more new species evolved.

In this process of diversification and increasing complexity, another remarkable characteristic of the present biosphere developed. The exploitation of organic material in any form has been pushed further and further—until it now seems that wherever there is organic material there is always some organism which exploits it for its own use. Parasites get on to and into other animals and plants, coprophages exploit the discarded droppings of other animals, and biting insects have developed a specific apparatus, of a mechanical and chemical nature, to exploit that most excellent source of food—the blood of terrestrial vertebrates. If an animal does not have the proper enzyme to break down some organic substance, such as cellulose, symbionts in vertebrate or insect herbivores do the work to the benefit of both. Practically the only notable exception is the anoxybiontic accumulation of dead plants under water, commonly resulting in the formation of peat and ultimately of coal or oil.

Incidence of Man

Relatively recently, only a few million years ago, there appeared the first signs of a new kind of animal with some exceptional qualities. It walked upright, developed a large brain and, contrary to what happens in other bipedal animals, the front legs did not gradually show atrophy but developed further into finely-designed functional arms and hands. Later it must have lost its hair cover and, whether developing in one or more places simultaneously, has gradually spread all over the world.

The species is in very many respects unusual, for it obviously depends for its survival not on strength or speed or fangs and claws, but upon its brain and upon communication, which makes large-scale concerted effort possible. Later it developed the cultural tradition which further emphasizes the exceptional place of this species in evolution.

We have some indications of the rate at which evolution occurs. It must be measured in hundreds of thousands of years, though individual changes, of course, can take place much more rapidly. Industrial melanism and resistance to insecticides develop in a few generations. But before a new species is established, a whole network of characters must be reshuffled and adapted, and that takes time—particularly if there are no violent changes in the environment. The longer an environment stays more or less stable, the greater will be the probability that diversity will increase, which implies a concomitantly greater exploitation of all details of the environment; hence the richness in species that we find for example in tropical rain-forests.

With the appearance of Man, considerable changes have taken place. The rate at which Man has evolved and spread over the Earth has made it virtually impossible for other species to adapt to the changes which followed this occupation. Man has disturbed the process particularly of the evolution of ecosystems which, we believe, until his appearance, was a more balanced one.

The action of Man on his environment is not completely new. Every animal has some influence on its environment. It feeds, builds nests, occupies space, provides food for other species, and finally dies. It will have direct or indirect effects on plants and other animals, which in return will affect it—often more or less profoundly.

Man is different, in both the quantitative and the qualitative aspects of ecosystem interrelationships. His technology has made the quantitative side excessive, while his chemical–industrial development has produced substances which are new—and which many species, including Man himself, cannot cope with physiologically, because of lacking the enzyme-system to detoxify them.

But not only is Man's individual influence much greater than that of any other animal. He has furthermore developed the techniques of medicine, and with the support of ethical convictions his numbers have been steadily increasing to densities which, locally, from the point of view of the biologist, are far beyond the carrying capacity of the locality. Whatever the exact definition of carrying capacity, it at least implies that the food available will sustain the population throughout the yearly cycle. This is not the case in any urban community of any real size—nor is it true for very large regions in the western world.

We can also put the matter in another way. The development of Man has produced a species which has introduced elements into the biosphere that are not in harmony with evolution as it was proceeding until a few thousands or even hundreds of years ago, and this creates problems that are both quantitative and qualitative.

Environmental Impact of Man

We must analyse the situation further and see what Man is actually doing to his environment—including the plants, other animals and the non-living elements. Some of these activities are in themselves normal from the ecologist's point of view: hunting, gathering plants for food and producing 'physiological refuse' are all acceptable. But they become a destructive factor in the natural environment if they exceed the permissible limits. These limits can be defined by stating that, below the limit, no irreversible effects are produced but that, above the limit, a permanent change in the relative abundance of species will result.

In addition, there are the non-biological influences which are mainly qualitative. The production of toxic substances (a useful neologism would be 'xenobiotic'), construction of dikes or dams, and the building of houses, are all such activities which do not fit in the biological complex. But there, too, quantity is important. A very low concentration of toxic substances is not always harmful but may even be a useful stimulant. A clear distinction between qualitative and quantitative cannot be given, but in assessing damage by or through some activity of Man, the two aspects should always be kept in mind.

Sewers have been emptying into rivers for hundreds of years. Only when the amounts are so large that the biological process of oxidation of organic material can no longer take place within the required time, do we speak of pollution. A small amount of sulphur dioxide in the air does not matter, but when the amount becomes such that is has a quantitative influence on corrosion and is bothersome to people breathing the air, we begin to object. Innumerable other examples can be quoted where it is quantities that matter, and quite often we find that it is interference with biological processes which makes pollution serious. And this is where we have to refer to some of the problems raised by Professor Tinbergen in his thoughtful paper.

The Problem of Human Population

The first question to be considered is that of the growth of the human population. Let me approach this from the viewpoint of a biologist.

No species can survive which does not have a reproduction rate that is higher than the replacement rate of physiological death. Every individual has a life-span which is limited, and it has to be replaced. To be sure that there are always enough individuals to exploit the possibilities of the species as near to the maximum as is feasible, there has to be a superabundance of young. In many cases this number is so high that most new individuals die before they attain sexual maturity: they are either eaten by predators or die through adverse environmental circumstances or by accidents.

Adherence to these 'laws of Nature' is essential for evolution and for the optimal functioning of ecosystems. Evolution is in essence composed of two elements: variability in genetic composition of individuals, and selective mortality. The variability is maintained by mutations and interbreeding within the population, while selective mortality eliminates a larger percentage of the less-adapted individuals than of those which fit better into the local environment. In this way the genetic composition of the population as a whole is constantly adapted to the qualities of the ever-changing environment.

In addition to all his other attributes (whether we like them or not), Man has the genetic capability to overcompensate for physiological mortality. In the beginning of his existence he was exposed to the selection powers of his environment, but this selection process brought him to a stage where he constantly grew in numbers and began to spread all over the world.

We have since added another element of importance for the quantitative aspects of our presence, and that is our concern for the individual. Under the guidance of our moral leaders, and with the support of our medical knowledge, we have developed the ethics of the care for the individual. We do not wish a human being to suffer, and we try to keep each individual healthy or, at least, alive for as long as possible. There is no ethically acceptable system for selection of, and no judgement possible to decide, who shall take part in reproduction and who shall not. (The death penalty for particularly heinous crimes is a problem apart, having no relationship to the argument developed here.)

This means that there is no longer a quantitatively significant differential mortality, which in turn implies that selection no longer functions, while the variability continues to increase, through mutation—probably at a slightly higher rate than before, due to mutagenic agents introduced by Man in the environment, and to increased interbreeding between different human populations. But serious though this may be in the long term, more important in the short term is the quantitative increase.

Obvious Biological Solution

The solution of the problem is, from the biological point of view, quite simple. As nobody could ever consider the possibility of wantonly increasing the mortality rate, the only solution is a reduction of natality. *Any argument against this principle misjudges the dire consequences of continued population growth.* If the human population continues to increase, we have but few options before us: starvation, large-scale wars, or catastrophic epidemics.

Whatever the agriculturists say, the rate at which we can produce more food is too low in comparison with the rate of population growth. Transportation of food in bulk from places of overproduction to where there is shortage is no solution; nor is the political system which we are developing towards more equal distribution of assets over the world. Famine looms ahead, and of course every effort should be made to try and alleviate the situation.

But there may be some other way out, such as a major war leading to quite large-scale deaths—not necessarily because of the use of nuclear weapons but because of the social disorder which is bound to follow. But even without a real war breaking out, the high population-density may at any time result in social disorder with the consequent outbreaks of epidemics. As the concentration of population increases, the possibility of epidemics increases too. We keep our hygienic system intact by a highly sophisticated organization based upon fossil fuels. When transportation is stopped and water supply interrupted, that system can easily break down.

This applies to all nations, although not to the same extent to each of them, and should be well considered by all leaders—whether political, religious or otherwise. Faith or hope or science will not increase the size of the world. Those who do not take this matter seriously are misled by the time-scale. They cannot imagine the rate at which the exponential graph is shooting upwards without any real sign of slackening off to stabilization.

In the meantime we are all clamouring for further economic growth. What does that mean and where will it lead us?

If we all want to partake in this process of economic growth, it will mean further increases in the rate of exploitation of non-renewable resources: metals, oil, coal, etc. If those who have enough could stop consuming and wasting, those who do not have enough could be better served. All economists agree that there is not enough to go round for a high standard of living all over the world. All the growth is based on increasing consumption, and in spite of the fact that energy consumption per unit of GNP is less than it was some time ago, the overall fuel consumption continues to rise.

We have for some time been told that nuclear fission energy will solve the problem, and latterly that nuclear fusion will give us a clean solution of all our energy problems. But the doubts about each of these have also been growing, and fusion at best still seems to be very far away. Meanwhile problems of disposal of radioactive wastes, and the possibility of misuse of the basic materials, hang over us like the sword of Damocles.

Metals are mostly still abundant and can be extracted if we spend more and more energy on refining poor ores. But the problem of supplying the necessary energy, and dealing with the waste materials and pollutions with which industry continues to saddle us, have not been effectively solved so far as one can judge. Certainly the pollution in some countries is quite appalling.

Economic development can go on as long as there is power to drive the machinery. We still continue to destroy the natural environment, and even increase pollution of air and water. So we get asked the questions: does it all really matter, and, if it does, how much destruction of the biosphere can be tolerated?

The answer could possibly be simply that we do not know enough to state any quantity. We cannot assess the right number of square kilometres of savanna or wetlands or tropical forests. We do not know in sufficient detail where areas could be reduced without destroying the viability of the community. What we do know is that in many places irreparable destruction has taken place already, and that the rate at which animals and plants are becoming extinct and their habitats are being destroyed by or through Man's activities, is not slowing down. What we must consequently expect is that if we do not curb the destruction of natural areas, we shall come to a point at which we shall see that we have gone too far and then it will be too late.

But this does not constitute a proper answer to the questions of whether it really matters, and of how far destruction can go in a still-viable biosphere; nor does the other extreme position which we could take by stopping all further extension of cultivated land or other destruction of natural vegetation. It would be utterly unrealistic to attempt to achieve anything so drastic, quite apart from the fact that no mechanism exists which could be used for such a line of action. So it is perhaps worth while to look at the problem from an entirely different direction.

Possible Economic Solution?

What we are trying to achieve is a sufficient quality of life for as many people as possible. For as long as Man has been evolving, this has been practically identical with quantities. We have the fundamental idea that the more we have of good things, the better they are and we are. This is so because it has always been practically impossible to have too much of anything. Comfortable homes, enough to eat and drink, and at a later stage more leisure, have always been in short supply. Only small, elite top-layers of societies have ever had enough, or even too much, of any of these things.

Now the rate at which the western culture has been applying its knowledge, based on scientific research and applied technology, has been ever-increasing. This has been possible because an ever-growing percentage of the population have been set free from manual labour, and more and more have been able to apply themselves to increasing our knowledge. As a result, living conditions have improved to such an extent that what was extreme luxury for 90% of the population a hundred years ago is now considered the normal standard for everyone. Reduction of the standard of living can only be achieved by force: only a negligible number of people will give it up voluntarily for the sake of others.

In the development of this high standard of living in the western world, no long-term or global planning was involved. No doubt many people think about the future, and some talk about 'one world', but in terms that are too vague. Besides, most such thinking is about only one aspect: food, or war, or transport, or education. Few people have attempted an integrated approach although, after the lead particularly of the MIT studies of early in this decade, others are beginning to try. The quantification still gives trouble but progress is obviously being made.

It appears that human behaviour is so unpredictable that economists cannot foresee future developments with much confidence and, although they can predict trends, the time-scale raises serious problems. The present stagnation in economic activities was not forecast two or three years ahead. The symptoms of slightly increased reactivation which we see now were not forecast with any predictive value even a year ahead, and no economist will dare to tell you what the situation will be two years hence, five years hence, or twenty years hence. But that is what we need if we are to plan ahead.

We are quite good at short-term planning. Far-sighted economists make it clear as to how we should structure our community if we are striving for a better distribution of wealth among nations. Of course there is no unanimity of opinion on such a difficult problem, but the suggested trend is obvious: use less energy and concentrate on economic regulation which will make it possible for the poor nations to have a stable income from their natural resources, that are not to be exploited on a short-term basis but should be part of a long-term development plan.

Beware Forest Desecration

At present the technologically underequipped nations are selling their natural wealth for short-term gains. Looming large among these assets, in parts of Asia, Africa and Latin America, are the tropical rain-forests. They are biologically extremely interesting, valuable as a resource, and not only the result of global climatic conditions but surely influencing that climate by recycling water into the atmosphere.

Large parts of the currently remaining rain-forests have been practically untouched by Man, or have been lived in by Man at a very low density that harmonized completely with the biological processes which are going on there. Where more systematic agriculture was practised in certain areas, again this was done in harmony with the biological possibilities of productivity. Patches were cleared, cultivated for a few years, and then abandoned. The area was soon overgrown and the forest largely recovered, obliterating most of the traces that Man had left. The secondary forest is not identical with the original vegetation, but the difference is mainly a question of relative abundance of species—as well as, to some extent, of dominance and physiognomy.

With the increase in human population, the rate of cyclic use in shifting cultivation has likewise increased and the forest and soil have less and less time to recover. This has resulted, all too widely, in a gradual overexploitation and concomitant reduction in productivity. To this is now added the large-scale cutting down of forest for timber to be exported and used elsewhere. This results in the complete destruction of the ecosystem. The soils are exposed to sun and rain for too long, humus is oxidized, minerals are washed out, and if the soils are lateritic they are so changed as to acquire a hard and unworkable surface. The vegetation that can subsequently develop in such areas is extremely poor and consists mainly of coarse grasses.

It is known that in much of the Amazon basin, cutting of the trees will lead to a soil which will lose its natural fertility within a very short time. The forest exists because there is an immediate recycling of all dead matter, which does not happen on agricultural land. The destruction of the forest and soil is, from the practical point of view, irreversible.

In other places, forest destruction will often lead to erosion, with all the disastrous results of upsetting the water-economy of a whole region. Recent studies in Java show us something of what lies ahead of us if the overexploitation of the hill- and mountainsides continues to spread.

Yet, in spite of all this, the destruction goes on. What has developed in millions of years, and what has been the natural basis for the existence of Man for tens of thousands of years, is disappearing in a few decades because at present we seem to have no proper alternative, at least in the minds of the politicians and big business. And soon the next problem will arise because there will not be enough fertilizer to keep the soils productive. Particularly phosphorus, which is as essential for life as any of the other minerals, may pose a serious problem because no minable reserves are at present known which will carry us much further than a few decades.

In some areas the original forest is being replaced by planted trees. This will reduce erosion, and produce fibre and firewood; but it does not produce food and it does not restore the original ecosystem.

The maximum sustainable yield of a forest is well known. It is the amount which can be taken without exposing the uncovered soil to sun and rain for any undesirable length of time. Removal should be limited to small tracts at a time, so that the surrounding plants can recolonize the area. There is enough experience and general knowledge to quantify these conditions in each individual area.

Considering some more temperate regions, it is interesting to note that in Europe—particularly in Greece, Italy and Spain—the forests have been mismanaged and most of the hills and mountains have been cleared of trees, with resulting erosion. But the Alpine population has been careful to keep its woods intact on the hillsides, and thus has saved its valleys. One could speculate on the social and cultural background of this. Apparently the mountain farmers, who know about soil, have had a greater influence on management than in the countries bordering the Mediterranean. Perhaps also the fact that people who live along the sea used timber for ships—particularly warships—may be of significance, in which case there could be some basic truth in the claim that the wars of Rome against Carthage 2,000 years ago caused the flooding of Florence some 10 years ago.

The practical necessity of keeping the tree cover of this Earth more nearly intact than we are doing is evident. In particular, emphasis on rain-forests is essential. They seem to be highly productive but that is largely an illusion. The very efficient recycling makes the standing crop relatively large. When depleted of the forest, the soils are found to be poor and are soon exhausted if used for agriculture. But besides, these forests harbour a wealth of species of plants and animals. They are part of the natural heritage of this globe, which we are responsible for. Moreover, many untapped resources of specific products remain to be studied before the economic value of these areas can be assessed at all precisely.

As regards the present manner of exploitation, it is clear that only very little is gained in the form of a short-term increase of incoming money. Therefore, from the biological and the economical points of view, this exploitation should be stopped.

Wetlands also Important

Another bone of contention are the wetlands—those bothersome areas of which you are not quite sure whether they are lakes or not and where, in many cases, changes in water level produce floodplains which cannot be used by Man either to walk or ride on, or to use a boat on. Hence the implicit contradiction in the term 'wetland'.

Nevertheless these wetlands are biologically of great importance. Generally of high productivity, they are a source of food for many animals, the most spectacular being commonly the birds. This applies not only to the resident species but, more particularly, to migrating birds—those truly international assets of Nature of which many species depend upon wetlands as feeding areas at appropriate intervals along their migration routes.

If we have any respect for other creatures on this Earth, those wetlands must be saved from destruction, pollution, or too much use for tourism.

Wetlands can often be made into fertile land with comparatively little work, or they may become planned extensions of harbours, which will bring wealth to the human population of the neighbourhood. Living as I do in the Netherlands, I am very much aware of the urge to do something to them. They are not neat, they do not fit into the orderly way the world should be, and I know that this alone is considered by some engineers as a reason to do away with them. But their biological value is too well known; and agricultural land being not always so urgently needed in that particular place, and the harbour, with a little more cost, being quite well situated a little further along, we should save them for posterity wherever possible. With wetlands in particular we should calculate our costs, not only on the value of today's estimate, but including future values of a dimension far larger than the local values alone and over a much longer period. The nature conservation values are bound to rise as more and more people become aware of the part played by Nature even today in our urbanized lives. Too many wetland areas have already gone, in many cases without our knowing even the percentage of the population of a number of species that were dependent upon a particular wetland, The quantitative data concerning wetlands are known, and any further reduction of them should be considered a serious infringement on all that has been said—even by politicians—about Man's respect for Nature.

Economics and Nature

I am not going to repeat all the arguments in favour of nature conservation, but rather to consider how the economic evaluation and the ethical claims of nature conservation compare with each other.

Specialists in development planning often claim that they have quantitative arguments for their plans to increase agricultural land, to dam rivers, to extend towns, to site polluting industries, and even to produce radioactive waste, because all of these will improve the standard of living of the area concerned. The biologists are accordingly asked to produce comparable figures to substantiate their claims for conservation.

Admittedly the claims of conservationists often lack the expected precision expressed in present and future monetary values. But in many cases that is not needed and, for instance, where watershed management is concerned, data are sufficiently precise. But the disappointing experience is that, even in these cases, the short-term profit is valued more highly than the long-term effects. I believe that the entire problem of rain-forest destruction falls into the same group. The facts are known, but the consequences are not accepted. All too often, suggested economic development 'has' to take place at the cost of natural areas, and this applies particularly to certain large-scale agricultural operations which are under way.

It is clear that Man has always tended to use the best soils for agriculture. This means that many areas which are still unoccupied are in just that situation because it has not been profitable to use them as agricultural lands. Most of the great cultures have developed in delta areas—where we find the most fertile soils known—although later in their development they have certainly also occupied higher plateaux, and even mountain ranges have become centres of culture.

It is particularly now, with the world-wide population growth, that other areas are sought. Many of these are 'marginal', although there is no doubt that the increased knowledge of agriculture and technology today makes it possible to exploit areas which formerly were considered unprofitable. New varieties of cultivated plants, application of fertilizers, irrigation mechanisms, and various other considerations, all help to make it economically profitable to exploit areas which hitherto could not be used.

But we must emphasize that something will also be lost. The diversity of Nature will be reduced, species will be pushed to the brink of extinction or actually over it, and possible knowledge of the working of Nature, which we must have in order to plan well for our future, may become unobtainable for ever. We will lose reserves of species—especially plants—which have not yet been exploited, but may have the potential of yielding important materials for medicinal purposes or improvement of vital crops. Even more subtly, we may exterminate some vital component of an ecosystem which can never be resurrected without it. Quantification is also a problem, particularly as we do not know enough. It is because we do not know, that we have to be particularly careful as to what we do.

There can scarcely be any places left where population pressure requires increased food production and yet high-quality space is available to augment the area under cultivation without destruction of natural areas. In all areas of increasing population pressure, the production per unit of surface will have to be augmented. Agriculturists claim that this can be done quite well and, compared with increasing the cultivated area, this argument has everything in its favour in the eyes of the politicians.

Constraints Foreseeably Severe

There is, however, another pair of questions that remains unanswered. What will be gained by increasing the agricultural area and for how long will a positive effect be noticeable? Soils remain fertile under continuous exploitation only for as long as certain minerals are available in them, and normally nowadays fertilizers are added to 'keep things going'. High yields, such as are required at the present time, need high doses of fertilizer. These fertilizers are produced at the cost of energy and must be transported from the factory to the land—often a considerable distance. Because we expect to need about three times as much food some 40 years hence—to make up for current shortages and feed the foreseeably doubling population—enormous amounts of fertilizer are likely to be required. And some think there may be another doubling of human population after that!

We need to know *inter alia* the ratio of recent increases in food production between that obtained through growing highly productive varieties, that obtained by increasing the area cultivated, and that which is due to improved methods used on the existing cultivated land. Further, we must know what influence all this has had on people's social life in the different areas, how irrigation schemes, etc., have actually worked out, and in particular what the actual impact on the food situation has been in relation to the increased population.

Only if we have these data per area and for each kind of soil and crop, can we know what the effect is of every effort to increase agricultural production. Only then can we determine whether this will make a real impact on the food situation for a considerable period, and only then will it be possible to compare these positive results with the negative ones of loss of natural areas. Conservationists do not think that all animals and plants are worth more than people. But they do believe that Nature has some value and therefore wish enough of it to be protected. It is no use trying to express the value of Nature in monetary terms, because the agricultural value is something different from the usual value calculated by the agricultural economists. Hunger or misery do not equal so many dollars.

In particular the time-factor should be brought into account. Our plans as a rule are for 5- or 10-year periods. Predictions for longer terms have very little value at the moment. But we do need them, because destruction of Nature is for as long as we care to think ahead!

The number of people who think Nature has some real value is increasing, and they can be found all over the world. If we see the interest taken particularly by African people in Zaire, in Tanzania, in Kenya, and also in some Latin American countries, we know that this is not just a wish of some well-fed rich people who still think of colonial times.

If we continue to let human populations grow, if we continue to squander energy on pleasant but unnecessary things, if we go on polluting our waters with waste that could quite well have been partly or largely recycled, and if we do not succeed in reducing the gap between rich and poor and then, besides, destroy Nature which we know is of unique intrinsic value, we shall be forcing a multiple insult on posterity. We should not go along with the cynical contention that we do not owe anything to posterity as it has done nothing for us: rather must we now seriously begin to think how the world can be managed in 30 years' time, or posterity will blame us for not having done what our knowledge showed us we should do and to a considerable extent indicated how.

Background Reading

Many of the ideas in this article have been taken from published works, of which the following are some of the most relevant:

CALDWELL, L. K. (1972) *In Defense of Earth*. Indiana University Press, Bloomington, Indiana.

CURRY-LINDAHL, K. (1972) *Conservation for Survival*. V. Gollancz, London, England: xiv + 355 pp.

EHRLICH, P. R. & EHRLICH, A. H. (1972). *Population, Resources, Environment* (2nd edn). W. H. Freeman, San Francisco.

FOUNEX REPORT (1972). *Development and Environment*. Mouton, Paris & The Hague: viii + 225 pp.

MASSACHUSETTS INSTITUTE OF TECHNOLOGY (1970). *Man's Impact on the Global Environment*. MIT Press, Cambridge, Mass.

MELLANBY, K. (1967). *Pesticides and Pollution*. Collins, London, England.

MESAROVIC, M. & PESTEL, E. (1974). *Mankind at the Turning Point: The Second Report to the Club of Rome*. Hutchinson, London, England: xiv + 210 pp.

MICHALJOW, W. (1973). *Protection of Man's Natural Environment*. Polish Scientific Publishers, Warszawa, Poland: 663 pp.

POLUNIN, N. (1972). *The Environmental Future: Proceedings of the First International Conference on Environmental Future*. Macmillan, London & Basingstoke, and Barnes & Noble, New York, NY: xiv + 660 pp., illustr.

RAAY, H. G. T. VAN & LUGO, A. E. (Eds.) (1974). *Man and Environment Ltd*. Rotterdam University Press, Rotterdam: 332 pp.

DISCUSSION (Session 1)

Pauling (Chairman, introducing)

Man has been here on Earth for some thousands of generations with essentially his present genetic composition and abilities. Only recently has he achieved the power to determine the nature of the world. Before that he was in essential equilibrium with the rest of the world and was not able to damage it in a really significant way. I feel that the effort which Man should be making now—the decisions that he ought to come to and act on— should be such as to make it possible for every human being to lead a good life.

There are, I trust, going to be tens of thousands, perhaps millions, of generations of human beings following us on Earth. The decisions made now will, without doubt, [help to] determine what sort of lives the people of future generations will be able to live. It is of the greatest importance from the standpoint of morality that the present generation and the immediately following generations of human beings should not rob future generations of their birthright. The exhaustible resources of the Earth should not be exhausted, and the environment should not be damaged in such a way that it would be impossible for human beings to lead good lives in the future. This is the theme of our Conference, which I believe will make its contribution to the achievement of the goal of a world in which every human being can lead a satisfactory and productive life, and in which each generation can pass on to the succeeding generation a world with the same capability.

Our opening speaker today is Professor Kuenen, of the Netherlands; but first let me ask if the three panellists for this initial session would come and sit with me here [on this dainty little platform] while Professor Kuenen is speaking. [He is President of the International Union for Conservation of Nature and Natural Resources and Rector Magnificus of the ancient and famous University of Leiden.]

Kuenen (Keynoter, commenting on Tinbergen's paper)

Unhappily our friend Tinbergen has not been able to attend the Conference in person but as a biologist I think it essential that his contribution [which had been circulated and is now printed above] should be touched upon before we start off on other discussions, because it is of such importance for the background of all. So let me indicate a few of the essential points, starting with the one that this whole problem of economic growth is so recent that until 1940 or even 1950 hardly anybody worried about it. But then [rather suddenly] human population growth together with industrial development made the conflict between our private little world as humans and the world as a biosystem continually more serious. To consider these matters it is necessary to predict— which Tinbergen says has always been difficult and has become even more difficult [in economics] during the last few years because of the unexpected recession. Assuming that the recession will be overcome, I feel it to be a very fundamental thing to consider that even the most eminent economists cannot forecast the development of the economic world at short notice and we have to live with the certainty that we do not know what will happen in a few years' time.

In view of this, the interesting thing is that we do know what will happen in a much longer range of time, although we have not the actual value of the time; for we are sure that if matters continue to develop as at present there will be a conflict between the natural environment and mankind that will be catastrophic. In the relationship of Man and his world, the use of energy looms very large—including the use of automobiles and the heating or cooling of buildings, with the large consumption of fossil fuels contributing to more and more pollution. The percentage of tolerable growth may be 5% or 4% or 1%, but in any case it is quite clear that, whatever the size of that value, there will be an increasing impact on the biosphere. Yet how far it will be possible to take note of the environmentalists depends, according to Tinbergen, to a large extent on the way in which they quantify their wishes, so that the economists can take note and say what they think can be done or should not be done.

As an example Tinbergen quotes three Dutch agriculturists who have said that if you have an optimal use of the world's agricultural soils, and an optimal use of the newest and yet-to-be-developed varieties of plants, and if you cultivate under optimal conditions, it should be possible to produce in the world about 30 times as much food as was done in 1970. The unhappy thing is that this has been read and repeated by a lot of thoughtless people forgetting the ifs—saying that there is no problem in agriculture, because we now know that the world can support 20 or 30 times as many people as at present and so what are we worrying about?

Two things are wrong. In the first place the ifs have been forgotten. In the second place the rate at which the population is now increasing would, if continued, bring us to such numbers rather soon. President Pauling was talking of thousands of generations, but only a few tens of generations will be needed to reach such numbers and lead to absolute catastrophe. [That is my considered opinion as a biologist; now to return to my own paper and conclude by saying]:

If the economists agree that there is a problem of nature conservation, and they do, then they must specify much more precisely where and how the production of food can be increased. They must also sit down and calculate—and it would take only a very short time—how much this destruction of natural areas will improve the situation for a growing population. For in many cases the agricultural increase is compensated within a very short time by the growth of population. If we saddle those who are going to come after us with a large population which is still not well fed, and with a diminished natural environment which will not give them the possibilities to live on this Earth, we have doubly insulted them, and I think we should be very careful before we do so.

Butler (Panellist)

Thank you Mr Chairman: I find myself at a little bit of a disadvantage and in rather unfamiliar circumstances because I have absolutely no criticisms of what was said by Professor Kuenen and what would have been said by Professor Tinbergen if he had been here. Therefore I will probably disappoint the Secretary-General by not starting an argument! [Laughter.]

Six years ago at a General Assembly of the Scientific Committee on Problems of the Environment (SCOPE), a Committee of the International Council of Scientific Unions [ICSU, the world's scientific summit], we identified two basic or primary causes of environmental problems. The first was the continuing increase of human population, both in localized areas and in the world as a whole. The second was the uncontrolled growth of technology, with the attendant consumption of energy and production of waste products. The problem of controlling or reversing these adverse trends is political and social, and it will be solved perhaps by new economic or ethical values; since I do not feel expert in these areas I will leave them to someone else. I would, however, like to say to Professor Kuenen, that in my view the thing which makes Man as a species so terrifyingly successful is not that he retained good use of his front limbs but that each member of the species when it is born has available to it the whole learning and experience of all other generations that went before it, so it starts off with a great advantage over all other species.

The only question that I might raise in this area of social science concerns the title of this symposium which was given by its Steering Committee—they talk about economic growth as though it is inevitable or at least desirable, and I wonder if this can be accepted. I might [also] challenge the Chairman of this Session when he talked about 'the Good Life' and ask him what he means by [that]. Is it made up of the kind of things which we see in North America, such as one man riding in a large car all by himself every day? Is it ten million people using hair driers, each consuming a kilowatt? Is it people eating and drinking to excess at the expense of their health?—and so on. Nevertheless I will pass over these questions and turn to something more scientific, where I feel a little more at home.

Professor Tinbergen said it was necessary to have a quantitative assessment of the problems which he mentioned, and I can comment on some activities that are going on and that might give us some measure of how bad the condition of the world is. One way of carrying out an audit of what is happening to material resources of the world is to study on a global scale the biogeochemical cycles of such important and essential elements as sulphur and nitrogen, phosphorus and carbon. For example, does the conversion of the world's fossil-fuel sources to carbon dioxide exceed the fixation of carbon dioxide into organic compounds by photosynthesis?

Another way of assessing the biological well-being of the world is to make measurements on a world scale of such well-known and acceptably important biological variables as the total biomass, the disappearance of species, the diversity of species, and the balance or equilibrium between different species. Also to give the world some forewarning of any undesirable trends, we can measure the amounts of pollutants in the environment and the amounts of pollutants in biota. We can only hope that people will then listen to the warnings which we give them and that the proper authorities will take the proper action.

Glasser (Panellist)

When I was asked by the Secretary-General, Professor Polunin, if I would wear my economic hat this morning he was referring to the fact that, as well as being an economist, I am a psychologist. But you will see in a moment I think that this has certain advantages. In my work in the developing world—in India and Pakistan and Bangladesh, and in Kenya and Senegal—over the past 12–14 years, I became increasingly oppressed by the thought that the human race has been proceeding for many generations on the wrong assumption as to the aims of life. We inherited a dogma from the Age of Reason to the effect that Man's fulfilment would come by increasing his power to consume. If that were true, then it would also have been true that those people who at that time were rich and privileged were also happy! Yet that was manifestly not true.

The people of the Age of Reason did not question this dogma, and nor did the Social Reformers of the 19th century—presumably because they knew instinctively that the answers would be inconvenient. Therefore, under the sponsorship of the OECD [the Organization for Economic Cooperation and Development], I decided [a few years ago] to conduct an investigation of a peculiar kind—peculiar because it appears never to have been done before. This involved going to a community of a pre-industrial nature—and there are still some left—and attempting to discover, by probing under the rational level of their minds, exactly what they really did want out of life. It is against the background of that research, to be published quite soon in a book, that I am making the following comments.

What seems clear is that consumption on its own does not produce fulfilment. That is one point. The second point is that it is the tragedy of the human race that they have discovered and applied the means of stopping people from dying and of keeping them alive long before being able to answer the question—which they have still not answered—of how to feed the increasing populations that have come on us as a result of these medical and allied advances.

The trouble with economic thought and teaching in the last hundred years or so, is that we have been taught to accept demands as given, not to enquire whether they are right or wrong but only to look—as economic engineers if you like—for the means of satisfying them.

Of course, as T. S. Eliot said very neatly, all desires are not desirable; and so I would say that the problem of our time—and we have not got a lot of time to try to solve it—is to answer a different sort of question. This is a philosophical one, namely, what are the patterns of needs which will provide fulfilment, and what kind of selection should we make in order to achieve the pattern which is most desirable.

Now I have the greatest admiration for Professor Tinbergen [by many considered the greatest living economist], and all that I would like to say now is perhaps to add one or two footnotes to what he has so admirably [brought forth]. On the subject of growth, I think it is useful to say that we ought to regard demands as being in two categories. First it is convenient to think of income demand, that is to say demand which can be satisfied from resources that can be renewed. The second type is capital demand, involving consumption of resources once and for all. It must be perfectly obvious to everybody here that we, as a race, have been for many hundreds of years, and particularly for I suppose the last 150 years, consuming our capital at a fantastic rate.

It must be remembered that, after all, civilizations have disappeared without trace, or with very little trace, and it is a fair assumption that they did so because they were unable to adapt in time. Our job therefore is to *adapt in time,* so as to transform our capital consumption into income consumption.

Here it is instructive to consider a very blunt and rather terrifying statement from the Government of Kenya, when it says in its current development plan: 'There is no conceivable rate of economic growth that will alone absorb all the people in search of work.' That is indeed a terrifying statement, because Kenya is only one of the many countries of the developing world which is in this position. And here is another from Kenya: they estimate that their population rate of growth is approximately 3.9% p.a., which is among the highest in the world, and they expect their population, which is at present approximately 13 millions, to become—and this is rather indicative of our problem, that they are not even sure of the likely increase but *think* it will reach as a minimum—approximately 28 millions by the end of the century, and may by then have gone as high as 34 millions. So at the best estimate in one single generation they will have to provide for more than double the population that they have now, and they have already said there is no conceivable hope of absorbing the increase into employment.

So may I sum up with concrete suggestions, such as we are asked to make? The first—and it can only be done by governmental action—I believe should be to convert people's horizons, to convert people's hopes and material aspirations and lead them towards attitudes which will make it possible to consume on an *income* basis and not on a capital basis. We must have energy budgeting. We must go in for technology that will encourage people to make things to last and not to throw away. [For the time may come when there is no 'away' to which to throw things. Already] you can go to almost any town in the world and see a mountain of cars—used cars—and even if you were to recycle them you would be unable to recycle two of the most precious attributes of mankind, namely time and work. You cannot do that—these have gone and you cannot recover them. As J. M. Keynes has said, 'bygones are bygones for ever'. I would also say that we should abolish consumer advertising, which encourages waste.

Magnússon, Gudmundur (Panellist)

I want to dwell almost entirely on the economic side of these questions. Some years ago the Club of Rome propounded that the world would end in a disaster—

there were so many limits to growth that could be found. In a way, they deserve and still deserve credit for this. They showed quite clearly that we should mind what is going on. Many of the problems that haunted them have been mentioned already at this Conference. What I would like to do is to strike a somewhat optimistic keynote that has so far been forgotten in this discussion.

First of all the measure of economic growth, as used by economists and adopted by others when they talk about economic growth, is quite crude. You do not include in it many important things that would change the picture and ought to be taken into account if it is to be used as an indicator of welfare. I could mention leisure, for instance. Leisure is usually not included in a measure of welfare. Yet it is *the* materialistic measure in a sense. If we should include leisure and the detrimental effects of growth, I am certain that the 4% figure mentioned by Tinbergen would be much smaller.

I would like also to point out that it is our experience (and it is straining the budget of every country) that the demands for health services, for instance, have been increasing very rapidly and that is not unnatural because when standards get higher you want better services. And that is what we have experienced in the last decade. The demands for a purer atmosphere and environment have grown, and I think this is a built-in stabilizer; when people get a higher living standard they will ask for a purer environment.

Another optimistic thing which I would like to point out is that if you take the [resources of raw materials] needed for production (I agree that one of the most critical is food supply), such as copper for instance, you can extract now very much more per kilo of crude copper than you could before, so that, in a sense, efficient means of production have added to supply.

We should also mention here that with planning and forethought you can get improved conditions, indeed optimal conditions, for production of food, and if you adopt these improvements in other areas you save a lot and increase production. Moreover you could produce less and keep the same living-standards, or even improve them, by producing more qualitative goods, which in a sense is often related to conservation. If you take fish, for instance, we have fished a lot in Iceland and maybe the price mechanism has been such that we have put the main emphasis on quantity rather than quality of the fish, whereas we could have got the same living-standard by using the catch better than we have done up to now.

Energy conservation—as you see, I am proposing some remedies to alleviate strains on resources—has been mentioned. One of the critical factors is to save energy, and some types of it are very capital-intensive. We can save by building better houses, and by filling up our large and wasteful cars [ultimately replacing them with public transport, or, better, simply bicycles]. Certainly we need to be far more wisely selective than hitherto in many of our actions and proposals.

We have somehow to solve also the problem of the haves and have-nots, although personally I do not think that the rich countries are exploiting the developing countries to the extent that is often claimed. On this I can quote Tinbergen who made a study that was designed to answer the question: would the terms of trade between the developed countries and the underdeveloped be worse for the latter if the rich countries used their full bargaining strength? He came to the conclusion that, if anything, the terms of trade were better for the underdeveloped countries than they would be if you took competition—pure competition—as the basis. So I do not think the rich countries are as bad as they are often supposed to be, even though everyone could surely be better off through planning production more efficiently and getting somewhere near the optimum conditions which Professor Kuenen talked about.

Pauling (Chairman)

Now with thanks to our keynote speaker and panellists we can open the Session for general discussion. This being the first such occasion, I ask that you rise before speaking because it is easier to hear if the sound is not blocked by the people sitting in front of you. Also please give your name quite clearly when asking your question. When I say 'asking your question', it does not mean that you need to ask a question; but we do need, in order that as many people as possible can participate in the discussion, to keep individual items short. If you want to direct a question to a member of the panel, please do so. That is what they are here for—to enter into a free dialogue. [Let us remember that all here are responsible people who have been especially invited to speak freely in their personal capacities and that we all want to hear as many *enlightened* views in lively debate as time allows.]

Schütt

I would like to address a question to Professor Kuenen, about his point that it will not be possible in the future, because of the transportation costs, for the United States to produce [for other countries. Is that true?].

Kuenen (Keynoter)

The answer to that is extremely simple: economists tell me that the bulk transport of subsistence food is just an impossibility in view of the costs which it would take over the distances involved and in view of the economic situation of the people to whom it would be brought—somewhere in the centre of, let us say, the S'ahel region where they are short of food. If you took the excess wheat that can be produced in the United States of America and transported it in bulk, so that it really could have an impact on the food shortage over a long period of perhaps 5 or even 10 years until the drought period had passed and the area recovered a little bit, the population would probably not know what to do with it and in any case would not be able to pay for it. Therefore it would not be a practical proposition, unless you started giving the food away and paying for the transport; but that of course is such a revolution of the fundamental thinking of all economists that I do not believe it would be considered at all seriously.

Schütt

It depends how much you have to pay for the resources which you are using.

Kuenen (Keynoter)

If there are resources, which in the Sahara there are not [and in America will not always be].

Magnússon (Panellist)

The point about people in the less-developed countries not knowing what to do with wheat, even if they could purchase it and pay the transportation costs, reminds me of a dilemma sometimes discussed by economists. It is often difficult to get people to consume what they need to consume, and one way of solving the problem could be to pay the opinion-leaders—usually the rich people in these developing countries—to consume it and then the others would adopt it. However, this may not be a very welcome proposition of enticing the United Nations to subsidize the rich people to consume the food in the beginning, so that it would never get to the vast majority.

[**Polunin** (Snr)

A jocular note: this talk of rich people in poor countries makes me think of the wag's definition of foreign aid as being the money which poor people in rich countries give to rich people in poor countries.]

Worthington

The other solution, if we are looking at the really long-term possibilities, is surely to move the people to where the surpluses are produced.

Kuenen (Keynoter)

My perspective in this, and I am under the impression that this is the perspective of the organizers of the Conference, is that we are not here concerned with the short-term solution of particular types of hardship such as the Sahel one and others. We are concerned with trying to modify Man's behaviour as a whole, so as to give him a chance of long-term survival; and I think we must not forget the point that a large proportion of the world's population have been educated, have been exhorted by demagogy of all kinds, to regard the position of western Man as their goal, without being told that western Man himself does not know where he is going and is not happy with his high consumption. As I have already tried to indicate, that level of high consumption just is not possible [to maintain indefinitely]. And unless Man's attitudes are changed in time, then the discontent of something like 3,000 million people will overflow and burst and quite possibly destroy the world as a whole.

Guppy

Professor Kuenen made a very important point which seems to be forgotten in many of these [comments], which is that when Man rose on his hind feet his forelimbs did not atrophy. Throughout all of these deliberations we are considering human beings entirely as a problem. They can also be an asset—provided they can feed themselves. Now and again one hears about the problems, for example of the Government of Kenya, of feeding people or finding jobs for people. Yet all of this comes, I think, from the emphasis which we place on investment in large-scale agricultural methods, in heavy manufactures, and so forth.

Now this is perhaps not strictly relevant to conditions in the Sahel or in Kenya, but none the less it is very interesting to note that on the BBC there was a programme in which a woman described how she fed a family of four entirely on green vegetables grown in four window-boxes. Now it may be that we are producing too many tractors and too few window-boxes. But what is needed in the Third World is a change of emphasis from thinking of how to finance large organizations, to how to lend small sums of money to individuals and small groups to enable them to have window-boxes and trowels and things like that.

Goldsmith

About this question of world food: the number of exporting countries with a food surplus was very high 30 years ago. Since then it has been falling all the time. The number of countries now with a net surplus is very, very limited, while 75% of world food imports come from the United States of America and Canada. Now according to Professor Kassas, who is shortly due among us, the United States is losing 2 million acres (810,000 ha) of land a year to urbanization and another 2 million acres to erosion, etc. If you calculate on that basis, before the end of the century the US will have lost probably 20% of her food-producing potential, as she has currently about 400–500 million acres of arable land. By that time she is expecting a population of nearly 300 millions instead of the present 215 millions—her population is increasing rapidly, mainly through illegal immigration from Mexico of some 1.5 million people a year. So by the end of the century, America on these calculations will cease to be a food-exporting country.

Even if you could increase food production indefinitely, which is a total illusion, there will come a time when your increased price of chemical fertilizers and other inputs, with diminishing proportionate returns, will bring you to a point where it is no longer economic to increase productivity and indeed is likely to be environmentally dangerous. So in countries that are not independent in the matter of food supplies, widespread starvation is to be expected. The notion that you can [go on getting] food from abroad is a very, very dangerous illusion.

Franz, Eldon

I would like to call your attention at this point to a very simple model consisting of men [and women and children] and their food. Given the factors of climatic variability which govern plant production—a point which Professor Bryson has made several times in the past, in fact in one reference I saw he cautioned a group of agronomists to stop worrying about optimization all the time—he was talking about a moving target, such as we have in the world food production, which is governed by vagaries of climate. Now the only way to make a stable situation with respect to supplying a man's simple food requirements in a model such as this is to have a big storage reservoir which buffers the variations in inputs from year to year. Yet the outputs from food storage and into the support of human population are going to need to be increased, while the simple fact is of course that there are currently no adequate supplies—no major storage of stocks in the world today. We are essentially living from hand to mouth. So we are very sensitive to any major changes in climate such as are all too likely to occur.

Fosberg

With reference to Mr Guppy's window-box version, I would like simply to call your attention to several books by a man named Robert Gregg who was a long-term resident of India and a student of the problems, economic and social, of that country. He recommended—and I think justified in his books—the substitution of the idea of cottage industry for industrialization. I would simply recommend that anyone interested in this problem should look up those books, which are small, very readable, and extremely interesting.

[**Polunin** (Snr)

On a not-so-jocular note which I really feel needs uttering (else I would not take even this single minute to do so), there is a tendency among many—though not our distinguished participants I feel sure—in population considerations to confuse projections with mere predictions. The oft-cited UN and other demographers' doubling of the world's human population by around the end of this century and doubling again 30 or so years thereafter, is *merely a projection* and should never be cited as a prediction. Many ecologists are convinced that it will not happen, and if it should do so the world would be one helluva place!]

Flohn

Regarding climatic variability and food production, I think it is generally agreed that we experienced, in the period between let us say 1930 and 1960, extremely small variability. Indeed this was an extraordinary period in [climatic history. On the other hand] since the early 1960s, we have experienced a lot a dramatic extremes in different parts of the world. And we have to realize that this is, at least in the last 500 years, the more or less normal situation. So we have to expect further variability, stronger variability in the future, and that means we have to expect simultaneous occurrences of droughts and other dramatic variations, resulting in variability in food production. Consequently I think we have to visualize even stronger reductions of food production in any year or perhaps two adjacent years as a very real problem.

Kuenen (Keynoter)

Although what we have been discussing latterly has little reference to the introductory papers, which [of course describe a not very] encouraging situation, we should not allow ourselves to forget [other] major threats to the biosphere [but] deplore how very little is actually being done by the politicians. They are the ones who have got to solve the problems, by *forcing* measures on people. [But at present they simply refuse to listen—shades of Paul Ehrlich's mule*.]

[At this point the recording failed completely and interventions were lost from the following, to whom our sincere apologies are offered: Worthington, Dodson Gray (David), Myers, Olindo, Dodson Gray (Elizabeth), Crabb, Kuenen, Bergthorsson, Glasser, Butler, Widman, and Pauling (who concluded the Session and then asked for written suggestions for consideration by the Resolutions Committee).]

* This refers to a story published in the first-ever issue of *Environmental Conservation*, and which we repeated so many times that we will not be guilty of doing so again here. But the moral is that many ecologists believe it will be necessary to have an ecodisaster that kills many millions of people before the politicians will pay the necessary attention to environmental problems.

Session 2

WITHER THE ATMOSPHERE AND EARTH'S CLIMATES?

Keynote Paper

Man's Increasing Impact on Climate: Atmospheric Processes

by

HERMANN FLOHN

Professor and former *Director, Meteorologisches Institut der Universität Bonn, Auf dem Hügel 20, 53 Bonn 1, West Germany.*

INTRODUCTION: ENERGETICS OF THE CLIMATIC SYSTEM

After a series of climatic anomalies from about 1968 onwards, with serious consequences on human welfare and economy, the problem of man-made or man-triggered climatic variability has reached general attention. For the time being we ought perhaps to avoid the term 'climatic change', which should be restricted to major changes (as between an ice-age and a warm interglacial period), although this is a matter of definition and of the time-scale under consideration. Rather do we prefer the term 'climatic fluctuations' for short-living (e.g. interannual) deviations, and the term 'climatic variation' for such changes as have been observed, using 30-years-averages, , since the beginning of instrumental observations—that is, after A.D. 1650.

Recent model simulations have led to a growing consensus that our climate cannot be considered as stationary. Studies in climatic history (National Academy of Sciences, 1975; WMO–ICSU Joint Organizing Committee, 1975) have revealed the fact that quite abrupt natural changes occurred between ice-ages and warm interglacial periods, evidence having been found that the transition between two quite different climatic patterns may occupy no more than a century or even less.

It is thus obvious that the more than 300-years-period of instrumental observations—first organized by the Florentine Academy and by the Royal Society of London after 1650—is much too short to show all possible excesses of climate. Consequently it is necessary to extend the records into much longer time-scales by using (and quantifying) 'proxy data'—such as weather diaries kept without instruments, notices of ice (especially sea-ice) and snow, tree-rings, lake levels, pollen profiles from bogs, and ice and sea-bottom cores with annual layering or radioactive chronology (C_{14}). The Icelandic Annals are among the most valuable of these sources (Bergthorsson, 1969).

How far is Man's increasing activity responsible (SMIC Report, 1971) for recent climatic vagaries? Has Man inadvertently interferred with the highly interactive climatic system (Schneider, 1976; WMO—ICSU Joint Organizing Committee, 1975), which consists of atmosphere, ocean, ice and snow, and soil and the biosphere (Fig. 1)? A large-scale Global Atmospheric Research Program (GARP) was organized, at first only for the purpose of extending the predictability of the weather and perhaps helping to answer these questions, but soon expanded to provide also a deeper insight into the physical basis of climate, and especially into the multitude of non-linear feedback mechanisms within the climatic system and its subsystems.

In recent years, impressive work has been done in designing physico-mathematical models of the large-scale atmospheric circulation and of climate, but still many physical interactions within the system are imperfectly understood (WMO–ICSU Joint Organizing Committee, 1975). The results of these computations should therefore only be considered as sensitivity tests of as-yet-incomplete models, not as simulations of the real climate.

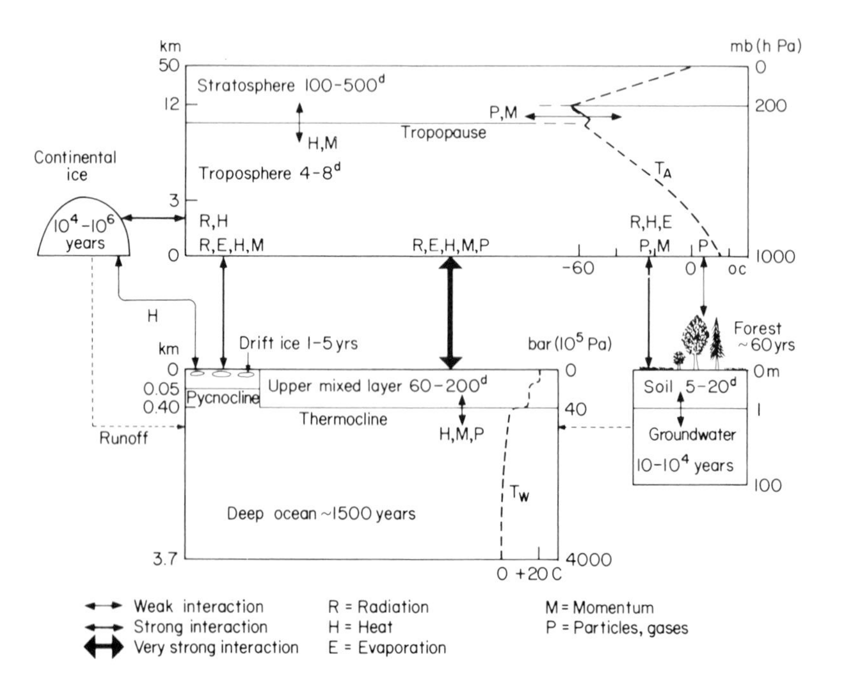

Fig. 1. Climatic system with characteristic time-scales and interactions.

In this present paper it is intended to look mainly at atmospheric and oceanic processes contributing to climatic variability, together with some consideration of the role of land-use conversion. However, due to the many interactions, some overlapping can hardly be avoided.

Energetics of Natural and Man-made Climatic Processes

A truly realistic simulation of climatic variations is now (and will remain for years to come) not possible. Thus we are forced to compare natural phenomena—such as volcanic eruptions or possible variations in solar radiation—with man-made effects (Table I) merely by estimating their

TABLE I

Causes of Global Climatic Fluctuations ('External' and 'Internal' refer to the Climatic System).

Natural causes	*'Man-made' causes*
External	
Δ solar constant	Carbon dioxide (CO_2)
orbital changes (10^4-10^5)	Fossil energy
volcanic dust	Nuclear energy
Internal	
Δ arctic sea-ice	Tropospheric dust
Δ antarctic sea-ice	(incl. man-triggered dust)
antarctic surges ($>10^4$)	Conversion of vegetation (10^2-10^3)
	(land-use, desertification)
	Impact on hydrological cycle
	(irrigation, reservoirs)

( ) = probable time-scale (in years).

energy contribution to the climatic system. As the climatic system is driven, not by the whole of solar radiation but merely by its regional (e.g. latitudinal) differences, only a small fraction of it is used in the weather- (and climate-) producing processes (Flohn, 1973*a*). These last are the production of 'available' potential energy by the pressure and temperature distribution and—equal to that—the final dissipation of the kinetic energy of the wind by friction (*see* Table II).

In comparison with that reference term, we try to estimate the energy contribution of some factors which are assumed to cause short-period climatic fluctuations (Table IIB). We consider first—disregarding some recent controversies—the stratospheric dust-veil after some of the largest volcanic eruptions, such as that of Krakatoa in 1883: a world-wide cooling effect lasts a few years, with a maximum near 1°C in the first year, equivalent to an energy-loss of about 160 terawatt (TW). Average cloudiness could change by 1–2% without being detected by surface observers or satellites; evaporation above the equatorial Pacific can be substantially increased by ceasing upwelling during an El Niño episode (as in 1972); satellite-observed snow-cover can vary, from year to year, by about 12%;

the extension of the arctic sea-ice has varied by about 20% during the last millennium.

All these 'internal' changes within the climatic system contribute (*see* Table IIB) an energy input or loss of between about 100 and 350 TW (WMO–ICSU Joint Organizing Committee, 1975), equivalent to global temperature variations of 0.6–2.0°C; such values are typical of minor fluctuations of climate as experienced during the last few centuries. They are also typical of the (still uncertain) role of solar events through modulations of the magnetosphere.

TABLE II

Natural Climatic Energy Parameters.

		Global scale Terawatt (10^{12} W)	(W/m^2)
(A)	Solar constant	173,000	340
	Net radiation, Earth's surface	52,000	102
	Production of available potential energy	1,200	2.4
	Geothermal heat	32	0.063
(B)	Photosynthetic processes	104	—
	Large-scale volcanic eruption	160	0.3
	Change of cloudiness (1%)	350	0.67
	Change of oceanic evaporation (equatorial Pacific)	~300	—
	Change of snow-cover (12%)	110	—
	Change of arctic sea-ice (20%)	100	—
	Large magnetic storms	100–200	—

Such events are the natural background against which we may judge the role of anthropogenic effects (*see* Table III) (SMIC Report, 1971; Bach, 1976). By far the largest effect is that of agricultural irrigation (Lvovich, 1969; Flohn, 1973*a*), which spans more than 2.1×10^6 km²; here the area of the reservoirs should be added—unfortunately there do not exist any statistics for the small ponds, but the larger ones cover about 0.4×10^6 km². If we estimate the total area at 2.5×10^6 km² and the additional evapotranspiration from both at 1 m² p.a.—this is a conservative figure in arid and semi-arid areas—we come to an annual water-loss of 2,500 km³ (Lvovich (1969) gives a figure of 1,700 km³, but for 1965), or the equivalent of about 150 TW.

TABLE III

Anthropogenic Climatic Energy Parameters.

	Terawatt (10^{12} W)		*Area*
Large-scale irrigation, reservoirs	~150		2.5×10^6 km²
Heat input by energy consumption	8	(15–20)	
Annual increase of CO_2	~1.6		World-wide
Tropospheric dust	~6		Mostly northern hemisphere
Savanna bush-fires, etc.	~3		~10×10^6 km²

The direct input of heat, due to consumption of fossil and nuclear fuels, is at present equivalent to 8 TW (SMIC Report, 1971; Flohn, 1973*a*). This input consists not only of 'sensible' heat, into which most other energy forms are finally converted: an increasing part of it consists of latent heat (evaporation from cooling towers, etc.), and a small part of chemical energy of compounds and of potential energy (e.g. in tall buildings). Due to the increasing concentration of population, industry and traffic (altogether less than 500,000 km^2), the greater portion of it is released in high concentrations (*see* Table IV), which are not negligible when compared with the natural net radiation—i.e. the balance between incoming and outgoing radiation at the Earth's surface, both in visible and infra-red parts of the spectrum (global average ~ 100 W/m^2, Central and Western Europe 50–70 W/m^2).

TABLE IV

Energy Consumption (EC) at Local and Regional Scales.

	Area (km^2)	*EC density* (W/m^2)	*Population density* (cap/km^2)	*EC intensity* (KW/cap)
New York, Manhattan	59	630	29,000	21.0
Moscow	878	127	7,300	16.8
Osaka–Kobe	742	56	5,800	9.7
West Berlin (building area)	234	21	9,900	2.2
Industrial Area NW Germany	10.3×10^3	10.2	1,100	8.9
'Boswash'-Megalopolis, USA	87×10^3	4.4	380	11.2
Donez Basin, USSR	53×10^3	1.4	145	9.3
Japan	366×10^3	0.71	270	2.7
14 eastern states, USA	930×10^3	1.11	97	11.6
Central + West Europe	$1{,}433 \times 10^3$	0.77	170	4.5
Africa, South America	48×10^6	0.004	11	0.4

A typical energy consumption density in large cities and industrial centres is 10–20 W/m^2, which is valid above areas of 100–1000 km^2, and in some cases (Table IV) even up to 10,000 km^2, which is still far below the synoptic scale (10^6 km^2). Thus direct heat input affects mainly the local scale (Landsberg, 1970), producing permanent urban heat-islands and, under unstable conditions, increasing convective activity. This is also true for intense-point heat sources, such as power-station parks, cooling towers or local fires. In such cases the heat input can reach 10^5 (Hamburg bombing fire, 24 July 1943, 6 hrs, surface more than 13 km^2) or even 10^6 W/m^2 (cooling tower, surface 700 m^2), with a substantial injection of water vapour.

While this heat is included, as a small but locally significant amount, in the global atmospheric heat-budget near the surface, the role of CO_2 (Bolin, 1975; WMO–ICSU Joint Organizing Committee, 1975) emanating from fossil fuels is different. Here absorption of terrestrial radiation in the 12–15

μm range leads to higher temperatures in the troposphere (and ineffectively lower temperatures in the stratosphere). This radiational warming is difficult to assess, because of its possible interaction with water vapour and its combination with dynamic processes. Taking these processes into account, a doubling of the CO_2 content alone should lead to a global warming of +2.9°C (Manabe & Wetherald, 1975). CO_2 concentration increased from about 290 ppm before massive industrialization (1880) to 327 ppm (+13%) as the most recent value (1975): this would have yielded a global warming of nearly +0.4°C, equivalent to about 60 TW. Owing to the inadequacies of the network of climatological stations—especially to the lack of long homogeneous series above large oceans and polar areas—this value is possibly not yet distinguishable as an entity. Local values up to 30 ppm higher have been observed, but due to a residence time near 6 years, the global distribution of CO_2 is—if seasonal variations are neglected—fairly uniform. An increase of 1 ppm (1–2 ppm annually in recent years) yields an energy input of 1.6 TW.

Recent investigations (Wang *et al*., 1976) have indicated that the 'greenhouse effect' of CO_2 is further enhanced by other man-made trace-gases, such as the halocarbons ('freons', with an atmospheric residence time of 40–70 years), N_2O (from chemical fertilizers), CH_4, and NH_3. If the further use of freons is prohibited, the combined warming effect of these gases will nevertheless reach about 50% of the CO_2 effect; if it cannot be prohibited, the combined effect may even double that of CO_2 alone. Due to the long residence time of these infra-red-absorbing gases and their fairly rapid mixing, they will soon take the leadership in the anthropogenic impacts on climate on a global scale.

The role of tropospheric dust has been frequently investigated with regard only to scattering—not to absorption—which then results in atmospheric cooling. This is inconsistent with the facts which indicate that nocturnal and average temperatures not only in industrial areas, but also in dust-laden arid areas, are significantly higher than in others. More recent models, including realistic absorption coefficients, correctly result in warming—at least in areas with high or normal surface albedo. The essential quantity is the ratio between the particle absorption and its backscatter; as the bulk of the aerosol particles are produced above land, where the surface albedo is higher than above sea, their warming effect predominates. Our estimate is based on the warming of a local or regional low-level dust layer (below 850 mb). The role of man-made vegetation fires, especially in the tropical savannas, should not be neglected, even if the estimates (Table III) are uncertain.

The result of all these estimates converges to a slight, but general, warming, each effect contributing essentially in the same direction. It should be mentioned, however, that the energy comparisons in Tables II and III are not quite compatible, at least regarding the efficiency time. While 'sensible' heat (and latent heat after precipitation) has only a 'residence time' of 1–2 days, due to infra-red cooling, the residence time and thus the efficiency of CO_2 and of the halocarbons is of the order of many years, i.e. 10^3-10^4 as much as that of heat (enthalpy).

It should be mentioned that Bryson (1974, 1975) still maintains a hypothesis of a predominating cooling, caused by tropospheric dust (with its residence time of 2–20 days). But most climatologists now agree to an hypothesis of increasing predominance of warming, and they give particular attention to the future role of CO_2 and other infra-red-absorbing gases.

There is no question that the impact of Man on the climatic system has now reached a level near to that of natural climatic fluctuations, and that we are on the fringe of anthropogenic climatic fluctuations on a global or at least a hemispheric scale.

MAN-MADE IMPACTS ON THE GLOBAL WATER-BUDGET

Even more effective are the man-made changes in the hydrological cycle between precipitation, evaporation and runoff. According to the most reliable estimates (Baumgartner & Reichel, 1975), the global amount of evaporation (E) and precipitation (P) is 496 × 10^3 km^3 p.a., equivalent to 973 mm p.a. E needs 76 W/m^2 or about 75% of the net radiation at the surface. At the continents P_L (E_L) yields 111 (71) × 10^3 km^3, while runoff R with 40 × 10^3 km^3 closes the budget. According to the estimate of Lvovich (1969), valid for 1965, Man uses about 2,850 km^3 or 7% of R, from which 1,800 km^3 is added to E_L. These figures (Flohn, 1973*a*) must be upgraded, over the past 10 years, by 20–30%; thus an estimated increase of E_L by 2,500 km^3 (= 3.5%) is not unrealistic. On a global scale these figures involve a redistribution of only E_L; similar figures for P_L cannot be given.

This increase of E_L, mainly from irrigation, is a fairly recent development, following an inadvertently slow reduction through many centuries. Since Neolithic time 4,000 (or more) years ago, Man has incessantly destroyed (SMIC Report, 1971) the natural vegetation of grasslands (by overgrazing), dry forests (by burning), and humid forests (by shifting cultivation), with the result that E_L has been substantially diminished. In mid-latitude forests the ratio E/P changes, after deforestation, from 52% to 42%: this loss by E_L of about 20% can be taken as a conservative minimum for tropical rain-forests (Amazon). Then more energy is available for direct heating of the air (sensible heat): deforestation means local warming, in contrast to local cooling after irrigation. Large-scale irrigation—e.g. the Punjab with 85,000 km^2, using an energy amount near to 13 TW for E_L—may thus also to some degree alter the regional climate, provided that irrigated areas lie close together and are not interspersed with arid land. An increase of E_L by 2,500 km^3 would need 150 TW of heat-energy which is no longer available as sensible heat. This is apparently above the critical threshold of natural climatic fluctuations, but due to the wide-scattered distribution of irrigated areas it would only be of local importance.

On a global scale, a change of 2,500 km^3 is insignificant (= 0.5%) compared with a global $E = P = 496 \times 10^3$ km^3 or E_S (oceans) = 425 × 10^3 km^3. The oceans will remain the great buffer of the water-budget (Flohn, 1973*a*), smoothing man-made variations on land, as long as their surface properties remain unaltered by chemical or biological pollution, and as long as the solar 'constant' remains really constant (which we still do not know). Table IIB contains a regional short-lived phenomenon, the suppression of equatorial upwelling in the Pacific during an El Niño episode: in this case solar energy is no longer used to warm the upwelling cool water and hence is available for evaporation—which has far-reaching climatic effects.

Sensitivity of the Arctic Sea-ice and Its Consequences

Taking a future warming with increasing E_L for granted (Fig. 2, cf. also Lvovich, 1969; Flohn, 1973*a*, 1977; Broecker, 1975; Schneider, 1976; Kellogg, 1976), would this not be a beneficial modification of our climate, and all the better if it could be supported and even accelerated? Unfortunately this is not the case, owing *inter alia* to the high sensitivity of the arctic air–sea–ice system (Flohn, 1973*b*; Budyko, 1974). This is demonstrated by its large variability during the past 5,000 years and its strong correlation with the position of the large-scale circulation patterns.

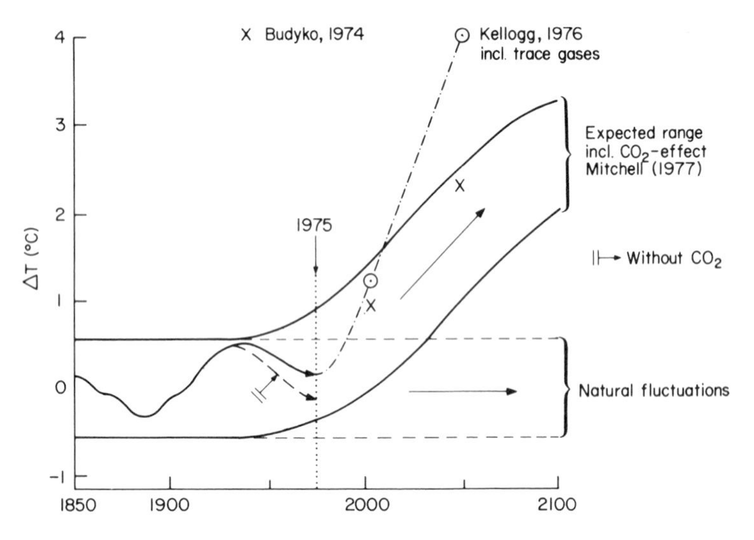

Fig. 2. Projections of global temperature (Budyko, 1974; Kellogg, 1976; Mitchell, 1977) assuming constancy of natural effects (*see* text).

The arctic sea-ice, with an average extension near 10.5 × 10⁶ km² (12 March–8 September), consists of individual ice-floes with an average lifetime (of an ice-crystal) of 5–6 years and an average thickness of 2–3 m, although varying between 0.5 and more than 6 m. This drifting ice decays from above during the melting-period (mid-June to end of August), while it grows from below during the rest of the year. This occurs in a shallow (~50 m) low-saline upper layer of the Arctic Ocean, maintained from inflow from the large freshwater rivers of Siberia and Canada. Thus the ice is very sensitive to any changes in the length and intensity of the melting- and growing-seasons, as well as to density variation in the stratified ocean.

Most probably its central core has been maintained since 1–2 million years ago.

In the northern Atlantic, the ice boundary receded before the time of the Viking expeditions (AD 800–1200) as well as 5,000 years ago up to northern Greenland, and spread southwards 200 years ago towards the Faeroe and Shetland islands. Thickness variations are documented, but doubtful; they should be correlated with the observed surface temperature variations which are 3–4 times as large as those in the mid-latitudes.

Historical evidence (Budyko, 1974; National Academy of Sciences, 1975) shows that hemispheric temperature variations of 1–1.5°C coincide with advances and retreats of the arctic sea-ice of up to 2,000 km in the Atlantic sector: this indicates its key role in the climatic evolution. If the CO_2 content of the atmosphere varies by 50% or more, together with the other above-mentioned warming effects, drastic changes in that region must be envisaged, which may perhaps even grow to a rapid and complete disappearance of the sea-ice.

Among specialists there is little doubt about the *possibility* of an ice-free Arctic Ocean: such an evolution would soon become irreversible. The high solar radiation during the polar summer would be stored in the ocean and lead to a rapid warming of surface waters and air, especially during winter. Since the temperature gradient from the Equator to the poles controls the position of the large-scale climatic belts of winds and precipitation (Flohn, 1973*a*), we would have to expect, together with an open Arctic Ocean, a shift of these belts by some 300–600 km to the north in the Northern Hemisphere, with severe consequences for the water-budget of densely-populated areas. Large increases of snowfall and snow-cover would also be expected along the relatively dry coasts of the Arctic Ocean, including northern Greenland and the Canadian Arctic Archipelago.

It should be added that any substantial diversion of fresh water from rivers running into the Arctic Ocean, for irrigation or other purposes, would accelerate this melting of the ice, due to the increasing salinity and density of the shallow upper layer of sea water.

ROLE OF LAND-USE CONVERSION

It has been pointed out (Sawyer, 1965) that large-scale weather phenomena are produced by differential heating only when this occurs on a 'synoptic' scale, i.e. over closed areas with a magnitude of several 10^5 or 10^6 km^2, and when the heating function varies locally by about 20 W/m^2 or more. This excludes local point-sources, such as power-stations or 'power parks': their effect is restricted to the local scale, e.g. to the frequency and/or intensity of showers and thunderstorms.

Inadvertently, Man has changed the surface conditions of the Earth through many millennia to a very considerable extent—by conversion of forests into arable land, by conversion of natural grasslands into pastures, and by many processes of devegetation. The need for firewood is, in many hitherto underdeveloped countries, one of the most powerful bases for this steady and often irreversible process, which extends and intensifies with increasing pressure from a rising population.

Estimating the areas which have been affected by this process since the beginning of agriculture, a total of 45×10^6 km^2 or about 30% of the surface of the continents is still conservative. The process had been started even earlier, when Man learned to change his environment with the use of fire, which may already have happened for hunting purposes in the Palæolithic age.

The destruction of the natural vegetation leads nearly everywhere to an increase in the reflectivity (albedo) of the surface and thus decreases the amount of (absorbed) solar radiation. Two examples may be given: the typical albedo of a tropical rain-forest is 0.12–0.14, while that of a humid grassland or cropland is 0.20–0.22; a typical albedo of a green (yellow) steppe is 0.20 (0.25), while that of a desert with bright, sandy soil ranges around 0.35. These values are given for the visible part of the spectrum (0.35–0.7 μm); in the near-infra-red (0.7–3 μm) part even higher values have been observed (Otterman, 1974).

Man's activity, e.g. protection of large farmland or grassland areas against herds of nomadic tribes, can thus create significant horizontal differences of the heat-budget. In some cases the bright surface of a desert may even be cooler than the darker surface of a desertic steppe; but one has to take into account also the local change in the evaporation, as indicated earlier in this paper. In a partially humid climate, any decrease of evapotranspiration of living plants is replaced by an increase in direct heating of the air (flux of 'sensible' heat from surface to air).

The role of surface albedo has been best demonstrated in the radiative model of Manabe & Wetherald (1967): increased surface albedo leads to a (nearly linear) decrease of the surface equilibrium temperature (about −1.2°C with an albedo increase of 0.01). The large-scale role of the surface albedo has been strikingly evidenced by satellite data (Raschke *et al.*, 1973): during July the Sahara and neighbouring deserts act as one of the main heat-sinks (cold sources!) of the northern hemisphere—similarly to the deserts of the southern continents during the southern summer. This unexpected, but convincingly verified, result motivated Charney (1975) to a model experiment, in which he could show that a drastic albedo increase results in a likewise drastic reduction of rainfall (which is prevented by increasing subsidence and heat-import necessary to replace the loss of energy to space). Because of the great significance of this result, it has been checked and verified by using several other models for the general (or local) circulation of the atmosphere (e.g. Berkofsky, 1976).

This process aggravates the existing circulation above the great deserts of the globe, in relation to the increasing man-made desertification process (Hare, 1976, 1977). It may have contributed, to some extent, to the very gradual desiccation of the margins of the desert since the last moist period, which ended around 4,000 years before present, i.e. after the beginning of the great civilizations of the past, such as the Old Empire of Egypt and the Indus Culture. However, this quite slowly spreading process of desertification has been overshadowed by the much larger natural climatic variations of the past; now it increases and accelerates practically before our eyes, intensified by the increase of population and its herds and also by some technological developments. The recent Sahel drought (1968–73) was basically a consequence of large-scale circulation anomalies (of natural origin), similar to earlier droughts (1941–43, 1908–13), and in the nineteenth and eighteenth centuries).

In a large-scale comparison, the Man-triggered degradation of the natural vegetation has locally quite different, even opposing, consequences (*see* above). These include:

(1) Reduction of absorbed solar energy due to higher albedo: cooling.
(2) Reduction of plant evapotranspiration: drying.
(3) Replacement of flux of latent heat by flux of 'sensible' heat: warming.

It should be mentioned that, under certain circumstances (except in mountains), a reversal of such degradation is possible, if strict protection of the vegetation is maintained; examples at average rainfalls of 70 mm (southern Tunisia), 150 mm (near Khartoum) and 250 mm (near Jodhpur), demonstrate this capability for recovery. Construction of large reservoirs and conversion of arid lands by irrigation leads to increased evapotranspiration and to (local!) cooling.

One of the largest conversion projects is now being carried out in the Amazon basin. In addition to the ecological consequences it should be mentioned that, from the climatological point of view, the area is particularly sensitive to such environmental changes: for the high rainfall is maintained by a large-scale convergence of water-vapour transport together with high evapotranspiration and the orographic barrier of the Andes. The regional circulation leads, during the southern summer, to a permanent high-level anticyclone (Kreuels *et al.*, 1975). This whole system might be altered if the regional evapotranspiration over an area of nearly 10×10^6 km^2 is drastically reduced; a significant decrease of rainfall and runoff of this (semi-closed) system cannot be excluded.

Until the beginning of our century, such changes of the land-use pattern went at a comparatively slow pace; with increasing population, especially in developing countries, and modern technology, the rate of change increases rapidly. As an example: the tropical rain-forest of Ivory Coast diminished from 1954–57 to 1966 from 75% to 53.6% (in absolute figures, by 28,000 km^2 in 10 years, according to Synnott, 1977).

Conclusions: A Scenario

Any attempt to speculate about the climatic evolution during the next century must be based on the assumption that the unpredictable natural causes of climatic variation remain without major impact. These are (Flohn, 1973*b*):

(a) constancy of the solar 'constant';
(b) no unusual frequency and/or clustering of heavy volcanic eruptions;
(c) no unusual advance of the antarctic ice-shelf; and
(d) no significant variation of the average global cloudiness.

The present situation in the field of climate modelling, and the multitude of (mostly non-linear) feedback mechanisms within the climatic system, preclude an early solution to problems concerning the prediction of climatic variations, even if we accept the above-mentioned assumptions without further discussion. In addition to this, the growth-rates of energy consumption, and of the CO_2 content of the atmosphere and likewise of other trace-gases, depend on many social and economic developments and on political decisions: they are also largely unpredictable.

Under such conditions we can only try to imagine what would happen if a further increase (not necessarily exponential) in man-made effects, due to increased human population, were extrapolated. This would give us a scenario leading into the twenty-first century, with its speed depending on the future political, economic and social evolution—cf. Budyko (1974), Broecker (1975), Kellogg (1976), Flohn (1977), Mitchell (1976, 1977), and WMO Technical Committee Panel of Experts (1977).

We should expect, around (or slightly before) the turn of the century, a climatic situation similar to the period 1930–60, with its unusually high temperatures. Later on we may expect a climate like that of the early Middle Ages (Viking period) or that of the 'climatic optimum' of about 6,000 BP: at least this evolution is much more likely than a return to the 'Little Ice-age' of between 1600 and 1850, with its extreme variability. Intensive studies of the historic patterns of climate in that period are urgently needed: whatever happened may indeed happen again (Bryson, 1974, 1975).

The last stage of this evolution—probably not before mid-century—might be the irreversible transition to an ice-free Arctic Ocean, with its unimaginable consequences on climate and economy, and probably in a short time (such as a few decades). According to the indications of deep-sea cores, such a pattern (with a very different climate in the Arctic) has not occurred in the past at least 150,000 years, and most probably not in the past 1–2 million years (National Academy of Sciences, 1975).

Any reasonable estimate of the climatic pattern that would accompany an ice-free Arctic Ocean can only be obtained from model studies, which have yielded up to now quite different results. Adaptation of a world population of 10–15 thousand millions (10^9) to such an unprecedented, true climatic catastrophe—especially regarding fresh water and food supply—seems inconceivable. Such an evolution could be avoided only if the energy problem could be controlled at an international level, and if the mesmerizing idea of unlimited growth—which necessarily leads to overshooting and disaster—can be overcome by an acceptable compromise. It is our generation which bears responsibility for a global-scale problem that will be facing our grandchildren: let us take care to match the challenge, and remember that there is all too little time left.

References

BACH, W. (1976). Global air pollution and climatic change. *Review of Geophysics and Space Physics*, **14**, pp. 42–74.
BAUMGARTNER, A. & REICHEL, E. (1975). *The World Water Balance*. R. Oldenbourg, Munich: 179 pp.
BERGTHORSSON, P. (1969). *Jökull*, **19**, pp. 94–101.
BERKOFSKY, L. (1976). The effect of variable surface albedo on the atmospheric circulation in desert regions. *Journ. Appl. Meteor.*, **15**, pp. 1139–44.
BOLIN, B. (1975). *Energy and Climate*. Secretariat for Future Studies. Stockholm: 55 pp.
BROECKER, W. (1975). Climatic change: Are we on the brink of a pronounced global warming? *Science*, **189**, pp. 460–3.
BRYSON, R. A. (1974). A perspective on climatic change. *Science*, **184**, pp. 753–60.
BRYSON, R. A. (1975). The lessons of climatic history. *Environmental Conservation*, **2**(3), pp. 163–70, 8 figs.
BUDYKO, M. J. (1974). *Izmenenija Klimata*. Gidrometeoizdat. Leningrad: 260 pp.
CHARNEY, J. (1975). Dynamics of deserts and drought in the Sahel. *Quart. Journ. Roy. Meteor. Soc.*, **101**, pp. 193–202.
FLOHN, H. (1973*a*). Der Wasserhandhalt der Erde: Schwankungen und Eingriffe. *Naturwissenschaften*, **60**, pp. 340–8.
FLOHN, H. (1973*b*). Globale Energiebilanz und Klimaschwankungen. *Bonner Meteor. Abhandl.*, **19**, 43 pp.
FLOHN, H. (1977). Climate and energy: A scenario to a 21st century problem. *Climatic Change*, **1**, pp. 82–8.
HARE, F. K. (1976). *Climate and Desertification*. Component Review for the United Nations Conference on Desertification. Institute for Environmental Studies, Toronto, Canada: 189 pp., illustr. (mimeogr.).
HARE, F. K. (1977). Connections between climate and desertification. *Environmental Conservation*, **4**(2), pp. 81–90, 5 figs.
KELLOGG, W. W. (1976). Effects of Human Activities on Global Climate. *WMO Bulletin*, **26**, pp. 229–40 and **27**, pp. 3–10.
KREUELS, R., FRAEDRICH, K. & RUPRECHT, E. (1975). An aerological climatology of South America. *Meteor. Rundsch.*, **28**, pp. 17–24.
LANDSBERG, H. (1970). *Science*, **170**, pp. 1265–74.
LVOVICH, M. J. (1969). *Water Resources for the Future* [in Russian]. Gidrometeoizdat, Moscow.
MANABE, S. & WETHERALD, R. T. (1967). Thermal equilibrium of the atmosphere with a given distribution of relative humidity. *Journ. Atmos. Sci.*, **24**, pp. 241–59.
MANABE, S. & WETHERALD, R. T. (1975). The effects of doubling the CO_2 concentration on the climate of a general circulation model. *Journ. Atmos. Sci.*, **32**, pp. 3–15.
MITCHELL, J. MURRAY, Jr. (1976). *Quaternary Research*, **6**, pp. 481–93.
MITCHELL, J. MURRAY, Jr. (1977). *Environmental Data Service Magazine*, March 1977.
NATIONAL ACADEMY OF SCIENCES (1975). *Understanding Climatic Change*. National Academy of Sciences, Washington, DC: xv + 239 pp., 52 figs.
OTTERMAN, J. (1974). Baring high-albedo soils by overgrazing: A hypothesized desertification mechanism. *Science*, **186**, pp. 531–3.
RASCHKE, E. *et al.* (1973). The annual radiation balance of the earth–atmosphere system during 1969–70 from Nimbus-3 measurements. *Journ. Atmos. Sci.*, **30**, pp. 341–64.

SAWYER, J. S. (1965). Notes on the possible physical causes of long-term weather anomalies. *WMO Technical Note*, **66**, pp. 227–48.
SCHNEIDER, S. (1976). *The Genesis Strategy*. Plenum Press, New York and London: xxi + 419 pp., 38 figs.
SMIC REPORT (1971). *Inadvertent Climate Modification. Report of the Study of Man's Impact on Climate (SMIC)*. MIT Press, Cambridge, Massachusetts: 308 pp.
SYNNOTT, T. J. (1977). *Monitoring Tropical Forests: A Review with Special Reference to Africa*. Monitoring and Assessment Centre of SCOPE, Report No. 5.
WANG, W. C. *et al*. (1976). Greenhouse effects due to man-made perturbation of trace gases. *Science*, **194**, pp. 685–90.
WMO TECHNICAL COMMITTEE PANEL OF EXPERTS (1977). Technical Report. *WMO Bulletin*, **26**, pp. 50–5.
WMO–ICSU JOINT ORGANIZING COMMITTEE (1975). *The Physical Basis of Climate and Climate Modelling*. GARP Publication Series No. 16, Geneva: xxiii + 265 pp., 96 figs.

DISCUSSION (Session 2)

Malone (Chairman, introducing)

We are particularly happy to welcome [as our principal speaker in this Session] one of the most distinguished contributors to climatology in the world, Professor Hermann Flohn, until very recently Director of the Meteorological Institute of the University of Bonn, Germany, who will address us on 'Man's Increasing Impact on Climate'. As panellists we have first the [outstanding] Canadian Environmentalist and Climatologist, Professor F. Kenneth Hare, now of the University of Toronto. As our second panellist/discussant we have—very naturally from our host country in view of its [prominence in weather reports]—the man who in a sense is responsible for them from here, namely the Director of the Icelandic Meteorological Office (Vedurstofa), Hlynur Sigtryggsson. As our third panellist we have an old friend from the University of Wisconsin, Professor Reid A. Bryson, the Director of their Institute for Environmental Studies, and I take the liberty of remarking that soon on the bookstands there will be a book, *Climates of Hunger*, by Bryson & Thomas J. Murray, published by the University of Wisconsin Press, that is very much apropos of our topic this morning.

I have two announcements to make on behalf of our Secretary-General, who apologizes for being called away on urgent business. For those of you who did not get the full text of Professor Flohn's paper—there was some problem in collating, due basically to the general strike—plentiful copies are now available and our Conference Secretary, Miss Angela Gibbs, will pass along and if you want one, please hold up your hand and she will come down the aisle and give you a full text of the complete, unabridged Flohn. The second announcement is that, as in our opening session, rather than sending notes of questions or expected times of interventions up to the Chairman, if you will please raise your hand and then wait for Mr Polunin Junior with the microphone to come over to you, we can get your name and message recorded and save you the problem of writing it and it will then be preserved for posterity. [This is now envisaged as the best procedure for the rest of the Conference as set by our President for the first session.]

By way of prefatory remarks I might say that Professor Flohn himself, in his brief intervention earlier this morning, advanced the crucial consideration that not only stress between Man and environment, but also the perception of stress between Man and his environment, has taken place precisely during a period of abnormally benign climate, so that we can expect the stress to be aggravated rather than attenuated.

The timeliness of this Session's topic, of 'Whither the Atmosphere and Earth's Climates?', results from a series of circumstances of which the following stand out particularly: (1) We now have for the first time in the history of meteorology a global observational technological capability. (2) We have the computational technology to handle these data. (3) Our understanding of the physical processes in the atmosphere has grown explosively during the past four decades. Moreover new conceptual formulations are beginning to appear which make it possible to employ the methods of statistical sampling and estimating theory in deducing the relevant properties of a non-linear dynamical system, so that the levels of climatic 'noise' can be identified and internal behaviour distinguished from external forcing functions.

In addition to the above three chosen circumstances we now have: (4) proxy data obtained from ice-coring and ocean-bottom coring, as well as fossil pollen counts, which have permitted us to reconstruct past climatic variability, while (5) impressive strides have been made in setting forth the hypotheses which relate climatic variability to orbital characteristics of the Earth and Sun, and (6) the elements of a scientific strategy for a global research assault on this problem have begun to emerge from the series of national and international conferences that have been held during the past few years. Finally (7) it is becoming increasingly clear that human activity is reaching a dimension which poses the potential for human interference in an inadvertent manner on climate, ranging from local areas to the entire globe. But that is really what Professor Flohn is telling us about.

Flohn (Keynoter, remarks in presenting paper)

It is a great honour for me to speak here of my paper in an abridged way, and to introduce it by saying that in recent years we have come to understand a lot of things which we did not understand when I started my own meteorological education, more than 40 years ago. The first of these is illustrated in my Fig. 1, namely that the climate depends on what we are now calling a *climatic system*, which consists of many subsystems that are *per se* quite different, each extending over a very large area and being very difficult to handle. This is because, within them, non-linear actions and feedback mechanisms having a positive or negative direction are going on persistently. Perhaps the most important and the most difficult thing to countenance is that these subsystems have 'memories' which are apt to be quite different in their time-scales, ranging from a few days within the atmosphere to several hundred thousand years with the great continental ice-sheets—particularly of the Antarctic and of Greenland. This means that we have to deal with all these systems together; and this aim, indeed, has not been reached yet. We have now a rather good perception of the physical processes that are going on, but to take them together with the variety of time-scales and space-scales involved goes far beyond the limit of the present computers and our ability to handle such a system quantitatively.

The second point is that the causes of climatic variability which we have seen, as Dr Malone pointed out, in the history of the last at least 100,000 years [though he did not mention a figure], we now know with very much more precision than some time ago, though the causes are very various. Thus we have natural causes which are, practically, not predictable, at least at the present time; and if you look at my Fig. 2 [Table I in this book] you will see some of them—including the solar constant, though we still do not know if it really is a constant. We may also deal with orbital changes; but as they are on a very long time-scale of some ten thousand years and upwards, these are uninteresting for us. We also have to deal with individual events of volcanic activity which can cluster to such an extent that the global heat-balance of the Earth and its atmosphere is greatly disturbed. Indeed most of us believe that this is the main origin of the climatic variability in the past, at least on short time-scales of the order of a few tens or maybe some hundreds of years.

Then we have a lot of internal changes within the climatic system, though here I should add an item which is so familiar to most of us that I have not included it in my figure, namely the air–sea interaction, which is in fact the most prominent of all these internal feedback mechanisms. But I did include the [so-called] antarctic surges—a little-known new item for research, which only a few specialists are aware of. Although apparently the time-scale in which these events occur is a very long one, we have insufficient insight into their mechanisms to say anything more than that.

The questions of the constancy of the solar 'constant', of the volcanic events, and of these antarctic surges—indeed of all of these natural causes of climatic changes—are now, and will remain for some time, unpredictable. We can only follow what has happened in the past; and as Professor Bryson has frequently said, what has happened in the past may happen in the future. So the past is the key to the future, and we must understand it—this is one of the prominent tasks of research.

In addition to the natural causes we have now the man-made causes of climatic fluctuations. And as Dr Malone has pointed out, we are just at the fringe of where these man-made changes reach the level that the natural changes have reached. Looking into the energy contribution of these natural changes, it is quite clear that the energy level of the climatic fluctuations in the past, caused by natural forces, are of the order of 100–300 terawatts (each of 10^{12} W), which is fairly small compared with the solar constant, but which is relatively large compared with the production of potential energy and the destruction of kinetic energy, namely with those energy sources that are driving our weather and ultimately our climate. This is of the order of 1,000 TW only. So we are here dealing with between about 10 and 30% of the amount of the forces within our weather and climate. With one exception, all man-made changes are small compared with that; but they are increasing, and they may reach rather soon the level of 100–300 TW. The most likely value of the sum of all these man-made influences is now of the order of between 15 and 20 TW, but there is a general feeling that this is the lower level of estimate, and that the correct figure may be perhaps much larger. One has to deal here also with the time-constant, but I cannot go into all the details.

The above-mentioned exception is, of course, the increase in the carbon dioxide content of the atmosphere. This increase is, from the beginning of the industrial age, up to now of the order of 13%, and we expect it will rise by the end of the century maybe to something like 25 or even 30% over the original value—reaching about 400 ppm instead of [the 'basic'] 290 at the beginning of the industrial era [and for quite some time thereafter].

Nobody can say whether this value will be reached at the year 2000 or 2010; but if economic growth and the use of fossil fuels goes on increasing, it will be unavoidable. Indeed it is the general feeling of climatologists that this question of the increase of the carbon dioxide content of the atmosphere and the concomitant so-called 'greenhouse effect' of warming the Earth's surface, is the key question of climatic change. But I must say that in addition to this we have also some other sources, even if it is not worth while to deal here with tropospheric dust, as it can be controlled to some extent at an industrial level. Indeed countries such as the United Kingdom and Japan and my own country, Germany, are more or less successful in this respect.

Rather should we deal with another problem which has already been stressed here very aptly by Professor Kuenen, as well as by our Secretary-General elsewhere, namely the conversion of natural vegetation into croplands, including grasslands very widely into arable lands, and the continuing devastating deforestation which I personally feel poses one of the most vital of all threats. I should emphasize that the destruction of the natural vegetation and its replacement by man-made vegetation types (including the processes of desertification) which is going on all the time under our very eyes, is extremely serious also from the meteorologist's viewpoint. For all these processes lead generally to an increase in the surface albedo, and that means to a change in the heat-budget of the Earth.

Regarding carbon dioxide, we have to deal with several storages or 'sinks', and one of these is the forests. We have latterly come to think that the storage capacity of the forests is not large enough in view of their rate of destruction, and that we should assume that the forests are no longer a sink of fossil carbon dioxide but actually a source of more carbon dioxide. What we do not know sufficiently is the interaction between the atmosphere, the ocean, and especially the upper mixed layer of the ocean and the deep ocean. However, the general idea of all specialists nowadays is that the exchange processes between the mixed upper layer and the deep ocean are much less active than we had assumed before, which means that practically only an exchange between the atmosphere and the mixed layer of the ocean is occurring. Thus the storage capacity of the oceans which store now something like 50% of all carbon dioxide will be limited, and will be further limited by any ongoing acidification and warming of the ocean; so we have to expect that in due course even the ocean may be a source of carbon dioxide because of the slow exchange with the vast depths of the ocean.

Projections of the future concentration of carbon dioxide in the atmosphere show that we are now slightly above 300 ppm and should countenance the possibility of something like a doubling [of this level]. If we start from the assumption that the use of fossil energy will be reduced immediately, then we expect reduction of the carbon dioxide input into the air. But this is quite unrealistic, as it assumes that control of economic activity and energy use can be effected on an international level with an optimum strategy. All other cases indicate that perhaps in the second half of the next century we have to expect an increase of atmospheric carbon dioxide not only by a factor of 2 but by a factor of 5 or perhaps even 8, which would have a very horrible effect from the viewpoint of climatic evolution.

We have lots of models describing the relationship between carbon dioxide increase and temperatures. The best is probably that of Manabe & Wetherald, published in 1975, and taking into account the dynamics of the atmosphere and also parts of the interaction with the ocean. It indicates that, with a doubling of the carbon dioxide content of the stratosphere, we would have some cooling which would be relatively modest in the lower latitudes. But in the polar and sub-polar regions there could be an increase of up to 10° Celsius. This would mean a complete change of the present situation and also of the climate which we have experienced in the last 100,000 years—with the assumption of a mere doubling of the carbon dioxide content.

What aggravates the situation is that quite recently we have found that there are several other trace-gases besides carbon dioxide which absorb in the infra-red and thus increase the 'greenhouse effect' of the carbon dioxide. Among those are nitrous oxide as a product of denitrification and of the use of fertilizers. Then there are the fluorocarbons or freons which are so frequently discussed nowadays; their present concentration is rather small but their lifetime is very long, being of the order of 30–60 years. So we have to expect a dramatic effect in addition to that of the carbon dioxide; indeed all these effects together may double the effect of carbon dioxide in the next few decades. Yet the most essential question in that respect, from the climatological point of view, is what would happen to the Arctic and Subarctic, and so I am very glad to have this Conference right here, where many data are available since the early settlements of the Vikings in Iceland, indicating the dramatic changes of climate in this particular region of the world.

We can only make useful projections on the basis of assumptions that the unpredictable, natural causes remain more or less constant. This would mean that there would be no change of the solar constant, no major clustering of volcanic activity, no antarctic surge, and—what is quite essential from the climatological point of view—no major change in the cloudiness of the Earth. Under those circumstances we can produce scenarios and conclude that a marked change of the polar climate, with a marked retreat of the arctic ice to such a level as it reached either a thousand years ago or maybe about 2 million years ago, before the beginning of the Quaternary ice-age, with a completely ice-free Arctic, would quite dramatically change the position of the climatic zones of the Earth—including the rainfall belts, the supply of fresh water, and of course the temperature. This is what we are visualizing; and indeed it is not impossible, if we are unable to control further economic growth and the use of fossil fuels for energy, that we shall have to expect a true climatic catastrophe in the form of a sudden transition to a completely different kind of climate.

We know—and I say this in all seriousness—of at least six and probably eight examples during the last 200,000 years of very sudden changes of climate occurring in the time-scale of from one generation to one century. So it is by no means new that such a sudden change could occur. Hence my impression—and I know that I speak for many of the climatologists who are studying the problem—that such an evolution could be avoided only if the energy problem can be controlled, which would also mean curbing the basic idea of unlimited growth.

Malone (Chairman)

Thank you, Professor Flohn. In case there is any doubt, I think the question of what an ecodisaster might be has been answered [with a superb example], though I would stress that Professor Flohn was careful to point out that it need not happen.

Hare (Panellist)

It is always an instructive pleasure to listen to Professor Flohn, and I have no desire to challenge this apocalyptic vision that he has put before you. We do indeed lead scary lives these days, and I think that it is particularly appropriate that this Conference should be taking place in the country which is probably the most vulnerable in the world, climatologically speaking. Iceland is in the very core of the region where natural climatic change has maximum amplitude, and it has been the country from which we have learned most about the past thousand years and the vulnerability of the human economy to climatic events. So it is entirely suitable that we should meet here, quite apart from the fact that it happens to be my favourite country.

I would like to confine my own remarks to calling attention to the professional weakness of meteorology, and I hope that economists and ecologists present will consider whether this also applied to their own disciplines. We were caught napping by the events—by the processes that Professor Flohn referred to. We neglected atmospheric chemistry to an astonishing extent; yet most of the significant changes that Professor Flohn is talking about are the work of trace constituents, in some cases present in the atmosphere to concentrations of less than 1 part per thousand million. None the less, those trace constituents turn out to be keys to the behaviour of the atmosphere in a thermal sense. We neglected this completely; it was not in our education. It still is not in the professional curriculum of most meteorological training schools. One major research organization of international scope not long ago deliberately dismantled much of its programme in atmospheric chemistry, apparently regarding it as irrelevant to the situation!

Secondly, the orthodoxy of the professions to which all of us sitting here in front belong has essentially been confined to one particular discipline; we meteorologists have taken the viewpoint of the physicist. This has turned out to be, as Professor Flohn's remarks indicate, absolutely inadequate to an understanding of the problem that we are confronted with. I think we have done this, not only because of our limitations, but also because physics as a discipline tends to create in the mind a certain arrogance. Physics is thought of (by physicists) as being the reduction of all problems to the simplest possible set of generalizations. Unfortunately the things we are discussing in this room are not susceptible to the simplest kind of generalizations, as this morning's session demonstrated.

So I think we have been wearing blinkers, and one of the transformations that we have to achieve—immediately, as I fully accept Professor Flohn's statement—is to solve our problems in the next 100 years or else [we shall have little if any chance of solving] them at all. One of the things we have to achieve is a fundamental re-ordering of our own discipline. I would feel a traitor to my colleagues for saying such a thing in this particular gathering, if it were not for the fact that I am going to be saying precisely the same thing in a lot more detail at 9 o'clock next Monday morning before the Executive Committee of the World Meteorological Organization. Physics is not enough. But then, all of us—whether we are ecologists or economists [or some other category of '-ists']—have found ourselves caught up in what it is fashionable to call a paradigm, or let us say just simply a way of thinking, which has turned out to be fundamentally inadequate. If this Conference achieves nothing else, I hope it will produce some other confessions of inadequacy from representatives of some other disciplines.

So with that pessimistic—or, rather, realistic—note, allow me to give way to the next discussant.

Sigtryggsson (Panellist)

My remarks will probably be short and perhaps not entirely to the point, but I must say that I am very much impressed by this rather sudden burst of knowledge which I have experienced, during the account just given, and its impact. As Director of a national meteorological service which must maintain a network of meteorological stations by means of which minute climatological changes have to be detected, I am somewhat concerned about the problems of observations which beset the theories we are discussing. In fact we are aware of errors and variations in the observations that may be of the same order of magnitude as the changes which we are observing. And these errors may be not at all at random. They probably affect both temperature and especially precipitation observations and reports, which in a windy country such as Iceland, with frequent snow, are apt to be too low by some 30–50% or even more. The inadequacy of these observations over oceans is also easily established. Efforts should of course be made to remedy these shortcomings. The existing theories are inadequate to explain even recent climatic changes. The change for the worse around AD 1600 remains unexplained, and so does the change for the better in the 1920s. I expect that we do have adequate theories for the effect of CO_2 and probably for the effects of the carbohalides or fluorocarbons also; but when we impose these effects on something that we cannot predict, it becomes extremely difficult to verify the predictions. That has to be done on the basis of the theory, which is also not a very satisfactory method.

Speaking about climatic changes in relation to Iceland, we have, during our history, experienced either the arctic climate, which has not been very pleasant for us, or the cool-temperate climate, which has been quite tolerable—apart from some hardships.

When I think about the future, and compare it with the pre-settlement climate of Iceland which was very good indeed around 3,000 to 4,000 years ago, I must say that I am less worried about the predictions of a warmer climate as just expressed by Professor Flohn—even if the consequences would be serious in some parts of the world—than about the possibilities of a colder climate as expressed at times by Professor Bryson. But this is of course rather egoistic. We must also think about the rest of the world; but I leave the expression of such thoughts to others.

[Climatic variations are now a well-established fact, but cannot be predicted as yet—neither with regard to time nor amplitude. The national and world economies seem to adapt fairly rapidly to improving conditions, with unpleasant consequences when conditions deteriorate. In view of the existence and non-predictability of climatic variations, it would seem advisable to make the adjustments of the economy to improve conditions more slowly than is done at present, and never to rely on optimum conditions continuing for any length of time.

[The World Meteorological Organization has the responsibility of encouraging and coordinating the national programmes of atmospheric observations, together with their exchange and processing on a regional and ultimately world-wide basis.

Thus it accomplishes a World Weather Watch.

[Many of the activities of WMO relate directly to this and other topics which we are discussing, while several of the resolutions passed by the WMO Congress in 1975 have a bearing on these matters, being concerned with: (1) WMO activities in the field of environmental pollution, (2) participation of WMO in the UNESCO 'Man and the Biosphere' programme, (3) the UN Environment Programme, and (4) climatic change. More recently the WMO Executive Committee at its session in May–June 1976 passed resolutions on (5) WMO activities related to energy problems, and (6) marine pollution.

[It should be evident from this selection of resolutions that WMO is actively engaged in furtherance and administration related to several of the main topics of this Conference. As an observer of WMO at this Conference and the Permanent Representative of Iceland to WMO, I am grateful for this opportunity to gain further insight into the problems which have been so ably discussed at this Conference and into their possible solution.]

Bryson (Panellist)

This meeting started with some reference to the future of the environment and with some comments on economics, especially with reference to the important role of economics and social studies. Now here on the platform we have some meteorologists/climatologists and atmospheric scientists, and you all know that meteorologists are forecasters. And you *all* know, as my colleague here admitted, that we are not very good forecasters. But there is one group that we, as professionals, really love, because we can beat them at forecasting, and that is the economists. We have not yet learned how to forecast at all well, but we have learned a great deal about what will not work—what will *not* give a good forecast. One thing that we know will not work in forecasting is to assume that the future history of some variable in which we are interested depends only on the past history of that variable and nothing else. The future of the economy, measured in dollars, krónur, or whatever, can not be predicted on the basis of the past history of dollars, krónur, or whatever, *alone*. The future history of the temperature cannot be predicted on the basis of the past history of the temperature alone. There are other factors involved. I suppose you might say that is what ecology is all about—the interconnectedness of the whole world and the idea that the course of one variable does affect the course of another.

We have had a discussion this morning on the future build-up of atmospheric carbon dioxide and its impact on climate. The future course of carbon dioxide is quite clearly associated with the future course of the consumption of fossil fuels, which in turn is very tightly linked to the gross world product, economically speaking. So here we are talking about an ecodisaster—if you want to call a sudden warming of the Earth an ecodisaster—but it is related to a resource crisis which is related to an economic disaster which is related to many other things. One usually thinks about putting such interlinked systems on a computer, but unfortunately computers do not think. They are very fast but very stupid. I believe that this consideration provides a call for all of us, as Professor Hare suggested, to reach beyond the confines of our own disciplines, to look at the messages of other disciplines, and to include those in a creative, individually thinking, sense. Then let the computers do the number-smashing. But let us put our creativity together soon because, as Professor Flohn has emphasized, the time is short.

In the few seconds I have left, and changing the subject slightly, I wish to defend myself. It has been said by my colleague from Iceland that the carbon dioxide increase suggests a warming, but Bryson has said 'It is going to get colder'. Please, I have never said that. What I have said is that, 'If certain things happen, then the northern hemisphere will get colder; if other things happen, it will get warmer'. But let me add one other caution here. Ecodisasters do not have to be dramatic. I have studied the past history of climate for 30 years. For a number of years scholars in general did not regard climate as being variable, because they did not find those small events in the past which had wiped out whole civilizations. When once we had developed some techniques, so that we could find these small changes in the past, we could see that it really did not take very much change globally or hemispherically to shift the rainfall patterns of the Earth so that whole cultures would disappear. Then it became fairly obvious that there have been dramatic effects of environmental change where the change itself was so hidden, so insidious, that the people at the time were not aware of it. There are some cases, of course, where past civilizations were aware of climatic changes. The Hittites said: 'Let's get out of here, it's getting too dry.' They were aware of it. This kind of thing has happened.

Now let me give you just a slight example with specific numbers. We have talked about small changes, such as half-a-degree, not being terribly important, but concluded that a big change would come if certain things happened. However, a 1°C change in the temperature this coming summer in the north-central United States will affect the gross income of the spring-wheat farmers (of whom there are not a very large number) by about a $130m dollars. When you consider how [relatively few] of them there are, this means a significantly different net income. Looked at on a world-wide basis, where the margin between feeding the world or not is of the order of a few per cent, little changes like this can be significant changes. They do not occur overnight but they occur at just the wrong rate—slow enough that the next generation of politicians do not have to answer for them, but fast enough that cultures do not usually cope. Let us put our brains together, transcend the disciplines, and face these problems in a creative way.

Buchinger

I would like to give you a little example [from a remote part of South America] to show how right Professors Flohn and Bryson are about the need for close study before action. Those of you who know the Andes will be aware that, where the plains meet the mountains, the precipitation may differ over a short distance from 200 mm to 4,000 mm per year. Now in the 200 mm region they were cultivating plants which were adapted to the dry climate, with irrigation. Then to improve the economy of the region a big reservoir was built which was supposed to provide energy and also plentiful water for irrigation. This reservoir was unfortunately put into the 200 mm precipitation area and, as you can well imagine, there is now much more water available and also precipitation, with lots of storms, and those plants which were adapted to the dry climate are absolutely incapable of growing there, so the entire economy has to be changed.

Myers

It seems to me that there we have a classic instance of a potentially significant ecological non-event. It is not so much a case of one side wins and the other loses. In the long run, or possibly the short run, we shall all either win or lose. But I would like to put this question to the panellists and especially to Professor Bryson, who I believe has occasion from time to time to take some of his findings and recommendations to Washington. I would like to know what happens when you get there, Professor Bryson: is the administrative system able to absorb, to listen to what you are saying, and the same with the governmental system? What is the reaction of congressmen and senators who operate within a planning time-horizon of just a few years to the next election, when what you are talking about is likely to happen either rather or much later? Do you find that the institutional structure in Washington is in some way deficient—to the extent that it cannot handle the situation and cannot address itself to problems of this nature and scale?

Bryson (Panellist)
You have really put me on the spot! But I am pathologically honest and so will answer bluntly.

My experience with presenting ideas about this topic to congressmen, to politicians, and to bureaucrats of the upper levels such as heads of divisions or secretaries of departments, has been that it is easy. There is no problem in getting them to see what the situation is, and no problem for them in considering a thoughtful reaction. Now I may be selective on who the politicians are: I do not talk to all of them. Obviously I talk to those who are willing to listen and talk to me. The problem has been with my colleagues, not the politicians. If you include everybody's idea and therefore give only general opinions, they say, 'Well, go away and make up your mind'. If you say 'This is our best judgement and I can give you odds that it is correct', then they are ready to act. But you cannot get from the scientists a unified, clear statement—except for the one that most scientists who talk to politicians make: 'Give us more money to do more of the same.' So I do not blame the politicians for not taking action when there is no clear-cut statement presented or clear action for them to take.

Miller
I would like to ask Professor Flohn if, in the second graph which he showed us just now, * the curve marked C describes the condition without nuclear energy? To my mind advocates of nuclear power development could make use of that reference. What is the context for this curve?

Flohn (Keynoter)
I took this from a report which was given by the International Institute for Applied Systems Analysis, Laxenburg, Austria, and it refers to the assumption that no nuclear energy shall be used at all; there is another curve dealing only with the present technology of nuclear reactors, namely light-water reactors. This is one of the crucial points of the energy discussion; my own point is that in the long run the continued use of fossil energy, regardless of how long we can use it, will reach at least as high a risk-level as nuclear energy.

Miller
May I ask a second question? What effect might be attributed to prospective releases of nuclear energy on the global heat-budget?

Flohn (Keynoter)
I do not think that this is much different from what we have now. Radioactivity, generally speaking, does not add to the heat-budget of the Earth. There are some estimates regarding the electricity-budget of the atmosphere, with some hints that this may change and perhaps also influence the overall rainfall-budget. But I do not consider this to be too serious, because rainfall and evaporation are mainly controlled by the heat-budget. It consumes something like 75% of the net energy available at the Earth's surface—a figure that can hardly be changed by Man's activity.

* In presenting his paper—not in the paper itself.—Ed.

Miller

Is it fair to say, Professor Flohn, that the ultimate release of sensible heat to sustain nuclear energy generation might very well cause a problem, but that this problem is several degrees removed from the immediacy of the CO_2 problem?

Flohn (Keynoter)

Here we tackle a rather difficult question of the answer to which I am not certain myself. I think that with all these different man-made effects we are not dealing merely with the energy contributions but also with what we may call 'the residence-time within the atmosphere'. For example, the residence-time of a carbon dioxide molecule in the atmosphere is of the order of 6 years. In contrast to this the residence-time of enthalpy (or sensible heat) in the atmosphere is of the order of a few days because of the long-wave radiation to space when we heat the atmosphere. The residence-time of a water molecule in the atmosphere is of the order of 10 days, while that of a halocarbon molecule is of the order of 50 years. If we take this into account it will change the figures I have indicated here, the long-living effects being more marked than the short-living ones. This means that the input of sensible heat plays in reality a role only at a local and not at a global scale.

Sigurbjörnsson

Professor Flohn was referring to the greenhouse effect of CO_2 which leads to the warming up of the atmosphere. However, there is another greenhouse effect of CO_2 which is well known to greenhouse operators: if you increase the CO_2 concentration in a greenhouse, you get increased photosynthesis—increased growth-potential—and thereby increased production of oxygen. Now, I wonder if increasing atmospheric CO_2 the way you describe will lead to increased rates of photosynthesis and increased production of crops and wild plants—whether this will, in effect, neutralize the effect we are talking about and the net result will be no increase in temperature?

Flohn (Keynoter)

This is indeed a very essential question, that has already been considered elsewhere with regard to the study of biogeochemical cycles. In that context we have indeed to ask, what is the greater effect: the increase of growth-rates of the living forests produced by the increase of carbon dioxide, or the rate of destruction of forests by Man? This has recently been studied with the perhaps surprising result that probably the rate of destruction of tropical forests by Man goes on at a faster rate than the possible increase of growth of the remaining (mainly boreal) forests.

Malone (Chairman)

I might say that within the past few weeks Professors Roger Revelle and Walter Munk from San Diego, California, have made some calculations which suggest that if you treble the biomass, you would just be able to take care of the increase in the CO_2 content [and arrive at a steady state]. They do not anticipate this happening but give it as the result of a simple calculation.

Glasser

If I heard correctly, Professor Bryson said that the increase of the amount of CO_2 was directly related to the burning of fossil fuels, and therefore was also directly related to economic growth. That would be rather puzzling, because it seems to me that if you follow a different method of economic activity whereby you use what I describe as income energy as opposed to capital energy, which is perfectly possible, you could have economic growth at a pace that may be different from what we have had in the last few generations, but you would still have it.

Bryson (Panellist)

There is a key word in what you said: we *could* have growth with an income economy, but that does not mean that we have had or do have it. My statement was based on the way things happen just at the present time, which certainly does not preclude the change to an income economy. But we are not there yet and we will not be there in this decade.

Laconte

Just a comment to sustain what has been mentioned by our two keynote speakers this morning, about the grave problem of deforestation. In the case of Java, the last survey made by the Dutch Government before leaving in 1947 showed that 27% of the island was covered with tropical forest. This has been published by Professor 'Otto' Soemarwoto, Director of the Institute of Ecology of Padjadjaran University in Bandung. Here we have a clear case which shows the kind of action that may be needed by developing countries. The Indonesian Government is possibly not in a position to give the necessary effort to make it possible for the people to use the forests in the manner which they want. Effort is probably needed at an international level to secure national action towards remedying this kind of problem, and one wonders if a better use of the satellite photographs, which show the same kind of thing happening in other countries, could not be made, with a suitable public relations effort, to convince public opinion and the governments concerned that some kind of action is urgently needed.

Malone (Chairman)

I might add that this very week there is going on in Tel Aviv a meeting of Working Group 6 of COSPAR (Committee on Space Research of the International Council of Scientific Unions) and they are considering precisely the question that you raised.

Kuenen

Just a short additional remark: the destruction of the tropical rain-forests has two causes, one of which is the increased population and the attempt to increase agricultural area, but the other is much worse—the extraction on behalf of you and me up here in the north, of timber from the tropical rain-forests to make [all] sorts of beautiful things.

Worthington

Relating to the influence of the 'greenhouse effect' and increased CO_2 on photosynthetic activity at ground-level, there is a factor which strikes anybody who travels by air over Central Africa or South America, namely the amount of smoke in the atmosphere during the dry season and its absence during the wet season. In Africa, particularly, almost everything which is burnable is burned annually over vast areas, and closed-canopy forest, which does not burn, tends to be replaced by savanna or grassland, which does.

I would like to know the effect of all this smoke on radiation which reaches the Earth's surface, and hence on photosynthesis and on the CO_2 content of the atmosphere. Professor Flohn referred of course to dust, including man-triggered dust, but I would think that smoke from this annual burning, which is increasing as a result of the destruction of the closed-canopy forest, is even more important.

Flohn (Keynoter)

The question of grass-burning in the semi-humid and semi-arid tropics has been discussed on numerous occasions in relation to the facts that were described so convincingly many years ago from Africa. We know that an area of the order of at least 10 million square kilometres is affected by this process every year, and I agree completely with the view that this greatly changes the heat-budget. We know from radiation measurements that atmospheric turbidity in the tropics increases substantially in the dry season, partly due to the accretion of water molecules to the hygroscopic particles. Here the turbidity is as great as in the most polluted industrial centres of the world, e.g. Tokyo or Los Angeles. But in the tropics this effect is spread over many millions of square kilometres, so that it is of a much greater importance to the large-scale climate. It leads to a local surface-heating during the dry season; this includes also areas where the remaining humid forests are being burned down, as is now widely the case in South America.

Bryson (Panellist)

I would like to add a few comments in response to this. You are right, there is a lot of that smoke; the best quantitative estimate that I have been able to come up with anywhere in the literature, or by calculating myself, is something of the order of 60 million metric tons of agricultural smoke put into the air every year. If you add to that the dust put into the air by the overuse of marginal lands in dry areas such as the Sahel, Rajasthan, etc., the air pollutant product of these mostly Third World countries is equal to the total industrial pollution of the world.

Last November I published a paper on the effect of this material on the hemispheric surface heat-budget. I found that to the present time the effect of man-made low-level particle pollution (about half industrial and half agricultural), is just about equalled in the opposite sense by the effect of increased carbon dioxide. Now this would give us a very interesting result if we were to switch to nuclear power immediately and relieve the industrial air pollution totally. What would then happen? We would have neither the carbon dioxide nor the industrial smoke part, but we would have still the agricultural pollution and its effect is to drive the temperature the other way.

What all of this means is that, as we do not know the whole answer, we should not walk on the edge of the cliff, where we do not know just how solid the ground is. We should preserve a broader safety-zone between where we are and the carrying capacity of the Earth because of such uncertainty. It does not mean that we are safe because we do not know the answers; the more uncertain, the more cautious, and the more prudent, we must be.

Nicola

I first had a question about atmospheric dust which has been well answered, so I have switched to two others. The first one is to Professor Flohn: as far as I have noted, you have not spoken about microclimatic studies being of use in agriculture, and whether from a practical point of view something could be done in that direction? I am speaking about the influence on insects, on the ground, and the albedo of cultivated areas.

The second question is whether any of you have any news about any current studies of abrupt orbital changes, and if anything is known about the influence of orbital changes, which could be very important. I think that some centuries ago there was such a case which at that moment completely altered the whole situation rather abruptly.

Flohn (Keynoter)

Regarding the first question, I am not sure how many areas we can handle in this way. Looking at the large-scale and indeed global question in which we should be mostly interested, only those effects which are operative over areas of the order of several hundred thousand square kilometres are effective enough.

Regarding the question of abrupt changes in the past, the orbital elements work in the time-scale of 20,000 years and more, and their variation from one century to the next is practically negligible. Recently, evidence was found that there are apt to be rather abrupt events with dramatic climatic variation—as large as between the last ice-age and the present time—occurring in the time-scale of a few hundred years or less. For example, the equatorial Lake Kivu in Africa, with a depth of more than 300 metres, has been desiccated and filled again in a period of about 300 years, and this is only one of the available examples. Very little can be said as yet about the causes of these abrupt changes; the most appropriate hypothesis is that of a clustering of large-scale volcanic events.

Schütt

I am concerned with energy use and its environmental effects, working as I do in the Energy Research and Development Commission in Sweden, which corresponds to ERDA in the United States. Future decisions concerning our energy system in Sweden will be taken at the parliamentary level. Sweden represents one of those industrialized countries that are using fossil fuels to a large extent. Approximately 70% of our total energy comes from oil. If we go for nuclear energy as much as we can afford until about the year 2000, which we have not yet decided on, we will be able to support ourselves as to approximately 30% from nuclear energy—not more. Yet we are one of those countries with quite a lot of uranium within our borders. We also have hydroelectric power, supplying at the moment about 15% of our energy, which is almost the maximum we can produce directly from water. So whenever we switch from oil to something else, we need some other energy source than hydroelectric and nuclear power. We do not have many possibilities: coal is one, that, from an environmental point of view, is even worse than oil. A second choice—though we do not know if it is possible or useful to us as yet—is wood or other biomass. This alternative would reduce our dependence on imported fuels. But such burning of wood, etc., would release carbon dioxide to the atmosphere, though something like the same amount would be fixed again during the following year, continually in the growth of new trees and shrubs.

With this background in mind I would like to ask a few questions. First of all it would be good to walk into an eventual carbon dioxide catastrophe with one's eyes open, even though we might not be able to do much towards avoiding it. But before we are willing to take any decision in the matter of reducing the use of fossil fuels, we must know much better than at present what will be the effects of the carbon dioxide emissions. We know that the carbon dioxide in the troposphere is increasing. But is this an effect also of deforestation? Or is it an effect entirely of the increased burning of fossil fuels? Or will the burning of fossil fuels be the basis of only a small increase in temperature and at the same time increase the fixation of carbon dioxide in photosynthesis, so that an equilibrium will be reached? I do not know the answer. Can [our galaxy of leaders on the platform help me in this predicament?]

The emissions from burning fossil fuels include not only carbon dioxide but also a lot of particulates, which it can be argued will work in the opposite direction by reflecting solar energy. So perhaps there will be a decrease in the Earth's surface temperature as an overall result? As long as we only fear but do not really *know* the total effect, we cannot do very much or advise at the political level [where the final control actions have to be taken]. So I will finish by asking this question: What action should be taken at the moment, from your point of view, and what resources would it be possible to use, to become more sure about these effects?

Malone (Chairman)

I will seek to answer some of your questions in my summation, but first call on Professor Hare.

Hare (Panellist)

In the wake of the last speaker, who has said almost precisely what I intended to say, I shall be very brief. May I point out that if part of the solution is to go nuclear, because going nuclear for electric supply does not increase the thermal loading at the base of the atmosphere, and does reduce the carbon dioxide load, then clearly going nuclear may well be an environmentally attractive option. But where in the world can you find an environmentalist to admit this at the present moment? Politically, the nuclear energy trade is in deep trouble, the public belief being that nuclear energy is more dangerous than any other form of energy production. Yet the statistics show otherwise, suggesting that coal-mining has perhaps the higher potential impact on human health and on things ecological. It certainly does not do anything, as you suggest, to alleviate the carbon dioxide loading; but may I just stress that, at the moment, any suggestion from this Conference that we should advocate going nuclear to a very much greater extent than at present—perhaps to 40% of electric supply by the end of the century—would be received with hoots of derision from the people who make environmental opinion. This is something that we have to face—and properly so.

Bergthorsson

To what degree is the CO_2 content the cause or maybe the effect of the warming up of the atmosphere and the oceans? If it is so that the warming leads to an increase of atmospheric CO_2 in some way or other, for example due to interaction of air and sea, this will, of course, affect dramatically all forecasts of future happenings. However, this does not change my opinion that Man has to be extremely careful regarding all the possible man-made causes of environmental change.

Malone (Chairman, concluding Session)

I believe the answer to your question is that the CO_2 increase so far has been calculated to be responsible for 2–3 tenths of a degree Celsius, but this is below the 'level of noise' of natural variability of climates. Thus while the increase in CO_2 has tended to produce warming of the atmosphere during the last 30 or 40 years, this effect has been more than counteracted by other causes tending to produce a net cooling. That is about the situation as it stands today.

Our Secretary-General has stressed the importance of making concrete suggestions to avoid the kind of ecodisasters we are discussing. The lessons taught to us by historians of the politics of science tell us that, when confronted with situations such as those which Professor Flohn has outlined so eloquently this morning, there are three requirements for a successful response.

First is the searching re-examination of human value preferences. We will hear more about that later this week, but I will simply illustrate this requirement by two decisions that emanated from the Stockholm conference. One is the resolution which states that natural resources are a shared human resource, and so there should be some equity about their distribution among all people. The second—and it was interesting that this should come from an intergovernmental conference—was a clear-cut recognition of our responsibility to future generations.

The second requirement for successful tackling of environmental problems is that new conceptual frameworks must be developed, and I think that was pointed out to us this morning by Professor Bryson when he touched on the need for a systems point of view in looking at these problems. [There is all the time strong] mutuality and interaction of the internal natural processes and human intervention plus external forcing functions. The whole matter of solar–terrestrial interaction is gaining a new perspective and a new sense of respectability. The ocean interaction linkage has emerged as an important element of this new conceptual framework within which we must address the climatic problem.

The third requirement in confronting environmental situations, and one that does involve the political process and also due human determination, is institutional innovation. In our case this requires replacement of international cooperation which is simply the aggregation of national efforts, by proper integration of the scientific work. Aggregation characterized the activities of IGY and IBP. True integration is accomplished by bringing together scientists of different nationalities, different cultures, different political persuasions, working shoulder to shoulder in the same laboratory or institute. I would say further that the co-mingling of inter-governmental and non-governmental institutions is a part of this process of fashioning new institutional instrumentalities. And I think the challenge here to the scientific community is to do as Professor Hare has suggested—to illuminate these issues before the politicians get to arguing about them at intergovernmental conferences where courses of action are determined on political rather than scientific grounds. There is clearly a need for these integrated operations of the kind that is represented in the International Institute for Applied Systems Analysis in Austria, of which the principal function must be to develop the kind of international environmental impact assessments that Professor Hare has suggested. Especially do we need these assessments for alternative kinds of energy strategies.

Such an institution will be effective only if it is coupled closely to the inter-governmental bodies which in the last analysis are going to make the decisions that will emanate from the kind of international impact assessments which the scientific community must be prepared to develop. We might further note that one of the characteristics of such an institution must be that it is not only integrated internationally but also integrated in a disciplinary sense. This will require lawyers, and our colleagues in prediction such as economists, psychologists, biologists, chemists—yes, and atmospheric scientists. We are going to have to fashion a new kind of intellectual elite or at least understanding which places a true value on, and properly rewards, that strange creature who wanders aimlessly in academia today, preoccupied with [seemingly obscure] interdisciplinary activities. If only we are able to fashion that kind of instrumentality, and couple it with a decision-making instrumentality, I believe there is hope that we can avoid the kind of ecodisaster which Professor Flohn has pointed out so beautifully.

Thank you all—especially our keynoter and our animated panellists, as well as many members of our audience for their patience and enlightened participation in the proceedings.

Session 3

WHITHER TERRESTRIAL ECOSYSTEMS AND HABITATS?

Keynote Paper

Whither Terrestrial Ecosystems? Preservation of the Habitat of Man

by

F. RAYMOND FOSBERG

Senior Botanist, National Museum of Natural History,
Smithsonian Institution,
Washington, DC 20560, USA.

INTRODUCTION

The subtitle of my paper, 'Preservation of the Habitat of Man', was chosen to place in sharper focus the general subject of this Conference, as what we are here for essentially is to discuss Man's future prospects on Earth. His future, if any, depends very largely on his relationships with the environment in which he finds himself, and from which he cannot escape. In this sense, his environment extends throughout at least the inner 100 million miles' radius of the Solar System, and perhaps farther out. We do not know if light and other radiation from the rest of the Galaxy have any significant effect on Man as a species. Perhaps they do. Starlight has at least a psychological effect; it also enables Man to see in times of darkness, to navigate his ships at sea, and to locate himself precisely on land. So perhaps the Universe is Man's real environment. In the narrowest possible view, the Sun must be a part of his environment, as he is utterly dependent on its radiant energy.

Of course, the segment of Man's environment of immediate concern to this Conference is the Earth, including its atmosphere. This is the part, the future of which Man, with his present and definitely foreseeable technology, is able to influence, and most portions of which he, indeed, does influence—sometimes profoundly.

'Habitat' is a term long used in natural history, defined in *The Oxford Universal Dictionary*, third edition, revised 1955, as: 'The locality in which a plant or an animal naturally grows or lives; habitation. Applied (*a*) to the *geographical area* over which it extends; (*b*) to the particular *station* in which a specimen is found; (*c*) but chiefly used to indicate the *kind of locality*, as the sea-shore, chalk hills, or the like.'

The third of these alternatives is the one almost always intended by modern users of the word, or at least by ecologists—which presumably most of us are, anyway in the broadest sense.

Even with this simple and logical definition of habitat, however, we are not without a few problems in the use of the word. Is the 2,000 miles (3,200 km) stretch of sea over which the Pacific Golden Plover [*Pluvialis dominica fulva*] migrates, between its summer habitat in Alaska and its winter habitat in Hawaii (Munro, 1944), also a part of its habitat? Is the east coast of the United States a part of the habitat of the Smew [*Mergellus albellus*], one of which strayed there recently from its proper habitat in northern Eurasia (Finch, 1976)?

The reason for these questions in our present context is to clarify our circumscription of the habitat of Man. There is much discussion as to what was Man's habitat at the time when he emerged as the recognizable species, *Homo sapiens*. One current idea is that Man's primordial habitat was in or near the boundary between forest and savanna in East Africa (cf. Cloudsley-Thompson, 1975). Of one thing we can be certain—Man's habitat a million or more years ago was far more restricted than it is at present. His developing ability to create microhabitats for himself, and to adapt himself in the matters of food habits and temperature tolerance, has enabled him to extend his total habitat quite vastly. He has enlarged the area which he effectively occupies, so that it includes most of the land area of the Earth. A few deserts and ice-caps, a few mountain peaks, and remnants of tropical forest, have, for the time being, been left unoccupied; but even these are within the capacity of his technology to occupy. It is mainly Man's economics that have so far retarded this occupation.

The only area really in question is the sea (and large lakes). And here, Tektite and similar experiments, as well as Oriental and Indonesian floating villages, suggest that it will only be a matter of time before permanent habitations spread to the sea-bottom, and surely to its surface.

An anecdote may show that even the question of the sea as a part of Man's habitat is merely a matter of viewpoint. A class of schoolchildren in Guam were from various islands in Micronesia. They were asked to write essays on their home islands. A boy who was native to the island of Ponape ended his essay with words to the effect that, best of all, his home island was surrounded by the sea, which made it easy to go from his island to others.

Are the sea, which certainly is where Man lives when he travels by ship, and the upper air, where he lives when travelling by 'plane, parts of Man's habitat? I think that for our purposes we must say 'yes'. Perhaps we can, for the present, exclude the hot interior of the planet. With this exception I think we may accept the entire Earth as the habitat of Man.

Thus the scope and meaning of my subtitle become clearer. We are concerned with the preservation of the Earth as a habitat for Man.

Man not Alone

Personally I do not like this latter emphasis. Several million other species share with Man the Earth as a habitat. I like to think that, in the cosmic scheme of things—if there is one—these other species might, along with Man, have some right to a share in our common habitat, rather than continuing to exist only at Man's discretion or sufferance. However, this does not seem to be a tenable viewpoint, given Man's basic instincts—a product of his evolutionary moulding. We shall come back to this question later, and I hope it may receive some attention in this Conference. Meanwhile, we will discuss the preservation of the Earth as Man's habitat, and his prospects of inhabiting it for the indefinite future. Presumably concern as to this point is the reason for most of us participating in this Conference.

Now, what are the characteristics of this planetary habitat of Man? Possibly its outstanding feature is its diversity. Depending on how you divide it, or, perhaps, regardless of how you divide it, the clearly different sub-habitats number, or numbered, many thousands. They are much more numerous than most people realize. Such obvious gradients as that from eternal snows on high mountains to steaming hot lowlands—from Greenland's ice-cap to the scorching Sahara or Death Valley—are familiar to us all. Other gradients are more subtle, but may be equally effective in determining the areas that are habitable by particular organisms.

The gradient from oceanic to extremely continental regional climates may not be too restrictive for modern Man, but may have had a strong effect on his remote ancestors. Day-length, varying from almost uniform in the tropics to enormously fluctuating in the Polar regions, exerts a profound effect on what plants grow in an area—including food-plants. Differences in the frequency of typhoons or hurricanes, as well as tornadoes, have an obvious effect on the habitability of different regions. Variation in the seasonality and amplitude of rainfall fluctuations even now influence the suitability of many areas as human habitats. Then there are the short and steep gradients such as those from the sea to dry land, from the forest to meadow or savanna, so-called 'edges', and from volcanic rock to limestone. One could go on and on with examples. Suffice it to recall that there is no area on the Earth where a combination of intersections of several or many of these gradients does not produce an environment that is different in some respect, however large or small, from all or most others.

The phenomenon of dispersibility, characteristic of all forms of life, enables the species of plants, as well as of animals, to reach and occupy those habitats that best suit them or satisfy their requirements. This in itself produces new habitat features, as it is a continuing process and the biotic components of a habitat are fully as important, and limiting, as the physical ones.

The result of the interaction of these myriad factors is a vast and shifting mosaic of habitats—major, minor, and micro—for Man as well as the other living inhabitants of the globe.

Species vary enormously in what is called ecological amplitude—that is, in the range of different potential habitats in which the species is physiologically able to live. Some are so finely adjusted to a specific set of conditions that their distribution is extremely limited, although they may be very successful within these limits. Such species are vulnerable to even small changes. Other species are able to tolerate a vast range of different environmental variations, either through genetic variability or through their phenotypic adaptability. No known species has the capability of adapting to all combinations or all extremes.

Within its range of ecological tolerance or amplitude, no species actually occupies its full potential habitat. All species—with the possible exception of ones that live in the most extreme conditions, and, of course, Man, who modifies conditions to adapt them to his physiological tolerances and also eliminates competing species—have their actual ranges restricted by competition for the resources of the habitat. Most species actually occupy only those parts of their potential ranges that are the most favourable for their particular requirements.

This, of course, is elementary ecology, and I may be criticized for boring those who are familiar with it already. However, all too many of the people whom this Conference aims to influence and educate are not ecologically trained or even versed. If they were, perhaps the Conference would be unnecessary.

Evolution of Man's Influence: Five Thresholds

We can, with our knowledge of human physiology, probably map with some confidence the potential total geographical range of prehistoric Man, or, more exactly, precultural Man. What part of this range was actually occupied by our remote ancestors we may never know, and we can only conjecture about what Man's competitors for environmental resources were then. Undoubtedly the competition was real and severe.

With the emergence of Man's intelligence, he gradually developed the use of weapons against the competitors, fire to increase hunting effectiveness, and tools, clothing and hunting implements to assist him in securing food and providing protection against climatic and other environmental hazards. He began to be able to modify his environment and to disturb and shift the relative equilibrium in which he had lived for ages as a hunter and gatherer.

This disturbance increased dramatically when plants and animals were domesticated and agriculture began. Large areas of relatively stable 'climax' or 'subclimax' biotic communities were cleared and shifted back to pioneer status, which is where most agricultural ecosystems are and must be. This process of accelerated change, which is still going on, results in the most pervasive and far-reaching modification that Man has ever brought about in his environment. It has totally and profoundly changed the landscape of most of the land surface of the Earth.

The sequence of changes outlined above may be viewed as a series of thresholds—the early ones being more or less conjectural, the later ones very real indeed (Fosberg, 1958). The earliest 'hunting and gathering' human populations most likely lived in the same sort of equilibrium with their environments as can be observed to characterize the relationship between other more or less omnivorous primates and their habitats under relatively undisturbed situations. Their effect on the environment was probably well within its 'resiliency' or capacity to recover from any injury thus suffered.

As hunting technology developed to the point where the hunters could kill their prey more rapidly than its reproductive capacity could maintain its normal population density, a *first* threshold was passed and a changed habitat resulted. This event probably coincided with what has been termed 'overkill'. This apparently resulted in the disappearance of the vast populations of large mammals and large birds that characterized most continents and very large islands in the late Pleistocene and early Recent periods. Although this may not have happened simultaneously everywhere, the general effect was a changed world in which game became harder and harder to find and other means of subsistence than hunting and gathering were favoured. With plant life also greatly affected, a new equilibrium was approached.

The use of fire in hunting was probably a part of the technology referred to above. However, the use of fire deliberately to clear land may have caused Man to cross the next, the *second* major threshold. Large areas changed from forest to open or semi-open land, perhaps leading to or making possible the development of agriculture. About this time, plants and animals were beginning to be domesticated—resulting directly in agriculture, with its profound effect on the landscapes and the life-style of Man. An open landscape in the more favoured regions, especially where water was abundant but not too abundant, fostered a sedentary pattern of human occupancy, production of surplus food, and the growth of population. In less favoured, drier or less fertile areas, pastoralism and a nomadic life-style frequently developed. Here human populations grew less fast, but herds of domestic ungulates increased and formed the surplus needed for a less precarious existence. The accumulation of surpluses led to larger aggregations of people, to cities, and to commerce.

This whole syndrome, based on production or extraction from the environment of more than was needed to support the producers, brought about the crossing of a *third* important threshold. It is questionable whether, except in limited areas, any significant equilibrium followed the passing of this division point. The existence of surpluses made possible the focusing of attention on the deliberate accumulation of knowledge and on its use in further—and increasingly rapid—development of technology, including those aspects which exert a strong effect on the environment. From this point onwards, environmental degradation became predominant in many, if not most, areas. Such evidence as the desertification of formerly fertile lands—as in North Africa and the Middle East—accelerated erosion in the Mediterranean regions, in China, and in India, and the spread of blanket bogs and moors in wetter areas such as northern and western Britain, suggests this (Pearsall, 1950; Bryson & Murray, 1977).

Expanding technology—especially in the fields of extraction and processing for human use of the inorganic resources of the Earth and the products of agriculture—and the expanding markets resulting from population increase, inevitably brought about the crossing of a *fourth* threshold, commonly referred to as the 'industrial revolution'. This was characterized by the utilization on a large scale of the mineral and agricultural resources of the Earth, for production especially of non-food items. The resulting increase in need for labour and concentration of human population in cities, depletion of readily available resources, the depopulation of rural areas and development of mechanized agriculture and 'agribusiness', and the appearance in aggravated form of the phenomenon of pollution—the accumulation of wastes more rapidly than the environment can absorb and convert them without serious degradation of the environment—have been with us for the last 150 years. Man has, during this time, acquired the ability to change his total habitat almost completely and very rapidly.

In the 1940s, with the explosion of atomic weapons and the introduction into the biosphere of abnormal amounts of radioactive materials, we have passed still another, *fifth*, threshold. We can now destroy ourselves and all other life on Earth in a [thousand-millionth] of the time which it took life to develop, or, alternatively, by the 'peaceful use' of atomic energy, we can do this in perhaps a millionth of that time. Of course, we can also refrain from doing so.

It may be noted—although I have given no precise figures for the time-intervals between these thresholds—that the periods between them became shorter on a roughly exponential curve. [The curve we can draw] indicates, essentially, the increase in Man's ability to alter his environment. If it were only that, namely his ability to change the character of the world he lives in, it might not be such a source of worry. The bothersome thing is that he seems to be unable to refrain from doing this even when the consequences are obviously going to be bad.

Population the Basic Problem

This exponential curve, if one stops to consider, seems to be characteristic of many, if not most, ongoing phenomena involved in environmental change—and at whatever scale. Basic to all is the exponential growth of the human population. The biotic potential of the human species remains unimpaired, while the normal constraints on population growth are rendered ineffective by medical science and public health advances, as well as by increasing availability of resources due to technology.

The normal population curve is asymptotic, the build-up of predators, disease organisms, and/or waste products, and the depletion of resources, bringing about a levelling off—or, in the case of 'pioneer-type' organisms, a sharp down-swing or even a 'crash', which at least for a time may render the population unimportant in the ecosystem in which it lives. In the case of a 'climax-type' organism, the population curve tends merely to flatten out and continue with only minor fluctuations for as long as the ecosystem remains viable.

Whether or not the human population curve will flatten out, as the optimists—at least those of them who think at all—seem to expect, or whether it will continue to rise exponentially until a crash occurs, is the underlying question facing conferences such as this one, and is the basic problem facing the human race. It can be stated differently according to whether the human species is considered a pioneer or a climax species.

Man has been behaving for the last few thousand years in typical pioneer fashion, both in his rapid exponential increase in numbers and in his effect on his habitat. This effect will, it appears, if he persists on his present course, eventually make his habitat unsuitable to support him further, and a crash will occur. This has, of course, occurred before on a regional scale, for example in areas of North Africa where we have evidence of former highly-developed civilizations where uninhabitable deserts now exist (Bryson & Murray, 1977). In the past, this phenomenon has usually been obscured in that the population crash has been brought about prematurely—that is, before the habitat became uninhabitable—by invasion by 'barbarian' hordes who have destroyed the advanced cultures that they did not understand. Such were the fates of successive highly-developed cultures in Middle and South America, down to and including the Inca, Chibcha and Aztec civilizations that were destroyed by the Spaniards. The Maya may have been an exception to this, in that their system may have collapsed from internally-generated causes as outlined above, rather than from warfare. To the best of my knowledge we are not certain as to what caused the end of the Maya culture, although many guesses have been hazarded (Carr & Thorpe, 1961; Thompson, 1966).

In our own situation we may, of course, be destroyed by warfare before habitat degradation is sufficiently advanced to bring about a crash. Almost certainly when the crash comes it will be accompanied by war. But in either eventuality, the wars will not be invasions from barbarian hordes from without (unless science-fiction predictions should prove correct and invasions come from other planets or outer space), as the human species now essentially belongs to one culture, although comprising the parts of our 'one world'.

It will be clear from the above that I am a pessimist: I have seen little reason to be otherwise. It should also be clear, however, from the very fact that I am here participating in this Conference, that I am not the kind of pessimist who throws up his hands and abandons all hope.

In this whole dismal picture there appears to me to be one mitigating circumstance that may, with help from us and from others like us, come to bear and give us a way out. This is the factor of human intelligence. Man is, as I see it, the first species in the history of the world that has been given an apparent choice as to whether he will behave as a pioneer or as a climax organism; but more on this later.

What I have offered up to this point has essentially been an outline of how Man, as a species, reached the present critical situation—wherein he may quite likely destroy himself, either by a sudden 'big bang' or more slowly, by causing or permitting his habitat to be altered, polluted and degraded, until it is rendered irreversibly unfit for his continued tenancy.

Some of the concepts that are necessary to my discussion have been introduced along the way—e.g. environmental resiliency, the availability to Man of a choice as to whether he behaves as a pioneer or a climax organism and, most important, the exponential acceleration of the rates of environmental change.

Vast Biological Diversity on Earth

I have emphasized the ecological importance of the enormous habitat diversity exhibited by almost all the major regions of the Earth. What I have not sufficiently stressed is their incredible biological diversity. This is manifested in the almost limitless range of morphological and physiological variation in the plant and animal kingdoms. What this really represents is the surviving portion of the innumerable array of genes and combinations of genes that have resulted from thousands of millions of years of evolution. In spite of the arrogant branch of biochemistry that has succeeded in isolating, identifying and manipulating a few of the molecules (DNA) that, apparently, are actually genes, and despite the dreams of 'genetic engineering', this array of genetic material is, I submit, the most precious and irreplaceable heritage that the human race possesses. For this genetic diversity is what ensures the perpetuity of life on Earth, and brightens any future prospects which we may have on other planets. It is the essence of all ecosystems, however we define or circumscribe them, as well as being the essence of human diversity.

Let us examine the present status of a few of the major ecosystems that go to make up the terrestrial world habitat of Man. How do they now compare with their 'normal' or recent historical condition? Can they still satisfy Man's requirements and continue to do so? First, perhaps, we might discuss briefly what are Man's requirements? What should his environment be expected to provide? At what level of satisfaction and fulfilment should modern Man expect to live? What will determine the level at which he will live? In what numbers can Man exist and still enjoy a reasonable degree of fulfilment of his potentialities? And for how long?

It is obvious—as everyone must agree—that Man must have breathable air, potable water, a certain limited range of temperature, food to nourish and energize his body, and a reasonable amount of living space. Much past and current thinking on these matters has accepted the assumption that, because these are the absolute essentials, they are the most important things and that, so long as they are provided either at minimum, adequate or excess levels, society is in good shape. Employment is accepted as a necessity, but, in all-too-many quarters, only in order to provide these requirements for subsistence (or, more aptly, existence). Accordingly, the question of why this cycle of 'work in order to subsist in order to work' is often dismissed as facetious, or is regarded as a philosophical matter that has no meaningful answer. The person who asks this question can expect to be regarded by many if not most of his listeners as crazy, or at least as an impractical dreamer. Of course a Man works to eat and to feed his family! Why else?

I submit that this answer places Man precisely on a level with domestic animals. They clearly enjoy the same physical satisfaction and pleasures that we do, at least in some degree.

I have no clear answer to the question of what are Man's other requirements, beyond those listed above. But I am completely certain that he has them—at least for normal life—that they are varied and complex, that they are fully as important, although perhaps in a different way, as those listed, and that Man's habitat must fulfil them.

Two obvious ones are beauty and variety or diversity. At least when he or she is a child, almost everyone shows great attention to, and has need of, beauty and variety. When a person becomes older, he may have had this educated out of him, or suppressed. He also has a curiosity that motivates him in many diverse paths. Most people have this, especially when young—a sense of wonder! (Carson, 1965). This is a response to aspects of their environment that may have nothing whatever to do with subsistence. Even grown men, and much more so children, are attracted to animals and plants. The thrill at seeing a fawn, a wild orchid, a delicate fern, at hearing the honking of a flock of geese flying overhead, a wood-thrush song, the wind blowing through the trees—these are all very real. The emotions evoked by beautiful and interesting things and experiences are strong, and furnish much of the motivation of Man. His reactions to such things are the clues to the meaningful aspects of life. The stimuli—or, if you like, the experiences themselves—are a vital part of what is supplied by Man's environment. These phenomena are as complex and diverse as are the myriad of habitats and components of these habitats that surround Man. He evolved in this beauty and diversity, and his life will be incomplete to the extent that he is deprived of contact with these elements.

I do not mean to lecture on the philosophy and psychology of human responses—apart from trying to indicate that the quality of the human environment cannot be measured by economics, or by how well fed the population may be. The uniformity of a vast and productive monoculture of maize or wheat produces a temporary thrill and satisfaction when first experienced. But very shortly it evokes a feeling of deadly monotony and an enervating boredom. Even the city, which is diverse in its way, becomes something to escape from—witness the congestion on the highways leading out of New York or London, Paris or Tokyo, on a week-end morning! The beauty and diversity of a more natural landscape are eagerly sought out—and, unfortunately, tend to be loved to death.

The beauty of the natural world is beyond measure—the diversity equally so. In addition to the capacity of Man's habitat to supply his physical and physiological requirements, its capacity to satisfy his emotional requirements is equally essential. As a part of the latter we must also insist on a continuing variety to satisfy Man's intellectual curiosity. To sum up: environmental quality includes, at the very minimum, the capability of satisfying the physical and physiological requirements—the requirements for beauty and diversity, as well as the quality of inherent stability or, perhaps better said, of maintaining a state of dynamic equilibrium.

DETERIORATION OF WORLD'S ECOSYSTEMS

Now let us look at the current conditions of a few of the world's major ecosystems. Let us compare their present state with what we know of their primitive states. Let us pay particular attention to the quality of resiliency, to their capacity to recover—to re-establish something resembling their primitive or former condition after being seriously disturbed. Let us accept the premise that any major ecosystem, before perturbation by human technology, has approached or arrived at a condition of maximum utilization, by its biota, of the resources of the regional climatic area that it occupies, and that such maximum utilization is the basis of the diversity and stability that the system manifested in its primitive state. An implication of this, which some of my colleagues might take exception to is that, making allowances for geographical isolation which may bring about excess speciation, the totality of biota (number of species) should be an indicator of the total abundance and diversity of resources in a major ecosystem. Of course, it is doubtful if anyone has come close to compiling even the number of species in any major ecosystem—except perhaps the Polar ice-caps and their associated nunataks and morainal deposits, which tend to be the simplest of all and among the poorest in resources and species.

Let us look briefly at one of the most significant, in human terms, of all major ecosystems—the tall-grass prairie, occupying major areas in central North America, in Argentina–Uruguay, and in the Ukraine and surrounding areas. This is one of the most suitable for human occupancy and highly productive of human food. Geographically one of the less diverse, and biologically in the middle range of productivity—at least in temperate latitudes—the tall-grass prairie has felt the heavy hand of Man to the extent that few unaltered examples of it still exist. Its biota and its ecological processes are of the greatest interest and importance to Man, yet today probably not a single complete, primitive, functioning example of [any substantial extent] remains. A few scraps may retain most of their original floras—with an additional increment of weeds and other exotics—perhaps a smaller fraction of their small animal faunas, few or none of their large mammals, and some of their birds (including prairie-chickens [*Tympanuchus* spp.]). Interest in saving these last from extinction has been responsible for the preservation of a few of the remaining small areas. Substituting for the prairie biota are a few agricultural monocultures—especially maize and other cereals, and soybeans—towns and cities, and gardens and parks; there are also some pastures, but these are mostly of exotic grasses and legumes, with cattle and other domesticates in place of the Bison [*Bison bison*], various deer [*Cervus* spp.], and other large mammals of yesteryear.

A shocking proportion of the topsoil of the prairies has been lost by wind and water erosion. Fertility is maintained by the pouring on of chemical fertilizers. Even the people no longer live close to the land. They are being crowded together in towns and cities, more and more replaced by the machines of agribusiness. The productivity of this modified ecosystem is still high, but its diversity is almost gone. Its capacity to support Man could be halved in a season or two by a really effective disease of maize, and at least seriously reduced by a disease of soybeans or one of wheat. Its beauty is already mostly gone. Compare it now with what was described by some of its authors and poets!

Another of the ecosystems that has been a principal habitat of Man, and that has been profoundly modified by his activities, is the broad-leafed deciduous forest. This may at one time have covered much of the North Temperate Zone, lying between the cooler taiga or boreal coniferous forest and the southern, warm-temperate and subtropical broad-leafed evergreen forest. Judging by the disjunct areas where it survived the Pleistocene glaciation, it must have been an extremely rich community, biotically. The Japanese and Southern Appalachian floras are among the richer temperate floras of the Earth. Not all, by any means, of the plants and animals that survived the ice-ages in these refugia were able to move back quickly and reoccupy the moraines and lake-beds that were left uncovered by the melting ice, and so the biotas of these areas were not as rich as those of the unglaciated parts; but one has the impression that some swamp-lands, now altered beyond recovery, must have been fascinatingly diverse, at least biologically.

The non-mountainous parts of this ecosystem had enormous potential for production of food-crops, and were almost completely brought under cultivation before anyone had the idea that any areas should be kept unchanged for future and possibly wiser generations. So now in the whole of North America and eastern Asia there are almost no virgin deciduous hardwood forests left—not even tiny samples. One or two small parcels still remain in the Smoky Mountain National Park that give us a glimpse of what was much more widespread. A few very old pieces of second-growth exhibit many of the features of the original forest. Most present-day hardwood forests have at some period or periods been clear-cut, and they are believed to have lost substantial thicknesses of rich topsoil. The forests of eastern North America have lost one of their principal dominant tree species, the American Chestnut [*Castanea dentata*]. The corresponding European forests are at present having the same thing happen—the Chestnut-blight Fungus [*Endothia parasitica*] is gradually wiping out the European Chestnut [*Castanea sativa*]. The main change, however, has been clearing and transformation to agricultural land—crop-land and meadow-land—and the development of suburbia with accompanying playgrounds and golf-courses. This has happened to practically all of the richest land and, at least temporarily, to much of the mediocre and even the poor land as well.

From the human viewpoint this change has been largely beneficial, at least where the land has been farmed by men who loved their land and developed sufficient understanding of it to leave it in as productive a condition as they found it—or better. Some of the finest landscapes on Earth may still be seen in good farming areas of this biome. Ironically, these conditions seem to persist where 'scientific agriculture', with its highly-developed mechanical tools, chemical fertilizers and presticides, has not yet become popular—where there are still domestic animals to produce manure for fertilizer and to pull smaller implements that do not tear up the soil so drastically. These things are, of course, on their way out except in a few pockets of 'Pennsylvania Dutch' and 'organic' farmers, who have resisted 'progress' and value their old ways. Even in the British Isles, where this farm, woods and meadow landscape reaches a high development, the hedgerows are disappearing, mechanization is taking over and suburbs are proliferating.

This is impoverishment, even though agricultural productivity may be increasing, at least temporarily. Heavy applications of chemical fertilizer do the same thing for the land that chemical stimulants do for the human athlete or the racehorse. But they do not put humus back into the soil. Should the crutch be taken away, the going will be painful and slow.

We do not really know what the eroded and devastated areas of North and Central China were like, but large areas were doubtless occupied by broad-leafed deciduous forest. Most of us who studied elementary geography probably remember pictures in our school-books of the stark, eroded hills of China, the horrible examples of overexploitation and bad land-use practices, which were said to be the results of overpopulation. Certainly these landscapes may be foretastes of conditions in other parts of the world if widespread present trends are allowed to continue.

Marginal Examples

Let us turn from these well-endowed, high-potential ecosystems to some of the extremely marginal ones.

The 'desert scrub' ecosystem, epitomized by the American Creosote-bush [*Larrea divaricata*] desert seemed, back in 1927 when I first explored it, to be pretty nearly eternal: a few roads and trails, a few patches of irrigated alfalfa along river floodplains, a prospect hole here and there, a railroad or two crossing the desert, but essentially an undisturbed, seemingly almost totally virgin system. Farther east from California, in Arizona, subsequent research has suggested that at least some of the desert scrub has been created by overgrazing over a long period of time. At least some of these shrublands seem to be degraded desert grasslands.

This is merely a reminder that this ecosystem is not uniform, either in physiognomy or history. However, nowadays even the vast desert areas are not being spared the heavy hand of Man. Places that 40–50 years ago came about as close as anything on Earth to being perfect wilderness, are now scarred and torn up by off-road vehicles—so badly that the devastation can be seen even from the stratosphere. Other places, and extensive ones, are carved up for 'development'—the construction of large numbers of houses. An extraordinary number of people have discovered the beauty of the desert and the emotional effect that it had on those of us who learned to love it back in the 1920s and 1930s. Suddenly there is a market for the desert—for bits of it with houses on them. For now people want to live out there and feel continuously this combination of peace, beauty and exhilaration. One wonders how soon they will awaken to the fact that they have not bought peace and solitude, but more of the human company which they have tried to escape!

Far worse than this is an ominous development that is destroying even the clear air of the desert. Huge coal-burning power-plants are being located in the desert, pouring enormous volumes of smoke into the air. Strip-mining is taking place to provide the coal for the power-plants, and other minerals that markets have been developed for. Even in Death Valley National Monument, ruthless money-hungry corporations are tearing up the landscape, mining for borax. And the desert landscape is very slow in healing itself when wounded. We are going to have to live with these scars for a long time. If these activities—off-road vehicle recreation (synonymous with noise-making), road building, irrigated farming, housing development, cement plants, air pollution, strip-mining, littering, and more—continue, increasing at the exponentially accelerating rates that we have observed during the past 50 years, we are going to deprive our descendants of a valuable emotional experience that we can still take advantage of, if we are interested enough to search out what desert wilderness areas still remain. If we have the foresight to withdraw large areas of desert—of the order of millions of hectares—from further 'development', and reserve them for our descendants to enjoy, we can assure them of the chance to experience what I have been talking about, namely the peace, beauty and exhilaration that some of us, at least, discovered long ago in these 'wastelands'.

Another major ecosystem that has for the most part been a marginal habitat for Man—especially modern Man—is the lowland tropical rain-forest. A few tribes of hunters and gatherers were normal animal components of this system, but few of them still survive. Yet the tropical rain-forest has, by far, the most species of both plants and animals of all comparable ecosystems. It also has probably the greatest standing crop of biomass of all ecosystems, although some of the temperate coniferous rain-forests and cloud-forests may be competitors for this accolade. An all-too-prevalent confusion between productivity and standing crop has led in relatively recent times to much unwise destruction of large areas of this magnificent concentration of the genetic heritage of the world. It seems hard for people to understand that a soil which can support a dense 50 m high forest can be other than a rich soil. The fact that this is a marvellously closed system of recycling of a limited supply of nutrients (d'Hoore, 1961; Went & Stark, 1968), and that its functioning depends on its dominant components being trees, is not obvious to the temperate-zone orientation of most of those who advocate the conversion of this natural system to agriculture, and its development as 'the breadbasket of the world' (Meggers, 1971; Fosberg, 1973).

Before the period of European expansion, the core areas of tropical rain-forest, where seasonality was not strongly evident, had generally resisted significant encroachment by all but the hunting and gathering tribes (Meggers, *et al.*, 1973). Sedentary or semi-sedentary peoples were mostly found along rivers, in coastal situations and at higher elevations, as well as in seasonal-forest ecosystems. These last tended to be converted to savanna under their influence.

Under European influence, changes began to come about in the rain-forests—at first around the peripheries and along main rivers. Later, when markets for rubber, bananas, dye-woods, gutta-percha, mahogany and other tropical timbers developed, penetration was more widespread. Plantations were established, avenues of penetration were pushed into the forest, and missionaries attempted to domesticate the hunters and gatherers. As medical and public-health measures were introduced into the tropics, population pressures began to build up. Increasingly, attempts were made to 'settle' in the jungle. Shifting agriculture, probably first practised in more fertile mountainous regions, spread into the lowlands. Most of these attempts failed or were only partially successful, because of the inherent sterility of most of the lowland soils (Sioli, 1973; Myers, 1977)—except for those which were flooded annually and those which had been formed on recent volcanic materials. Malaria and yellow-fever as well as other diseases were, until recently, fairly effective deterrents. Change, in the core areas of rain-forest, was, until this century, and even until post-World War II times, fairly slow. This biome was, quite rightly, regarded as a not very favourable habitat for Man—especially for European–American Man.

In this century unrestrained population growth, as well as aggressive commercial enterprise, brought more and more attention to the 'undeveloped resources' of the tropics. More and more, the forest was hacked away. Savanna ecosystems expanded at the expense of the marginal areas of rain-forest. Some of the changes were successful, and reasonably stable agriculture, mostly based on tree crops, began to be established. Enormous mechanical devices were invented to clear and subdue the jungle, and tropical timbers began to be exploited on a vast scale—both for lumber and for cellulose. At the exponential rate that these activities are still increasing, it is now being said that the tropical rain-forest will be mostly a thing of the past before the end of this century. And most of the hundreds of thousands of species of organisms forming it and depending on it will be lost with it. To my mind this will be the greatest and most tragic ecological disaster in the history of the world. The greatest part of the products of hundreds of millions of years of evolution will have disappeared—many of the components without even being scientifically studied and described. Some people say that what Man does not know he is losing will not make any difference to him. Perhaps so, but I do not subscribe to this theory which seems to me a non-starter.

The above examples are entirely typical, in their range of degree and kind of alteration, of the major land ecosystems that I know. They were selected not only to show a spectrum of kinds and of changes, but also to demonstrate the treatment of habitat by peoples in an assortment of technological and economic levels.

It becomes obvious that, even in cultures that have a reasonably good environmental record—Britain, for example—the realities of the economics of overpopulation will force, indeed are already forcing, changes that lower environmental quality. Changes in availability of technology are enabling cultures that have never shown any regard for their habitats to speed up exploitation and destructive processes to the point where, in a few years, these processes may stop for lack of environmental resources to feed on.

It is not possible in a paper of this length to detail the world assortment of human habitats and the range of ways in which the multitude of cultures are going about degrading these habitats. It is safe to say that ruthless exploitation enormously exceeds, and will continue to exceed, any effort even to maintain the quality of habitats, let alone to move towards improving them or reclaiming what has been degraded. The trend seems to be almost universally, exponentially and inexorably downhill.

Conservation Hopes and Activities

A pinpoint of light in this dark picture is the growing interest in the protection of natural areas and in the saving of endangered species, although one must stress interest rather than claim much activity as yet. About the most notable accomplishment of the movement to save endangered species has been to give us a shocking view of how serious the situation is. Of course there have been a few successes with individual species—all, to the best of my knowledge, of birds and large mammals. These are encouraging, but they are dimmed by such realizations as that 10% of the vascular plant species of the continental United States are endangered or threatened, and that between two-thirds and three-quarters of the vascular plant species of the Hawaiian Islands are either already extinct or rare, threatened or endangered (Fosberg & Herbst, 1975). We can only guess at the figures for other tropical countries, but there is no doubt that they are comparable or in some cases even worse. For destruction in the tropics tends to be much more complete and irreversible than in temperate regions.

The only bright side to this situation is that, even in the tropics, where human poverty tends to be vastly more extreme than in temperate lands, there is a growing and audible concern in this field. The people who have come to a realization of what is happening are becoming vocal, and in some cases are working effectively in the right direction. Their motivations are various. There is concern that potential economic or medicinal plants and animals may disappear. There is concern about loss of genetic resources. There is concern about loss of environmental diversity and resulting vulnerability to environmental catastrophes. There is concern about the loss of aesthetic quality and opportunity for experience of beauty and intellectual excitement. These are all understandable and important human-centred concerns, and are adding up to an effective force.

Perhaps the most surprising and significant development, however, is a concern for Nature as such, and for our co-inhabitants of the ecosphere—the other species of organisms. Compared with the magnitude and momentum of greed and selfishness in utilizing the living resources of the world, the realization that perhaps other species have a place in the scheme of things and a right to life above and beyond their potential usefulness to

Man, is a mere pinpoint of light. The concept seems to be one that many of us are born with, but that seldom survives our education. If we could, in some way, preserve and foster this idea, the other concerns would automatically be taken care of. We might then place the same value on living Nature that is now placed on money and other tangible wealth. Dreaming? Of course I am! If I were not capable of this sort of dreaming I would have little basis even for these glimmerings of hope and would not be here.

General Degradation on Earth

Now I would like to get to the point that I want to put across. There seems no doubt as to the almost universal degradation of our habitat. Any of us with ordinary sense must deplore this. Any of us who say 'All is well, Science will take care of us', are to be pitied for the kind of optimism that is epitomized by the fable of the ostrich with its head in the sand.

I intend to use some rough words. I make no apologies for them and simply suggest putting the shoe on if it fits.

The kinds of problems that we are discussing at this Conference can be approached at a series of different levels. Whether by coincidence or not, these correspond somewhat to increasing degrees of human aggregation, and to stages in the development of what has been called civilization.

At the first, lowest level, one can think only in terms of the individual—either himself or one or more others. There probably never was a time when this characterized an effective life-style.

A more natural and logical level is that of the family. This may have been sufficient in the early hunting and gathering phase of human history. A somewhat rational, rather than strictly instinctive, behaviour came into play when this stage was reached historically. The extended family is still an effective unit on a purely short-term basis. But it is inadequate in a world where there are no longer limitless territories and limitless resources for all. Warfare probably came into being with the advent of this level, historically (or, rather, prehistorically).

The tribe or organized community was the normal pattern for many millennia. Again, so long as resources were abundant, this was an adequate frame of reference for human society. It, however, quickly depleted the resources of its effective range of activity. So long as the tribe could move on without impinging on the territory of another tribe, all was well. When such impingement occurred, warfare or some other form of conflict frequently resulted. In modern terms, economic competition may be substituted for warfare; but the effect on the environment may be equally devastating, though in a different manner. One has only to inspect the environs of almost any tropical village or small town to see what I mean. Environmental poverty is very visible there.

Inevitably, historically and in modern terms, tribes and communities formed alliances in one manner or another. Also, increasing size of cities poses similar problems. Governments had to evolve in these states and cities. The aims and objectives of such alliances and governments are to promote the welfare of their particular communities—usually at the expense of the welfare of their neighbours, and almost always to the detriment of the common environment. Rulers of such states or cities historically have developed ideas of conquest, either by peaceful means or, more often, by war. So nations came into being.

The nation is the level of organization at which we currently function, and at which most of us think—regardless of how much lip-service we may pay to 'one world', 'humanity', 'mankind', and such concepts. Nations are perhaps the ultimate expressions of territoriality that have so far been evolved by Man. Alliances have proven ephemeral, lasting only until they proved inconvenient to one or more of the allied parties. Of cartels and multinational corporations I will have more to say later.

The obvious purpose of a nation is to promote the interests of its citizens, or, more usually, of its influential citizens or ruling classes. One might think that, predominant among the interests of the citizens, or even of the ruling classes of a nation, would be the maintenance of the best possible habitat for them to live in on a permanent basis. After all, there is no place elsewhere to which they can go and remain effective, functioning citizens of their own nation (except in special cases).

I have visited the majority of the larger nations of the Earth and many of the smaller ones. In almost none of them have I seen much in the way of intelligent and effective, long-term concern for the habitat of their citizens. Lip-service in unlimited amounts, ill-conceived manipulation or exploitation of the environment for short-term advantages, and frantic activity to convert as much as possible of the resource-base into money in the shortest possible time: these are the patterns which I have seen almost everywhere. Exponentially increasing degradation is the rule. And these nations and their leaders are not at all content with devouring their own entrails. The more successful ones among them are shipping their destructive agents, their 'know-how', their machines, their pesticides, and their bad habits, abroad to gain as much of their neighbours' resources and wealth as possible. They not only help in the devastation and pollution of other nations, but are apt to take the lead in these nefarious ways—especially when the other nations prove slow or 'unprogressive' about such matters.

As though these processes were not efficiently destructive enough, depending on the initiatives and ingenuities of individuals and ordinary corporations, supra-national or trans-national organisms—cartels and multinational corporations—have come into being. These are organized by those who have a great deal of the money in the world and are out to get the rest of it, and to do all of the things that are done under national industrial economic systems—but more effectively, while evading any restraints that may be put on their activities by national laws and tax and tariff barriers. Probably if the full story of the activities and influence of these entities were ever put together, they would be found to rank next to modern warfare in destructiveness. Indeed, some of them may even make use of warfare as a means of advancing their interests. Their role in changing the face of the Earth is a major one, unrestrained by either national pride or national conscience.

One should perhaps not be too critical of 'developing' nations for their haste to adopt the destructive practices that they see as having produced the 'high standard of living' of the 'overdeveloped nations'—with its smog, its carcinogenic air and water, its noise and its ugliness. The will-o'-the-wisp of industrial prosperity is so attractive that they do not even notice the accompanying phenomena. These are 'catastrophes that take place slowly'—usually the very worst kind—but are not so obvious as typhoons and earthquakes.

We see on every hand these days the 'developing nations' short-cutting the slow development of technology, adopting blindly the worst features of the life-styles of the overdeveloped nations—with or without their good features. We hear much of the 'right' of the developing nations to the fruits of modern technology as a part of the rights of Mankind. And the over-developed nations are all too ready to accommodate them—for a price, either in economic advantage or in ideological advantage (perhaps the same thing, ultimately).

It is an interesting fact that, along with the knowledge of how to destroy the world, some of the industrial nations have, all too tardily, learned a few of the principles and techniques of at least slowing down some of the destructive processes that they have fostered. We actually know enough to be able to retard greatly some of the industrially-related threats to our habitat. And we are beginning to apply some of these—with spectacular results where they are given a fair chance.

Although a few of the more far-sighted citizens of Third World nations are very well aware of the dangers we are all facing, this cannot be said of many of their leaders and people in positions of power. We hear in international forums biting criticism of those who, after getting rich through exploitation of their own national resources, would like to impede others from doing the same thing. Most international attempts at environmental protection founder in this sea of protest and righteous indignation—perhaps territoriality is a better term.

What the protestors seem to want is not only the right to the benefits of modern technology, but the right also to learn the hard way—to make all over again the mistakes which the industrial nations have made and which have destroyed much of the best in their habitats. This, in my scale of values and definitions, is nationalism at its worst. It is not merely nationalism, it is stupidity. If there is such a thing as national commonsense, it must very largely consist of the ability to observe the mistakes of other nations and avoid repeating them. There are people in every country that I have visited who desperately want to see their countries avoid what they observe in the wake of advanced technology. Unfortunately, few of these are elected to office or otherwise placed in positions of leadership. This is sad, not merely for these nations but for the entire world.

The national level of organization of human society has clearly not only failed to consider the ecological welfare of the people, but, by its built-in territorial behaviour-patterns, has gone far in leading or pushing us towards the point where the ecosystems in which we live, our habitats, and the world ecosystem, will no longer be able to satisfy our needs. To this extent our society has failed, and does not seem to be a satisfactory framework for thinking or action.

And What of Man?

The next-higher level of approach to the problems we are considering is that of the human species, *Homo sapiens*. It is only as a species that we have enjoyed continuous tenure of our world ecosystem. Individuals come and go, families mostly disappear finally, and tribes, states and nations rise and decline—even cultures are likewise transitory. But hopefully the human species may hold an indefinite tenure.

If we look at humanity as a species, we must consider its habitat as a whole. Destruction or attrition in one part of this habitat reduces the quality of the whole. Impoverishment of the biota of the Magdalena Valley by profligate use of pesticides is not just the concern of the Colombians. It lowers the diversity of the world, which is the habitat of all of us. The Japanese have just as much right to protest a threat to the Grand Canyon as have the Californians, or even the Arizonans. It is a part of the riches of the habitat of Man.

If we continue to behave as a pioneer species, two things are clearly predictable. We will modify our habitat so that it will no longer support us. Then a population crash, from whatever direct or immediate cause, will follow. If this comes by atomic war it may wipe out the species. If it comes by famine or disease, a few seem likely to survive to continue the existence of *Homo sapiens*. If this happens, and a few wretched survivors are left in an unfavourable and hostile environment, a severe process of natural selection will inevitably ensue. The products that evolve from this may be very different from what we know as humans. Very likely we would not care very much for them; for surely the decent, kind and gentle ones will not survive to inherit what remains of the Earth.

What of the habitat itself? If present trends continue—and there is every reason to think they will do so—we shall have a dreary world. Its organic diversity will be reduced first to a fraction of the existing species, and those remaining will be mostly cultivated and weedy. If Man disappears, most of the cultigens will disappear with him. Then the weeds and insects that have been able to evolve general resistance to pesticides, and the descendants of such domesticates as are able to survive without Man, will inherit the Earth. These will repopulate an eroded, sterile possibly radioactive Earth. If they survive they will serve as the genetic material from which a new evolutionary diversification may originate. Perhaps in a thousand million years or so, maybe less, a new dominant, intelligent species may evolve. Hopefully it will not be one that fouls its nest and destroys its own habitat, and, hence, itself!

The alternative to this dismal prospect is that *Homo sapiens* may forego the evils of nationalism, may have the intelligence to bring his population increase under control, may give his love of beauty and variety a chance to call the shots for a change, and may place sufficient value on Nature to set aside adequate (and adequate is big) samples of all the important major ecosystems to provide stocks to repopulate and rehabilitate the already devastated Earth. Under intelligent management and loving care, even the most degraded ecosystems may be brought back to productivity and the capability of providing satisfaction for a rehabilitated humanity—one that will have chosen a 'climax' pattern of behaviour. This will mean that its individuals will leave their habitat a better place than they found when they were born into it. If as individuals we each go on as usual and leave it to the others to change their life-styles, there will be no 'environmental future' for us. We will have demonstrated our unworthiness and we will follow the dinosaurs into extinction!

References

Bryson, R. A. & Murray, J. T. (1977). *Climates of Hunger*. University of Wisconsin Press, Madison, Wisconsin: xv + 171 pp.

Carr, D. & Thorpe, J. (1961). *From the Cam to the Cays*. Putnam, London, England: xii + 190 pp.

Carson, R. L. (1965). *The Sense of Wonder*. Harper & Row, New York, NY: vi + 89 pp.

Cloudsley-Thompson, J. L. (1975). Environment and human evolution. *Environmental Conservation*, **2**(4), pp. 265–9.

d'Hoore, J. (1961). Influence de la mise en culture sur l'évolution des sols dans la zone de forêt dense de basse et moyenne altitude. (Abidjan Symposium.) Pp. 49–58 in *Tropical Soils and Vegetation*. UNESCO, Paris, France: 115 pp., illustr.

Finch, D. W. (1976). Report on Northeastern Maritime Region.... *American Birds*, **30**, p. 69.

Fosberg, F. R. (1958). Preservation of Man's environment. *Proc. Ninth Pacific Sci. Congr.*, **20**, pp. 159–61.

Fosberg, F. R. (1973). Temperate-zone influence on tropical forest land-use: a plea for sanity. Pp. 345–50 in Meggers *et al.* (*q.v.*).

Fosberg, F. R. & Herbst, D. (1975). Rare and endangered species of Hawaiian vascular plants. *Allertonia*, **1**, 72 pp.

Meggers, B. J. (1971). *Amazonia, Man and Culture in a Counterfeit Paradise*. Aldine & Atherton, Chicago & New York: viii + 182 pp.

Meggers, B. J., Ayensu, E. S. & Duckworth, W. D. (1973). *Tropical Forest Ecosystems in Africa and South America: A Comparative Review*. Smithsonian Institution Press, Washington, DC: viii + 350 pp., illustr.

Munro, G. C. (1944). *Birds of Hawaii*. Tongg, Honolulu, Hawaii: 189 pp.

Myers, N. (1977). Garden of Eden to weed-patch. *NRDC Newsletter* (Washington, DC), **6** (1), pp. 1–15, illustr.

Pearsall, W. H. (1950). *Mountains and Moorlands*. Collins, London, England: xv + 312 pp., illustr.

Sioli, H. (1973). Recent human activities in the Brazilian Amazon region and their ecological effects. Pp. 321–34 in Meggers. *et al*. (*q.v.*).

Thompson, J. E. S. (1966). *Decline and Fall of the Maya Civilization*. University of Oklahoma Press, Norman, Oklahoma: xv + 328 pp.

Went, F. W. & Stark, N. (1968). Mycorrhiza. *BioScience*, **18**, pp. 1035–9.

Discussion (Session 3)

Gill (Chairman, introducing)

I would like to begin Session 3 by extending the appreciation of Canada and Canadians to the patrons and sponsors of this Reykjavik Conference and especially to Dr and Mrs Polunin for the guidance they have given us in organizing this most urgent exchange of concepts concerning our environment. This Session is devoted to the preservation of the Earth's terrestrial habitats as the home of Man, and its keynote address will be given by Dr Raymond Fosberg, of the Smithsonian Institution, Washington, DC [—probably the world's largest scientific outfit and certainly among the best, whose head was on the Steering Committee of this Conference]. Now many of us first became aware of Dr Fosberg's work through his writings on island ecosystems. However, his [interests and influential contributions] extend far beyond that sphere—not only in space but also in time. Dr Fosberg [in all his continuing vigour] has been working on many of the world's ecosystems for almost 50 years—think of that, nearly half-a-century! This wide experience has equipped him particularly well to speak on this [most vital and] urgent topic of the preservation of the habitat of Man. But not only Man; it is gratifying that Dr Fosberg recognizes that, to preserve Man's habitat, we must likewise preserve our shrinking natural ecosystems for the benefit of those several million other species that share our world as well.

It will, I think, be borne out this afternoon that Dr Fosberg is a pessimist, or perhaps a realist—or would a pragmatist be a more appropriate term? However, he is also a futurist—one who holds that the human organism, although it has now evolved to a level where we are able to destroy the world ecosystem, may yet continue to evolve to a higher level where we just *might* be able to reverse the universal trend of habitat degradation.

I agree with Jan Tinbergen that, because of the incredible complexity of the real world, the major problem that we—as ecologists, environmentalists, or whatever else you may wish to call us—must face is that of quantifying the biophysical environment, even at the microscale. And of course at the macroscale we have even further to go. But before we are able to begin affixing figures, we must give birth to new perceptions, and one of the great values of this kind of conference, and of the address which you are about to hear, is that it stimulates people like myself to think in macroscale or even global terms. Up to now my own ecological research, and my own ecological awareness and perception, have been on a microscale or at best a mesoscale, and meeting with people like Dr Fosberg and our panellists is helping me to enlarge my perceptions and to recognize that we need really a world view of Man's habitat before we can begin properly the vital job of restoring and ultimately preserving it.

Fosberg (Keynoter, remarks in presenting paper)

I am very much flattered to be here, speaking to such an audience as this. I do not think I will tell you very much that most of you have not heard before or thought out for yourselves, unless it be that in 1958 I published a short paper entitled 'Preservation of Man's environment', and I have regarded this occasion as an opportunity to enlarge upon and update the ideas set out in that paper. I must add that I have not found it necessary to retract or change any of the ideas there expressed, but am giving you here a statement of self-evident principles, observations, and perhaps obvious conclusions, with no claim whatever to originality. [Where my paper is] undocumented, I am perfectly willing to take full responsibility for the statements that I make, without falling back on authorities. But although the object has been to place in sharpened focus the subject of this

Conference, unfortunately my paper has more of the gloom that we were talking about this morning.

I do not in general attempt to give figures, for example of the possible duration of the five thresholds that I see in the evolution of Man's influence on his habitat and the world. In fact I am not altogether in accord with what was said this morning about the necessity for quantification. I think quantification would be a good thing if it could be done reliably; [yet] on the scale which we are talking about I am perfectly certain that it cannot be done reliably, and that is why I will not indulge in [attempting] it. However, the stages separated by those thresholds obviously decrease in length of time and may be roughly arranged on an exponential curve, which seems to be characteristic of most phenomena of man-influenced environmental change. Thus deforestation, pollution, loss of diversity and many other current phenomena you will find fit this exponential pattern all too well. [It seems to me] that the exponential curve which we observe in human population increases merely the exponential portion of the normal population curve resulting from less and less restrained population growth. I do not think Man is likely to escape the full population curve indefinitely, but rather that we are now in the exponential phase of it. I think a good analogy may be pointed out, namely that Man perhaps is playing the role of a pioneer organism. The population curve of a pioneer organism is very well known: after the exponential increase there is usually a catastrophic decrease or crash.

In my paper of 1958 I pointed out that perhaps Man was behaving as a pioneer organism but perhaps he was the only organism in the history of the world that had a free choice as to whether he would behave as a pioneer or as a climax organism. By climax organism I mean one that occupies its habitat and leaves it in as good a condition as it was found to be on entry: in other words, an organism that is capable of sustained occupation [at least of the habitat under consideration]. I think all the evidence points to the fact that Man continues to behave as a pioneer, and to occupy a pioneer role, and the consequences of this are evident if we look at history, or what we know of it. Thus on a regional scale, there have been cultures—very highly developed cultures—which have reached the crash period and then vanished, leaving very little evidence of their existence except perhaps for some ruins, perhaps a few accounts in old literature. Of course it was not all as simple as this. I think that most crashes or ends to cultures in the past have been accompanied by wars, invasions by barbarian hordes, and so forth. I guess we are at the point now where we do not have any more barbarians to invade us but this does not mean that if we reach the stage of a crash it will not be accompanied by wars—particularly wars between ourselves. These can be fully as devastating as Attila and his hordes proved to be in Europe.

It seems evident to me that such a war between ourselves would result in the destruction of our present culture, and possibly in the destruction of Mankind completely. Obviously I am pessimistic, as Mr Chairman pointed out; but I am not the kind of pessimist who simply throws up his hands, otherwise I probably would not be here. There is one possible mitigating feature that I see in this dismal present situation that we are in, and that is the factor of human intelligence. This could enable Man to choose a different role—a different course from the one which he seems to be following at the present time. One of the things to be emphasized here is the importance of biological diversity. We have millions of species of organisms living in this world. This represents a pool of genetic resources that has developed over more than a thousand million years, and I think it is unquestionably the most important single resource that we have on the Earth. But of course we are busily destroying it, and that which we lose we will never regain.

What must the human habitat supply—the requirements that we have a right to expect from our environment? The commonly accepted ones are breathable air, potable water, livable temperature-range, adequate food, and space in which to live. Perhaps also now—although this was not always so important—we need energy to fuel the cultural machine that we have created. These are usually regarded as the essentials, and unfortunately I think that most writers and planners accept this as the pool of essential necessities, so that their plans and projections rest on this. [The assumption is that] if these things are satisfied, at least at a minimum level, the situation is in hand; there is nothing to worry about. But I think acceptance of this point of view places humanity on exactly the level of domestic animals, which have the same requirements. I think that we have far greater needs than these bare essentials, [including] beauty and variety or diversity in our habitat.

In most of the major ecosystems that I have examined in various parts of the world [over a period of nearly 50 years], the trend at least to some extent is universally, exponentially and inexorably downhill. In other words our habitats are almost all becoming more and more [limited, depleted and generally] degraded. In this situation there is perhaps one relieving feature, and this is something that has emerged on a significant scale only very recently—the widespread and more and more vocal interest in the preservation of examples of our natural environment and in the preservation and protection of endangered species of plants and animals. This is still mostly talk, although some progress has been made and there are several people in this room who have contributed materially to the preservation of natural areas; indeed I think that their activities are some of the most significant that we have going on at the present time. The individual interest on the part of people in these things is the most encouraging development that I am aware of, and although it has not had much effect as yet compared with the magnitude of the problem, it is getting under way and perhaps it will behave in the same exponential fashion that our other phenomena mentioned have behaved! [Applause.]

Gill (Chairman)

Dr Fosberg has given us some blunt and very straight talk, and I think you will agree that our panellists have their work cut out for them. I would like to introduce them in the order in which I plan to call on them. Sitting on Dr Fosberg's right is Dr Raymond Dasmann, from [Switzerland and] the United States, who is perhaps the leading synthesist in ecological writing today. He has been with the IUCN for some years and [is now on his way to take up a Chair in] the University of California at Santa Cruz. He will comment on Dr Fosberg's paper from the standpoint of an ecologist. To Dr Dasmann's right is Mr Perez Olindo, from Kenya, for many years Director of their famed National Parks and now Science Secretary to the Kenya National Council for Science and Technology, in their Ministry of Finance and Planning. He will comment on Dr Fosberg's paper from the standpoint of a field biologist. On the far right is Dr Andrew Macpherson, from Canada, who is Director-General of the Western and Northern Region of Environment Canada. He will comment on Dr Fosberg's paper from the standpoint of a resource manager.

Dasmann (Panellist)

[Unfortunately] I am not in the position to criticize Dr Fosberg's paper in any way because I agree with everything he said! I was sitting here thinking of how to be more optimistic about the future, because Norman Myers over there [from East Africa, and our Secretary-General and many others who are present] does not like ecologists to be doomy and gloomy: he wants them to be cheery and beery. Actually there are some good signs, and we can take particular heart in the enormous growth of interest in the environment and in conservation which Ray Fosberg and I have both seen happen, as we both were involved in trying to preach a conservation message when it was something that just did not get into the newspapers at all. During the 'sixties in particular, and following on into the early 'seventies, we saw a great upswing of interest in the environment. Oddly enough this distressed one group of people more than anyone else, and that was the ecologists. The ecologists were really upset at this upsurge of interest in ecology [albeit in the popular connotation], and for a while it seemed that the principal concern of ecologists was rushing around defending their discipline against public support [laughter].

In the United States the peak of interest in ecology and conservation seems to have been reached, but I think there will be another upsurge as things get worse. In Europe I believe such an upsurge is just beginning. You see really remarkable things happening these days in Europe where, for example, the Ecology Party in France can scare the major parties to the point where they are actually considering some environmental measures. The spread of environmental interest in the developing world has been considerable, [rising] from practically zero when for example Perez Olindo started in Kenya with the National Parks, to really quite a reasonable level of interest at the present time. That is one encouraging trend.

The other trend which I see as particularly hopeful is the fact that in Europe and in the industrialized world generally, many countries have reached a level of no-growth population while others are rapidly approaching it. You do not have to go through a long transition period to stop growing: you just have to cease having more babies than there are people dying, in order to balance the two. It can happen virtually overnight, and in Europe this does seem to be taking place. And as the developing world is not the real cause of the trouble whereas the developed world *is*, I think the fact that population growth is ceasing in the industrialized world is really quite a cause of good cheer. But if the level of economic demands can be reduced on the lines we were discussing this morning, I think we could be even more cheerful.

I am also concerned about two things that have been said so far in [this Conference], and one is the appeal for quantification—as though people were greatly convinced by figures. The only figures they are really interested in are those for example on a pay-cheque, or those they have to fill out on income tax forms. Figures indicating quantities are not necessary to get people moving. And then I am concerned about the appeal for more and more research. I like research—part of me is still scientist—and I like to see the frontiers of knowledge advance. But the idea that more research is going to influence the political process very much is, I think, misleading. People start moving when their hearts and minds are engaged, and their feelings are [roused]. That is what happened in the 'sixties with the environmental movement. It will happen again, but not by accumulating information or doing more research. It will happen when we get out and convince others, selling our cause to them by talking heart-to-heart to them to get them moving again. If we succeed in this you will see the political picture changing very rapidly.

Olindo (Panellist)

In the developing countries we started practically in the [wilderness], from the point of view that we did not know very much about Nature; so to shield against obvious mistakes in this area we decided in Kenya that perhaps the best thing we could do for the national environment was first to search for what looked familiar. So what we did was establish National Parks as far as possible in representative ecosystems. These were representative in so far as our eyes could see and the little tests we could make could prove. From our East African experience in Kenya we came to a point where we felt that if we could succeed in setting aside up to 15% of the total land area, we might be making a good start.

We did not feel brave [enough] to talk too openly about this strategy, for a number of reasons. However, we are convinced that if 15% of the total land area turns out to be too big a proportion maybe to give complete protection to, there will always be time to use it or to put it to some alternative use. In pursuit of this objective we have suffered a number of setbacks and a bit from political comments to the effect that our major goal seemed to be to convert the entire countries of East Africa into international parks. [This unfortunately generated a measure of resistance to the concept of nature conservation.]

I was talking to a friend just last week and he expressed the view that Mankind takes action when there is imminent danger. So perhaps the role of the pessimist is justified, because if there is enough call of the dangers that are approaching us, reaching the larger population of the world, there is every reason to believe that corrective action to counteract what is to come would be considered if indeed humanity does not wish to be swept away by the current of destruction.

From the little knowledge I have of human expectation, each individual wants to survive and will do everything possible to achieve this goal—which brings me to the point of sharing knowledge. We in the developing countries are led to believe that there is a lot of knowledge around the world which has not yet been put to work, being stored in books, in theses, and generally in libraries. We want to go on a search for this knowledge, to identify gaps in it, and to encourage research to fill those gaps, putting this vast store of knowledge to work to save Mankind—to save the Earth for ourselves and for those to come. We are making a move in the direction of controlling population by restricting the rate of natality. I accept and fully agree with Professor Kuenen on his approach to that great issue of population explosion. However, I differ from Professor Kuenen on the concept of sharing of resources, as I believe that in our effort to look for solutions to the impending problems that may lead to disaster there is a sufficiency today of resources to share, if [only a transport] system can be found and designed and if the economics of the matter are accepted by those who have more resources than others.

This morning there was [reference to] the question of energy consumption. In many developing countries the majority of people depend on firewood as a source of energy, and I need not overstress the impact of this demand on terrestrial ecosystems. But what can we do about it? They will need to continue to use firewood in order to survive—to provide a means for cooking their food and so on. Looking at the crises that have been created recently by the restrictions on the movement of fossil fuels, the rocketing prices, and the inability of many developing countries and young economies to cope, I think we have reached a turning point [and should] look again at our life-style, [and consider the practicabilities] of sharing and change. In many parts of the world there is very little that can be shared, while in other parts there is a lot that is wasted. When we look at all these happenings, one pointer comes to surface: the need for evolving an international code of ethics which may later be developed into an international legal code for the sharing of the limited resources on Earth which, I believe, as we started the day by feeling, can give a reasonably full though simple life to the people we have on Earth.

This morning we commented briefly on the restrictions [imposed by] increasing human populations in developing countries. I do not want to be a spokesman for all, but should say that in Kenya until now there is a tax incentive for having several children. This morning a friend suggested to me that perhaps this tax incentive should be shifted from children to the number of trees that a person can grow. I fully agreed with him, because it would be a move in the right direction.

Macpherson (Panellist)

My personal experience [has been largely] limited to a rather fortunate country, and indeed to a rather fortunate part of it—Western Canada, where I think we are above the second threshold of Dr Fosberg's very interesting analysis. To his concerns about degradation of habitats one might also add the problem we have in protecting habitats from natural ecodisasters, and one quite striking example of this is the Elk Island National Park outside the city of Edmonton, Alberta, which used to be part of the general parkland zone of that part of the country and has been protected from fire for a long time. But now, due to agricultural and other activities going on around it, instead of containing the array of plants and animals which it supported when it was established, the Park is largely part of the immense Boreal Forest Zone which otherwise starts a considerable distance north of its location. An [allied] problem is the damage that is done to delta areas and riverine habitats downstream of impoundments on rivers, due to protection from the spring flooding and disturbance. This is a problem that concerns us greatly in western and northern Canada now that we have several new hydroelectric projects in the planning or construction stages.

The prairie—the tall-grass prairie which extends also into western Canada—I quite agree with Dr Fosberg is very difficult indeed to find remnants of, though one turned up by the International Biological Programme CT section was inside the bounds of a city, whereas all the surrounding rural parts had been ploughed up and had totally vanished. But quite apart from the problem of identifying and setting aside special areas, seems to be the lack perhaps in most countries of the institutional arrangements for establishing and looking after such areas. The IBP programme turned up a number, identifying very useful and interesting candidates for reserves, but in my country at least we have had great difficulty in institutionalizing these reserves and even deciding what sort of body might preside over them and regulate their use. The problem is not only institutional but also attitudinal. The people who depend on wildlife for recreation in Canada have a saying that 'ducks don't vote', and that is really part of the problem: instead they have to be voted for.

There is one great improvement as Dr Dasmann said: the era of elites who defended the biosphere has been replaced by one of popular enthusiasm for environmental causes. Of course there are some questions which I think of asking technocrats and resource managers such as myself, including the perceptions of the public and whether they are really reliable and well-founded. We recently completed a very interesting exercise in western Canada with an enquiry into the consequences of building a gas pipeline down the Mackenzie Valley, and the social perceptions involved. The enquiry moved through the north and 'did' most of the major Canadian cities in the south, and I think as a consequence of that one might generalize and say that the environmental consciousness goes from a high spot somewhere in the middle of the [remote countryside] to a low in the most sophisticated urban concentrations. As a matter of fact the native people's conception of the realities of relationships of animals to their habitats and ecology in general was remarkably good, and many of them saw the land, the untouched environment, as a sort of savings bank to which they could turn when they needed resources to support them. The urban people on the other hand seemed to be very much out of touch with Nature and even squeamish about it, showing a general failure to understand its realities. Thus a neighbour of ours, sitting in her backyard writing a letter to the local paper about field hunting, had a caterpillar drop on her notepaper and immediately telephoned the exterminator, prevailing on him to come and spray the whole neighbourhood—which was surely unrealistic!

For the future, perhaps we are getting into a condition of having a better-rounded information base. Perhaps this will eventually relieve and submerge the ignorance that surrounds the issue of environment and the narrow thinking that characterizes it. Undoubtedly the public attitudes that are the forerunners of improved institutional arrangements will have to be cultivated if there is to be a light at the end of the gloomy tunnel, both nationally and internationally.

Gill (Chairman)

Let us now open Dr Fosberg's address and the comments of the panellists to speakers [from the floor]. As before, please signal for the microphone and, when you get it, state your name quite clearly and where you are from, and also please stand up so that we can all see who is addressing us on the panel.

Sigurbjörnsson

I was glad Dr Fosberg announced that he is a pessimist, otherwise I might have missed the point [laughter]. His paper was very gloomy and I am surprised that he did not put some bright spots into it, because I find a lot of bright spots even in the topic he was talking about. I would like to make three points. The first concerns the erosion of genetic material. I agree that there is a great danger in losing a lot of the existing genetic material and valuable plant species [that we now have to choose from]. Certainly we must collect and preserve germ-plasm—especially gene-complexes of agronomic value concerned with the cultivated crops, and we must preserve the old or obsolete cultivars that have been threatened by monocultures. But the bright spot here is that on the single-gene level we have methods of creating artificially [through induced mutations] any single gene that now exists in plant populations, and possibly also those that have occurred during evolution and later disappeared, or even those that have never occurred during natural evolution. This method is available, and it has been demonstrated that this can be done, though I should emphasize that it is possible only on the single-gene level—not for gene-complexes. So I agree that we have to be very careful of losing too many of the genotypes and gene-complexes which now exist.

My second point concerns fertilizers. The implication has been that chemical fertilizers are something bad, at least to the extent that they are like a stimulant given to an athlete, and I want to state that chemical fertilizers can and do improve our environment, as may be seen quite clearly from examples here in Iceland. I will cite one instance. In the central parts of Iceland we have fairly large areas of black desert sand which has been a source and still is a source of erosion and dust-storms which cover that part of the country. About 25 years ago, some of this area was seeded with exotic seeds and fertilized, and then the area was abandoned after about three years. But 20 or 25 years later this sandy desert has turned into a beautiful 'natural' grassland, with some shrubs and other native vegetation, and with birds nesting. Some of this land was ploughed last spring and found to contain a lot of humus, which had [developed following treatment with] the fertilizer. Here I believe that chemical fertilizers have resulted in some improvement in our 'natural' habitat.

The third point which I want to make concerns the many references that have been made to the beauty of untouched Nature, and I certainly agree that Nature has beauty that has to be preserved. But I do not like to look at our Earth—my country and others—only as a national park that tourists come to enjoy. I think Nature is part of our home and a part of the environment from which we have to derive our living. I also believe that human habitation can improve Nature in many ways, although it can of course destroy Nature. Certainly I agree with many of the points you have made in that respect; but there are some bright spots. Yesterday, by coincidence, I was travelling along a narrow fjord in Iceland and I was looking across the fjord. There is a mountain there, bleak and barren like most of our mountains here in subarctic Iceland. But along the coast there were three farms with bright-coloured houses, red roofs and green cultivated land, and I thought

'Thank goodness for some human habitation!'. For the farms really improved the landscape and without them the prospect and all Nature would not have been nearly so beautiful. To our men from the Canadian Rockies I must say that when I compare the two mountain ranges, the Canadian Rockies and the European Alps, I prefer the European Alps primarily because of their human habitation. I like the alpine huts, I like the farms, and I like the animals and the orchards. I think that in these cases human habitation improves Mother Nature.

Myers

I would like to comment briefly with regard to what Dr Fosberg was saying about tropical rain-forests. I fully support his view that if these forests go under within the foreseeable future of the next four or five decades (as is likely if present patterns of exploitation persist, let alone accelerate), it will surely be a major ecological tragedy—one of the greatest in the history of our society.

I would like to comment also on an angle which was raised by Professor Kuenen this morning and by Dr Dasmann earlier this afternoon. Last autumn I went to Central America to have a look at their tropical forests and I found that their rain-forests are being cut out so fast that within another 5–8 years at the present rate of exploitation there will be hardly anything left. These forests are being cut down not so much to supply timber products as to make way for artificial pasture-lands for cattle. During the last 10 years the number of cattle in Central America has more than doubled. Yet during those same 10 years the amount of beef eaten by the local inhabitants has actually diminished. So where is all the extra meat going to? It goes north to the United States and Canada, where it is fed into the hamburger chains and other fast-food organizations that can get their beef from Central America more cheaply than they can get it from domestic sources such as Texas.

This means that the American consumer who goes out in the evening looking for a hamburger—and looking for the best quality at the cheapest price—may unwittingly be supporting this process which is going on in Central America. He might well be a man who belongs to the Sierra Club and other organizations, who sends off $10 a week to the World Wildlife Fund; so what he contributes with his right hand he may well be taking away with his half-dozen left hands.

Thereafter I went to Indonesia and visited the forest concessions of several major timber corporations—Weyerhaeuser and Georgia Pacific from the United States, Mitsubishi and Mitsui from Japan, Bowater from London, and others. I talked with these people and they say they are exploiting the hardwoods in order to supply luxury specialist timbers for the developed world—by which they mean veneer, fancy plywood, fine furniture, sophisticated panelling, and other stuff of that sort. Once again the market demand for these hardwoods is coming not from within the developing world but from far outside—from Europe, from North America and from Japan. At the same time these advanced nations are supplying, through the multinational corporations, the capital, the skills and the technology to enable these developing countries to exploit their forest resources at unsustainable rates.

I then flew home to Nairobi, and before I left there again six months ago I arranged to have a new house built and told the architect to make it nice and furnish it well. When I got home again I found that it was indeed a nice house—parquet floor, fancy furniture, fine panelling. Mr Chairman, maybe the upshot of all this is: by all means let us weep over the fate of what is happening to the tropical forests, but while doing so maybe we should not use too many [sheets of] Kleenex.

Rudolph

I want to say something about an area in the world that maybe is a little different from most which you think of and that is the great Antarctic Continent and the waters which surround it. Right now the oceans of the Antarctic Continent are being fished for Krill [*Euphausia superba*] at an extreme rate, in a system that is little understood, without knowing what the consequences of this fishing will be. The future may hold a similar picture for the continent itself, as there is nobody *in situ* to speak for it and protect it. So I wonder if any of the panellists have suggestions as to how one protects an ecosystem that has no resident [human] population to look after it, and whether it is worth protecting such an ecosystem anyway?

[With interspersed comments that the Antarctic was a world heritage and should be treated as such, and on the Mackenzie River pipeline being stopped for political rather than ecological reasons, the Chairman ruled that, with time getting short, this question should be held over for the 'marine panel' (Session 5)].

Royston

A comment: I now find myself in the most unexpected position of standing as defender of the multinational corporations, though over the last three years I have been consistently cast in the role of their attacker. But in view of some of the comments that have been made today both from the platform and from the floor, I think it would be well to recall that the multinational corporations are probably at least no worse than other corporations, and in a number of particular instances and especially in their environmental behaviour, are probably better than most corporations. This is because on one hand they have extreme 'visibility', being closely dependent on the host nations, while on the other hand they have large reserves of skilled manpower which enables them to behave in an environmentally sensitive way. One example—if we are looking for gleams of light—of the many instances which I could quote of multinational corporations operating in an environmentally sensitive way, was British Petroleum building the Forties Field Pipeline across Scotland. They studied every square centimetre of the ground along the proposed path, making sure that the pipeline was put down in a way which was both ecologically and sociologically the best run. They came across a particular patch of ground and when their ecologists looked at it they found a completely new ecosystem including subarctic flora that had not been detected before. On their own initiative [and at considerable cost], they deviated the pipeline around the area and it was declared a 'trust' area. There are many instances of this nature—of multinational corporations operating in an ecologically sound way. It does not mean to say they are angels—they certainly are not—but we do I think need to retain a little balance in this judgement.

The same thing with the developing countries. I was a bit put off by the idea that developing countries only want to industrialize and then, when they have got the fruits of industrialization, will start thinking about environmental consequences. This might have been the case five years ago, but now as we have found—and we have recently been working with 30 or so developing countries, surveying attitudes of some 12,000 decision-makers and policy-makers in such countries—the great majority of those say: 'We have learnt from you; we want to grow, yes, but we want to grow in a way which is environmentally sound. We want clean technology, clean industrialization; we want to make sure that growth is compatible with the physical and cultural environment.'

One sees springing up in the developing countries many instances of this being translated into practice. One of these in fact was already mentioned by Mr Olindo, and in the Philippines they are actually offering tax incentives or tax rebates on property tax for people who plant trees. This is a very real policy which is working in a country that has already suffered in the past from massive deforestation and the tremendous consequences of that deforestation for flooding, for the drying up of springs in summer, and for the loss of fertile topsoils. [Presumably as a result of having seen all this and of understanding the reason, the Filipinos are now some of the most environmentally-conscious people to be found in the developing world.]

So I think that we have to see the positive side as well as the negative side, and to avoid sweeping generalizations. But there is one comment that I would like to make, perhaps linking the above with the specific subject of the paper: moves are taking place very widely to protect habitats and terrestrial ecosystems, particularly through the rapidly-growing use of environmental-impact assessments. These are now coming in in Europe, where there are already a number of countries that demand such assessments as a matter of law. Within the European Economic Community it is going to come into law in the [ten] countries involved. In one developing country after another they are introducing environmental-impact assessment procedures into the planning process, as an integral part of the planning to make sure that the industrialization plans do not have a damaging effect on the sustainability of the ecosystems. They are also using this technique to get a broad involvement—not only of all the ecologists and other specialists who have something to say about the impact of a particular industrial development, but also of local communities.

One of the things I have been most struck by in a positive way [in this Session has been the] talk of beauty. In fact, in one of the provinces of the Philippines they have an eight-points development programme and one of those eight-points is for the beautification of the province. One section of the human ecosystem which is most concerned about beauty is the ordinary citizen in the village—in the [heart of the] community. When he gets involved through the environmental impact-assessment process, he can introduce those values into the planning process. So I see very many gleams of light. Perhaps it is rather like someone looking at the dawn, seeing a glint of light on the horizon, and saying 'The sun's coming up'. But someone else who is standing next to you is looking up over ahead and says, 'Oh no, it is still the middle of the night'.

Glasser

Several people have mentioned during the course of the discussions that there is a demand, unreasonable by implication, for quantification of the dangers which have been so aptly aired by, among others, Dr Fosberg himself. I think that we ought to pay a lot of attention to this demand, and I would like to put this in the form of a [suggestion]. People are apt to be scared of two things: one is the price of what they want, and the other is the fear of what might happen if they got it. If we could evolve a method whereby we could convey to constituents of all the politicians in this world that there is a danger of falling off the edge of the precipice which people have talked about, then we might get somewhere. One detects in the modern world a feeling that people, as they will not abandon their desires for higher consumption, are being prepared for entry into the nuclear age with a vengeance—in spite of the dangers residing there. I would suggest that perhaps one of the useful outcomes of this Conference would be to work out a method whereby people—meaning the common man—could be made aware of these [grave matters], so that these aware people in turn could influence the leadership groups in their countries [towards action for avoiding dangers looming ahead].

Gill (Chairman, concluding Session)

Thank you all for making this a rather successful Session. The topic has been [essentially the preservation of] Man's habitat, and I would like to put this in historic perspective a little bit. This Conference is taking place in 1977, and yet if we look back in the Old Testament in the book of *Isaiah* we see that he, too, had concern for Man's habitat even way back then. He said something like this: 'Woe be to them that lay house to house and field to field, so that there be no place where Man may be left alone on the face of the Earth.' So perhaps there is nothing really new in our concern today about Man's habitat, but what is new of course is our concerted effort [to do something about the situation] rather than having one individual [concerned] such as Isaiah.

Session 4

WHITHER FRESH WATERS AND THEIR BIOTA?

Keynote Paper Part I

Whither Freshwater Biota?

by

LETITIA E. OBENG

Chairman of Soil and Water Task Force Division of Environmental Management, United Nations Environment Programme, PO Box 47074, Nairobi, Kenya; formerly *Director, National Institute of Aquatic Biology of Ghana.*

INTRODUCTION

The subject of 'Whither Freshwater Biota?' implies a general and directional change in freshwater fauna and flora which is not wholly favourable and which might need checking. It also lends itself to a number of interpretations, although I prefer to consider it to refer to the effect of human activities not only on freshwater biota but also on the aquatic environments with which they are associated and which support them. It is accordingly relevant to examine these activities as causes and agents of change, and to consider the impact which they make in our particular context.

It is not my intention to present a scientific paper with new data and information on unknown freshwater systems; nor do I believe that this very serious environmental subject should be considered as only academically interesting. There has already accumulated over the years, through the conscientious work of many distinguished scientists of whom some leaders are in my audience, a vast amount of knowledge on the subject of fresh waters. I feel that we owe it to them that any further consideration of subjects such as the present one, and in a forum of this kind, should attempt to build an overall picture towards a holistic understanding of fresh waters and what we do to them—in order to assist in giving practical guidance for better environmental management. For this purpose, the consideration of a fair amount of information which is already well known is unavoidable, and my presentation should be seen against such a background.

As water is such a key resource for life, freshwater systems, represented by rivers, lakes, and other non-salty surface waters are subjected to extensive demands, and the changes which are made affect the biota. Extraction of water from stationary water-sources may cause changes in the biota. Modification of water-sources, such as is effected by the damming of rivers, may hold the water flow in check and increase the local quantity of water which then becomes available for enlarged fisheries, hydropower generation, navigation, agriculture and recreation; such modification has an impact on the biota.

In addition to being sources of water-supply, rivers and lakes are often deliberately used for disposal of human and domestic wastes, municipal sewage and industrial effluents—including heated waste-water. They also receive drainage water from storms, floods and irrigation works. These kinds of interference with freshwater resources, which also introduce various materials into them, tend to affect and change the quality of the water and also affect the biota.

Other changes, indirectly due to human action but normally not within easy control of Man, may also cause quality deterioration. In watersheds where, because of geological characteristics, the rate of chemical weathering is slow, some lakes and streams do not have the necessary buffering against acidity from atmospheric precipitations. Wright & Gjessing (1976) report that a number of lakes and streams in Scandinavia and North America, which have received highly acidic precipitations (weighted average pH 4.0–4.4) over several decades, are seriously threatened.

Quite often, the introduction of exotic species, deliberately or accidentally, has produced major biotic effects on fresh waters and their biota (e.g. Courtenay *et al.*, 1974; Courtenay & Miley, 1975).

Wherever interference caused by Man or other agents results in physical, chemical or biological change of the water-system, some or the entire range of biota may be affected. To put these changes in an appropriate perspective, we should bear in mind that, under 'normal' circumstances, freshwater plants and animals are adapted to specialized, complex aquatic environments with specific chemical and physical characteristics, which distinguish them from brackish and salty waters.

Biotic Tolerance of Disturbed Habitats

The finer characteristics and nature of freshwater bodies tend to differ according to their location. Temperate-zone rivers, for instance, show marked differences in faunal diversity and populations from tropical ones. In some instances, variations exist between the source, the body and the mouth of the same river, and the differences are reflected by the fauna—and also, of course, by the flora. Consequently, the response and reactions of the biota to human interferences with rivers and other water bodies may be severe and widely varied.

In freshwater systems, as elsewhere, the fauna and flora have tolerance limits for ranges of physical factors. In reaction to a low content of dissolved oxygen, for instance, such faunal components as nymphs of mayflies (*Ephemeroptera*) and stoneflies (*Plecoptera*) tend to avoid waters containing little dissolved oxygen—except in instances of animals which are specially equipped to meet this condition. An example of such adaptation is the larvae of *Eristalis tenax*, which are provided with breathing tubes for atmospheric air and can accordingly live in water having an extremely low oxygen content. Chironomid larvae contain haemoglobin which holds atmospheric oxygen, while some aquatic pulmonate snails and mosquito larvae surface regularly to take in air, and so normally such groups can survive in low-oxygen-tension environments. There are also fishes which are structurally adapted for life in waters with very low oxygen contents, and which survive perfectly well under such conditions. At the other extreme are some, such as *Clara lazera*, which has a highly vascular organ and has been recorded as able to live out of water for 30 hours.

Among species of a genus, the adaptation may be even more finely adjusted than among higher taxa. *Simulium damnosum* larvae are generally found on surfaces and areas exposed to turbulent waters, but *S. unicornutum* lives in a wide range of ecological niches. Whatever the reasons, these differences in adaptation show the importance of specific environmental conditions for the biota.

There are limits to the adjustments that the biota can make for existence in conditions of 'abnormal' levels of such factors as suspended matter, acidity, alkalinity and salinity, as well as to concentrations of chemicals and accumulated organic compounds. And in a world where increased human populations and complicated life-styles produce enormous quantities of complex wastes, and water resources are subjected to extensive modifications through 'development', the freshwater systems and biota exposed to these wastes and ecological changes face a fearsome hazard.

The Problem of Human Wastes

We know that deliberate discharge of allochthonous substances and especially of human wastes into rivers and streams is an old story which goes far back into human history. By the very nature of human physiology, we have to expect the production of body wastes; disposal of such wastes has always been a nuisance for the human race. In biblical times, the ancient Israelites were specially instructed on the subject by Moses, as described in the book of *Deuteronomy* xxiii (12–13). His advice, even by present-day interpretation, was ecologically sound.

Also, in their time, some ancient civilizations forbade the discharge of human wastes into rivers. In some African cultures, where certain rivers were regarded as sacred for the cleansing of the soul, it was taboo to foul them with human waste.

But although such action was in consonance with conditions that existed then and is to this day still practised to a limited extent or in modified forms, a different attitude to the disposal of human waste became necessary with the wide adoption of the water-closet system which is reported to have been invented around 1600. The use of water for carrying waste from towns and cities directed further attention to rivers for waste disposal. With the boom of industries, rivers and streams became the obvious and regular waste- and effluent-disposal channels, and soon became murky and developed unpleasant odours. Deliberate pollution of rivers became a common practice, and rivers that had previously been suitable for drinking-water very soon could not even support aquatic life.

This was the fate of many rivers associated with cities and towns all over the world. The Tiber received the filth of Rome and the Thames received the London sewage; the Rhine and the Seine, the Mississippi and many other rivers, large and small, were loaded with wastes from the regions through which they passed, while lakes near towns were similarly defiled. All over the world, every country must have had its own polluted river, lake, swamp or lagoon. The problem of polluted waters is therefore an old headache which developed in response to an older but unavoidable headache!

With the increase in world population, with concomitant large-scale urbanization, and with the resultant increase of human waste and sewage, the discharge of untreated, or only partially treated, waste into surface waters was to be expected. For, what does one do with large quantities of waste-water—if one does not send it back into rivers and lakes?

The changes that have occurred in some freshwater ecosystems and their biota as a result of these practices of waste disposal began long ago and got progressively worse until recently, when rehabilitation programmes began to restore the quality of some rivers. In some cases they have been remarkably effective, although apt always to be costly.

Municipal sewage often carries both domestic and industrial wastes and, in some cases, storm and street-drainage waste-water as well. Sewage from domestic sources has a heavy load of organic matter in suspension or in colloidal solution, and is composed largely of proteins, fats and carbohydrates. The most serious impact of pollution by organic matter is due to its decomposition (including putrefaction) by Bacteria and other micro-organisms which results in oxygen depletion. Delivered in small doses, the organic matter can usually be adequately broken down aerobically. But in the quantities in which sewage is all-too-often discharged, the effect on the oxygen level is such that stretches of streams and even large rivers may be deprived of adequate quantities of oxygen and become devoid of fishes and invertebrates until recovery is effected. Oxygen consumption of fishes is substantial, and Heilbrunn (1952) has shown that in this respect a trout, resting at 14.7°C, may approach that of a sheep or even a horse!

Domestic sewage also contains Bacteria, Protozoa, viruses, parasitic worms and other living organisms. The coliform Bacteria are by far the most abundant and they give indication of faecal contamination of water. Some Bacteria, such as *Salmonella typhii* and *S. paratyphii*, *Shigella dysenteriae* and *Vibrio cholerae*, are pathogenic. *Entamoeba histolytica* is a serious protozoan pathogen. Eggs and other stages of helminth parasites such as hookworm, schistosomes and cestodes are expelled with faeces or urine. This category of contaminants may perhaps not affect the overall faunal balance of freshwater bodies but they contribute to an unsatisfactory water-quality.

Toxic Industrial Wastes

Organic and mineral wastes also come from many industrial sources, including food processing, oil and fat refineries, soap industries, laundries, textile and dye plants, tanneries and breweries. A large proportion also comes from the cleaning-up involved in the processing of various articles in daily use, such as sugar, coffee, fruits, meat in slaughter houses, fish, vegetables and dairy products.

The toxic organic materials that are widely discharged into fresh waters include phenolic and other wastes from gas and coke manufacture, from tar and wood distillation, from synthetic chemical and resin plants, and from plastic and dye factories—also chlorinated hydrocarbons. Unlike the organic wastes previously mentioned, these substances are resistant to microbiological decomposition and are potent poisons—and yet the processes which produce them often fulfil important roles in development.

Still more of a problem are apt to be the toxic inorganic compounds—the acids and alkalis which in quantity disturb the normal pH regime. They include wastes from synthetic preparations such as rayons, from the manufacture of batteries, and from tanneries. Equally serious in their effect on streams are ammonia, free chlorine, and salts of the heavy-metals—especially lead, copper, zinc, chromium, cadmium, silver, and mercury. They are usually by-products of processes in plating, photographic works or paper-mills. Copper sulphate kills trout, Algae and even sewage Fungi. In general, these toxic substances may destroy not only fish and other larger organisms but also kill Bacteria, so leaving streams sterile and upsetting the self-purification mechanism. According to Jones (1969), fishes can detect even low concentrations of toxic substances and commonly avoid them.

Impact of Agricultural Wastes

Apart from municipal discharges, streams may also become polluted with agricultural wastes from piggeries and dairies. Due to the heavy application of insecticides, pesticides and herbicides, run-off waters may also carry excesses of such substances to surface and ground-waters, and, by upsetting phosphate and nitrate balances with waste fertilizers, they may encourage eutrophication. There are reports on the enormous effort that has gone into the rehabilitation of some Scandinavian lakes, and at least hopes of success in saving some freshwater ecosystems. An encouraging example of what *can* be done is afforded by the London reaches of the River Thames in England, where, following World War II, there were said to be no fishes living; but with subsequent extensive clean-up operations a large number returned (Wheeler, 1969).*

In its heyday, DDT was used in large quantities. From the USA it is reported that, from the 4,500 tons made in 1944, the quantity rose to 50,000 in 1955, and to much more later on. The toxicity of DDT to fish and other

* Before going to press, we heard from the British Museum (Alwyne Wheeler, *in litt*. February 1978) that the total of different species recorded then numbered no fewer than 91!—Ed.

aquatic fauna is well known. When used in *Simulium* control, it not only removes the target organisms from a stream but also removes many non-target and natural predator organisms. Other chlorinated hydrocarbons, including Methoxychlor, Dieldrin and Toxaphane, are not only toxic to fish and other aquatic fauna but also persistent in Nature. Persistence of Toxaphane for 9 months in Lake Columbia has been reported. The organophosphate insecticides, though less toxic, are apt to be equally harmful. Herbicides such as 2–4D, when used on aquatic plants, need to be applied at concentrations which are also toxic to fish and other fauna. These chemicals, all in their way useful for development purposes, cause undesirable side-effects to which attention should be directed if surface waters and their biota are to be protected.

Another impact is bad land-use in catchment areas of rivers, which often results in erosion and excessive transfer of silt and suspended matter. In large quantities, silt affects the transparency of water and sediments may interfere with bottom spawning-sites and feeding-grounds of fish.

Thermal Pollution, Mining and Radioactive Wastes

Thermal pollution is another cause of disturbance of freshwater systems and their biota. Steam power-plants, some industries and atomic power-plants discharge heated water that has been used for cooling purposes. The temperature of the water may change local conditions and destroy various organisms, but it appears that fishes generally escape by migration.

Some kinds of suspended materials, such as fine coal particles, lead and gypsum, from mining waste-water, cause the death of fishes by clogging and destroying their gills, and by asphyxiation due to the development of a coagulated mucus-film on the body.

As regards radioactive wastes as causative agents of biotic upset in fresh waters, it appears that more intensive investigations are required to assess the impact, since there appears to be some controversy as to the gravity of the effect of released radioactive wastes on human health and on aquatic ecosystems. Preston (1975) believes the direct discharge of liquid radioactive waste to be of negligible consequence as compared with waste discharge due to the reprocessing of more conventional fuels. Other observations have been made by Osterberg (1975) on the basis of studies of the impact of discharge from eight reactors on the Columbia River when it was used for cooling, and it was receiving 1,000 curies a day of radioactivity from nuclear plants. The radioactivity could be traced to the mouth of the river and 350 km out to sea—also in sea-cucumbers (holothurian echinoderms) down at a depth of 400 m on the sea-bed. However, the discharge of long-life plutonium and its possible inclusion in geochemical cycles of aquatic ecosystems leave room for concern and due caution, especially as there are so many insufficiencies in our knowledge and understanding of such wastes.

Adverse Effects on Biota

In summary, the quality of water is impaired as a result of the introduction of the substances which we have discussed, and many others. Hynes (1960) put these substances, etc., into five categories: poisons, suspended solids, organic substances which cause deoxygenation, non-toxic salts and heat. They disturb the characteristics of the environment to which the biota are adapted, and some destroy fauna outright. Thus lead and zinc have a profound effect on worms, leeches, crustaceans, molluscs and fishes, although according to the same author they may not affect mayflies and stoneflies. Others, such as salt, destroy plants by upsetting water-balances, while still others, such as silt, may interfere with feeding- and spawning-grounds and depress photosynthesis.

All these effects can disrupt or actually remove populations, even if only temporarily. And this gives cause for concern over the long-term impact on aquatic organisms, especially as there is still a good deal to understand aeout the intricacies of freshwater systems and the reaction of their fauna and flora to adverse environmental factors.

When we consider the changes which result in the decrease or increase of sizes of freshwater bodies, we find influences on the biota. Some activities, including engineering works, may produce effects which are irreversible. The mere straightening of a river can destroy all shoreline flora and fauna. Deepening the bed has a number of effects: the bottom flora and fauna are removed, the bed is physically altered, and the feeding and spawning- and breeding-grounds, and places of attachment for fauna in the stream-bed, are destroyed. The change in water-flow also has its effects, and it may encourage the increased growth of particular plants and animals.

Damming a river has drastic physical, chemical and biological consequences, some of which are permanent. Thousands of dams have been built in recent decades. The effect of accumulated and almost stationary water in dams has an impact on the biota, although the effect may differ in different ecoregions. In the tropical regions, perhaps because plant and animal growth is so much faster than in temperate zones, weeds flourish. The fern *Salvinia* on Lake Kariba, and Water-hyacinth (*Eichhornia crassipes*) on Lake Brokopondo and many other bodies of fresh water, are notorious examples.

Example of Volta Lake

Let me illustrate the faunistic changes which result from physical modification of a river with the example of the Volta Lake, Ghana, which is the artificially dammed body of water that I know best. The dam held up about 165 km^3 of water in a few years and spread it over an area of 8,500 km^2, with a depth-range of 19–75 m. The immediate effect was the loss of the truly riverine species of all groups—perhaps due to the exceedingly high load of suspended matter, or low oxygen-tension, or sometimes straightforward drowning.

Because they were easily observed, the effects of the change on the fish population are the best documented. Sixty species were recorded within the first five months after the closure of the dam, but some were absent locally—especially from the deep, more lacustrine southern end of the impoundment. The momyrid population, typical of the Black Volta and previously abundant in the Volta River, was reduced, and this group has not really established its importance in the new, enlarged and deepened lake since the loss of the rocky rapids and other suitable riverine substrates. For other species, including the characid *Alestes nurse*, survival depended on migration northwards. The citharinidae and some *Synodotis* spp. were also absent. *Ctenopoma* spp. disappeared from the catches in the south. Those species which preferred calm waters, such as *Hepsetus odoe* and the cichlid *Hemichromis* spp., became well established. *Hydrocynus* sp., a previously very common Black Volta predator, unlike the equally carnivorous Nile Perch (*Lates niloticus*), has never become abundant in the calm lake. Provided with abundant food in the form of clupeids, *Lates* has become a powerful lake species, but some *Tilapia* species have disappeared. There has been much change in the fish population of the Volta Lake (Denyoh, 1968). On the other hand, in Lake Kariba, which is in a different geographic zone, Harding (1964) has reported that all known fish species survived the flooding.

As regards the invertebrate fauna, it would seem that similar adjustments occurred. The Cladocera and Copepoda seem to have been highly selective in their distribution. The most easily-sampled planktonic Algae showed the Bacillariophyceae (diatoms) to be more abundant in the south and the Myxophyceae (Cyanophyta) to occur more in the northern, river-like areas. A number of invertebrates, including buliniid snails which had not previously been recorded, became established on the aquatic weeds which developed. The mosquito *Mansonia africana* was a newcomer. The Odonata also flourished. The Simuliidae, and especially *Simulium damnosum*, although previously a major group on the rapids, completely disappeared. Some of the plant communities were also new to the Volta basin, but flourished and became a nuisance in some areas (Obeng, 1968).

Similar observations on invertebrate faunal changes in man-made lakes have been reported from India (Hussainy & Abdulappa, 1973), Australia (Williams, 1973) and other regions. Mitchell (1973) has described the impact of the enlarged Lake Kariba on aquatic plants.

Dams are an impediment to fish migration and a nuisance to other fauna, often upsetting the whole balance of populations above them—unless special facilities are created for them as, for example, passage for fishes (McConnell-Lowe, 1973). Besides the many major dams that have been built, there are probably millions of small impoundments throughout the world. We still have to understand them better and learn more fully the nature of their effects on faunal changes and freshwater biota. McConnell-Lowe (1973), summarizing the available accounts of the impact of dams on fishes, concludes that much more information is needed—especially on fish ecology in the tropics—to assist with the planning of new lakes.

The reduction of water resources also has an important effect on biota. Dams quite often reduce the downstream flow of rivers, with disastrous impact on fauna and flora especially in estuarine regions, a famous example being the effect of the Aswan High Dam on sardine catches in the nearby parts of the Mediterranean Sea.

Other Practices and Their Effects on the Biota

Another practice which has particularly serious effects on freshwater fishes is the introduction of exotic species. It may be accidental, as with the entry of the Sea Lamprey (*Petromyzon marinus*) into the Great Lakes of North America, or deliberate. The stocking of new lakes and old ones is quite common. *Lates niloticus* found its way into Lake Victoria and is believed to have caused quite an upset among the indigenous fishes. There are cases of exotic species altering the whole ecology of lakes (Zaret & Paine, 1973).* The introduction of the cichlid *Cichlia ocellaris* into Gatun Lake in Panama produced population changes which affected a wide range of trophic levels. The disappearance of the endemic species of *Orestias* in Lake Titicaca as a result of the introduction of trout (*Salmo* sp.) into the in-flowing streams is another example (Villwock, 1972). Many other cases of the introduction of exotic species, and accounts of the destruction of indigenous freshwater fauna and flora, confirm the impact of gross interference with natural environments.

It is clear that introductions in the long run do not always turn out as desired. The Common Carp (*Cyprinus carpio*) from North America, which was introduced into the milkfish ponds of S.E. Asia, is an example. But even such cases have not stopped the practice of introducing fishes, and the Chinese Carp is now being liberally used as an exotic fish.

There is also the converse of this practice when Man indulges in general as well as specific over-fishing and produces disastrous effects on the biota. Quite often the reason is bad management of fishery practices, such as the use of unsuitable mesh-size for fishing.

Because of the close interrelationship of the various freshwater communities and their mutual impact on one another, the maintenance of ecological balances would seem to be the most reliable mechanism for protecting them and assuring a sound future for them.

Finally, there is the matter of reducing evaporation by the application of thin films of organic substances such as alcohol. The practice has been found effective and can be of importance especially in water-short areas. Although it appears that there is no adverse effect on fish, concern has been expressed over difficulties experienced by some aquatic insects to emerge through such films.

Conclusion

These and all the other points that have been made about the effects of the activities of Man on fresh waters raise a fundamental question of priorities. The subject of this paper implies concern over the future of freshwater biota, and there can surely be no doubt as to the importance of this matter. The influences to which fresh waters and their biota are subjected are legion—quality changes as a result of organic and inorganic toxic wastes, increases in suspended matter and turbidity, temperature

*A striking example is furnished by the wanton killing of large bass and other depradations by the S.E. Asian Walking Catfish (*Clarias batrachus*) in Florida, as indicated by papers published in, or submitted to, *Environmental Conservation*.—Ed.

fluctuations, and changes in the concentration of nutrient materials, are only a few of them. Even human manipulations to increase available water-supplies have some adverse repercussions on fauna. And yet the activities which cause these effects may be, in most instances, justified in terms of attempting to create satisfactory conditions for Man or improve on existing ones. Killing Algae in recreational lakes may satisfy swimmers and bathers. Reducing evaporation of waters may be of crucial importance in arid areas. Spraying flies with DDT and other insecticides may remove a threat to health. And these operations are often given priority consideration. In some areas, in the minds of some people, the question of freshwater flora and fauna, apart from fishes, does not seem to exist—perhaps justifiably, the availability of water, especially for human and domestic needs, being the overriding consideration.

At the UN Habitat Conference in Vancouver in 1976, and at the Water Conference in Mar del Plata in 1977, there was no doubt as to the importance attached by governments to water. Discussions stressed water for drinking, domestic use, power, industry, agriculture and navigation. There was some mention of fisheries but none of 'freshwater biota'. Nor would freshwater biota have been mentioned by a layman or even by a narrow-minded 'non-natural' scientist. I have personal experiences of friends taking such a view. When I was setting up the Institute of Aquatic Biology after the Volta Dam was built in Ghana, I would often be asked whether I honestly believed there was 'anything' in the water—that is, apart from fish—to merit the establishment of a full-scale national research institute.

How then, in these circumstances, do we justify concern over the biota? The theme for this Conference, 'Growth Without Ecodisasters?', emphasizes 'growth'—that is, 'development' without destruction of the environment. Water, with land and air, constitute vital resources of the environment on which Man's continued existence on this planet depend. The continued viability of water resources as sustaining elements of the environment is equally important to development towards a better life. Satisfactory water resources provide a basis for fundamental human needs—food, clothing, suitable settlements and good health. Fresh waters, essential in this respect, are complex ecological systems, and the inhabiting fauna and flora are both vital components. Without them a water-body is described as 'dead'. Innumerable planktonic organisms, larger Algae and other plants—all members of the biota—clean up and refreshen the fresh waters and the oceans. The myriads of freshwater organisms, ranging from microscopic Bacteria all the way to large fishes, form a complex of interrelationships and balances—the basis for the continued viability of our own and Nature's waters.

There is every justification and indeed dire need to understand fresh waters better and better, in order to assure that they are treated in a satisfactory manner to guarantee their continued viability. The responsibility quite clearly falls on scientists to provide guidance to assist in reducing toxicity of wastes, finding alternatives to present waste-water disposal systems or effecting their improvement, and generally for the protection of freshwater resources and their biota—in order to preserve them for continued human development and for the future of a healthy environment.

References

Courtenay, W. R., Jr & Miley, W. W. II (1975). Range expansion and environmental impress of the introduced Walking Catfish in the United States. *Environmental Conservation*, **2**(2), pp. 145–8, 4 figs.

Courtenay, W. R., Jr, Sahlman, H. F., Miley, W. W. II & Herrema, D. J. (1974). Exotic fishes in fresh and brackish waters of Florida. *Biological Conservation*, **6**(4), pp. 292–302, 2 figs.

Denyoh, F. M. K. (1968). Changes in fish population and gear selectivity in the Volta Lake. Pp. 206–19 in *Man-made Lakes* (Ed. L. E. Obeng). Ghana Universities Press, Accra, Ghana: 398 pp., illustr.

Harding, D. (1964). Hydrology and fisheries in Lake Kariba. *Verh. Internat. Verein. Limnol.*, **15**, pp. 139–49.

Heilbrunn, L. V. (1952). *An Outline of General Physiology* (3rd edn.) Saunders, Philadelphia & London: xiii + 818 pp.

Hussainy, S. O. & Abdulappa, M. K. (1973). A limnological reconnaissance of Lake Gorewade, Nagpur, India. Pp. 500–7 in *Man-made Lakes, Their Problems and Environmental Effects*. American Geophysical Union, Geographical Monograph **17**.

Hynes, H. B. N. (1960). *The Biology of Polluted Waters*. Liverpool University Press, Liverpool, England: xiv + 202 pp.

Jones, J. R. E. (1969). *Fish and River Pollution*. Butterworths, London; England: viii + 203 pp.

McConnell-Lowe, R. H. (1973). Reservoirs in relation to Man-made lakes—Fisheries. Pp. 641–55 in *Man-made Lakes, Their Problems and Environmental Effects*. American Geophysical Union, Geographical Monograph **17**.

Mitchell, D. S. (1973). Aquatic weeds in Man-made lakes. Pp. 602–12 in *Ibid*.

Obeng, L. E. (1968). The invertebrate fauna of aquatic plants in Volta Lake in relation to the spread of helminth parasites. Pp. 320–5 in *Man-made Lakes* (Ed. L. E. Obeng). Ghana Universities Press, Accra, Ghana: 398 pp., illustr.

Osterberg, C. L. (1975). Radiological impact releases from nuclear facilities into aquatic environments—USA views. *IAEA–SM 198/58*, IAEA, Vienna, Austria: pp. 33–9.

Preston, A. (1975). The radiological consequences of releases from nuclear facilities to the aquatic environment. *IAEA–SM 198/57*, IAEA, Vienna, Austria: pp. 3–21.

Villwock, W. (1972). Gefahren für die endemische Fischfauna durch Einburgungsversuche und Aklimatisation von Fremdfischen am Beispiel des Titicaca-Sees (Peru-Bolivien) und des Lanao-Sees (Mindanao-Philippinen). *Verh. Internat. Verein. Limnol.*, **18**(2), pp. 1227–34.

Wheeler, A. (1969). Fish-life and pollution in the lower Thames: A review and preliminary report. *Biological Conservation*, **2**(1), pp. 25–30, map.

Williams, W. D. (1973). Man-made lakes and the changing limnological environment in Australia. Pp. 495–500 in *Man-made Lakes, Their Problems and Environmental Effects*. American Geophysical Union, Geographical Monograph **17**.

Wright, R. F. & Gjessing, E. T. (1976). Changes in the chemical composition of lakes. *Ambio*, **5**(5–6), pp. 219–23.

Zaret, I. M. & Paine, R. T. (1973). Species introduction in a tropical lake. *Science*, **182**, pp. 449–55.

Keynote Paper Part II

Enlightened Management of Aquatic Biota

by

ARTHUR D. HASLER

Professor and Director, Laboratory of Limnology, University of Wisconsin, Madison, Wisconsin 53706, USA; sometime *President of the International Association for Ecology (INTECOL)* and *Director of the Institute of Ecology;* formerly *President of the Ecological Society of America* and *Chairman of the Societas Internationalis Limnologiae Fifteenth Congress.*

Introduction

I am honoured to be invited to address this Conference. In this presentation I will draw upon my research experiences at the University of Wisconsin and elsewhere, and attempt to relate them to the problems, both actual and potential, of drainage-basin ecology in general.

It is not only man-made lakes which have been changed by additions and manipulations, but natural lakes as well. An appropriate example is Lake Wingra, a small, shallow lake in Madison, Wisconsin. My colleagues (Baumann *et al.*, 1974) reviewed the history of man-induced changes during this century and observed:

> Wingra, like most lakes, has never been managed as an ecosystem, but rather individual problems have been attacked one at a time. To diversify the angling and [to] reclaim fish, new species were introduced. Little thought was given at the time as to whether these species could thrive or to whether they would negatively influence native species. Marshlands were drained for park and land development, and the reproductive success of higher predators was severely reduced. Carp were first introduced into our waters, then became a nuisance and were actively removed. Again, the secondary effects of carp removal were not anticipated.
>
> To be effective, management plans must consider the entire fish community and the total ecosystem. The historical perspectives described in this paper serve as illustration of the complex interactions and response capabilities of a total

> ecosystem. An integrated whole ecosystem approach is essential in establishing ecologically sound management. Although resources are often not available to make an overall ecosystem study, sound management would seem to require consideration of as many interactions and secondary effects as time, funding, and contemporary knowledge and techniques, permit. As always, hindsight is better than foresight. . . .

This site became the focus of a multidisciplinary study by the University of Wisconsin as its contribution to the Analysis of Ecosystems Project of the Eastern Deciduous Forest Biome of the International Biological Programme. The site offers an opportunity to assess the relative importance of urban versus natural wooded and prairie areas, being bordered on the north by the city of Madison and on the south by the University of Wisconsin Arboretum.

In the course of the Lake Wingra study, Kitchell *et al.* (1974) developed a model of fish dynamics based on principles of physiology, population biology, and trophic ecology. All parameters were for the Bluegill (*Lepomis macrochirus*), and simulation results were compared with data derived in the field and laboratory. This model provides simulation of seasonal changes in a natural population and, as an example of a potential future use, enables predictions to be made of the effects of a perturbation such as thermal enrichment. It has been tested against real data comprising 100 man-years of research effort taken from literature sources and against the comprehensive data from the Lake Wingra study.

Elegant predictive transfers of an improved model were made to Perch (*Perca flavescens*), that hold great promise (Kitchell *et al.*, 1977):

> The model can be used to generate tables of maximum consumption rates for various fish sizes and water temperatures. If water temperature, mean fish-size, and population density in the ponds, are monitored, daily food needs of the fry populations can be estimated, using the model-generated consumption tables. If food abundance is also monitored, it should be possible to predict when reduced food abundance might dictate the need to thin the population or supply additional food. The consumption tables may be useful tools which can provide a basis for management decisions on a time-scale [that was] not previously feasible. They could aid in determination of optimal density as well as contribute to an understanding of fish production in ponds. If growth-rates of the fish are determined, the model can be used to calculate actual consumption rates that were realized.

Moreover, simulations are remarkably accurate in predicting total body-load of PCBs in Perch after several years of exposure.

Natural and man-made eutrophic lakes world-wide grow harvestable crops of macrophytic vegetation that could be used for agricultural mulch, fertilizer or animal feed. Research on the role of the ecology of aquatic plants is meagre indeed, and no criteria are available on how much of the crop can be harvested and when. As a management tool, harvesting appears to offer unexplored potential. Harvesting could contribute to the development of lake strategies of management, ecosystem biology and concepts of lake ontogeny.

Because of the intensive IBP study of Lake Wingra, it has become an ideal site for a definitive experiment, so a model has been prepared which will predict the effect on the biotic cycle of the withdrawal of nutrients from the lake in a 90% harvesting operation. This experiment will be especially valuable in another way, namely as a test of the predictability of the model. The results will show how nearly the effects were assessed beforehand, when compared with actual findings of the post-treatment study. Details of the model (Loucks *et al.*, 1977) and of the design of the experiment are available from the Center for Biotic Systems, University of Wisconsin.

While the technique of systems analysis developed by engineers, economists and ecologists has limitations, it provides a systematic and logical approach to planning for and management of an ecosystem. It is therefore a far better approach than the hit-or-miss policy of the past, where individual problems were attacked one at a time.

A review of the losses and gains from the construction of man-made lakes points up the urgency for evaluation of the effects of these constructions. The technique of systems analysis is developing into a useful tool for predicting the effects of projects in which many interacting factors are involved.

In Wisconsin, a dam was recently proposed for the Kickapoo River—essentially for the purpose of flood control, but as a side-benefit to serve also as a recreation area. Based upon a systems study by the University of Wisconsin's Institute for Environmental Studies (1974), a model of the Kickapoo was used to predict the features of this proposed impoundment. The prediction was that the new lake would become eutrophic, with frequent nuisance blooms of Algae and massive growths of rooted aquatic plants. In short, the analysis and model predicted that the water quality would be poor for recreation; now because of this evaluation the construction has been delayed and the dam may never be built. Rather, is it recommended that smaller flood-control efforts may be sufficient to perform the same purpose as a single large dam.

Rehabilitation of Stocks of Salmon

Canada has persistently rejected proposals to build dams on the great Fraser River, largely because they would damage the Salmon (*Oncorhynchus* sp.) fishery—a serious disadvantage of the dams on the near-by Columbia River in the United States. In the view of the Canadians, the loss of a fishery would outweigh the advantages of hydroelectric power. Decisions to build the dams on the Columbia River had been made in spite of the warnings of fishery biologists, because the demand for hydroelectric power and irrigation water were valued more highly than the stocks of Salmon which spawned there. Now the problem is, how to salvage this fishery.

Permit me to describe an undertaking in our research programme which may lead to a partial rejuvenation of these Salmon stocks in the Columbia River. My example concerns the 'guideposts' on or emanating from the land which are used by Salmon to find their way home from the sea. What we have known for many years is that Salmon have the capability of returning from the sea to a precise home-stream. But there was no good explanation of what the cues or 'guideposts' were that directed them and helped them to identify this home-site of each individual fish.

When I first began to investigate the homing behaviour of Salmon, my principal goal was to satisfy a scientific curiosity and explain, by new facts, the mechanisms used by Salmon to find their way home. One of my working hypotheses was an ecological one, namely that the home-stream of a Salmon carried an odour—a fragrance which was different from the odours of all other streams. I reasoned that because the soils and plant communities of each catchment basin were different, then each stream was unique, and bore a characteristic bouquet—as do wines from grapes raised on different soils. My hypothesis proposed that the young Salmon became imprinted to this odour before they migrated; hence, after attaining sexual maturity at sea, they returned to this area and recognized their home-stream odour, having retained it in memory since their departure for the sea or open lake.

My students and I conducted a number of experiments in the laboratory and in the field (Hasler, 1966, 1975) which provided circumstantial evidence that this hypothesis might be valid. First, we demonstrated that fish could be trained in the laboratory to discriminate between two different streams of olfaction (Hasler & Wisby, 1951). The next question was to determine if adult Salmon need their noses to find their way in the river. To answer this, we captured 300 homing Salmon and displaced them down-stream, below a fork in the Issequah River. Of one-half of these we plugged the noses with cotton, while the other half served as controls, being appropriately marked. The normal Salmon returned correctly to the stream from which they had been displaced, while those with plugged noses were confused about the confluence and made mistakes.

While this evidence suggests strongly that a stream odour is unique and serves as a cue for homing Salmon, I felt it was essential to conduct the following experiment as a more conclusive test: young smolting Salmon, raised in a hatchery, would be imprinted with an artificial odorous chemical which they had never experienced in the history of their evolution. They were to be marked and released to swim to sea. Upon their return as mature adults, a synthetic imprinting chemical that could be smelled by Salmon at low concentrations would be dripped into a tributary downstream from the original imprinting site, to determine if they could be decoyed into a tributary of a river which was different from that of their birth.

One of my students, Warren J. Wisby, in a 1954 doctoral dissertation, undertook the task of screening various chemicals to find a suitable synthetic substance which we could use in testing this hypothesis. The chemical should have the following characteristics: (1) to be detectable by olfaction at low concentrations, as low as 10^{-6} mg/l; (2) to be neither a natural repellent nor an attractant; and (3) to have reasonable chemical stability.

Morpholine fulfilled these specifications, but we were unable to perform field tests during the ensuing years because of our location so far from Salmon territory. Yet such an experiment would be definitive in resolving the hypothetical aspects. Then, over 20 years later, in collaboration with my students and associates, it was possible to conduct these experiments on Coho Salmon (*Oncorhynchus kisutch*) which have been stocked and now flourish in near-by Lake Michigan.

We performed this experiment in Lake Michigan during 1970–74. Several thousand Coho Salmon smolts, raised in the Wisconsin hatcheries, were exposed, i.e. imprinted, to the synthetic chemical morpholine. An equal number of fish were not exposed (controls). Both groups were appropriately marked and stocked directly into Lake Michigan without previous exposure to any natural stream odours. During the spawning migration 18 months later, morpholine was dripped into a river near the location where the fish were released, and a census was made by electrofishing and examination of anglers' creels. Replication of the experiment over three years in Oak Creek (Fig. 1) and in the Manitowoc area has demonstrated that the exposed fish, as opposed to non-exposed fish, returned to the simulated home-stream to spawn in significantly greater numbers than to any other stream (Cooper *et al*., 1976; Scholz *et al*., 1976).

It is clear that an odour is the basis for identifying the 'home' water; in this instance, it was a simulated home-stream which had been scented by

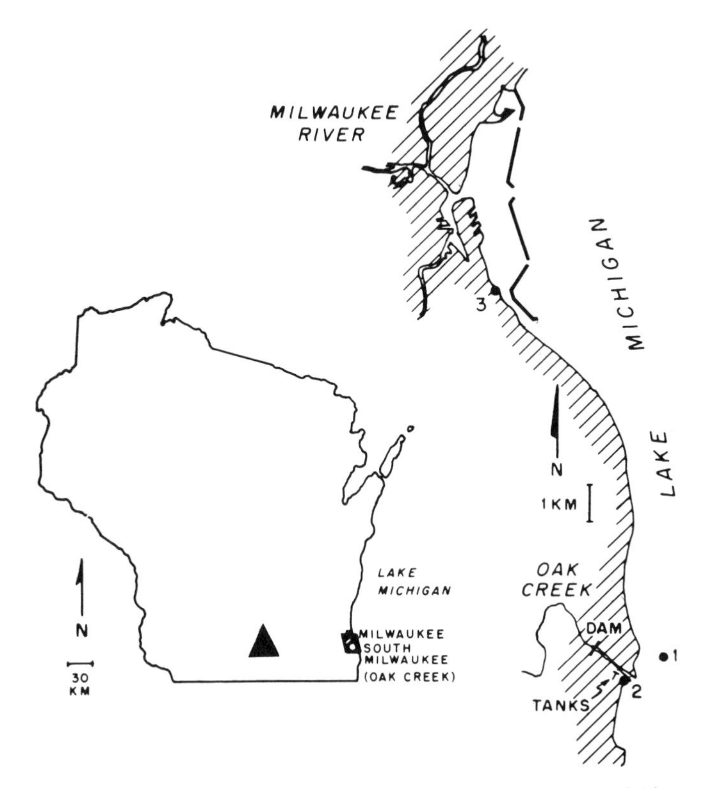

Fig. 1. Salmon imprinted to morpholine in a hatchery and released in Lake Michigan (Sites 1, 2 and 3) returned to morpholine-scented Oak Creek and were harvested there. (The location maplet on the left indicates the position of the large-scale sketch-map on the right.)

the imprinting synthetic odour. In addition, we fastened ultrasonic transmitters to some of the homing Salmon and displaced them 5 km out into the lake. Our tracking boats were then able to follow them with hydrophones. Meanwhile, we dripped a small quantity of morpholine in another streamlet. When the imprinted fish reached the streamlet, they stopped short and milled around. Control fish, however, swam right by. If the morpholine was not dripped into the water, imprinted fish also swam by without hesitating.

This research exemplified the linkages between the land and water of a drainage basin. Here Salmon rely upon the fragrances from the landscape's soil and vegetation to provide the 'guidepost' to their spawning ground upon returning from sea or lake.

Moreover, this example is used because a series of experiments designed to test a theory yielded, in the end, facts which not only satisfied our initial scientific curiosity, but had, as well, application to a human need—food—because artificially imprinted Salmon may now be attracted at sexual maturity to a point of easy accessibility to a stream near the coast where the protein-rich adults may be harvested. Moreover, ripe eggs can be readily salvaged and shipped to hatcheries where Salmon can be raised to provide stock to maintain a population of fish in an area where streams have been: (a) dammed with non-passable barriers, (b) warmed by cooling water, (c) polluted beyond biological restoration, or (d) rendered inhospitable to Salmon by eroded sediments.

An instructional manual of procedures is available (Scholz *et al.*, 1975). Already they are applied in the Great Lakes, the Columbia River, Japan and New Zealand.

In considering the management of our biota, I emphasize that Salmon are essentially a 'free' resource. At sea they need no stalls. There is no manure to haul and no one is required to feed them, treat them for domestic diseases and herd them; instead they deliver themselves at our 'back door' for harvesting, free of middleman charges. Their food—shrimps and small fish in the sea—is supplied free of charge by Nature. A series of interacting ecological systems comprising vegetation, soil, rivers, estuaries and ocean have been utilized in the evolution of a species that has become an important protein-source for Man (Fig. 1).

Too little is known about the migration of fishes in tropical and subtropical rivers and reservoirs. Presumably management of stocks of fish could be more responsibly and effectively carried out when such knowledge becomes available, and so I urge the prosecution of research of this type on fishes in other parts of the world, to determine if some of the migratory species can be saved for posterity.

Aquacultural Research

As in the example of Salmon in the temperate zone, tropical species may need to be reared through sensitive early stages in order to maintain a viable fishery or to develop an alternative source of food through aquaculture. According to Allsopp (1975), a recognized authority on tropical fisheries:

> The prime problem is the inadequate supplies of juvenile fish, generally referred to as 'fish seed', which are only seasonally available from natural sources. . . . For all species, expanded fish production is limited by supply shortages. To ensure adequate, reliable and year-round supplies of fish seed, research on standardized techniques of induced breeding of selected adults must be applied to the important cultivated species.
>
> Fundamental studies of the physiological and environmental factors governing the reproduction of these target species are needed in the localities of their cultivation. Additionally, studies of the nutrition of brood fish and juveniles and the behavioural responses of fish to aquaculture conditions are needed. Research efforts may be more productive if coordinated work is done on fewer species.
>
> For some fish, induced breeding has been achieved by the control of water levels, flowage, or physical manipulations. Even the technique of induced breeding by injection of hormonal extracts needs to be standardized for wider application to various tropical species. This will permit fish breeding and production throughout the year in tropical fish-ponds.
>
> Stripping, fertilizing, and artificial incubation of ova, may be possible for many indigenous species. When research makes their reproductive maturation processes better understood, the process of selective breeding for desirable characteristics and true domestication of wild species can proceed.
>
> Research is further needed on simple designs of efficient hatcheries suitable for rural conditions, as well as on efficient larval rearing and distribution.
>
> Efficient feed supplies is another problem area. There is need to standardize practices for various ecological conditions and fish species by increasing production of natural fish-food and satisfying the nutritive requirements with artificially prepared feeds.

Floating Cages

Harvesting fish by hunting and capture could be augmented materially by the use of floating cages in man-made lakes of the tropics. An Algae-eating native species of fish (e.g. of *Tilapia*) could be cultured and raised in these cages in an intensive programme.

This technique, used in south-east Asia, could be an extremely useful method of converting abundant Algae in lakes to fish protein, although at times supplementary artificial foods would need to be fed to the fishes.

Experimental Methods Applied to the Field

A major emphasis of limnological research in recent years had been away from the classical, descriptive study and towards experimental limnology which has been augmented by the development of modern apparatus and automation. In these studies ecologists have attempted to apply, as precisely as limiting circumstances permit, the methods of the experimental laboratory to the field.

At the outset in our laboratory, interest developed in the possibility of using small bog-lakes as experimental models for limnological studies.

Bog-lakes have a colloidal, brown-coloured humic stain which reduces the penetration of light and energy that is essential for the phytosynthesis of Algae. Upon contemplating a solution to this problem, we noted that limed bog soils had a clear-water drainage, and we demonstrated in the laboratory that the colloids could in fact be precipitated by hydrated lime (calcium hydroxide) (Johnson & Hasler, 1954; Stross *et al.*, 1961).

For a field test of this finding, a small hourglass-shaped lake (Fig. 2) was located for the experiment. By means of an earthen dam we separated it into two, enabling us to treat one twin with hydrated lime (Fig. 3) and observe the other as a reference. Within the first summer the transparency in the limed part increased by 60%, and in the second year the well-lighted zone had deepened from 2.7 to 7 m. The *Daphnia* population was found to replace itself every 2.1 weeks in the lime-treated lake, compared with 4.6 weeks in the untreated twin; therefore, both the size of the population and its rate of energy transfer were increased by liming. Introduced Trout (*Salmo irideus*) flourished where none could live before. In addition, further species of plankton and of rooted aquatic plants appeared and began to thrive.

In another experimental lake we (Likens & Hasler, 1963) sought to determine if chemical elements were leaving it through the emergence of insects living in the deep water. We introduced ^{131}Iodine into the bottom of a small chemically stratified lake where the larvae could absorb the iodine. Light-traps on the shores of the lake captured midges as they emerged. They were found to contain the radionucline iodine, hence providing evidence that insects can transport ions to the surface and on to the land. *En route* to the surface, the pupating midges were evidently eaten by surface-dwelling fish, as their bodies were found to contain ^{131}I. This experiment gave us an example of uphill migration of nutrients.

Many field studies (Hasler, 1973) support the view that eutrophic conditions (excessive algal growths) are caused and maintained by steady inputs of phosphorus from the drainage basin. A definitive field experiment (Schindler, 1974), modelled after our Peter and Paul lakes experiment and using twin Canadian lakes, showed conclusively that algal blooms could only be maintained with continuous additions of phosphorus to one twin-lake but were absent from the reference or control lake. Upon cessation of treatment, the blooms subsided as the phosphorus accumulated in the sediments.

This sequence—(1) formulation of a working hypothesis, (2) assembling examples from monitoring, (3) testing the hypothesis by experiments in Nature, and (4) duplication and verification of results by different methodologies—can thus help to answer for us a puzzling ecological question. In this case the causes, consequences and correctives of man-made eutrophication are now well understood and society has benefited.

Ecologists have shown that their science can identify the causes of eutrophication, can point to ways of solution and can carry out corrective measures which are effectively restorative. Because of ecological knowledge, therefore, man-made eutrophication is no longer a doomsday issue (Vallentyne, 1974).

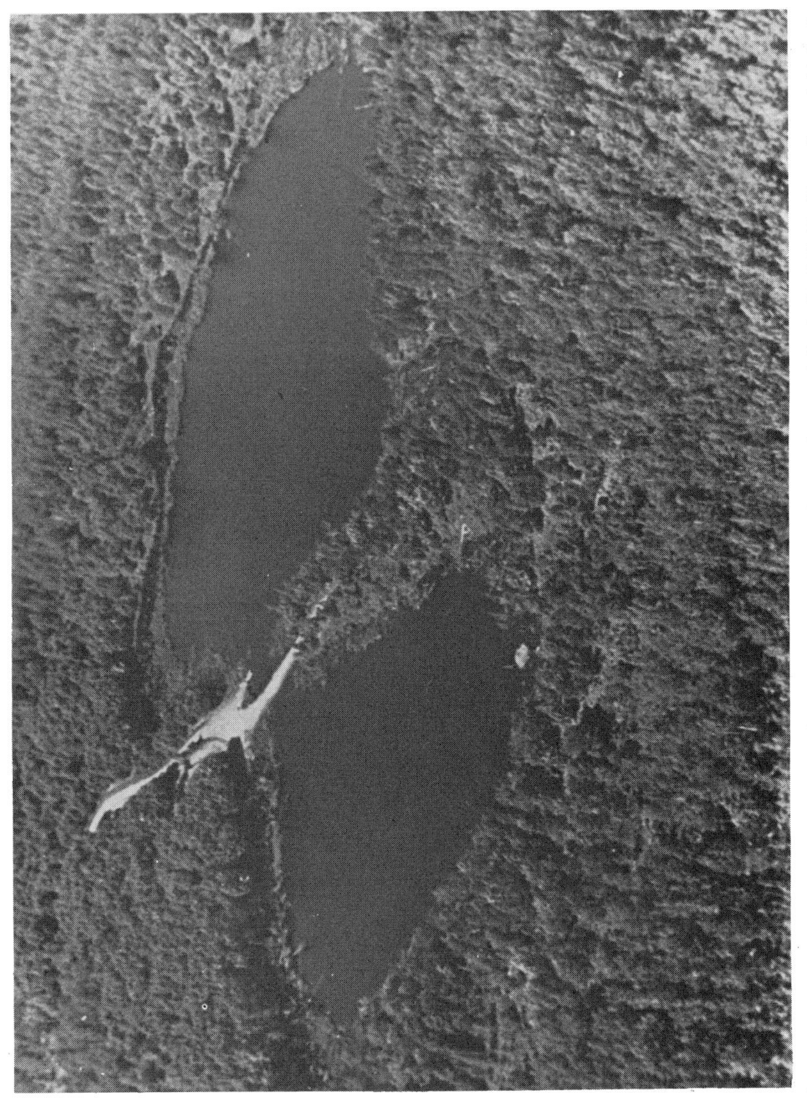

Fig. 2. Twin bog-lakes Peter and Paul, divided by an earth barrier. Reference lake is Paul (left). Experimental lake is Peter (right) which was treated with hydrated lime, $Ca(OH)_2$.

Fig. 3. Applying lime to experimental bog-lake. Clearing of bog pigments occurred, followed by increased production.

Chemical Control or Utilization of Nuisance Species

Allow me to review a few ecological problems where ecologists could perform an important service in management of aquatic communities.

There is an increasing demand for chemical treatment of lakes and streams to rid them of an existing nuisance species (e.g. Carp, Water-milfoil [*Myriophyllum spicatum*] or Water-hyacinth [*Eichhornia crassipes*]). Natural aquatic ecosystems are too complex to assume that a nuisance organism can be managed responsibly by resorting to chemical control.

In agriculture, pesticides have been reasonably successful in controlling pests because the crop to be protected consists of a single species such as Maize (*Zea mays*) in which the task is to eliminate all pests without harming the crop. In natural aquatic systems, by contrast, there is no community consisting of a single species. Hence, this agricultural practice is misapplied.

The ecologist who understands the complexity and diversity of ecosystems and their manifold interactions should question the wisdom of adding a poison to that complex system. Even if the toxin were species-specific, the sudden elimination of that species would be likely to produce other problems. Moreover, when herbicides are used, the decomposition of the dead plants (e.g. Water-hyacinth or Water-milfoil) releases phosphates and nitrates, etc., which will help to re-fertilize an already fertile environment.

In agriculture, comprehensive control is gaining acceptance because it makes available an array of weapons for combating pests—such as crop rotation, sex attractants, release of predators, parasites, or irradiated males, and breeding of disease- or pest-resistant strains of food species.

Integrated or comprehensive pest-control rests on the view that a crop and its pests form an ecological system, so no single control measure is considered a panacea. Natural controls are recognized as potentially powerful agents and are manipulated as required, whereas pesticides are considered as weapons, to be used only when necessary and in the smallest quantity that will prove effective.

New methods are now being developed to utilize aquatic vegetation. Automized machinery can cut, harvest, and de-water the plants, pressing from them a high-quality protein-rich product which constitutes a useful animal feed. Such harvesting is a better alternative than applying poison!

Species as a Resource

A great array of aquatic habitats and natural communities needs to be preserved for future study. We must provide examples for continuous comparison of natural communities with artificially simplified communities such as the world is bound to see more and more of with its still increasing human population.

Nature has developed ecosystems over thousands or millions of years; hence, great care should be exercised in saving and protecting representative examples of each. It must be recognized that other generations will follow ours; therefore, we have a responsibility to pass on to them untampered natural ecosystems and to place them on deposit for the future. Each instance where a population is poisoned leaves us with fewer potential study-sites for understanding natural systems.

As the demand for intensive management of lakes, streams and estuaries increases, we should make long-range plans which require complete bans on the use of chemical toxins in representative and selected habitats throughout each continent. There should be many natural areas where chemical toxins are never used, but where Nature's surpluses are harvested on a sustained-yield basis.

Intensive management in aquaculture, as in agriculture, is growing, and chemicals play an important role along with other tools. However, endangered species and local populations must be protected, while the introduction of exotic species should be indulged in only with great caution. Mistakes may be impossible to correct.

In conclusion, I quote from my presidential address to the First International Congress of Ecology, held in The Hague in 1974:

> What can ecologists accomplish through these multi-institutional and international organizations that will augment our understanding of ecosystems? In the past we have made only modest use of scientific organizations in reacting responsibly to the need for better management of our resources, but the need increases; hence we must strive to be of greater service [with] instances of applying ecological knowledge [leading] to reasonable cooperation [that] can become models upon which greater ones can be based. A new ecological ethic is gradually emerging which will help to create a responsible concept of world order in which concession and agreement will replace opposition and disagreement.
>
> Perhaps in the area of ecology rests a key to cooperative attitudes among nations. Sharing the products of common ownership, such as the fish of the oceans, will encourage each nation to act responsibly and in concert on other problems confronting society.

References

ALLSOPP, W. H. L. (1975). *Tropical Aquaculture Research Problems in Southeast Asia.* International Development Research Centre Report, University of British Columbia, Vancouver, Canada: 12 pp., mimeogr.

BAUMANN, P. C., KITCHELL, J. F., MAGNUSON, J. J. & KAYES, T. B. (1974). Lake Wingra, 1837–1973: A case-history of human impact. *Trans. Wis. Acad. Sci., Arts Lett.*, **62**, pp. 57–94.

COOPER, J. C., SCHOLZ, A. T., HORRALL, R. M., HASLER, A. D. & MADISON, D. M. (1976). Experimental confirmation of the olfactory hypothesis with artificially imprinted Coho Salmon (*Oncorhynchus kisutch*). *J. Fish. Res. Board Can.*, **33**(4), pp. 703–10.

HASLER, A. D. (1966). *Underwater Guideposts—Homing of Salmon.* University of Wisconsin Press, Madison, Wisconsin: 155 pp.

HASLER, A. D. (1973). Causes and correctives of Man-made eutrophication. Pp. 141–63 in *Ökologie und Lebensschutz in Internationaler Sicht* (Ed. H. Sioli). Verlag Rombach, Freiburg, Germany: illustr.

HASLER, A. D. (1975). How the Salmon comes home. Pp. 184–91 in *Science Year, the World Book Science Annual.* Field Enterprises, Chicago, Ill.: illustr.

HASLER, A. D. & WISBY, W. J. (1951). Discrimination of stream odors by fishes and its relation to parent stream behavior. *American Naturalist*, **85**(823), pp. 223–38.
INSTITUTE FOR ENVIRONMENTAL STUDIES (1974). *Environmental Analysis of the Kickapoo River Impoundment*. Institute for Environmental Studies, University of Wisconsin, Madison, Wisconsin: Rept 28, 116 pp.
JOHNSON, W. E. & HASLER, A. D. (1954). Rainbow trout production in dystrophic lakes. *J. Wildl. Mgmt*, **18**(1), pp. 113–34.
KITCHELL, J. F., KOONCE, J. F., O'NEILL, R. V., SHUGART, H. H. Jr., MAGNUSON, J. J. & BOOTH, R. S. (1974). Model of fish biomass dynamics. *Trans. Am. Fish. Soc.*, **103**(4), pp. 786–98.
KITCHELL, J. F., STEWART, D. J. & WEININGER, D. (1977). Applications of a bio-energetics model to Yellow Perch (*Perca flavescens*) and Walleye (*Stizastedion vitreum vitreum*). *J. Fish. Res. Board Can.*, **34**(10), pp. 1922–35.
LIKENS, G. E. & HASLER, A. D. (1963). Biological and physical transport of radionuclides in stratified lakes. Pp. 463–70 in *Radioecology* (Ed. V. Schultz & A. W. Klement, Jr). Reinhold, New York, N.Y.: 746 pp.
LOUCKS, O. L., PRENTKI, R. T., WATSON, B., REYNOLDS, B., WEILER, T., BARTELL, S. & D'ALESSIO, A. (1977). *Studies of the Lake Wingra Watershed*. Center for Biotic Systems, Institute for Environmental Studies, University of Wisconsin, Madison, Wisconsin: Rept 78, 36 pp.
SCHINDLER, D. W. (1974). Eutrophication and recovery in experimental lakes: Implications for lake management. *Science*, **184**, pp. 897–9, illustr.
SCHOLZ, A. T., HORRALL, R. M., COOPER, J. C., HASLER, A. D., MADISON, D. M., POFF, R. J. & DALY, R. I. (1975). Artificial imprinting of Salmon and Trout in Lake Michigan. *Wis. Dept Nat. Res. Fish. Mgmt Rept*, **80** (Univ. Wis. Sea Grant Advisory Rept, **414**).
SCHOLZ, A. T., HORRALL, R. M., COOPER, J. C. & HASLER, A. D. (1976). Imprinting to chemical cues: the basis for home-stream selection in Salmon. *Science*, **192**, pp. 1247–9, illustr.
STROSS, R. G., NEESS, J. C. & HASLER, A. D. (1961). Turnover time and production of the planktonic Crustacea in limed and reference portions of a bog lake. *Ecology*, **42**(2), pp. 237–45.
VALLENTYNE, J. R. (1974). *The Algal Bowl: Lakes and Man*. Dept of the Environment, Fisheries and Marine Service, Ottawa, Ontario: 186 pp.

Discussion (Session 4)

Worthington (Chairman, introducing)

May I call this fourth Session to order now and start by introducing the people on the platform. My own name is Barton Worthington. Basically a hydrobiologist, I have spent too little time in the field recently and too much time shuffling papers. On my right we have our principal speaker, Professor Arthur Hasler, elected Member of the National Academy of Sciences of the United States, Professor of Limnology at Wisconsin, former President of INTECOL (the International Association for Ecology) and also former Chairman of SIL—that is, the Societas Internationalis Limnologiae. On my left we have Professor Aristeo Renzoni, Professor of Hydrobiology and Fish Culture at the University of Siena in Italy, then Dr John Knox, Associate Professor in the Department of Geography and the Institute for Environmental Studies of the University of Wisconsin at Madison, who is specially concerned with the influence of climate and land-use on water problems, and finally, at the end, Professor Edward Goldberg, Professor of Chemistry at Scripps Institution of Oceanography, University of California, who is concerned with problems of marine pollution and author of various important works on the oceans. You will see [some intrusion of marine and estuarine interests on our panel for special reasons] that have no disadvantages.

I would like to draw attention to five points about the future of fresh water and its biota which it seems to me might give a [useful] bit of focus to our comments.

The first point concerns the rate of development of freshwater resources and the ecological impact of that development. There is no need to labour the importance of water to all forms of life, and for us especially the domestic uses, agricultural uses, industrial uses, and transport and yet other uses. Water is also an essential component of every plant and animal [and constitutes the habitat of many, so there is great] scientific interest in limnology—the science of lakes and rivers, reservoirs, and so on. Yet in spite of the all-pervading importance of fresh water as a natural resource, I think we should remind ourselves of its smallness: only about 0.01% of the water in the world exists as surface fresh water, with roughly 0.3% existing as ground-water. All the rest of the world's water as you know is in the oceans or the ice-caps, and we hope will remain there. But even this tiny resource of surface fresh water is mostly in the tropics, and is so far very little used. Thus only 2% of fresh water available in Africa and 3% in Latin America is as yet developed for human use.

Simple data of that sort led the United Nations Water Conference, held in Argentina in February of [1977]—I was present and heard a good deal of the discussion—to emphasize the need for accelerated development of freshwater supplies. Yet when you couple this demand with the avowed intent of providing all people on the globe with clean water by 1990, which emerged from the United Nations Habitat Conference [held in 1976] at Vancouver, it is quite clear that accelerated development of the freshwater resources is what we must expect and what the world must have.

My second point is that a great deal of the development of freshwater resources of the world consists of restraining the free flow of water through engineering works. Admittedly some of this involves making it run more quickly towards the sea through drainage of swamps, for example, but very often it is again held back further downstream by another dam or reservoir. This multiplication of reservoirs right down to the size of [those behind] small farm dams which are numbered in millions, with for example round half-a-million in Africa today, has been much criticized by some environmentalists who say 'Leave our streams alone' or 'Don't flood swamps and agricultural land by making dams'. But at the risk of being thought of as one of the 'cheery and beery' ecologists, I myself feel that most of the holding back of water from its loss to the sea is good conservation. Of course it raises problems; but its effect in general is to increase this enormously important, very small, freshwater resource. Even the Aswan High Dam of Egypt, which has been much criticized for its ecological effects, if you look closely at it and all its implications, is really not a bad bit of conservation.

A third point is that hydroelectric development has the benefit of producing energy without any reduction of the resource—for example of fossil fuels. It is surprising therefore that only 6–8% of the water power which exists in Africa, South America and Tropical Asia has yet been tapped [or otherwise used]. But used it will be all over the world, and the impact of this kind of development on the environment [as a whole and on the autecology of individual biota] is worth thinking about pretty closely.

My fourth point relates to the interdisciplinary approach to water development. The more one looks at these problems, the more does it seem essential to relate hydrology with biology, with engineering, with agriculture, with medicine and health, with sociology, with recreation, and of course with economics, and to look at each project from all these points of view. Here I would mention an event which is to take place a week from next Saturday, namely the first meeting of a new Water Research Committee, which is being set up jointly by ICSU (the International Council of Scientific Unions, involving its former Water Committee which was composed only of unions of ICSU) and the International Engineering Organization UATI. This marriage between engineering and the more basic sciences represents a considerable advance in the interdisciplinary approach to water problems.

My last point relates to conservation of water areas—waters that are specially important for reasons of their uniqueness, for example in physical or chemical characteristics, fauna and flora, or large investment in research effort. Such places provide scientific benchmarks, and a lot of work has already been done in making a world list of some hundreds of key freshwater areas—mostly lakes, rivers and large reservoirs—throughout the world. Though this documentation is not yet finished [and indeed can never be to entire satisfaction owing to ongoing changes] it has already saved quite a number of important key areas from undesirable development. This is a form of conservation which this Conference might very well push forward.

Now unfortunately Dr Obeng, who has written the first paper that you have before you, is not able to be with us until later in the Conference [so I will just draw attention to its most salient points* before asking Professor Hasler to present his paper].

Hasler (Co-Keynoter, presenting papers)

With the documents before you as basic material for study, I am going to speak from illustrations to give you a first-hand view of two of the items. [My object in slipping in these two examples is to illustrate how, in the context of] the interrelationships of land and water, [a combination] of basic research, systems analysis and multidisciplinary studies has enabled us to attack some of the problems, identify the causes, suggest methods of improving the situation that has gone bad through Man's activities, and project into the future some better ways of handling certain of our problems.

My first example will be on the topic of eutrophication. With the invention of the flush-toilet, resulting in the placing of effluents of Man, which are high in phosphates and nitrates, into rivers and freshwater lakes, the Algae of the environment receive this additional nutrient supply. We observe them responding by growing in leaps and bounds and producing nuisance scums on the surface. The other evident

*Omitted from this account in the interest of brevity.—Ed.

response is the increased growth of macrophytes—the large aquatic plants which also require phosphates and nitrates.

With the decomposition of the excessive plant growth we get objectionable, putrid odours developing [and a concomitant deterioration of] oxygen conditions in the lakes receiving these excess nitrates and phosphates from sewage. The deep-dwelling species of fishes such as lake-trout [*Salmo* and *Salvelinus* spp.] disappear, as was first observed at the turn of the century in Lake Zürich in Switzerland and has been repeated many times in various parts of the world. This is one of the symptoms of this disastrous disposal of Man's wastes. Another cause of eutrophication is the run-off from agricultural land which, in the spring of the year, can bring into the lakes rather plentiful sediments that are rich in phosphates, nitrates and manures that have been used to fertilize the land. Yet in Lake Washington, for example, when action was taken by the community [following sound scientific advice] to divert the sewage away from the lake, by 1968, only four years later, the lake had returned to almost normal conditions.

So here again we have a successful example of eutrophication not being any longer a doomsday story: the scientists knew the causes and recommended remedial action, and when this was taken by the community the results were wholly favourable. The study clinching this was done in Canada just a couple of years ago, in which twin lakes were treated artificially, both being divided in the middle. On one side of the division nothing was done [and this acted as a control], but to the other side phosphates were added and within a short period this treated part became overgrown with scums of Algae whereas the control part remained clear; yet when the treatment ceased on the other side of the division it again reverted. So here is a recent experimental example to show that one can take remedial action by knowing that phosphorus is the critical thing which can be manipulated, and we have another fine instance of success. [Further instances may be cited, including that of the Kickapoo 'flowage' outlined in the paper, and the most striking study of elucidating the homing of Salmon, which is described in detail in the paper.]

The work of homing of Salmon offers the possibility of restoring stocks by applying a new technique through bringing the fish in to places where you want them—such as areas below dams or places at the mouths of streams where they can be harvested and reclaimed as food. [There must be plentiful other instances where], with knowledge, basic research, and some understanding of the drainage-basin situation and of the lake involved, by employing a systems approach we are going to be able to solve often pressing problems.

One of the pessimistic aspects of these examples is the case of eutrophication: it took only something like five or six or at most 10 years for the lakes to become eutrophic after sewage was introduced into them, but to resolve the problem by public action took up to 25 years—even after the scientists had found the causes and suggested remedial methods there was a long period of delay. In the case of the Salmon situation there was a delay of some 30 years between having the idea and getting the results in, and these delays in getting action, considering the rapidity with which things are destroyed, constitute one of the most discouraging aspects [that tend to dampen our often] hopeful and optimistic approach. [Applause.]

Worthington (Chairman)

Thank you very much, Professor Hasler, for two admirable examples of fundamental research leading to practical applications—and I emphasize the word 'fundamental' as being basic to the practical research leading to such opportunities. We have not yet heard much about the quality of water in conservation, concerning which Professor Goldberg is [pre-eminently expert, though mainly in marine situations].

Goldberg (Panellist)

I certainly cannot enter the area of Professor Hasler; I admire his work but it is way above my knowledge. However, I am glad to address myself to this problem of the quality of waters—especially as the stepping-stone to this was given to me today by several persons who discussed the problem of quantification and advanced the idea that perhaps we do not have to quantify our researches, our concerns. My argument is that we must do this.

The story I would like to tell you is that of chloroform in potable waters. My first concerns about this were raised in the early 1960s, when a book came out about *Social Indicators*, which received very little [attention]. In this book it was pointed out that the life-expectancy of US citizens after the age of 1 appeared to be decreasing, and the authors of the book suggested this might be [because of additional anxiety or other] stresses upon human beings in the United States. As a consequence of this, many of us have been looking for the last decade at what materials might be stressing the human body. One general group that we have long been especially worried about are the organic compounds that contain chlorine, whether they be the large molecules such as DDT or the industrial chemicals such as polychlorinated biphenyls [PCBs], chloraphenes, benzines, or yet others. But then we became disturbed by the smaller molecules, the so-called halocarbons and chlorinated methanes.

Recently my attention has been directed to chloroform. It has been identified as a carcinogen, and we asked the question, how is the human being exposed to chloroform? There are three general ways in which we looked at the chloroform exposure problem: one is through inhalation in the atmosphere. Another one is by the ingestion of foodstuffs, and there is a very curious and foolish one here—the use of chloroform as an additive to toothpastes and cough lozenges, which somehow make the mouth tingle or give it a nice feel. The third means of entry of chloroform to the human body is through drinking waters in some cities—especially after treatment with chlorine to clean the waters and rid them of pathogens and organic materials. So we had three sources of entry of chloroform to the human body: atmospheric inhalation, ingestion through food and medicines, and finally through potable waters.

We had identified the most important source of chloroform for many people as being potable waters as a consequence of chlorination. But quite curiously, we could not identify the sources—and still cannot, by the way—of chloroform in the atmosphere. We are off by a factor of ten. We suspect that we know the natural sources to a large extent as being the ocean surface, but we are totally unaware of how the chloroform gets into the atmosphere. Of course this is very much our concern, yet how do we do the bookkeeping on chloroform? How may we turn off valves, so to speak, if it is introduced by human activity and we come to the point where we must reduce the leakages to the environment?

My conclusion and my concern is that we must quantify—we must quantify the addition of pollutants such as chloroform to the environment, we must recognise the sources, and we must identify the [channels of] entry to the human ecosystem and to other ecosystems—such that if we can pinpoint a number which says the human system can accept no more, that the ecosystems are in danger, we can turn off the valves.

Knox (Panellist)

One of the things I think it is important to note in the papers by Professor Hasler and Dr Obeng is the emphasis on natural variability in the sensitivity of biota to pollutants—particularly to their quantity and magnitude in fresh waters—and I would like to make a few remarks in that vein.

When we think about variability, the two obvious categories are of natural variability related to climatic variation and then of anthropogenic effects. Most people, or anyway many, do not perceive accurately the significance of climatic variations. This morning we heard about how unusual the characteristics of the first half of this century were, climatically speaking, and how the patterns of the last two decades may be more representative of the long-term average; since 1950, approximately, we have had high variability in climate as a common characteristic, and one of the things that we have observed is that this has greatly affected the magnitude and frequency of stream-flows—particularly the occurrence of extreme events such as floods and droughts with low flows. So our whole concept of randomness that dominates much of our planning and perception of hydrology in general is a bit out of [gear] in the sense that randomness is a very poor assumption. This becomes particularly important if what climatologists are telling us is correct, namely that the period which dominated the [first half of this century] may not be a good one to base our predictions on, but rather that this high variability of the last decade or so may be a more [widely normal] situation. To give you a specific example, one of the traits that we have been monitoring in the Upper Mississippi Valley in the central midwest United States, the surface run-off for one year, may [account for anywhere from 20% to 50% of the total rainfall in years] having different climatic circulation patterns. Now this variation is fairly representative of the kind of thing that I am concerned with here, because pollutions by nutrients such as nitrogen and phosphorus of course are very much related to the flow of surface run-off, and when we are dealing with these kinds of variations in a non-random recurrence we can see that we have possible major problems of variations in expected and actual yields.

The second category of variability that I mention is the anthropogenic factor, and there are several subdivisions here that we can think of, agriculture of course being one that we hear a lot about. Some of our studies in Wisconsin, where we have only been farming the landscape for 150 years and before that had a covering of natural forest and prairie, show that the conversion to agricultural land increased the frequency of recurring floods—say that recur once in a couple of years—by an order of four- to five-fold, with a much greater proportion of the total run-off of the watershed going to surface run-off and much lesser amounts retained for base-flow and addition to the ground-water-table. At the same time the sediment delivery from the water has increased many-fold. During urbanization—another type of land-use that we should be thinking about in areas actively undergoing urbanization—sediment deliveries may be at 50,000 times the natural rate. Of course they go way down after a watershed has been fully urbanized, but in general we see the same tendency as in most anthropogenic impacts—variability of flow is increased and this puts greater stress on the system, so that other parameters such as temperatures [and floods] generally go up.

Concerning quantification we have heard a lot of negative remarks but I would like to support Professor Goldberg as I feel it is very important and that we should not back out of it. This is particularly necessary I think in stream systems where the changes, the responses in particular to external as well as internal factors, are of the non-linear, discontinuous types of response. We know very little about these. In many cases there are thresholds. We do not know where those thresholds are, but when we surpass them we get a major metamorphosis in the watershed—often accompanied by an episode of trenching or alluviation in the channel.

In terms of things that we should be concerned about when looking ahead, and some of the trends that we see in tackling specific problems, the suggestion of our Chairman, that we might have more dams, more development in general of rivers, indicated that there is an adequate supply of fresh water which is very little utilized;

yet one of the problems is that its distribution in the world is far from uniform. It is true that it is little utilized, but in places where we really need it, it is often a very limited resource, and when we see things going on such as mining the ground-water-table, which is being practised in many areas today and is indeed mining (as the water is not being replenished at the rate that we are using it but is part of the palaeo-environment), this kind of activity must have a great impact eventually on freshwater resources.

One of the other things that we ought to think about is the elaborate [planning of often limited water resources in attempts to meet future needs, which is] being proposed by various countries today, and which of course will have great impact.

Lastly I should comment that the losses of soils, with concomitant pollution of streams, is immense, though there does seem to be some progress in correcting this. But because the systems are so interdependent, with practically everything inter-linked, I think we must expect to see both positive and negative responses. A particular example is the control of sediment from watersheds. If we shut this off, we can expect some erosion in the system by removal of storage sediment that has accumulated through accelerated run-off from land-use, because there is a very delicate balance between the amount of water flowing in a stream and the sediment that is delivered by it, and if we change conditions to improve land-use we are going to get some negative responses in some cases. Indeed when I think back over the last few years in areas that have very elaborate soil conservation programmes, extreme climatic events have produced some of the highest pollutions from sediment that have ever been recorded.

Worthington (Chairman)

Thank you very much Dr Knox; now Professor Renzoni, from the other side of the world.

Renzoni (Panellist)

After the very pessimistic view given to us by Dr Fosberg and the eminent first panellist of this afternoon, I would like to introduce a small note of optimism. During the last six years, in my laboratory in Italy, we have been studying the concentration of mercury in the biota (fish and freshwater mussels) in two rivers of an area that is particularly rich in cinnabar deposits, where mercury has been mined in large quantities over the last 100 years. Recently, the world demand for this metal has decreased sharply and the one remaining Hg plant has installed depuratory equipment. The results were quite evident in our sampling: over the last three years we have detected a mercury reduction in fish and in mussels of about an order of magnitude. Almost the same thing has been occurring in an area of the Tyrrhenian Sea where a chlori-alkali plant was polluting the water, biota and sediments. After repeated inquiries, analyses and fines, the factory finally installed depuratory equipment, and here too the mercury concentration in molluscs, crustaceans and fishes has been greatly reduced.

Since 1976, 18 Mediterranean countries have been participating in a monitoring programme for heavy-metals and common pesticides under the sponsorship of a FAO/UNEP common project. The first results seem to show that, except for certain small areas (such as harbours, sea-inlets, and so on), the anthropogenic contamination of the Mediterranean is not so high as we thought, at least for these kinds of pollutants.

Finally, regarding the question of pessimism or optimism, I would like to call your attention to a statement which Commandant Jacques Cousteau, the French submarine engineer, made about 1968: he said at that time that the Mediterranean Sea would be dead within ten years. Almost a decade after that dramatic forecast the Mediterranean, even though polluted in some areas, is far from dead, and the fish catch for several species has increased, though not in terms of fish catch per boat. So even though people like Cousteau can be extremely useful [in alerting public opinion, especially when they have a wide following], their statements should be taken with a grain of salt. [Applause.]

Worthington (Chairman)

Maybe the Mediterranean would not be quite so clean today if Cousteau had *not* made that statement. [Laughter.] Now we have had some interesting contributions from fundamental and applied biology, and from various aspects of water chemistry and physics and geography if I may call it so, yet we still have half-an-hour for contributions from the floor.

Goldsmith

I want to go back to the question of quantification, for I, too, feel that quantification is not a way of solving our problems. I want to look at this question of pollution—pollution of drinking water, pollution of food, pollution of the air we breathe—as I think one of the best indicators of pollution is the cancer rate. The cancer rate now practically amounts to an epidemic. In America and most other industrial countries, 25% of the people die from cancer. As people are especially susceptible to carcinogens when they are very young—maybe 100 times more susceptible then than in later life—and as many of us are in the age-group that is most prone to get cancer, it is likely that in 20–25 years the cancer rate will have increased still further—perhaps up to 50% or 60%, becoming generalized in industrial populations. This is the view of the Director of [the] National Cancer Institute, who has said that we are sitting on a time-bomb with regard to cancer.

Now this being so, and it is being accepted more and more widely that 80% or 90% of cancers are due to environmental factors, in particular to pollution, we can say that the whole class of nuisances which we are putting into the environment are responsible for about a million deaths a year. In America [every year] some 700,000 [people] are being told they have got cancer and some 400,000 to 500,000 are dying [of it]. You can say that pollutants between them are killing this number of people. They are guilty, and in any court of law would be found guilty. Yet if I happen to be a man with a factory oozing lead or mercury or any of the [other horrors] you may name, I could produce endless 'experts' to show that my particular pollutant is not doing the damage. All the people producing these pollutants are able to do the same thing. As a result, pollutants are very rarely classified as carcinogenic.

We are putting into the environment 2 million chemical compounds, of which about 3,000 have so far been examined for possible carcinogenic properties—3,000 out of 2 million! Of these 3,000, some 1,000 have been shown to be carcinogenic in certain animals, but of those 1,000 only 30 have been classified as carcinogens for Man. Why is this so—probably because it is too difficult to prove carcinogenicity in particular cases? It cannot be done by quantitative means. But let us look at the problem still further. According to the SCEP report on pollution, *Man's Impact on the Global Environment*, which I am sure many of you will be familiar with, synergic effects more often exist than not among the different chemicals which we put into our environment. Therefore these things will [be quite likely to] affect us differently in different combinations [as well as concentrations]. An obvious example is DDT. If you put it in water it is virtually insoluble, but if you put it in oil it becomes some 10,000 times more soluble. Think of millions and millions of different effects of this sort: it is not just 2 million substances that you ought to examine, but all their possible combinations. It may also be that some of the 2 million different substances are effective at levels which are not even measurable by current techniques. There is no evidence that high levels of particular chemicals are more dangerous, are more likely to give cancers, than very low levels which cannot even be measured over a long period of time.

Quite apart from the above considerations, there are different vulnerabilities of different peoples—particularly different age-groups—and tremendous differences in personal vulnerability. Moreover levels can change—even while you are measuring them; [for example] the heavy-metals in barracudas vary according to the season.

Now if you put all these things together you will see that it is impossible to carry out the necessary number of measurements. And there is another factor: when you are dealing with 700,000 people who [may] get cancer every year in America, a 1% increase in the cancer rate [may mean] 7,000 extra cancers, and you have got to look at terribly small amounts of possible carcinogens that could have a massive effect. And remember the samples you need to look at are absolutely massive: there is no point in just looking at 1,000 white mice or something. You have got to look at maybe 100,000 or 500,000 and will be dealing with experiments which may cost 5 or 6 hundred thousand dollars—just to test one single chemical for carcinogenicity! Even then you will not have tried the thing out on humans; so now you see the whole thing is no longer a problem that can be solved by scientific methods.

Worthington (Chairman)

[A voice having been heard to call out 'But statistics are a form of quantification.'] Can I call on our medical friend [from the University of Oxford] to comment on this?

Juel-Jensen

I was very interested in Mr Goldsmith's comments. He is obviously worked up about this whole matter [laughter], but I am afraid has lost his sense of proportion. It may be desirable to prevent a few people from getting cancer: I am not altogether convinced of that. People already live too long. If you are a physician, much of your work is to do with trying to patch up worn-out bodies when you can do jolly little for them. It may well be right to stop many of the infectious diseases, and it may well be, as Dr Worthington [implied], that the Aswan Dam was a success [on balance], even though it has raised the incidence of schistosomiasis in the Egyptians to 95%. It will certainly do away with a good many people, and that will help to solve the local population problems.

As for your carcinogens, maybe some of those substances are carcinogenic—but does it really matter? You cannot extrapolate from what happens in mice to what happens in Man. It is a terribly dangerous assumption. This is true for many medicaments as well. Let us cite the example of penicillin, which should have been damned and banned as a dangerous substance, never to be used in Man, if the original animal experiments had been carried out on guinea-pigs. [The wags say that] you need only *show* a guinea-pig a penicillin bottle and it will topple over dead [laughter]. So, therefore, be cautious before you say of any substance that has had suspicion cast upon it, that it is a possible carcinogen and should be done away with. Life is too short, and it would be far too difficult to live if we were to be cautious all along the line. Let us preserve a sense of proportion. By all means tell people that if they smoke, every cigarette will shorten their life-span by five minutes. They can then make up their own minds. But concerning some of the remoter possibilities we should use our common sense.

Fosberg

As I seem to have been the one who started the argument about quantification, perhaps I had better clarify my viewpoint a bit. I regard quantification as a method that you use to accomplish certain things, and if it is an appropriate method, by all means it ought to be used. I do not regard quantification as the solution to all the scientific or other problems in the world. Indeed I think we are more and more coming to realize that ecosystems and even components of ecosystems are enormously complex, and that they are probably not amenable to investigation by simple quantification. We can quantify only single factors. If we are dealing with a million factors in a system, we are [surely] not going to be able to quantify all of them. Therefore, the result of our investigation, if we do depend on the quantification of the few that we can so treat, may be quite unreliable. But by all means quantify if it solves your problems.

Miller

I am greatly concerned about freshwater habitats and whither they go, and especially as I am from an area of the world [the southwestern United States] where irrigation is very important. There has been diversion, for economic purposes, from the streams and from the lakes, which have been the habitats of native species. We have, because of this, encountered problems of endangering the natural populations of plants and animals in those waters, and there are now measures being taken throughout the United States for protecting endangered species lest we eradicate some of the gene-pools of those kinds which we have not yet learned how to adapt to Man's uses for Man's benefit. Nevertheless we have learned that we ought to be cautious.

I have in mind one particular instance where there is a population of very tiny fish: the adults are about the size of the last joint of your little finger. The population in Devil's Hole, Nevada, is in a unique hot-spring environment, ranges in the order of 200 up to 700 individuals throughout the season, and then drops back again because of attrition. But this population has existed for thousands of years [and appears to be] a remnant. [In 1952 the chasm and pool of Devil's Hole and a tract of land around it was set aside by presidential order, as a protected site, part of the Death Valley National Monument. Later, owners of nearby land began pumping groundwater from several wells, thereby drawing down the water level in the habitat pool. In litigation brought by the United States, a District Court enjoined pumping that would 'lower the level of the pool below the level that was crucial for spawning'. This ruling subsequently was affirmed by the Court of Appeals (U.S. Ninth Circuit) and just recently by the United States Supreme Court. In its unanimous decision the Supreme Court held to the lower court's assertion that 'the pumping had lowered the water level in Devil's Hole, that the lower water level was threatening the survival of a unique species of fish and that irreparable harm would follow if the pumping were not enjoined'. It found with the District Court that if the injunction did not issue 'there is grave danger that the Devil's Hole pupfish may be destroyed, resulting in irreparable injury to the United States'.* Thus goes the business of protecting water habitats. Does not this suggest that society is becoming sensitive to the importance, if not the rights, of our fellow creatures?]

Another instance is the pollution of Lake Tahoe, which straddles the border between California and Nevada, [with] drawdown of good water there for domestic and industrial purposes and the use of the Lake as a sump for domestic wastes. There, too, the courts, and the community through the courts, finally decided that there had to be pure water, and tertiary treatment plants were [required]. But then there was a decree that no water should be returned from these treatment plants into the Lake itself but that it all had to go over the mountains and

*Cappaert *v*. United States (Nos. 74–1107 & 1304) 426 US 128, Supreme Court Reporter v. 48, Lawyers' Edition 2nd vl., pp. 523–539. Decided Jn 7, 1976.

out into other drainages in order to keep the Lake free from eutrophication.

These are examples of what is going on, and there is something that the other speakers did not cover—namely the use of water for transport [—and not only on the surface. Thus] in southern Nevada there is a coal-burning power-plant which gets its coal through a pipeline. The coal is ground into a powder and is carried in a slurry through the piepline from Utah into Nevada, where it is converted to electricity. [Power goes thence] by wire to southern California. The figures for consumption of coal I cannot give you but the requirements of water are of the order of one ton of water going into the slurry to carry one ton of coal. Now if that is the case we have got an innovation, because in the current designs for improving the use of energy in America and using coal instead of imported oil, there is a great move to have these power-plants activated by coal. [In] the process of doing that [we] move coal from such areas where it has not been produced in quantity up to now. [With no previous means] for transport, [we] want to use the slurry system in pipelines, and this from an area where there is not now an adequate supply of water to satisfy [even] the needs of irrigation. So it is very likely that the designs for using water as a medium of transport will [be limited and in contention].

Worthington (Chairman)

[Thank you, Dr Miller. In connection with your last point] it might interest you to know that the [proportion] of biological water in the hydrosphere, in plants and animals [including Man], is 0.0001%—a calculation that was made not long ago.

Thorhaug

To add a cheery note to some of the rather gloomy remarks that have been made in the last half-hour, I would like to comment on both the terrestrial and the aquatic ecosystems. [In 1976] we held a symposium on Restoration of Major Plant Communities in the United States, ranging from Alaskan high-altitude and high-latitude ones to Florida and the Caribbean. There are many kinds of [desirable plant species whose populations] are now capable of being restored, ranging from the Arctic to tropical regions and from high mountain to desert or marshland and aquatic systems. Much of this work has recently been published in Dr Polunin's Journal *Environmental Conservation*. Biologists should note the kind of individual management [of] which Dr Hasler talked about with fish. Botanists are turning their attention to restoring plant populations in ecosystems in various places. I think this is part of the aquatic/terrestrial situation that should be examined more carefully for its potentials of mitigation, restoration and plant resource management.

[**Polunin** (Snr): You don't always have to restore artificially: many biota migrate adequately on their own—Alwyne Wheeler, of the British Museum, now reports I believe 91 species of fishes as having been captured in recent years in the London reaches of the Thames—and we still have our Principle of Ecosystem Completion to fall back on.]

Kuenen: [Bravo to that; the resilience/plasticity of life is absolutely stupendous. But] I would like to come back once more to this problem of quantification. I was very glad to hear what Fosberg said about it because I thought it was putting things right which were going wrong. Certainly we must try to quantify where we can, but my point in opposing [wholesale] quantification was that the economists pretend that they quantify well enough and now require the biologists to do the same. My point was that the quantification of the economists was no good at all, because they were not looking at the long range of time—only at the short time. But particularly with the example of the chloroform, nobody could doubt the absolute necessity of quantifying in that sort of case; so I do not think that we have really been talking [quite the same language]. Certain things must definitely be quantified, whereas others—such as beauty, which we all appreciate, or misery, which we all want to abolish—are not quantifiable in the way that economists try to quantify matters by using dollars or francs or whatever happens to be [convenient].

Knox (Panellist)

As one of the advocates of quantification I would agree with the last comment: some items obviously need to be quantified and some others do not. But one of the things I think we must be careful about is allowing several people to say that we do not need quantification. For if this gets picked up in the popular press, we might be confronted with statements that we do not need quantification any more, which could be a bad thing.

Goldberg (Panellist)

I would like to answer both Professor Kuenen and Mr Goldsmith about why we concern ourselves so with chloroform. Is it just whimsy that we pick chloroform out of the air to examine? The answer is definitely 'no'. What has been done is to look at the potential exposure-levels of toxicants on the basis of their production—chloroform by the way is one of the most-produced organic chemicals in the world—at their persistence in the environment, at their toxicity, and at their bio-accumulation, so it is not from whimsy that we pinpointed chloroform.

Now as to quantification—and I think this is a sort of tempest in a teapot—[evidently] most of us agree that there are certain areas in which we do need to have numbers. But one area that concerns me the most is that of transforming our scientific bases into regulations at a government level. And one of the most direct and I think persuasive ways [to get politicians to listen and even act], is to have numbers in respect of your data. Much of our information on the regulation of the release of chemicals to the environment is terribly weak. But where we can back it with numbers it makes the conversion to social action much easier.

Goldsmith

Now I just want to go back to my point that I do not think quantification is [widely practicable; indeed it] gives a spurious air of precision. If you look at the studies of the last few years—take the limits to growth—I think they would have been far better without any figures, because anybody who tells you he knows for instance how much copper there is in the world is above all a liar. Nobody knows how much copper there is in the world. And take the case of oil: the estimates of the reserves in Canada are thought by some to be 15 times too high. The oil companies pretended there was more around, but why? Because they wanted to justify the sales of oil to the United States—they were mainly American companies. Yet it appears that in America the opposite is occurring: the oil companies pretend there is far less oil than there really is, because they want to justify the import of Arab oil on which they make more money. So in fact both sets of figures appear to be wrong, and if you look at all the figures you come across you will find many of them equally wrong. We just do not have any means of quantifying these things properly.

To return to Professor Goldberg's argument, it is not a question of figures. If you are driving a car towards a cliff, I think it is an absurd notion to start putting out a lot of instruments to determine at what point you have got to stop the car and go into reverse. How can you measure it? We are moving in the wrong direction. We are moving in the direction in which we are wiping out the environment on which we depend for our sustenance and very existence: we are wiping it right out. At what point we have to turn back seems an irrelevant question. [The same with those of] whether there is 30 years' worth of copper, whether there is 40 years' worth of oil, or whether we can still go on behaving the way we are doing without getting generalized cancer. All these things are irrelevant. We are moving in the wrong direction, and the sooner we start moving in the right direction, the less painful will be the transition. We just cannot wait until the last minute. Quantification gives us an air of precision which is totally unjustified and it is largely irrelevant to measure these things in any case.

Worthington (Chairman, concluding Session)

We have had a long day but have finished our main discussion I think in good time. Yet I would just like to make two points before closing. One is that I do not myself press for water control in its many forms. But while the world's population curve is in that upward sweep, we are bound to have a great deal more water development and water control than has yet taken place, and the resource is there for the job. It is no good trying to stop that sort of development, but there are many ways in which we can turn it to better effect than if we do nothing about it. For example, in connection with schistosomiasis, there are ways in which the transmission by snails can be broken during the planning and management of irrigation schemes. And our medical participant did not mention that there was at least 90% infection *before* the Aswan Dam was built, and the initial increase was predicted in advance. So I make my last point which is that, considering how we have spent all the day discussing ecological systems and attempting predictive exercises in conservation and ecology of the future, I am surprised that we have scarcely heard the expression 'systems analysis'. This I have found a bit of a relief; but I hope that in the rest of this Conference we shall not forget the tremendous advances which have been made in mathematical modelling drawn basically from engineering and applied now to biological systems in a large and often successful way. So let us close by thanking you very much Professor Hasler, panel members, and all contributors. [Applause.]

Session 5

WHITHER THE OCEANS AND SALT SEAS?

Keynote Paper Part I

Diagnosing the Future Health of the World Ocean

by

EDWARD D. GOLDBERG

Professor of Chemistry, Scripps Institution of Oceanography, University of California at San Diego, La Jolla, California 92093, USA.

INTRODUCTION

There exist today a growing number of threats to the integrity, stability, and possibly even the continuity, of marine ecosystems. Human impact on these systems is increasing both in intensity and diversity. One consequence is that, despite continuing scientific research and even a modest growth in research efforts, our apprehension as to the medium- and long-term effects of such impacts is, in many areas at least, tending to increase rather than to diminish.

Human actions having an impact include mining and dredging of the sea-bed, constructions and other changes on land near shore which affect critical habitats of animals and plants in coastal zones, man-made changes in régimes of freshwater run-off from land, and above all pollution and over-fishing. In the two parts of this paper we examine particularly the last two actions as examples of what we have seen in recent years, and what we must continue to expect unless very drastic changes are made, and soon, in the management régimes of land- and water-use—including, of course, salt-water-use.

By 'pollution' we mean here the addition, by human agency—whether intentionally, incidentally or accidentally—of substances or energy such as to cause harmful consequences for various uses of ocean space and its resources, including consequences for human health, fisheries and amenities. Similarly, by 'over-fishing' we mean the conduct of fishing operations in such a way as to prejudice present or possible future values of marine living resources, taking account of the synergy of fishing pressure and deleterious environmental changes, whether they be natural or man-caused.

The ability of human society to affect and even manipulate the nature of the ocean environment is well documented (e.g. Goldberg, 1976). Surface waters contain more lead now than they did a century ago. The creatures of the sea are tainted with measurable amounts of the pesticide DDT and its degradation products. Artificial radionuclides, primarily from the detonation of nuclear weapons, are evident in all ocean waters. Some changes have jeopardized the continued use of marine resources, while other alterations have benefited mankind—such as the creation of artificial reefs and the consequential development of novel fishing areas. So far such catastrophes as the Minimata Bay mercury poisoning or the Torrey Canyon release of oil* have been limited both as to region and to short intervals of time.

But what of tomorrow? How and where might the marine environment be vulnerable to degradation by the actions of Man? The continued development of technological societies—disposing of materials rather than conserving them, and shaping the physical character of the environment to their needs rather than adapting those needs—creates all manner of dire disturbances to our surroundings.

The identification of significant problems in the future decades will continue to suffer from all of the difficulties we have in predicting activities of human society. The governing parameters are too many and their interactions are too complex. Still, the continued development of strategies to identify areas of concern should help us to maintain acute awareness of the powerful influence of Man upon the marine environment.

Fortunately, a mood of protectionism has entered political thinking as a consequence of the knowledge that our fuel supplies are not limitless and our water supplies are not always available in usable forms. But more and more energy is being sought by a world population that is still increasing in size and needing more and more water to satisfy its industrial, agricultural and social needs. The non-energy-related wastes of Man are now produced at a rate of about three thousand million tonnes per year. This annual flux might better be considered in geometric terms—it would form a cube a whole kilometre on edge. Much of this waste eventually enters the marine domain. In amount, it may have a doubling-time of merely decades, although only a part would suffice to threaten ocean resources. Herein we will attempt to identify that part and consider where the undesirable impacts might take place.

THE POISONS

Of the many thousands of substances mobilized by Man to the sea, only a very few challenge its integrity. The deliberate and unintentional releases of petroleum have killed many seabirds, but such events appear to be

*Later, in the spring of 1978, by far the biggest-ever oil-spill from a wrecked tanker occurred on the northern coast of France, 'desecrating hundreds of kilometres of shoreline'.—Ed.

short-lasting. Our beaches are soiled with thousands upon thousands of cigarette filters and plastic bottles which offend the sensitive eye. We have yet to measure the impact upon marine organisms of their increased body-burdens of anthropogenic halogenated hydrocarbons. But with what materials should our concerns be, and can we justify these concerns?

The marine discharge of artificial radionuclides from facilities used in nuclear fuel-cycles—especially reprocessing plants—pose formidable problems, in my opinion. The largest operating reprocessing plant is at Windscale, situated in the United Kingdom on the border of the Irish Sea. Although only a minute fraction of the total amount of waste is disposed to the sea, the potential challenges to human health from such an activity are already evident from official publications (e.g. Hetherington, 1976). Recently, the highest degree of individual public exposure to radiation is a consequence of exposure to radionuclides accumulated on beaches. The population which could suffer the highest external exposures are the salmon fishermen working the Ravenglass Estuary. In the year 1974 alone, the most highly exposed individual would have received 7% of the recommended dose-limit proposed by the International Commission on Radiological Protection (ICRP). Such doses accumulate in the body through the years. Similarly, the levels of caesium isotopes in fish and shellfish caught in the Irish Sea led to estimates of consumption by heavy eaters of fish of 14% per annum of the ICRP dose-limit. I wish to emphasize that these estimates are very conservative and the true dangers are probably less than indicated. Still, the UK has just entered the nuclear fuel-cycle and other countries, currently without reprocessing facilities, are considering their construction in coastal areas with consequential discharges to the sea.

What is the capacity of the ocean systems to accommodate these nuclear wastes? A knowledge of the amounts and types of radionuclides that are being discharged is essential for a description of the present situation and for a prediction of future levels. Although the call for such a bookkeeping has been made over the last two decades, no international agency has been able to provide the scientific community with the data (Goldberg, 1977). It is my contention that the concerned scientists must assume the role of detectives and ferret out this information in any way possible. At present the United Kingdom is the only major nuclear country which has made available its release data, although these do suffer substantial inadequacies.

What other substances might jeopardize the oceans? Let me present to you one collective of chemicals whose members might act in concert against the well-being of living systems—the low-molecular-weight halocarbons. These chemicals, which include carbon tetrachloride, chloroform, the fluorochloromethanes, trichloroethylene, tetrachloroethylene, trichloroethane, and methylene chloride are produced all together in annual amounts of the order of megatonnes (each of 1,000,000 tonnes). Present environmental levels are of the order of nanograms (g^{-9}) per litre in sea water and nanograms/m^3 in the atmosphere. Their rates of production in general are increasing at the present time. They are markedly persistent in the atmosphere and in natural waters.

The chlorofluorocarbons have been postulated to alter the ozone levels of the stratosphere through their involvement in photochemical reactions that reduce ozone. But in the marine environment they might play another undesirable role—the interference in methyl transfer reactions that figure so importantly, in fermentation processes. Yet up to the present time their levels of toxicity, either individually or collectively, to the involved microorganisms have not been measured. Maybe there is an intuitive sense among interested biologists that these levels are far too low to have an impact upon the health of marine life. Still, the teratogen dioxan is toxic at levels of parts per million millions or less. I submit that an assessment of these chemicals as environmental pollutants is called for.

Clearly, other man-mobilized chemicals may damage ocean resources—including heavy-metals, petroleum components and plasticizers. Their identification as pollutants, it is much to be hoped, will be made by analytical methods rather than by a catastrophe. The protocols to identify potential pollutants have been proposed (NAS, 1975), but actions to implement such proposals are yet to be taken.

The Ocean's Vulnerable Zones

With the aim of defining where Man's impact will be most intense, the oceans may be subdivided in several ways: the northern-hemisphere oceans and the southern-hemisphere oceans; the coastal ocean and the open ocean; the surface ocean and the deep ocean, etc. Each division offers some insight as to where unwanted episodes might occur.

If the gross national product (GNP) of a nation is accepted as a measure of its pollution potential, i.e. the number is proportional to material usage, then clearly the northern-hemisphere ocean will receive a far larger amount of Man's wastes than its southern counterpart. So far, the major pollution problems have occurred in the developed nations of the Northern Hemisphere—the destruction of non-target fish-eating birds by DDT; the methyl mercury poisoning epidemic in Japan; the increasing burden of artificial radionuclides in living matter of the Irish Sea, and so on.

However, the developing world south of the Equator shows indications of inheriting from technologically advanced societies the capability of jeopardizing their marine resources. For example, although there are restrictions upon agricultural use of DDT in most large countries of the Northern Hemisphere, there is increasing use of this biocide in Latin America (Central and South America), Africa and Asia—primarily in cotton production and in malarial control (Goldberg, 1975). Indeed the present-day overall usage appears essentially equivalent to the highest usages of northern-hemisphere countries in the mid-1960s when use in other countries was minimal. There is no sign of let-up in its overall applications, and there is no surveillance of any undesirable effects.

Multinational corporations are transferring some of their manufacturing activities southwards in an attempt to find lower internal and external costs of operation. In developing nations, labour costs are often lower and the concern for the environment is overriden by their need for economic development. There are many instances of this industrial drift to the south being mediated by multinational corporations, including the transfer of much of the US pharmaceutical industry to Puerto Rico and the dramatic development of petrochemical industries in the coastal areas of Portugal which to a large extent will service northern Europe.

There is a continuing shift in potential pollution problems from the Northern to the Southern Hemisphere. As a result of this situation, there should be an increased surveillance of pollutant levels in the living and non-living resources of the sea-waters south of the Equator.

Marine Time-scales

For the pursuit of both scientific and societal problems, the world ocean can conveniently be divided into the zones designated as 'coastal' and 'open'. The coastal ocean constitutes about 10% of the total oceanic area and includes estuaries, lagoons and many marginal seas. Examples include the North Sea, Chesapeake Bay, the Sea of Japan and the Persian Gulf. Their properties are strongly influenced by their sea-floors and also by the adjacent land-masses. The coastal ocean receives direct injections of continental materials through rivers, glaciers and the atmosphere, as well as by the constructions of Man (such as domestic and industrial outfalls and ships).

The time-scales enjoyed by sea-waters and their constituents in the coastal ocean, i.e. before mixing with the open ocean, undergoing sedimentation to the sea-floor, or entry into the atmosphere, range from months or less to decades. Estuaries exchange their waters with the open ocean in such periods of time. The persistence of lead in the biologically productive waters of the Gulf of Mexico is less than a month (Bruland *et al.*, 1974). Indeed, these short time-scales are in part related to the high rates of biological activity. The open ocean areas, with few exceptions such as some productive equatorial waters, are the marine deserts. Of importance to the consideration of pollution problems is the fact that nearly all of the marine organisms consumed by Man spend much of their lives, if not all of their lives, in the coastal waters.

The open ocean constitutes about 90% of the world ocean and is in general unaffected by the properties of the sea-floor and by the continents. In addition, the persistence of chemical 'species' in its waters ranges from centuries to hundreds of millions of years. Sodium, one of the principal elements in sea-water, remains in the open ocean for perhaps over a hundred million years before accommodation in the sediments. Aluminium compounds, on the other hand, following entry to the open ocean from continental weathering processes, spend about 100 years or less before becoming precipitated to the sea-floor.

It is with this background that we can now consider the scientific time-scales in comparison with the societal ones, i.e. the periods in which Man can react to potential problems or catastrophes in the marine environment.

The Times of Societal Response

The concerned scientific community or the unexpected catastrophe alerts governmental bodies to regulate the releases of polluting substances to the environment. The possibility that the promiscuous releases of artificial radionuclides to the oceans could jeopardize public health through the return of the contaminants to society in seafoods was recognized in the early 1950s by scientists in many countries. As a result of their deliberations, sovereign nations formulated regulations upon the dissemination of radioactivities to our surroundings. Guidelines to permissible releases have been prepared by such international agencies as the International Atomic Energy Agency and the International Commission on Radiological Protection.

But how rapidly can nations react to a tragic event? Two recent incidents suggest that the time-scales are of the order of a decade or less—times comparable to those of the persistence of chemicals in the coastal zone. Perhaps the textbook example is the Minamata Bay incident in which over 100 Japanese fishermen and family members fell victim to methyl mercury chloride poison as a result of the release of mercury to this enclosed marine zone. The symptoms of the first patients were evident in 1953, but it was not until 1959 that mercury was identified as the toxic element. In 1963 the compound methyl mercury chloride was found to be the substance of concern. Discharges were halted in the mid-1960s, a decade after the first indications of a catastrophe. The mercury was first discharged to Minamata Bay in the early 1930s.

Similarly, the rise and fall of the use of DDT in the Northern Hemisphere spanned several decades. Extensive use followed World War II, but it was not until the late 1950s that environmentalists, such as Rachel Carson, were emphasizing the deleterious impact of the biocide upon non-target organisms. Mounting evidence by ecologists emphasized the apparent depopulation of fish-eating birds such as the Osprey (*Pandion haliaetus*) and pelicans (*Pelecanus* spp.). The DDT was bioaccumulated in the fish; the birds feeding upon the fish produced thin-shelled eggs, many of which broke during incubation or otherwise failed to produce healthy young. As a consequence of such impacts, the use of DDT was banned by the USA and many European countries in the early 1970s, some 40 years after its introduction and about 10 years after its identification as an environmental hazard.

Governments may respond to the protective mood for the environment in periods of decades after the problem has been posed—times of the order of the retention of chemicals in the coastal waters of the ocean. It thus seems that the highly vulnerable coastal zone can be managed in principle for the forthcoming years.*

* Fundamental to this is appropriate research into effects of pollutants and monitoring of their concentration in marine organisms, of which a pioneering large-scale example is the 'Mussel Watch' (involving also oysters) of the coasts around the United States which is headed up by Professor Goldberg himself and of which the first detailed descriptive paper has now been published in *Environmental Conservation* (Vol. 5, No. 3, pp. 101–25, 10 figs, Summer 1978).—Ed.

Dealing with the open ocean is a more formidable task. The vast expanses of the ocean fall into this domain, and through mixing and exchange with coastal waters reduce the levels of pollutants in them.

On the other hand, the long residence-times of chemicals in the open ocean may lead to the formation of a toxic 'broth' through the accumulation of Man's wastes—a broth no longer capable of assisting in the purification of coastal waters.

Although the possibility of the long-time build-up of toxic substances in the open ocean with very small annual increases in their open-ocean concentrations appears unlikely, it is not inconceivable. There is taking place today a gradual but continuous build-up of halogenated hydrocarbons, both of high molecular weight (such as the polychlorinated biphenyls) and of low molecular weight (such as the chlorinated methanes). They enter the open ocean *via* the coastal waters, being introduced from the continents in sewer outfalls and rivers. They can enter the open ocean directly following atmospheric transport from the continents. But can they act in concert to destroy a member or members of marine communities at levels that might be reached in the future?

To add to the disturbing nature of this last thought is the sense that there are members of the marine biosphere that are especially prone to any unusual stress, be it a temperature change, an increase or decrease in the salinity of the water they are accustomed to, or a pollutant. Is it possible that the increased body-burdens of strontium-90, DDT, lead, and petroleum, acting together, can lead to the demise of a marine organism?

There is strong motivation to address ourselves seriously to these possibilities and not to disguise them as portents of doom. The open ocean, like its coastal counterpart, is not immune to the impact of Man. But furthermore worrisome is the fact that its great volume makes removal of a toxic substance, identified by a tragic event, impossible by the technologies of today or of the foreseeable future.

REFERENCES

BRULAND, K. W., KOIDE, M. & GOLDBERG, E. D. (1974). The comparative marine geochemistries of lead 210 and radium 226. *Journal of Geophysical Research*, **79**, pp. 3083–6.

GOLDBERG, E. D. (1975). Synthetic organohalides in the sea. *Proc. Roy. Soc. Lond.*, **189B**, pp. 277–89.

GOLDBERG, E. D. (1976). *The Health of the Oceans*. UNESCO Press, Paris, France: 172 pp., illustr.

GOLDBERG, E. D. (1977). A time for accounting. *Marine Pollution Bulletin*, **8**, pp. 49–50.

HETHERINGTON, J. A. (1976). *Radioactivity in Surface and Coastal Waters of the British Isles 1974*. Ministry of Agriculture, Fisheries and Food, Directorate of Fisheries Research, Technical Report FRL 11, Lowestoft, England: 35 pp., illustr.

NAS (1975). *Assessing Potential Ocean Pollutants*. National Academy of Sciences, Washington, DC: xx + 438 pp., illustr.

Keynote Paper Part II

Productivity Aspects with Special Reference to Fisheries

by

SIDNEY J. HOLT

Adviser on Marine Affairs, Department of Fisheries, Food and Agriculture Organization of the United Nations (FAO), Via delle Terme di Caracalla, 00100 Rome, Italy, and *Professor, Environmental Studies Board, University of California at Santa Cruz, Santa Cruz, California 95064, USA.*

OVER-FISHING

In a recent article, McCloskey (1977) has argued that, notwithstanding provisions in United States legislation for sustainable forest- and land-use, and the existence of governmental agencies whose supposed responsibility is to ensure implementation of these provisions, the implied policies are not in fact being generally pursued. Much the same might be said regarding the conduct, management and trends of international fisheries, and of some major national ones.

In the years immediately following World War II, 'over-fishing' had been detected in a very few areas and of a few marine species—some bottom-living fishes in the North Sea and the northeastern Pacific, and the large baleen whales of the Southern Hemisphere. The signs noted were (1) reductions in the 'catches per unit fishing effort' (the effort being measured as total time of operation of vessels, brought in each case to a common standard), and (2) diminution of the average length or weight of fish of a given species in the catches, which was taken to indicate a reduction in the average ages of fish in the population. In certain cases reductions even in the total catches were noted, notwithstanding continued fishing effort at the same or even higher intensity, and the respite which the stocks had been given by the diversion of fishing vessels to other war-time uses that had allowed depleted stocks to recover at least partially (Holt, 1969). Some earlier attempts to reach agreement on restraining the post-war build-up of fishing fleets and their power were soon abandoned. In 1947, for example, the British Government proposed that the total tonnage of trawler fleets operating in the North Sea should be held at the immediate post-war level, which was only 40% of the pre-war level; Britain had the biggest pre-war fleet, and this proposal was not acceptable to other European countries that were seeking to increase their share of the North Sea catch.

In the following quarter-century, the attempt to bring some sort of order into the growth of fishing power took the form mainly of the establishment of a considerable number of regional fisheries bodies, the members of which were governments whose vessels operated in a particular region or were those of countries that were coastal to the region. Many of these bodies were set up by treaties negotiated directly among the countries concerned; others were established under provisions of the constitution of FAO. Their powers and responsibilities varied; at most they could agree on recommendations to governments for certain management measures to be taken by them in concert. Such measures usually concerned the establishment of what are now called 'total allowable catches' (TACs), the sizes of meshes of nets and the minimum sizes of fish that could be landed; in a very few cases the amount of fishing could (in theory at least) be directly controlled (by agreement, of course) and some measures for enforcement adopted—such as inspection of gear and exchanges of national shipboard 'observers'. Also in a very few cases the regional bodies, or the governments of their members, were empowered to negotiate the national allocations of TACs.

Over the same period, global fishing power increased steadily and so did world catches. By 1970 it had become a commonplace observation that catches had for decades been increasing faster than the human population and faster than most other kinds of protein production; this was used in some quarters to justify even further efforts to increase fishing capacity. Yet, equally, by that date it could be said, on the basis of careful analyses by FAO staff and their collaborators, that the end to expansion, at least of so-called 'conventional' fisheries, was in sight (Gulland, 1970). It was true that there had been several earlier predictions of upper limits which had quickly fallen down as these 'limits' were exceeded in the real world. Dickie (1975) has likened the process to 'a gigantic roulette game in which new players continually joined and left their winnings on the table in the hopes of one more lucky play'. But there was 'growing nervousness among the managers and administrators responsible for running the game'. Yet *were* they responsible? Few governments controlled the growth or even the direction of fishing industries; their servants could hardly be held fully responsible for what was beginning to happen.

It turned out that 1969 gave the first clear sign of a break in the statistical trends—a decline of 3% in the total catch as compared with 1968. Two years of recovery were followed by an 8% decline from 1971 to 1972; a further partial recovery for two years was followed by a 2% decline from 1974 to 1975. The 1970–71 peak catch level of 61 million tonnes has only recently been regained (Table I).

The above figures do not include catches of whales. These were of the

TABLE I.

World Marine Catches (excluding Whales)—(Millions of Tonnes).

	All fishes and shellfish	*Excl. Anchoveta*	*Annual % increase (excl. Anchoveta)*	*% of catch reduced to meal and oil*	
				all	*excl. Anchoveta*
1938	18.8	18.8			
1948	17.7	17.7			
1958	28.0	27.2	5.6		
1961	37.0	31.7			
1962	40.1	33.0	4.1		
1963	41.2	34.0	3.0		
1964	45.4	35.6	4.7		
1965	45.6	37.9	6.5		
1966	49.4	39.8	5.0		
1967	52.7	42.2	6.0		
1968	55.9	44.6	5.7	42.9	28.5
1969	54.4	44.7	0.2	41.4	28.6
1970	60.9	47.8	6.9	43.5	28.0
1971	61.0	49.8	4.2	41.8	28.7
1972	56.2	51.4	3.2	36.3	30.4
1973	56.7	55.0	7.0	32.8	30.3
1974	60.2	56.2	1.0	34.8	30.1
1975	59.5	56.2	0	35.1	31.3
1976	62.2	58.8	4.6	35.8	31.1
Average:					
1948–58			5.6		
1958–68			5.0		
1968–75			3.3		

order of 2–3 million tonnes annually in the 1930s, 1–1.5 million tonnes annually in the early post-war years, but declining steadily to less than 0.5 million tonnes by 1970, and considerably less thereafter as more stringent catch regulations were introduced in a last-minute attempt to hold catches down to sustainable levels (Holt, 1976*a*). Thus the slowing down of fishery growth, as manifest in total catch levels, was somewhat more distinct and perhaps began earlier than Table I indicates.

Unfortunately, no global figures are available for fishing effort and the trends in catch cannot therefore be interpreted unequivocally as indicating the limits of the resources. This is revealed only by detailed studies of the kind described in Gulland (1970). At that time it was judged that catches could be increased to two or three times the level then attained (about 55 million tonnes) provided better management measures were also adopted. Less than a decade later, the FAO secretariat had to report to the FAO Committee on Fisheries, meeting in April 1977, that the potential was, if anything, less than 100 million tonnes (FAO, 1977*a*)—a figure which had already been foreseen in 1973 at the Vancouver Conference on Fishery Management and Development.

The break in 1969 can be attributed to fluctuations in the catches, mainly by Peru but also by Chile, of the Anchoveta, *Engraulis ringens*, combined with a general slowing of the rate of increase in catches of other species. The Anchoveta in 1968 accounted for over 20% of the world catch of marine fish and shellfish. Its peak was reached in 1970 (13 million tonnes), followed by a dramatic decline to less than 2 million tonnes in 1973—from which level it is gradually recovering. The massive 8% drop in world marine catches from 1971 to 1972, compared with an average 6% annual increase over the previous 23 years, was thus caused essentially by the collapse of the Anchoveta stock.

If we now look at the catches of everything except Anchoveta and whales (Table I, column headed 'Excl. Anchoveta'), another phenomenon is seen—a longer-term decline in the rate of increase. In addition, it appears that the annual rate of increase of catches begins to fluctuate more and more, until between 1974 and 1975 it becomes, for the first time, zero.

Before looking more deeply into the nature of these trends, it is interesting to examine the uses to which the marine catches are put. Basically they are either used for direct human consumption in fresh, frozen, canned or cured form, or they are reduced for the production of fish-meal and oils. The oils are refined and enter human diets in various ways; the meals, which are more voluminous and valuable, enter livestock feeds and reach the human diet as eggs, and as the flesh of domestic fowl and pigs, or to a much lesser extent as cultivated freshwater fishes. This process, and the international trade in reduced-fish products, led to a bias of effective consumption vastly in favour of the richer countries—regardless of where the fish are originally caught, and by whom (Holt & Vanderbilt, 1974).

The separation of uses is not complete—especially in recent years when the flesh of demersal (bottom-dwelling) fishes, mechanically filleted from the bone and frozen, has been going directly to human consumption, while residues are reduced to meal. On the other hand, an unmeasured but not insignificant amount of fish recorded as canned goes to feed pets rather than people. In addition, a significant amount of the actual catch is in some fishing operations discarded at sea as 'trash fish', but in others may be retained and processed in some way. These factors should be borne in mind when interpreting the last two columns of Tables I and II; nevertheless the trends revealed there are basically valid indications of the direction of industrial development.

The proportion of total catch going to reduction generally increased after World War II. Since 1968, there appears to have been a tendency for this proportion to decline but, again, this is a distortion caused by the wild changes in Anchoveta catches which are practically all reduced to meal and oil. If Anchoveta are excluded, we see that nearly one-third of the remainder goes to reduction and that the percentage thus processed continues slowly to increase.

Ecologically different types of fish tend to be treated in different ways, as Table II shows (from FAO (1975) and FAO yearbooks of Fishery Statistics—Catches and Landings, and Fishery Commodities). Since 1961,

TABLE II

Utilization of Catches, by Type (Fish Only, Excluding Anchoveta).

Year	% small pelagic fish (all fish)	% demersal fish (all fish)	% large pelagic fish (all fish)	% of small pelagic fish reduced	% of demersal fish reduced	% of all fish reduced	Food consumption, fish equivalent		
							incl. Anchoveta	excl. Anchoveta	% increase
1961	29.7	30.4	39.9	42.2	4.3	14.2	29.8	28.0	
1962	32.1	33.4	34.5	45.7	4.1	16.0	31.0	29.1	3.9
1963	28.5	29.7	41.8	42.7	5.0	13.6	32.2	29.8	2.4
1964	36.7	35.0	28.3	38.2	5.7	18.3	34.2	30.9	3.7
1965	37.3	36.1	26.6	42.0	6.1	22.6	34.8	32.2	4.2
1966	36.2	36.2	27.6	42.1	7.4	22.8	36.9	33.7	4.7
1967	35.8	37.5	26.7	42.5	6.8	26.5	38.3	34.8	3.3
1968	34.5	39.0	26.5	41.9	7.5	29.4	39.9	36.1	3.7
1969	33.1	39.7	27.2	37.9	7.4	28.8	39.6	36.4	0.8
1970	32.3	40.7	27.0	34.4	7.5	27.3	43.6	39.2	7.7
1971	32.0	40.0	28.0	31.8	7.3	28.4	44.0	40.7	3.8
1972	31.3	40.5	28.2	44.7	7.6	29.4	43.2	41.6	2.2
1973	33.0	39.1	27.9	52.0	9.7	30.1	44.9	44.2	6.3
1974							47.1	45.8	3.6
1975							45.8	44.6	−2.6
Average:									
1958–68									3.7
1968–75									3.1

from 60% to 70% of all the catches of small pelagic (open ocean) species *including* Anchoveta have been reduced to meal and oil. Anchoveta are excluded from most of the columns in Table II, from which also are excluded shellfish (crustaceans and molluscs) as well as whales. All the bivalve and gastropod molluscs taken are used as human food (many of these are cultured, and recorded total weights give no measure of edible weight), as are 90% of the cephalopods, the world catch of which is about one million tonnes. The fishes may for this analysis be split, roughly equally, into three groups: (1) the small pelagic fishes, from lower trophic levels, used very largely for reduction; (2) the demersal fishes, from higher trophic levels, which mostly are consumed directly; and (3) the larger pelagic fishes, also from higher trophic levels, which are practically all consumed directly. After a decade during which just over 40% of small-pelagic-fish catches were being 'reduced' for other than direct consumption, there seems to have been some reduction (1969–71); but now the percentage so utilized has begun to climb to higher levels than those attained before.

While a much smaller proportion of the demersal fish than of other categories is reduced, the percentage has increased fairly steadily over the past 15 years, from 4% to nearly 10% (Table II). However, the demersal catch itself has risen faster than this, both absolutely and relatively to the other ecological groups.

In order to determine the trends in the supply to mankind of food from marine sources, a conversion must be made back from fish-meal to live-weight of livestock or to protein equivalents. In 1975 I used (Holt, 1975), on the basis of information available to me at the time, a ratio of 11:1 for the conversion live-weight of fish to live-weight of chicken. In 1974 we had used ratios of 10:1 (Holt & Vanderbilt, 1974) and, more generously, 5:1, in our analysis of the disposition of dietary benefits from the ocean, as between 'developed' and 'developing' countries. On the other hand it appears from FAO (1975) that an appropriate ratio for available protein, as between fish consumed directly or fed through livestock, might be nearer 2:1.

There are, however, big differences in protein yield as between, for example, frozen fish and canned fish, between 'white' and oily fish, between utilization efficiency in developing and developed countries, and between feeding meal to chickens or pigs and to freshwater fishes. One could attempt to allow for these variables in compiling total figures, but the present state of the available data does not seem to justify such refinement. In the last column of Table II I have used a single conversion-ratio of 3:1, which I believe may be somewhat generous with respect to the efficiency of fish-meal utilization. The quantities given are in equivalents of whole fish. Trends of annual rates of increases are to be compared with an approximately 2% rate of growth in human population over the period considered.

Lastly, to place these figures in perspective, figures provided by FAO (1975) indicate that, in the period 1964–66, the *direct* contribution of fish and shellfish to total human food-supply was 4.6% of the total protein and 14.0% of animal protein. These percentages include catches or aquaculture of freshwater and anadromous (ascending rivers to spawn) fishes, which all together contributed 24.4% of the fish for direct consumption; the strictly marine contribution was, therefore, 3.5% and 10.6% of total and of animal protein, respectively. If we now add to the direct ocean contribution that part of the non-fish animal protein which is of marine origin, we find that the ocean contributes in one way or another 4.1% of total protein and 12.7% of animal protein, respectively. The relative consumption of marine products varies very greatly, however, between areas of the world, between different economic groups of countries, between coastal and land-locked states, and, within countries, between town-dwellers and the rural and coastal populations and also between different income groups.

The trend of relative increase in demersal-fish catches shown in Table II contradicts the commonly expressed opinion that, as mankind exploits the living resources of the sea more and more intensively, increasing effort is directed to the organisms at the lower trophic levels. It is true that in recent years important fisheries have been developed for, e.g., the Capelin (*Mallotus villosus*)—an important food of Cod (*Gadus morhua*), fin-whales, and

TABLE III

Catches of Some Fish Species Showing Trends—(Millions of Tonnes).

Year	Herring—*Clupea harengus* (N. Atlantic)	Sardine—*Sardinops caerulea* (Eastern Central Pacific)	Sprat—*Sprattus sprattus* (N.E. Atlantic)	Mackerel—*Scomber scombrus* (N. Atlantic)	Capelin—*Mallotus villosus* (N. Atlantic)	Sand-eels—*Ammodytes* spp. (N.E. Atlantic)	Alaska Pollock—*Thelagra chalcogramma* (N. Pacific)	Polar Cod—*Boreogadus saida* (N.E. Atlantic)	Haddock—*Melanogrammus aeglefinus* (N.W. Atlantic)	Cod—*Gadus morhua* (N. Atlantic)
1938	1.75	0.51	0.16				0.19			2.08
1948	2.23	0.17	0.11				0.19			2.09
1950	2.04	0.32	0.16				0.15			—
1958	2.55	0.11	0.23				0.35			2.56
1959	2.80	0.05	0.26				0.45			
1960	2.64	0.04	0.25				0.51			
1961	2.47	0.04	0.30				0.47			2.96
1962	2.79	0.02	0.35		0.01		0.58			3.01
1963	2.93	0.02	0.45		0.04		0.68			2.96
1964	3.54	0.03	0.55		0.04		0.92			2.68
1965	4.05	0.02	0.18	0.29	0.28	0.14	1.05	—	0.25	2.76
1966	4.01	0.02	0.21	0.62	0.51	0.18	1.23	—	0.21	2.88
1967	4.09	0.03	0.15	1.05	0.51	0.21	1.73	—	0.12	3.20
1968	3.89	0.03	0.17	0.97	0.62	0.20	2.20	—	0.10	4.00
1969	3.33	0.03	0.20	0.94	0.85	0.11	2.55	0.13	0.07	3.66
1970	2.32	0.04	0.23	0.66	1.51	0.19	3.06	0.24	0.05	3.14
1971	2.14	0.05	0.30	0.74	1.58	0.40	3.59	0.35	0.05	2.85
1972	1.92	0.05	0.33	0.78	1.95	0.36	3.21	0.17	0.03	2.74
1973	1.98	0.06	0.51	1.02	2.05	0.31	4.62	0.08	0.03	2.54
1974	1.57	0.08	0.64	0.98	1.91	0.53	4.90	0.13	0.02	2.81
1975	1.52	0.12	0.98	1.09	2.24	0.44	5.03	0.06	0.03	2.42

seals—and for sand-eels (*Ammodytes* spp.). These have, however, been more than counterbalanced by declines, and in some cases virtual collapses, of fisheries for some small pelagic species which had previously been very important. Thus, the story of the Anchoveta, which itself followed an earlier eclipse of the sardine fishery of California, has been repeated with respect to other clupeoids—including the Herring (*Clupea harengus*). Some of the 'species fisheries' which have undergone massive changes over the years are indicated in Table III.

These examples are drawn mainly from the industrialized fisheries of the Northern Hemisphere. Developing countries are not, however, immune to effects of drastic man-caused changes in stocks. Thailand provides a good example. A trawl fishery for demersal species began in the Gulf of Thailand in the 1960s, in part as a result of effective technical assistance from the Federal Republic of Germany. As catches increased, catches per unit effort fell. Now for three years the total marine catch by Thailand has been lower than in preceding years, despite continued fishing effort (Table IV).

TABLE IV

Marine Fish Catches by Thailand (mainly in the Gulf of Thailand).

Year	*Millions of tonnes*
1963	0.23
1964	0.50
1965	0.53
1970	1.35
1971	1.47
1972	1.55
1973	1.52
1974	1.32
1975	1.35
1976	1.43

Apart from the special case of the large whales, the earlier examples of over-fishing of demersal stocks—flatfishes and some gadoids—seemed to derive more from changes in age- and size-structure of the populations than from changes in reproductive performance. This in theory could be corrected by increasing the size at which fish were first liable to be captured (by protecting 'nursery grounds' and by increasing sizes of meshes and hooks) and by restraining growth in fishing effort, or even reducing it. Later cases of collapse of pelagic fish, with shorter lives, have also involved reductions in reproduction and in survival of young fish to commercially useful size. Consequently, where mathematical models based on earlier experience have been applied crudely to the new situations, those models have broken down. In being used to predict or determine catch-levels, they have indeed encouraged the deployment of excessive fishing effort. When combined with statutory management objectives which not only define a 'maximum sustainable yield' (MSY) but also suggest that it is 'wasteful' *not* to take that yield, the consequences have been disastrous. These consequences are now stimulating a reappraisal of objectives, examination of the validity of existing models, and a review of our scientific ability to make predictions.

An additional factor in this reappraisal is the growing evidence of substantial interactions between the different species which are subject to fishing. This, of course, is not unexpected, but until rather recently was not obvious from the data. Some good examples come from studies of marine mammals. Thus the premature growth-rate, and hence the age of sexual maturity, of Sei Whales (*Balaenoptera borealis*) in the Southern Ocean changed before they were substantially exploited, but after their competitors—the Blue (*B. musculus*), Fin (*B. physalis*) and Humpback (*Megaptera nodosa*) Whales—had been depleted. At the same time the antarctic seals increased greatly in number and total biomass. The growth and reproductive rates of North Pacific Fur-seals (*Callorhinus ursinus*) appear to have changed in response to the impact of fishing on the Alaska Pollock (*Thelagra chalcogramma*), which is important in their diet.

Detailed studies of changes in North Sea stocks over a 20-year period have suggested that interaction between species may now be significant; in particular, high reproductive success of some demersal species has coincided with reductions in some of the pelagic species, though there is no direct evidence of a causal relationship. The Cod (*Gadus morhua*) in the North Atlantic as a whole does not seem yet to have been drastically affected by the growth of the Capelin fishery; the Cod stocks yielded catches of 3–4 million tonnes annually through the 1960s, but have been declining gradually since, whereas the Capelin catch grew from practically nothing in 1961 to 2 million tonnes by 1973 and has remained at that level since. It seems to me possible that the Cod is now suffering from the effects of very high catches in the late 1960s, peaking at 4 million tonnes in 1968, combined with stress on its food-supply. In other cases, over-fishing may not alone be the cause of decline or collapse, but of this stress coincident with an adverse natural environmental change at a time when the stock is small and less stable than usual because of a reduction in the number of year-classes present in it; the Anchoveta may be such a case.

The observation of interactions, especially between predators and their prey, poses new questions regarding management objectives. How are the values of Fur-seal skins and Pollock catches to be balanced, and how can the high monetary value of a relatively small catch of a predatory fish that is esteemed as human food be offset against a higher catch of its smaller prey, for reduction to meal? Such questions have arisen most recently and most strongly with respect to the potential catches of Antarctic Krill (*Euphausia superba*—the small shrimp-like crustacean which is the main food of the baleen whales, and of many antarctic seals, fish and birds)—from the Southern Ocean. Estimates of this potential vary greatly but are often quoted as of the order of 100 million tonnes per annum, which is the same order of magnitude as the estimates of the limit of world catch for all conventional fish species.

Already vessels from eight countries are conducting experimental trawling for Krill and are studying methods of processing and marketing it. It seems certain that substantial catches of tens of millions of tonnes would have large effects on the recovery rates of the depleted baleen whales, which are now protected, as well as on the other antarctic predators. There is no regulatory body for fisheries in the Southern Ocean, and judging from all previous experience it would take many years to establish one (10 years in the case of the International Commission for the Conservation of Atlantic Tunas (ICCAT)), even if it were clear—which it is not—with whom lies the initiative for establishing such a body and what could be an acceptable boundary for its area of responsibility. This latter is a crucial question, as the Krill is distributed partly within and partly outside circumpolar lines that are drawn at a distance of 200 nautical miles from the Antarctic continent and from Southern Hemisphere land-masses and islands, and it is far from clear which, if any or all, Antarctic Treaty powers will claim jurisdiction in the area, and, if so, when and with what effect.

The Consequences of Changes in the Law of the Sea and of Claims to Extend National Jurisdiction

It is well known that, following a rash of claims and declarations in 1976 and the early months of 1977, most nations have now extensive jurisdiction over 200-miles-wide Exclusive Economic Zones (EEZs) or their equivalents, or have announced their intention of claiming such jurisdiction. In nearly all cases the claims are a source of current dispute between neighbours. While the justification commonly given is on the basis that they will facilitate conservation, they are in fact more directed to securing a redistribution among nations of fish catches and/or economic benefits from fishing. They reflect scepticism as to the likelihood of the UN Conference on the Law of the Sea (UNCLOS) reaching agreement that would effect an equitable redistribution, and also reflect a recognition that regional fisheries bodies have not in general been successful in placing effective restraints on fishing effort—especially where they have been established in areas adjacent to developing countries, but which have become exploited in the past two decades by northern-hemisphere fishing powers.

In this situation even long-established regional bodies, such as the International Commission for the Northwest Atlantic Fisheries (ICNAF) and the North-East Atlantic Fisheries Commission (NEAFC), have been destabilized—North American countries have announced their withdrawal from ICNAF, and, in the north-east Atlantic, TAC (Total Allowable Catch) decisions are in abeyance because of disputes concerning the fisheries policy of the European Economic Community (EEC).

It has so far always been necessary, in practice, to negotiate simultaneously agreements on overall uotas and on national allocations. For example, countries engaged in pelagic whaling have not been able to agree each year on the proportions of total catches which they could each have until the International Whaling Commission (IWC) had come to agreement on the total catches, and vice versa. The *status quo* has not provided a very stable basis for such negotiations, and it seems that even if EEZ (Exclusive Economic Zone) claims are not related to needs of people for food, they might at least provide a stable negotiating base, provided boundary disputes are resolved. There are, however, a number of problems to which the Law of the Sea Conference has certainly not devoted sufficient attention, polarized as it has been on questions concerning sea-bed minerals and rights of states in coastal zones. These problems have been succinctly summarized by Saetersdal (1977) with reference to the north-east Atlantic. In that area, most stocks of fish are 'shared' by at least two and sometimes by a dozen nations; they occur almost wholly or mainly within 200 miles of some coasts. When an overall TAC has been agreed (on a basis yet to be determined), the question will arise as to how it is to be subdivided, with respect to both the fishing nation and the coastal zone. If 'historical' data are to be used, over what time-periods are past national catches to be averaged? Further, for various practical reasons, fishing has been concentrated more in some parts of the area of stock distribution than in others, so a particular stock might have been exploited mainly in one or a few zones. The distribution of fishing has not in any case been constant, but nor has the distribution of fish. Environmental factors have caused both year-to-year variations and medium- and long-term trends of change in geographic distribution.

Saetersdal (1977) reminds us that fishing has itself changed the distribution of the fish, and to varying degrees that depend on the type and intensity of the fishing. Moreover, different life-stages of a species may occupy parts of the total area, and we would then be forced to try to weigh the contribution to stock 'productivity' in one EEZ that happens to be a spawning or nursery area against that in another EEZ where adult fish feed. No guidelines for such decisions have yet been devised, and there are very few areas where sufficient is known about the biology of the fish to permit reasonable discussion.

Marine Fisheries Management

A beginning has recently been made, after several years of relative stagnation, in elaboration of a useful economic theory of marine fisheries management. This stagnation could, I suggest, be attributed, at least in part, to the fact that open and serious economic discussion has been effectively banned from international fishery bodies, while all attention has been directed to regulation of output but ignoring input and questions of economic efficiency. Recently, however, Clark (1976) has detailed the implications of taking account of the rate at which future values are discounted in cost–benefit analysis as it might be applied to renewable resources. Anderson (1975*a*, 1975*b*) has shown how an approach might be made to the negotiation of allocations of yield on the basis of economic equity in what he calls 'biologically and technologically interdependent fisheries'.

There is also the concept of 'surplus' to consider. In the language of UNCLOS, this is an amount of fish which may be allocated to a non-coastal state when a coastal state has decided how much it will take from an area. This concept derives from the one to which I referred earlier, namely that a resource which is not 'fully utilized' is a resource 'wasted'. I have argued (Holt, 1976*b*) that this latter concept is not a valid one and need not reiterate the arguments here; it suffices to note that it serves the interests of the more powerful fishing nations. It is remarkable that those who have maintained that it is 'wasteful' not to take a maximum sustainable yield (a Chairman of the IWC said only a very few years ago: 'The whales will only rot in the sea if we do not catch them') do not understand that it is 'wasteful' to expend fuel-oil and other natural resources in limited supply by striving for catches at MSY (Maximum Sustainable Yield) level when one could often secure, say, 80% of that level with half the fishing effort, and hence with considerably greater total net profits (Holt, 1977).

The economic and ecological desirability of permitting catches only up to levels somewhat less than the calculated MSY is reinforced by the very great uncertainties in the latter calculations, for the many reasons exemplified in this paper. Such considerations have led a number of scientists to attempt a formal reformulation of management objectives. One such reformulation was made at a workshop convened at Airlie, Virginia, in 1975, sponsored by ICSU, the World Wildlife Fund, the Smithsonian Institution and others; the essentials of the workshop proposals have since been formally accepted by IUCN and UNEP, but they are not yet contained in texts before UNCLOS. These proposals cover the following principles (Talbot & Holt, 1975):

(1) Ecosystems should be maintained in such a state that both consumptive and non-consumptive values to humanity of those ecosystems can be realized on a continuous basis.
(2) Options for different use by present and succeeding human generations should be ensured.
(3) Risk of irreversible changes or long-term adverse effects of exploitation should be minimized.
(4) Decisions should include a safety factor to allow for limitations of knowledge and for inevitable imperfections of management institutions.
(5) Measures to conserve one resource should not be wasteful of another.
(6) Survey or monitoring, analysis and assessment, should precede planned use, and accompany actual use, of a resource, and the results should be made available promptly for critical public review.

It will obviously not be easy to put these principles into operational form, although a start can be made with some of them. For example, the IWC has already incorporated a safety factor of 10% in its quota-setting procedure; the question remains of whether this factor is big enough in the light of the existence of enormously greater uncertainty than ±10% in MSY estimates. With respect to principle 6, the IWC set a zero quota for Bryde's Whales

(*Balaenoptera brydei*) in the Southern Hemisphere in 1976 on the grounds that insufficient was known about the stock to determine an appropriate quota. Unfortunately, the IWC does not take any collective responsibility even for planning, let alone conducting, the necessary research, and it was left for the country which wanted to catch those whales anyway to issue itself a large 'quota' of scientific permits (an act provided for explicitly in the International Whaling Convention) to catch and process 4 million dollars' worth of Bryde's Whales.

That 'non-consumptive values', both economic and ecological, must be recognized, has been accepted internationally at least by the scientific community concerned with marine mammals, for which such values are evidently high (FAO, 1977*b*). It remains for such recognition to extend to governments and intergovernmental bodies; the United States, through its Marine Mammal Commission, is taking a lead in this respect (Botkin & Sobel, 1977).

The single act which could perhaps give us more time to learn how to manage fisheries along sound ecological lines concerns principle 5 (above). By giving equal consideration to restraints on growths of fishing power and effort as is given to setting catch-quotas, application of most of the other principles would probably be ensured. The extent of present 'waste' of fishing inputs is illustrated by the example of shrimp-fishing by Mexican-flag vessels in the Gulf of California, where over two decades the same annual catch (20,000 tonnes) was obtained—early in the period by 400 small vessels, and later in the period by 800 larger and more expensive vessels. After 1973, when the number of large vessels exceeded 800, the total catch began to fall, there being still no quota regulation. If an MSY objective were sought, 20,000 tonnes would be taken by about 800 over-powerful vessels, when in fact nearly 20,000 tonnes could be taken, sustainably, by not more than 300 vessels.

Principle 2, concerning options for future use of resources, has so far been given very little attention with respect to its practical application. It seems to me that some new kinds of institutional arrangements to this end may be called for. One suggestion is that international-resource management institutions should, in making decisions regarding types or levels of exploitation and allocations of present benefits, necessarily consult, and heed, a designated person, tribunal or office, which might be described as a guardian of the presumed interests of future generations. Such a guardian would be empowered and enabled to collect data and evidence, and authorized to intervene in the deliberations of the managing institutions (Bruce & Holt, 1977). As far as marine resources, living and non-living, are concerned, the guardian should, it seems, have a special responsibility with respect to the areas beyond national jurisdiction. This is precisely the area to which very great attention has been paid by UNCLOS for the sea-bed minerals—leading to proposals for a new type of supra-national institution, the International Seabed Authority—but which has been largely ignored for the living resources.

It is true that there has been talk of the special problems of so-called highly migratory species (with mainly the tunas in mind—at least one-third of present catches being beyond 200 miles from land), but no fundamental changes have been proposed to the outdated concepts of freedom of high-seas fishing modulated through voluntary regional agreements. The idea that the off-shore living resources should be regarded as a 'common heritage of mankind' just as much as the deep sea-bed resources are, is being tentatively advanced by a few people, but it is very far from international agreement in principle, and the institutional consequences have barely been discussed.

At last the UNCLOS is, however, beginning to take more seriously the need for continuing flexible comprehensive instruments for ocean management, beyond merely the signing of one or more treaties. This was pushed, at the 1976 session of the Conference, by the Delegation of Portugal (1976). However, it should be obvious at this time that not only is there a great deal to be done to improve the management of fishing, if continuing benefits are to be expected from the living resources of the sea, but that a move towards integrated management of different uses of ocean space and other resources is now required.

There are growing ecological and physical interactions between fisheries and mariculture, oil, metallic ore and other hard-mineral extraction, energy extraction, navigation, communications, waste disposal, recreational industry, and other ocean uses. The institutional implications of these interactions have been ably analysed by Borgese & Pardo (1976), whose proposals have been presented in a number of fora. Closely linked with the discussion of arrangements for management of multiple use of ocean space is the question of the procedures and means for settlement of disputes between states and also perhaps between public and private legal factions and persons engaged in ocean use. This question, too, has arrived rather late in the deliberations of the UNCLOS—mainly through the initiative of its President, Ambassador Shirley Amarasinghe—but is now occupying a focal point in the debate.

References

Anderson, L. G. (1975*a*). Analysis of open-access commercial exploitation and maximum economic yield in biologically and technologically interdependent fisheries. *J. Fish. Res. Bd. Canada*, **32**, pp. 1825–42.

Anderson, L. G. (1975*b*). Optimum economic yield of an internationally utilized common property resource. *Fishery Bull. USA*, **73**(1), pp. 51–65.

Borgese, E. M. & Pardo, A. (1976). *The New International Economic Order and the Law of the Sea*. International Ocean Institute, Msida, Malta, Occasional Paper No. 5, 318 pp.

Botkin, D. B. & Sobel, M. J. (1977). *Optimum Sustainable Marine Mammal Populations*. US Marine Mammal Commission, Washington, DC: 83 pp.

Bruce, M. & Holt, S. J. (1977). *A World Guardian for the Future*. International Ocean Institute, Msida, Malta: 6 pp. (mimeogr., obtainable from 2nd Author, FAO, Rome, Italy).

Clark, C. W. (1976). *Mathematical Bioeconomics: The Optimal Management of Renewable Resources*. John Wiley, New York–London–Sydney–Toronto: xi + 355 pp., illustr.

Delegation of Portugal (to UNCLOS) (1976). [MS not available for checking].
Dickie, L. M. (1975). Problems in prediction. *Oceanus*, **18**(2), pp. 30–35.
FAO (1975). *Report of the FAO/NORAD Round-table Discussion on Expanding the Utilization of Marine Fishery Resources for Human Consumption*. Svany, Norway, 25 August to 4 September 1975: 47 pp. (mimeogr.).
FAO (1977*a*). *Review of the State of Exploitation of the World Fish Resources*. Document COFI/77/5 and supplements, FAO, Rome, Italy [not available for checking]: mimeogr.
FAO (1977*b*). *Report of the ACMRR Working Party on Marine Mammals*. FAO Fisheries Report No. 194, Rome, Italy: 43 pp.
Gulland, J. A. (Ed.) (1970). *The Fish Resources of the Ocean*. Fishing News Books, London, England: xi + 255 pp.
Holt, S. J. (1969). The food resources of the ocean. *Scientific American*, **221**(3), pp. 178–94.
Holt, S. J. (1975). Marine fisheries and world food supplies, Pp. 77–96 in *The Man/Food Equation* (Eds F. Steele & A. Bourne). Academic Press, London & New York: xv + 289 pp., illustr.
Holt, S. J. (1976*a*). *Statistics of Catches of Large Whales*. Contribution to Scientific Consultation on the Conservation and Management of Marine Mammals and their Environment, Bergen, Norway, September, 1976. FAO, Rome, Italy: 26 pp.
Holt, S. J. (1976*b*). *UNCLOS, MSY, FAO, Regional Fisheries Bodies and Whales*. Contribution to joint meeting of the Club of Rome and the International Ocean Institute on 'Reshaping the International Order', Algiers, 25–8 October 1976: 7 pp. (Reprinted in *The Center Report*, **1**(9), November 1976, Environment Liaison Centre, Nairobi).
Holt, S. J. (1977). *Some Implications of Maximum Sustainable Net Yield as a Management Objective for Whaling*. Contribution to meeting of IWC, June 1977, Canberra, Australia. [Published in 28th Report of the International Whaling Commission, pp. 191–3, 3 figs, 1978.]
Holt, S. J. & Vanderbilt, C. (1974). *Materials for an Appraisal of the Dependence of Nations on the Living Resources of the Sea*. International Ocean Institute, Msida, Malta: 29 pp. (mimeogr.).
McCloskey, M. (1977). Enlightened management of renewable resources: hope or reality. *Sierra Club Bulletin*, **62**(4), pp. 15–18.
Saetersdal, G. (1977). *Problems of Managing and Sharing the Fishery Resources under the New Ocean Regime*. Doc. COFI/77/Inf 11, FAO, Rome, Italy: 4 pp.
Talbot, L. & Holt, S. J. (Eds) (1975). *The Conservation of Wild Living Resources*. Report of a Workshop at Airlie, Virginia, February and April, 1975. Offset document available from Department of Fisheries, FAO, Rome, Italy: 51 pp.

DISCUSSION (Session 5)

Jakobsson (Chairman, introducing)

It is a special pleasure for me to call this [fifth Session] to order. The topic is of great interest to everyone in Iceland, and should draw the active attention of the participants of the host country. The panellists [as usual] have been asked especially to look into the future in their [respective fields], but we in Iceland have always had a tendency not only to look into the future but [first] to look into the past, [in order to have a firm basis for] our future perspective and planning.

Whenever we have had [the time], we have sat down and written a saga book or two as you all know. I am not going to read one of these saga books in my introductory remarks, but would only like to remind you that we had lived in this country for 1,000 years before we discovered the resources that we should really base our livelihood on. We had lived here for 1,000 years before we started fishing really, or at least to any [major] extent. The development of the fishing industry was, of course, our industrial revolution.

I think that when we say 'Whither the Oceans?' we could just as well say, as far as Iceland is concerned, 'Whither the Icelandic Nation?' or 'Whither Iceland?'—because we are so closely linked to the living resources of the sea around this island. Now until very recently one had looked upon the oceans as almost limitless, especially for dumping wastes. The living resources have also tended to be looked upon as limitless. I think it was Thor Heyerdahl [who is unable to be with us owing to a prior commitment on a raft in the Middle East] who first drew my attention to the fact that the ocean is just a landlocked lake that could easily be polluted and easily be destroyed as far as living resources are concerned. Even if this lake covers more than two-thirds of the surface of Earth, it is still in some terms, relatively speaking, a [limited] lake. And similarly for the living resources of the ocean, these are not nearly as limitless as we would think—[especially] when one realizes that most of the [area of the] oceans [constitutes a virtual] desert, and that the really productive tracts lie within the coastal zones which are again the most sensitive to pollution.

In the very last decade we have developed methods of assessing the abundance of these living resources, and this of course has made their rational management possible. [Such developments would have] changed the picture very much if it had not been for the character of the ownership of these resources, because, so far, most of the living resources have been a common property of all nations, and even where we are dealing with common properties *within* nations, these are always very difficult to manage and, first, to obtain agreement on appropriate management actions. This we have learned by bitter experience through the history of the international organizations which have been responsible for such management.

Following these short remarks I would like to introduce to you the author of the first part of the keynote paper, who I will shortly call on to present it. I refer of course to Professor Edward D. Goldberg, who we are very happy to have with us. An internationally renowned geochemist, Professor Goldberg is Professor of Chemistry at the Scripps Institution of Oceanography, University of California, and his main field of interest is marine pollution. [The author of the second part of the keynote paper, Dr Sidney J. Holt, of FAO in Rome, is unable to be with us owing to a prior assignment in the Southern Hemisphere.]

As panellists we have with us here Nicholas Polunin, Junior, who is a research student in the Department of Zoology at the University of Cambridge, England [having graduated from the one at the even more ancient sister University of Oxford], and who is interested in aquatic ecology—especially as regards tropical marine ecology in shallow waters [which to me spell coral reefs]. Then we have Dr

Anitra Thorhaug, Associate Professor in the [Department of Microbiology,] University of Miami, Florida, whose field of interest is [marine] primary production. We also have with us Professor Emanuel D. Rudolph, Professor of Botany and Director of the Environmental Biology Program in the Ohio State University [and Chairman of the Board of Governors of the Arctic Institute of North America], whose field of interest is antarctic marine resources. Finally on the platform we have a countryman of mine, Dr Jakob Magnússon, [of the Marine Research Institute here in Reykjavik], who has a wide interest in fisheries, and who has been working with FAO in many parts of the world—including the Philippines and Mexico—to mention [only] some of his past.

Goldberg (Co-Keynoter, presenting papers)

I feel sort of inadequate here without my friend Sidney Holt, who has written the other part of the keynote paper, although I hope the panel can take over for him in expressing his views [and filling in where I do not do so]. There is, however, a certain relationship between his presentation and mine. He is concerned with the positive resource of the oceans, namely the fish and the shellfish. My concern is with the oceans as waste space—a somewhat negative concern—the [question] being to what level, or to what extent, can the oceans accept the discards of Man without loss or limitation of his use of their resources? [Thereafter Professor Goldberg gave an account of his paper, with added examples, and also to some extent of that of Dr Holt.]

Thorhaug (Panellist)

As Dr Holt's expert paper covers so much of the field of fishing, I thought I would make a few remarks beyond what Professor Goldberg said in terms of biological resources of the ocean and whither they go. Although, as our Chairman has pointed out, most of the ocean is a desert, nevertheless one-third of the total primary productivity of the world occurs in the [collective] ocean, and most of this is in the coastal zones—particularly in the estuaries, where primary productivity gets as high as 2,500 grams dry-weight per square metre per year. The other areas of intense primary productivity in the ocean are those of upwellings [carrying nutrients], of which you are probably all aware—for instance the El Niño off Peru, which [produces] about 500 grams of fixed carbon per square metre per year at the surface. The Antarctic Ocean, as indicated by Professor Rudolph, also has upwelling zones producing about 100 grams [of fixed carbon annually].

So basically we have a lot of area in the ocean where very little happens, and in terms of pollution and effects we could actually dump a lot of materials into these desert-like oceans and not [do much harm]. On the other hand, if we dump just a bit into an estuary a great deal happens, so analogy of its limits should help us to understand the limits also of the ocean. Most productivity in an estuary is concentrated in a narrow zone along the shorelines, and as one goes out towards the centre the productivity falls. Also in an estuary, Man's impact [occurs] right at the edges, and it also [occurs] at the edges of the ocean: that is where we dredge and fill and make our harbours and marinas, that is where we put our sewage, and that is where we develop and make land out of the ocean's bottom or the estuary's bed. So I think the analogy of the ocean as an expanded estuary may give us the perspective also of 'Whither go the oceans?'.

There are two general regions of the oceans that are particularly vulnerable [to pollution, namely]: the tropical part where I have been working and the [polar parts] where Professor Rudolph has been working. We have found from our very detailed studies of a series of pollutants in Florida and the Caribbean that many of the standards [established for] aquatic and marine resources in the temperate zones simply do not extrapolate [into other regions. So] it is my feeling that you cannot extrapolate from the simple, primitive tropical marine environment, where life [supposedly] originated, to the more complex. [And even if you can, perhaps, extrapolate in this manner, you cannot do it] backwards, from the complex to the simple; yet that is what we have been [attempting]. Take the instance of thermal standards, which allow us to dump materials that are 5 – 10°C above the temperature of rivers and temperate marine areas into which they go: these standards simply do not apply in the tropics, where three degrees above ambient [can wreck] the whole system. This bears back to Professor Flohn's hypothesis in terms of atmospheric carbon dioxide and a general warming.

The over-fishing problem has been described by Dr Holt, and Professor Goldberg has just handled much of the pollution problem, but I would like to put in a few biological remarks about pollution. My first is that many of our standards currently are built on what is called an LD50—a lethal dose for 50% of the population. What one does is put a whole lot of animals in a tank and throw in the variable—DDT or radioactivity or whatever it is—and watch the time until 50% of them drop dead. Or one may determine the percentage that have died in 96 hours and [so, by trying various strengths,] determine the permissible limit for the pollutant under investigation. But this does not really take into account many of the long-term and genetic effects.

[As an example of the kind of thing that is happening we may cite the observations over a period of about 10 years of] a group led by Professor Eugene Odum, who recently found that male alligators like to bathe in the heated waters of Parr Pond during the winter, but for some reason the female alligators do not like to bathe in the warmer water. So what happens is that the male alligator becomes sexually ready to mate earlier than the female, because this wintering at a higher temperature has speeded up his responses. So he goes charging around looking for female alligators, none of whom [is at all responsive] to his advances. Then about a month later the female alligators all become sexually ready and they go charging around looking for the male alligators which are by this time totally indifferent, lying in the hot water. [Laughter.] So what is happening there after a period of many years of the male alligators wintering in the warm waters, is that there is an uncoupling of the reproductive cycle, and these are the kinds of long-term effects which unfortunately take a lot of money [and effort] to look at. We are simply not aware [of many of these possibilities and make no provision for them in setting] our standards for permissible limits of dumping of many pollutants. I am not sure what we can do about this except keep on looking very carefully.

From the viewpoint of a biological scientist, things in the ocean are not all gloom and doom, and I would just like to mention a few of what I consider extremely exciting developments in marine pollution research in the last 10 years. First of all we do have a lot of anti-pollution devices—the US Environmental Protection Agency of course want zero emission, which is clearly not possible, but due to [EPA's desire] the technologists have invented all kinds of devices to keep cadmium, mercury and other things from going into our estuaries, whereas 10 years ago of course there was none. And secondly we are [making use of] a lot of waste products. There was a conference two weeks ago in Miami on the uses of waste heat, and some of these are quite exciting. Germany is using waste heat for agriculture and greenhouses quite extensively. Aquaculture in Japan and other countries is stimulated by the use of waste heat from power-plants. My own suggestion which people for some reason are not listening to very much is that in Miami we have a 178-mile radiator canal system which cools our nuclear plant. Now we also have more old people [living] in Miami Beach than [perhaps anywhere else] in the world, and it just occurred to me that it would be perfect multiple use to bring them down to the radioactive warm mineral springs and make a spa.

Aquaculture in general—not just aquaculture in hot water—is very promising for the future, but I think people up to now have had the wrong idea, [as it has involved mostly] high-priced fisheries' products such as species of lobsters and shrimps, and expensive fishes such as salmon and trout. But there are many efforts all over the world where really a lot of aquacultural activity is proving quite successful. Other improvements include the new devices for getting dolphins out of the tuna nets, and much more knowledge about life-stages and where the life-stages take place in our important commercial fisheries—which incidentally are mostly in the estuaries that are being impinged upon.

Marine parks and preserves are a fairly recent [development]. We have known about land reserves for a long time, but within the last 10 years a system of marine preserves has been set up all over the world. [Dr Thorhaug thereupon showed slides illustrating some of her notable marine restoration projects, particularly through underwater planting of sea-grasses as described in *Environmental Conservation* and elsewhere.]

[We should encourage] multiple uses of the marine environment [and avoid allowing] our big industries to develop on coasts and especially in estuaries where land is cheap but great ecological harm may be done]. Towards this the State of Florida is now finally having scientists come in to plan ahead for power plants and sewage plants. The legal ramifications, too, are very pleasing. In the last 10 years we have had marine lawyers coming in—people who are capable of drafting refined laws that really are going to control and enforce conservation—whereas before this, many of the legal ramifications were left to chance or to the Corps of Engineers or to other bodies with [little conscience and less] expertise.

The degradation that has happened to the marine environment in the last 50 years I think is counterbalanced by a whole series of innovations which are only some 10 years old and many of which are still in experimental stages. However, I think a change in ethics is occurring from that of exploration to that of a somewhat more altruistic viewpoint regarding waste materials. Basically I think that if we trained our children on dry toilets we would not have this subconscious attitude that the minute we put our excrement into water it disappears and we do not have to worry about it any more. That is my solution to the whole problem. [Applause.]

Magnússon (Panellist)

I will restrict myself to [the subject of] Dr Holt's paper, which in my opinion is very informative on what is going on in the world of fisheries. So I wish to make some points in this general connection rather than to discuss the papers as such.

It [is clear] from Holt's presentation—and I agree with him in this—that there has been a great effort, but in vain, to limit the expansion of fishing fleets and/or intensity. It is also [pertinent] that fish stocks continued to be depleted while the discussion on the necessity of restricting the fisheries was carried on. I want, however, to point out that fisheries resources—unlike some other resources, such as mineral oil—can be restored. This has been done and is going on, e.g. in Iceland, with considerable success, and may throw some optimistic light on the case: even though the conditions of many fish stocks are rather bad today, they can be restored. But in Iceland this could not be done effectively until we had gained full jurisdiction over our waters.

The fact is—at least in this part of the world—that the fishing fleet is too big and too effective for rational utilization of the fish stocks. Almost the same quantity of fish could be taken by a very much smaller fishing fleet. But there has been a continuing increase of investments in bigger and more efficient vessels. [Nevertheless] we are taking less food in the form of fish or fish products from the ocean than we could and should do if we avoided over-fishing. At the same time we are wasting more and more materials deriving from sources which are *not* restorable, such as fuel oil and iron for the fishing fleets. It is truly a very serious matter that we do not get from the fish stocks the optimal yield of food because of over-fishing, as is at least very obvious in the Northern Hemisphere. One could talk at length about the causes and consequences of over-fishing, but I am not going into details as many are already given in [Dr Holt's] very interesting paper. Yet I think it is of crucial importance in connection with the [main theme] of this Conference that *these resources can be restored*.

Considering the interaction of species, it certainly can be said that fishing has affected Nature considerably. This is a complex matter, which has attracted relatively little attention lately [when it] appeared that, from depletion of some stocks, others were benefiting.

It has been pointed out earlier that most present-day fishing is of species that are rather high up in the food-chain. But there should be a move towards gaining more food from the ocean by trying to utilize species at an earlier stage in the food-chain. In his paper, Dr Holt touches on this problem, e.g. in connection with fishing of Krill in the Antarctic and how intensive fishing of Krill might affect the [maintenance] of other living resources in the region. But much more research is needed to say anything definite about this matter.

I think one of the main points in this connection, and which has a bearing on the topic of this Conference, is that our goal should be to obtain the maximum [sustainable] yield from living resources of the ocean as determined and controlled by a firm and rational management. At present this is not the case so far as fisheries are concerned in a great part of the world. We cannot achieve our goal by destroying the resources, [which is what over-fishing is doing all-too-rapidly].

Dr Holt described in his paper how this problem has been dealt with in the various international bodies. But until now it has not been as successful as the scientists had hoped, as has been pointed out already by our Chairman. Politicians and administrators have kept on arguing, year after year, about the advice of the scientists, while one stock of fish after another became depleted—at least in the Northern Atlantic. But what is of the greatest importance is that these resources are restorable. In many countries where fish stocks have suffered from over-fishing, great effort is now being given to 'bring back the fish', and here in Iceland we are doing this with considerable success.

Jakobsson (Chairman)

I think that before we come to the Antarctic and talk about Krill, we will keep to the warmer waters.

Polunin, Jnr (Panellist)

Without referring to specifics it seems hard to say anything that is general, yet true, about the marine world, but perhaps I should begin by emphasizing that we must remind, and continue to remind, ourselves of how little we really know about the oceanic environment. I often like to rationalize my interest in ecology in general by saying that Man is rapidly modifying, and in some cases completely destroying, systems of which he knows very little. The question is whether a being who makes a choice between two such fundamental systems—namely the natural environment of the oceans and that which Man would replace it with, and which would contain species that were more amenable to his fishery policies, when he has [virtually no understanding of the real situation or knowledge of the species concerned]—is really worthy of his specific epithet '*sapiens*'?

Clearly to be realistic the problems are enormous. One of the most important of these, and here I resurrect one of the themes which was touched on yesterday, is science itself. Science is clearly incapable of providing the precise information which is often required in the short time available. [Indeed to continue Edward Goldsmith's] analogy, it seems that we really are rushing towards a precipice—and that science alone will take too long to establish exactly why we are going over this precipice and how long it will be before we do so. I think the upshot of this analogy is two-fold: in the first place more scientists should concern themselves with generalities rather than with the fine minutiae of their fields of interest, and to support this contention we might note that enormous numbers of scientists are simply concerned with very specific academic problems. Secondly it seems clear that a lot of scientists, and no doubt mostly those who are already interested in generalities rather than academic minutiae, should be far more vociferous [than they tend to be] in their outpourings about environmental concerns. In a restricted sense, this viewpoint can be interpreted as being an anti-academic one, but it is merely sounding a note of severe caution. For let us remember that it is a community of highly competent, but extremely self-interested, physicists who have been perhaps the dominant force in the development of nuclear technology.

The fish species and fisheries tables which Dr Holt produced give us some cause for hope, in that we can see how in the past certain fisheries have been seriously depleted and yet have been replaced to some extent by other species which have come to the fore. But this does not necessarily imply that, by the exploitation of natural fisheries alone, we can increase the total yield. In order to increase the total yield it seems that we must either look to sources of natural primary or secondary productivity that are at present underexploited, or we must create our own culture systems.

In the matter of unexploited natural resources I can give an example of something which Dr Thorhaug has already spoken about, and that is the so-called sea-grasses which form extensive 'meadows' on shallow shores [for example in the Indo-Pacific regions]. I do not want to refer in particular to the restoration of the beds of sea-grasses because *inter alia* that is evidently a very costly kind of operation, but to call attention to the possibilities that sea-grass communities hold from the point of view of exploitation for human benefit, especially in the tropics where there is widespread protein deficiency. Sea-grasses are fed upon by two extremely useful groups of animals; one of these is the Green Turtle [*Chelonia mydas*] and the other is the Dugong [*Dugong dugon*]. In both groups the species mainly concerned are endangered—possibly seriously endangered and threatened with extinction—and certainly at present occupying only a tiny portion of their original ranges. Yet in many regions very large areas of sea-grasses—extremely productive natural communities—are waiting to be exploited.

The problems of such exploitation are again enormous: in the first place, relatively little is known about the ecology of these basic animals which utilize sea-grasses. Secondly, whenever one gets into a project of this kind, one immediately runs into problems of commercial ventures; people only put money into projects that look like becoming profitable, and so go into such things as oyster farming and so on. Fisheries which can yield a lot of money, and which are going to be sold essentially to these people who do not really need the protein, are unlikely to benefit directly those who really are in need of that extra protein.

[Lastly] there are the problems of getting across to people the need for conservation, and this means not only governments but often also traditional peoples who have always been dependent on the sea, and who yet find it difficult to accept the value of self-restraint—particularly when they have obtained the modern means of exploitation.

Jakobsson (Chairman)

Now we will turn to the Antarctic, where we have hopes of increasing the harvesting of the sea.

Rudolph (Panellist)

In the paper before you, there is one paragraph that mentions the antarctic marine problem, and Dr Magnússon has referred to it. I would like to approach it in a rather different way and talk about it in terms of the ecological situation. I am sure you are aware that the antarctic continent is a vast polar desert with very, very simple terrestrial systems. We tend to think of the marine situation around the continent as being very rich, as indeed it is, being probably the most productive area in the oceans as a whole. Now the interesting thing about this area is that over 50% of the invertebrate marine productivity is by one species, or possibly a few species, but primarily one. This is Krill (*Euphausia superba:* Crustacea], and the estimated productivity of that organism—and I have a figure which is [10 times as big as] the one given in the paper—is somewhere over 1,000 million tons a year. Now that is a vast amount of productivity, but if you think about it, the Krill is the key link in a large number of food-chains in the antarctic seas. For instance the Crab-eater Seal [*Lobodon carcinophaga*], which has the largest productivity of any seal in the world, depends entirely on this one resource for its food, as do many of the whales. We all know that the stocks of some of the whales—especially the Krill-eating baleen whales—have been severely depleted [in recent years], and so we may also presume that Krill is probably more abundant than it was, say 10 years ago.

It should also be noted that penguins [various spheniscid birds] and some flying birds depend to a great extent on Krill as a source of food. So what we have in the antarctic seas is a potentially weak link in the [food-web], because if this one species was depleted severely it would reverberate throughout [other] food-chains and affect many, many other organisms. The tragic fact is that we do not know really the amount or much else about this resource; [worse still], we do not know what is allowable in terms of harvesting. All we do know is that there are at least six nations now fishing this resource, and many others studying its possibilities. Yet there is no control at all. We know what the whale situation is: even with the International Whaling Commission, whale stocks have been depleted considerably. So it does not necessarily mean that, when once you have a commission [for a resource], it is going to protect that resource. All I can say is that we ought to be really concerned with knowing [far more] about the potential antarctic marine ecodisaster which might happen if we do not understand the situation in terms of ecological [possibilities. Consequently] I make a plea that somehow we have a resolution which says that somebody ought to be looking at the Krill situation officially and making recommendations [for proper action] before we end up in an ecological disaster.

Buchinger

I would like to take up [Professor Rudolph's] remarks and remind you that in 1964 at the Mar del Plata Conference on Renewable Natural Resources, the Argentinian and Chilean delegations suggested that, before exploiting any new natural resources [such as a species of plant or animal], an ecological study should be made, and that a certain amount of money coming from [any resulting] income or sale should be spent on continued ecological surveys, so that sustained-yield management would be possible in the future. So maybe this Conference could reiterate this recommendation, which unfortunately was not approved by the Conference on Renewable Natural Resources but which we still believe [to be sound].

There is another recommendation which we hope might be coming out of [our present Conference] with regard to Antarctica. There may be some differences of opinion as to how many miles of sea each country should have—as you may know, in Argentina we are for 100 miles whereas some are for 200 miles and others are for 12—but I think it is extremely [unfortunate] that the [antarctic] continent, which is now set aside for scientific research, should not have any [territorial waters] at all. So I would like you to consider this as a recommendation and approve of an antarctic unperturbed sea-shore which might be fair for all concerned. [Applause].

Miller

I also speak on behalf of the Southern Ocean bloc and wish to say that Dr Buchinger's comments are I think very appropriate. Moreover we should be encouraged by her history of performance, as in 1959 she was a member of a small committee in the IGY [International Geophysical Year] Symposium in Buenos Aires where she and I and several others put together recommendations for conservation of the animal and plant life of the Antarctic. This got adopted by the Scientific Committee for Antarctic Research (SCAR) and became the strong base for the periodically re-elaborated provisions of the Antarctic Treaty. We should not be discouraged to think that our words are wasted; sometimes they go a long way.

It may be true that birds and fish do not vote, at least not in our forums, though when I was in the Antarctic I [had the distinct impression that] the penguins were voting, though I do not know how it was recorded. Now I would like to further what Dr Buchinger has proposed and ask that a resolution be drawn up by this Conference to the effect that the Law of the Sea Treaty establish an exclusive economic zone for the Antarctic Continent and the unclaimed islands of the Southern Ocean, like those proposed for other maritime lands, to reach 200 miles from the Antarctic [continental] shores and [adjacent] island shores. I believe that it would be perfectly in order and probably a very helpful step to make this proposal.

I am greatly concerned that, as I understand it, the Law of the Sea Conference (where I have been an observer at some of the sessions), [has no provision] for a maritime country that does not utilize the resources of its own economic zone. The terminology seems to imply that when a maritime country does not harvest its own resources, these must be put up for harvesting by alternative countries—possibly under contract. It seems to be the intent in the Law of the Sea that the food resources be 'utilized' by Man. I am wondering, in the case of the Antarctic Continent, and in fact any country, whether there should not be some provision that it is up to the good thinking and good management of whatever country to utilize also its resources for non-consumptive purposes, including observation of the sequences of energy transfer, behaviour and economy of undisturbed populations, and biotic succession in sheltered habitats—or indeed for whatever purposes, scientific or otherwise, it may desire. I think that if we are going to honour any kind of sovereignty, then the sovereignty ought to extend as far as we feel our scientists and managers [can usefully] go.

Fosberg

I would like to ask a couple of questions of Professor Rudolph, though probably the answers are obvious to experts in this field. The first is about this matter of Krill in the Antarctic. Last night we were having some discussion and it was pointed out that, since the whales have been depleted, there must be an overabundance of Krill. But I wonder if this is necessarily true. Many populations of animals and even plants do not respond in this way—that is, they may be at an equilibrium or maximum and not be able to produce a super-population [in response to changed conditions however beneficial these may appear to be].

My other question is, what is the primary organism on which the Krill live? There must be perfectly enormous amounts of Krill, from all that we hear, but this must mean that there are many times that bulk of something else in these cold waters and I would like to know what it is.

Rudolph

I will answer that. First of all about any increase in numbers of Krill, I think we are really very ignorant. All that we do know is that in the last five or six years there has been an increase in the number of Crab-eater Seals and also in the number of penguins, so you then imagine that the Krill have increased. But obviously there is a limiting amount, and it is limited [in turn] by the primary producers, which are mostly diatoms [and other phytoplankton], of which there are vast quantities. The largest primary productivity measured anywhere in the world has been [in Antarctica] in the Weddell Sea, at least at one period. This of course does not mean that it is sustained all through the year, [for example when] it is frozen over, but even on the sea-ice there is a great deal of primary productivity by diatoms [as in the Arctic Ocean].

Dasmann

I would like to refer to Professor Goldberg's paper and his remark about the reaction-time of governments to information that, for example, a certain type of poisoning is potentially hazardous. The delay between receipt of such information and appropriate action to control emission of a harmful substance is considered to be not too long for the protection of the coastal zone. But in the case of the countries which have banned DDT, a 10-year period was involved from the time when the danger was first pointed out. There had been an enormous amount of activity by environmentalists and environmental organizations who spent great amounts of money in taking the issue to the courts, fighting it through the courts, and pressuring the governments constantly at all levels. Now in the areas where this problem exists today in the tropical world and in southern countries generally, there are usually no environmental organizations, or only very weak ones. The mechanisms for doing this type of court battling in order to bring pressure to bear on governments, such as exist in the United States, Europe, Japan, and so on, are just not present in the less-developed countries. So the situation may be far more risky than it would appear at first glance, and the likelihood that something will be done in time is not very great—unless there is increasing pressure from international organizations.

Gaekwad

I have two questions. Professor Goldberg, as well as one or two of the distinguished panellists, warned us of pollution of the ocean by mercury and lead and so on, but no one as far as I could catch mentioned or warned us about any possibility of oil pollution, or how grave it is.

The second question arises out of the last speaker's point about DDT. Now as you know, DDT in underdeveloped countries or developing countries like mine is a very useful instrument and has been employed now for a couple of decades—especially in antimalarial campaigns. [So what should we do?]

Professor Goldberg mentioned that there is evidence that bird-life has been a victim of DDT. But we have our bird life still flourishing. There is still no sign of a 'silent spring', and if a country like India uses so much DDT, why has it not had any evident effect? Or have you any evidence that it *has* affected countries like India?

Goldberg (Co-Keynoter)

Let me first address myself to your problem of oil pollution. I consider it is serious but in the time allotted to me I could not go into it in detail. A colleague of mine in the United States once argued that the increasing levels of man-mobilized petroleum hydrocarbons in fish did stress human health and might have genetic effects.

I think his arguments are in continued need of assessment, bearing as they do on the problems of multiple stress. You as well as I carry a higher body burden of lead, of DDT, of PCBs, than did our grandparents. How about the additional burden of petroleum products that enter our bodies through the consumption of seafoods? So I think you have posed a very serious problem. The consequences of oil pollution are being looked at by many people and, luckily, assessments and evaluations continue.

Now with respect to the second problem of DDT dissemination in your country as well as many others: India is, by the way, one of the big producers of DDT in the world today. Your argument that to your knowledge there is no impact upon fish-eating birds or upon other non-target organisms may be [valid, or else it may still be too early to say]: I simply have not looked at the available data put forth by scientists in your country who have [studied] the situation.

Macpherson

We have some evidence on the subject of the damage to birds from DDT in the Southern Hemisphere. For many years the temperate-zone Peregrine Falcons [*Falco peregrinus*] have shown a decline in most of North America and around the world, while our arctic populations have remained quite stable; but these populations have overflown the wintering grounds in the southern United States and the Caribbean, and have wintered in South America, yet they are now showing greater pesticide levels and some decline.

Jakobsson (Chairman, concluding Session)

We have overstepped our time a bit and I thank you for a very lively discussion. [There are still some hands being raised and so, if enough of you are interested, we could continue this discussion after the Workshop [*see* Preface, page xvi] tonight. But I will take it there is insufficient urge to make this worth while unless I am warned by the Secretary-General or enough of you by the end of the next session.]

Session 6

THE PANDOMINANCE OF MAN

Keynote Paper

The Pandominance of Man

by

STANLEY P. JOHNSON

Head of Division and Adviser, Environment and Consumer Protection Service,
Commission of the European Communities,
200 Rue de la Loi,
Bruxelles, Belgium;
formerly *Liaison Officer with International Organizations,*
International Planned Parenthood Federation, London, England.

THE PRESENT SITUATION

The world's population of human beings now exceeds 4,200 million. In mid-1975, according to the United Nations' *1975 Demographic Yearbook* (United Nations, 1976), world population reached 3,967 million—an increase of 77 million in the preceding year. That figure represented an annual growth-rate of 1.9%.

Breaking down world population by regions, the UN *1975 Demographic Yearbook* shows that, according to the mid-1975 figures, the 2,256 million people living in Asia constituted at that time more than half (56.9%) of all the people in the world. Each of the remaining regions had a much smaller proportion of the total, with 473 million (11.9%) living in Europe, 401 million (10.1%) in Africa, 324 million (8.2%) in Latin America, 255 million (6.4%) in the Soviet Union, 237 million (6.0%) in Northern America, and 21.3 million (0.5%) in Oceania.

On a regional basis, the most rapid population increase in the world is taking place in Africa. Rates of population growth are 2% or more in 40 out of 47 countries or areas of Africa for which data are available, while 10 of these are experiencing levels of 3% or more. Taking the African continent as a whole, the annual rate of natural increase is around 2.6%. In Europe, on the other hand, 26 out of 37 countries or areas report an increase of less than 1%. In fact three of these—the Federal Republic of Germany, the German Democratic Republic and Luxembourg—report a negative rate of natural increase or actual reduction in population. Countries or areas of

North and South America exhibit a wide range of growth levels, both high and low. The rate of natural increase in the United States, for example, is 0.6%, whereas in Mexico it is 3.5%. In Asia and Oceania (apart from Australia and New Zealand), rates of natural increase are generally high. Thirty-three countries or areas in Asia report rates of 2% or more, and 14 of these have rates of 3% or more. In the Soviet Union, rates are less than 1%.

Looking at natality, we find that eight of the 10 highest crude birth-rates in the world are found in Africa. Among countries or areas with 200,000 or more population, the highest birth-rates are in Senegal and Niger (about 55 births per 1,000 population), followed by Zambia (52). At the other end of the scale, we find that the lowest *net reproduction rate* is reported for Finland (0.709), followed by the German Federal Republic (0.710), Luxembourg (0.733), Switzerland (0.822), German Democratic Republic (0.822), Netherlands (0.843), Belgium (0.878), USA (0.879), Sweden (0.899), and Austria (0.907).

Looking at life-expectancy, we find that female infants in Norway enjoy the longest life-expectancy in the world—77.6 years. Thirteen other countries or areas—Sweden, Netherlands, France, Canada, Japan, Denmark, Iceland, Switzerland, Puerto Rico, Byelorussian SSR, United States, England and Wales, and Hong Kong—also record a female life-expectancy at birth of more than 75 years. Most African countries or areas show an expectation of life at birth of less than 50 years (19 of them less than 40 years, 20 of them from 40 to 49 years, and only nine of over 50 years). In the Americas, Europe, the Soviet Union and Oceania, however, the majority of countries or areas record an expectation of life of 60 years or more. All the countries or areas of Europe fall into this group. Twenty-seven of the 42 countries or areas of the Americas are also in this category, along with five of the six countries or areas of Oceania for which data are available.

It is estimated that, in mid-1977, the 25 most populous countries or areas are (World Population Data Sheet, 1977):

1.	China	(850 million)
2.	India	(623 million)
3.	USSR	(259 million)
4.	USA	(217 million)
5.	Indonesia	(137 million)
6.	Japan	(114 million)
7.	Brazil	(112 million)
8.	Bangladesh	(83 million)
9.	Pakistan	(75 million)
10.	Nigeria	(67 million)
11.	Mexico	(64 million)
12.	Federal Republic of Germany	(61 million)
13.	UK	(56 million)
14.	Italy	(56 million)
15.	France	(53 million)
16.	Philippines	(44 million)
17.	Thailand	(44 million)
18.	Turkey	(42 million)
19.	Egypt	(39 million)
20.	Spain	(36 million)

21.	Poland	(35 million)
22.	Republic of Korea	(35 million)
23.	Iran	(35 million)
24.	Ethiopia	(29 million)
25.	South Africa	(26 million)

Projections

With the increasing awareness of the importance of demographic factors in planning for social and economic development, work on demographic projections has been given ample attention in recent years, for example in United Nations Publications (cf. United Nations, 1976). This aspect was re-emphasized strongly at the world Population Conference in Bucharest in August 1974. The United Nations has been a major contributor of projections of total population and its structure by sex and age for the world, regions, and individual countries or areas; its latest work in this field, namely the World Population Prospects as Assessed in 1968, has been widely used and has provided an objective basis for studies of demographic factors in planning for food, housing, health, education, labour force, etc.

The revision of these projections was undertaken in 1973 in order to provide more up-to-date information which would be used in the studies needed for assessment of the United Nations Second Development Decade. The revision was also planned in such a way as to serve as a basis for preparing revisions of the projections of urban–rural population, economically active population, agricultural and non-agricultural population, school enrolment, and households and families, which are to be prepared by the United Nations and the specialized agencies, including ILO, FAO and UNESCO.

The 1973 revision by major area, region and country has the year 1970 as a basis and covers the period 1970–2000. It has several characteristics of which the following may usefully be pointed out here:

(1) Quinquennial projections were prepared by sex and every 5 years of age on a country-by-country basis, except for very small countries or areas whose total population is less than about 250,000 (for which only the total population was projected).
(2) Projections by sex and single years of age for population between the ages of 5 and 25 were provided, along with the quinquennial sex–age distributions, for all the individual countries or areas except for the above-mentioned very small ones.
(3) Projections by single calendar years were prepared only for the total population of all the individual countries or areas.
(4) Projections were prepared in four variants, namely 'high', 'medium', 'low' and 'constant fertility trends', for the world, the more developed and less developed regions, 8 major areas, 24 regions, and all the countries or areas.
(5) A number of selected demographic indicators were calculated for all the regions and countries or areas, except the above-mentioned small ones, along with the sex–age distributions of population. They include, among others, average annual rate of growth, crude birth- and death-rates, percentage of the population by major categories, 'dependency ratio', child–

woman ratio, median age, general and total fertility rates, gross and net reproduction rates, and life-expectancy at birth separately for males and females. Average annual rate of growth is available for the above-mentioned small countries or areas also.

Tables I–IV present the results of total population estimates and annual rates of growth by regions during the period 1965–2000, according to the above-mentioned four variants. Figures are given in thousands, rounded to the nearest ten (as also in the text). The world population of 3,289 million in 1965 is expected, according to the 'medium' variant, to increase to 6,494 million in the year 2000. The implied rates of growth indicate that the world population may be growing at an almost constant rate of about 2.0% annually until the middle of the 1980s, and then the rate would gradually decrease until it reaches 1.7% by the end of the century.

During the 35-year period under consideration, the population of the less-developed regions is expected to increase from 2,252 to 5,040 million, with a rate of growth which would remain virtually constant at about 2.4% until 1985 and then gradually decrease to 2.0% p.a. On the other hand, the population of the more-developed regions may sustain an almost constant annual rate of about 1.0% throughout the remainder of this century, thus increasing from 1,037 million in 1965 to 1,454 million at the turn of the century. As a result of these differing rates of growth, the projected population of the less-developed regions is expected to be about three-and-a-half times the projected figure for the developed regions at the end of the century, whereas the ratio in 1965 was only a little over two to one.

According to the 'low' and 'high' variants, the range within which the population of the less-developed countries would probably fall by the end of the century is from 4,523 to 5,650 million. The 'high' variant implies an acceleration of population growth at a rate equal to 2.6 or 2.7% p.a. until 1985, followed by a gradual decline in the rate of growth to 2.4% p.a. during 1995–2000. The 'low' variant envisages a more moderate rate of population growth which would gradually decrease from 2.3 to 1.6% p.a. by the end of the century (Tables II and III). In the 'constant fertility' variant, the rates of growth given in Table IV indicate that if the populations of the less-developed regions maintain their 1965 levels of fertility, they would have an accelerated increase in their annual rates of growth, from 2.6 in 1965–1970 to probably 3.4 in 1995–2000, and their total population may reach 6,369 million by the end of the century.

Among the world's major areas, the largest addition to the population during the 35-year projection period is expected in South Asia, which contains almost one-third of the world's population. As the data in Tables I to IV show, the population of this major area is anticipated to increase from 981 million in 1965 to 2,354 million in the year 2000 according to the 'medium' variant, and it may still reach 2,617 million if the assumption of the 'high' variant materializes.

The major area of next importance with respect to addition of population numbers, according to the 'medium' variant, is East Asia where, although the pace of growth is expected to be moderate (from 1.8 in the beginning of the projection period to 1.1 at the end), the absolute increase will be very

TABLE I

Total Population Estimates and Annual Rates of Growth by Regions, 1965–2000. Medium Variant.
Source: UN Population Division.

	Total population (in thousands)					*Annual rates of growth* (%)				
	1965	1970	1980	1990	2000	1965 –70	1970 –5	1980 –5	1990 –5	1995 –2000
World total	*3,289,002*	*3,631,797*	*4,456,688*	*5,438,169*	*6,493,642*	*2.0*	*2.0*	*2.0*	*1.8*	*1.7*
More-developed regions	*1,037,492*	*1,090,297*	*1,210,051*	*1,336,499*	*1,453,528*	*1.0*	*1.0*	*1.0*	*0.9*	*0.8*
Less-developed regions	*2,251,510*	*2,541,501*	*3,246,637*	*4,101,670*	*5,040,114*	*2.4*	*2.5*	*2.4*	*2.1*	*2.0*
East Asia	*851,877*	*929,932*	*1,095,354*	*1,265,343*	*1,424,377*	*1.8*	*1.7*	*1.5*	*1.2*	*1.1*
Mainland Region	700,076	765,386	901,351	1,042,864	1,176,176	1.8	1.7	1.5	1.3	1.1
Japan	97,950	103,499	116,347	125,330	132,760	1.1	1.2	0.8	0.6	0.6
Other East Asia	53,851	61,046	77,656	97,148	115,442	2.5	2.4	2.4	1.8	1.6
South Asia	*981,046*	*1,125,843*	*1,485,714*	*1,911,819*	*2,353,841*	*2.8*	*2.8*	*2.6*	*2.2*	*2.0*
Middle-South Asia	664,868	761,809	1,001,046	1,279,761	1,564,963	2.7	2.8	2.5	2.1	1.9
South-East Asia	249,349	286,925	380,367	491,775	607,709	2.8	2.9	2.7	2.2	2.0
South-West Asia	66,829	77,109	104,302	140,283	181,169	2.9	3.0	3.0	2.7	2.4
Europe	*444,642*	*462,120*	*497,061*	*532,636*	*568,358*	*0.8*	*0.7*	*0.7*	*0.7*	*0.6*
Western Europe	143,143	148,619	158,214	168,679	179,266	0.8	0.6	0.6	0.6	0.6
Southern Europe	122,750	128,466	140,059	151,605	162,674	0.9	0.9	0.8	0.7	0.7
Eastern Europe	100,060	104,082	112,392	119,607	127,277	0.8	0.8	0.7	0.7	0.6
Northern Europe	78,689	80,953	86,396	92,745	99,141	0.6	0.6	0.7	0.7	0.7
USSR	*230,556*	*242,612*	*270,634*	*302,011*	*329,508*	*1.0*	*1.0*	*1.2*	*0.9*	*0.8*
Africa	*303,150*	*344,484*	*456,721*	*615,826*	*817,751*	*2.6*	*2.8*	*3.0*	*2.9*	*2.8*
Western Africa	89,546	101,272	133,406	180,059	240,158	2.5	2.7	3.0	3.0	2.8
Eastern Africa	86,448	97,882	128,757	173,639	233,245	2.5	2.7	2.9	3.0	2.9
Middle Africa	32,318	35,893	45,785	60,449	80,214	2.1	2.4	2.7	2.9	2.8
Northern Africa	74,520	86,606	119,385	163,230	214,404	3.0	3.2	3.2	2.9	2.6
Southern Africa	20,318	22,832	29,387	38,450	49,730	2.3	2.5	2.7	2.6	2.5
Northern America	*214,329*	*227,572*	*260,651*	*299,133*	*333,435*	*1.2*	*1.3*	*1.5*	*1.1*	*1.0*
Latin America	*245,884*	*283,253*	*377,172*	*499,771*	*652,337*	*2.8*	*2.9*	*2.8*	*2.7*	*2.6*
Tropical South America	129,854	150,660	203,591	272,495	358,447	3.0	3.0	2.9	2.8	2.7
Middle America (Mainland)	56,961	67,430	94,706	132,387	180,476	3.4	3.4	3.4	3.2	3.0
Temperate South America	36,000	39,378	46,731	54,783	63,266	1.8	1.7	1.6	1.5	1.4
Caribbean	23,068	25,785	32,145	40,107	50,148	2.2	2.2	2.2	2.2	2.2
Oceania	*17,520*	*19,370*	*24,025*	*29,639*	*35,173*	*2.0*	*2.1*	*2.2*	*1.8*	*1.6*
Australia and New Zealand	14,015	15,374	18,785	22,659	26,214	1.9	2.0	2.0	1.5	1.4
Melanesia	2,452	2,767	3,585	4,743	6,107	2.4	2.6	2.8	2.6	2.4
Polynesia and Micronesia	1,053	1,229	1,657	2,237	2,853	3.1	3.1	3.1	2.6	2.3

TABLE II

Total Population Estimates and Annual Rates of Growth by Regions, 1965–2000.
High Variant.
(Less-developed regions only)

	Total population (in thousands)					*Annual rates of growth* (%)				
	1965	1970	1980	1990	2000	1965 –70	1970 –5	1980 –5	1990 –5	1995 –2000
Less-developed regions	*2,251,510*	*2,563,561*	*3,378,768*	*4,424,950*	*5,650,426*	*2.6*	*2.7*	*2.7*	*2.5*	*2.4*
Mainland region	700,076	785,095	983,009	1,183,317	1,369,757	2.3	2.3	2.0	1.5	1.4
Other East Asia	53,851	61,046	78,845	102,115	123,424	2.5	2.5	2.7	2.1	1.7
South Asia	*981,046*	*1,126,115*	*1,518,153*	*2,032,456*	*2,617,382*	*2.8*	*2.9*	*3.0*	*2.7*	*2.4*
Middle-South Asia	664,868	761,993	1,024,890	1,363,525	1,742,573	2.7	2.9	2.9	2.6	2.3
South-East Asia	249,349	286,925	387,315	522,096	677,570	2.8	3.0	3.0	2.7	2.5
South-West Asia	66,829	77,197	105,947	146,835	197,239	2.9	3.1	3.3	3.1	2.8
Africa	*303,150*	*345,818*	*466,366*	*648,854*	*905,702*	*2.6*	*2.9*	*3.3*	*3.4*	*3.3*
Western Africa	89,546	101,705	136,590	190,624	269,314	2.5	2.8	3.3	3.5	3.4
Eastern Africa	86,448	98,203	131,361	182,218	256,970	2.5	2.8	3.2	3.4	3.5
Middle Africa	32,318	36,013	46,754	63,457	88,626	2.2	2.5	3.0	3.3	3.4
Northern Africa	74,520	87,027	121,883	172,708	236,900	3.1	3.3	3.5	3.3	3.0
Southern Africa	20,318	22,871	29,778	39,847	53,892	2.4	2.6	2.8	3.0	3.0
Tropical South America	129,854	151,266	208,241	288,203	394,822	3.1	3.2	3.3	3.2	3.1
Middle America (Mainland)	56,961	67,498	96,505	138,609	196,659	3.4	3.5	3.7	3.5	3.5
Caribbean	23,068	25,851	32,754	41,915	53,842	2.3	2.3	2.5	2.5	2.5
Melanesia	2,452	2,771	3,645	4,963	6,625	2.4	2.6	3.0	3.0	2.8
Polynesia and Micronesia	1,053	1,230	1,737	2,472	3,337	3.1	3.4	3.6	3.1	2.9

Source: UN Population Division.

TABLE III

Total Population Estimates and Annual Rates of Growth by Regions, 1965–2000. Low Variant. (Less-developed regions only)

	Total population (in thousands)					*Annual rates of growth* (%)				
	1965	1970	1980	1990	2000	1965 –70	1970 –5	1980 –5	1990 –5	1995 –2000
Less-developed regions	*2,251,510*	*2,522,681*	*3,136,625*	*3,819,836*	*4,523,382*	*2.3*	*2.2*	*2.0*	*1.8*	*1.6*
Mainland region	700,076	752,802	855,508	945,776	1,034,638	1.5	1.4	1.0	0.9	0.9
Other East Asia	53,851	61,046	76,468	92,659	107,712	2.5	2.3	2.0	1.6	1.4
South Asia	*981,046*	*1,121,456*	*1,438,771*	*1,785,862*	*2,119,009*	*2.7*	*2.6*	*2.3*	*1.8*	*1.6*
Middle-South Asia	664,868	758,481	967,173	1,191,467	1,403,391	2.6	2.5	2.2	1.7	1.6
South-East Asia	249,349	286,062	369,499	461,531	550,240	2.7	2.6	2.4	1.9	1.7
South-West Asia	66,829	76,914	102,100	132,864	165,378	2.8	2.8	2.7	2.3	2.1
Africa	*303,150*	*343,596*	*448,006*	*582,872*	*734,159*	*2.5*	*2.6*	*2.7*	*2.4*	*2.2*
Western Africa	89,546	100,928	130,536	168,751	210,587	2.4	2.5	2.6	2.3	2.1
Eastern Africa	86,448	97,637	126,633	165,633	211,152	2.4	2.6	2.7	2.5	2.3
Middle Africa	32,318	35,766	44,757	57,033	71,306	2.0	2.2	2.4	2.3	2.2
Northern Africa	74,520	86,470	116,964	154,130	194,285	3.0	3.1	2.8	2.4	2.2
Southern Africa	20,318	22,795	29,117	37,325	46,829	2.3	2.4	2.5	2.3	2.2
Tropical South America	129,854	150,035	198,648	257,832	325,152	2.9	2.8	2.7	2.4	2.2
Middle America (Mainland)	56,961	67,136	92,831	127,219	167,641	3.3	3.2	3.2	2.9	2.6
Caribbean	23,068	25,762	31,713	38,814	47,677	2.2	2.1	2.0	2.1	2.0
Melanesia	2,452	2,765	3,533	4,579	5,786	2.4	2.5	2.6	2.5	2.2
Polynesia and Micronesia	1,053	1,213	1,632	2,179	2,733	2.8	3.1	3.0	2.4	2.1

Source: UN Population Division.

TABLE IV

Total Population Estimates and Annual Rates of Growth by Regions, 1965–2000.
Constant Fertility Variant.
(Less-developed regions only)

	Total population (in thousands)					*Annual rates of growth* (%)				
	1965	1970	1980	1990	2000	1965 –70	1970 –5	1980 –5	1990 –5	1995 –2000
Less-developed regions	*2,251,510*	*2,559,001*	*3,381,131*	*4,583,220*	*6,368,737*	*2.6*	*2.7*	*3.0*	*3.2*	*3.4*
Mainland region	700,076	780,941	991,228	1,275,390	1,673,559	2.2	2.3	2.5	2.7	2.8
Other East Asia	53,851	61,573	82,445	113,879	156,700	2.7	2.8	3.2	3.2	3.2
South Asia	*981,046*	*1,126,074*	*1,515,875*	*2,100,924*	*2,988,562*	*2.8*	*2.9*	*3.2*	*3.5*	*3.6*
Middle-South Asia	664,868	761,904	1,023,084	1,414,629	2,012,112	2.7	2.9	3.2	3.5	3.6
South-East Asia	249,349	287,050	387,272	537,323	762,368	2.8	2.9	3.2	3.5	3.5
South-West Asia	66,829	77,121	105,509	148,972	214,081	2.9	3.0	3.4	3.6	3.7
Africa	*303,150*	*344,496*	*456,620*	*622,901*	*872,798*	*2.6*	*2.7*	*3.0*	*3.3*	*3.4*
Western Africa	89,546	101,272	133,360	180,901	252,231	2.5	2.7	3.0	3.3	3.4
Eastern Africa	86,448	97,882	128,711	174,009	241,750	2.5	2.7	2.9	3.2	3.4
Middle Africa	32,318	35,958	45,603	59,449	79,683	2.1	2.3	2.6	2.9	3.0
Northern Africa	74,520	86,606	119,719	170,143	247,424	3.0	3.2	3.4	3.7	3.8
Southern Africa	20,318	22,779	29,227	38,399	51,710	2.3	2.4	2.7	2.9	3.0
Tropical South America	129,854	151,523	209,966	295,754	420,972	3.1	3.2	3.4	3.5	3.6
Middle America (Mainland)	56,961	67,485	96,413	140,425	206,814	3.4	3.5	3.7	3.8	3.9
Caribbean	23,068	26,041	33,725	44,540	60,115	2.4	2.5	2.7	2.9	3.1
Melanesia	2,452	2,767	3,612	4,886	6,798	2.4	2.6	2.9	3.2	3.4
Polynesia and Micronesia	1,053	1,229	1,733	2,478	3,544	3.1	3.3	3.6	3.6	3.6

Source: UN Population Division.

high. Sizable increases in total population are also expected in Latin America (from 246 million in 1965 to 652 million in 2000) and Africa (from 303 million to 818 million) during the same period. It is also to be noted that, for the year 2000, the 'constant fertility' estimate of total population is higher than the 'medium' estimate by 27% in South Asia, 42% in East Asia (excluding Japan), 17% in Latin America and 7% in Africa.

We may summarize the world's demographic situation in the following manner. It took from the beginning of Man to about AD 1830 for the Earth's population to reach its first 1,000 million human inhabitants, another 100 years to add the second 1,000 million and only 30 years (between 1930 and 1960) to add the third. Some time in 1975 the 4,000 million mark was passed.

If we assume a continuation of present levels of fertility, the world's population would reach about 8,000 million or nearly double its present total, by the end of the century; 6,369 million (or three-quarters of these people) would live in the developing countries. Even with declines in fertility—the medium variant—the world's population may reach 4,457 million in 1980, 5,438 million in 1990, and 6,494 million by the end of the century. In this event the population of the less-developed regions would total 5,040 million in the year 2000.

Implications

Food

In the developing countries, rapid population growth has stimulated a growing concern over the twin problems of hunger and malnutrition. *Per caput* food production in these countries remained practically unchanged between the mid-1950s and mid-1960s, in spite of substantial increases in total output. Much hope has been placed in the so-called 'green revolution', and it still seems possible that, with new agricultural technology, providing for a country's minimum food needs will not necessarily have the attributes of a chronic crisis; but many qualifications need to be made even to this cautious statement. Agronomists are keenly aware of the need for proper husbandry of the new strains of wheat and rice, of the importance of the right applications and combination of inputs, and of the need for credit and marketing facilities. Many financial and administrative constraints have to be eliminated before yield-increases that have been achieved in certain parts of the developing world can be sustained, and improved upon, over wider areas. Huge social problems will increasingly arise with the transformation of agriculture in the less-developed countries. Land-tenure patterns, distribution of agrarian incomes and farm holdings by size, and rates of rural–urban migration, are likely to be significantly modified. Rapid population growth seems certain to increase the number of landless, expand the subsistence or disadvantaged sectors of society, and raise the administrative burdens and social costs of absorbing urban arrivals.

In both developed and developing countries the continued intensification and extensification of agriculture to meet the needs of growing populations create major problems of environmental pollution, with concomitant soil and water deterioration. Increased use of pesticides and chemical fertilizers may have serious ecological consequences. Large-scale irrigation schemes, the reclamation of marginal land, the clearing of forests, etc., may pose severe threats to the stability of natural ecosystems.

A population growing annually at:

1% doubles itself in 70 years
2% doubles itself in 35 years
2.5% doubles itself in 27 years
3% doubles itself in 23 years
3.5% doubles itself in 20 years

By far the greater part of the additional demand for food over the next 15 years will be due to increases in population, particularly in the developing countries. The population-growth factor will require an increase of two-thirds in food supplies over the next 20 years in the developing countries merely to maintain existing nutrition-levels and patterns of consumption. The demand for food in the developing countries as a whole by 1985 will be about 140% greater than the 1962 level. At least two-thirds of this increase will come from population growth. Another third will result from assumed higher *per caput* consumption as income levels rise. To meet projected demand, the present trend of around 2.6% p.a. increase in food produced in developing countries and retained for domestic consumption would need to be stepped up to 4.3% p.a.

As Dr Norman E. Borlaug put it in his Nobel Peace Prize lecture: 'The green revolution has won a temporary success in Man's war against hunger and deprivation; it has given Man a breathing space. If fully implemented, the revolution can provide sufficient food for sustenance during the next three decades. But the frightening power of human reproduction must also be curbed; otherwise, the success of the green revolution will be ephemeral only.'

Education

The number of children enrolled in the primary schools of the less-developed countries rose 150% during the 15 years from 1950 to 1965, and the percentage of all children 6–12 years old who were in schools rose from less than 40% to more than 60% in that period. This marked increase in enrolment ratios reflected in large measure the value placed on education by people of all classes and income-groups in the developing countries.

It is estimated that the child population in the developing regions during the 1970s will continue to constitute 40–45% of the total population, and sometimes more. The number of children of school age (5–14) is expected to increase from 560 million in 1965 to over 1,000 million by the end of the century, even if fertility is substantially reduced. This means that almost intolerable burdens will be placed upon the educational system, even to maintain the present unsatisfactory school attendance. It seems that even with a major allocation of resources to education, i.e. construction of schools, training of teachers, etc., the absolute number of illiterates—already more than one-third of all the adults in the world—is likely to increase rather than diminish. In fact, between 1960 and 1970, the absolute total of illiterates increased by 70 million because of the rise in population.

Health

It has long been obvious that progress in science and medicine has destabilized any equilibrium which the human race may formerly have enjoyed, and that this tremendous problem has to be solved if mankind is to survive in reasonable fettle.

Public-health technology applied on a mass scale in the developing countries has indeed reduced death-rates dramatically. Yet the level of personal health services for the individual and the community varies widely and, in general, remains far below the levels of the more developed regions. High fertility forces health ministries to run fast in order to stay in the same place—let alone improve services. For the next 20 years at least, the demand for health services will outrun the supply by such measures as doctor/population ratios and number of hospital beds. Rapidly growing population combined with growing aspirations makes this inevitable. The age and geographical distribution of the population also affect the health services. In a high-fertility community, the primary stress on health services will be the care of mothers and children. The problems of medical treatment for infants are substantially greater than the problems of treating young adults, and hence care of the young requires a higher doctor/population ratio than the care of people aged 15–45. The levels of personal health services are usually much higher in urban than in rural areas, both in terms of numbers and quality.

Rapid population growth and its accompanying rapid urbanization are injurious to the communal health, although it may be difficult to isolate the effects of crowding from other conditions such as poverty, poor nutrition, poor housing, and pollution. These problems, of course, are in themselves partly a consequence of rapid population growth.

Urbanization

It is estimated that the world's urban population, which was 33% of the total world population in 1960, will be 46% in 1980 and 51% in 2000. At the beginning of the twenty-first century, the urban population of Latin America may include 80% of the region's total population, making that region more urbanized than Europe, while the urban population of North America is projected to reach 87% of the total. The percentage of urban population during the 1960–2000 period will grow from 23 to 40 in East Africa, from 18 to 30 in Africa as a whole, and from 18 to 35 in South Asia. In the year 2000 it is projected that 62% of the world's urban population will be living in the regions now described as 'developing'. The corresponding figure for 1960 was only 41%. In 1900, probably no more than one-quarter of the world's population lived in urban settlements.

High rates of population growth and rapid urbanization are inevitably linked. Today, when the population growth-rates of urban centres often exceed an annual average of 6% and sometimes reach 10% or even more, many developing countries have reached a critical period, and others are on the threshold of transformation from primarily rural to primarily urban societies.

It is characteristic of the urbanization process that the more it progresses, the more significant is the part represented by large cities among the total urban population. The over-concentration of population in certain centres, and subsequent overcrowding in squatter settlements and slums in developing as well as developed countries, creates social problems and environmental deterioration. Pollution of air, water and land, and damage to the ecological balance of the environment, have become major problems of urban settlements in developed countries, and are beginning to affect developing countries as well.

Housing

At the beginning of the 1960s the United Nations Organization, in its proposals for its first development decade, estimated that around 10 dwelling units per 1,000 inhabitants had to be built each year in the developing countries in order to offset obsolescence and to cope with the foreseeable urban population growth throughout the decade. Yet in most developing countries, less than two houses per 1,000 inhabitants are actually being built each year.

In 1960, the housing deficit in Latin America was about 20 million units, while in Asia and the Far East it was about 22 million in urban areas and 125 million in rural areas. In industrialized countries the average rate of construction has been below the suggested targets. Today a considerable proportion of the world's population is poorly housed and is likely to remain so, as the resources that are at present being allocated to the housing sector are modest compared with the rising demand for housing that is generated by the rapidly growing population. With the world's projected rate of population growth, 1,100–1,400 million new dwelling units would need to be built before the end of this century.

Employment

The world's labour force in 1970 was estimated to be over 1,500 million—an increase of about 200 million since 1960. By 1980 the total may have reached nearly 1,800 million. During the 1970s it will be necessary to absorb an increase of some 226 million in the labour force of the developing countries, with the bulk of this total coming from South-East and East Asia, where the labour force is expected to increase from 804 million to over 970 million or by roughly 20%. Although smaller in absolute terms, the foreseeable increase in the other less-developed regions is relatively even greater: 32 million (23%) in Africa, and nearly 30 million (32%) in Latin America. In the industrialized countries the anticipated increase in the labour force during the 1970s is about 56 million (11%).

Within the next three decades there could be more than 1,000 million new workers in the developing world, with all the attendant problems. Already 300 million new jobs will need to be provided during the Second United

Nations Development Decade (1 January, 1971–31 December, 1980), to make good the backlog of 76 million already unemployed and to provide for the expected 226 million growth in the labour force. Moreover, these figures make no provision for 'disguised' unemployment.

Of equal significance is the large number of young workers who will swell the world's labour force during the 1970s. There will be an increase of some 21 million workers in the 15–19 age-group, and this group will represent in 1980 some 15% of the total labour force in Africa, 15% in Latin America and 12% in Asia, as compared with 9% in Europe and 8% in North America. The problem of finding useful and productive employment for young people in the developing countries will be one of the most urgent and challenging for governments in the years to come.

Environment

The environmental impact of population growth differs between regions and even between countries within regions. In the developing countries the ills caused by poverty leave little room for concern about, or expenditures on, the environment in ways that now preoccupy many people in rich countries. The environmental problems of the less-developed countries arise much more from rapid population growth combined with lack of technology than from rising incomes and the presence of new technology. Rapidly rising populations bring unwelcome pressure to bear on resources—e.g. extension of cropping into rain-fed areas that are at best suitable only for grazing; also random deforestation, etc. The very rapid spread of new technology, e.g. the 'green revolution', designed to feed rising population, may have secondary effects on the environment to a degree and extent that are as yet unknown.

The effect of rapid population growth on economic situations and prospects clearly has environmental consequences. The contamination of water supplies by sewage, rampant and unplanned urbanization, and health and disease problems, in city and countryside—all reflect the very lack of development that is at the root of so many poor countries' problems.

In the more industrialized countries, and in those countries where the process of industrialization is merely under way, environmental problems manifest themselves rather differently. It is not easy to distinguish the extent to which industrial pollution, problems of solid waste, marine pollution, etc., are a function of population increase. Certainly, rising standards of living and growing affluence, combined with the force of technology, exert a 'multiplier effect' such that even relatively modest increases in the populations of so-called developed countries may have a quite disproportionate effect on the 'human ecosystem' as a whole. It might be argued that, environmentally, the growing populations of Europe, North America, Japan and the Soviet Union constitute a truly global threat that is not posed by the populations of Africa, Asia and Latin America; for it is in the so-called 'advanced' industrial nations that growing populations make the most impact upon terrestrial ecosystems and upon the scarce resources of the biosphere. These nations are the rich nations, whose technological achievements are the highest, and it is their patterns of production and consumption which, when allied with steadily increasing populations, have led to many of the present acute environmental problems.

Per Caput Income

In aggregate terms the rate of population growth in the developing world renders difficult, if not impossible of attainment, rapid increases in *per caput* income. For example, with a GNP (Gross National Product) which is growing at an annual rate of 5%, and a stationary population, it takes 12 years to double *per caput* income. If population growth is rising, say, at 2.4% a year (the projected average rate for the less-developed world), it will take 27 years to achieve a doubling in *per caput* income. If the annual growth in national income is 4% rather than 5%, doubling of *per caput* income will take 43 years; and as two-thirds of the world's population live in countries where *per caput* income is less than $300 p.a. and often less than $100, even twice the present income is hardly likely to represent a satisfactory fulfilment of economic expectation. As the World Bank indicates in its paper on Population Planning (published in 1969):

> . . . despite its limitations, one of the best available measures of economic progress toward these goals is the growth of *per capita* income. This is the growth of national income, adjusted for growth of population. Thus the relationship between the growth of a nation's income and that of its population is fundamental to the improvement of human welfare. While neither the causes nor the effects in this relationship are fully understood, one central fact is clear: the higher the rate of population growth, the more difficult it is to raise *per capita* income. Today the world's population is growing much faster than at any time in history. This simple fact led the Pearson Commission to say, in 1969, that 'No other phenomenon casts a darker shadow over the prospects for international development than the staggering growth of population'.
>
> The problems created by the large numbers and high growth-rates of population concern both the world as a whole and individual countries. Both more and less developed countries confront such universal questions as the ultimate size of population the world can sustain and the rate at which the limit will be approached. The earth can undoubtedly support substantially more than the 3.6 [thousand million] people now living on it.* But there is a great doubt about its ability to sustain unlimited numbers at decent standards of living, which a majority do not have even now. [Cf. Hopkins, 1971; Ehrlich & Ehrlich, 1972; Johnson, 1973].

Actions

What can be done about the population explosion in the developing countries? The first thing to make clear is that the only people who can properly do anything about the population problem of the developing countries are the people of those countries themselves. If there was any underlying theme in the International Development Strategy for the Second United Nations Development Decade, adopted by the UN General Assembly in October 1970; it was that the primary responsibility for the development of the developing countries rests on themselves. This is as true of popu-

*This referred to several years ago, the figure being currently well in excess of 4,000 million.—Ed.

lation as of any other field of activity. In fact, in this matter of population, there has been very considerable progress over the past few years—to the extent that the total number of governments with policies to reduce population growth-rate is now over 30 (Nortman & Hofstatter, 1976).

Within northern Africa, for example, Tunisia led the way by adopting in 1964 a long-term objective to reduce population growth to 1% p.a. by the year 2000. Egypt followed, in 1965, with a national policy to reduce population growth-rate and, in 1969, specified that the target was to reduce the crude birth-rate by 0.1% p.a. for 10 years. The demographic target set in Kenya's five-year Development Plan (1974–78) is to reduce the population growth-rate from an estimated 3.5% to 3.25% at the end of the period of the Plan.

Within Asia, Iran's official policy is to reduce the population growth-rate from 3.1% to 1.5% over 20 years, in order to facilitate socio-economic growth. China told the United Nations World Population Conference in 1974 that 'in order to realize planned population growth, what we are doing is to develop medical and health services . . . and strengthen our work in the maternity and child care, so as to reduce the mortality rate on the one hand and regulate the birth-rate by birth-planning on the other'. In South Korea in March 1976, a Population Policy Deliberation Committee was established and charged with developing a comprehensive population policy aimed at reducing population growth. In the Malayan peninsula the goal of the national policy is to achieve a 2% annual population growth-rate by 1985. In the Philippines, according to the four-year plan (1974–77), the target is to reduce the birth-rate from an estimated 43.2 per mille in 1970 to 35.9 in 1977. In Bangladesh, the first five-year plan (1973–78), adopted after independence in 1971, assigned the highest priority to the Population Control Programme, equal to that of food production. The ultimate target is a stationary population of less than 150 million within the next 25–30 years. In India, the current goal is to reduce the crude birth-rate to 30 per mille by the end of the fifth five-year plan (1974–79), and to 25 per mille by the end of the sixth. In Pakistan, the goal is to reduce the birth-rate to 35 per mille by 1978.

In Latin America, Colombia's Minister of Health stated in January 1975 that the rapid population growth was an impediment to development, and that the people had the right to plan their families. In El Salvador, in July 1974, the government announced the establishment of a National Population Commission to implement the objective of improving the quality of life, while in September 1973, President Echeverria, of Mexico, marked the shift in Mexico's position on population from pro- to anti-natalist. The Mexican Population Policy Law, issued as of 7 January, 1974, established a National Population Council empowered to provide family planning, educational and health services in the public sector and to monitor the private sector in order to stabilize population growth, improve the utilization of the human and natural resources of the country, and preserve the dignity of the family. In Trinidad and Tobago, the official aim of the government is to reduce the birth-rate to the vicinity of 20 per mille by the end of the tenth year of the programme.

Other governments provide assistance for family planning without having an official population policy. Within Africa, these include Algeria, Sudan, Benin, Nigeria, Rhodesia, South Africa, Tanzania, Uganda, Zaire, etc. Within Western Asia, these include Afghanistan and Iraq. Within Latin America, these include Bolivia, Brazil, Chile, Costa Rica, Cuba, Ecuador, Honduras, Panama, Paraguay, Peru, Venezuela, etc.

Family Planning

This must be a familiar enough term to any Western audience. In an organized and institutionalized way Britain, for example, virtually invented the idea. The essential philosophy is that a married couple (or indeed any other kind of couple) should have the number of children that they want, and no more. So as not to have more children than they want, the husband or wife (possibly both) practise some form of contraception. The methods, which there is no need to go into here, are familiar. Family planners tend to advocate what they call the 'cafeteria' approach: anything will do as long as it works, from abstinence at one end of the spectrum to the latest hormonal preparations and vasectomy at the other.

The bulk of the effort in the developing countries has so far been concentrated on the extension of family planning programmes. National Family Planning Associations work in 120 countries. Where governments have official family planning programmes, either inside or outside the health service, the Family Planning Association will work alongside and try to dovetail its efforts into the government's own programme. Where governments do not yet have a family planning programme, the private Family Planning Association can bring pressure and influence to bear and can hope to demonstrate to the government that family planning is something which women want and governments should provide.

Reasons for Need to Limit Births

Now it is conceivable that if family planning services, i.e. organized contraception, was widely and cheaply available, so that all couples had as many children as they wanted and no more, then we might arrive at a rate of population growth that was acceptable and tolerable and did not impose excessive burdens on the economy and social structure of countries and on the environment. But the trouble is that, all over the world, people tend to want (or anyway get) too many children from a demographic point of view. There is, of course, endless speculation as to why this is so. Some argue that the collective consciousness in developing countries has not yet woken up to the fact that more babies are surviving than used to be the case, and that it is no longer necessary to have, say, seven children in order to have a net gain of two—or even to have two survive and to keep the family number stable.

Others point to deeply-ingrained patterns of behaviour: children are needed to revere dead ancestors, and the more the better; two sons at least may be required to open up the father's skull on the funeral pyre; children may provide support and succour to their parents in countries where social security services are virtually non-existent; agricultural systems in much of the developing world require a ready availability of labour at certain times of the year, even if there is a good deal of unemployment or underemployment at other times.* Consequently, the probability of a population policy being successfully implemented solely through the provision of family planning services—as a non-priority matter—is remote. This is not to say that there may not be important additions to human welfare from family planning programmes. No doubt, from the point of view of the individual mother and family, the provision of family planning can be seen as a boon and a relief from the burden of constant childbearing. But it will not bring the birth-rate down very much.

All this explains why such countries as India, Pakistan, Indonesia and Egypt, which, on broad socio-economic grounds, so desperately need to bring their birth-rates down—i.e. those countries which have official population policies and defined demographic goals—have decided to do more than merely provide family planning services: they have decided to publicize and advertise them.

States' Arsenals of Devices

The idea is that the State has at its disposal an arsenal of devices to influence people's desires and inclinations. The word often used to describe this process is 'motivation'. The instruments of motivation may be human—the family planning worker wobbling her way on a bicycle through the villages of Taiwan bringing the good news—or they may be mechanical, including movies and mass-address systems, television programmes, and huge billboards posted in the streets. Or again, the instruments of motivation may have to do with incentives or disincentives, privileges or penalties.

There is some evidence that this wider approach is working. Thus there have been limited successes in achieving modest reductions in the birth-rate in countries such as Taiwan, Singapore, Hong Kong and Korea, but it is not clear what relation there is between these successes and the deliberate implementation by the State of population policies.

There are those who argue that a falling birth-rate is a consequence of a general rise in the level of economic and social development. They continue the argument by saying that, so far from investing large sums of money in population programmes that will have no effect—as you cannot change, from the outside, people's fundamental attitudes to something as important and intimate as procreation—you should put this money into 'conventional' development programmes. The standard of living rose in the West, they say, and that itself induced a decline in fertility. When people wanted to have fewer children, they found a way. This is known as 'the theory of demographic transition'. The idea is that, after a certain period of time, another category can be added to the ones mentioned earlier: after falling fertility and low mortality, you reach low fertility and low mortality.

*Among further reasons we might cite the teachings of certain religions and the financial benefits offered by still rather many governments—the cash bonuses paid per child being reputedly sometimes alone sufficient to support the entire family.—Ed.

There is evidence that what happened in the West will be repeated in some developing countries. The peripheral areas and Asia, as mentioned above, have come within the orbit of the Western Industrial System, and Japan has seen a very remarkable check in population growth with its development as a major industrial nation.

The relationship between population growth and economic development has, of course, long been a matter of dispute. From the environmental point of view, the theory of demographic transition presents certain dangers. For if population decline only comes about because of increases in industrial growth and *per caput* income, we may in fact come up against a new set of constraints—namely, those related to carrying capacity and the ability of the Earth to meet the demands made by the industrial process itself.

Revised Theory of Demographic Transition

This explains the importance of the so-called 'revised theory' of demographic transition. Dr Norman Uphoff, of the Center for International Studies at Cornell University (Uphoff, 1977), points out that a

> number of Third World countries have already begun to reverse their rate of population growth at *per capita* income levels of $150, $200 or $300 in conjunction with strategies of development that stress not so much the expansion of a modern industrial sector starting out with advanced technology, but rather the development of agriculture and the rural areas, using throughout most of the economy production techniques that are appropriate to the existing factor endowments, particularly their abundant labour.

Citing particularly the cases of China and Sri Lanka, Uphoff suggests that the

> equivalent of the demographic transition can be achieved at *per capita* income levels well below those observed historically in Europe and North America. The more equitable development strategy, taking intensive agricultural production and labour-using technologies as its base rather than capital-intensive industrialization, appears able to provide the socio-economic equivalent of what occurred earlier in Europe and North America.
>
> However, the development strategy followed by these early developers, with its unequal distribution of benefits concentrated in the urban–industrial sector, meant that average *per capita* incomes had to reach, say, $600 to $1,000 before the rural majority began to receive some improvement in income, welfare and security that would be conducive to reducing fertility.
>
> With national policies that have more widespread impact, these improvements can be brought to rural families—who in the Third World countries produce the large majority of children—while national *per capita* income levels are still relatively low, certainly lower than the transition level noted for Europe and North America.
>
> This question of the *distribution* of benefits and of consequent decline in fertility points to the basic weakness in the strategy linking anti-natalist programmes with GNP growth-maximizing efforts. Even small reductions in the fertility rate of the rural majority will contribute more to bringing down national population growth-rates than will achieving much larger drops for the urban middle and upper classes, who constitute a definite minority.

Chinese Model

The case of China is of particular interest. As part of its policy (noted above) to 'realize planned population growth', the government strongly advocates late marriage and the two- or three-child family. It provides an active programme of contraceptive services, including sterilization, as well as abortion on request. The delivery system, as Freedman & Berelson (1976) note—including services and supplies in the traditional sense as well as motivational activity—is integrated not only into the health structure, from locality to central hospital, but also into the entire administrative, economic, social and political structure of the country as well. This integration is a key and distinctive feature of the system, enabling the government to launch mass campaigns that reach down to the household in a systematic and potentially powerful way—in the manner of national Chinese campaigns in other fields, rather than family planning programmes in other countries.

Freedman & Berelson (1976) point out that committees on planned births are organized in China at all administrative levels—ranging from village or urban neighbourhoods to commune or production units, and to counties, urban wards, and so on. Usually chaired by an official of the political party in power, the party apparatus participates actively in the implementation of the programme. A powerful force for family planning appears to be the peer influence and pressure for conformity to fertility goals—exerted, among other ways, through the vast network of small primary groups in which the whole population appears to be organized, both at place of work and in home areas. Since 1971, targets have been set for cities and counties (reportedly, growth-rates of 1% and 1.5% p.a. by 1975 for cities and countryside, respectively), and the targets flow down to the allocation of births within the smallest units (roughly 20–40 households) on the basis of group decisions. Where the target requires allocation of eligibility for pregnancy, the guidelines reportedly apply first priority to newlyweds, next to those not yet having completed family size (recommended size: two children in the cities, three in the countryside), then to those with the longest birth-interval (recommended interval: five years).

The rationales for such a campaign include support and strengthening of the revolution; availability of women for labour and learning; release of young people's energy for national service; maternal and child health; and facilitation of social and economic development. The common theme of 'practising birth-control for the revolution' provides substantial social support; in many other developing countries the family planner feels socially isolated and deviant, whereas in China he is made to feel that he or she is contributing to the common welfare. And 'the use of Maoist phrases to sanctify birth-control clearly indicates that the campaign now has unequivocal support at the highest level'. Moreover the propaganda, pressure and service go on within a social system that has reportedly improved markedly in the areas of social security, small industry in rural areas, income equalization, women's status, and education—all of which presumably lead towards fertility reduction.

Thus, according to these accounts, there is a strong policy commitment to the 'planned birth' programme, reinforced through the political structure, and strongly pursued through all available channels—on its face, the strongest such effort in the world today.

What has such an effort accomplished in such a setting? Freedman & Berelson (1976) state that 'the Chinese themselves have been modest and circumspect in their statements, perhaps in part because they appear not to have statistical evidence on a national basis. However, the United Nations Demographic Division has taken all of the cumulative evidence sufficiently seriously to estimate that the birth-rates are lower than was earlier believed. The United Nations estimated the birth-rate for China as 28 [per mille] for 1965–70, and estimate a rate below 27 for 1970–75. These estimates, prepared as part of the background material for the Bucharest Conference [on human population], appear unbelievably low to many demographers, but even if the rates were five points higher, they would be substantially lower than those in the other large developing countries and low by the criterion of previous estimates for Chinese fertility.'

Conclusions

The recent experience of China and certain other countries gives some grounds for guarded optimism, as far as both environment and demography are concerned (Brown, 1976). *For the evidence appears to be that fertility declines can be achieved without increasing the general environmental surcharge.* This is probably the best news of the decade; but it is news, of course, which has profound implications not merely for the internal organization and structure of Third World countries but also for programmes of international economic assistance (many of which now include an important population and family planning component).

Whether the revised version of the theory of demographic transition is also of domestic relevance to western countries which are already experiencing low or even negative rates of population increase, remains to be seen. The task in these countries is in some ways particularly hard. Not only do they have to maintain the pattern of low fertility; they now have deliberately to seek a lower standard of living than hitherto—one that is more in conformity with both the new political, economic and ecological realities and with the difficult but important concept of 'global equity'. So even for the West there may be an important message inside the 'bamboo tube' as we try out new patterns of living, new social structures, and 'softer' technologies, etc., that are more appropriate to the circumstances of today.

One of the most remarkable realizations of the last few years is the increasing recognition that the growth of population is an international responsibility. In the first seven years after it was set up, for example, the cumulative resources of the United Nations Fund for Population Activities grew to $280 million (Salas, 1976). The World Bank has been devoting steadily growing resources not just to population and family planning but to the so-called 'alternative development schemes' which may in fact (as has been argued here) result in improved diffusion of the population and family planning message and quicker reductions in fertility. The International Planned Parenthood Federation (1976) is continuing important work in the private sector.

Will national efforts to reduce fertility, coupled with increasing and better-directed programmes of international assistance, be sufficient to deal with the problems of population growth in an orderly and humane way? Does the concept of 'pandominance' include also Man's ability to control himself and the proliferation of the human species? Today there is no easy answer to these questions. Ten years from now, the evidence one way or the other may be a great deal clearer. My own view is that if the ability to control his own fertility escapes him, then the impact of Man's pandominance in all other spheres must in the long run be pernicious, with little doubt ending in disaster for the world at least as we know it.

[Not only does the total demomass of Man with little doubt exceed by far the biomass of any other species of land animal, plant, or microorganism, but the unique human brain gives this unprecedented monoculture a capability of almost endless possibilities. So Man has come in a relatively short time, explosively yet seemingly inexorably, to colonize virtually the entire Earth and conquer Nature. In the process he has become the world's great pandominant—to the extent that such activities as large-scale irrigation schemes, 'reclamation' of 'marginal' land, severe overgrazing, clearing of forests, etc., depletion of irreplaceable raw-materials, wasteful use and depletion of water resources, and pollution *inter alia* causing climatic changes, are posing severe threats to the stability of natural ecosystems and even undermining the very life-support system. We have seen that the environmental problems of the less-developed countries arise much more from their rapid population growth combined with lack of technology, than from rising incomes and the presence of new technology; yet somehow we have to steer this human population-swarming away from the proverbial crash-course.]

References

Brown, Lester (1976). *World Population Trends: Signs of Hope, Signs of Stress.* World Watch Institute, Washington, DC: Paper No. 8, 40 pp.

Ehrlich, Paul R. & Ehrlich, Anne H. (1972). *Population, Resources, Environment: Issues in Human Ecology* (2nd edn). W. H. Freeman, San Francisco, California: xiv + 509 pp., illustr.

Freedman, R. & Berelson, B. (1976). The record of family planning programmes. *Studies in Family Planning*, Population Council, New York, 7(1), pp. 1–40.

Hopkins, J. (1971). *Rapid Population Growth: Consequences and Policy Implications*. National Academy of Sciences, Washington, DC: 696 pp., figs and tables.

International Planned Parenthood Federation (1976). *Family Planning in Five Continents*. IPPF, London, England: 32 pp.

Johnson, S. P. (1973). *The Population Problem*. David & Charles, Newton Abbot, England: 231 pp.

NORTMAN, D. & HOFSTATTER, E. (1976). *Population and Family Planning Programme: A Fact Book*. Reports on Pollution, Family Planning; New York, NY: Vol. 2 (8th edn), pp. 1–104.

SALAS, R. (1976). *People: An International Choice—the Multilateral Approach to Population*. Pergamon Press, Oxford, England: 154 pp., illustr.

UNITED NATIONS (1976). *1975 Demographic Yearbook*. United Nations Sales No. E/F.76 XIII, New York, NY: x + 1118 pp.

UPHOFF, N. (1977). Accelerating the movement to smaller families: Hope not income is what matters. *Development Forum*, **5**(1), pp. 4–5, illustr.

WORLD POPULATION DATA SHEET (1977). Population Reference Bureau Inc., 1337 Connecticut Avenue NW, Washington, DC 20036: 1p.

See also:

The Past and Future Growth of World Population—A Long-range View. *Population Bulletin No. 1*. United Nations, New York, NY: Sales No. 52.XIII.2.

Framework for Future Population Estimates, 1950–1980, by World Regions. *Proceedings of the World Population Conference 1974*. United Nations, New York, NY: Vol. III, Sales No. 55.XIII.8, pp. 283–328.

The Future Growth of World Population. United Nations, New York, NY: Sales No. 58.XIII.2.

World Population Prospects as Assessed in 1963. United Nations, New York, NY: Sales No. 66.XIII.2.

World Population Prospects as Assessed in 1968. United Nations, New York, NY: Sales No. 72.XIII.4.

DISCUSSION (Session 6)

Gaekwad (Chairman, introducing)

May I call this sixth Session to order and say how extremely flattered I am to have been asked to chair this meeting on a subject which is so important to the world today. I am no expert, [nor even a] scientist. At the moment I belong to what is considered the most despicable tribe of *Homo sapiens*, popularly known as politicians [laughter], being a member of the Lower House of Parliament of India.

You may well [wonder] what I am doing here—a good question indeed—but why not ask Nicholas Polunin? Why am I chairing this Session [laughter] followed in by that great organizer, with so many endearing qualities—but, mind you, he is a bit crafty, too? Actually may I say this, with no humility attached to it, that he could not have made a better choice. [Laughter.] I have three main qualifications: firstly, that I hail from the second most populous country; secondly, as a former Minister in Government, I, for four years, held the portfolio of health and family planning—which meant, unlike you, facing and solving the practical problems arising out of pious resolutions passed at Conferences like these [laughter]; and thirdly, I now belong to a 'species' of human beings made extinct by Man himself (I was, unfortunately or fortunately, born a prince and not long ago was called a Maharajah, but am now a poor, plain 'mister'). These are my three qualifications, but mainly I make my opening remarks as somebody who has had to implement, as I said, pious resolutions passed by Conferences such as these.

In India, you may know, we were accused for a long time of not implementing the population programme. Moneys were being poured in by some of our friendly countries and we have constantly been accused of not using them properly, of misusing them, or not using them at all. Do you know—I am sure most of you know—that only a couple of months ago, the government in power, that had been in power for 30 years, fell because of the family planning programme; or, rather, because of the excesses which were committed [in its name] especially over the last two years.

But to say that we have not vigorously pursued the population control programme—popularly known as the family planning programme—is far from the truth. Yet today as you know, with all this progress, we have about 623 million people, [so roughly] every sixth person you meet on the face of this Earth ought to be an Indian. I believe every [fourth] ought to be a Chinese. [In population] we produce one Australia per year or one South Africa every two years, and we produce nearly 70 Icelands every year. So coming from a country like India, I know what pressures of population mean—not only preventing the possible birth of a certain number of children, but also providing food, education, health services, housing, employment, etc., and this has been a task which we have tried to fulfil as far as possible.

Perhaps at the end of the Session when I sum up I may make a few remarks about these challenges. But let me first of all introduce to you [our keynote speaker] Stanley P. Johnson, Head of Division and Adviser, Commission of the European Communities, Brussels, Belgium—a deeply-committed demographer who has produced several [highly pertinent] papers and books. To assist this Session we have three eminent panellists: Nicholas Guppy—a botanist, a traveller, a lecturer, and I have read a few articles written by him; I was only wondering whether that exotic [group of fishes], called Guppies, did not come from his family? Then we have Professor Donald Kuenen, Rector Magnificus of the University of Leiden, and also President of the IUCN, who you all know very well. We also have Dr Bent Jeul-Jensen, the first Medical Officer to the University of Oxford, [who also] looks after our Secretary-General's health, so he must be quite important [laughter].

Johnson (Keynoter, presenting paper)

I do not think a Conference like this can afford *not* to remind itself of the pandominance of Man if you take it in the sense of the sheer proliferation of the human species. The figures of course are quite familiar: I do not want to bore you with them now, as I have given them in my paper. They rise to something over 4 thousand millions at present; but it is interesting to add here that a few years ago female babies in Iceland had the [world's] longest life-expectancy at birth. That figure has just been overtaken in Norway, where female babies now have a life-expectancy of 77.6 years at birth.

The United Nations, who have been involved in this population business for 25 years, from time to time make projections about the population future, though it is important to point out that these are only projections of what would happen if certain assumptions about fertility and mortality were fulfilled. [They are not predictions or even forecasts.] In 1968 they published the first basic set of projections, and for World Population Year in 1974, these projections were revised. The projections are based on four different sets of assumptions: the so-called 'constant fertility variant' (i.e. what would happen if present trends remained unchanged), a high fertility variant (which already presupposes some reduction in fertility), a medium variant, and a low variant.

In my paper I have spelled out in some detail what the various projections made by the United Nations would give in terms of total population and also populations by regions in the various key years between now and the end of this century. The figures are given in four tables, [and in the text I comment on their significance].

Professor Paul Ehrlich and others have asked, what good are these projections? Everybody [enlightened now] knows that only an imbecile really believes the world's population will reach 8,000 millions by the year 2000. This is because everybody [realizes]—I am paraphrasing the argument—that there is no way in which the demands that such projections would pose for health, housing, urban facilities, education, employment, and on the environment, could conceivably be met by any kind of system which we know today. And of course food is at the basis of it all. Therefore we are, if you like, presented with a very stark choice. Either there will be a general inability to meet the requirements posed by these increasing populations, or we can make the choice of organized and sustained efforts to control fertility. The latter is now being pushed, and after some 15 years of family planning effort we have roughly 30 countries with deliberate policies to reduce the growth-rate of their populations.

Now we ask ourselves what is being achieved, and is it in any way commensurate with the need to control fertility—as any analysis of the implications of the projections would require? There is some evidence that the propagation of population and family planning programmes is working, as there has been limited success in achieving modest reductions in the birth-rate in countries such as Taiwan, Singapore, Hong Kong and Korea, though it is not clear what relation there is between these policies and the successes. So there is some evidence that what happened in the West will be repeated in some developing countries.

The danger from an environmental point of view is that the demographic transition of population decline following increases in industrial growth and levels of *per caput* income may come up against a new set of constraints, namely those related to carrying capacity in the ability of the Earth to meet the demands made by the industrial process itself. You raise levels of growth of income high enough to bring self-sustained fertility decline, only to create at the same time a new set of environmental problems or exacerbate old ones!

A possible way out of this dilemma may be indicated by a number of Third World countries that have begun to reverse their rate of population growth at *per caput* income levels of 150 or 200 or 300 US dollars, i.e. far below the conventional levels at which self-sustained fertility decline is to be expected. This has been effected in conjunction with strategies of development that stressed, not so much the expansion of a modern industrial sector starting out with advanced technology, but rather the development of agriculture and rural areas, using, throughout most of the economy, production techniques that are appropriate to the existing conditions—particularly where there is abundant labour. This, I think, is an extremely important [and encouraging] observation if it can be sustained.

We have [meanwhile] to recognize that the recent experience of China, and of certain other countries but mainly China, does give us some grounds for guarded optimism as far as both environment and demography are concerned. For if the Chinese experience is in fact something to go by, it would appear for the first time that fertility declines can be achieved without increasing the general level of environmental surcharge, which would be splendid news indeed!

It is also encouraging to note the growth of international assistance for family planning and the recognition of population as an international responsibility. I have been involved since it started with the United Nations Fund for Population Activities, and in seven years the resources of that fund grew from zero to $100m a year. The International Planned Parenthood Federation and the World Bank are [also active] in this field. Will national effort to reduce fertility, coupled with increasing and better-directed programmes of international assistance, be sufficient to deal with the problems of population growth in an orderly and humane way? Does the concept of pandominance include the concept of Man's ability to control his own fertility? I think it impossible [as yet] to give answers to these questions. [But we must] urge that population be not shelved as a 'forgotten' subject merely because World Population Year has ended. Conferences such as these [on Environmental Future] really should take every opportunity to remind the world that this is one area on which all other solutions depend. [Applause.]

Gaekwad (Chairman)

Thank you Stanley Johnson: [from what you have said and] other important aspects dealt with in your paper, which I would highly commend as very interesting and thought-provoking, it is clear that this [most vital matter of human population] is not a problem for any one country but a global responsibility. And as we go on each month, each year, I do find countries waking up to this fact that it is no longer their problem only, but indeed a global one. I do not want to come between our distinguished panellists, so I shall at this stage merely warn them that the time allotment for each is five minutes [though in case of need I understand we might manage to squeeze] a few more minutes [laughter], and there is always a possibility of a return later if time allows.

Guppy (Panellist)

Stanley Johnson has emphasized one of the most crucial issues: it is not just the question of there being a very great growth in human population, but that this growth is concentrated in [expanding] cities. You may well ask, where do these city-dwellers get their food from, and of course they are asking the same question themselves—particularly in the Third World, where vast urban shanty areas are mushrooming with population increases that may be of the order of 6% per annum, thus exceeding anything in any other region.

This growth has sprung from social disorganization in the countryside, [which situation] has been very extensively examined. But how does this social disorganization proceed? Usually it arises from the modernization of agricultural and other production in the wrong manner, accelerated by mistaken quantification of what is needed. Quantifications are extremely useful, but the first thing you have to think of is what are you quantifying and what is its *significance*, and all these statistics from Third World countries show that what they quantify most is *per capita* income. If you want to increase *per capita* income you go out to increase industrial production by heavy investment in all sorts of industries, which are usually sited in cities. Let us take for instance a big shoe factory in a city such as Lima, Peru. No more shoes are going to be produced in the countryside because the city-made shoes are cheaper and smarter. So shoemakers in the country are put out of work. Multiply this by all the other industries, and what you get is social disintegration in the countryside, followed by the workless flocking to the cities, and then social disintegration in the cities as well. So this emphasis on quantifying *per capita* income is one of the factors that has led to a totally wrong investment attitude—even with the best will in the world—in countries which want to solve their own problems.

So I would like to suggest that what we need is a different set of statistics—the statistics of self-sufficiency. Before countries compete over *per capita* income, let them compete to show the world that they are self-sufficient and not improvident beggars. Evidently this is what China has decided to do, and this fundamental change to a self-sufficient attitude on the part of the individuals and groups has replaced their former hopelessness and helplessness.

It is my belief that, though thinking in terms of reducing populations is obviously very important, what is equally significant is that, given the chance, people can help themselves. The problem of these vast, growing populations at present is their absolute helplessness. Until we learn to invest in very small individual concerns and in the countrysides—and not just in mammoth hyperproductive programmes—we can never reverse this process.

Kuenen (Panellist)

I have read the paper by Stanley Johnson and the first sight has extinguished all hope I had for the future. If you look at the figures given there, you see absolutely no way out. But then, like so many of us, even if without any hope himself, he may be an optimist who tries to look for difficult points and that of course is, in a sense, an essential aspect of this whole Conference. Although we [may feel] that there is no real solution, we keep on saying that we try to believe that there is one, and [Johnson] gives a number of examples where changes have taken place—where birth-rates are going down, and such like [rays of hope]. Now I should warn you that in my own country, Holland, where we are very much aware of what overpopulation is, and where, according to fairly intelligent people, the population is about twice the size that it should be, we have had an extraordinary reduction in childbirth in the last 10 years. Everybody said that at least we are intelligent enough to worry about what is going on and acting in the right way. But last year the reduction stopped and now the birth-rate is going up again in Holland.

My belief is that it is a very dangerous symptom of what may happen if we do not know *why* birth-rates go down, and one of the things [which make me] rather doubt this [demographic] transitional theory is that nobody can tell me why industrialized people in the last 100 years have been reducing their birth-rate. I do not see that there can be any probability that developing countries, which have totally different social structures, totally different original techniques, and totally different ways of moving towards the more technological approach, will then in one-tenth of that time automatically reduce their population. But we must realize that the sociological technique which has been applied in China has brought some very impressive results. Indeed it is extraordinary that a country about which we know so little, and which systematically tries not to let us know what is going on, is the one which gives us some hope that we may be able to induce other countries to do the same.

From Johnson's paper it seems to me quite obvious that the reduction in birth-rate in China is a part of the political system. Therefore I believe that only if we get our social scientists to find out why the birth-rates go up or down, and try and decrease them accordingly, can we then [tackle] the problems of the environment. Although the [scourge is the dual one of] both population and pollution, unless you do something about this surge of people it is of no real use to wonder about our future because there will not be one.

Juel-Jensen (Panellist)

I have little to say, other than that I think Stanley Johnson's paper is admirable. I would, however, take issue with him on his statements concerning *per caput* income. Why are we so obsessed with the annual income of a person; it tells us comparatively little. Income does not necessarily create happiness. I have my doubts, for instance, whether the average car-worker in Cowley, which is a suburb of Oxford, where I live, who has his two cars, colour television, and perhaps buys £30 or £40 (sterling) worth [of goods] every Saturday in the supermarket, is any happier than the peasant living in a remote part of the highlands of Tigray in northern Ethiopia. In many ways, he is much less happy. So income is not necessarily an indication of happiness.

It is tremendously important that the population of the world should not grow [further, but] the limitation in birth-rate [should be] implemented for the right reasons. One of the reasons why people do not have more than one or two children in the West—in the First World or the Second World—is that if they do, they cannot afford a colour television and they cannot afford that [extra] motor-car. In the process [of enrichment] they have also lost something which you still find in the Third World, and which is of great value, namely the family unit, which still will look after the old in the family. It has been my privilege as a doctor to look after both very poor people in the Third World and some in the First World. One's headache in the First World is to get help for looking after the old, because the young have opted out; some are far too keen on colour television and motor-cars, whereas in the Third World the poor usually will look after the old.

It is not enough just to limit the number of people; you must also be very sure that the quality of life of those who follow us is of a high standard. The First World and the Second World are responsible for many of the things that have gone wrong in the Third World—for instance, the shanty towns that have been referred to. Transistor radios are thought to be desirable things, but it is questionable whether they, or even shoes, are really desirable, and it may be that it is more important to try to put things into reverse.

We have all made some terrible blunders. A very close friend of mine, who in some ways was in the same boat as our Chairman (although he came from another continent), said that the biggest mistake he ever made (at a stage when he was Minister for Communications and Public Works in an East African country) was to build a very large teaching hospital which produced doctors in the capital of that country. For it swallowed up 90% of the total budget for health in the country and has done so up to date. In the course of about 15 years the medical school has produced 43 doctors, none of whom will work in the countryside, because it is much more profitable to be in a big town. Clearly the advice that was given was well meant but totally improper, and my friend's view is (and I would share it) that the same money should have been spent quite differently: on a lot of dressers and a few health officers and probably no doctors at all.

Gaekwad (Chairman)

Now we come to question-time: ladies first.

Dodson Gray, Elizabeth

I would like to speak on behalf of the women of the world, being amazed at how one can discuss population problems for an hour-and-a-half and never mention the word 'woman', though I am sure other women in the room have much the same reaction. Part of the problem is our use of what may be called 'male generic language'—'Man' or 'man' when you mean 'people', 'humans', or 'men and women'. You can talk at length about reproduction of Man or the pandominance of Man—and somehow never think that you are not even mentioning the word 'woman'. Now I would like to fill in a bit of how I see that women are involved—in both our population problem and its final resolution. I understand that in China they do have a change in the role of women, and it has been suggested by David Korten at Harvard that the crucial factor in changing the fertility rate may lie in the status and role of women.

I would ask you what motivation there is for the women of the world to stop having children, when their meaning in life is defined in this and so limited by the social conventions of a given country that there is [little] else that they can do? Housework is boring in any culture, and one of the things that makes it at all interesting for women is to have children. Indeed I see no way that women are going to give this up without having some more attractive role substituted, bringing an alternative meaning into their lives. In wrestling with human reproduction and population increase, you have got to think about the women—their feelings, their identity, their role, and the change in their status. You have also got to look at the health impacts on women of various forms of birth-control.

Women in the United States are getting fed up with birth-control measures directed mainly at them. Why? Because all of these methods are apt to have adverse health impacts on women. [Dr Royston] was telling me of a vivid instance of such differential health impacts of birth-control measures between the two sexes. In a region of India, vasectomies were being offered to men and tubal ligations to women. But it was primarily men who could read, and printed notices were posted calling attention to potential adverse health effects of vasectomies upon men—but nothing about the more serious possible risks of tubal ligations for their wives. You can guess how many more tubal ligations there were than vasectomies in the region!

We know that India has pursued vasectomy rather vigorously, and I am under the impression that part of the backlash against Prime Minister Indira Gandhi in the recent election came from the fact that the men did not care for having *their* fertility dealt with by using force. I would like to pursue that [laughter]. To a friend of mine who is a rather active feminist in the United States, I was saying something about the genuine concern for population control throughout the world, and she said, if *they* [the men] think that *we* [women] are going to stop having babies because *they* have decided it [creates] problems, they have another think coming [laughter], and this is the final issue it seems to me; women as I know them in the developed countries (and I suspect in the underdeveloped also), are really wearying of having the major decisions of their lives made by men, sitting in Conferences like this, or in boardrooms, in which there are no women representatives—men making decisions about the use of nuclear power, which will impact upon their children's lives, birth defects, and all kinds of things. In short, women are coming to want a say both in defining the problems involving them *and* in how these problems shall be resolved.

Johnson (Keynoter)

I just want to say two things very briefly. First, [we have been] using 'man' in the sense embracing women. Secondly, I do not think it is fair to extrapolate your 'women's lib' ideas to the women of the developing world. I would much prefer to hear a woman from Ghana with 10 children tell me that she wants to make these decisions herself than hear you say so on her behalf.

Cunningham

I have just been observing and taking notes on what has been said. But although I have no direct expertise on the topics under discussion, I have one remark to make from the developing or Third World country [of Jamaica whence I come. Surely] one of the reasons why population growth in some developed countries has been reduced to some extent, can be attributed to the fact that various opportunities [for time-absorption] are available in those countries. This is so because the capacity and potential exist in developed countries for people to be preoccupied with professional careers and enjoy a lot of other facilities of all sorts. There are also more amenities to keep the people in the developed world detached from family life than in the less-developed world.

We in the developing or underdeveloped countries have a lot of time on our hands—with, for example, 25% unemployment in Jamaica. There are a lot of people who sit back and relax, or go to the beach and waste time, and it is easier for them to have social interaction there. [It would seem logical to assume that the more time people spend with one another in non-productive pursuits, the more will their chances of resorting to making babies be increased. On the other hand, it would not be illogical to assume that people with less time on their hands would be less inclined to indulge so heavily in the occupation of starting babies.]

It would seem to me therefore that if we were to concentrate on making people in the underdeveloped countries more preoccupied, giving them more things to do in terms of jobs, with better job opportunities, more cultural activities, and greater recreational facilities, we might then be able, in the long run, to see the world's population taking a new turn—downwards.

Glasser

I am much disturbed by what might be called the functional approach to this overwhelming problem [of population]. When the atom was first split, it was said that Man was taking a step into the dark—that he had broken through the last frontier. I wonder now whether that was a true assessment. For we have come to another frontier, [dividing social systems and organization from the abiding spiritual goals of life as reflected in our] attitudes to child-bearing and to the family. These inherited attitudes rest upon an extremely complex and delicate structure of emotions, evolved over [some] 40,000 years, concerning the very purpose of life. Now of course to the technologist or the economist, [such subordination] is unwelcome as it conflicts with the functional objective of relating demand with supply. But I suggest to you that we are opening a door—perhaps a final one in this case—to a very dangerous situation, because if we destroy this very delicate conceptual framework of ideas with regard to the purpose of life itself, we will, in emotional terms, take away the very foundation from Man's life.

In 1924 a novel, *WE*, was published by the Russian writer Zamiatin, that so far as I know is still banned in the USSR. It portrayed a totally controlled society of the future in which, if you wanted to have sex, you would need a special permit—rather like our having to present a chit to get our lunch. Now this sort of conception of a completely functional, mechanistic approach to life I suggest we must look at very carefully indeed, and perhaps a Conference like this ought to have a discussion as to how the purpose of life itself, and the value-systems of communities, can be related in a gentle and congenial fashion with this problem of input and output in terms of resources. [For at all costs we must avoid the soulless world of Zamiatin.]

In the book that I referred to yesterday, I have discussed this aspect as a fundamental issue, and what I suggest we ought to address ourselves to in all parts of the world—not only in the underdeveloped as they call it, but in the developed as well—is resuscitating and safeguarding this inherited value-system as the foundation for all policy, and to make only policies of adjustment that are congenial to it instead of vice versa.

My final point is that I think the argument that the birth-rate falls when income rises tends to be misunderstood. Anyone who has experienced backward societies will be likely to expect the [basic consideration to be] their attitude to large families—to the desirability of large families as a kind of insurance policy against the unknown, against the dangers of disease and hardship, war, pestilence and other accidents of life—in addition to supporting the stream of inherited values already referred to. The reason why birth-rates tend to fall when incomes rise, is that the income itself provides a kind of insurance policy against the future. Therefore I suggest that if all the approximately $30,000m of aid or loan money which goes [each year] from the rich countries to the poor, were to be diverted away from capital investment to the creation of welfare incentives for population reduction, you would achieve not only the reduction in the birth-rate which is desirable but thereby remove the 'population drag' on GNP growth. [As a result, you would begin to have a net increase in indigenous capital growth, an uplift in local pride and identity, and a strengthening of local value-systems. All would gain, and the environment would recoup its ability to fulfil 'income demand'.]

Goldsmith

One of the reasons for the increase in the population in the Third World is, probably, the disruption of traditional social systems, because what actually provided security for the individual was his family and his community, and when once his family and community became disrupted, he then needed the insurance you were talking about. So his population started growing, [as surely,] the sort of security that may be provided by material goods is not a patch on the security provided by the traditional family and community. Therefore the best way to cut down population, if this interpretation is correct, is to stop interfering with the functioning of the local traditional social system.

To return to the problem of the role of women, I am a member of the Ecology Party in Britain, and in elections we get mainly the votes of the women—not because we have more sex-appeal than our competitors, but because women are more concerned about the future than men are. And this was our argument: when I went into houses as a candidate for the local County Council in Wadebridge, Cornwall, I said 'What sort of a future do you think your children are going to have? Are they going to be living in skyscrapers a mile high, with nuclear power-stations, rubbish dumps, and motorways all around them, eating plastic cornflakes for breakfast—is this the sort of life?' And they would reply, 'No, thank God, I don't have any children', or 'Thank God I'm almost dead already', or 'I'm emigrating to Australia'—and almost everybody agreed.

There is the point which I think is most important and which Stanley Johnson raised. [It follows the Guest Editorial by] Paul Ehrlich & John Holdren in *Environmental Conservation*, which was [reprinted] in *Development Forum* and in *The Ecologist*, [and contended] that the population of the world is not going to go on increasing—that it will not support [more than about 6,000 million] people. How are you going to double the population of Bangladesh? It is just not remotely conceivable; rather is the population going to start falling, with the death-rate going up. Aid programmes are largely unsuccessful because nobody knows that people are starving until it is too late: [governments and even United Nations outfits] do not admit it.

I worked for Environment Canada in 1975 and in their document on perspectives for the next 10 years they suggested that about 300 million people will die of starvation in [that time]. Of course you do not [actually] die of starvation; you die of diseases to which you have a greater susceptibility when you are underfed. My feeling is that we are going to lose many more people [than that] within the next 10 years—very many more. We are facing now a massive demo-catastrophe such as the world has never seen. We could lose anything up to a [1,000 million] people—a quarter of the population of the world — in the next 10 years [and some say even more]. All you need is for a few things to go wrong at the same time [or in rapid succession]—bad climate and drought, bad transport problems in America [so that] you cannot get the food over—and you may lose [1,000–2,000 million] people.

How can we mitigate this horror, for that is what it is? First thing, I disagree with [our Chairman], with due respect; I do not think we must regard this as a global problem. [Rather do I feel with] Jay Forrester, who is one of the world's wise men, and think, that if we are going to solve problems it is not going to be on a global level [but] at a [local one]. [For we can no longer export our problems—Mexicans illegally by the million to the United States, pollution by building high chimneys so that the winds carry it to devastate the fresh waters of Scandinavia, for example.]

Malthus was wrong, because he did not expect us to develop the technology to control populations; but we have not controlled our populations by technology, we have controlled them by exporting. We have exported our surplus people to all over the world [and likewise our goods and pollutions]. That is the only method we have found of solving this problem. We dispose of radioactive waste by diluting it or by diverting it somewhere else. But unfortunately there is ever-less water to dilute it in, and ever-less other places to divert it to. When a problem becomes global, export becomes impossible, and we are now reaching the situation where the possibility of exporting problems when they were local, is over. The problems are now global—we can no longer export them, but have to deal with them ourselves. I really believe that it is not just on a national level that we have got to solve our problems, but even at a village level.

To solve problems at the local level means that you have got to realize that you are responsible for them, and will have to pay the consequences of not solving them. If you overgraze your land, erode your soil, or produce too many children, you have got to [risk starvation]. This is the fact we have all got to face. The time is over for permissiveness, liberalism, do-gooders, people weeping over individual problems—that is over. If you are going to solve any of these problems at all, you have got to be tough. There is no alternative to this. You go to a village in India and say, 'Here is your problem: you've got so many people, so much land, you've got to solve it yourselves because nobody's going to do it for you. The only way to solve it is to look after your land—not to overgraze it, not to over-crop it—and to control your population by whatever means you decide. It's no longer a question of something that fits in to your local culture or cultural pattern'.

The sooner people [get tough and] do this, the sooner are we going to solve our problems; at the moment, not only do we not solve them, often we do not even know they exist. The chief scientific adviser to the British Government told Mr Heath when he was Prime Minister that Britain had no population problem, [and this fallacy has been repeated by government spokesmen since. If] Scotland breaks away from England there will be about 1,000 people living in each remaining square mile, making England one of the most grossly overpopulated countries in the world. Yet the Chief Scientist, [originally exported from South Africa, and now a Lord, seemed] incapable of realizing that Britain was overpopulated. Why? [Perhaps] because he [presumed that] the population could continue to be exported elsewhere; and he thought the problem could be exported, he [may not have realized] that it even existed!

So it is time to take our responsibilities seriously, to stop exporting problems, but to deal with them at the smallest possible level—that of the village. [Applause.]

Malone

In response to Mr Guppy's request for more germane statistics, I wonder if he would see merit in the proposal of Professor Emile Benoit, of Columbia University, that we should develop a new concept of economic productivity which would be based, not on increased output per unit of labour and/or capital input, but on increased output per unit of environmental damage expressed in terms of pollution and/or utilization of non-renewable resources. This could take into account various negative externalities that are neglected in our present productivity concept.*

Guppy (Panellist)

[Although Professor Benoit's proposal] would involve a very good shift of emphasis, I think it would still not solve the prime cause of the problem which is that, in our production, we are always trying to increase efficiency and use better methods of pollution control. [This is all very well in its way; but it] is a capital-intensive approach that does not get to the human roots of the problem, and [involves] treating human beings as a sort of surplus, to be discarded if possible. What makes the world population problem most important is that, because of it, people are regarded as a sort of waste matter, which somehow has to be fed—darn it!—instead of being used as fuel or dumped in the ocean in sealed containers, or something rational like that! But these millions around the world's great cities are in fact living on *surpluses*, produced by a small part of the population which is very highly productive because of modern methods. And my feeling is that a lot of environmental problems [stem] from the fact that we are using just such methods—which are often short-term-cheap but long-term-dear—because they devastate the environment [sooner or later].

In advocating an increase in window-box technology, [my hope would be] of just getting ordinary humans to be productive, and finding ways of persuading them that there is growth to be made by themselves—actually in their own countryside—which is important. Of course the possibility of doing this varies from place to place, but in the wet tropics very much could be done on these lines, [instead of always trying to solve problems of production at a heavy-industry level or the level of very large-scale agriculture, and devastating the environment in consequence].

In South-East Asia very large populations, often of what we might consider an urban density, live and are supported in [quite local] agricultural regions where soil and conditions are suitable. Their life-styles may not be luxurious, but they have evolved agricultural skills and social systems and satisfactions which the rest of the world could emulate, though the way in which they are handled would vary from place to place. I think it is a question in the first instance of education, because a South-East Asian agricultural productivity could not be created, say, in the relevant parts of Amazonia, as the local people there have not got the traditional skills of knowing how to handle land. So in such cases as Amazonia you probably do in fact need large-scale governmental intervention to create, at one go, the sort of irrigation systems that have been built up in Asia over centuries [or even millennia]. In a sense this makes the Amazonian-type problems easier, when once they have been properly analysed [and tackled], because they will resolve into quite large projects.

* Professor Benoit, having been unable to attend the 2nd ICEF on its revised dates, sent us a prepared statement on this theme. Unfortunately, shortness of time in the Conference and of space in these Proceedings, have prevented us from using it, but we hope to publish it as a 'short communications' in a forthcoming issue of *Environmental Conservation*.–Ed.

The danger is that it is all too easy for accountants to handle [large-scale] projects which cost for example $5m and employ 200 people, but that is usually not the best way to spend those $5m [—especially when you have thousands of people standing around and needing to be occupied as well as fed]. We need more accountants to start looking at what individuals require in order to increase their self-sufficiency—which might be just simple implements or small parcels of land, that they could buy if lent a mere $200 each. If we could do more of this sort of thing, we could bring hope to the countrysides within existing social systems, and stop the flow of workless people into the cities—which [flow] is the crucial danger-area of the population explosion. I think this would do more to minimize social disintegration and damage to the ecosystem [—at least if it could be coupled with education and local forest conservation and utilization—] than any setting of firmer controls on existing high-exploitative methods, which are usually aimed at supplying export needs. Thus instead of large-scale clearance of forests we would develop further the existing zones of productivity.

Malone
I am sure you realize that Professor Benoit's argument would support your window-box economy.

Thorhaug
I direct this question to Stanley [Johnson]. From a biological point of view it seems to me that what Man has done is uncouple his natural system by this technology which we call 'medicine'—first by increasing the birth-rate and then by improving infant [survival]. A 'natural' species would have done this over a very long period of time, which would probably have resulted in an increased life-span and an increase in the age of sexual maturity. The way that most people have dealt with this has been to bring in yet more medicine [in the form of] some kind of birth-control method, such as pills or something else 'unnatural' to our system. It appears to me more ecologically sound to deal with population control in the Chinese way, which is to extend the age of the parents at the birth of the first child. Now in Scandinavia, [where you indicate there tends to be the longest life-expectancy in the world] they have been doing this for many years, as also in Germany and other countries. This seems to me much less likely to upset the social morality of the village tradition: by extending the age of the parents at the birth of the first child, one could still have almost the same number of children, but the result in terms of the number of people alive at any one period on Earth (which is really what we are [concerned with]), would change dramatically.

A very simple biological example to illustrate this is an Indian in South America, where the first child comes at 13. If the first child came instead at 26, you can see that in a couple of generations there would be a dramatic [decrease in the total] number of children. [This could benefit the ecosystem, but in any case the exportation of modern] medicine to non-European settings introduces a powerful new device into the ecosystem. In human, other animal, or plant systems, when you introduce a new device, there is the possibility of uncoupling the whole ecosystem, and I am afraid that is what we have done with our exportation of medicine—especially without exporting the other side of the problem simultaneously, namely family planning and contraception.

Johnson (Keynoter)
What you say is absolutely right. Demographically speaking the average length of the generation is crucial. At the moment we have an average generation-length of around 22 years or something like that (which means the average age at which a

woman has her children). Certainly, if you push that up, you can make appreciable differences to the rate at which a population grows. But I very much doubt whether that by itself will be enough, without bringing down the total completed-family size as well.

Juel-Jensen (Panellist)

What Messrs Guppy and Goldsmith said I think is [also] absolutely right. You want not only appropriate technology, you also want appropriate medicine, and this I would underline in reply to the last speaker, because your country, Ma'am, as mine, has an awful lot to answer for. We have exported an awful lot of rubbish, and this includes quite irresponsible elimination of diseases without knowing or realizing what the consequences would be—just because, intellectually, for a medical man, it is terribly satisfying to get rid of, say, smallpox or plague or something like that. But the whole pattern and meshwork is very delicate. Medicine has much to answer for, because it has upset that delicate balance. Yet it is not only medicine, but also the United States and England and Western Europe, that have much to answer for. The kind of life that the majority of people lead in the States, and in England [and widely elsewhere], is not necessarily enviable. [Personally] I would rather live in a poorer community in the highlands of Africa.

Worthington

I do not often find myself in full agreement with my friend Goldsmith [laughter], but I would like to comment on his thesis of [local solutions to problems]. It is perfectly true that the world is much too large a unit to consider in relation to population dynamics, and I personally like thinking about islands as microcosms of the world, so to speak. If you look at islands, you will see a lot of interesting things, [biologically and otherwise]. Consider this island we are on, described as being on the margins of human habitability. Here we have a population which looks to me pretty much in balance with its natural resources, [at least provided the foreign] trawlers really keep their distance. Look at Malta, which is comparable in some respects; it provides another example of accomplishment in dealing with the population problem, as far as one can see. The instance is well documented, because it was a British colony for so long, and we are a great people for counting heads—so that one knows of the big increase in population, and then of its levelling off in spite of Malta being a Catholic island. It has balanced its population against its resources. Now some jokers say this has happened in the same period as the withdrawal of the British Navy [laughter]. But as [an earlier] commentator said in this discussion, it is not the man but the woman who controls the babies [laughter].

Now look at two other examples—Singapore and Hong Kong. Both of them are on the way to controlling their populations, which have increased to the limits of their urban resources. And look at the UK, which has exported so much rubbish; I do not quite include the doctors among the rubbish, by the way [laughter]. But as an island, Britain is now rubbing up against the limits of its resources, with the population levelling off after an explosion during the last century. It can no longer export its human rubbish, let alone other rubbish, and now at last we see its population [in incipient decline, yet with no room for further importation]. We could go on with a lot of other islands which have sorted out their problems locally; but could the whole world follow suit?

Kuenen (Panellist)

Just one remark about Dr Thorhaug's comment concerning the biological aspect of all these matters: I do not want to be facetious, but woman is not a machine to produce babies. One point is that you must be very careful not to postpone the age of the mother at the birth of her first child until too late, because [we are informed] that, as the age goes up over 25, the first baby is in increasing danger of not being as perfect as it would be at an earlier age of the mother. So I am a bit afraid of postponing the first birth for too long, although I must of course agree that from the quantitative aspect it is a very efficient method. But I think the total process of reproduction should be considered, which includes the men.

Schütt

Sweden is a very small nation; out of each 1,000 people [in the world] you meet only two Swedes, but you will meet both women and men. I find it hard to understand whether this [vital matter of population] is a national problem—a problem in the small—or a global problem, and so whether the solution should be sought in the small or in the large. But of course [our wise colleagues are] right when [they tell us that] if we do not know the solution in the large, we should try to force a solution in the small, because if we wait and get no solution in the large, the fall will be much greater.

Gaekwad (Chairman, concluding Session)

I think we have had a lively and absorbing Session, and must thank you all for bearing with the Chair and obeying the Chair. There is not time for more than a few comments, but I think Goldsmith got me a little mixed up on the [topic of] global responsibility. What I was really referring to—and I entirely agree with him that charity must begin at home—is that, because a country does not face a problem at a given stage, it does not mean they must not think ahead of it. That is what I meant by global responsibility. I was not suggesting that rich nations must fund a programme of the developing nations with population problems.

About this question of the feminine angle, I think we must all write to the compilers of the *Oxford* [*English*] *Dictionary*. Because the definition in the old days when we said 'man' included woman, the spelling of 'woman' includes 'man'. I think we should change that. [Laughter.]

Professor Kuenen talked about delaying the birth of the first child,* but this has created great social problems in India. If the daughter of a family is not married by the time she is 17 or 18, it is then difficult for the parents to find a husband for her. Secondly, it becomes an economic problem, because they have to support an unmarried daughter. These and many more are local problems. But what is good for the States need not necessarily be good for India, and [we can have different ways of tackling] these problems arising out of what we have concluded here today—that population control is a necessity. Let each country then deal with their problem independently, because local conditions I feel must be taken into consideration before evolving a national policy—not an international policy. There [can surely be] no two opinions on this.

Finally an announcement: the film on Iceland will be shown at 6.30 right here this evening, and thank you again for bearing with me. [Applause.]

* To us it seems a pity that he did not also talk about what in an editorial in *Environmental Conservation* in 1975 we called 'The Beirut Syndrome'—of people killing one another in increasing numbers in various parts of the world, which he himself had expressed the opinion would suffice in time to mitigate the human population explosion.—Ed.

Session 7

DANGEROUS DEVEGETATION AND MONOCULTURES

Keynote Paper

Dangerous Devegetation and Monocultures

by

STURLA FRIDRIKSSON

Head, Agronomy Department, Agricultural Research Institute, Keldnaholt, Reykjavik 110, Iceland; Chairman of the Genetical Committee of the University of Iceland, Ingolfsstraeti 5, Reykjavik, Iceland.

INTRODUCTION

The vegetative composition and cover of our Earth has been extremely varied ever since plant life began to colonize and conquer the dry land several hundred million years ago. Plant species evolved with varying genetic structures, composition, and morphological complexity, and species of terrestrial plants have expanded over a wide range of topography. The habitats thus acquired extend frm coastal lowland to the summits of high mountains, and the plants have colonized a great diversity of substrates, forming the main basis of a series of successions and various types of communities: swamps and grassland, brush and forests.

But the Earth is not a stable system; volcanoes erupt, spreading lava and ashes over extensive areas, while its crustal plates have drifted apart, separating continents, the land has subsided, or high mountains have been elevated. The landscape has also been altered by the eroding action of temperature changes and the effects of wind and water (Fig. 1). Rocks have disintegrated and the mineral particles have undergone multiple cycles of sedimentation and erosion.

Sunlight has reached the surface of the Earth in varying amounts, and the atmosphere has been unstable in its composition of gases and its offering of water. The Earth has undergone drastic climatic changes which have had major effects on the global range of vegetation. There has been intensive plant production during hot and moist periods, whereas the vegetation has dwindled during dry and cold periods. An elevation of the ocean level as a result of the melting of polar ice has submerged large areas of swamps and coastal vegetation and has separated land-masses.

It is obvious that during these different epochs of flourishing and diminishing vegetation there have also been great differences in the availability of food for herbivorous animals, which were reflected in the sizes and numbers of these animals. Although coal deposits seem to indicate a surplus of plant material during the Carboniferous period, later, during glacial periods, food was scarce for larger animals and many species became extinct.

This brief historical account of vegetative development should indicate its extreme variability and demonstrate that Man cannot look upon the Earth's vegetation as a stable phenomenon.

Many of the natural changes acting upon vegetation are, however, so exceedingly slow in comparison with Man's relatively short life-span, that they may go unobserved; but it should be remembered that such changes are constantly taking place, irrespective of Man's existence and activities. It seems probable that the majority of the species ever formed were exterminated by natural forces long before Man appeared on Earth. Yet even erosion must not be regarded as a totally destructive factor influenced by Man; on the contrary, it has been the cause of formation of the richest areas of cultivation on Earth, and is thus a natural phenomenon countering the uplift of mountains and upbuilding by volcanoes.

It is now fully recognized that during Man's relatively short existence he has been able to accelerate the extinction-rate of both plant and animal species and also alter his environment more drastically than any previous species on Earth. It is moreover becoming increasingly apparent that Man has the power to change life on Earth quite markedly, and make irreversible alterations in the world's vegetation (Wagner *et al.*, 1973).

Influence on Various Biomes

Man's modification of the environment began with his ability to use fire to burn underbrush and forest. This he did for the purposes of clearing his neighbourhood and maintaining good hunting facilities, for increasing the areas of grassland and improving pasture for his livestock, for obtaining a more fertile soil for sowing, or even as a device of war. Thus the natural forests have, increasingly, been severely reduced.

When Man began to use timber for construction, tools and firewood, the forested areas were still further encroached upon. With the destruction of the forest there has been a spectacular increase in soil erosion in many parts of the world, which has accelerated the extinction-rate of many plant and animal species that previously occupied the forested areas. Allow me to deal with some examples as follows:

(1) The mid-latitude deciduous forests grew in an area of variable seasonal climate, with abundant rainfall, moderate temperatures, and soil of relatively high nutritive value. The diverse topography offered suitable

Fig. 1. Erosion in central Iceland. The grazing has gone and, with it, all semblance of fertility. (Photo: Sturla Fridriksson.)

Fig. 2. Large-scale tree girdling has destroyed wide areas of *Nothofagus* forests in Tierra del Fuego, Argentina. (Photo: Sturla Fridriksson.)

conditions for a large number of species and a wide range of complex communities. With the destruction of wide forest areas, many of the most sensitive species disappeared or became extremely rare, as the remaining forest was often not sufficient to protect animals from the hunter or to support the necessary maintenance-population. As an example, it may be mentioned that the Central European forests have been drastically reduced during the last 1,000 years, with the consequence of major changes in flora and fauna, but with improvements in agriculture which have made possible the support of a large human population. The deciduous forests of other continents have been similarly treated (Darby, 1956)—cf. Fig. 2.

(2) As a general rule the coniferous forests are not easily converted into arable land and have thus not suffered as much as deciduous forests from Man's activities. In these coniferous forests, however, Man finds the greatest source of lumber, which is in heavy demand for the rapidly expanding human population. These forests are therefore in increasing danger of being overexploited (Børset, 1976).

(3) Tropical forests constitute the most complex ecosystems, containing thousands of species of animals and plants, with year-around sunlight and abundance of warmth and rain. Growth is thus rapid, but many of the soils are thin and can be easily destroyed.

In tropical forests, Man has practised shifting cultivation, for which the trees in a strip of forest are burned to prepare a seed-bed for planting various crops. Such areas are then cultivated for a few years before the people move to a newly-cleared strip, abandoning the cultivated tract for a length of time before the cycle is repeated. However, if there has been an intensive burning, the soil may gradually deteriorate, and the strip may not revert to forest but be converted into low-productive grassland or worse. This has been the fate of forests in many tropical areas, where up to 40% has been lost and, with them, numerous species have disappeared (e.g. Bartell, 1956; Polunin, 1960; Ehrlich *et al*., 1977).

(4) The grassland and savanna, which are biomes of regions with rather low rainfall, have been highly valuable to Man and provided pastures for grazing animals. The grassland has, however, often been severely mis-treated. In the native condition there was harmony between the available grass production and the herbivores, but with the advance of Man's agricultural activities, a heavy load of livestock has often overexploited vast grassland areas. Sheep and goats have been especially destructive, depriving the vegetation of the most sensitive species and leaving behind an open sward with unpalatable shrubs—indeed often turning the land into desert (Fig. 1).

In Neolithic times, Man started to cultivate the fields for grain by preparing a seed-bed for sowing. This was usually repeated at the same site, using a single crop or monoculture. The techniques of ploughing, seeding and harvesting have been practised ever since. In the cultivation of a single annual crop, the land must be ploughed regularly and carefully weeded. This leaves the soil unprotected during part of the year, and thus it is very susceptible to erosion. In Europe and Asia, the cultivation has been most widely of wheat and barley. In the New World the monocultures have also included maize, cotton and tobacco, which have had a rather disastrous effect on the soil in some districts of the United States.

Marginal Lands

Although Man has caused great damage to the more densely vegetated tracts, changed vast forest areas into cultivated land, and exploited the grasslands, the greatest damage may have been done to 'marginal' lands, where the conditions for growth are such that the ecosystems are highly vulnerable. There has been a steady creep of cultivation upwards—for example in the mountains of the Himalayas, the Andes and the East African highlands—forced by population pressures from the bordering farmed regions. The hillside vegetation has been cleared and the heavy rains can easily wash the soil down the slopes, increasing silt and causing severe floods in rivers of the lowlands. Similarly, the marginal lands bordering the deserts and the polar regions are badly affected by overgrazing and/or acceleration of deforestation.

In Patagonia and southern Chile, large-scale wood-burning and girdling of trees has destroyed wide forest areas when the land has been cleared for sheep-grazing, There were few grass species to replace the trees, and the sward was unprotected and open to the eroding effects of wind and water. In these districts the return of the forest is exceedingly slow, and complete removal of the woods will therefore have serious consequences for the soil and remaining vegetation (Fig. 2).

Similarly, there has been a major change in the native ecosystems in Iceland since Man discovered the country and settled here some 11 centuries ago. It is now estimated that half of the original vegetation cover has been lost, mainly due to intensive sheep-grazing (Fig. 3). The rate of deforestation in Iceland has been faster than in other European countries, and its effects more serious, due to rapid leaching of the porous volcanic soil and the vigorous erosion effect of the harsh North Atlantic climate (Fig. 4).

As the Icelanders were extremely isolated until the close of the nineteenth century, they had to rely to a great extent on the produce of the land. The vegetation imposed effective limits on the number of inhabitants who could live in the country, and thus the size of the human population declined in accordance with the deterioration of the land. It was becoming critically small when modern agricultural techniques and the use of fertilizers, as well as increased fishing, reversed the trend towards modern prosperity (Fig. 5). But fortunately it is realized that the natural resources and space for cultivation are limited. Man's position in such a closed 'island ecosystem' can in many respects be compared with Man's status on Earth (Fridriksson, 1972).

So far, the tundra vegetation of the arctic regions farther north has been only slightly altered by Man's activities; but with increasing human population and its need of natural resources in northern territories, the life of the tundra will be endangered. Increased traffic, or any other injury to the insulating top layers of the soil, will cause thawing of permafrost and upset the natural balance between water and vegetation. And the damage may be

Fig. 3. Former pasture from which about 60 cm of rich lava-derived topsoil with the entire grassy sward has been eroded almost everywhere as a consequence of overgrazing by sheep and subsequent erosion. (Photo: George A. Petrides, Michigan State University.)

Fig. 4. Soil erosion near Bláfell in southern-central Iceland, with typical landscape in background. The boggy depression in the middle-distance has probably been enriched by soil washed down from the eroded gully in the foreground. (Photo: Sturla Fridriksson.)

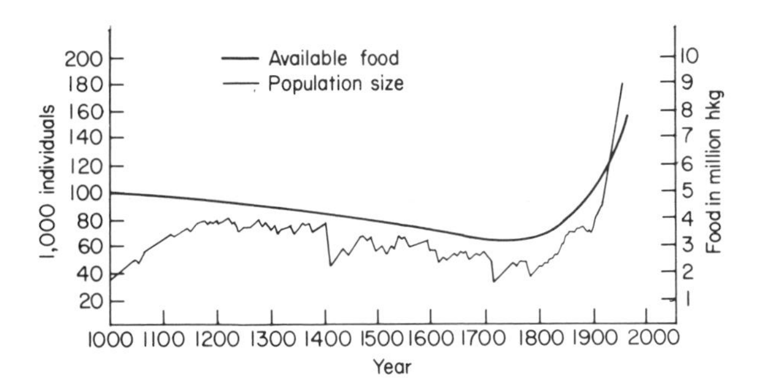

Fig. 5. Graph showing trends in the human population of Iceland compared with the amount of available food at various periods from the time of settlement (AD 874–930) to recent. (After Fridriksson, 1972.) The present population is about 218,000 (President Eldjárn, *voce*).

even more serious than in milder climates, as the recovery from any ecological disturbance in the tundra is extremely slow (Committee on Geological Sciences, 1972; Rickard & Brown, 1974).

Misuse of Vegetation

When Man started cultivating plants and keeping livestock, he began to affect Nature's ecological equilibrium on a large scale. He transferred both plants and animals from their native habitats into new environments. This enabled him to control his food supply and to establish food sources practically throughout the Earth; but by doing so, he often drastically upset the prevailing homeostasis with new intrusions.

In distributing his useful vegetables and crop-plants, Man has also been a distributor of weeds, and although much effort and expense have gone into controlling their spread, many have become a great nuisance in their new environments.

Man has introduced livestock into new territories where they have competed with local animals, and so new pests and diseases have been brought in to native populations. Man has thus diminished the livelihood for the native animals and even exterminated some of them. He has drastically altered much wild vegetation by letting his livestock graze it too heavily, and thus he has exhausted the soil's fertility (Fig. 3). However, his intensive clearing of forests and cultivation of fields for open-tilled crops has probably posed the greatest threat of all to the natural environment (Fig. 2).

By practising monoculture and using only one or two crop species in farming extensive areas of land, Man has caused tremendous changes in the composition of local plant communities and their dependent ecosystems practically all over the world. The old communities, with their diverse plants and animals, have been wiped out very widely. The new crops are attacked by pests such as specific insects that may be the food of particular predators; in these and other ways the uniformity in crops also simplifies the whole web of life. There is always the great risk in monoculture that the gene-pool of the crop species concerned may become reduced so markedly that the species may lose its adaptability to changing environment, and it may thus be completely wiped out by droughts, temperature changes, or wild competitors.

Similarly, such monoculture crops provide ideal conditions for attacks of pests and disease epidemics. A case to remember is the potato blight in Ireland, caused by *Phytophthora infestans*, which wiped out the crop in two successive years and caused widespread depopulation by death and emigration in the 1840s. And every year there is a tremendous reduction of wheat yield due to rust and other diseases in the USA and Europe, although fungicides are applied heavily. Such monoculture crops are also highly vulnerable to any attack by insects, as the balance provided by other organisms preying upon them has commonly been destroyed.

Furthermore, the minerals drawn from the soil by the single crop are relatively uniform and, with the uniformity of conditions of the rhizosphere imposed by the single crop-plant, tend to simplify the soil's microbiota. The application of fertilizers necessary to maintain the crops, and, in addition, extensive use of irrigation, often upset the nutritive balance in the soil. Salt may accumulate in irrigated soils, resulting in lowered crop-yields due to toxicity, which may have dramatic consequences. Thus as a result of such salinization, millions of hectares of irrigated land have been abandoned over the centuries (Eckholm, 1976). The deserted lands of ancient civilizations remind us that irrigation is not a constant benefactor to agriculture of the arid districts.

Man is today becoming increasingly aware of the Earth's limited space and precarious food supply for the expanding human population, and that all arable land is needed to produce the necessary crops for his sustenance. Man must, therefore, show a special responsibility towards the soil and use all his scientific and technical knowledge to secure its fertility with proper management and control. The problems of degradation of the land are no longer restricted to small farms or local communities. The cultivated areas that are deteriorating by bad management are so extensive that they are bound to concern us and ultimately affect all mankind.

Man is now recognizing these serious trends, and important actions have been taken at both the local and international levels to conserve the living world and its natural resources (Nicholson, 1970). The International Union for Conservation of Nature and Natural Resources (IUCN) was founded in 1948 for this purpose, and its sister organization, the World Wildlife Fund (WWF), was established in 1961 to arouse public awareness of Man's grave threat to Nature and to support conservation programmes financially. A step, taken by IUCN, in this direction is the publication of *Red Data Books* that list the endangered species. A recent survey of endangered plant species in Europe, for example, shows that there are 1,389 species of higher plants which are either rare or threatened on a world scale in that area alone (Lucas & Walters, MS.). It is also estimated that perhaps 10% of the world's flowering plants are in some way threatened (Lucas, 1976; Lucas & Synge, 1977), while some 50 species and subspecies of birds have become extinct during the past 300 years (Greenway, 1958).

It is evident that the intensive use of land and increased human population will limit the space that is available for many organisms, and will eliminate others. Monoculture has accelerated this extinction of dryland organisms. But the word monoculture might also be used in the broad sense for a uniform cultivation of fishes in lakes and rivers. Now, modern Man is on the verge of starting the management of marine flora and fauna, and he will have to ensure that the marine biota and surrounding ocean will not be abused to the same extent as terrestrial life and its environments have been.

Monoculture has encouraged the breeding of high-yielding strains, but at the same time intensive breeding has resulted in loss of genetic material, which may have originally existed in the ancestral races of domestic animals and cultivated plants. It has also left the pampered crops quite dangerously susceptible to pandemic attacks of diseases that could be quite devastating. From an aesthetic point of view, any extinction of life-forms is always regrettable. But it has repeatedly been demonstrated how difficult it is to predict what may be the ecological consequence of such a loss to the living world.

There are still large areas of temporarily-used lands on our Earth, and even some that are almost unexploited. These will be in increasing danger when more space is needed for food production to supply the rapidly expanding population. Today Man cultivates about 1.4 thousand million ha, or approximately 11% of the Earth's land surface. This is less than half of the surface that is considered potentially suitable for cultivation, which is estimated to be 3.2 thousand million ha (Brown *et al.*, 1975); however, the unused land is in general much poorer than that which is currently being cultivated. As natural resources—fossil fuels, fresh water, and minerals such as phosphorus—diminish, it becomes increasingly difficult to maintain a level of food production that keeps in step with the current rates of population growth. It seems, however, that much of the yet uncultivated arable land will have to be taken into use in the next 25 years. If it becomes necessary to double the food production from what it is today, a similar area of land must be taken into cultivation to that which has taken the whole history of Man to prepare (Brown *et al.*, 1975).

Is it inevitable that, in the near future, Earth must be dominated by the monotonous simplicity of short food-chains, with monocultures of a few essential crops and a monopoly of Man? Can we visualize vast areas of ocean filled with a single culture of Algae and wide tracts of land cultivated with a single crop of grain supplying food for future Man? A sounder and happier way of life for the human race would be to strive for control of its population size, and adjust this to a more moderate rate of food production—thus securing the existence of other living beings, with all their diversity, so that they also may share and enjoy the richness and beauty of our planet which should also be theirs.

[Allow me please to conclude by reciting a poem which I have composed for our Conference.]

The Old Web of Life

The Sun keeps sending light from outer space
and paints the mountain tops with golden stroke,
as lakes and rivers catch reflection rays
of landscape which has disappeared in smoke.

And down the mountain ran a river clear—
so fresh—and winding like a silver coil
which moistened trees, but now it has a smear
of frothy soap and greasy spill of oil.

And there the caterpillar used to creep
in search of fruit upon a cherry tree,
but at that moment it was put to sleep
and poisoned with a spray of DDT.

And thus the worm became an easy prey
and food for skylark with a tender voice,
that filled the air with music far away
but was not heard because of traffic noise.

A falcon caught the skylark during flight—
a skill that only birds of prey acquire—
but then itself an instant later died
by getting tangled in a rusty wire.

The falcon was the game of Man with mind,
who mastered Earth and all its wealth he spent;
but by his footprints later one may find
a pile of dust, a broken instrument.

References

BARTELL, H. H. (1956). Fire, primitive agriculture, and grazing in the tropics. Pp. 692–720 in *Man's Role in Changing the Face of the Earth* (Ed. W. L. Thomas, Jr), University of Chicago Press, Chicago, Ill.: xxxviii + 1193 pp., illustr.

BØRSET, O. (1976). Introduction of exotic trees and use of monocultures in boreal areas. Pp. 103–6 in *Man and the Boreal Forest* (Ed. C. O. Tomm). Swedish National Science Research Council, Ecological Bulletin No. 21, 153 pp., illustr.

BROWN, A. W. A., BYERLY, T. C., GIBBS, M. & PIETRO, A. SAN (Ed.) (1975). *Crop Productivity—Research Imperatives*. Conference sponsored by the Michigan Agricultural Experiment Station, East Lansing, Michigan, and the Charles F. Kettering Foundation, Yellow Springs, Ohio: viii + 399 pp.

COMMITTEE ON GEOLOGICAL SCIENCES (1972). *The Earth and Human Affairs*. Canfield Press, San Francisco, California: xii + 142 pp., illustr.

DARBY, H. C. (1956). The clearing of the woodland in Europe. Pp. 183–216 in *Man's Role in Changing the Face of the Earth* (Ed. W. L. Thomas, Jr). University of Chicago Press, Chicago, Ill.: xxxviii + 1193 pp., illustr.

ECKHOLM, E. P. (1976). *Losing Ground*. W. W. Norton, New York, NY: 223 pp.

EHRLICH, P. R., EHRLICH, A. H. & HOLDREN, J. P. (1977). *Ecoscience: Population, Resources, Environment*. W. H. Freeman, San Francisco, California: xv + 1051 pp., illustr.

FRIDRIKSSON, S. (1972). Grass and grass utilization in Iceland. *Ecology*, **53**(5), pp. 785–96, illustr.

GREENWAY, J. C., Jr (1958). *Extinct and Vanishing Birds of the World*. American Committee for International Wild Life Protection, Special Publication No. 13, New York, NY: 518 pp.

LUCAS, G. Ll. (1976). Conservation: Recent developments in international cooperation and legislation. Pp. 271–7 in *Conservation of Threatened Plants* (Eds J. B. Simmons, R. I. Beyer, P. E. Brandham, G. Ll. Lucas & V. T. H. Parry). Plenum Press, New York & London: xvi + 336 pp., illustr.

LUCAS, G. Ll. & SYNGE, A. H. M. (1977). The IUCN Threatened Plants Committee and its work throughout the world. *Environmental Conservation*, **4**(3), pp. 179–87.

LUCAS, G. Ll. & WALTERS, S. M. (M.S.). *List of Rare, Threatened and Endemic Plants for the Countries of Europe*. Compiled by the IUCN Threatened Plants Committee Secretariat, at the Royal Botanic Gardens, Kew, Surrey, England: [vii] + 97 + 166 pp., mimeogr.

NICHOLSON, E. M. (1970). *The Environmental Revolution*. McGraw-Hill, New York, NY: 366 pp., illustr.

POLUNIN, N. (1960). *Introduction to Plant Geography and Some Related Sciences*. Longmans, London, etc., and McGraw-Hill, New York: xix + 640 pp., illustr. [new edition in preparation].

RICKARD, W. E., Jr & BROWN, J. (1974). Effects of vehicles on arctic tundra. *Environmental Conservation*, **1**(1), pp. 55–62, illustr.

WAGNER, K. A., BAILEY, P. C., & CAMPBELL, G. H. (1973). *Under Siege*. Intext Educational Publishers, New York, NY: xii + 386 pp., illustr.

Discussion (Session 7)

Misra (Chairman, introducing)

Perhaps the most spectacular change of the face of the Earth that has been brought about by Man, and mostly within the [geologically very] short period of the last 200 years, has been the widespread removal of the natural plant cover from the land. This has been done, by and large, for growing crop-plants and trees in monoculture—mostly for short-term economic gain. Modern Man must be educated at all levels to realize the dire consequences of wiping out the natural flora and fauna and destroying the vegetal cover [on which life depends].

It is argued that the yields of organic matter from monoculture are in no way inferior to what one might expect from the primary productivity of the natural vegetation of any area, and accordingly that the state of gaseous balance of the atmosphere is not disturbed by widespread cultivation. But the modern technologies of agriculture and silviculture employ fertilizers, pesticides and herbicides, besides exposing the soil to erosion, to loss of nutrients and useful microbes, to desertification in semi-arid areas, etc.—which together give us diminishing returns for the investments made therein. These activities undoubtedly change our environment for the worse.

Management of our vital resources of land, water and vegetation presents an interlinked package programme. Devegetation and monocultural practices, on the other hand, threaten our existence much as does an armaments race, [though more insidiously]. Both of these activities need to be controlled, and the large investment made on them should be partly utilized in educating the public on these matters, through the science of ecology. [From within the wisdom here assembled] I hopefully look forward to the valuable suggestions which I can foresee being presented at this Session, and to recommendations which we may adopt at this Conference [—hopefully for the world's benefit].

Let me first introduce our keynote speaker Dr Sturla Fridriksson, though as a leader among Icelandic scientists [and author of the standard book on *Surtsey*, the volcanic island of substantial dimensions that rose out of the sea some years ago off the south coast of Iceland], he can scarcely need introduction to any of us. He has worked extensively on the reclamation of grasslands by breeding new varieties of grasses and also by fertilization in the volcanic areas of Iceland, and will [elaborate on some aspects of our vast topic of the grave dangers of] too much devegetation and monoculture of crops and plantations.

As panellists we have an obvious choice in Dr Fosberg, whom you already know. Next to him is Dr Tan Koonlin, from Malaysia, which is one of the countries that has suffered most from deforestation. I especially admire her knowledge of the agricultural geography of tropical lands, [and only regret that her distinguished husband, Professor Robert Stanton, is not able to be with us on these revised dates]. Next to Dr Tan we have [on the platform] Dr W. Ted Hinds, of the Ecosystems Department of the Battelle Pacific Northwest Laboratories, Richland, Washington [and formerly of our co-sponsor ERDA in Washington, DC], who has worked for many years on the vulnerable habitats of North America, and from whose experience of revegetation from monocultures we shall surely benefit [witness, for example, his exhibit alongside Dr Gossen's in the vestibule outside this theatre].

The panellists are asked to supplement the paper [by looking into the future as far as possible] and adding observations from their various viewpoints. After they have finished their 3–5 minutes each, the Session will be opened for general discussion. As for myself, whose privilege and pleasure it is to conduct this Session, I am Emeritus Professor of Botany at Banaras Hindu University, Varanasi, India, being a plant ecologist whose work has latterly been mainly with IBP and now MAB.

By way of introduction of our topic I should emphasize that agriculture, practically all over the world, has claimed, for growing farm crops, and these usually as monocultures, by far the largest total area of land which has yet been devegetated. Likewise, silviculture of timber species has replaced multispecies natural forests very widely during the past century. Besides these two monocultural practices, we find that mining, construction of roads and airports, human settlements and reservoirs have also claimed a heavy toll of natural landscapes. When we add to these yet other processes of devegetation such as arise from fire and war, it appears that no [major area of] landscape has remained untouched by Man; yet natural vegetation is among the most precious resources of Man. It undoubtedly moderates the extremes of climate and cushions the landscape against soil erosion and rapid runoff of water, whereas clear-felling and bulldozing of the land have undoubtedly caused landslides in hilly areas.

The scale of human operations in relation to space and time assumes great importance in the matter of devegetation. Further, the amount of vegetal cover and the period for which it is available in relation to the topography of the terrain, are most important factors in the cropping of herbs, shrubs and trees, or their replacement systems. Shifting cultivation in the lush forest vegetation of the tropics (as in India) did not disturb the ecosystems of the olden days when population pressure was low, as the rotational interval on the spot used to be of the order of 20–25 years and the area opened for raising crops was small, which altogether permitted regrowth of the forest. But nowadays this may happen every 2–3 years over increasingly wide areas without giving the wounds a chance to heal. This is resulting in bald hills all over the northeastern hill states of India and widely elsewhere. The slash-and-burn practice, 'Jhuming' as it is called, is the most devastating way in which the primitive tribes raise their crops on the forest-enriched soil—for 2–3 years during which time the soil is completely washed away by the torrential rains of the monsoon tropics.

Unfortunately village populations in much of the undulating countryside of India have started practising mechanized agriculture, growing the so-called improved [or 'miracle'] crop varieties which need heavy inputs of water, fertilizers, weedicides and pesticides, at tremendous costs of fossil fuels but with ever-diminishing returns. Such monocultures become vulnerable to diseases, floods and drought. Additionally, the resulting soil-erosion, -salinity and -alkalinity lay larger and larger areas to waste. This kind of desertification drives people to devegetate and cultivate whatever remaining land is available in their neighbourhood. Increases in population and rises in expectations add further to human misery.

What applies to the croplands is also true for silvicultural monoculture. *Pinus radiata* [Monterey Pine] crops raised at Kodaikanal by the British have withered on account of disease and failure to regenerate. *Gmelina arborea*, raised at tremendous cost in Bihar and Assam, was destroyed by outbreaks of pests. Very often the political exigencies set a chain-reaction towards these local ecodisasters.

Defoliation of the beautiful forests of Vietnam is a well-known case to be mentioned here. It seems that the more a country becomes economically better off, the more destructive of its poorer neighbours it tends to be. The contractors of Japan and USA have destroyed much of the Indonesian forests to replace them with plantations of coffee at the bidding of the native governments. And this was done in the face of the lessons that ought to have been learned from the earlier experience in Ethiopia.

Roughly 10% of our land is devegetated for construction or for road or rail and air transport, with human settlements now occupying a substantial proportion of fertile land in most parts of the world. And as we produce xeric conditions, irrigation channels and reservoirs become necessary—which in turn submerge further significant amounts of the [more and more limited available] land.

Recently a very rich coalfield has been discovered at Bina [and Singrauli] in northern India, where the forest is being devegetated on a large scale for operating the mines. Similarly, quarrying for stones, bauxite, iron ores, and other economically useful materials, is destroying virgin forests in many areas. Such competitive demands on the land are accelerating the process of desertification in modern times.

It is thus evident that devegetation and replacing monocultures lead among the world's ecodisasters. But how can we stop or minimize their effects, taking into account the need for economic development especially of the tropical countries? Can we make agriculture and silviculture obsolete? It appears that low-energy technology (such as microbial processing) for food production from natural products such as leaf-litter is the only answer. It has recently (1974) been estimated that the world's primary productivity per year amounts to 172×10^9 tonnes. If it could be suitably processed into food, only a fraction of this amount would feed the entire growing human population of the planet. Further, we are told that all the world's requirement of protein could be met by processing 1% of the fossil oils that are being consumed today. Does mere cultural inertia stand in our way of taking these steps?

It appears that a multipronged attack is necessary to stop devegetation and minimize the adverse effects of monoculture, for which it may be useful to consider the following ameliorative measures:

(1) Transform capital into income consumption.
(2) Change the life-style into one of conservative uses of goods and materials.
(3) Encourage institutional arrangements for monitoring of natural resources and maintaining consultancy services.
(4) Promote cottage industries and clean technology.
(5) Adopt a constructive approach towards rehabilitation of Nature in order to increase the resilience of ecosystems.
(6) Perform all the above on an appropriate scale of time and space, harmonizing quality of life and quality of environment.

Fridriksson (Keynoter, presenting paper)

I am honoured by the privilege of participating in this Conference which we all hope will stimulate better management of our environment and our natural resources. To begin with I wish to point out that my paper may to some extent not reflect the views of scientists working in highly productive parts of the world, but rather of the ones who are coping with the problems of Man's survival in a climatic area which by many is considered to be marginal by present-day standards.

As has already been mentioned at this Conference, our land is on the border of that which can be used for agriculture, and marginal ecosystems are extremely vulnerable and sensitive to any changes in climate, to unwise exploitation, to erosion, to pollution, and to overpopulation.

In earlier days it was generally believed that the Earth's resources were so great that Man could have little influence on them—that although he utilized some of the vegetation and [grew more] and fished, he would not be affecting the great wealth of life and its environment. Today we are aware that, with modern technology, Man's influences on ecosystems are apt to be so drastic that there is a tendency to blame the activities of Man for all natural changes in the environment. Indeed it is very difficult now to imagine the world without Man, and hence to visualize the changes that would be taking place on Earth in his absence, though we hope it will not come to that in the near future!

We must remember that various changes are constantly taking place irrespective of Man's present activities. There have been great variations in life-forms of both plants and animals and also of [dominance] during the [hundreds of] millions of years since life started to evolve and colonize the Earth, while habitats have [multiplied and ecosystems have] increased in complexity or have deteriorated. Species have evolved and multiplied, but most of these were exterminated by natural forces long before Man appeared. There was intensive plant production during hot and moist periods such as the Carboniferous, whereas the vegetation dwindled during drier and colder glacial and other periods, when food became scarce for larger animals and many species became extinct. The landscape has also been changed drastically by volcanic activities, uplifting of mountains and variation in ocean levels—to mention three of the main agents of geomorphological change. In addition, rocks have been disintegrated by temperature changes, while the effects of wind and water and mineral particles have undergone multiple cycles of sedimentation and erosion—which accordingly must not be regarded as a totally destructive factor engendered by Man.

Fosberg (Panellist)

There is very little that one can quarrel with in the speaker's paper. It is a very sound assessment, and all I can do is perhaps add a little bit from my own experience and knowledge. Most field-crops belong to pioneer species (in the successional sense) and require open habitats. This is necessarily so, but does introduce many problems—such as weeds, erosion and so forth. This pioneer culture is not appropriate in some climatic areas and regions—for example, temperate agricultural practices introduced into the tropics have created problems and introduced instability. Monocultures have inherent susceptibilities to various pests, diseases, mineral deficiencies and man-induced [debilities] that are due to interference with nutrient balances, as was pointed out in a way by our speaker. Therefore they require extraordinary care and attention from their cultivators. Also, pioneer monocultures have an inherent susceptibility to erosion and to such soil degradation as is brought about by exposure to sunlight. Moreover, run-off is likely to be increased and ground-water bodies are accordingly less well replenished than hitherto. These pioneer monocultured crops are also frequently shallow-rooted and in rainy climates the nutrients may be leached down, so that they accumulate below the reach of the roots of the crop plants.

In very rainy areas of the tropical and temperate regions, this loss of nutrients may be at least partly offset by planting tree crops. However, tree-crop monocultures are even more vulnerable to pests and diseases [than are most herbaceous crops]. For where life-cycles and cultural cycles are long, it is not easy to control the troubles by rotation or fallowing. In the tropics and also in many temperate areas, the solution to some of these problems may be crop diversification of several sorts. Thus in wet tropical areas I have seen, for example in [Sri Lanka], a highly stable and highly productive culture which I call tree gardens, where the interfluves, in a mosaic with wetland rice-culture, are covered by a mixture of food-producing and other useful trees growing around the houses of the rice cultivators. They have coconuts, breadfruits, jackfruits, cassava, betel-pepper, areca palms, caryota, mangoes, bamboos, cashews, etc.

In temperate regions such as Europe and the eastern United States, for example, at least before the era of agro-business (agriculture employing big machines over vast areas), the diversification was effected by meadows and a fine mosaic of plots or small fields of different crops, which may be rotated in a short-term régime. What is accomplished in either of these systems is to increase the stability of the system and to lessen its susceptibility to disasters of various sorts—such as disease, insects, climatic vicissitudes, soil depletion, etc. Chemical cycles are made more complex. The erosion and leaching effects of rainfall are lessened, and the uncertainties of the food-supply of the people are reduced. In the case of the Ceylonese example described, the mosaic of rice and tree culture gives the added advantage that nutrients which are washed out of the interfluves are caught in the marshes and utilized by the rice crops. This complex seems to approach the ideal of a permanent agriculture about as well as any scheme that I have seen in operation. It is based on an effective system of recycling of nutrients and very efficient use of water.

I would like to add one more thing that was brought to mind by the reference to shifting agriculture in tropical forests. I think we must distinguish in such a reference between upland and lowland tropical forests. I think that the shifting agriculture probably originated in tropical uplands and there is at least a body of thought which maintains that it is not a bad system. In fact some geographers maintain that it may be the only possible successful system in a tropical upland. That is open to argument, but there is certainly something to be said for it. However, in the tropical lowlands, where the soil is almost everywhere extremely poor, there is practically nothing to be said in favour of shifting agriculture. You can bring a new area of tropical rain-forest under cultivation and depend on getting a crop or even sometimes two crops, but after that you are finished. The total nutrient supply in the lowland tropical forest has been shown to be up to at least 70% contained in the vegetation itself—in the plant biomass. You cut and burn that, or else cut it and remove it for timber, and the nutrients which are left are rendered soluble, so that the first good hard rain that comes along washes them away. You can catch enough for the first crop and then you are finished. And I believe that such dreamy schemes as bringing the Amazon region under cultivation and so forth should definitely take this fact into account, and think the whole thing through before deforesting the Amazon basin and creating a desert or weed-patch.

Tan (Panellist)

I have noticed that throughout this Conference whenever speakers refer to the tropical lands they invariably think only of the rain-forest. As we all are very much aware, there is a great deal of deforestation, and this happens not only because we need to deforest for our own development but also because of the insatiable demand by the developed world for our products. Malaya [where I come from]—not Malaysia but Malaya, the peninsula that juts out down from Thailand—was the world's largest producer of rubber and palm-oil although we are only about 20% larger than Iceland. Now of course we are joined by Sarawak and Sabah and have become also the world's largest producer of tropical hardwoods, while from time to time we are also the world's largest producer of tin. This is quite a remarkable achievement but you should realize what it has done to our country, in that large areas are being converted to various forms of grassland (often dominated by the pernicious lalang or Spear-grass, *Imperata cylindrica*) [many of] which are unstable in our humid climate.

I also draw your attention to another form of ecological system which is still very much in existence in Southeast Asia, and is still relatively little-known and under-exploited, namely the wetlands, or what we call swamplands. Professor Kuenen called them bothersome areas—neither land nor water—and indeed they do not conform to what most people [think of as part of] the tropical environment. The swamplands of Southeast Asia are I think just as representative of the tropical environment as the rain-forest, and yet little has been done about them as I say because we just do not know what to do with them or how to exploit them without drastic alteration of their ecology—despite their [potential] productivity. So far it has been in other parts of the world that swamplands have been considered productive, though still chiefly in terms of animal life—particularly to attract and feed the birds and waterfowl. But in fact in Southeast Asia we have identified such areas which are relatively productive in terms of supporting fairly high densities of rural population, requiring no undue modifications of their natural state to provide sustained productivity of a relatively high order. These are areas where the concept of conservative agriculture can be fruitfully applied for economic development.

In one respect peat swampland is very distinctive in that the dominant vegetation is exploited in a manner which seems to contradict one of the themes of this Session. I would describe it as a form of monocultural silviculture which is a unique departure from the usual dryland agricultural systems that we are familiar with. These areas have sustained coastal and fluvial communities for centuries—perhaps even for over a millennium in Southeast Asia—and we have been delightfully surprised to find some fairly sophisticated cultures and village technologies still existing there. Last year we organized a symposium in Kuching, Sarawak—the First International Sago Symposium, of which the proceedings are being [printed] by the University of Malaya. These cultures exploit trees as the source of their staple food—namely the starch-bearing palms of many species including some that are collectively known as sago palms. They are very much underexploited [and would seem to represent useful] food and commercial crops for the future.

Hinds (Panellist)

Dr Fridriksson's paper gave an excellent perspective, but he was very modest about Iceland's considerable understanding of what to do with a devegetated landscape. Such landscapes can be a source of exasperation to a plant ecologist, and what to do about them has a long history. So I suggest that for the next two or three minutes I outline a few of the ecological principles which we need to keep in mind when walking over cobbles and sand that ought to be grass and flowers.

Revegetating a landscape requires the making of two choices. First you have to decide whether you are going to [aim at] an immediate or a long-term landscape stability. This is not a simple decision, and it is beyond the scope of this panel's discussion. It is an ethical problem, among other things, and has many ramifications. Maybe we will be able to come back to that. Secondly, revegetation requires you to decide whether you are going to use native species or exotic species. The search for a magic plant that can be broadcast indiscriminately, has occupied for a long time agronomists and others who are involved in revegetating landscapes. But it will not work that way. One of the things that an ecologist finds pretty universally true is that it is easier to modify the habitat to satisfy the species that are there, than to develop new varieties: in other words, take advantage of the physiologies that are already available.

Monocultures, as Dr Fosberg mentioned, have problems of their own, and as monocultures are particularly inviting as an immediate revegetation approach, you should keep these problems in mind. This is particularly true as revegetation with monocultures has a long history of using annual species which compound the problem [already at the end of their first year].

Let me bring up for your consideration four generalized guidelines that we ought to bear in mind when we consider a devegetated landscape.

First, I think we should recognize that revegetation is an agronomic exercise, not an act of ecological penance. We cannot replace what was there before. [Nature may be able to restore her] ecosystems but we cannot [do more than help her].

Second, revegetation probably should not begin until the proper handling of the water-balance has been assured. The requirements will vary in different circumstances, of course, depending upon whether you are working for example with slight slopes and deep soil, or rocky slopes, or bogs with permafrost. But without prior control of the water-balance you are in trouble already.

Third, revegetation probably should not depend on monospecific plantings even within a single life-form—[for reasons that have already been made clear and yet others which could be enlarged on].

Fourth, and last, revegetation probably should not rely principally on exotic species. But although these guidelines are almost platitudinous, they are not yet reflected in many legislated requirements.

The topic this afternoon is certainly one with a long history, but I think we can profitably generalize by saying that we now know far more about it than we [actually use in practice]. The reason for [this reluctance to apply ecological understanding] might very well be a central issue of this whole Conference, namely that we are involved in a long-term process but we think as a short-term species.*

Bergthorsson

I hope you do not mind if I draw your attention to an ecological problem of the country you are staying in during this week. It is the problem of growing grass [under conditions of] climatic variations which are particularly strong due to the [proximity] of the polar ice. The sunshine is relatively strong today and the day is long, and you have probably noted that the grass is green and rich—which grass is one of our main natural resources. But the temperature was below freezing-point in most of the country this morning and here in Reykjavik the minimum temperature of the grass was 5.5 degrees below freezing, while it was snowing in the North. The situation is, however, not bad this year; but you can see that a rather slight cooling of the climate could be important. It is impossible to discuss this matter in any detail in a few minutes, so I have taken the opportunity to leave on a table [outside this] hall some copies of an Icelandic farmers' periodical containing a paper on this problem with a summary in English. I would, however, like to mention briefly a few points from this paper.

First the annual temperature, tabulated for [the past] 75 years, shows a sudden warming after 1920, then a stable and warm climate until 1965, as mentioned by Professor Flohn, and after that a cold period. This emphasizes the fact which Professor Bryson and others have brought into the light, namely that the long-term variations of climate are surprisingly great compared with the variations from one year to another.

Secondly, you can see in the tables the figures for the annual hay-yield per hectare for the first 75 years of the century and observe a very close correlation between this hay-yield and the temperature. At the same time the hay-yield is of course closely related to the fertilizer applied per hectare, which is also tabulated for every year. A [simple] mathematical model makes it possible to explain more than 90% of the variation in the hay-yield, assuming it to depend only on temperature and fertilizer. In this cold climate the precipitation seems to be of slight importance, as could be expected. But the computed relationship between hay-yield and temperature is tabulated, and you can see that there is a certain critical temperature of −0.6°C [mean] annual temperature when total failure of the hay [crop] is to be expected. During 130 years of observations this critical temperature has never been reached, but two years in the same decade had +0.9°C compared with the normal [average] temperature of 4.2°C during 1931 to 1960. This confirms the well-known statement that Iceland is on the borderline of the habitable world as regards agriculture.

* **Polunin**, Snr [noted but not uttered]: This could be another instance supporting our idea of having some time in the future a conference on the theme of 'Man as a Pioneer Species'].

Furthermore it seems that the winter temperature greatly influences the hay-yield, as a certain temperature deviation in the preceding winter is just as important as the same temperature deviation during the summer. This is probably due mainly to the extensive winter-kill of grasses in cold years. A forecast of the yield is thus possible in the Spring, when the fertilizer is applied. So the point I want to make is the possibility of *compensating the long-term climatic variations by varying the amounts of fertilizer.* If in the warmer periods we use a maximum amount of fertilizer, we lose the opportunity to increase [effectiveness of] the fertilizer when the cold sets in. This could result in an ecodisaster in hay-yield during a cold period, and so this notion of keeping the production constant irrespective of reasonable climatic variations might be something to think of in our efforts towards [attaining] economic equilibrium.

Royston

I would just like to add a footnote to the very interesting statement made by Dr Tan about the sago palm [situation]. It [could be important] to note that there are a number of groups of plants around the world which grow in fair abundance but are not much used by mankind because they tend to have a high starch content and a low protein content. Sago comprises one of these [groups of species] and of course Manioc (*Manihot esculenta*, or Cassava) represents another. It is interesting to note that in Brazil at the moment 5% of all the gasoline used is, in fact, alcohol derived from Manioc, and the plan is that within 5 years there will be a 20% blend and within 10 years all gasoline used in Brazil will be made from Manioc by fermentation.

This is a development which is going to provide not only a great contribution to solving the fossil fuel problem and the problem of CO in the atmosphere which we were talking about yesterday, but also an enormous contribution—social and economic—to the 30 million people living in destitution in the north-east of Brazil, through the setting up of small-scale and labour-intensive cultivation and processing units.

Allow me to extend this to a more general point bearing on deforestation. It is one thing to tell people to keep their hands off the trees. I think it is a much better thing to show people how, by planting trees and increasing forest cover, they can not only do a great deal for the retention of soil, the retention of water, the increase of the fertility of the land, the amelioration of the climate, the provision of shade, and all manner of other things, while at the same time doing something for their own direct material and economic good—by providing themselves with food, providing themselves with fibre, providing themselves with timber, and providing themselves with a whole series of other things that can come from this direct use [of land].

In the Philippines, which were heavily deforested, one of the main objectives of their reforestation programme was to show the villagers the great economic advantage of going in for massive reforestation with commercial trees, and the example that has been cited of the sago palm was I think very good. We must tackle this very severe deforestation problem in the world where our Chairman has pointed out [in a pungent aside] that the future of Man depends on the future of trees. But we need to tackle this in a positive way by planting more trees, and the way to do this is to show people both in the short term and the long term that their well-being depends on tree planting.

Olindo

The question of deforestation occupied our thinking yesterday. It has also been covered this afternoon. I would like to suggest a possible [solution to this increasingly grave problem]. The destruction of tropical forests is undertaken in many countries as government policy and financed either by the USA through FAO or through aid-programmes or by the World Bank. Now if the few financing organizations could agree, as a matter of policy, not to finance any project whose environmental impact or effects have not been [expertly assessed and approved], only then could one say that a major step towards sorting out this problem has been taken. If external finance that is used deliberately to destroy tropical forests under any guise whatsoever could be stopped, then those forests would continue to grow. When tropical forests have been destroyed as a matter of policy, they are replaced in many countries [by planting] monocultures of mainly temperate-derived species of trees, such as pines, for the express purpose of supplying foreign tastes in [distant] countries several years after untrained subsistence farmers have been sent into the area, have destroyed the tropical forests, and have attempted to cultivate the land. By the time the plantation is established, the soil has often been eroded away and we are really dealing with a desert condition! I am fully convinced that we have the [means] to solve this problem if we decide to and organize ourselves to that end.

Gossen

I agree with Dr Hinds's comments that it is not ecologically desirable to revegetate solely by using exotic species. Our experience in revegetation in the Canadian subarctic has revealed that applying a mixture of exotic and indigenous native grass species provides the necessary plant cover. Of course, in permafrost tundra soils it is essential to establish a rapid plant-cover to preserve the thermal régime. Now with respect to exotics—and when I say exotics, I mean genetically-selected ecotypes of native species—these grow quickly but die off in about two or three years, when the slower-growing native species take over. [Hence, not only is the thermal régime preserved, but the system eventually reverts to practically its original state.]

Fosberg

I can make one comment here in regard to the employment of native species. It has been found in the prairies of the central United States, especially as a result of research by agronomists and ecologists, that the original climax grass species, if employed for revegetation of agricultural land that has been depleted, are far more productive than any mixture of exotic species that have been tried out.

Fridriksson

I would like to point out that the pictures which I showed to you were taken of Icelandic soil in its deterioration. There is, however, a great movement in Iceland to control utilization of vegetation, and our Soil Conservation Institute has been able to stabilize the eroding soils in many areas; that particular one you saw in the picture has possibly already been checked. I do not want you to leave this room with the impression that nothing is being done about this problem in Iceland!

We had our 1100 years' anniversary of settlement of Man in Iceland three years ago, and there was a substantial sum of money set aside by the government to pay back some of the debts that the nation owes to the country—in the form of revegetating or putting a new vegetation on many of the badly eroded lands. [Applause.]

Misra (Chairman, concluding Session)

Because there is no time left for me to summarize the presentations [or deal further with these vast and vital topics from the platform], I suggest that we continue to discuss them informally during the tea-break, [meanwhile closing this Session with my warmest] thanks to my colleagues and to those in the audience who have participated so keenly and effectively in the proceedings.

Session 8

EXTREME URBANIZATION EFFECTS AND DANGERS

Keynote Paper

Effects and Dangers of Extreme Urbanization: The High-density—Low-rise Layout as a Possible Planning Remedy

by

PIERRE LACONTE

Director for Expansion, Université Catholique de Louvain, 13 Avenue G. Lemaître, 1348 Louvain-la-Neuve, Belgium;
President, World Environment and Resources Council.

Formulation of the Problem: The Dangers of Extreme Urbanization

In what was probably his last paper published during his lifetime, the late Constantinos A. Doxiadis (1975) formulated the main danger of extreme urbanization in the following way:

> As soon as commercial forms of energy and machines enter our settlements, they start to expand at a much greater rate than does the growing settlement's population and economy. As the growing settlements need to add a lot of industry and many big buildings, their expansion is not directed towards mountainous areas but occurs chiefly on the plains. It is here that the best soils usually lie: thus urban growth means profligate elimination of agricultural land.
>
> This is not an immediate problem for tomorrow, in terms of days; but it is a very big problem in terms of years and decades, as agricultural land covers a very small percentage of the global surface and we cannot afford to lose it. The danger is very great and we must act quickly if we are to avoid it. (*Ibid.*, p. 13.)

Let us elaborate somewhat on the facets of the urbanization process:

Population Growth

The first and foremost facet is demographic: we are in the middle of a population explosion; only conjectures can be made about its future, as was indicated by several other contributors to this Conference.

Industrial Growth

The second facet is industrial growth; the industrial growth of the nineteenth and twentieth centuries has generated a need for space to accommodate its own activity. More recently, it has led to a huge complex of administratve activities, strongly prone to geographic concentration. Thus the concentration of employment in industrial and service centres has fostered urbanization and increased the physical size of the city.

Expanding Transportation Networks

At the same time, technical progress and cheap energy have changed the 'uni-speed city', which was formerly limited by the walking distance to fulfil necessities and enjoy amenities, into an expanding transportation network.

Urban Sprawl

The variety of transportation systems provided in cities has been a further cause of space consumption which in itself led to further sprawl.

Increasing Mobility

The mobility involved in technical progress was further supported by progress in telecommunications, which also helped disseminate settlements. The notion of Daily Urban Space (DUS) more adequately reflects today's urbanization than does the word 'city' (Berry & Kasarda, 1977).

Rural Decline

For the rural world, urbanization is associated with the idea of 'modernization'. The city is presented to the rural world through 'the media' as an attractive way of life and a means for the country-dweller to escape from the cultural constraints of the village. All over the world, villagers have flocked into cities—never to return.

Estrangement of City Dwellers

Because of its function and its way of life, the city has become a melting-pot of strangers, where social controls have been replaced by anonymous police-forces which, despite their ever-growing staffs, have not been able to check the rise in crime. Much should also be said about the influence, on the psychological and physical balance of city-dwellers, of a daily life cut off from Nature (Rapoport, 1977).

Functionalist Development of Cities

The 'actors' behind city development have tended to become anonymous and powerful private or public organizations whose way of using the city is more concerned with their own growth-objectives than with a concern for the well-being of the citizens. The tower offices and super-blocks symbolize the capital-intensive 'organization city'. The high-rise apartment blocks are another expression of building economics playing against the preferences of the people concerned, and particularly against families with children (Adams & Conway, 1974).

Technology versus Nature

The technology which enabled cities to grow, also made them independent from Nature and its cycles, even blurring the distinction between day and night, and between the seasons of the year. The city's effluents are rejected into non-urbanized areas wherever possible, contaminating the water resources, underground space, and atmosphere in ever-expanding areas (Purves, 1977).

Dependence of Urban Populations on Technology

The same technology, which tends to make us independent from Nature, also makes us dependent on its own reliability. Mechanical failures will never be excluded, but the larger the urban systems become, the more will mechanical failures have linked consequences. The two recent power failures in New York are only one type of example. The risks of nuclear accidents, and of large-scale blackmail through criminal handling of technical equipment on which entire cities depend, are increasing proportionally with the degree of technical integration of urban systems.

Urbanization versus the Life-support System

Returning to the point made in the quotation from Doxiadis which opens this paper, there is a *grave threat to the life-support system* in continued urban expansion. This should be treated quite apart from concomitant deforestation and vulnerable monocultures (Polunin, 1972). Little will be said here about the carbon cycle and build-up of atmospheric CO_2, or about the nitrogen, phosphorus, and other cycles, as these are treated at length in other reports presented at this Conference.

The above aspects collectively constitute the link between urbanization, human ecology, and the environmental future of our planet (Duvigneaud, 1974; Polunin, 1974).

This inevitably simplified picture of the dangers of extreme urbanization brings us to the question: are there no alternative patterns of human settlement? Indeed, one of the factors of extreme urbanization, i.e. cheap energy, may soon disappear and bring about a major change based on a compelling incentive towards containment of urban space.

Role of Technology

What about the role of technology? The forces which organized the modern city used technology in a way that led to complexity, capital-costliness, mass-production of build-up space and high energy consumption. Schumacher (1976) advocated the use of technology towards tools for *efficient* small-scale production without undue complexity, and with modest capital requirements. This would necessitate a new attitude of groups towards (*a*) a return to middle-of-the-road technology, (*b*) a closer relation with Nature, and (*c*) smaller-scale communities.

To summarize the formulation of the problem, the population growth and the way it settles in urban areas affect the planetary ecosystem. It is now being realized that we live in a finite world, and that these processes cannot go on for ever. Therefore, the problem arises as to how one can, within the framework of an ever-expanding demography, achieve the following objectives:

— maintain a maximum area of land for agriculture by adequately concentrating the urban housing areas;
— in so doing, bear in mind the need to keep to a human scale in favour of small, self-contained towns or villages rather than large, anonymous, and continuous, urban space;
— bear in mind the need to save resources, particularly energy, and therefore avoid reliance on energy-consuming devices—for example, by trying to reduce the automobile traffic within a small-scale community, by avoiding reliance on energy-consuming elevators, and by adopting a type of architecture which relies on natural lighting and ventilation.

We shall now proceed to go further into the practical aspects of these proposals by presenting, in the second part of this paper, an inductive approach to a practical planning answer to these grave problems.

The High-density—Low-rise Layout as a Possible Planning Remedy

Urban settlements are proliferating ever farther into the countryside. How can these settlements be laid out in a way which would, at one and the same time, save land and avoid high-rise building concentrations and the effects of living off the ground? This could be called the high-density–low-rise alternative, and takes its inspiration from the traditional pattern in continental European cities, from the denser parts of the English 'garden cities' (Letchworth, Welwyn Garden City, Hampstead Garden Suburbs), and from the less dense parts of recent developments, such as Kévin Lahti in Finland, Reston in the USA, La Verrière-Maurepas in France, and Louvain-la-Neuve in Belgium. This last development will now be taken as a case-study.

Overall Description of a New University Town

Principles of Layout at Louvain-la-Neuve

In 1968 the academic authorities of the Catholic University of Louvain in Belgium became engaged in transferring the University to the Plateau of Lauzelle on an ex-urban site situated about 30 km to the south of Brussels. They chose to redevelop the ancient university in a form integrated into a new town rather than as an autonomous university campus. The goal was to foster town-and-gown interaction and the service role of the university towards government and industry.

The long-term planning objective was to have a balanced community of 50,000 people (with a maximum of 15,000 students). The shorter-term objective was to have a resident population of some 13,500 people by 1980 (comprising 8,000 students and 5,500 residents made up of university staff and people occupied in business or trade but wishing to live in Louvain-la-Neuve). These were the only bases for overall planning. Otherwise, considerable flexibility was to be maintained for developing the entire new town around a strong linear backbone. The master-plan was approved in 1970 and the first buildings were put into use in 1972.

The university decided to use some 350 ha. (a little less than 900 acres) for its own and the new town's development. Within these 350 ha., *high-density–low-rise construction* would enable a maximum of 50,000 people to be housed. The maximum distance within the urbanized area would be less than 2,000 m.

The general layout is based on a long pedestrian main street along which the community facilities are located. Access to the buildings is by external roads, underpasses, and a new underground railroad station which was inaugurated in 1975 and puts Louvain-la-Neuve at less than 30 minutes of transport time from the centre of Brussels.

The design of the residential areas excluded detached single-family houses, so development is based on attached double-family houses, maisonettes, and apartments in low buildings. In 1977, some 8,000 people were already living in Louvain-la-Neuve.

The university's open spaces and sporting facilities are meeting-places for the whole population, and are linked by pedestrian routes. The master-plan allows for the parallel growth of urban services and population according to the response of the housing market.

Economic Constraints

The university was able to acquire its 900 ha. through a state loan at a low interest-rate (3.18% p.a., including capital payment). From the very beginning, the university had to play the role of leader in the development of the new town, although it had never attempted anything of the kind before. Its policy was to retain ownership of the land and sell long-term leases (from 50 to 99 years) to private builders who were to implement the master-plan.

Adjacent to the immediate areas that are being used for university development and housing, 140 ha. (340 acres) of the acquired 900 ha. have been set aside for industrial research and development activities. The basic objectives of this aspect of Louvain-la-Neuve are:

(1) To encourage scientific interchange between the university and industrial research, while maintaining a total independence of the university from industry.
(2) To attract non-polluting firms using advanced technology.

(3) To diversify the social mix of the population of Louvain-la-Neuve by creating employment opportunities within walking distance of the residential area.

Some 16 firms (all specializing in research and development or in technology-orientated production) have chosen the Louvain-la-Neuve location, six of them during the year 1976. They brought in about 1,000 jobs. To ensure a good interaction between academic research and industry, a policy of strict selection was applied. This meant turning down applications from enterprises having nothing to do with research or advanced technology.

The close proximity (less than 1 km) between most of the resident firms and the social facilities, shops and restaurants, of Louvain-la-Neuve, which are used by their personnel, encourages contacts more effectively than any organized procedure.

Relationship of Layout to Energy Concerns

The layout and development options of Louvain-la-Neuve have resulted in the following features of energy-saving interest:

Layout

The high-density–low-rise option has resulted in minimizing motorized transportation within the urbanized area, the distances never being greater than about 1 km. In this manner up to 50,000 inhabitants can be housed together in an area not exceeding 350 ha., and within a radius of about 900 m. This is the basic prerequisite for encouraging non-motorized relations.

The road design is based on a main road with *culs-de-sac*, such that the resulting network encourages non-motorized trips from one point to another of the development site. If, for some reason, people want to use their cars, this can be done but it will involve a longer trip. It is expected nowadays that all parts of a city should be accessible by car.

Railway access is provided near the middle of the town. This encourages trips to the outside, i.e. mainly Brussels, to be made by train rather than by car—thereby saving energy. The new station is a terminal one that is linked with the main line of Brussels–Luxembourg–Basle, and beyond the station is a 'non edificandi' area. This will allow for the creation of a loop which would serve other settlements that are now served by terminal railways, and would mean in the future a more efficient railway service.

After two years of operation, the number of railway passengers in both directions is 100% higher than had been expected by the railway company. Thus, service to the public has doubled and is now provided by using twice as many trains with four carriages each—rather than the double-carriage exploitation that was originally envisaged.

The low-rise option (i.e. an average of three storeys for the university buildings) induces a limited employment of elevators, these being provided mainly for the use of the handicapped and for removals. Small apartment buildings provide no such elevators at all.

A single-level transportation network minimizes infrastructure investment, for example in overpasses. Except in the centre of the town, the layout provides for a separation of traffic through the *culs-de-sac* system and pedestrian paths with occasional underpasses—but without systematically providing construction on artificial floors, which is expensive and requires costly lighting, maintenance and surveillance.

The option of dividing the project into a large number of small buildings designed by some 50 different architects involved employment of a number of separate contractors. This has had an unexpected overall result in vastly increasing the competition between contractors and in discouraging cartel-formation between the different contractors. In some cases, the cost has been 30% below the architects' estimate. The increased cost of coordination (borne by the university) and occasional bankruptcy of small contractors did not counteract this fundamental advantage.

The emphasis has been on small groups of 'town houses' (Fig. 1). It has been found that encouraging the layout of groups of seven town houses, each of 120–200 m^2 of floor-space including attic, was a particularly economical way to build because it could tap the large market of building equipment for large single-family homes. Thus, if the group of houses is larger than seven in number, one enters into the market of large projects. One is then induced, for economic reasons, to build larger and larger units, which become inhuman in their sheer scale and cause problems of vertical transportation.

The overall density of buildings and population achieved by attaching units together, even if they have small individual gardens, is as great as with high-rise buildings separated by collective empty spaces.

Although not intended to save energy but rather to allow families to have an attic which is the 'memory of the family', this systematic use of roofs is a potential support for solar heating devices. Up to now these have not been used but, in view of the increased market for solar roofs, it may well become practical to install solar heating devices in the years ahead.

Sewage and Water Management

Although dual sewerage and rain-water disposal networks have become standard in several countries, this system is not yet employed at all widely in Belgium. But in Louvain-la-Neuve, such a system has been systematically imposed and used.

It was felt that the water taken on the site should be given back to the ground-water resources, and so a separate sewage system was installed from the start. The storm-water is conducted to a small artificial lake that is located at the lowest point of the site and operates as a storm basin. In dry periods, the lake's level is kept constant by the water coming from the sewage treatment plant which is located near the lake. The waste goes by a separate sewage system to the sewage treatment plant. When once it has been treated, the sewage water is used to complement the water input of the lake. The bottom of the lake is not sealed, which circumstance allows natural infiltration to take place and therefore recharging of the aquifer as the local water reserve.

The specific object of this project is to concentrate all the rain- and storm-water in a 6 ha. artificial lake, which provides an amenity at the same time as a regulation storage for water and, through infiltration, a potential refilling of the underground water resources. At the same time all the domestic and industrially-used waters are sent through a sewage pipe to a sewage treatment plant located downstream in the valley.

The fact that any used water is treated and does not include rain-water means that a reduced quantity of water is treated. This represents an important saving, as the participation in the treatment costs is proportional to the quantity of water treated and not to its degree of pollution.

Heating

The high-density–low-rise concept has an immediate result, in the case of Louvain-la-Neuve, of protecting the plateau from the unpleasant northern winds and improving the microclimate, which in itself is a source of energy-saving. This is also a feature of old towns where a 'curtain' of continuous low-rise buildings acts as a barrier to the winds and thus improves the local climate.

After five years of operation, a noticeable improvement has taken place at Louvain-la-Neuve. The buildings retain the heat, as terraced houses with common walls form an aggregate which diminishes the disposition of heat—as compared with isolated apartment blocks or detached houses.

The high-density construction allows the distribution of heating from a centrally operated heating-plant, which at present burns natural gas. As developer of the site, the university has concluded a long-term agreement with the gas distribution company and has therefore been able to impose a ban on fuel-operated individual heating. Even for uses not connected to the urban heating-plant, the main heating source is natural gas.

The fact of having small building-units involves an improvement in the natural heating and cooling situation. It also allows greater flexibility in future use of appropriate technologies as they become available on the market, e.g. heat-pump, heat storage, etc.

Management and Maintenance Costs

The scaling down to small units operated individually is not *per se* a source of economy but, in fact, it has been noted that the care which people take of their own individual house within a small group is conducive to improved care and effective voluntary maintenance as compared with reliance on maintenance provided by large organizations.

The variety of design of individually conceived spaces allows increased flexibility for future uses and therefore tends to reduce the

Fig. 1. One of the small groups of 'town houses' characterizing the residential layout of Louvain-la-Neuve. (Photo: Pierre Laconte.)

Fig. 2. The kind of low-rise layout in which suitable plants, if properly tended, flourish and attract birds and butterflies. (Photo: Pierre Laconte.)

rate of obsolescence of the buildings and the need for renewal. The saving in road investments, together with the linear pattern of development, allow a 'generative' way of growth which can, according to economic circumstances, accelerate or stop altogether without undue debts for capital repayments.

Conclusions

Beyond the case-study of Louvain-la-Neuve, the high-density–low-rise type of layout has its economic advantages—including saving of energy *per caput* and lowering the cost per square metre of used space—which would be worth considering for new developments in any country. It is well fitted to urban extensions (*see* Epstein, 1976) as well as to new settlements. Its very flexibility also allows the best use to be made of local materials, local craftsmanship and local techniques.

In terms of social policy, the high-density–low-rise layout can obviously not *per se* determine life-styles and change attitudes or behaviour in relation to consumption, but it does create the physical framework for informal contacts between people and their contact with the surrounding grounds—including some with Nature, as plants will flourish in such areas if wisely chosen and properly tended, and birds will flock to such *milieux* (Fig. 2). In the context of already existing cities, such an arrangement allows increased harmony through preserving similar scales among the existing buildings, as well as growth of trees and desirable new types of construction. It links urban amenities to the urban history, which has perhaps more to teach about the home for Man than a few decades of monofunctional urban space-allocation.

However, such kinds of layout rely on smallness and competition between a wide range of contractors. They also employ unsophisticated techniques and a relatively small scale of infrastructure, which is rather far removed from that of the dominant building economy of recent times.

While such implementation requires specific skills in terms of developing good governmental relations, the involvement of administrative aspects with academic interests in Louvain-la-Neuve has already engendered several comparative research projects on the institutional prerequisites for a more humanely-built environment.

Let us now go back to our main theme of extreme urbanization, to see how this example fits into the constraints formulated at the end of the first part of this paper. The high-density–low-rise concept as applied in Louvain-la-Neuve, and as in many old cities, has the following consequences:

— it allows the concentration of a medium-scale community into a small area surrounded by urban or rural space;
— it fosters the kind of human qualities which one can find in a small town rather than in a large metropolis;

— it relies very little on elevators and not at all on air-conditioning, for example—thus contributing to energy-saving policies.

Will the foregoing cancel out the dangers of extreme urbanization? Frankly speaking, they could not by themselves cope with the problems arising from a multi-million aggregated population. Thus although they can be a means of accommodating extreme urbanization, they will never counteract the effects of its very reality.

References

Adams, B. & Conway, J. (1974). Social effects of living off the ground. *Habitat International*, **2**(5/6), pp. 595–614. [Pergamon Press, Oxford, England.]

Berry, B.J.L. & Kasarda, J.D. (1977). *Contemporary Urban Ecology*. Macmillan, New York, NY: xiii + 497 pp., illustr.

Doxiadis, C.A. (1975). The ecological types of space that we need. *Environmental Conservation*, **2**(1), pp. 3–13, illustr.

Duvigneaud, P. (1974). *La Synthèse Ecologique*. Doin, Paris, France: 270 pp., illustr.

Epstein, G. (1976). Planning forms for 20th-century cities. Pp. 157–66 in *The Environment of Human Settlements* (Ed. P. Laconte). Pergamon, Oxford, England: Vol. I, ix + 311 pp., illustr.

Polunin, N. (1972). The biosphere today. Pp. 33–64, in *The Environmental Future* (Ed. N. Polunin). Macmillan, London & Basingstoke: xiv + 660 pp., illustr.

Polunin, N. (1974). Thoughts on some conceivable ecodisasters. *Environmental Conservation*, **1**(3), pp. 177–88.

Purves, D. (1977). *Trace-element Contamination of the Environment*. Elsevier, Amsterdam, Netherlands: xi + 260 pp.

Rapoport, A. (1977). *Human Aspects of Urban Form: Toward a Man–Environment Approach to Urban Form and Design*. Pergamon, Oxford, England: viii + 438 pp., illustr.

Schumacher, E. F. (1976). Patterns of human settlements. *Ambio*, **5**(3), pp. 91–7.

DISCUSSION (Session 8)

Jones (Chairman, introducing)

Having just arrived this morning on the 'milk flight' from New York and being already 20 minutes adrift, I feel I should introduce the panel [without more ado]. At the far end we have Mr Peter Stone who I am sure is still a geologist [at least at heart, though of course] he was the person who was ultimately responsible for a great deal of the successful publicity which surrounded the United Nations Conference [on the Human Environment in 1972] at Stockholm. Next to him we have Professor Reid Bryson, who is the Director of the Institute for Environmental Studies at the University of Wisconsin, Madison. He is a [celebrated meteorologist] and microclimatologist who I hope will tell us something about the microclimate of large cities [as part of] the subject of this Session [—unless he feels he did it sufficiently at the first ICEF in Finland in 1971]. Next to him is Mr Ralph Glasser, from London, England. He is a psychologist and I am sure he will tell us something about the psychological aspects of urban development problems. Perhaps he will [also] tell us something about what I [can only] describe as a social mutation which seems to be taking place in young people, some of whom apparently think that milk [simply] comes in bottles and is found at the side of the door in the morning. When I gave a lecture to a group of students in a high school in southern Ontario about water pollution, and about some of the amenities we are losing through water pollution—such as swimming in lakes—I was quite horrified to discover the majority of the students looking at me in amazement: they really did not think anybody ever swam in lakes, but that in fact you swim in swimming pools, and lakes are for fish and industrial wastes and things like that! This could be described as a social mutation and is certainly a psychological phenomenon.

Next of course we have Dr Pierre Laconte, who is the main speaker and will tell us [particularly] about the case-study of the University of Louvain [extension in Belgium]. But before calling on him I would like to explain that from my point of view as an engineer the problem associated with urban development has to do with the generation of wastes and how to deal with those wastes. So I would like to use my Chairman's prerogative to share with you a little secret which I share with some of my engineering students, about how engineers have traditionally treated wastes, and which I suspect may be new to some of you at least when presented in this form.

Traditionally we used to throw all the wastes from a city or other settlement into the nearest body of water, [whether it was salt or, originally, fresh]. Waste can be generalized in chemical terms as $(CH_2O)_x$ N,P, which is some kind of an organic complex with nitrogen and phosphorus, and it used to go straight into the water and cause some odour and discomfort on hot sunny days. Then the engineers stepped in and [proposed] the sewage treatment plant that is supposed to be a panacea. It was designed, of course, to convert the organic carbon complexes to carbon dioxide, water, and nitrogen and phosphorus. They then proceeded to release this material into the water, having spent millions of dollars building a plant and lots of money operating it. With the addition of sunlight, Nature has a capacity of putting together again those same elements in the form of Algae which, if you look at the chemical formulation, appears awfully like what we had in the first place, all this being accomplished after a great deal of investment.

My point in mentioning this is that it is the kind of waste-water treatment thinking associated with urban development which engineers have employed for a very long time, and it is only within the past 10 years that we have started to think about other things which should be removed. Clearly, if the nitrogen and phosphorus had been removed from the effluent, we would have broken that cycle; we would therefore not have produced the Algae and we would not have produced essentially the same material that we were treating in the first place.

The next question that arises is: what are we going to have to be treating for in the future? Clearly we are now starting to look at such things as heavy-metals. We have gone through a series of arguments concerning separation of storm-water from sewage. In Roman times there were separate sewers, though in fact it was only the very wealthy people who could really afford to connect to the drains for their domestic waste. Subsequently, in Elizabethan days, people used to dispose of their domestic wastes out of the upper-storey windows, and this led to a number of the cries and the sounds and smells of [old] London. After that people [at least in England] became fairly civilized and decided to put it all (storm and sanitary wastes) into one pipe and call this a combined sewer. This waste in London was discharged into the River Thames, until Elizabeth I, looking out of her bedroom window one morning [so the story goes], saw what a frightful mess the river was in. So she set up a Royal Commission which looked into the condition of the Thames. It was then that sewage treatment plants began to be constructed to perform the function which I described earlier.

After this we came to the conclusion that in rainy periods we were increasing the magnitude of the problem by mixing up rain-water with sanitary wastes, which of course are the result of urban development. So a big argument ensued and indeed continues in a large number of older cities in the world: whether to conduct separately the sewage to treatment plants and the storm-water to streams, or leave them together in the combined sewers. Many cities have in fact made a major investment and commitment to separate the sewers. I think you will hear about the University of Louvain, where they are actually separating their systems. Many cities have gone to a great deal of expense to make sure that the storm-water—which is thought of as being relatively clean—should be discharged directly into the rivers, creeks and water-bodies. The sanitary sewage is then treated in efficient waste-water treatment plants.

Now many engineers are starting to complete that cycle once again, and are starting to worry about what is contained in storm-water—what it is that washes off the roads, what particulates, what heavy-metals, what forms of hydrocarbons, and what are the effects of these materials? Do they not warrant serious consideration and treatment? I think many engineers are coming to the conclusion that they do warrant serious consideration. So, strange as it may seem, many engineers are reverting to the conclusion that maybe it was a good idea to have combined sewerage systems after all. Of course, some method of trapping the waste during high-flood periods should be devised, and provision should be made for treatment after the storm is over. This would allow treatment to take place at a fairly uniform rate throughout the low-flow–high-flow fluctuation.

Other urban problems, particularly in the colder climates such as are experienced in Canada, include the application of de-icing substances, such as salt to highways during the winter time. This practice results in a number of quite serious problems. It provides little pools of very concentrated brine from which wild animals are apt to drink. It also ultimately gets into the ground-water supply and of course it has a disastrous effect upon the roadside vegetation. But when one considers the alternative of not de-icing the highways at all, society—and ours in particular—throws up its hands in horror and questions what would happen if we did not have bare pavements in winter and had to drive on ice. This is something that our society in Canada in particular is not ready for.

There are many other facets of urbanization, and I am sure other speakers will tell you about some of these. The air pollution problems of course are well known to us, and so are transportation problems associated with urban development—not to mention the noise pollution (as we are coming to call it) which is particularly serious to visitors. If one goes to Athens, for example, one finds the noise level offensive; yet the people of Athens would suggest there is no noise pollution. In Greece this is their way of life, and the noise which comes from the tavernas and the motor-horns, which offends us, does not offend them; it is again a perceived problem. So with this introduction which has taken longer than I intended, I will pass directly over to Dr Pierre Laconte, who will present a summary of his paper which I hope you all have.

Laconte (Keynoter, presenting paper)
[Besides giving a summary of his paper, Dr Laconte showed pictures of the 'new' University of Louvain and its urban surroundings. He concluded]: I have passed my time-limit but want to say just one last thing: What you have seen here indicates that it has been possible for a body which has no specific power, and which is not a governmental institution as such, to spend an amount that now totals approximately 13,000m Belgian francs. This has been accomplished by a group of people who, in adverse circumstances, had the will to create a new environment based on ecological principles as outlined in the paper which you have received. [But what of the future?] The residents have formed an association which is defending its identity, upholding the master-plan against any infringement which could come either from the municipality or from the university. They will gradually assume control, which will also solve the maintenance problems. The University has done all it can to create a new town, which now has to be taken over by the people in order to create a viable community. [Applause.]

Stone (Panellist)
Just for once, hearing a keynote speech, I have almost nothing to say [except that it all] sounds very sensible. I cannot think of anything that raises my personal hackles, and I cannot really see [anything to comment on] except for the fact that, by not mixing intellectuals and students with working-class people in industry, you have got a very big sort-of-Baedeker university element, and a lot of people disapprove strongly of Baedeker universities—by which I mean universities in beautiful places that are set apart from the real world and are in a kind of ivory tower, or a concrete realization of an ivory tower. Apart from this it would seem that [Dr Laconte and his colleagues] have made an excellent job of it, and I am particularly thrilled to hear that [they] used 50 architects. I think lots of people feel that giving one architect—as we very often do in Britain and many other countries—the licence to wreak his own particular kind of havoc on the scene, can be a bad idea; to avoid this is an excellent precedent, and I hope a lot of other people will follow it.

Laconte (Keynoter)
I am in complete agreement with your remark. The [time-limit for] presentation did not allow me to speak about that aspect but it is our idea that the town should have a minority of students and people related to the university. At present, 30% of the residents are not connected with the university, and the industrial research park has more than 15 firms and 1,000 jobs. So the working population is gradually coming in. Land has been leased to a kind of 'council homes institution', much as in England, in order to create social housing in the middle.

Bryson (Panellist)
I probably should not be saying anything, not being either an urban planner or an architect or an engineer—just a mere climatologist—but those of you who know me know that I never pass up a chance to speak.

My reaction is not 'Wow, isn't this neat!' but, as a citizen of the world thinking of Mexico City—rather than of this beautiful little town to house elite university people—I think also of Calcutta, and I think of Hong Kong and its suburbs. I think of the real cities of the world, not of these beautiful little towns. And I also think of the 13,000m Belgian francs. How much per person is it, and could that amount be spent on repairing a totally absurd urban life of, let us say, the inhabitants of Calcutta? No way! How many of the 4,000 million people plus, now on Earth, live in cities—live in cities which Man invented but never figured out how to make viable? How much of the politics of the world is dictated by the fact that people in cities can get together and collect a mob and dictate the politics which then say that the people out in the country who are feeding them have really no voice. When the relief grain that the US sent to India—whether it was proper or not is not the question—but when it was sent to India it never reached the [country areas], because the political crowd was in the big cities.

Now this is a beautiful thing [they have created in Belgium] and I do not want to denigrate in any way what has been done there. Yet would it not be nice if the unworkable cities of the world could be taken apart, and the people moved out and spread into that kind of environment where they could live a life with some kind of beauty around them? But it cannot be, there is not that much money in the world! And this brings us back to the problems that we were talking about the other day. Climatic change, fuel shortages, overpopulation: these things say that we have to think of the cities as a number of time-bombs, not as opportunities to use our intellectual capacities to build nice little exemplars of what could be done if there was enough money in the world—say $10,000 or $100,000 per person—to spend on making a beautiful environment.

The real cities of the world have totally changed their environment, and have modified the environment of areas much larger than those cities themselves. I remember back in the 1930s when there was a great depression, I lived in the city of Detroit. During that depression one-quarter of the houses in the city were standing empty—because there was no work and the people had returned to the farms and rural communities from which they had come to man the factories of the cities. And now I think of that big city and how, if there is another depression, the city will be bigger than it was and they can not go back to the family on the farm the way they did before, because the ties are not there, and the families are not there. There is not that much of a rural population any more.

It seems to me that while we have heard about what can be done if there is enough money, we should ask what is practical, what is necessary, and what are the dangers—if we do not do something about the big cities in an efficient way because we cannot afford to do it in the best way, [we shall be in for trouble]. What is the best that we can do before the cities explode when the fuel is gone? How does the food get to the cities if the railroads are gone? We can build new railroads—but where, and how quickly? What do we do in the meantime, because of the slow response?

René Dubos thinks that there are already adaptive changes in the population of the city of New York—that there is already selection for humans who are capable of living in a totally polluted environment. But could we then take these people who are adapted to the city and move them back into that horrible stuff known as clean air, and in the country? What would it do to their lungs? I think we have got bigger problems than how to house a university, even though I enjoyed greatly [the paper] and I enjoyed thinking about what it could mean in terms of new buildings on our own campus. Moreover, there are some examples here of what can be done without planning, [given particular] approaches and attitudes; but I think the real problem of the cities is a vastly different one.

Glasser (Panellist)

I have no comment to make of a technical nature on the admirable introduction we have heard. Like [those of Professor] Bryson, however, my thoughts dwell upon what the city means, upon its significance in terms of what I talked about in an earlier session—including the purpose of living and what these habitations and settlements are really for. We ought to remember that our concepts of the city, even though we all use the same word, are not often the same. The city as we know it is far from what it was two or three thousand years ago. In biblical times, for instance, most cities were about the size of what I would describe as the 'network township'—a settlement of less than 6,000 inhabitants. Each consisted of a collection of people—mainly craftsmen and traders—who provided a servicing and trade organization for the surrounding agricultural economy and for the life of the rural dweller. It was convenient in those days to centralize those services. Cities grew up to serve the industry of the country and not, as at present, the reverse.

As Professor Bryson rightly points out, the town-dweller draws in sustenance from the country and from the areas of primary production in general. To a large extent he is divorced from the problems of the outer world of which he is largely ignorant. That divorce is what concerns me most of all. This line of thinking, as I pointed out yesterday, inevitably brings into focus other factors that are highly explosive in the modern world. In olden days the word 'townsman' was a term of abuse! The townsman was assumed, rightly, to be out of touch with the needs of the production cycle and likewise with the attitudes, the ways of thinking and the spiritual posture of the countryman, and also with the demand for maintenance of a natural balance that the countryman felt bound to observe, or else denied at his peril. We see the reflection of that in our own day. The farmer does not trample down his own crops; but when townsmen roam about the countryside—and many of them do so carelessly—they are surprised and slightly hurt when admonished for it. They do not understand. The old links have been broken. But they *should* understand! For the countryman that understanding maintains an organic coupling which provides a continuous reminder of his identity, its sources and the models of behaviour which he must strive to emulate.

Again the countryman, when *he* transfers to the town, loses that coupling, is aware of the loss, and seizes upon the false refuge of the townsman's isolation. He experiences, too, a similar decay of certainty about his identity goals, for he loses the inherited frame of emotional reference—the continual reminders about behaviour and self-discipline—that his forebears possessed, enmeshed as he is in the smaller and emotionally manageable community. It is a feature of human nature that if you do not understand your neighbour, and if you do not even understand who you yourself are—if your own identity is not clear to you—you care less and less about what happens to anyone else. Why should you, unless constant contact and interdependence compels you to do so? And why should they care about you? We retreat, then, into embattled watchfulness—into a savage indifference that is so typical of the metropolitan existence.

If, in a modern city, you see a man fall down in the street, or knocked down by an attacker, you almost certainly do not know who he is; he is nothing to do with you, nor will his absence mean anything to the economy or emotional balance of your life. And the general climate of attitudes encourages you 'not to get involved' (as the cant phrase goes). But if that happened in a village, you would know at once who the victim was; you would know instantly that something had to be done about it, because his absence from the economy would have its impact on everybody else. In contrast in a city you build up an indifference which is extremely destructive—not only upon the life which the other people in the city live, but also because its influence on everybody else is to induce a total indifference and a total lack of responsibility. So it is no surprise that the massive urbanization which is a feature of all parts of the world today, leads to a destruction of much social discipline and most of the social values which we [in towns] have grown up to feel unnecessary.

Returning to the man who takes his livelihood out of the soil, he knows very well that he must not overdo things in terms of extraction—in terms of the overuse of the soil, and so on—whereas for the town-dweller [practically the only real] discipline is the economic one of price. Nevertheless there appears to be an irresistible attraction of urban life leading to its continual expansion. As a bizarre result of this we now have the new-vogue expression of 'urban renewal', whereas it seems clear that we should favour the reverse process, and address our minds to rebuilding, and sustaining, the type of small, intertwined community that is internally knowable by all who live in it. In such a township of about 6,000 people, which is perfectly feasible economically with the aid of adjusted technology, everyone's life is perforce joined, in a direct and instantly recognized fashion, with that of everyone else. Consequently each person in such a community is obliged to maintain a posture of *total* responsibility—the attribute of true maturity.

Laconte (Keynoter)

I was probably wrong to quote an overall figure for expenses, as that figure is very impressive when compared with the usual budget of a university. But it is only a drop in the ocean if you compare it with the total amount of urbanization investment in Belgium, and I would like to point out that the kind of layout with which we have experimented at Louvain-la-Neuve uses less road-space and allows for closely-linked low buildings. It is about 50% cheaper as regards building and maintenance costs than a layout resulting from the dominant economy, which leads to high repetitive buildings and abundant road-space.

My point is that, as urbanization is still going on, we should note that there are means of coping with this phenomenon which are both cheaper and on a more human scale [than those that are usually practised]. This remark is [particularly] valid for developing countries. When I see the kind of huge road projects that are [being] built all over the developing world, I think of the layout alternatives which are cheaper in terms of infrastructure, building costs, energy consumption, and loss of agricultural land, whilst also being on a more [appropriate scale for] Man. I believe this indicates a direction in which we should carry out further research. It will not be easy, as the implementation of this kind of idea is *against the dominant economy*.

At every stage of our project we saw that whatever we were proposing was against the administrative culture of our country. Each time we had to lobby and fight to convince people that our cheaper and smaller-scale solutions were workable. It is a fact that, in general, large public and private administrations prefer large-scale, standardized and impressive achievements. The cost is of little importance [to them]. But for our project, from necessity and by philosophy, we were cost-conscious from the start in keeping the scale of things small. The last time the Minister of Public Works came to see us he said, in a moment of truth at the end of a meal given in his honour: 'Please give more work to my engineers, ask for more roads', and we said 'No, we don't need more roads than the ones we already have'. So that briefly explains the background to the money spent. However, in making this point, I do not wish to disagree with Professor Bryson but merely to put the matter into perspective.

Jones (Chairman)

Now you have heard all kinds of viewpoints. You have heard of a microcosm in Belgium. You have heard Professor Bryson say [in his inimitable style] 'Yes, that's great; but it's atypical'. You've heard about the community structure [and why there is apt to be too much infrastructure and road-building]. But what questions or comments do you have to make [from the floor]? Please raise your hands, be recognized, and speak to the [Chair *via* the] microphone. If what you have to say is in the form of a question, I would like to direct it specifically.

Royston

I wish to make a general comment and then ask a question of Dr Laconte. Like most people here I guess I am totally unqualified to enter into a discussion of this particular topic—except, perhaps, for the fact that I did live for 10 years in Crawley New Town [England], so I have some personal experience of living in a community which in some respects did not look so very different from some aspects of Louvain-la-Neuve. At the moment, in Geneva, I live in an apartment block which houses 8,000 people under one roof, and this is another form of extreme urbanization. My general comment is that I like enormously what is being done at Louvain-la-Neuve, because it does introduce two elements that I think everybody recognizes as tremendously important in urban development. One is diversification; we see that our urban settlements are apt to be inadequately diverse. We need to get in the community or town a good mixture of different social classes, of different human activities—the residential, the commercial and the industrial—as I experienced them in Crawley New Town. Now you have Academia added in as well, and the quality of life in those communities will be directly related to that diversity. The sustainability of those communities as [such will also be] related to that diversity, and here the connection with natural ecosystems is quite clear.

My following question is about size, which is obviously tremendously important. Whether we are talking about Louvain-la-Neuve or about communities in developing countries, or indeed anywhere, the size issue is vital, because obviously the problems of food supply, of transportation, and particularly of alienation, come from excessive size. One has only to look at the social indicators of communities of different sizes to see how sheer scale is important in [conditioning, if not actually determining,] the quality of life and in the behavioural patterns of the people living there. But there is another reason for size being important, and that is, of course, because it is being realized now that a city is not something in isolation: a city and the countryside surrounding it form in fact one unit. The quality of the city is determined by the quality of interaction between that urban unit and the rural environment in which it operates, and the quality of that interaction is determined by the size of the city and in the same sort of way that, [for a living] cell, the [supply by] diffusion through the wall [becomes inadequate for sustenance] when the cell grows beyond a certain size. Thus the rural/urban balance is very much determined by size, and I think that the sort of size which [Dr Laconte and his colleagues] are planning is just right for that type of operation.

As one sees in other new towns, the concept is of building up the new town not with a central commercial area, but as a series of communities of 5–10 thousand inhabitants per community, each one containing industry, commerce, schools, churches, and being in direct contact with the countryside. This is the sensible development, and if we can see this sort of development taking place, then we have a basis for preventing that flood of people to the cities—to those great and in many places most horrible slums such as surround Calcutta and Bombay and Caracas and other areas in the world.

Although we like to idealize the village life or consider it as idyllic, in many rural communities the village life is in fact dull, the individual is oppressed by having his friends and relatives peering down his neck all the time, and there is not the excitement, the diversity, the opportunity for individual fulfilment, that towns have to offer. So he goes to the city, even though he might have a full belly in the countryside—which is doubtful—and even knowing that he will probably have an emptier belly in the city. And so I think it is a question of size and diversity, and of relationship between city and countryside. This is something for which I heartily applaud [Dr Laconte's] efforts, and although it is a response to a particular set of problems in a specifically Belgian environment, the basic principles I think are universal and I would not go along so much with Professor Bryson's view that this is irrelevant in terms of the needs of the developing countries.

Just one question at the end. I would like to ask [Dr Laconte] whether [he has] established any sets of social indicators on this project, so that [it is possible to] monitor the various dimensions of the quality of life of people living there? I think that one of the great faults of many master-plans is that they spring straight from the brains of some master-planner who cannot possibly imagine the manifold dimensions of human well-being. The best he *can* do—if he is not prepared to bring the people into that planning process from the beginning, which is probably the best way—is to set up [what amounts to] a monitoring system involving the people who are living there, to find out how they are reacting. [For it can be important to know] how the quality of their life is being influenced, for example by that particular assemblage of concrete, bricks, steel and glass, and try to adapt the plan progressively. One note of concern: I was a bit dismayed to hear Dr Laconte say that he did not want to change or deviate from the master-plan. I hope that there will be built into it the possibility of flexibility, so that you *can* deviate from it in response to demands from the people living there for an urban environment which is going to maximize their well-being.

Laconte (Keynoter)

I just want to talk briefly about the kind of flexibility we have tried to incorporate. I do not know if people are familiar with the type of master-plan which is described as 'generative'—which means that it can be modified at any point or even completely stopped. Growth [with such a plan] takes place in accordance with a strongly stabilizing backbone. The main line is indicated on the map, and the extension of the buildings takes place along each side of the main line or main street. This means that buildings which are to remain for a very long period are built along the main street. Whatever increase is needed to meet the new demand can always take place in any direction—providing, of course, there is still space. This enables changes to be made at every step, which is a kind of generative planning. The inhabitants of Louvain-la-Neuve have accepted the master-plan, in that they do not wish to change the principles of density or to build over the countryside; but, with a linear pattern of development, growth can be accelerated or halted as required by circumstances and demand.

I would also like to bring to your attention the case of the new town of Letchworth, [in England,] which has enjoyed a very interesting achievement. In 1963 a large developer (a hotel group) wanted to take over the town, as it was aware that only one-third of the land had been used for buildings. The sense of community was so strong among the inhabitants that they united successfully to defeat this takeover bid. The story of Letchworth has been told by C. B. Purdom in his book entitled *The Letchworth Achievement: From the Establishment of the First Garden City to the Recent Dramatic Defeat of a Take-over*. It is my hope that within the next 20 years the same fine example of civic togetherness will be demonstrated by the Louvain-la-Neuve project.

Magnússon, Gudmundur

First I want to say that I found this all very interesting, particularly because the plan of our university [of Reykjavik] site is on the drawing board now and maybe we can use some points from this discussion even if we do not have to move to another town or to the countryside. For instance, if you take this street which goes into the town from here, it is moving back and forth on the drawing board and we are now struggling with the authorities [to determine] what to do.

I would like to know something about things associated with a question [posed by Dr Laconte]. Even if they were more or less obliged to go from old Louvain, how does this work out for the people who are going to move—for the professors, [the other staff], the students, and so on, and for communications between the university people and others? The beautiful pictures which we have seen do not tell us that story, which it would be very interesting to know. This is also related to what is happening in [other] Scandinavian countries, for there the movement has been started to decentralize institutions. They have the experience in Sweden that this can be done, but it is very expensive. We have had a committee here in Iceland which has proposed that all the engineers [should] go to Akureyri and the theologians to some other part [of the country]!

Laconte (Keynoter)

As a further answer to your question, we come again to a point raised by Mr Glasser. The fact that the university had to move has had a kind of catalytic effect: the whole community, instead of being concerned with internal arguments, suddenly realized that the institution was in danger and reacted accordingly with great enthusiasm. Fortunately, the engineering faculty was the first one to be transferred, and engineers, when faced with such a situation, react by trying to find practical means to solve the problem. Sociologists, on the other hand, might have been tempted to try to analyse the reasons for the situation being less than perfect! [Laughter.]

As to the town-and-gown relationship within the new town, this is a long story, as it results from the relations with the municipality and with the various [local] administrations. Political and administrative culture is always on the side of the dominant capitalist economy, and each time you want to do something a little different, you immediately have these people against you. You are then faced with the task of lobbying to try to bring them around to your point of view. The university community has had to do this, and although the process has not always been pleasant, a lot has been learned about how to deal with conservative bureaucrats.

Glasser (Panellist)

I must admit that I am always very depressed when I hear entirely admirable people such as Dr Royston, for example, and Professor Bryson on my right, speak, from the point of view of sophists, of conditions which they rightly deplore and feel repelled by. When, for example, they talk of Calcutta, and by inference of other places of that kind—which anyone who knows them, as I do, must feel very unhappy about—as we walk through [and see] all those wretched people living in terrible conditions, our hearts go out to them and we say how awful, why don't we build them better houses so that they can live in better conditions, give them decent employment, and so on and so forth. Instead [we should be] saying: 'Why are they here? Why have they been attracted by the media—by the powerful metropolitan influences—to leave their own environment for another one which cannot [properly] absorb them?'

Now Dr Royston, if I may say so—without any disrespect at all, because I know how this is—we are the products, you and I and almost everyone in this room, of an upbringing which is highly sophisticated; we are highly educated people, for better or worse, and we say: 'How can we possibly live in a village, where life is dull, with people breathing down our necks, and so on'. What we do not realize, I am afraid, is that this is not [necessarily] the way *they* feel, and they may say: 'You see, some of the people who are discontented will always give you a sort of respectable reason for going'. But I would only ask you to reflect upon this and try, if you can, to see it from the point of view of those who want to remain because they still have faith in their own inherited values, [or, at the worst, cannot visualize any alternative].

Franz, Kristi

While creating new communities, such as the project under discussion, does solve habitat problems for some, we still must deal with those who are left in the less desirable habitat of the cities. There the few living species are generally limited to [occasional] trees, patches of grass, and many, many human beings.

It is useless to put money into housing, transportation and employment, without also dealing with individual people, who, as Mr Glasser pointed out, have lost the feeling of responsibility, have lost their dignity and pride, and have become indifferent towards their surroundings in [the struggle] to survive. They need to be recognized as being able to be responsible for their environment and be assured of input and control of their own destiny.

Let me give an example of how this has been done in one community—Athens, in the State of Georgia—through the federally [financed] Model Cities programme. The racial situation had prevented people from having equal public input and control over their own lives, so the Model Cities programme provided for citizen input and citizen participation in the planning and carrying-out of community programmes—rather than someone else coming in and determining people's needs for them. Individual persons were listened to as their hurts, feelings, needs, ideas and efforts were expressed [and described], and a coalition of many citizens, rather than a few powerful politicians, determined the directions and futures of the community.

We need to return to the village concept in our cities—with grassroot growth and development, with citizen input and control, and with indifference and hopelessness towards one's environment replaced by responsibility and opportunity for improving the human habitat.

Jones (Chairman)

That sounds like anticipatory democracy which could I am sure be the subject of another entire [session].

Dodson Gray, Elizabeth

I would like to give an example of how much the built-structures in which we live can influence the quality of human life that we can sustain.

Many years ago my husband and I were doing graduate work at Yale University, and the university owned two large apartment complexes for graduate-student families which stood side by side on the slope of a hill, separated only by a high fence, so that each had a quite defined area for its own life. The architecturally more attractive building had been designed by the Dean of the School of Architecture; the other was like many similar buildings in the same city.

However, as places for young children to live in, the two apartment complexes provided sharply contrasting environments. The [Dean] had provided for all the children in his buildings a single play-area at one corner of the property. The result was that only a small number of families had apartments from which mothers could supervise their children at play. With all the children competing in a single play-area, and little adult supervision, that play-area became an absolute [shambles].

The built-structures of that apartment complex assured that it would be such a jungle.

Meanwhile, across the fence, were other similar children of similar families living in built-structures which provided a very different environment for young children. These were much more conventional apartment buildings, and had been subdivided into smaller communities of about 16 families who looked out on to each courtyard, so that mothers could supervise their children while they were doing household chores. Families who shared this sort of smaller-scale built environment overlooking smaller numbers of young children sharing a play-area developed very positive social relationships between the parents and the children. It was a very positive experience for us all, while across the fence next door the layout of the buildings prevented a very similar group of adults and children from developing a good climate among the young children who used their larger and more isolated play-area.

My conclusion is that structures do influence how much you can do. Also, I think there is clearly a need for the people who live in structures to help design them, because often women and children, for example, know what are the features which will contribute to a life of nurturing, and they will know this—which it is difficult for men, who do not live those lives of nurturing, to be able alone to plan for.

Crabb

I believe the whole idea of Louvain-la-Neuve in substantial part has been to give an ancient traditional university, due to the peculiarities of the situation described, a spontaneous reproduction of the Catholic University of Louvain, which has been going since 1425. It is one of the very traditional European universities, among the oldest, and consequently you have the university in the midst of the town which still shows traces of its mediæval layout and origins. Therefore I believe the idea was, when the split came, for Louvain-la-Neuve to be not a reproduction as such, but nevertheless to reproduce the ambiance of what I believe is now called Louvain-la-Veuve, namely the old part that remained after the split. [Laughter.] Also in terms of trying to reproduce the traditional university community, I have the impression that there has been a change in the mores of the university community. What strikes me particularly is that, both at Louvain-la-Neuve and at the old University of Louvain, the students do not stay around the university during the weekend. [Both are then almost] completely deserted, and Louvain-la-Neuve has no students except for the foreigners. So I wonder whether trying to reproduce this ready-made traditional university with new bricks and all does [not] run into the problem of whether the traditions on which you are basing [your planning] are indeed going to continue to serve your purpose?

Jones (Chairman)

I think that is an appropriate note on which to turn back to the panel members if they wish to make a final one or two minutes' comment. It has been an interesting afternoon and we are well past the time when we should have got out of here.

Stone (Panellist)

All I can say is that I have great admiration for those of my colleagues who managed to get from this delightful Baedeker university to Calcutta. I think it is a terrific trick [laughter]. Reid Bryson did the first step: I did not realize [even he] could do that. Mr Glasser was even better: he got into abstract theories [laughter] though I am sorry it did not inspire me to comment on towns. I have only got one particular point to make on his remarks about why people go to places like Calcutta. I think a lot of people do not realize that, in fact, ghastly though Calcutta, Bombay and such places may look to us, they still are better than people's villages, because [in the towns people] do in fact get better medical services, and a chance of some sort of a job, which is simply not the case in [most] villages.

Bryson (Panellist)

Now I can really defend myself, [though] for a while I thought I was 'way off base'! The topic of discussion is 'Extreme Urbanization', and our concern is how we can make a viable environment—a sound environment in the future—and I wish to say that I was not talking about Calcutta as being worse than the surrounding villages, or Mexico City as being worse than its surrounding villages. I was talking about the question of whether they will be viable in the future. Are they viable in the absence of fuel for transport? Will they be viable as they grow still larger [in a manner that seems to be considered inevitable and is encouraged by some highly unfortunate religious teaching]? Consider that Mexico City has one-fifth of all Mexicans living in it: can it grow to have 90% of all Mexicans in that one city? Will it survive? Are there strategies for opening it up so that, as in a Japanese city, the space between the vehicles in a used-car lot is used to plant some flowers or [grow] some food? What are the strategies to make these cities viable?

Glasser (Panellist)

To sum up, I must remain at issue with my esteemed colleagues. I conclude my remarks by emphasizing my concern with the increasing difficulty which the individual encounters in his necessary search for a desirable identity within a supportive community. That true community, in the totally interpedendent fashion I have mentioned, cannot be 'provided' as an artefact—an additional welfare service like any other, or an infrastructural additive to the consumers' shopping list! It must grow out of the diurnal needs of life—how you earn your living, how you constructively respond to the continual demands of life's necessarily reciprocal relationships—and that cannot be done, or the requirements even correctly perceived, in an urban environment where people are conditioned to an essential isolation. [I reaffirm that] if the world is to be saved at all, the way forward is by means of the 'network townships' of 6–7 thousand inhabitants each that I have already mentioned—internally knowable, self-regenerating communities whose economic constituents use adjusted technology in balance with local resources. These townships, grouped in regional clusters, would be served by regional centres of higher technology and higher education. Only in this modest fashion, in scale with himself and emotionally manageable, will Man rediscover himself, and keep his demands and desires in consonance with his abiding identity aims and with the environment. If we refuse this option, it will soon cease to be available at all; and] nothing will stand between Man and global disaster.

Laconte (Keynoter)

Just three short remarks on what has been said by people in the audience and on the panel. Someone mentioned model community experience. I would like to know more about successful experiences, as I understand that most results have proved negative. I have been interested in the recent studies of David Harvey, who wrote the famous *Social Justice in the City*, which attacks the problem by looking into the banking and mortgage systems, by investigating the factors influencing a person's choice of housing, and by trying to explain why there are 50,000 empty homes in New York. It would seem that this is not by chance but due to links between various policies.

We all know that the problem of extreme urbanization does not lie in the cities but in the countryside, which people leave because they no longer feel any attraction in living there. The question is, what can be done to improve the urban extensions? We can, of course, improve the implementation of the suburbanization process by creating small entities, which would form a kind of densely-built space rather than an overall sprawl, leaving no [clear] distinction between countryside and city. These small clusters would be potential communities.

This concept is valid for any country. A high-density–low-rise and low-energy-consuming pattern, which would nevertheless avoid becoming urban sprawl in the American sense, has a certain validity which could be universal. However, the problems are political and related to many factors.

Lastly, I would like to answer Mr Glasser's remarks. The problem of the empty city at night or at the weekend is basically that of our few decades of experience in functionalism with regard to architecture and planning. This problem does not arise provided you do not have split-functional spaces. It arose less frequently, for example, in the old cities. I believe that one of the positive discoveries of the latest trends in planning has been that the ideas of Le Corbusier, who originated the monofunctional use of space, [neglected to allow for the fact] that they were dealing with human beings in the cities and not just with functions.

Jones (Chairman, concluding Session)

I am sure you would like me, on behalf of the panel, to thank everybody who took part in these [lively and enlightened exchanges], and I am sure you would like to join me in thanking the main speaker, the panel and the other participants. [Applause.]

Session 9

WHITHER THE LIFE-SUPPORT SYSTEM?

Keynote Paper

Whither the Life-support System?

by

EUGENE P. ODUM

Director, Institute of Ecology, University of Georgia, Athens, Georgia 30601, USA; Tyler Laureate for Ecology, 1977

&

ELDON H. FRANZ

Institute of Ecology, University of Georgia, Athens, Georgia
new address: *Environmental Research Center, Washington State University, Pullman, Washington 99164, USA.*

Introduction

The human population of the globe has been increasing according to some superexponential law. One possible form of this relationship is:

$$N = \frac{1.79 \times 10^{11}}{(2026.87 - t)} \times 0.99$$

where N is the total population size, and t the time in years AD. This expression was derived by fitting a model, which assumes the rate of increase to be a weak monotonic increasing function of N, to 24 estimates of the human population of the world ranging over the last 2,000 years (Foerster *et al.*, 1960). The time, $t = 2026.87$, was appropriately identified as 'Doomsday'. The assumed relationship between the rate of increase and N was logically justified on the grounds that an increasing population would reduce environmental hazards to survival by forming coalitions to combat them. By thus invoking the 'principle' of 'adequate technology', the authors concluded, sardonically, that 'our great-great grandchildren will not starve to death. They will be squeezed to death'. It is somewhat chilling to note that current population estimates are higher than would be predicted from the above equation. This particular version of 'Doomsday' need not concern us, however; our food and life-support systems seem likely to give out first.

The concept of 'life-support system' has been given impetus by the popularization of the 'Spaceship Earth' metaphor (Boulding, 1966; Ward, 1966), by efforts to design bioregenerative systems for manned spaceflights (H. T. Odum, 1963; Cooke, 1971) and, perhaps most of all, by the dramatic failure of Apollo 13, the only one of the Apollo series of journeys to the Moon that had to return to Earth without completing its mission. The failure was not due to any flaw in propulsion or guidance systems, but to a malfunction in the life-support module. When this happened, the only mission possible was for survival, reminding us as inhabitants of Spaceship Earth that if life-support is threatened, nothing else matters.

That natural ecological systems function as the Earth's life-support 'module' is, of course, fundamental ecological knowledge which provides the basis for our understanding of Man's relationship to the biosphere. As Woodwell (1974) has put it, 'We live as one species within a biosphere whose essential qualities are determined by other species'. It is those other species—especially plants and microorganisms—that serve, as it were, the bioregenerative needs of Man on Earth. Furthermore, a homeostatic biosphere capable of supporting Man and his machines, with living organisms providing the essential controlling mechanisms, evolved long before Man himself appeared on the scene—a perspective inherent in the *Gaia* (Greek for 'Mother Earth') hypothesis, as advanced by Lovelock (1972) and Lovelock & Margulis (1974).

Perspective

In addressing the question, 'Whither the life-support system?', attributes of the system, as well as of Man and his relation to it, provide suitable approaches. A selection of important considerations includes the following:

Stability

Stability of the life-support system is desired to guarantee that essential resources will always be available. While ecologists are not in full agreement on many aspects of stability theory, both intuitive and experimental wisdom indicate that functional complexity at the ecosystem level enhances stability because such complexity provides many alternative methods of resisting or recovering from perturbation. Unfortunately, in pursuit of his own self-interest, Man tends to reduce the complexity of the life-sustaining environment, and thereby weakens stability.

Carrying Capacity

A workable definition of carrying capacity is the maximum population that can be sustained in a habitat without degradation of the life-support system. On a local scale, increase in air and water pollution, malnourishment, crime, and so on, indicate that carrying capacity is being exceeded in some parts of the underdeveloped world and in some large urban districts of the industrial world.

The Tragedy of the Commons

The 'tragedy of the commons' involves exploitation of a resource with some positive utility gained by an individual and any associated negative utility shared by all. When coupled to the desire for economic gain, each individual is locked into a system of pursuing self-interest to the ruin of all (Hardin, 1968). As a large part of the Earth's life-support system, particularly the atmosphere, rivers, forests, lakes and oceans, are in the general category of 'commons' (i.e. in common ownership), the overall result of the conflict between private and public good is a degradation of the life-support system and ultimate reduction in carrying capacity—unless, of course, mankind acts in a positive, organized way to preserve the priceless 'commons'.

ASSESSMENTS

So far, or at least until very recently, the activities of Man, as we have already noted, have had local adverse effects on life support but no clearly discernible effect at the global level. Thus, climate in the vicinity of large cities has been definitely affected by the intense metabolism of the city, but, as yet, no overall effect on global climate has been detected. However, the marked increase in atmospheric carbon dioxide over the past 40 or so years indicates that the massive burning of fossil fuels and extensive deforestation are combining to add CO_2 to the atmosphere at a rate that overtaxes the homeostatic regulatory capacity of the global life-support module (Bolin, 1977; Stuiver, 1978). If the trend continues, then most climatologists and meteorologists agree that climatic changes will occur, and that such changes will be mostly detrimental to Man's interest (Bryson, 1974).* Much of the difficulty of forecasting future climate is due to the possibility that several effects may, at least for a time, offset one another (Rasool & Schneider, 1971); increase in atmospheric turbidity could offset the heat retention effects of CO_2, for example.

Whatever the trends, human population growth, industrial development and intensive energy consumption have reached the stage where Man can no longer take life-support for granted; he must now seriously consider ways and means of preserving and, perhaps, repairing the life-support system.

E. P. Odum (1975, Ch. 2) has suggested partitioning ecosystems of the biosphere into four compartments according to their energy sources: (1) solar-powered natural ecosystems, such as oceans, lakes, forests and grasslands; (2) naturally subsidized solar-powered ecosystems, such as tidal marshes, flowing streams, and upwelling coastal zones; (3) man-subsidized agroecosystems; and (4) fuel-powered, urban-industrial sys-

* Cf. Session 2 (including discussion) above.—Ed.

tems. Category (1) functions to provide basic life-support, while ecosystems in category (2), although not as extensive in global areas as those of category (1), are especially valuable because their higher power-level (energy-flow per unit time per unit area) increases their capacity to assimilate wastes from categories (3) and (4). Agricultural systems, of course, provide the food component of life-support for Man and, as we shall discuss later, they may or may not contribute to other aspects of life-support—depending on the kind and amount of subsidies that Man uses to increase yield.

The urban–industrial systems, which can be equated with the geographer's 'Standard Metropolitan Districts', are consumers rather than producers of life-support functions, and thus are metabolically parasitic on surrounding natural or semi-natural ecosystems for all life-supporting necessities. As currently managed, our cities and industries are fuel-powered rather than solar-powered. Comparatively little of their high energy output is 'fed back' for waste treatment and recycling, leaving the burden of this work largely to Nature. Although occupying only a small portion of the Earth's surface—perhaps no more than 2% of land area—the power level is so intense, and the waste products are so poisonous to life, that fuel-powered, urban–industrial ecosystems can quickly saturate the carrying capacity of adjacent life-supporting habitat. Let us now review the status and future prospects of those of the above compartments that together comprise most of the Earth's biological life-support module.

Agroecosystems

Agroecosystems occupy about 15% of the land area of the globe, with both the extent and intensity of use likely to increase in the coming decades as the rising world population places increasing demands on food production. Agroecosystems now come in two contrasting types: (*a*) the labour-intensive system where human labour and animals provide the 'energy subsidy', and (*b*) the energy-intensive type where fuel, chemicals and machines augment and intensify the sun-powered photosynthesis.

Impressive yields, especially of staple grains bred for mass monoculture, have been chalked up by energy-intensive agriculture which is most efficient on a per-man-hour basis; for example, one farmer in the USA feeds 48 other human beings—far more than in any other country. On the other hand, many labour-intensive systems have a higher efficiency on a calorie-to-calorie basis—that is, they produce more food per unit of energy subsidy expended. Where crops and land-use are diversified and waste products are used or recycled, a less energy-intensive agroecosystem also places less strain on other life-support systems; in fact, such agriculture may make a positive contribution to life-supports in ways other than food production. In contrast, where fuel energy, machines and chemicals replace human and animal labour, the energy-intensive agroecosystem comes to resemble more and more the fuel-powered, urban–industrial system in terms of deleterious impact on the life-support environment. Leakage of toxic chemicals and nutrients from intensely cultivated regions is increasingly worrisome, and soil erosion, once thought to be 'cured' by modern technology, persists or even increases (Carter, 1977). Cultivation of high-protein crops that require heavy fertilization (Bhatia & Rabson, 1976) will tend to increase further the deleterious impacts on life-support.

Industrialization of agriculture in many countries has not occurred because of a shortage of labour, but because of the availability of cheap and abundant fuel—especially petroleum—and a growing emphasis on 'agrobusiness' (where crops are not so much thought of as for life-giving food, but as commodities with high economic value in the manner of manufactured goods). Energy-intensive agriculture, coupled with an agrobusiness economy, in fact, tends to create chronic unemployment if the industrial sector, which is also increasingly automated, is unable to provide jobs for displaced farm labour—a trend that has been called by Feder (1971) 'The Rape of the Peasantry'. Also, the monoculture nature of fuel-subsidized agriculture threatens the stability of the agroecosystem, making it more vulnerable to perturbations, both natural (weather, disease) and man-made (market fluctuations).

Because of these evident problems with industrial agriculture, and in view of scarcity and rising costs of fuel, many expert panels who study, report to, and advise, governments are recommending: (1) a shift to less energy-intensive agriculture in developed countries, and (2) improvement in technology for enhancing existing labour-intensive systems in developing countries (*see* Pimentel *et al.*, 1973). If world opinion does move in these directions, the global life-support system will surely benefit!

The Oceans

The oceans are by far the largest and most important component of the global life-support apparatus. If it were not for the oceans' ability to stabilize the atmosphere, including their capacity to absorb CO_2, the global climate would already have been changed by Man's activities. While some of the stabilizing feedback regulators in the sea involve purely physical–chemical processes, the living components of the sea are absolutely vital to maintaining homeostasis. Thus, any threat to marine life is also a threat to global life-support. Without seafood, protein starvation would now be epidemic, not just chronic, and now the oceans' 'carrying capacity' for pollutants is reaching a limit along many sea coasts. The oceans are the ultimate 'commons' that must be protected by international law, reinforced by a realization by each nation that it is emphatically in its own self-interest to abide by international agreements that may now be necessary to preserve the integrity of the whole oceanic system.

Forests and Wetlands

Forests and wetlands contribute to the stability of the life-support system to a greater extent than is indicated by the area occupied by these ecosystem types. Forests, for example, now occupy only 30% of the land area of the globe, but constitute *ca* 90% of the biomass (i.e. total amount, or weight, of living matter). For this reason, forests are important carbon 'sinks' and climate stabilizers. Because large trees and associated organic soils take up and store large amounts of CO_2, forests, like the oceans, are a buffer against rapid changes in the atmospheric CO_2 content. In arid regions, woody vegetation also tends to reduce albedo (reflectance of solar radiation) and thereby enhances rainfall, although there is controversy as to the degree to which forest removal has contributed to man-made desertification in such regions as northern Africa.

Forests also function as moderators, or 'governors', of the downhill flow of water in the hydrological cycle. As solar energy powers the uphill part of the cycle, the return downhill flow provides energy for work that is of vital benefit to mankind. The immense economic value of such work of Nature is seen, not only in the potential for hydroelectric power, but more generally in the capacity of a forested watershed to provide high-quality water for urban–industrial areas at a very low direct cost to Man. The monetary cost of getting such water artifically by recycling waste-water or desalinizing sea-water is hundreds or even thousands of times as great, because expensive fuels would have to be used to power such substitute recycling systems.

Up to now, most domestic and industrial wastes have been discharged into bodies of water (rivers, lakes or shallow coastal waters). As the capacity of these very efficient, but area-limited, natural treatment systems is overtaxed, interest will naturally turn to disposal of treated wastes on land. With proper management and restraints, forests can function as living filters for degradable wastes and noise. Again, if such natural or semi-natural tertiary treatment systems are not available, the economic cost of completely artificial waste treatment would be prohibitive.

In contrast to the values we have been discussing, the value of forest products, such as lumber, paper and chemicals, is well recognized and need not be detailed here. Forest products are important in the economy of any country that is fortunate enough to have forests and to manage them wisely (as, generally speaking, is the case in North America and Scandinavia).

Finally, wood which today is used as fuel, chiefly in poor countries, will undoubtedly increase in value as a renewable source of energy and industrial chemicals throughout the world as other fuels and chemical feedstocks become depleted.

The point to make is that direct commercial and recreational values are really 'bonuses', and that the primary value of a healthy forest cover is life-support. Inevitably, there has to be 'trade-off' if one wishes to maximize for yield. If a natural, large-biomass, diverse and multiple-age forest is converted into an even-age monoculture with short-rotation cutting, then the yield of fibre will be increased, but some life-support functions will be decreased—owing to the reduction in standing crop and to the soil disturbance and heavy fertilization that would ultimately be needed to maintain a high yield. Preserving at least the present 30% natural-forest coverage world-wide would seem to be a worthwhile goal for international planning. This would mean, of course, maintaining a 50–60% coverage in regions with wet climates (to balance areas that are too dry for forests). In our opinion, short-rotation silviculture (i.e. 'tree-farms') should be considered a part of the agroecosystem and should be practised on land that is suitable for agriculture in general. Such a policy would reserve hilly and mountainous land, thin soils, floodplains, and other 'non-arables', for natural forests that are capable of self-maintenance with minimum management effort by Man.

Wetlands (estuarine, riverine and lacustrine, and also marshes and swamp forests) have a very special value in maintaining global cycles of carbon, nitrogen, sulphur and other vital elements, because of the anaerobic sediments (in the waterlogged, reduced zone) that constitute a unique feature of most wetlands. In this zone, microbial transformations 'gasify' many important nutrients, such as nitrates and sulphates—and, of course, organic matter. The evolution of methane and CO_2, and of ammonia and other gaseous nitrogen compounds, thus recycles important elements into the atmospheric pool. Recent studies indicate that methane plays an important role in maintaining the stratospheric ozone layer that provides the protective shield which is so vital to life on Earth.

Wetlands, of course, have many other values on a local or regional scale—as, for example: (1) water-storage reservoirs that help maintain ground-water levels, (2) filters for agricultural or industrial chemicals, (3) scenic open space, and (4) sources of food and wildlife. Wetlands should be preserved wherever possible for the very good reason that their multiple-use value is usually greater on an area basis than any other kind of landscape that might replace them. In other words, existing wetlands should *not* be converted to something else unless it can clearly be shown that the 'something else' contributes more to life-support and the quality of human existence.

Mitigation

Fortunately, the broad patterns of degradation of life, with profound implications for ourselves, are now being recognized throughout the world. Adverse trends can be reversed; land can be reafforested; polluted waters can be restored to health; marshes can be rebuilt (Thorhaug, 1977); and man-made deserts can be reclaimed (Lowdermilk, 1960).

Reciprocal Design

The time has perhaps come to think in terms of what we like to call 'reciprocal design', in which the engineer and the ecologist work together to create a harmonious coupling of the urban–industrial and the life-support systems. In most countries the first response to pollution problems involves a regulatory approach; limits to waste discharges into the 'commons' are set, to be adhered to if a firm or municipality is to receive a legal permit to operate. As a second step that would complement the regulatory process, reciprocal design would place the emphasis on cooperation rather than confrontation between the producers and the consumers, as it were.

Everyone, then, would have a common goal of maintaining a stable and high-quality living-space.

To take an example, Industry A creates a highly alkaline waste-product containing heavy-metals, toxic organics, degradable organic matter and inorganic salts. Reciprocal design would require that the industry design the whole process, including in-plant treatment facilities, in such a way as to reduce pH, remove the metals and other toxins, and reduce the concentration of degradables. The responsibility of the applied ecologist who is hired to work with the industry would be twofold: (1) to inventory the assimilative capacity of the surrounding waters, forests, marshes, or other environmental components into which the treated effluent must eventually be discharged, and (2) to establish self-designing semi-natural buffer ecosystems, such as holding ponds filled with vegetation in which the effluents will reside until concentrations reach low enough levels to be 'safely' discharged into the general environment.

There is much to be said in favour of the concept that every industrial complex and every city needs to have its own assimilative green-belt life-support system—not only to reduce pollution-stress locally, but also to protect the global life-support system. Thus in planning any new development, 'strip cities' (that is, continuous urban development along a coast or river) should be guarded against. Rather should high-energy-consuming areas be separated from one another by wide green-belts of croplands, forests, marshes, and so on. Here is a challenge for architects, planners and engineers to expand their scale of thinking and use their imaginations at new levels!

'Market Failures'

What might be termed the 'shortsightedness' of current economic systems stands as the greatest obstacle to rational planning for the preservation of life-supporting ecosystems. Economists are in general agreement that the market system, which is quite efficient when it comes to allocating man-made goods, often fails to function in the public interest when it comes to allocation of natural resources. Such a 'market failure' is especially evident when it comes to life-support 'goods and services' which, by and large, are accorded no value (the so-called 'free goods') or are grossly underpriced in comparison with man-made goods and services. Water, a product of the natural hydrological cycle, is a good example of an 'underpriced' life-support commodity. Gertel & Wallman (1960) have calculated the economic yield per unit of water for various uses, finding that the monetary return is much higher when water is used to make plastics or mine coal than when it is used in agriculture or for people to drink. In the market system, allotment to food and drink thus loses out in competition with use in manufacturing. Theoretically, as water becomes scarce the market would eventually act to correct the situation, as the rise in prices of food and drinking-water would increase allocation to these uses. However, before the 'correction' could occur, irreversible damage might be done to the life-support system and the economy. It would be much more prudent to recognize these shortcomings of the market system and move to correct the obvious 'failures' before it becomes absolutely necessary.

As yet, there is no consensus among economists, or anyone else for that matter, on how to correct these market failures. Westman (1977) has reviewed suggested remedies that involve pollution taxes and/or charges for the use of the 'commons'. While these 'quick-fix' measures would help, they would not be at all adequate. One approach which we have suggested (Gosselink *et al.*, 1973) is to extend economic cost-accounting to include the work of Nature as well as the work of Man. By using energy quality as a common denominator, one can first assess the life-support or other useful work that is going on in a natural area in terms of energy-flow. Then, as money and energy are counterflows, one can convert energy-flow (corrected or weighted for quality) to equivalent monetary units. When we did this for a productive coastal estuary, its total monetary value to the public (but not to an individual 'owner') was 40 times as great as the generally assessed economic value based on harvested fishery products obtained from the estuary.

Because life-support is really 'priceless', many thoughtful resource scientists feel that we should not try to put a monetary value on it, but, instead, recognize that large areas of the solar-powered natural environment are essential parts of the life-support module of Spaceship Earth, and take such areas out of the market entirely. This, of course, is what we do when we set aside national forests, wildlife refuges, parks, or industrial green-belts, to be in the public domain and no longer for sale. Even when this is done, political and economic pressures from the market-place can still degrade life-support functions—as, for example, witness the pressure in the USA for wholesale clear-cutting of National Forests.

Recoupling Man and Nature

Perhaps the only long-range solution to human shortsightedness is to build world opinion for a need to recouple the four ecosystem types that we have been discussing—instead of trying to solve our problems by piecemeal management of each system as if it were a separate, unrelated entity. Mankind has evolved to exert increasing control over his environment and thus achieve greater freedom from its constraints. Much good for us had come from this 'uncoupling' strategy, but all of the feedback signals we have enumerated in this paper point to the necessity of reversing this trend if we are to continue to survive and thrive as saturation levels of population and pollution are approached. We must be plugged back into the system. Our role is precisely that of comparator in a cybernetic device (Wiener, 1948), and we are suited for such activity as our very nature is determined by feedback (Paiget, 1971; Potter, 1971). As Holling (1969) has noted, however, the signals are weak. For thousands of years our interface with the life-support system was a 'way of life', but now that interface is occluded by the expedient of 'making a living'. For thousands of years tribal chiefs regulated their collective impact on the fragile Sahel ecosystems, adjusting to the vagaries of climate. The drive to produce cash-crops for foreign exchange then changed all that, and unchecked populations exceeded the carrying capacity of their fragile environment (Wade, 1974). The tragedy is that the inherent wisdom and common sense of local people with respect for carrying capacity and local control of environments has been replaced by well-meaning, but faulty, outside advice.

Education is an essential element of this process of using wisdom, but it has too frequently been confused with indoctrination, and too often fragmented. We find C. P. Snow's *Two Cultures* engaged in a game of cat-and-mouse—the social scientists deferring solutions of certain political problems until technical advances are achieved, and the natural scientists deferring the solution of certain technical problems to politics. The concept of local people dealing with local problems is also essential. Little wonder, then, that these three: education, integration and regionalization, are mentioned with increasing frequency as we seek to maintain our life-support systems (Ashby, 1976; Engel, 1977; E. P. Odum, 1977).

Conclusions and Summary

The natural solar-powered ecological systems such as oceans, forests, wetlands, lakes and rivers, *plus* man's agroecosystems, function as the Earth's life-support module. Human population growth, industrial development and intensive energy consumption have reached a level that is not only causing local deterioration of life-support carrying capacity but is also beginning to have global effects, as indicated by the rise in the CO_2 content of the atmosphere. Man can no longer take life-support for granted, but must seriously consider ways and means of preserving, and repairing where need-be, the life-support system.

In addition to 'quick-fix' solutions (for example, point-source pollution control), the following long-term strategies are presented and discussed: (1) a shift to a less energy-intensive but more energy-efficient agriculture; (2) international agreements for the protection of the biological integrity of the oceans; (3) preservation of at least 30% forest coverage world-wide; (4) preservation of wetlands and other special ecosystems that have a high capacity for assimilation of wastes; (5) development of a new technology for 'reciprocal design' to produce a more harmonious coupling of the urban–industrial and life-support systems; (6) correction of the 'market failures' inherent in current economic procedures that now tend to increase the inequity between individual and public benefits, so creating 'the tragedy of the commons', and (7) promotion of educational efforts world-wide to establish public opinion for the need to 'recouple Man and Nature' and adopt more 'holistic' approaches to solving societal problems.

References

Ashby, E. (1976). A second look at doom. *Encounter*, **46**(3), pp. 16–24.

Bhatia, C. R. & Rabson, R. (1976). Bioenergetic considerations in cereal breeding for protein improvement. *Science*, **94**, pp. 1418–21.

Bolin, B. (1977). Changes of land biota and their importance for the carbon cycle. *Science*, **196**, pp. 613–15.

BOULDING, K. (1966). The economics of the coming Spaceship Earth. Pp. 3–14 in *Environmental Quality in a Growing Economy* (Ed. H. Jarrett). Johns Hopkins Press, Baltimore, Maryland: pp. 3–14.
BRYSON, R. A. (1974). A perspective on climatic change. *Science*, **184**, pp. 753–60.
CARTER, L. J. (1977). Soil erosion: The problem persists despite the billions spent on it. *Science*, **196**, pp. 409–11.
COOKE, G. D. (1971). Ecology of space travel. Chapter 20 in *Fundamentals of Ecology*, 3rd edn (by E. P. Odum). W. B. Saunders, Philadelphia, Pennsylvania: pp. 498–509.
ENGEL, G. L. (1977). The need for a new medical model: A challenge for biomedicine. *Science*, **196**, pp. 129–36.
FEDER, Ernest (1971). *The Rape of the Peasantry*. (Anchor Books) Doubleday, Garden City, NY: 318 pp.
FOERSTER, H. VON, MORA, P. M. & AMIOT, L. W. (1960). Doomsday: Friday, 13 November, A.D. 2026. *Science*, **132**, pp. 1291–5.
GERTEL, K. & WALLMAN, N. (1960). Rural–urban competition for water. *J. Farm Econ.*, **45**, pp. 1332–44.
GOSSELINK, J. G., ODUM, E. P. & POPE, R. M. (1973). *The Value of the Tidal Marsh*. Center for Wetland Resources, Louisiana State University, Baton Rouge: 32 pp.
HARDIN, G. (1968). The tragedy of the commons. *Science*, **162**, pp. 1243–8.
HOLLING, C. S. (1969). Stability in ecological and social systems. Pp. 128–41 in *Diversity and Stability in Ecological Systems* (Eds. G. M. Woodwell & H. H. Smith). US Dept of Commerce, Springfield, Virginia: pp. 128–41.
LOVELOCK, J. E. (1972). Gaia as seen through the atmosphere. *Atmos. Environ.*, **6**, pp. 579–80.
LOVELOCK, J. E. & MARGULIS, L. (1974). Atmospheric homeostasis by and for the biosphere: The Gaia hypothesis.. *Tellus*, **26**, pp. 1–10.
LOWDERMILK, Walter C. (1960). The reclamation of a man-made desert. *Scientific American*, **202**(3), pp. 54–63.
ODUM, E. P. (1975). *Ecology; The Link Between the Natural and Social Sciences*. Holt, Rinehart & Winston, New York, NY: 244 pp.
ODUM, E. P. (1977). The emergence of ecology as a new integrative discipline. *Science*, **195**, pp. 1289–93.
ODUM, H. T. (1963). Limits of remote ecosystems containing Man. *Amer. Biol. Teach.*, **25**(6), pp. 429–43.
PAIGET, J. (1971). *Biology and Knowledge*. University of Chicago Press, Chicago, Ill.: 304 pp.
PIMENTAL, D., HURD, L. E., BELLOTTI, A. C., FORSTER, M. J., OKA, I. N., SCHULES, O. D. & WHITMAN, R. J. (1973). Food production and the energy crisis. *Science*, **182**, pp. 443–9.
POTTER, V. R. (1971). *Bioethics*. Prentice-Hall, Englewood Cliffs, NJ: 205 pp.
RASOOL, S. I. & SCHNEIDER, S. H. (1971). Atmospheric carbon dioxide and aerosols: Effect of large increases on global climate. *Science*, **173**, pp. 138–41.
STUIVER, M. (1978). Atmospheric carbon dioxide and carbon reservoir changes. *Science*, **199**, pp. 253–8.
THORHAUG, A. (1977). Symposium on restoration of major plant communities in the United States. *Environmental Conservation*, **4**(1), pp. 49–50.
WADE, N. (1974). Sahelian drought: No victory for Western aid. *Science*, **185**, pp. 234–7.
WARD, Barbara (1966). *Spaceship Earth*. Columbia University Press, New York, NY: viii + 152 pp.
WESTMAN, W. E. (1977). How much are Nature's services worth? *Science*, **197**, pp. 960–4.

WIENER, N. (1948). *Cybernetics*. John Wiley, New York, NY: 228 pp.
WOODWELL, G. M. (1974). Ecosystems and world politics. *Bull. Ecol. Soc. Amer.*, **55**(1), pp. 2–5.

Discussion (Session 9)

Miller (Chairman, introducing)

This morning our subject is the [orientation of the] life-support system [of Man and Nature, than which nothing could well be more vital and absorbing. But as my predecessors in this Chair have all been more or less world figures, I will have to introduce myself and do my best at justifying the International Steering Committee's choice before presenting in turn] the speaker and panellists. I am Richard Gordon Miller of the United States, where I live in the far west. The work I do is in [or for] the Foresta Institute for Ocean and Mountain Studies—a small foundation concerned with the natural environment and its biota, and with research and education in field ecology and environmental studies. For the past 18 years we have been working with graduate and undergraduate university students, and with high-school and [other pupils], in the study of natural history, field ecology and [the maintenance of] environmental quality. I have been a director of research, of youth programmes and of summer camps, and have explored in South America and Antarctica, besides being a Nevada rancher. I am trained as an ethologist and vertebrate zoologist, particularly in ichthyology and fisheries biology, and have an abiding curiosity in the fish fauna of the waters of the world.

Conferences such as this on the environmental future help me in my work towards improving curricula, teaching methods and administration in schools. They also help me in planning 'out-of-school education' programmes to promote public awareness, while encouraging research and engendering support relating to the state of environmental quality.

Our subject today, of the global life-support system [and where it is going], emerges as the most crucial concern of mankind. But how we can define it differently from the other subjects of this Conference, I am not sure. Our keynote speaker has a definition and we will hear it. I would like to suggest that, throughout the Conference, this life-support system be treated as an underlying theme relating to the whole of Planet Earth, which we must keep before each [session of this free international] forum.

This system is expressed at different levels throughout the natural world, and in ways that we ought to consider. First, an unnatural expression of such a system is the orbiting capsule for Man in Space, incorporating all the necessary requirements for sustaining life, at least on a short-term basis. You are familiar with the elements that go into that kind of system, and will know of the planning required to provide energy, food and cycling of nutrients and wastes, to make an artificial life-support system complete. Right from the Algae providing food through photosynthesis, it is all based on Nature, which has done it much better than any man has yet devised. The natural systems have come as the product of evolution, by trial and error and change over so many ages that Man cannot hope for an adequate mimicry or substitution in his short span.

I am evolutionist enough to feel that there is no substitute which we can provide through manipulation of genes and environs beyond some temporal satisfactions for us. It seemed once as though it would be very clever of us if we could create changes approaching what Nature has accomplished over a long period, [but now I at least am convinced that instead] we must be clever enough just to save the systems of Nature as they are.

I would like to call your attention to other possible descriptions of life-support systems. Besides the artificial kind just mentioned, there are examples in Nature and in our daily life. In my own community in western North America a river system supported a Paiute Indian tribe, whose economy had evolved with the river's fishery resource. In 1912, after ages of success in this fishery, the United

States Government entered the scene and constructed a dam on the river, diverting water to another valley for irrigation agriculture. This dam deprived the large and abundant anadromous trout of access to its spawning grounds upstream, and reproduction of fish came to an end. In a mere few months the [basis of the] fishery was exterminated.

Recently, after 50 years of deprivation, the Indian tribe brought a claim against the Government, asking $27m in compensation for economic loss. In other words, [ruination of] their life-support system had deprived them and they were seeking recourse. I was very intimate with the subject because I was brought in to help to determine what was a fair share. But in settlement of the claim for somewhat less, a measure of the value of a life-support system was recognized. Now continuing litigation seeks to restore natural water conditions, and militates towards stabilizing the economy, the ecology, and ultimately the natural life-support system. Thus whether it be on lands of American Indians in Nevada or in waters of Iceland and its cod fishery, or indeed on the whole global system, we must think of life-support as being a natural phenomenon which we must be very careful to maintain.

We have as usual a panel to respond to the speaker, who is Dr Eldon H. Franz, of the Department of Botany of the University of Georgia—also in the United States. He is a member of the Executive Committee of their Institute of Ecology, which is known very widely for the style and [exemplary] manner in which it has approached ecological problems. [As co-author of the paper, Dr Franz is presenting it in the absence of Professor Eugene P. Odum, for reasons which are at once compelling but happy. For, as indicated on the cover of the version of their paper which is before you, 'Owing to Professor Odum's very recent receipt of the Tyler Award for Ecology, which his friends and admirers applaud but which has necessitated his participation in ceremonies ranging from the White House, Washington, to Pepperdine University in Malibu, California, the main drafting of this paper for our Conference had to be done in the end by his "right hand", Dr Franz. The final paper for publication will, however, as promised and announced earlier, be by Eugene P. Odum & Eldon H. Franz, with possible inputs also from Professor Howard T. Odum and/or Dr William E. Odum. There has not been time to pre-edit or otherwise treat the present version before circulation' at our Conference. As it is a rather new and little-publicized prize, we might add that the Tyler Award for Ecology is among the most valuable in the world, and perhaps the most valuable given annually, being at least until recently ahead of the maximum amount given for any of the Nobel Prizes and little if at all less desirable in other ways. So with our President present we can surely feel trebly honoured at this selection of one of our own chosen members!]

Our panellists who will tackle the main theme of this Session from different angles are Dr Norman Myers, from Kenya but known around the world as a celebrated wildlife research worker, writer and photographer, who will be getting his eye in for the next exacting chairmanship. Then we have Mr Erico Nicola, from Switzerland and the Netherlands, who has been chief of the meteorological services of the Netherlands Air Force in Holland as well as in Indonesia, and has been director of meteorological and aerological stations in the Swiss Alps. [A board member of our sponsoring Foundation for Environmental Conservation, he is a critical] student of human–environmental condition (or I suppose you could say incondition or lack of condition), which he observes from the vantage-point of his delightful estate on Lake Geneva through gatherings of enlightened people both young and old whom he invites for free discussion. Finally we have Miss Leila J. Baroody, from Saudi Arabia and the United States, where she is a graduate student in ecology and environmental planning at Cornell University. She is a qualified pilot, having learned to fly in order to undertake aerial photography to study the changing patterns of vegetation on the Earth's surface, and her father is Saudi Arabian Ambassador to the United Nations in New York where, it may be recalled, he put forward last year a proposal for a one-per-cent-barrel levy on oil from all OPEC countries for environmental purposes which we understand is still before the UN General Assembly.

Franz, Eldon H. (Co-Keynoter, presenting paper)

It was a singular honour for me to have been asked by Eugene P. Odum to be here instead of him this week, and I might say an even greater honour to have had Secretary-General Nicholas Polunin approve my participation with you in this way today. In dealing with the problem of 'life-support', there are at least two extremes to which one can go in coming up with a definition of the 'life-support system'. At one extreme we can ask 'What are the minimum requirements for life?' I find that, frankly, to be uninteresting. A far more interesting approach [seems to be] to ask what is it about this Earth of ours which, unique to our knowledge in at least the visible universe, has permitted the elaboration of life-styles or ways of life that we have observed. One of the interesting things about our life-support system and our relationship to it is that this relationship is dynamic. Throughout human history we have observed the process of the human population adapting [itself]. Our major question at this point [of evolution] is whether or not we can continue to adapt [ourselves], in the face of increasing unadaptiveness of our life-support system, to satisfy our fundamental requirements for life.

We have heard of many scenarios concerning the nature of ecodisasters, but I have to say that my own personal favourite is one which was contrived by [Foerster, Mora & Amiot] in a little paper [published in 1960] in *Science*, and I think that paper still bears re-reading by all of us because of the very interesting way in which the arguments were couched. In particular we have at several times during this Conference discussed the spurious precision of equations, and [these authors'] approach is to provide us with an empirical formula that fits estimates of human population size from around the year zero up until about 1958. Utilizing this empirically-fitting formula, they do the next obvious thing for anybody who has such a tool in hand—extrapolate it. They extrapolate it back into the past and also into the future.

One of the very curious things about this particular formulation of the population-growth model is that they did not assume negative feedback based upon an increasing population-size but rather a positive feedback. For they believed, or at least argued in this paper, that an increasing population would actually result in a reduction of environmental hazards to survival, because the people engaged in interacting with one another would form coalitions to combat those hazards. They therefore invoked what I believe is properly called the Principle of Adequate Technology, which is that, regardless of the hazards that Man of the human kind faces—and I hesitate whenever I say 'man', because I know somebody is going to nail me for that [laughter]—we would always remain innovative enough to continue the process of adaptation. These authors point out, somewhat sardonically at the end, that actually the pessimists among us—the Malthusian pessimists—were quite wrong, that in fact if their model is extrapolated, we see that the human population does not starve to death but rather gets squeezed to death. And I have heard that somebody has calculated the rate at which the mass of human population would be expanding into space at the time that this particular ecodisaster is approached, and it [is quite fantastic].

I believe that we have come quite a long way in recent years in terms of our perceptions of our real relationship to our life-support system, and I believe that we are beginning to reach a consensus that the next 30 years [will be] crucial in the entire history of mankind's adaptive [evolution]. There is certainly no basis to assume that population growth will continue unabated, and surely no basis to believe that [the above-cited authors'] model will continue to be an adequate representation of our population's growth for very long into the future. Interestingly, though, when I calculated the particular value for population that was predicted by this model for the year 1977, the actual estimate which I was able to obtain was in excess of what was predicted by the model.

I think it is important for a conference like this to develop the concept of two basic kinds of approaches to the problems of adaptation. I will refer to these as 'hard' and 'soft' approaches. Now these are terms that have been battered around a great deal. We talk about the 'hard sciences' and the 'soft sciences'. And I raise this point at this time because [rather chronically] during the Conference people have been asking about the necessity for quantification. Well, it turns out that in terms of 'soft' understanding of systems, we may have intuitive feelings about the way in which [something] will behave—based upon our understanding of its internal workings, but without being able precisely to place a numerical value on various thresholds for its different kinds of behaviour. And I think this is very pertinent to our discussions here, because establishing the limits of the carrying capacity of this planet in terms of life-support is, at least at the stage of consensus that I think we have now, a strictly academic procedure. We need to begin, with all possible speed, to bring human population growth deliberately to a halt.

My basic reason for this [conviction] is that every person I know who really looks at the Earth's life-support system in a holistic way has come to the conclusion that the overall trend is one of degradation. This is the opposite of the overall trend which we can document through [the evolutionary span], which trend has been a process of life building upon life. At the present time this is in [a state of] reversal. We are essentially drawing upon the capital which has been built up by life throughout evolutionary time and the ultimate [outcome] of withdrawing that capital of course is the destruction of the life-support system itself.

There are many perspectives that one can take, and in drafting this paper we selected a few more. I certainly think that human health and well-being in relationship to environment are also extremely important, while there are others that are too complex to discuss [in the time available]. Stability is certainly an essential one. I have looked at stability in a theoretical way at various levels of integration and I find that basically we have no adequate concrete understanding of what it means. However, I think it is useful to view stability, at the level of life-support systems, as a property which simply results in the availability of resources to sustain the human population, [or in general the *status quo*].

A [further] major idea is that of carrying capacity, which we define as the maximum population that can be sustained in a habitat without degradation of the life-support system. One of the key words here I think is 'habitat', because we must recognize that the Earth is a mosaic of linked systems and each unit in the mosaic has a different capacity for life-support; consequently we cannot think about an average density of the human population without having some idea of the way in which that population is allocated to specific units.

In looking at Man's relationship to his environment, particularly in the context of this Conference theme of 'Growth Without Ecodisasters?', I believe that the 'tragedy of the commons' is an essential perspective that has arisen simply because we have, over the past several hundred years, become very economic animals, and the motivation for gain, driven I guess by that perhaps essential [animal] attribute of greed, results in our exploiting 'the commons' to the ultimate general detriment. The difficulty, in terms of our economic systems, is that, whereas individually we may be able to gain from this exploitation, the negative aspects are essentially shared by all of us.

One of the real revelations of the last seven years I think has been within the agricultural sector, and [the realization of agriculturists] that there is an impending trade-off between their ability to continue to supply the food needs of [an increasing] world population and [the seemingly inevitable] degradation of the life-support system itself. This is not merely ironic of course but could lead to an early ecodisaster. One of the conventional-wisdom types of solutions to this, that seems to be cropping up again and again, is that we need to begin to shift from energy- and capital-intensive agriculture to more labour-intensive agriculture. Among the people I have talked with recently about agriculture have been American experts who have returned from the People's Republic of China, and we could [speak protractedly about] that situation and its many ramifications; but the one that has most impressed me is the consensus among them that the kind of agriculture which is practised there is more efficient in terms of numbers of calories made available to human populations per unit of land area than the high technological agriculture of a country such as the United States.

Forests are among our most important components of life-support. Probably 90% of the terrestrial carbon is in the forest compartment—comprised [mainly] of trees [and shrubs], their leaves, stems and roots. In theoretical/ecological terms, one of the best measures of stability of systems that has been identified is the size of a storage compartment, and in terms of biospheral stability the forests must certainly themselves constitute an extremely important element of the overall stability of life-support systems.

My recent discussions with personnel from several forest experiment stations in the United States, has indicated a growing concern about the renewability of the forest resource. It is of course obvious that, whenever forest products are removed from a unit of landscape, a [part] of the capital which was previously established in the system, in terms of the nutrient stocks in the soil, will be removed with the trees. This problem can either be ignored or it can be examined experimentally and in fact simulated. Some of the best simulations that I have seen of this process have suggested that, for certain kinds of forest ecosystems, the number of short-term rotations that can be sustained without loss of the ability of the site to support further forest growth is of the order of 9 thirty-years rotations.

In approaching problems of life-support, I think one of the most important questions to be asked is—what kind of life and for whom? Here I would like to quote from a statement made by Chief Seattle of the Danawish Tribe to the United States President, Franklin Pierce, in 1855: 'The whites, too, shall pass—perhaps sooner than other tribes. Continue to contaminate your bed, and you will one night suffocate in your own waste. When the buffalo are all slaughtered, the wild horses all tamed, the secret corners of the forest heavy with the scent of many men, and the views of the ripe hills blotted with talking wires, where's the thicket? Gone! Where's the eagle? Gone. And what is it to say goodbye to the swift and to the hunt? The end of living and the beginning of survival.'

In mitigating the assault on our life-support system, I think it is extremely important to draw together several different kinds of approach, and I find that these approaches, in some sense, are now almost becoming a part of [one after another] paper which I see written—in identifying, over and over again, the needs to deal with education, to integrate across the hard and soft sciences, and to begin to deal with local problems at the local level in terms of regionalization.

And before I close I would like to make a few comments about education. I believe that we need to open up the channels for education from the tribal level right back up to those of us in academic life, [and not ply our trade always in the ongoing direction]. A good example [supporting this thesis] is that recently in Georgia a so-called growth industry has been the hydroponic production of tomatoes. Now if you want to know about the hydroponic production of tomatoes, you go to the Agricultural Extension Office and they give you a brochure that tells you everything you need to know about the hydroponic production of tomatoes. The unfortunate thing about this whole process is that the products do not taste like tomatoes. [Laughter.] And it is increasingly ironic to me that in a nation like my own, the ability to grow a tomato that tastes like a tomato is increasingly concentrated in a minority of the population—people who read *Organic Gardening and Farming* magazine. So I think we do need to reverse the flow of education. This will enable us more effectively to perceive the real needs of people at the local levels, so that we can begin to modify our institutions [towards making peoples'] lives better.

One of the discouraging aspects of the Stockholm Conference on the Human Environment was, to me, that the less-developed countries' initial response to the call for need for concern about terrestrial life-support systems [seemed limited to concern for] their share of capital gains. But now I think in the past seven years I can begin to discern a trend away from this—a tendency for them to say: 'Listen, we know about our local systems; we have some ideas about how they operate. We don't want anybody to come in here and show us how to run them [without concern for the environment].'

In the final analysis the Earth is after all a living system, and I think it is time for us to begin to manage it as if it were a living system and not simply an economic system. It is often said that technological solutions to problems result in more technological problems, many of which may not even have solutions. As a matter of faith, I believe that if we begin to manage the world as a living system, we will discover that the solutions to problems which we can find will solve essentially more than one problem.

Miller (Chairman)

Some people believe that chairmen, like economic problems, if ignored will go away! But I cannot withdraw and give way to our panellists without thanking you, Dr Franz, for your excellent description of the problems we are in.

Myers (Panellist)

I should like to address myself to one particular aspect of this problem of the life-support system and where it is going, and that deals with the planetary spectrum of species. The species problem is a category of its own in so far as, whenever forms of environmental degradation take place, such as pollution or other forms of stressing of ecosystems, it is generally possible, if we so decide, to reverse the process. If we want to use the atmosphere as a garbage [dump] for 20 years, and then we want to clear it up, quite often we can go a long way [towards doing so]. But when a species goes under, that is it for good and all, and it is an impoverishment, not just for present-day society, but for all generations to come.

Now I am not just talking about the great whales and cats but about all species on Earth, and a figure which has been used fairly frequently recently suggests that there are [some] 10 million species on the planet with us. Of those 10 millions it is thought that 4 millions exist in one limited series of biomes on Earth, namely the tropical moist forests. And, as we have heard several times in the last few days, if present rates of disruption and destruction persist in those tropical moist forests, there will be a little left of that particular [facet of the] life-support system by the end of the century except in severely degraded form.

Supposing we are optimistic and say that by the end of the century in the tropical moist forests we lose only 1 million of the 4 million species. If you do a bit of arithmetic on the back of an envelope, I think you will see that we are being very hopeful if we say that right now we are losing each day only one species. Now let us ask ourselves, would it really matter all that much if we lost 1 million species out of 10 million by the end of the century? Of course it would matter a great deal, but would it be so very catastrophic? Is this a number one problem or is it lower down in our hierarchy of priorities?

There are various ways of looking at this, [even without considering] the aesthetic or the ethical angles, but, let us do it from the standpoint of what the species spectrum contributes to diversity and stability in life-support systems. Obviously they contribute a lot of diversity—but whether this contributes [in turn] to stability or not is still an unresolved question. The tropical moist forest has more diversity than any other biome, but it is by no means the most stable, especially when it is subject to the kinds of assaults and disruptions that Man is able [and all too prone] to inflict on it. Supposing we look at matters this way: at one end of the gradient you take away one species out of the 10 millions. Does that really make much difference? Does the planet creak a little more as it circulates each day? Presumably not. It is so marginal that it would really make hardly any difference. It is regrettable to lose a species, especially when it is unnecessary, but if one species goes, what does it matter? Go to the other end of the spectrum and if you lose 10 millions, that will be the end of all life on Earth. If you lose 9 millions you would really be in very big trouble. Go back nearer the other end of the spectrum: if we lose 1 million, are we in really bad trouble or will we still survive?

Let us break down the question a little bit further and try to look at what practical steps we should be thinking about for the [coming] years: if we wish to maintain the species spectrum or most parts of it, and we recognize we cannot maintain all of it, we should start taking some agonizing decisions about which ones we allow to fall over the side of the boat, and which ones we direct special efforts to [maintaining].

For instance, does it make much sense to keep on spending twenty thousand dollars a year on the Mauritius Kestrel [*Falco punctatus*]—13 of them left, I believe—when there is a very similar kestrel, much more numerous, roaming far and wide on the continent of Africa? Should we maybe stop giving quite so much attention and so many resources in terms of skills and funds to the cranes, almost all of which are in various degrees of trouble? Should we not give more attention to the passerines and insects, which have the capacity for much more rapid speciation and maintaining the evolutionary process, than is the case with the slow-breeding K-selected species such as cranes? Ants, which make up maybe half of our 10 million species, have a tremendous capacity for speciation by virtue of their fast turnover of generations each month. Maybe we should give more attention to those creatures, many of which are unidentified, in contrast with some which are evolutionary dead-ends and are not going to throw up others in millions of years.

These are the sorts of questions which I think ecologists and conservationists, [as proponents of] a predictive science, should be addressing themselves to, and indeed directing some pretty hard-nosed attention to.

Nicola (Panellist)

May I first express our great respect and thanks to Professor Eugene Odum, author of the leading textbook which you all certainly know, *Fundamentals of Ecology*, and to Dr Eldon Franz for his thought-provoking presentation. The very broad aspects of the subjects in the theme of our present Session make it impossible to present in five minutes more than a few comments on certain aspects. So, in agreement with our Chairman, I will limit my remarks by starting with a quotation from Odum: 'Mankind has evolved to exert increasing control over environment to achieve greater freedom from it. This pattern of increased regulation and uncoupling now must be reversed for us to function effectively in preserving the life-support system. We must be plugged back *into* the system.' This means first that mankind does belong, and is a part of, the greater life-support system. But if we belong to it, we also have a function in it. If such is the case, the question has to be raised now, near the end of the twentieth century, as to *which functions in Nature* we, human beings, should have, being the greatest, and, why not, the cleverest, predators in our closed life-support system (our resources being limited).

Humanity will have to adapt itself to radically changed circumstances; up to now we have taken food and energy from our environment, practically without restriction. This being no longer possible, Man has to find, not only his own necessary limits, but also the way to integrate himself 'functionally' in the life-support systems to which he belongs. This he will have to do in the manner of all predators, if he is to avoid impoverishing the biosphere dangerously, including improving the life possibilities (taking without giving is bad politics!).

Thus I would like to suggest that, next to the very urgent problems of environmental conservation, energy, pollution, food, and above all the population growth of mankind, we should also consider from a more objective point of view the overall complex of ecosystems which constitutes the biosphere, and within it the demomass of human beings at the right scale.

Humanity long thought that we, on our planet, were in the middle of the stage. Yet we should remember the struggles of Copernicus and Galileo. Are we still so short-minded in science as to put ourselves *above*, instead of *in*, the beautiful life-systems of our planet?

Do we not need an important change of thinking in our research approach, evaluating the multi-directional dependence of the activities, and the place, of Man in Nature? Ought we not to orientate ourselves in a more objective and dynamic way, for the benefit of future generations, in an era which we could call the Ecological Age?

Miller (Chairman)

Thank you, Mr Nicola, for your very thoughtful comments, and I want the audience to remember the questions you put forward—together with other questions about Man as a predator and about the regulating functions of a predator.

Baroody (Panellist)

As the baby of the Conference, as I have been dubbed, I would like to begin by saying that the proceedings thus far have been an extremely fascinating experience for me, and I am very honoured to be able to attend and observe the workings of such a group as are assembled here today. In the sense that I am young and I have had limited experience, I feel a little bit overwhelmed at times both by the calibre of the participants and the deluge of information that has confronted me at the Conference. On the other hand, I feel very much encouraged and stimulated by these same two factors. Although I do not feel qualified to delve into a technical discussion of the fine paper that was presented by Dr Franz, I hope you will permit me to make several comments and suggestions regarding both the proceedings of the Conference and the paper.

Various themes have been brought forward and reiterated during the course of the programme up to this point. There have been presentations of various problems—population growth implications, toxic chemical concentrations, etc., that could potentially reach [ecodisastrous proportions]—and there have been some suggestions for possible courses of action. The idea that [remedial action] should be carried out at a local level—even the village level, where necessary—is one of the most recurrent and fundamental themes. But I wonder, are there not some refinements to this idea?

Regarding the paper before us, several critical problems have been presented: for example, the dilemma of labour-intensive as opposed to energy-intensive agricultural systems. What can be done about this shift in emphasis from labour- to energy-intensive systems? The carrying capacity—a term which was again defined this morning—with which most of us are familiar, was also briefly introduced. How can environmentally uninclined people be made aware of the implications and dangers of exceeding the carrying capacity? What measures might be taken to ensure effectively that the carrying capacity of developing countries is not exceeded [as a result of] meddling by developed countries? In other words, how can we stimulate the formation of an environmental awareness among peoples around the world? What types of environmental education programmes are possible and [also] viable? Do we need a three-phased process of communication, with interactions—for example, an environmental technician from one country, an interpreter from a second country, and an implementer from that same country? In effect the interpreter would serve as a liaison between the technician (I am attaching labels to people, and there may be other implications associated here) and the implementer, in an effort to avoid cultural and technological misunderstandings. What kinds of regulatory solutions regarding the pollution of the environment by national and multinational corporations could be developed? Is regulation a necessary approach, or can men and women be depended upon to develop [remedies] by mentally acceptable solutions to these problems on their own initiatives without regulations?

I realize that some of these questions are related to the future agenda topics but they are all interrelated of course and it is to be hoped that some suggestions for courses of action can be brought forward at those times. In some instances we have been given the 'whats' and the 'whys', but the specifics of the 'hows' and the 'whens' are still lacking, as I see it.

It is clear that we are living beyond our natural means; let us not wait for the volcano to erupt. Let us take positive constructive action as the Icelanders did in their evacuation of the Westermann Islands, for example, in the face of imminent [further] volcanic eruption.

The feeling I get in talking with other conference participants on an informal basis, is that we need more of an emphasis on the submission, discussion and partial refinement of possible courses of action, as well as follow-up strategies before the last day of the Conference is upon us.

Miller (Chairman)

We have some good questions before us and I think we should be stimulated to consider carefully what we should do [about them] between now and the end of the Conference. Do we wait for the last day to take any decisions, or to panic into some kind of resolutions? The same thing goes for the Doomsday possibilities: do we just plan, or do we actually look at the 'when' and the 'how'? As the last speaker has indicated, we need to have some kind of *action*–[stemming from due] contemplation of what we are [getting] into.

We cannot think of the possibility of evacuation if we are facing a global kind [and scale] of disaster. We cannot think of who to sue, because there will not be anybody to stand responsible any longer. In our questions and answers this afternoon, we might consider which are [likely to be] the first elements to collapse; what indicator species are in trouble? Where are the vulnerable spots—the 'hot spots' on the face of the globe? And if we are going to get into that kind of thinking, maybe we can see where we can do some salvaging on an immediate level. The floor is now open for questions or comments.

Goldsmith

I refer to Dr Myers's [comments]. I regard conservation as a whole as being no more than a holding action; it is a very necessary action, but only for a limited period [of time]. The reason is that, in the face of continued demographic and economic expansion, there will soon be very little left to conserve.

I have visited most of the national parks in Africa and India, and everywhere I found the problems to be very much the same. Pressures to transform wilderness into agricultural areas to provide food for starving people are irresistible. Take the case of Kaziranga in Assam. Illegal immigrants, at least when I was there, were pouring over the border from Bangladesh in search of work, land and food. It is very difficult to tell people that they must starve so that we may preserve a few hundred Indian rhinoceroses, so one knows in advance that the parks' land-area can only contract. The same is clearly true of Serengeti [and Tsavo in Tanzania,] and many other national and other parks and reserves surely throughout the world.

The pressures to turn wilderness into cash are also strong, even when people are not starving. With industrialization, wilderness inevitably disappears. In Great Britain, for instance, with all our pious talk of conservation, we have only left 270,000 acres [109,000 ha.] of land to relatively undisturbed Nature. That is the extent of our nature reserves and most of this land is in Scotland. In England there is [a total of] less than 100,000 acres [40,500 ha.] of nature reserves, and this is mainly made up of scraps of land that nobody wants—land, in fact, that for one reason or another is difficult to turn into cash. So our priority must be to stop further demographic and economic expansion. Conservation is only of use if we accept this principle; it can then limit the damage and start reversing present disastrous trends.

There is another point I would like to make. It seems to be conventional wisdom among environmentalists that the western world must stop expanding so as to release resources to permit economic expansion in the Third World. This was basically the theme of Mesarovic and Pestel's second report to the Club of Rome. I cannot accept this thesis. If economic expansion is a bad thing for us, then it must be [bad] for the Third World too. This reminds me of a story: a little boy was shown a picture of Christians being eaten by the lions in the arena in Rome. The little boy burst into tears at this horrible sight and his mother naturally tried to console him the best she could: 'You must not cry Johnny', she pleaded 'that was a long time ago, Christians are not thrown to the lions any more.' Johnny was not impressed: 'That is not what I am crying about, mummy. It's that poor little lion over there. He didn't get one.'

In a sense we are all reacting like Johnny. Economic growth has devastated our environment and destroyed our society, and for no purpose other than to increase our consumption of plastic buckets and electric toothbrushes and other equally useless consumer goods. Yet our reaction is inexpiably to deplore the fact that the people of the Third World have been deprived of the opportunity of doing the same thing.

Tan

I would like to relate the remarks on restitution made by yourself, Mr Chairman, and Dr Franz's disturbing contention about the Stockholm Conference in relation to the developing countries. Has Dr Franz wondered why the developing world should still, now, keep asking for their so-called fair share of the world's resources? A study of the history of economic development in the developing countries and some perception of the abiding self-interest of the developed world, so strongly evinced in their current protectionist economic policies, should provide one of the answers. I would further illustrate this with our experience in a problem that is giving us increasing concern, to which I think the USA, reinforced by the USSR, have shown lamentable disregard. I refer to the future of the Straits of Malacca, which is an ancient, international waterway. It is being polluted inadvertently by ourselves, and we are enacting legislation to control this, but I have been told that the worst form of damage is committed by international shipping, namely by bilging into its crowded waters.

When the three nations which border the straits tried to cooperate with one another, in spite of political differences, to control the needless abuse of the straits that is now evident and can further be anticipated—e.g. from oversize oil tankers and military vessels—the USA and USSR were most vociferous against any exercise of control that interferes with their 'rights' of passage. But the straits lie within our national borders (we need only exercise the 12-mile [19 km] limit), and we need them for more fundamental purposes: they have long provided the three countries with their main source of protein, and, for substantial numbers of rural people with no alternative livelihood, their sole source of income. We have seen how seriously depleted our estuarine and maritime fisheries have become in recent years—every oil tanker is an ever-present symbol of impending massive destruction, and two collisions resulting in major pollution have already taken place in recent years.

I am consequently wondering how your message on environmental conservation can be appreciated by your politicians and military strategists who do not think at all the way you do about ecological responsibility—a responsibility which they not only do not regard as important, but which does not even figure in their arithmetic. I wish the super-powers would fight their wars on their own territories, and let our people carry on their peaceful lives. My example of how, in practice, ecological responsibility is thrown overboard whenever other interests appear to be threatened, makes nonsense [also] of the contention that all the developing world wants is [only] to have a share of the world's resources [, although I see nothing amiss in that desire].

Sigurbjörnsson

I think it is rather easy for a western ecologist to say 'Let's de-industrialize, and let's not allow the developing masses to compete with us Westerners in getting the good things of life', but I do not think the 500 or so million people who are said to be undernourished, and maybe starving, will be worrying about plastic buckets or electric toothbrushes. I think we owe them a little bit more, and I do not think we can abruptly say 'let's de-industrialize'. We must not forget about the progress that still has to be made in the technology for growing more food.

I sometimes wonder whether we, when talking about an impending ecodisaster, are maybe forgetting about a human disaster; or should we choose between an ecodisaster and a human disaster? One thinks of the scenario that Dr Fosberg was describing after the ecodisaster, when the insects take over. That is not [necessarily] a very bad situation. I do not think the insects would miss us. [Laughter.] I think they would probably populate a really nice, naturally intact Earth. Maybe what we should be worrying about is the world *homo-disaster* and not an ecodisaster, because we should be thinking about the well-being of Man on Earth, and of course in that respect we have to consider the environment. Let us remember that we live here and have to derive our living from this Earth.

Polunin (Jnr)

I would like to make three points. First of all, Dr Myers referred to species extinction rates as though these might be possible in linear terms, i.e. you might get, say, one or two species going a year. [Yet] as far as tropical species diversity is concerned, [a major] point is that very large numbers of species depend on other species; therefore one should probably expect logarithmic rates of extinction to pertain to tropical situations, where the more species you lose, the more tendency there will be for further ones to go.

Secondly, in all these deliberations there has been very little mention of traditional peoples. Yet there are traditional peoples in both terrestrial and marine coastal environments, and these people are in the important position of living in a traditional way—perhaps often in the kind of way we would envisage as being ultimately the most viable way for us to live. [Both for their sakes and for ours, the lives of] such people should be taken into consideration when we talk about what *we* should do.

Thirdly, Miss Baroody spoke of the problem of changing from energy-intensive to labour-intensive industry. I think one of the most important difficulties here is, effectively, that of seizing power in some way from increasingly centralized governments and increasingly internationalized industrial concerns. If we are going to try to bring human activities down to a more local level, [the power of such concerns poses] very real problems, and perhaps we can think about this when relevant meetings come up.

Guppy

First of all, Dr Myers was discussing the propriety of concentrating conservation measures on saving *certain* species of animals rather than others—or, presumably, plants. The important thing in doing this is that we probably are saving their habitats as well. This increases the benefit, and the significance, of concentrating on what may seem unlikely species to invest in from the evolutionary point of view. He also raises the question of conserving ants and other insects. There is no conceivable way in which Man could take over the work of the honey bee if it were exterminated. This shows how dependent Man is on quite inconspicuous components of the [ecosystems to which he belongs].

It seems to me that the most important thing [to realize] in all these discussions is that a great psychological change still has to take place among ordinary people, so that we can get them to accept the ideas that *we* have accepted. It is [now] a decade since the Club of Rome prescribed that we must cease heavy industrial investment and concentrate investment on agriculture, education, and population control, and this same formula is basically what we are all propounding. But [at the risk of sounding presumptuous I now say that] the time *has* arrived when what we say strikes home to the public.

I have here a clipping on the latest Harris survey of US opinion, and it is quite remarkable reading. It says:

> The American people have begun to show deep scepticism about the nation's capacity for unlimited economic growth, and they are wary of the benefits that growth is supposed to bring. The latest Harris survey shed some significant and sometimes startling light on current American thinking. By [votes of] 79% to 17%, the public would place greater emphasis on teaching people how to live more with basic essentials than on reaching higher standards of living. By 76% to 17%, a sizeable majority opt for learning to get pleasure out of non-material experience, rather than satisfying needs for more goods and services. By 59% to 33%, a majority stressed putting real effort into avoiding those things that cause pollution, rather than finding ways to clean up the environment as the economy expands. 82% over 11% would concentrate on improving those modes of travel we already have, and only 11% would emphasize developing more ways to get places faster. By 77% to 15% the public comes down to spending more time to get to know each other better as human beings on a person-to-person basis, instead of improving and speeding up our ability to communicate with each other through better technology. By 63% to 29%, the majority believe they

would be better served if emphasis was put on learning to appreciate human values rather than material values—rather than on finding ways to create more jobs for producing more goods. By 65% to 22%, the public would choose breaking up big businesses and getting back to more humanized living, over developing bigger and more efficient ways of doing things. By 64% to 26%, most Americans feel that finding more inner and personal rewards from their work is more important than increasing the productivity or work-force.'

All this should encourage us to hit really hard in our resolutions.

Miller (Chairman)

Thank you, Mr Guppy, [though I should say that the] kind of responses you get are an indication also of how a survey is worded. It must have been a good survey—perhaps in a forum such as this! [Laughter.]

Altenpohl

I refer to the panel speech and intervention, respectively, of the two ladies this morning, and come back to their remarks that action is needed. In a way this is a follow-up of the Stockholm Conference, but there are very few NGOs (Non-Governmental Organizations) that would really try to take action. I happen to be a member of [such an organization]. It is a subsidiary of, or a daughter organization of, the YMCA. They were present at the Stockholm Conference and, ever since, have held seminars in one country at a time, such as Egypt or Yugoslavia, and have dealt with problems of what they call ecology [—using the term in the popular sense]. It came to my attention that more than three-quarters of all the problems which they deal with are 'punktuell'—[concerning very large cities or other wide areas in which] industrialization causes real problems, such as the acid rain in Scandinavia.

I do not want to underestimate the [magnitude of these problems but to emphasize that counteraction] can be best taken on a country basis, or a regional basis, or a city basis, or an industrial basis in one country. I personally believe that the industrialized countries such as West Germany and some others should set up a school to train 'goodwill ambassadors', mostly with [some suitable] technical training (especially as biologists or ecologists), and send them to the countries in need, for example in Africa or the Middle East. They should be people who are patient and listen, and will ask 'What do you need? What can we do for you? How can we arrive at the common denominator where we can do something which helps you?'. In France we have the Ecole Nationale Administratif, and in Germany we have a lot of money spent for all kinds of so-called Entwicklungshilfe—foreign aid—though not this one.

A final remark: countries such as Romania, Yugoslavia and Bulgaria have gone through a [period of] very rapid industrialization in the last 20 years. To their great regret they now find that they have real trouble environmentally. In one of those meetings of that YMCA group, participants from these countries were so alarmed, I remember, that one lady professor from Belgrade had tears in her eyes. She said 'Our beautiful river is all messed up by a red mud and industrial waste, though it was a clear mountain stream a few years ago', and so on. Thus even the 'eastern' countries are interested in putting things right and knowing what to do about these environmental problems. But whether it is in certain parts of Yugoslavia, or in Mexico City or [Cairo, Tokyo or Hong Kong, New York or London,] these problems, which can reach the proportions of substantial ecodisasters, are mostly local ones.

Buchinger

I would like to contribute an example of something which was not stressed enough yesterday, namely the means of communication between what you call the developed countries and the underdeveloped ones. One of the panellists said that there is a need of an interpreter between the two groups [—meaning, I presume, a way of communication—so] as an example I would like to talk about the International Working Meeting on Environmental Education which was held at Foresta Institute in Nevada [a few years ago]. Some 40 of us were invited from all over the world: from India and Pakistan, from Argentina and Venezuela, from England and various African countries, and so on—people who had already a way of understanding both civilization and culture. It was recognized as being not quite easy [to adapt rapidly when one goes] from one place to another. As I recall, everybody had been trained in his or her own country but had done some graduate work in Europe or the United States or England. The outcomes of this Meeting were to find a common means of communication and make recommendations that were understood by, and useful for, all those who were present. These same recommendations and practices then had to be presented and explained carefully to the authorities in each home country, who had not the same background.

These follow-up meetings, adapted to the local conditions and other circumstances, were extremely successful. So when [Dr Altenpohl] mentioned the possibility to train people as 'goodwill ambassadors' who would go to different countries with the talent and power to transplant suitable programmes into new surroundings, this seems to me a useful approach towards saving our life-support system. At least such people will be perceptive of environmental problems and receptive of ideas for action to combat them.

[With several others signifying their desire to speak, the debate was adjourned at this point for coffee before the next Session.]

Miller (Chairman, resuming Session after adjournment)

We are fortunate in this topic of 'Whither the Life-support System?' that it is [practically] all-pervading in the whole concern of global environment, and so anything you may have wished to say [when we adjourned yesterday should be] pertinent here as well.

One of the things that I am particularly concerned with as a person is the subject of environmental education, though there has not been a specific designation for it here in these sessions—perhaps because it is another of those widely-pervading themes, on which, moreover, there is a world conference coming up this fall in Tbilisi, Georgia, USSR. Sponsored by UNEP and UNESCO, it is an official conference at which delegations from governments will [deliver previously agreed 'set pieces' and little if any 'open' debate is likely to be allowed]. It is the outcome of a plan for having grass-roots discussions in various countries, to be followed by discussions on a regional basis throughout the world in the second year. In the third year comes a world conference at which the [means of execution] developed in these regional areas will be put forward [and discussed]. We are not sure just what the national delegations will accomplish, but we are counting on them to reflect and use the expertise of their various countries, and relying [for input] on environmental education people [who have a truly global viewpoint].

Olindo

The matter to which I wanted to draw attention yesterday is the gigantic UN Conference on Science and Technology [for Development that] is being planned for 1979. As I am very actively involved in the preparatory process leading to that Conference, I can state definitely that the 'Group of Seventy-seven', representing the developing countries, is going to demand the transfer of technology from developed countries to developing countries.

At the present time we are concerned about the inadequacies of a large number of those technologies, documentation being available on their negative impacts on the natural environment. Yet in 1979 and subsequently, decisions—far-reaching decisions with far-reaching implications—are going to be made to transfer exactly these same technologies which we are concerned about to fresh grounds in the [so-called] developing world. This is why I hold the different view that scientists should simply discuss scientific matters and leave them at that point. I agree with Maurice Strong that interactions at all levels should be encouraged, and should be quickly undertaken, so that scientists can widely make public [all pertinent] information about the dangers that are so prevalent with certain technologies. [This could help us] to avoid the prospect of these same dangers being transferred into developing countries which may genuinely not be aware [of impending problems from some technologies which they are ready to accept but which hold dangers that for them are totally unknown or at least] are not yet part of their way of life.

Dasmann

I would like to follow-up on Mr Olindo's comments and also on comments made last night about this subject. I wonder whether Mr Olindo will agree with the suggestion which we have had here that ICSU would be the logical body to sponsor [a suitable scientific meeting to discuss and give guidance on these matters]. I am also interested in knowing how the Second International Congress of Ecology, which is to be held in Jerusalem in 1978, might fit into this scheme. The first [such] congress was held at the Hague a few years ago. Would this forthcoming Second Congress in any way fit the need for a preparatory meeting for the UN meeting, or would we need a separate meeting? I am not sure just what is needed. I have a feeling that if we wait until 1979 and hold a concurrent scientific meeting it will not do any good, [any more than did the Stockholm side-shows], but that the meeting has to be held early enough to feed into the UN meeting in order to inform the scientists [and others] before they go there.

Olindo

I would like to respond. I think there has been a slight inadequacy in our methods of arranging meetings like this one, because in many cases we select people who think in like manner, whereas to have an interplay and an impact on 1979—or on any other meeting for that matter—I think there will need to be a mixture of influential politicians, scientists and those people who actually implement the decisions, meeting more frequently. Then we can begin to have a favourable impact on the world we want for tomorrow [and thereafter]. I think that if the scientific community can remove the atmosphere of confrontation from the 1979 Science and Technology Conference and bring about an atmosphere of a [real] meeting of minds between the developed and the developing countries, so that the needs of the peoples of the world rather than the needs of [particular] countries can be seen and accepted, and solutions accordingly worked out, the problems that the economic and growth systems of the developed countries impart to the developing countries might also be solved. But if the [proposed] ICSU gathering were to be a meeting mainly of scientists, and then after that one expects to go and address politicians on a subject in which they have not had any input, naturally they will resist!

Miller (Chairman)

Very good: I would only like to add a thought to that. In such a meeting, in order to get a basis—a common basis—for discussion, we have to find the points of agreement, and it would be good to review the points of agreement and then go from there to explore where we can get together and where we can understand better what the others' problems are.

Kuenen

Two remarks on this matter. In the first place, some of us feel that this famous [ICSU] Conference should not be in competition with the UN but some time before, so that we avoid the unpleasant situation we had in Stockholm where there were two or three other conferences going on at the same time, in competition. It certainly cannot have been Mr Strong's idea to have that repeated, and whoever goes, and whatever is going to be done, someone should take care of that aspect. I still believe ICSU is the organization [that could best] do it, and you need an organization to do it and [make sure that it is] carefully planned, as Mr Olindo has made quite clear.

About the INTECOL Congress, I may remind you that at the Hague the applied ecology got kicked into the corner. There were quite a number of contributions which had something to do with applications—at least half of the ecologists there were very much aware of their obligations in all sorts of general matters. But it was very much a scientific congress, and those who have seen the programme for Israel will see that, although there is a slight indication of application there, the main matter is still a scientific congress.

So unless you fundamentally change the set-up of the INTECOL Congress—which I am not at all sure would be a good thing to do, because I still rather believe in scientific congresses—it will make no or very little [applicational] impact. I would very much prefer to leave that alone and have an ICSU Conference on Applied Technology *sensu lato*.

Miller (Chairman, concluding Session)

As you know, the UNEP has a Governing Council which meets and develops policy for it as a UN service organization in which 80 countries participate. Some of the environmentalists have now been recognized by their governments at home and placed on the delegations that go to those Governing Council meetings. This is something that you, as individuals in your own country, might well explore—maybe you can be appointed to some of those delegations that your governments send, and bring in your comprehension to their substantial benefit [while enlivening their proceedings].

I want to thank those who showed up to participate in the conclusion of this Session, [and, once again, Dr Franz and the panellists and speakers from the floor for their widely excellent contributions to a stimulating Session at which much has emerged, including some points that seem to be new].

Session 10

AGRICULTURE AND WORLD FEEDING ALTERNATIVES

Keynote Paper

Ecological Constraints of Global Food Production*

by

GEORG BORGSTROM

Professor of Food Science and Human Nutrition, and of Geography, Department of Food Science, Michigan State University, East Lansing, Michigan 48824, USA

INTRODUCTION

Food production is rarely analysed from the ecological point of view, while even more rarely are the constraints placed in clear focus. Yet these matters have to be moved to the centre-stage of the current debate around the food issue.

Early in his history, Man tended to concentrate his feeding on the carnivorous level, relying extensively on the gathering of molluscs (shell-mounds all around the world bear witness to this) or on the trapping and/or hunting of wild animals, whether terrestrial or aquatic. A less prevalent pattern, but one that was singularly important to several groups, was recourse to the insect class—not only locusts and ants, but also innumerable others, and including larvae as well as full-grown forms (Bodenheimer, 1951). Only after methods had been evolved for breaking up the cellulose encasements of plant cells—by crushing, parching, fermenting or other devices—did Man get reasonable access to the nutritive riches of the plant kingdom. This opened up the herbivorous alternative, which, owing to the pressure of growing human numbers, has become the dominant pattern in major sections of the human species. Abel, the herder, and Cain, the tiller, may be chosen as symbols of the two roads taken by mankind in its quest for food.

As it is not feasible in this context to trace the historical development of these two evolutionary channels, attention will merely be focused on two

* *Presented, in the absence of the author, by Dr Raymond F. Dasmann.*

decisive turning-points. What for a long time was a harsh confrontation between these two feeding forces got its ecological atonement in the creation of balanced agricultural ecosystems, typified by the all-round European 'prairie' farms and New Zealand dairy-farms. This was achieved through ecologically well-balanced plant rotation schemes coupled with some kind of livestock production. In functional terms, dairy-farms were particularly effective as regards the protein–nitrogen cycle, although gradually the heavy outflow of minerals through, for example, urban milk deliveries, became a crucial drain on this viable cycling-system. One key factor in the creation of such balanced farms, combining plant and animal production, was the incorporation of pulses.*

The second major evolutionary route is symbolized by exclusive crop production, finally ending up as large-scale monocultures. Early steps in this direction were the tropical plantations for cash-crops†, such as sugar, banana, rice and others. Their counterparts in temperate and subtropical regions were grain crops such as wheat, maize and soybeans—all of them increasingly moved out of sensible crop-rotation schemes. Wet rice cultivation on a more modest scale, however, has held the scene unchanged by associating with nitrogen-fixing blue-green Algae as representing the indispensable attachment with the nitrogen cycle.

In the well-to-do world, crop production has gradually dissociated itself from animal-raising. Protein was the chief but largely unrecognized lever in this break-up. The first step in the process came early in the present century when farmers started buying supplementary feed for their livestock—to begin with, forage (chiefly hay) but, in time, predominantly protein. This was the starting point in the undermining of the ecological awareness of the farmer. In Europe in particular, but also in Japan and North America, oilseed cakes and meal, brought in from outside, started to blur the picture. Both farmers and politicians lost touch with the land as the prime basis of agricultural production. They blotted from their minds the vast expanses of land which were out of sight but on which they depended. Trade statistics and money accounts were hazardous substitutes. Fundamental to a clearer grasp of the ecological consequences is the fact that protein crops are the most acreage-demanding; protein was not only the key factor, but, in all likelihood, also the driving force. This is where macroecology enters the world food scene on a major scale, resulting in recent years in absurdities such as

— most European countries as well as North America providing their livestock with far more fish protein than was channelled to human food;

— more dehydrated skim-milk in several European countries fed to livestock than to Man, or in recent years the anomalous massive storage in Europe of non-fat milk-solids (almost wholly dependent on net importation of feed protein); and

* Edible seeds or plants of the pea family (Leguminosae, hence often called 'legumes').—Ed.
† Cash-crops are crops that are cultivated primarily for their commercial value and hence for sale (as opposed to ones cultivated for subsistence, etc.).

— 3–4 million metric tons* of milk protein being either wasted or at the best used as animal feed.

The separation of livestock and crop production had profound ecological repercussions. A kind of animal production emerged that might be better characterized as a secondary industry than an agricultural operation.

The blotting out of ecological perceptions is most clearly mirrored in the fact that food issues are almost exclusively analysed in completely other parameters than ecological ones. Viewed in the long-time perspective, three considerations dominate: food supply, farm income and release of labour. In particular in this last area false data are profuse, omitting as they do the large and growing subsidiary manpower that is required to uphold modern agricultural production. This distorted manner of defining the world food issue is lacking any relationship to basic biological factors and to the balance between food and people.

The Lost Dimension

For far too long, biologists have acquiesced in playing the 'numbers game' of statisticians, demographers and traditional economists. There is an urgent need to introduce new gauges of the true feeding burden exerted by Man. This can no longer be satisfied by counting heads. It is almost a truism that it takes so much more land, water, fuel and other resources to satisfy a European than a Chinese; yet few efforts have been made to measure this disparity in biological terms.

It is obviously true that we have a reasonably accurate picture of the number of humans to be fed, as to both countries and continents, but this simple arithmetical way to approach the task of appraising Man's role on Earth has very obvious limitations, and frequently leads to erroneous conclusions and distorted perspectives. Nevertheless, far too often such computations are encountered in scientific treatises as well as in numerous textbooks on economic geography, agriculture, nutrition and related disciplines.

The present misunderstanding of Man's true biological role on the globe is one major reason why we were caught in what is termed the population explosion and its tragic consequences. The failure to place Man in proper relationship to his environment is also the chief explanation for the many fallacious concepts disseminated about the potentialities of the globe and the many easy-going diagnoses of Man's present dilemmas. The all-important, indeed basic, biological dimension has been missing. In reformulating our picture of the world in terms of food, protein again takes the central stage.

'Man's Biosphere'

For numerous reasons the human biomass cannot merely be interpreted to mean the human numbers as composed by age, sex, marital frequencies

* Whether or not they are so indicated, the tons used in this paper are metric tons = tonnes.—Ed.

and race. It needs to be given a far broader connotation to signify, in general terms, the Man-controlled biomass or, more specifically, the living world on which Man is depending for his food. For this purpose he marshalled vast domains—much larger than sheer human numbers might imply. The living mass, active within his sphere at all levels is, in effect, enormous. The immediate biomass in terms of plants and animals, directly exploited by him, needs a more precise identification and quantitative appraisal. More distant spheres of influence, such as forestry and wildlife, will not be discussed here except for a few illustrative examples to bring Man's subsistence into clearer focus. The vast realm of microorganisms in the soil, in the rumen, and in the intestinal system, will be touched upon only briefly and on a few crucial points to elucidate the ecology of food production. In addition, microbial operations that are active in the degradation and recycling processes have increasingly moved into Man's biosphere.

Population Equivalents

Various methods have been devised to allow a direct comparison between Man's livestock and himself, whether these animals serve him as beasts of burden or traction, or to provide food. The most valid comparison on a commensurate basis can be made by acknowledging the primacy of protein as the key food. Man and other animals are gauged as to the pressure which they exert as reflected in their total intake of primary plant protein (Borgstrom, 1964; 1965 and 1972*a*). The computations are based on a man of a standard weight of 70 kg as a unit and assuming a disposal of 70 g of protein per day. Measured by this human gauge, the total livestock as protein consumers are converted into population equivalents (PE-units).

In this manner one arrives at data which measure the living mass directly belonging to Man's realm or operating under his control within the wider biosphere. One then finds that the globe currently (1977) is inhabited not merely by 4.2 thousand million humans but, in order to maintain their present nutritional standard and to retain the type of agriculture now prevailing in various parts of the globe, the green plant cover must carry a feeding burden far in excess of that figure. Domestic livestock accounts for 16.8 thousand million PE-units (1977). Estimates swirling around as to the ultimate limit of Earth's feeding capabilities, arriving at figures of 15, 25, 40, and up to 147 or even 900 thousand million people, are mostly computed with little recognition of the fact that the world in protein terms already carries a feeding burden of some 21 thousand million.

Ratio of Livestock to Man

Many ecological lessons can be drawn from the data in Table I. The world ratio of livestock to Man is now 3.9, but differences between continents are considerable. Overpopulated Asia has a ratio which is only slightly more than half of the global value, while Europe, second in line as to population density, surprisingly holds closer to the world average in this ratio of livestock biomass to Man's. Europe, however, does not produce

TABLE I

The 'Continental Biomass' (in thousand million PE-units), 1975.

	Man	*Livestock*	*Total*	*Ratio livestock/Man*
Asia	2.3	4.8	7.1	2.1
World	4.0	15.8	19.8	3.9
Europe	0.48	2.22	2.7	4.7
Africa	0.41	1.89	2.3	4.6
USSR	0.26	1.54	1.9	6.2
North America	0.24	1.66	1.9	7.0
Latin America	0.33	2.97	3.3	9.1
Oceania	0.021	0.59	0.61	21.1

more than approximately half the feed required for its livestock. Taking this into account, its ratio drops to 2.3—in effect, close to that of Asia. It is also noteworthy that the livestock of Europe almost equals the direct human pressure of entire Asia, of which the feeding load (of Man *plus* livestock) is less than 2.7 times that of Europe, although its human population is about five times as large. It is also noteworthy that the livestock of Europe almost equals the direct human pressure of all Asia.

The USSR and North America carry an almost identical feeding burden of around 1.9 thousand million PE-units each, with ratios that are not far apart. New Zealand and Australia, the chief partners of Oceania, are well known for their cattle and sheep. Less recognized is the fact that the biological pressure (of Man *plus* livestock) amounts to 600 million; yet their total population is one-quarter of one year's growth of the human family, rendering a ratio of livestock to Man of 21.1.

The ecological constraints of the African continent place it on a surprisingly low level, close to the world average. The broad disease-ridden zone across the continent explains this. The regional variations are considerable (Table II), with elevation and latitude as key factors.

TABLE II

Regional Differences in 'Man's Biosphere' in Africa (in millions of PE-units), 1970–71.

	Man	*Livestock*	*Ratio livestock/Man*
East Africa	95.5	635	6.1
West Africa	111.2	350	3.1
North Africa	82.0	328	4.0
South Africa	22.2	216	9.7
Central Africa	34.8	105	3.0
Total	345.7	1,634	4.7

The ecological constraints of individual countries in the world are well mirrored in the ratio of livestock to Man (Table III).

TABLE III

'Man's Biosphere' in Selected Countries (in millions of PE-units), 1973–74.

	Man	*Livestock*	*Total*	*Ratio livestock/Man*
South Korea	35.2	18.0	53.2	0.51 (0.35)*
Japan	108	87.8	195.8	0.81 (0.25)
Indonesia	134	110.1	234	0.82
Egypt	37.5	48.8	86.3	1.3
Jamaica	2.14	3.42	5.56	1.6
Philippines	42.7	76.0	118.7	1.8
Nigeria	60.4	133.2	193.6	2.2
Italy	55.0	147	202	2.7 (1.2)
Spain	35.1	105.5	140.6	3.0 (1.9)
UK (Britain)	56	202	258	3.6 (1.5)
West Germany	62.1	217.8	279.9	3.5 (1.8)
Turkey	39.2	172.0	211.2	4.4
France	52.3	278.1	330.4	5.2
USA	211	1,540	1,750	7.3
Canada	22.3	147.2	169.5	6.6
Mexico	57.2	299.2	356.4	5.2
Poland	33.6	225.6	259.2	6.7
Brazil	103	1,261	1,324	12.2
Argentina	25.6	500.5	527.1	19.5
Australia	13.2	337.3	350.5	25.6
New Zealand	3.0	140	143	47.1

* Figures in parentheses indicate the ratio when those fed through importation are discounted.

The acceleration of Man's pressure on the biosphere stands out as the most crucial aspect. In the period from 1948–52 to 1976, the world ratio rose (Borgstrom, 1965 and 1972*a*; 1973*a*). Livestock expanded by 7.23 thousand million PE-units, with a concomitant addition of 1.71 thousand million people, rendering a ratio for the increase in livestock to Man of 4.21 (Table IV). Europe shows the biggest relative increase in the livestock

TABLE IV

Livestock Increase in Relation to Man (in thousand million PE-units), 1948–52 to 1976.

	Livestock			*Man*			*Increase*
	1948–52	1975–76	Increase	1948–52	1975–76	Increase	*ratio*
World	9.59	16.82	7.23	2.50	4.21	1.71	4.2
USSR	0.78	1.62	0.84	0.18	0.26	0.08	10.5
Europe	1.47	2.22	0.75	0.39	0.48	0.06	12.5
North America	1.13	1.66	0.53	0.17	0.24	0.07	7.6
Latin America	1.95	2.97	1.02	0.16	0.34	0.18	5.7
Asia	3.41	4.80	1.39	1.37	2.32	0.95	1.4

sphere, but chiefly by buying feed from outside the continent. Asia has shown a growing squeeze over many years, and a falling ratio. Latin America is also in constant and more swift decline, but from a much more favourable ratio.

Mention is frequently made of the expectation, based on the oft-quoted United Nations projection, that the world will be adding almost 1 thousand million humans in the next 10 years, but, in view of the projected simultaneous increase in livestock, this should read more than 5 thousand millions (when counted in PE-units).

Livestock Categories

The regional analyses need to be supplemented by an accounting for livestock by category (Table V).

TABLE V

Global Biomass: 'Man's Biosphere' as to Categories (in thousand million PE-units), 1976.

Cattle	8.5	Horses, mules, asses	1.2	Total ruminants	11.1
Man	4.2	Sheep	1.4	Non-ruminants	9.5
Pigs	2.4	Buffaloes	1.2	—excl. of Man	5.3
Poultry	1.7	Goats, camels	0.6		

Despite widespread mechanization, the world's horses still account for a protein intake that corresponds to that of 720 million people, thus constituting a consumptive force exceeding that of India's entire human population. The Americas, with 570 million people, represent as consumers only about one-quarter of the intake of the pigs (as measured on a global scale). Cattle represent an intake of primary protein which is 3.6 times that of the direct intake of the humans in the population-rich Asian continent, or 52% of the total feeding burden of all livestock within its 'human biosphere'.

The ruminants constitute an invaluable beachhead in the plant kingdom, vastly expanding Man's feeding basis by their capacity to utilize cellulose (1), by strictly economizing with nitrogen (2) and, most importantly, by microbial synthesis of protein in their rumen (3). They are, in effect, parasites on their own microflora. It is, therefore, not surprising that ruminants are more numerous than any other group, representing 11.1 thousand million PE-units as against 9.5 thousand million for non-ruminants, of which Man accounts for 4.2 thousand million (Table V). In Nature, too, ruminants are the most numerous among the larger animals, being represented by deer, buffaloes, antelopes, giraffes and others. An intriguing aspect of the microbial activities of the rumen is the grazing balance between Protozoa and Bacteria, which is also reported to prevail in soil. This is a microecological matter, the understanding of which could be essential to future protein production.

How each human in the 'Satisfied World' exerts a biological pressure on the primary production of the order of twice that of each in the 'Hungry World', is evidenced in Table VI.

TABLE VI

Livestock and Poultry in the 'Satisfied World' (SW) and in the 'Hungry World' (HW) (in thousand million PE-units), 1976.

Total		*Man*	*Livestock*	*Ratio livestock/Man*
9.4	SW	1.3	8.1	6.2
11.6	HW	2.9	8.7	3.0
21.0	Total	4.2	16.8	4.0

Ratio Ruminants/Non-ruminants

World-wide, the ratio of ruminants (R) to non-ruminants (NR) is 1.2, but when excluding Man it is 2.4, as indicated in Table VII, which also gives this ratio by continents, etc.

TABLE VII

Ruminants (R) and Non-ruminants (NR) Excluding Man, by Continents, etc. (in thousand million PE-units); also Ratio R/NR (1970–71).

	R	*NR*	*Total*	*Ratio R/NR*
World	10.3	4.3	14.6	2.4
Asia	3.3	1.5	4.8	2.2
China	0.68	1.09	1.8	0.63
India	1.44	0.05	1.5	28.9
Latin America	2.0	0.84	2.8	2.4
Europe	1.06	0.84	1.9	1.3
Africa	1.37	0.26	1.6	5.3
North America	1.01	0.43	1.4	2.4
USSR	0.92	0.38	1.3	2.4
Oceania	0.50	0.02	0.52	25.0

The ratio R/NR for individual countries reflects in a similar manner the degree to which it is dependent for its food on the broad base that the ruminants provide for human survival. Table VIII gives such data for selected countries.

TABLE VIII

*Ratio of Ruminants/Non-ruminants (Excluding Man), Based on PE-units in Selected Countries During 1970–71.**

	World 2.4		
Nepal	110.2	Japan	0.54
New Zealand	37.6	*China* }	0.62
India	28.7	Malaysia }	
Kenya	25.0	Poland	0.74
Australia	24.7	South Korea	0.77
Sudan	15.1	Mexico	0.99
Pakistan	14.7	Spain	1.1
Tanzania	14.3	Czechoslovakia }	
Madagascar	13.5	Italy }	1.2
Argentina	6.8	West Germany }	
Colombia	4.3	North Korea	1.3
Turkey	3.5	Philippines	1.7
Nigeria	3.4	Egypt	1.9
Thailand	3.2	Brazil	2.0
Indonesia	3.1	UK	2.2
Iran	3.0	France }	2.3
USA }	2.6	Canada }	
Ethiopia }		USSR	2.4
Morocco	2.5		

* Note that order decreases downwards in the left-hand column but upwards in the right-hand column.

The ratio, whether high or low, reflects the ecology, economy or preference of the country, and there is no dividing line between rich and poor nations. China, with a high level of waste recycling through pigs and poultry, and with little leeway for any alternative, has a low ratio; India, on the other hand, having chosen to recycle its waste *via* ruminants to obtain supplementary protein *via* milk, has a high ratio. The intermediate ratio for the USA and Canada reflects a high degree of biological affluence which allows the use of feed-crops for pig and poultry production. A similar ratio for some European countries mirrors economic affluence, allowing them to buy such feed from abroad.

It is most revealing to compare the animal production of China and India with that of the USA—in particular, pork in China and milk in India (Table IX). The production level for dry village cheese in India, chiefly from buffalo and approaching the cheese output of the United States, is most notable, because the ecological parameters of the two countries are so widely different. China produces 1.6 times as much pork as the USA, but could never do that on its model.

The Vegetarian Option

The all-vegetarian claim that thousands of millions of additional humans ould be fed by stamping out livestock production is questionable. Through

TABLE IX

Comparison Between the Animal Production of China, India and USA (in millions of metric tons), 1974–75.

	China	*India*	*USA*
Meat	15.5	0.8	23.0
pork	9.6	0.05	5.7
Milk	5.2	25.0	52.3
cheese	0.17*	1.52†	1.59

* Inclusive of sheep- and goat-cheese.
† From cows and buffaloes.

grazing, ruminants utilize much land that could not be effectively used in any other manner. They are also capable of using much agricultural waste which otherwise would be lost.

Pigs and poultry, together totalling 4.1 thousand million PE-units (Table V), have, through intensified production largely in the well-to-do world, become Man's true competitors. Historically this was not the case, as they both started by being partially waste-consumers. Some USA prairie farms still retain a residue of this pattern by allowing the pigs to feed on the droppings of grazing cows. Dairy farms in New Zealand and Europe were commonly based on feeding skim-milk and whey to hogs. Hog farms were once common around major USA cities, the hogs being fed on kitchen swill. Chickens were formerly found around most farms, supplementing their feed in summer with weeds and insect grubs. It is only recently that large-scale, high-efficiency production was started by feeding both pigs and poultry even more meticulously than our own children. Their feed is supplemented with vitamins, minerals (inclusive of trace-elements) and amino-acids, in well-researched doses. In China and to some degree in India, pigs are still the prime scavengers, including the utilization of human faeces. Their pigs and poultry represent only a limited pressure on primary production, and somewhat reduce the PE-units as calculated above in terms of the true feeding burden.

In an effort to get the current discussion on the competition between livestock and Man into some kind of realistic framework in global ecological terms, I have made the following calculations. (The figures are approximations based on the distribution of the various categories of livestock between the 'Hungry World' (HW) and the 'Satisfied World' (SW), and on data concerning the use of feedstuffs in leading SW countries.

By cutting out pigs and poultry as competitors, we could gain approximately 1.0 thousand million and 0.7 thousand million PE-units, respectively. By refraining from the use of feed-crops in the raising of meat, and to some degree reducing dairy cattle on feed-crops, possibly another 0.8 thousand million could be saved. This gives a total of 2.5 thousand million PE-units.

In the HW, beasts of burden are indispensable and account for about one-third of the cattle. To this should be added all buffaloes and 80% of the world's horses, mules and asses. The total for beasts of burden comes to 3.86 thousand million PE-units (Table X).

TABLE X

Beasts of Burden (in thousand million PE-units), 1976.

	Total	*Used as beasts of burden (estimate)*
Horses, mules, asses	1.2	0.96 (80%)
Buffaloes	1.2	1.2
Cattle (HW)	5.3	1.7 (33%)
		3.86

A modest gain of 2.5 thousand million PE-units out of a world total of 21 thousand million (1976) would thus be feasible, although it seems even less significant in the light of an expected added livestock burden exceeding 4 thousand million PE-units in the next 10 years. In any case it is a far cry from the 16 thousand million which a complete abandonment of all livestock held for food purposes would imply.

New Population-density Concept

These population-equivalent data have been used to formulate a new population-density concept, in which the total biomass within Man's sphere of action is related to soil resources—both of tilled land and cultivated pastures. This approach offers many new insights and evens out considerably the seemingly large differences between countries. Each PE-unit has at its disposal between 8 and 18 acres (800 and 1800 m^2), which is a far smaller amplitude than conventional *per caput* figures.

Quite another example may be brought in to illustrate the value of the PE-unit concept, in particular as regards the ecology of protein production (Borgstrom, 1973*a*, p. 170). New Zealand, with 3 million inhabitants, is frequently referred to as a country the size of Italy or the UK, yet only carrying less than one-twentieth of their respective populations. When analysed in terms of population equivalents, taking into account the livestock fed through importation from other countries or from ocean resources (fishmeal), which are significant in both the UK and Italy, there is in effect only a minor difference in the feeding burden between these three countries (Table XI).

Protein Transfers

The basic prerequisite to understanding the macroecology of Man is, on one hand, to relate it to Man's intake of protein, but, on the other, to

TABLE XI

Comparison of New Zealand, Italy and UK (in millions of PE-units), 1973–74.

	Man	*Livestock*	*Total*	*Trade-borne*	*Ocean-borne*	*Fed from own feeding base*
New Zealand	3	137	140	—	2	138
Italy	55	147	202	66	4	132
UK	56	202	258	113	19	126

coordinate this with the origin of the protein production. This connection has been woefully missing in most analyses, and this has resulted in a distorted notion of the degree of self-sufficiency. Trade and fisheries have rarely been recognized in other terms than tonnage or money. The present author introduced several years ago the procedure to calculate, in terms of tons of protein, the quantities contributed *via* importation and from the aquatic harvests (Borgstrom, 1965 and 1972*a*; 1973*a*, 1975*a*). In order to make these data ecologically meaningful, they were taken one step further and, for each individual country (on the basis of its current state of agricultural techniques, employing registered yield figures), the acreage was computed which would be required to raise the amounts of protein acquired from the outside in the most efficient (acreage-saving) way. This gave the *trade acreage* and the *fish acreage*, respectively, which together were called the *ghost acreage* of a country. Some data of this kind are found in Table XII, in percentage of registered tilled land. This same procedure can be used for an individual farm, region or state.

TABLE XII

Ghost Acreage (in percentage of tilled land), 1974–75.

	Trade acreage	*Fish acreage*	*Ghost acreage**
Japan	365	348	713
Netherlands	562	75	637
Switzerland	31	194	625
Belgium	403	54	457
Italy	187	23	210
UK	159	39	198
West Germany	100	47	147
Czechoslovakia	103	10	113
Portugal	88	18	106
Spain	41	42	83
East Germany	56	28	76
France	37	7	44
Poland	17	25	42
Hungary	18	13	31

* Composed of 'trade acreage' and 'fish acreage' combined.

The lack of this ecological dimension recognizing the pressure exerted in distant lands, has led to innumerable fallacies. Singapore and Hong Kong are not the economic masterpieces they are generally claimed to be, nor is Japan feeding so many people from so little land. The same is true of several European countries, in which the notion still persists that their respective farmers are to be given credit for feeding the nation. Such data reveal to what degree Europe, as a whole, as well as individual countries, led by Japan, have lost touch with their geographical base and moved out of their ecological context. One further aspect, frequently not recognized, is the massive influx of additional nitrogen into soils through fertilizers and manure and, indirectly, through oilseed cake. The manner in which subsequent sewage is apt to place great strains on water recipients is also widely overlooked.

Still more neglected is the biological upheaval resulting from the European emigration of 1850–1950—the biggest-ever in history (Borgstrom, 1973*b*, 1975*a*). Not only did Europe lift off one-fourth of its population, primarily to the Western hemisphere, but, more importantly, it vastly expanded its feeding base by a doubling of its tilled land and more than a trebling of its pasture land, this being done chiefly in favourable temperate latitudes. The concomitant push into Siberia, only in latter years reaching an advanced stage, was less rewarding owing to more adverse climatic conditions. This is the basic factor explaining why the Satisfied World now has three times as much tilled land per inhabitant as the Hungry World.

This migration resulted in hunger decisively vanishing from the Western scene, and we should take serious cognizance of the fact that it happened late—not much more than five generations back. But this large-scale operation moulded the global trade-patterns in a decisive way. Another significant but equally little-heeded fact is that it was only then that transcontinental railroads and transoceanic shipping became a reality—in effect a basic prerequisite to the feeding of multi-million cities—further contributing to the disappearance of ecological awareness.

This grand land-grabbing operation had its repeat in the oceans during the post-World War II era, resulting in two-thirds of the ocean catches moving into the well-to-do world. Half the catches of fish were, at the peak years of 1970–71, channelled as fishmeal into the feeding troughs of the affluent countries.

Protein is the key item in both these phenomena of long-distance hauling from continents and oceans, constituting what amounts to the creation of major global survival-bases. Europe is little aware of the two huge 'ghost continents' on which it depends, one based on net importation and the other on the oceans. Computed on the basis of the current levels of yields per hectare in each individual country, this amounts to a 'trade acreage' of 74 million ha. and a 'fish acreage' of 25 million ha. (1973–74).

In order to transfer these data into ecologically more meaningful terms, the 'ghost acreage' of *each* European country has been computed. Added together, they amounted in 1973–74 to 99 million ha., i.e. 69% of the tilled land of Europe—for earlier data *see* Borgstrom (1965 and 1972*a*).

Trade Balance

It is frequently said that the United States is feeding the world, although such a contention is seemingly oblivious of the fact that international trade never accounted for more than one-tenth of what mankind is eating. As a matter of fact, more than half of the US cereal export is for purposes of feeding domestic animals. Although maize holds half that share, attention is mostly focused on wheat, of which one-third goes to the well-to-do countries and is even used for stock feed. To this should be added the soybean which, with the exception of some 5%, moves to support the protein sanctuaries of the rich world. The ratio of (animal) feed protein to (human) food protein in these United States exports is 2.8 to 1.

In the minds of both experts and the public, trade has far too often become synonymous with aid. The chief recipients of the net trade in cereal grains are in effect Europe, Japan and the USSR. China and India together have only in a few odd years received more than Japan. Several European countries are the recipients of as much as, or even more than, either China or India. It rings very hollow when we pompously declare that it is 'high time the developing world learns how to take care of itself'. This admonition would be far more appropriate if addressed to the affluent world—in particular when we recognize that most oilseed in the form of cake or meal (92%) moves to underpin its protein empires. More than half originates in the protein-short world. The ratio of feed to food protein in the world market is 6.7 to 1, and there is in effect no more telling indicator than this of the dominance of domestic animal feeding, chiefly in the Satisfied World, over human feeding.

Much too little attention has been devoted to these anomalies. The entire debate has got caught in the semantic trap of identifying grain-deficit countries and failing to establish which countries are on balance or truly deficient (and as a consequence are depending on large additional acreages for support). Europe (both West and East) then holds the lead, followed by Japan, constituting in effect heavy burdens on the world household. In contrast to general belief, neither China nor India belongs in this category. Very few countries in the Hungry World are in this sense truly a burden to the world household. Taiwan, South Korea, the Caribbean, and parts of the Middle East and of North Africa, are the main selected beneficiaries within that vast poverty-realm of the poor world.

Per caput figures for net import of wheat bear this out (Table XIII). China figures at the bottom of the list, with some 6 to 7 kg, *per caput*, followed by Mexico (8.9) and India (9.8). The 'ghost acreages' of China and India are, however, almost nil, as both sustain a countervailing export of vital agricultural commodities. India competes with Africa in providing Europe with peanut protein to the tune of 6 million metric tons per year, which is adequate to supplement the cereal diet of 250 million people. It is high time that we moved out of this outmoded pattern, which was shaped in the bygone grand period of colonialism, when populations were very much smaller.

Right up to World War II, human food and animal feed moved in the world market from the food-short world to the then already overfed world,

TABLE XIII

Net Wheat Import (*in kilograms* per caput) *in Selected Countries, 1974–75.*

Country	kg	Country	kg
Netherlands	189.5	Japan	50.5
Libya	144.0	Belgium	49.4
Guadeloupe	137.1	Sri Lanka	48.1
Algeria	122.5	Poland	47.8
Kuwait	117.0	Czechoslovakia	47.6
Israel	116.5	Iran	46.0
Cuba	105.0	Venezuela	45.6
Jamaica	99.5	Tunisia	44.5
Trinidad	97.5	South Korea	40.6
Chile	91.7	Costa Rica	39.7
Martinique	91.0	Syria	35.2
Singapore	85.0 (rice 65.3)	Hong Kong	33.4 (rice 70.0)
Barbados	82.5	Italy	24.4
Egypt	78.0	Brazil	23.6
Pakistan	74.3	Dominican Republic	23.6
East Germany	74.0	Bangladesh	21.7
Norway	67.8	Haiti	13.9
Grenada	64.8	Philippines	11.8
Morocco	61.2	India	9.8
Switzerland	56.0	Mexico	8.9
UK	55.4	China	6.1
Saudi Arabia	50.8		

and this pattern has, despite aid deliveries, returned as the increasingly dominant pattern. Yet we have the audacity to declare in response to the developing world that long-distance feeding must come to an end. This is, in the long run, unquestionably true; but to do this abruptly after almost two centuries, during which the affluent world has wholly relied on this device—and to its almost exclusive own benefit—is foolhardy. Besides, it is politically both venturesome and imprudent, in particular since we have exhibited the myopia of monopolizing and giving top priority to an accelerated build-up of our own nutritional oases.

The Soviet Switch

The province of Scythia, i.e. Ukraine, was one of the key grain-baskets of the Roman empire. It is therefore no surprise that Russia, ever since czarist days, and later the USSR, has been a net exporter, chiefly of wheat—except in years of droughts, which have been a recurrent phenomenon on that scene. Since the decision was made in the 1950s to break with the bread-and-potato-line—the two staple items for centuries in the diet of the country—more than 650 million PE-units were added (up to 1973–74) to the feeding burden of this giant, placing the USSR as top-ranking country in the world as regards milk production. The increase in PE-units amounted to 68% within 25 years.

There is also clear evidence of considerable gains in most other animal products (Table XIV). The USSR has further reached the global top position as wheat producer, despite its climatic and political adversities. Meanwhile wheat is reported to be increasingly used for animal feeding.

TABLE XIV

Soviet Agricultural Production 1948–52 to 1973–74, with Selected Comparisons (in millions of metric tons).

		1948–52	*1961–65*	*1973–74*
Cow's milk:	Europe	92.6	135.7	158.9
	USSR	33.2	63.8	87.3
	USA	52.3	57.0	52.4
USSR:	all meat	4.53	9.32	14.2
	beef	2.03	3.47	6.13
	pork	1.34	3.29	5.29
	poultry	0.28	0.75	1.35
	eggs	0.68	1.58	2.92
Wheat:	USSR	35.8	63.8	96.8
	Europe	41.1	59.4	86.4
	USA	31.1	33.0	47.6
Potato:	Europe	130.2	138.3	129.5
	USSR	88.6	81.6	94.5
	USA	10.7	12.4	14.5

The USSR has, within the past quarter-century, become a mighty biological force which can no longer be safely supported within its own ecological limits and now is resorting to the same biological evasion as Europe initiated towards the end of the preceding century. Joining the prairie corporation—i.e., like Europe and Japan, importing from North America—appears to be an indispensable recourse, but this is heavily taxing world protein resources.

From the biological point of view, this is an impressive accomplishment, but it is bound to have major ecological repercussions. It is therefore *no surprise* that the USSR reports on one hand considerable protein deficits in their feeding requirements (1), and on the other have appeared on the world scene for large supplementary purchases, primarily of feed commodities (2), and also have started up plants for the manufacture of petro-protein (3). But the big question to the world is: Can the world household support this big addition to the world's feeding burden, almost like adding a new India within 25 years?

The Peruvian Crisis

In 1970 the Peruvian and Chilean catch of Anchoveta (*Engraulis ringens*) reached its peak at 13.1 million metric tons (mmt). That constituted 38% of the total fish-catch of the Pacific Ocean and 19% of the global ocean harvests, thereby placing Peru as the world's top fishing nation, surpassing the two fishing giants Japan and the USSR, and harvesting in the Pacific 22% more than these two together extracted from this basin. The potency of this protein input, chiefly in the feed-protein balance and its macroecological significance, has been analysed in detail (Borgstrom, 1972*b*). It is therefore no surprise that the 1972 crisis greatly upset the feed balance, primarily of Europe and North America. Due to disturbances of the Peruvian current coupled with over-fishing (the standing stock had been reduced by half), a drastic decline took place in the Anchoveta catch (down to 1.96 mmt in 1973), and a corresponding shrinkage of fishmeal deliveries (Caviedes, 1975).

No less than one-third (0.94 mmt) of fish protein that had been moving into the world's feed markets was lost through this Peruvian *débâcle*. It was somewhat compensated by increased production of fishmeal in South and South-west Africa, Angola and Europe. More than one-fifth of the landings in Africa went into fishmeal. But the Western world then transferred part of its long-distance protein dependence from the SW Pacific at least temporarily to the North American prairies, by increased purchases of soybean. The feeding value of the fish protein is, however, higher than that of soybean.

No example more clearly brings out the almost clandestine breakdown in the awareness of ecological relationships. A number of European countries, as for example the Netherlands and Hungary, channelled far more fish protein into animal production than into the direct feeding of fish and fish products to its human populations. Two Scandinavian examples illustrate the degree of this ocean dependence as well as its lop-sided channelling into the human-support system (Table XV).

TABLE XV

Fishmeal Production from Ocean Catches of Norway and Denmark (in millions of metric tons), 1970–75.

	1970	*1971*	*1972*	*1973*	*1974*	*1975*
Norway						
total catch	2.86	3.07	3.20	2.97	2.65	2.55
made into fishmeal	2.10	2.21	2.30	2.14	1.85	1.76
fishmeal (%)	73	72	72	72	70	69
Denmark						
total catch	1.23	1.40	1.44	1.46	1.84	1.77
made into fishmeal	0.89	1.0	0.99	1.07	1.45	1.40
fishmeal (%)	72	71	69	74	79	79

Re-creation of a Rural Ecosystem

Due to the long-distance hauling of feed—particularly protein—animal production has been created around the globe with little consideration for ecological constraints. In principle, it has become feasible to raise all categories of livestock almost anywhere. Major animal-feed companies are quite capable of starting such enterprises, independently of regional climatic conditions. They may, in other words, move in and disrupt traditional agricultural production that has evolved through the interplay of prevailing ecological circumstances.

Just one telling example: due to the arid conditions of the Iberian peninsula, sheep and goats were for centuries the main livestock. Cattle, when gradually introduced, were restricted to meat-producing grazers. Pastures adequate for milk were limited, so milk was produced in and around the cities by bringing in feed. As a result, manure odours became a dominant feature in many Spanish cities. Thus it was only by linking its animal production to the North American prairie (by importation of feedstuffs therefrom) that it became feasible for Spain to accelerate milk production. It is worth noting that the post-war increase of milk production in Spain corresponds in volume almost exactly to the decline of Swedish milk production in the same period. Swedish dairying is in close harmony with ecological conditions, and in this case trade exchanges should have been in far better conformity with a sound ecological balance than long-distance hauling of feed over the oceans.

The almost complete dissociation between crop and animal production has created around the globe foci for producing eggs, broilers, pork, beef and milk, which are totally out of their ecological context. This is currently happening on a major scale in the Middle East, where neither feed nor adequate water is available. Ground-water depletion will soon assert itself, as regional production evolves.

As energy costs continue to climb, the 3- to 8-times larger transport volume of feed as compared with the ultimate products will, however, become a cumbersome burden in most parts of the world. Animal production will gradually be forced to move back into the position of an intermediary link in a restored rural ecosystem, based on recycling.

An entirely different aspect emerges when we analyse the dried-milk imports as compared with the milk production of importing countries. Several tropical countries import many times more than they produce. Emergency deliveries may well be justified, and even regular trade may be called for, but danger lies in the distorted ecological understanding that ensues. Reports and development programmes do not take cognizance of the vast dairylands on which they ultimately depend. Frequently the protein balance is misjudged when this dimension is overlooked. To judge these matters in monetary terms or even trade tonnage without taking into account the livestock, the land and the water behind these figures, is persistently resulting in false conclusions both as to the potential of global milk production and the feeding capabilities of countries or regions. It is no coincidence that milk basically is a privilege of some 700 million people of the temperate latitudes. The poor, hungry parts of the world with more than two-thirds of the world's human population account for only one-fifth of the global production of milk. The single major exception is tied to the ecologically better-adapted Water Buffalo (*Bubalus bubalus*).

TONNAGE—A MISLEADING GAUGE

The productivity of agricultural lands mirrors to some degree ecological factors, whether these be favourable or restraining. Yet, certain basic concepts of yield also need reformulation. Tonnage is an inadequate gauge that needs to be amplified and to take into account nutritional parameters. On such a basis, there is a major difference in productivity in terms of yield between one hectare growing sugar-beet as compared with one carrying grain, potato or legumes, and really little freedom of choice. The United States could, in effect, depend on its maize acreage to provide adequate calories and convert most of its agriculture into wonderful parks, such as are desperately needed for recreation. Growing only sugar-beet in the north and sugar-cane in the south would provide the USA with far more calories than today's total world consumption. The energy approach to this problem offers little substance in an appraisal of the world's food-producing capabilities, yet such simplified energy computations constitute the basis for a great many of the easy-going, over-simplified remedies which are advocated, and which contend that the world could feed from 50 to 157 thousand million people. The ensuing massive energy subsidies are not taken into account.

The increase of world food production has been most notable in calorie-rich crops (Table XVI). Cereals, sugars and fat (*via* oilseeds) have taken the lead, and to a pronounced degree outside the Euro–American sphere. The western colonization of Africa, Asia, Latin America and the Caribbean, in fulfilling food requirements in these regions, involved a switch from a varied, relatively protein-rich diet to a monotonous, sugar–fat menu. This pattern has prevailed and is almost universal in the new African countries and Latin America.

TABLE XVI

Increase in Global Crop-production (in millions of metric tons), 1948–52 to 1973–74.

	1948–52	*1973–74*	*% increase*
Cereals	691.9	1,347.1	94
Roots and tubers	317.2	562.3	78
Pulses	29.2	46.7	59
Oilseeds	31.6	118.2	276
Sugar	37.8	79.0	105
Population (thousand million)	2.50	3.94	58

Higher yields have, as a general rule, been attained by sacrificing the carbon–nitrogen (C/N) balance of 17:1 and bringing it above the crucial limit for growth and maintenance. There is a transition area up to a C/N ratio of 20–21 above which protein deficiency develops (Nicols, 1967). Animal feeding as a rule adhered to this principle, and most feed-mixtures stay within this limit, although those for ruminants afford a notable exception. Aquaculture has, in testing various aquatic organisms, established the same rigorous limitations. Rapid growth is only assured when the C/N ratio comes down close to 10:1 (Russel-Hunter, 1970).

Nitrogen fertilizing normally assures higher yields, but the protein level can only be sustained either through late dressing (after the plants' final structure has been fully created) or *via* continuous input through such a nitrogen-fixer as the blue-green Algae in wetland rice cultivation.

The Cereal–Legume System

Pulses (legumes) are nowadays documented as having been crop-plants as early as cereals in most civilizations and have, right through history, been indispensable as supplementation to cereals in human food and presumably also in animal feed. Legumes partially supplement the aminogram (amino-acid composition) of cereals, but more importantly allow adjustment of the C/N ratio down to the crucial 17:1 level. The bean plants aid in fixing nitrogen and hence in fertilizing the soil.

Ecologically tied in to protein production are various systems of survival strategies such as intercropping and transhumance. Beans have been grown intertwined with maize by several American Indian cultures.

The use of legumes as human food (after due preparation), therefore, early on became the base of the diet of millions of cereal-eaters. Brazil, China and India account today for almost half the world's legume (pulse) crops. Latin America, Asia and Africa, together account for 76%. No other issue is more vital and yet so neglected. Many hundred millions of humans depend on legumes for their protein balance. In Africa, pulses are far more important than these figures indicate, as wild pulses are not registered statistically; many such wild legumes, including trees, provide protein supplementation.

Two features are ominous in this context. Legumes have, in the past 25 years, lagged behind in the global drive for more food. Secondly, the 'green revolution' has unfortunately intruded on the pulse acreages of several key beans and thereby inadvertently aggravated nutritional imbalances—most conspicuously in India (Table XVII). Cereals have widely taken the place of beans.

TABLE XVII

India: Acreage Changes During 1961–65 to 1973–74 (in millions of hectares).

	1961–65	*1973–74*	*Increase*	*Changes (%)*
Wheat	13.4	19.3	+5.9	+45
Other cereals	80.0	80.8	+0.8	+ 1
Total cereals	93.4	100.2	+6.8	+ 7.3
Pulses	24.0	21.9	−2.1	− 8.7
Chick-peas	9.3	7.3	−2.0	− 2.2
Dry beans	7.0	7.9	+0.4	+ 5.7
Miscellaneous other pulses	3.3	2.5	−0.8	−24.2
Population (millions)	462	580	+118	+25.6

Most analyses of the global food production, particularly in recent times, have overlooked the fundamental role of legumes in crop and animal production. The crop-rotation schemes, as they evolved in Europe, were firmly anchored to legumes as a link that was indispensable to sustained production. Only when clover and alfalfa reached New Zealand and Australia, and were incorporated as a crop in the wheat-belt of the humid pampas of Argentina as well as in many parts of the North American prairie, did these regions become dependable for animal production.

Again, the settling of the pastoral nomads of Soviet Kazakhstan only became successful when alfalfa was introduced and irrigated to provide supplementary feeding. This basic relationship was widely broken later on when nitrogen fertilizing moved in on a major scale—but at an ecological price of losing this cost-free support and depriving Man of the free-of-charge microbial underpinning. This might be one factor in the diminishing crop-yields. The net return is, in relative terms, dropping further, being reflected in the growing losses of nitrogen to adjacent waters which it enters as a pollutant.

We need greatly improved knowledge of the role of organic nitrogen compounds in the soil and of their possible input as plant feed—e.g. *via* amino-acids, which are the basis of the strength of symbiotic relationships.

The Cassava Syndrome

Crucial to nutritional ecology is the calorie-rich production of the humid tropics which is dominated by cassava. No plant products can fill in the protein gap that its use leaves; even legumes can only partially compensate. So West Africa early resorted to dried fish, which was (and still is) mixed with cassava flour (Annegers, 1972).

With cassava there are, however, additional complications. When it is not carefully prepared, the hydrocyanic acid in it is detoxified through the formation of thiocyanates. The methionine is taxed in this process, thus reducing the content of the amino-acid which stands out as the prime limiting one. Besides, the total available protein is reduced. There are several studies showing the crucial role of this dilemma both in Brazil and West Africa.

Transhumance, as practised by several nomadic civilizations, constitutes an alternative way of balancing out the diet. The recurrent Fulani trek to the south to sell meat to the protein-short, calorie-excessive southern region of Nigeria is a good case in point. The dried beef, *charque*, of Brazil is another good example.

Soil Frontiers

The whole matter of utilizing the vast expanses of the forest lands of the humid tropics has been continuing along a largely disastrous course. The proportion of carbon to nitrogen in tropical forest litter is out of balance, and nitrogen is in short supply. Plants and animals, including the decay-producing organisms in the soil, all compete for the very limited supply of protein matter. Decaying plant residues provide ample energy-source but little, if any, protein for growth. Seed-producing plants do not do well under these conditions. Starchy crops flourish but are not grain-producers. One conspicuous exception is rice, which lives in symbiosis with, and is fed by, nitrogen-fixing blue-green Algae. The tropical rain-forests of Indonesia have a layer of nitrogen-binding microorganisms on the leaves—another of Nature's ways of circumventing the carbon–nitrogen barrier.

It is, therefore, not merely a coincidence that inhabitants of these tropical regions are those who suffer most seriously from protein-deficiency diseases. It is most significant that the original populations of the tropics were concentrated along the shores of lakes, streams and oceans, where fish or shellfish provided this imperative supplement. The explosive increase in diseases originating from insects, Fungi, and Bacteria, following in the wake of clearing such forests and starting crop or animal production, has generally been attributed to the enormous diversity and wealth in species-numbers of these areas, all of a sudden going in search for new vectors or hosts; but it could equally well be related to a drastic readjustment of the C/N ratio in favour of nitrogen, protein and growth.

Proteinaceous crops require major inputs of nitrogen to the soil, but under such conditions applied fertilizers tend to be rapidly leached out or washed away. It is generally cheaper to import food than fertilizers to such places. Studies in Brazil further indicate that only by continuous import of protein concentrates is it feasible to sustain animal production. The alternative is strictly-controlled pasturing on certain feed grasses within fenced lots.

This whole matter widens into the necessity of reviving interest in soil life, and tying it to the same C/N parameters as life in general. This becomes particularly significant in view of the need to find alternatives to the mounting nitrogen inputs, and is intimately coordinated in particular with the phosphate balance. The entire mineralization process is basically dependent on microbial activity, but it is equally vital to sustain the nitrogen fixers with organic nutrients. The drop in the humus and other organic content in so many of our intensively-cropped lands has consequences that we have not heeded in time. It should also be recognized that no less than one-seventh of the world's soils are organogenic and will rapidly 'burn away' unless major countermeasures are taken.

Non-fertilized plots within field experiments with nitrogen, are reported by old-timers as today visible for miles by their yellowing caused by nitrogen shortage. In corresponding experiments performed 25–30 years ago, such plots could not be immediately identified. It is high time we got a more comprehensive picture of the overall interferences by Man in the intricate balance between the nitrifying and denitrifying microbial systems of the soil. Is it possible that we have stamped out or heavily impeded Nature's complementary nitrogen fixation, or inadvertently built-up the denitrifying systems with inevitable losses to the air?

Of special interest is the fact that some wood-boring insects carry cellulose-digesting organisms. The most thoroughly studied case is that of the termites, which harbour Protozoa that digest the cellulose of the wood which is ingested by the termite. In turn, the termite ingests the product of the protozoan digestion. To what degree these insects could be marshalled for a second Man-controlled cellulose beachhead in the plant kingdom remains to be seen.

Another route would eventually open the lignin potential as an energy source for protein production through the use of lignin-decomposing Fungi, marshalled by the wood-eating insects in their 'gardens'. The nitrogen-binding capabilities of several insects could also open up new avenues for insects as food, if this has not already happened among human insect-eaters. There are also several indications that traditional free-feeding poultry depend on the contribution of grubs, larvae and insects. Moreover, several insects are known to carry active nitrogen-fixers in their hind-gut (Toth, 1946).

Such fields of research have been little pursued, and might well place at least some insects as integral parts in restructuring the urban ecosystem. An indispensable feature of this would be the creation of new food-producing centres in lieu of conventional waste-disposal systems.

Yield and Photosynthesis

Baffling misconception prevails as to the role of photosynthesis in food production. Far into the realm of key textbooks and professional papers, food is identified with photosynthesis, in complete disregard of the fact that much more is indispensable to make food—namely the massive microbial input to collect and fix nitrogen in order to assure the basic C/N ratio as the rigid parameter for all life. In addition, the whole range of other key elements, led by S, P and Ca, are required—all with their particular major cycles not yet well mapped ecologically in nutritional terms (Borgstrom, 1964).

Let us add a few brief but pertinent points. Generally the exuberant flow of solar energy is contrasted with the pitifully small fraction of it that is actually involved in photosynthesis. But leaving aside the modest proportion of organic plant matter that ultimately ends up in human or other food, and forgetting the root in its rhizosphere and other unharvested parts of the plant, the photochemical reaction itself is highly efficient—around 30%. But far more important is the fact that about 82% of the energy reaching the Earth's surface is used in the evaporation of water, some 25–30% thereof being connected with plant transpiration. Without this vital cooling function, the Earth would hardly be liveable; our field crops and forests would be almost ablaze without this mechanism. The Hydrological Cycle has the triple function of being: (*a*) Nature's big distillation machinery (through its desalination), constantly renewing the freshwater reserves; (*b*) a large global refrigeration machine; and finally (*c*) the mechanism for the transfer of energy which is basic to climate.

The Water Phase

The most potent ecological force in protein production is the disproportionate amount of water required in producing animal protein as compared with plant protein. This aspect is rarely recognized in discussions on the dietary balance between animal and plant products, or in the formulation of agricultural development programmes. In broad figures the amplitude is considerable, being between approximately 1,200 litres per person per day for a strict vegetarian diet, and 10 times as large (12,000 litres) for an average Western diet. The USA diet currently requires 16,800 litres daily, of which animal products account for no less than 13,200 litres (Borgstrom, 1969 and 1971).

Some 30,000 litres are required for the production of 1 kg of beef. An increase of 30 g of beef accounts for 600 litres—almost equal to the *per caput* USA daily take from the average faucet (currently 650 litres per person per day). Several conclusions can be drawn from this. One is that there is a wide disparity in water requirements between, for example, cereals, legumes, nuts and other plant products, as against meat, milk and eggs—these latter two being slightly below the needs of meat. Introducing animal production as a replacement or supplement, as frequently happened in development programmes, not only affects land and energy requirements but also taxes, often harshly, available water resources. As a rule-of-thumb, protein from animal resources demands, kg for kg of protein, from 3 to 6 times as much water compared with plant protein. For this reason, grazing pressure in marginal arid regions multiplies the destructive forces of human numbers.

Water is unquestionably the most limiting of all factors in world agriculture. Yet the 1900s might truly be called the century of irrigation, as since the turn of the century, the world's irrigated acreage has increased more than four-fold, to reach some 240 million ha. A further doubling is anticipated before the year 2000. The International Hydrological Decade has shown even this irrigated area to be wholly inadequate, and has established another 270 million ha as practically indispensable, with a price-tag of some $2,300 thousand million. Further tapping of ground-water will be called for, but must in the long run be put in reasonable line with replenishment. So far the exploitation has in many regions been proceeding at rates which exceed those of refill, as for example in Australia, South Africa and the US south-west.

The competition of industry for water enters as another decisive force. Many regions cannot afford to return water to the Hydrological Cycle *via* crops when it can be recycled for industrial and urban use—a dilemma that South Africa, Australia, and many sections of the arid and semi-arid United States, are facing. Industry and cities are on a direct 'collision-course' with their food survival base (cf. Vallentyne, 1972).

An expansion in the range of 270 million ha as stated above would require massive desalination of ocean water. Already the energy demands are prohibitive; but even more critical is the salt factor. Each American would cause the accumulation of at least 450 kg of salt per day on such a basis. Each 100 mm/ha would require the removal of 35 metric tons of salt to produce such irrigation water from ocean sources (one acre-foot, 42 tons of salt). Production of a ton of wheat creates 35 metric tons of salt, and production of a ton of rice creates three times as much. Man is already in an ominous battle on the salt front in most areas of perennial irrigation, and a growing percentage of so-called irrigation water has to be used for salt removal. Using desalinated water for the irrigation of 100 million ha, Man would be faced with the task of disposing *each year* of no less than 16 thousand million metric tons of salt accumulated in the 'manufacturing' of this water.

Making deserts bloom is one of technology's masterpieces. Yet at the same time Man has created five times as large an acreage of deserts as existed heretofore, or some 1.2 thousand million ha now in all—whether through negligence, ignorance or sheer pressure of numbers in Man and livestock. It is food for serious thought that Man, through his intervention, has created almost as much desert land as he put under the plough for tillage. This transcendence of ecological limits is an ongoing process; thus no less than 60–70% of currently-used pasturelands are threatened as to their future through overgrazing, thereby jeopardizing the plant cover.

The blessing of irrigation is, in a similar manner, menaced by ongoing salination resulting in gradually declining yields and finally eliminating cropping. US and Soviet experts have reached the conclusion that 40–60% of the currently irrigated lands exhibit declining yields owing to the mounting salt content of the irrigation water. In many areas, extensive and costly drainage systems have been or are being installed. No less than 30% of the irrigation water in the Nile Valley is no longer available for boosting crop production but is required for the removal of salt. A further irony is the fact that increased use of fertilizers elevates the salt content and brings the risk levels closer in time.

Energy Aspects

As late as 1870, 95% of mankind's energy budget within the cultivation sphere consisted of the input of human and animal work. The massive input of fossil energy largely belongs to this century and came into full force only after 1940. Thus it represents only a brief recent episode in Man's lengthy history (Borgstrom, 1974, 1975*b*). This study will not review the profound changes that have taken place through the growing energy-subsidies which are required in modern agricultural production. The collection of solar energy, enhanced by growing yields, has persistently required increased inputs of fossil energy (solar energy collected in previous eras through the same photosynthetic process). The ratio between energy output (in human food or animal feed) in relation to energy inputs, has gone from a level of 50 to 25 *via* 5–10 down to 1 or occasionally below. Also in this field, animal production emerges as a considerable burden, giving ratio-levels mostly below 0.5. Long-distance fishing is the most extreme case, with figures down to 0.05. Shrimping, in particular in the Mexican Gulf, shows a still lower figure of 0.006. This is partially connected with the very high discarding rate of catches, involving 7–10 tons of fish for each ton of shrimp harvested.

Table XVIII summarizes some data on the energy costs of protein, produced in alternative ways. Frequently the statement is made that larger

TABLE XVIII

*Energy Input (MJ = Megajoule) per kg of Protein, and Energy Ratios in Food Production.**

	MJ	*Energy ratio output/input*
Shrimp (Mexican Gulf)	3,450	0.006
Fish (long-distance)	489	0.05
Shrimp (Australia)	366	0.06
Eggs	353	0.14
Pigs	317	n.a.
Broiler (meat)	290	0.10
Bread (white, sliced, wrapped)	243	0.53
Milk (UK)	208	0.37
Cattle (beef)	185	n.a.
Rice (intensive USA)	143	1.29
Potato (UK)	96	1.57
Wheat (UP in India)	87	1.7
Cereals (UK)	64	1.9
Maize (USA)—1970	62	2.6
Wheat (UK)	42	3.4
Rice (Philippines)	33.5	5.5
Subsistence farm (India)	11.3	14.8
Peasant farm (China)	3.6	41.1
Petroprotein	195	
Methanol (microbes)	170	

* After Leach (1975*b*).
n.a. = not available.

energy subsidies are compensated by the acquisition of highly-coveted protein. This has been given as a rationale for the big energy inputs in long-distance fishing or large-scale animal factories (feedlots, broiler establishments and similar concerns). Doubt is, however, being raised as to the validity of this reasoning when the true energy-costs are ferreted out, and in particular when energy prices zoom. Judged in terms of the developing world, it can be concluded that the affluent world has shaped agricultural and fisheries operations in such a manner that they are beyond the reach of the poor world (Pimentel *et al.*, 1973; Steinhart & Steinhart, 1974). Western food-systems such as those of the US and UK require, per person, three times as much energy as the poor world disposes of for all purposes (Leach, 1975*a*, 1975*b*).

This has truly a macroecological, leading-factor effect on the world scene. Copying our format throughout the world would earmark some 80% of the current global energy account. It is highly dubious if even the affluent world can afford to continue the present energy extravagance on the food front.

In order to bring this aspect of the food issue into a better ecological focus, this author is calculating the energy acreages, i.e. the tilled land required to raise, on a particular farm or totally in a country, the amount of energy used. In order to fulfil his daily needs, each US citizen depends on the annual output from 0.6 ha of tilled land (+0.3 ha into net export) *plus* 1.4 ha of grazing land *plus* 0.7 ha of forest land; but 2.5–3 ha are needed for energy if it is to be procured through maize, sugar or forest plantations.

The Mineral Balance

If one could conceive of a world without chemical industry and its related huge plants, 500 million humans would be deprived of nitrogen and consequently of their protein. The energy relationships are, however, very unfavourable, as the annual production capacity of 1 million tons of nitrogen requires an input into such a plant of at least 1 million tons of steel. No less than 3 million tons of coal, calculated as energy equivalents, are required for energy per year. This is a most crucial factor for the energy-poor parts of the world, further aggravated by transportation costs for what by the year AD 2000 would annually exceed the weight of the present human family (~200 million metric tons).

In many regions of the world the use of fertilizers has not increased yields over those once provided by Nature; rather, it is a stop-gap measure designed to prevent further yield reduction through ongoing mineral depletion, currently accelerated by expanding urbanization. Only gradually, in limited areas, and under climatically favourable conditions, has the original fertility of the soil been restored, with possibly chances created for a raised production-level—as in parts of Western Europe, southern Scandinavia, and the eastern United States.

The future of crop and animal production needs to be analysed in ecological terms as to the rate of mineral depletion—not merely in terms of fertilizer inputs, but also as to the sustained mineralization rates *via* microflora. The forests as run-off regulators and controllers of erosion are parts of this picture. Clear-cutting emerges as a major mineral-depleting operation, and the losses must be compensated for—whether the land is going to be used for reforestation or for agricultural tillage.

Land Resources

Around the globe a ferocious battle is raging to secure land. Agriculture finds itself in the midst of a tug-of-war with forestry, hence between the meagre food rations of the hungry millions and their basic needs of fuel on one hand. But it is also in harsh confrontation with the demands of the expanding cities of multi-millions and the huge requirements of industries and transportation systems. Big swaths of land are cut out for transfer of electricity. Recreation and wildlife reserves are threatened or gradually eliminated by the crying needs for space and food, in many instances clandestinely. There is a growing discrepancy between land surveys and actual conditions. The bureaucracies in the capitals frequently fail to register what is already taken—even destroyed—by overgrazing, by land-hungry settlers, or by fuel-short masses.

Some spokesmen for what is labelled progress are so totally unaware of the indispensability of tilled land that they look in all seriousness towards the day when all surface traffic is placed underground and agriculture has been moved out to ocean platforms, with terra firma exclusively devoted to bases for supersonic air traffic and housing! According to this school of thought—and they are serious—all food will be made synthetically, presumably also on floating installations to circumvent pollution. This is a telling example of the total absence of even the most basic insight into Man's dependence on the biosphere and, for that matter, also as to the prerequisites for life on our space-ship.

In endless variants the statement is made that mankind is only using half of the world's arable land; a doubling is consequently claimed to be within reach. Such statements have no relationship whatever to the harsh realities of the globe.

Far too little attention is paid to the basic soil features of continents. Yet soil-less areas are fairly extensive in the tropics, with the African continent having a far greater share than others (Table XIX), while other areas are ice-bound or flooded or too steep to bear soil or at least to cultivate.

TABLE XIX

*Soil-less Areas of Continents (in Percentage of Land Area)**

	%		%
Africa	29	Eurasia	6
Australia	14	South America	2
North America	11	Total world	20

* After L. D. d'Hoore, 'The classification of tropical soil', pp. 7–28 in *The Soil Resources of Tropical Africa* (Ed. R. P. Moss). Cambridge University Press, Cambridge, England: 226 pp. (1968).

Man manages to harvest each year only 850–900 million ha out of the 1,470 millions that are being tilled. Diseases, floods, droughts, and other adversities, account for the losses. Prior to the year 2000, 300 million ha are anticipated to have disappeared through erosion, salination and diseases. Urbanization has already taken a toll of 400 million ha, half of which was farmland. The acceleration of urbanization makes it likely that an additional 300 million ha of tilled land will be taken for this purpose before the turn of the century.

The United States needs to recognize its ecological constraints. The acreage restrictions through its Soil-bank (withdrawing, from tilling, cropland that was prone to erosion) were a godsend to the semi-arid sections of the dry prairie, and were in part crucial to averting a repetition of the 'dust-bowl' catastrophe of the 1930s. Reportings of record crops in recent years overlook the fact that close to 25 million ha were added to the tilled land in 1974–75. Such a gain corresponds to the total tilled lands of France and the United Kingdom combined, or to twice those of Italy. Almost no other country in the world could have made such a jump in so brief a time. This dubious advance has hidden or masked critical adversities in US crop production, and little is being said about the high price that continues to be paid in soil erosion. (Since 1974, the United States has suffered the biggest such losses since the crisis of the 1930s.)

New Strategies

Within the limited sphere of food production, top priority must be given to channelling into human consumption the huge amounts of protein (two-fifths of the fish-catch of the oceans, one-quarter of the milk protein, and two-thirds of the oilseed protein) which are now going into animal feed. Second in line of urgency should be a restructuring of the urbanization process to incorporate a recycling of the large amounts of organic waste into human food (such as fish, mussels, poultry, quail and others) or animal feed (such as yeast, Algae and other types). This urban waste will gradually become a third important source of food, supplementing agriculture and fish harvests. The current experimentation with non-conventional foods will attain its greatest rationale within such a framework, and thereby move from its present tactical sphere into meaningful strategic operations.

On the whole, food and people can only be brought into reasonable balance through coordinated efforts. For this purpose I have developed (Borgstrom, 1973*a*) the 'Hexagon of Survival' (Fig. 1). A strategy, long overdue, needs to be formulated along such coordinating lines. Several action spheres must be harnessed together, but primarily the following six,

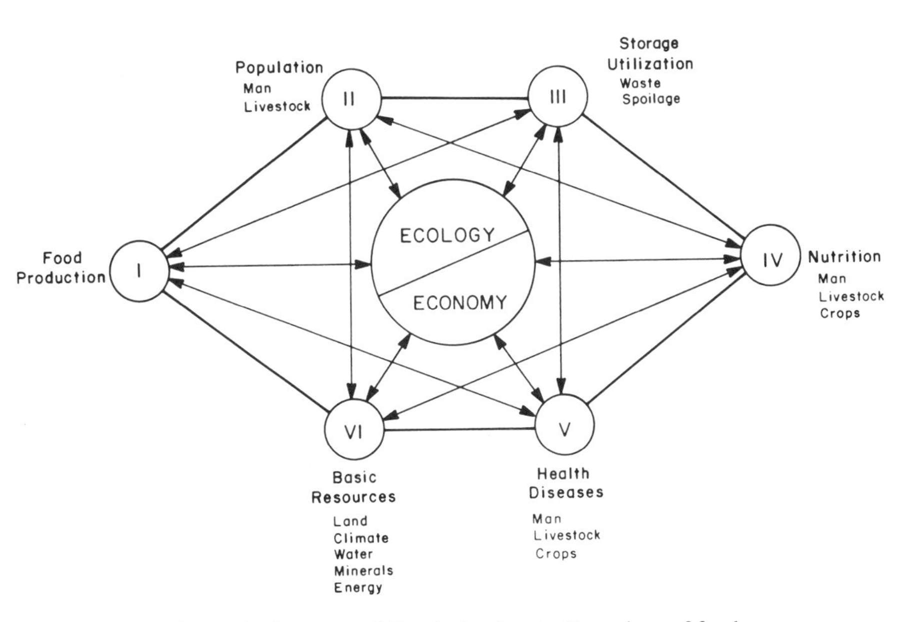

Fig. 1. The Hexagon of Survival—the six dimensions of food.

represented by the nodal points of the Hexagon of Survival: (1) food production, (2) population control, (3) better storage and utilization (both for human food and animal feed), (4) nutritional requirements, (5) disease control, and (6) resource appraisal (water, energy, land and minerals). Each measure taken needs to be linked or related to all the others—something which has been done only on rare occasions. There is considerable reason to maintain that sight of the overall goal—food for all—was frequently lost in the struggle to achieve more immediate advantages. Such short-range gains are quickly dissipated by countervailing forces which in the long-range result in the opposite, namely less food.

Waste and spoilage together remain the most neglected aspects of the food issue, though seemingly this does not belong to the realm of ecological aspects of protein production. Indirectly, however, protein is also here a cardinal factor, for the growth of pests and diseases is, by and large, dictated by the same C/N ratio that rules production. The insect and microbial menaces to grain, legumes and nuts, are, for this reason, far larger than are those tied to fats and sugar. Many more and deeper studies should be made of this aspect in relation to shipping and storage, once again opening up new vistas of fundamental ecological significance. Rodents, birds and other pests need in a corresponding manner to be brought into research and investigated as to their interference in the carbon and nitrogen cycles and in the basic balance between these two.

Ecology and economy are the two balancing wheels holding this system together—a responsibilty which has thus far not been too well discharged by either group of proponents.

References*

Annegers, J. F. (1972). *Geographic Patterns of Diet and Nutritional Status in West Africa*. Ph.D. dissertation, Michigan State University, East Lansing, Michigan: xi + 344 pp., illustr.

Bodenheimer, F. S. (1951). *Insects as Food*. Junk, The Hague, Netherlands: 352 pp.

Borgstrom, G. (1964). The human biosphere and its biological and chemical limitations. Pp. 130–63 in *Global Impacts of Applied Microbiology* (Ed. M. P. Starr). Almquist & Wiksell, Stockholm, Sweden and J. Wiley, New York, NY: 572 pp., illustr.

Borgstrom, G. (1965 and 1972*a*). *The Hungry Planet*, Macmillan, New York, NY: ix + 487 pp. [2nd rev. edn. (1972)], illustr.

Borgstrom, G. (1969 and 1971). *Too Many–An Ecological Overview of Earth's Limitations*. Macmillan, New York, NY: xiii + 368 pp. [2nd rev. edn. (1971), xiv + 400 pp.], illustr.

Borgstrom, G. (1971). *See* preceding reference.

Borgstrom, G. (1972*a*). *See* Borgstrom (1965).

Borgstrom, G. (1972*b*). Ecological aspects of protein feeding: the case of Peru. Pp. 753–74 in *The Careless Technology* (Ed. M. T. Farvar & J. P. Milton). Natural History Press, New York, NY: ix + 1030 pp., illustr.

* Basic statistics were taken from FAO Yearbooks for Agricultural Production and Trade, as well as from FAO Fisheries Yearbooks. All tables are computed by the author on the basis of these statistical data.

BORGSTROM, G. (1973*a*). *Focal Points*. Macmillan, New York, NY: xii + 320 pp., illustr.
BORGSTROM, G. (1973*b*). *The Food–People Dilemma*. Intext Publishers, North Scituate, Massachusetts: xii + 140 pp., illustr.
BORGSTROM, G. (1974). The price of a tractor. *Ceres*, Nov.–Dec., pp. 16–19.
BORGSTROM, G. (1975*a*). The food and population dilemma. Pp. 43–51 in *The Man–Food Equation* (Eds F. Steele & A. Bourne). Academic Press, London, England: xv + 289 pp., illustr.
BORGSTROM, G. (1975*b*). Food and energy in confrontation. *Proc. Amer. Phytopathol. Soc.*, pp. 31–6 [not available for checking].
CAVIEDES, C. N. (1975). El Niño—its climatic, ecological, human, and economic implications. *Geogr. Rev.*, **65**(4), pp. 493–509.
LEACH, G. (1975*a*). The energy costs of food production. Pp. 139–63 in *The Man–Food Equation* (Eds F. Steele & A. Bourne). Academic Press, London, England: xv + 289 pp., illustr.
LEACH, G. (1975*b*). *Energy and Food Production*. International Institute for Environment and Development, London & Washington: 137 pp., illustr.
NICOLS, H. (1967). *The Limits of Man*. Constable, London, England: 283 pp.
PIMENTEL, D., HURD, L. E., BELLOTTI, A. C., FORSTER, M. J., OKA, I. N., SHOLES, O. D. & WHITMAN, R. J. (1973). Food production and the energy crisis. *Science*, **182**, pp. 443–9.
RUSSEL-HUNTER, W. D. (1970). *Aquatic Productivity*. Macmillan, New York, NY: 306 pp., illustr.
STEINHART, S. & STEINHART, E. (1974). Energy use in the US food system. *Science*, **184**, pp. 307–16, illustr.
TOTH, L. (1946). *The Biological Fixation of Atmospheric Nitrogen*. Hungarian Museum of Natural Sciences, Budapest, Hungary: 116 pp.
VALLENTYNE, J. R. (1972). Freshwater supplies and pollution: Effects of the demophoric explosion on water and Man. Pp. 181–99 and following discussion in NICHOLAS POLUNIN (Ed.), *The Environmental Future*. Macmillan, London & Basingstoke, and Barnes & Noble, New York, NY: xiv + 660 pp., illustr.

DISCUSSION (Session 10)

Myers (Chairman, introducing, following announcements by Secretary-General)

Thank you indeed Professor Polunin for putting us in the picture in so many different directions, [with your welcome announcements of plans to circumvent the general strike, reassurances about the recovery of yesterday's feared platform casualty, details concerning excursions and arrangements for the Lord Mayor's reception tomorrow and the State Luncheon on Saturday, and many other matters big or small]. I am very glad to have the opportunity to chair this Session on 'Agriculture and World Feeding Alternatives', in which we are going to lead off with a paper by Professor Georg Borgstrom, of Michigan State University. I am looking forward to this paper because, like other writings of Professor Borgstrom, it will be full of facts and, even more important, it will be full of relationships between facts. Professor Borgstrom himself cannot be here [on this postponed date], which is a pity, but his paper is to be presented by Dr Raymond F. Dasmann, whom we have all heard about on a number of occasions, and whom we have listened to here during the past few days, and I personally look forward to the numerous wise and penetrating punctuations which I am sure [he] will slip into his presentation of this paper. [Dr Dasmann has just relinquished the post of Senior Ecologist at IUCN and is on his way to take up a Chair of Environmental Studies in the University of California at Santa Cruz.]

[Our panellists are four, and they will comment from different standpoints: Dr Tan Koonlin, from Malaysia, who is an Agricultural Geographer *cum* botanist, and whom you have had the instructive pleasure of hearing before; Dr Björn Sigurbjörnsson, Director of the Icelandic Agricultural Research Institute which is located here in Reykjavik; Dr Thomas B. Starr, of the University of Wisconsin's Center for Climatic Research, who will address this topic from the standpoint of the impact of climatic changes upon food-producing capabilities; and Dr Björn Jóhannesson, a distinguished soil scientist who has worked internationally under the aegis of the UN Development Programme but is now active mainly in his native Iceland.]

Dasmann (Presenting paper for Keynoter)

Originally [Secretary-General] Polunin asked me if I would assist Georg Borgstrom in preparing a paper for this Conference. It was my expectation that he would send me his rough notes as he was pressed for time, and it would be necessary for me to put them in some final form. But in fact shortly before I left Switzerland he sent me the completed paper, [a copy of] which you [now have], and I thought nothing would be gained by my changing it in any way. So it has fallen on me to present in this short period of time some rather complicated material. What has emerged from my digestion of the paper is a compact mass of generalities, whereas the particulars in the original paper are of principal importance. [So] what I am going to do is to present some of the main themes of Borgstrom's paper along with a few comments of my own.

Professor Borgstrom has three major concepts to emphasize. One is the dissociation of modern agriculture from ecological reality. Another is his concept of population equivalents referring to domestic animals which I think changes our concept of population densities and carrying capacities. The third is his concept of 'ghost acreage' which he has been advancing for some time and which will help to answer the question of who is feeding who on this planet.

To start with the first concept: food production is rarely analysed from an ecological viewpoint and [still more rarely] are the constraints placed in clear focus. Yet these matters must be moved to the centre stage in the current debate about world food and population. Failure to place people in proper relation to the environment is the chief reason for many fallacious conceptions [that are still being] disseminated about the potentialities of the globe and also for many easy-going diagnoses of humanity's present position. The biological dimension has been missing [all too widely].

Protein is a 'key element' in reformulating our picture of the world in food terms. Easy access to protein is by the utilization of animal products. Early in history people were hooked in on a carnivorous level. Hunting, and gathering of shellfish and insects, may have preceded more sophisticated methods of breaking down cellulose to obtain access to the nutritive riches of the grasses, legumes and tubers—calorie-rich and protein-rich members of the plant kingdom. Ruminant animals in particular have constituted an invaluable beachhead in the plant world by expanding the feeding-base of people through their capacity to utilize cellulose as well as by strictly economizing with nitrogen through microbial synthesis of amino-acids in their rumen. They are in fact parasites on their own microflora and microfauna and can thereby use otherwise nutritionally valueless areas and plant species. Early in time people recognized the value of ruminants and became herders. Whether it was before or after this that they learned techniques for breaking down plant cellulose and became tillers of the soil, is a matter still under debate. Eventually, however, a creative synthesis of animal breeding and plant production formed highly stable, rotational farming systems with effective promotion of the protein-nitrogen cycle.

One key factor in the creation of balanced farms was the incorporation of legumes, with their nitrogen-fixing symbionts, in the farming system. Crop rotation systems—as they have evolved, for example in Europe—were firmly anchored to legumes as a link that is indispensable to sustained food production, and only when legumes, such as clover and alfalfa, were extended to the New World, did those areas become dependable for animal production. Farm surpluses thus became possible, and milk, meat, and other farm products, could be used to support the cities. Indeed cities could not exist without such surpluses. However, the outflow of minerals to expanding cities became a crucial drain on the viability of farm cycling-systems. So farmers started buying additional feed—first hay, but then a little later also protein supplements for their livestock. This was the starting-point of the undermining of the ecological awareness of the farmer.

In the well-to-do world, crop production is now gradually disassociating itself from animal-raising. Protein was the chief but largely unrecognized lever in this break-up. The dissociation became so complete that ruminants (which can synthesize protein) were fed on protein-rich oilseed cake, and even on fishmeal and milk which are potential foods for people. Gradually, also, the concept of land as a place where one lived was replaced by the concept of land as a resource to be exploited for monetary gain by people who live, or who seek to live, somewhere else. This led in its most extreme form to monoculture—with its extensive use of chemical fertilizers, of pesticides, and of the whole gamut of associated activities.

With his second point Professor Borgstrom has helped to change our concept of population and pressure on the environment by his idea of feeding burdens in terms of human equivalents of domestic animals. This pressure he measures by comparing the amount of plant protein required to support a human being with the amount needed to support a domestic animal, whether it be poultry or swine, or cattle or horses.

Statistical methods for measuring food requirements usually leave out these relationships of animals to people. The human biomass must be given a broader connotation to include the living animals on which people are directly dependent for food. It has to be realized that, in order to maintain the present nutritional standards [of mankind], and to maintain the type of agriculture that now prevails, green plants must carry a feeding burden far in excess of that of human numbers. The world has a population of 4.2 thousand million people, but when their necessary domestic animals are included, we find that the plant kingdom on Earth is supplying protein to the equivalent of 21 thousand million humans. Current estimates of the Earth's carrying capacity are mostly computed with little recognition of the fact that the world, in terms of protein, is already supporting these 21 thousand million.

Mention is frequently made [of the expectation] that we may see an additional thousand million people [on Earth] in the next 10 years, but in fact we should say that in the next 10 years you will see an additional 5 thousand million human equivalents. Arguments as to whether the world can support 8 thousand million people should be changed to whether it can in fact support a population of 42 thousand million [including] protein-eating domestic animal equivalents. When viewed in this way, the differences in population densities between areas and continents seem less marked. Asia, with the greatest human population, has fewer livestock [on a *per caput* basis of] approximately two population equivalents in animals to each human being. Oceania, with the lowest human numbers, has by comparison 10 times as many livestock equivalents per person—21 to 1—as Asia.

Increases in human population are usually accompanied by decreases in livestock population—a process that has been noticeable in the last 10–20 years in Latin America. Whereas Asia has nearly 10 times as many people as North America, it has roughly only three times as many human population equivalents. This could lead the vegetarians among us to leap to an immediate but perhaps false conclusion. The all-vegetarians' claim that thousand of millions could be saved by stamping out livestock production is questionable. Grazing animals use land that would otherwise not be productive of food. Domestic animals such as pigs and poultry, are capable of using agricultural waste-products that would otherwise be lost for human consumption. Nevertheless some gains in food could be made, primarily by stopping the feeding of potential human foods to domestic animals, and Borgstrom estimates that this might be 2.5 thousand million population equivalents. But there are now 16 *plus* thousand million livestock human population equivalents, and nearly 14 thousand million of those are feeding on areas or on plants that cannot support people directly with present economy and technology.

The third major concept that is useful and central to Borgstrom's paper—and I may say there are many other minor themes which I just cannot go into—is his concept of 'ghost acreages'. We have to understand, in order to understand human ecology, the protein intake of people, both in its amount and in its place of origin. On examination of this protein in relation to its place of origin, we find that the rich countries of the industrialized or developed world are not self-sufficient in food in many cases, but are in fact supported by the poor countries of the hungry world. Borgstrom has analysed food imports and oceanic fisheries harvests in terms of the equivalent in land acreages that would be required to produce the corresponding amount of protein, using the technology available in the particular country concerned—that is, without [employing] any new agricultural techniques. Just how much land would it take to produce the protein equivalent of what we now get by harvesting the sea, or from imports from foreign countries? This results in his calculations in two figures: the 'trade acreage', which is calculated from protein imports *minus* protein exports, and the 'fish acreage', which is calculated from the amount of land needed to produce the protein obtained from the ocean fisheries. These two together form what he calls the 'ghost acreage' of the country—the statistically invisible area that in fact is [helping to support] the people.

Investigation of these figures shows that countries which are often given credit for feeding many people from little land are in fact not doing so. Japan, the Netherlands and Switzerland are notorious examples. Japan, with a 'ghost acreage' that is over 700% greater than its agricultural acreage at home; Netherlands with over 600%; and Switzerland in the same range. Europe as a whole, as well as Japan, have lost touch with their geographical base; they have moved out of their own ecological context. Europe of course enormously expanded its land-base by occupying the Americas and Oceania, so that they have far more acreage per person of European-originated population than the rest of the world.

This land-grab of earlier centuries had its repeat in an ocean-grab after World War II, so that now two-thirds of the oceanic fish-catches go to the already well-fed and well-to-do world. Half of the world's catch of oceanic fish in 1970–71, which was the peak year in oceanic fish production, was channelled into fishmeal that for the most part went to feed, not the people, but the domestic animals of the affluent world. This takes place on a planet where [a substantial proportion of] people suffer from protein hunger.

The result of this activity is that two major global survival-bases [have been] created, in Europe and North America, and these are supported by two 'ghost continents'—'ghost acreages' [based on] fisheries and [agricultural] imports. The idea that the US is feeding the world does not hold up. The movement of food in world trade is only a small percentage of food production anyway. But half of the US cereal export is for livestock feed for the affluent, well-fed world. The chief recipients of world net trade in cereal grains are Europe, Japan and the Soviet Union. China and India together, with [the largest grouping] of the world's people, have only, in a few exceptional years, received more of the world's cereal grain-trade than Japan.

It therefore rings hollow when we declare that it is high time the poorly-fed world learned to take care of itself. In fact Europe—East and West— and Japan, are the major burdens on the world's food household. In comparison, China and India have virtually no 'ghost acreages', and [consequently] are no burden at all. In fact, India competes with Africa in providing peanut protein in Europe—6 million metric tons a year, which is enough to supplement the cereal diet of 250 million people. If we are really interested in feeding the hungry people of the world, the means are there, the food is there, but we have to sort out our ideas on what we do want to do [and how to do it].

We produce food [nowadays] for the purpose of economic profit, or for use as a political weapon, and we probably find these aims incompatible with feeding the hungry. Poor people cannot compete with rich people in buying food, so perhaps the first step that the rich countries can take is to get off the backs of the poor. But Professor Borgstrom points out that this should not be done too quickly—please—or the sudden release of pressure might topple them over. They are used to supplying food; their whole economy is based on shipments of food to the rich countries, and they cannot quickly change over.

Professor Borgstrom points out the new direction in which we should go to get out of this dilemma. The first priority he gives is channelling the huge amounts of protein from livestock food into direct human consumption. This amounts to two-fifths of the present ocean fish-catch, one-quarter of the milk proteins of the world, and two-thirds of the oilseed protein—an enormous amount which should go directly to people. In other words, our pampered pigs, battery chickens and feed-lot cattle, in the wealthy world, need to be restored to their proper ecological place. Secondly, he calls for a restructuring of our technological system to incorporate organic waste into food for fish, shellfish or poultry, or into livestock feed in the form of yeast, Algae, etc. Food for people can only be brought into balance by coordinated efforts in what Borgstrom calls his 'hexagon of survival', of which the six sides are food production, population control, better storage and use of foods and feeds, nutritional requirements, disease control and resource appraisal. To make this system work, ecology and economics must be brought together [and henceforth work in unison].

Myers (Chairman)

Thank you for an informative, interesting and widely illuminating paper, presented in a style which we could expect of Ray Dasmann [whom many of us consider the foremost general ecological writer of our day]. I would like to interpolate here one comment with respect to what we have just heard about the relationships between the developed and the developing worlds—the 'haves' and the 'have-nots'. I am thinking of trade relations—where some of the food originates, who has to pay for it, and at what sort of price. I sometimes think that one definition of a conservationist might be a person who goes into his coffee shop in Manhattan or London or Geneva and jumps in the air with delight when he finds he has to pay a dollar for a cup of coffee. When that day comes, the food-production systems and other agricultural systems in developing countries in the tropics, such as in my own country of Kenya and also Brazil, might be a good deal changed.

Tan (Panellist)

I am impressed by this paper [which] gives some idea of some of the numerous ways of dealing with this subject of feeding alternatives, though I think [the author has] not quite perceived the intricacies, at least from the perspective of changing the food habits [of people living] in tropical areas. I am concerned about this because very often in the [less-developed] countries traditional diets are changed to incorporate substitute components which have to be imported into the country, and this deprives us of the opportunities for developing our own protein sources more intensively. It also diverts a lot of our [limited] capital and resources to procuring something that is not really beneficial to us.

It happens that the developed world sometimes feeds us, not because we need it but because propaganda has made it appear that to be fed in the way that Western Man is fed is the proper way, whereas the ways that we were used to are primitive or disagreeable or dirty or whatever prejudiced adjective might be used for it; for example sago worms are despised by some of the local people, because the westerner finds this type of food repugnant and unacceptable. That kind of reaction colours our own attitudes to many traditional foods which our older generations consume without prejudice [and with well-being].

A further complexity in our efforts to improve our food sources is that we are also being encouraged in our husbandry to adopt animals into our farming system which are not compatible with our environment; for example, cattle are being perpetuated as *the* animals for our grassland, [and it can be said that we are] getting grassland at the rate at which we are deforesting; in Malaya we have destroyed a million hectares of our forest land in the last five years. In fact there is in the traditional system an animal that is conservative of the environment, and that does not require all those high-energy and feed inputs which are associated with the temperate cattle breeds. This is the Water Buffalo, which in Southeast Asia is depended upon extensively for draught power. Unfortunately only when it gets too old [for such work] is it killed off for meat, and that has given buffalo meat a bad reputation. When beefs are compared, people of course prefer the beef of cattle, and that kind of prejudice has unfortunately crept in various insidious ways into our perception of developing alternative food-sources. So we have been needlessly destroying our own well-tried traditional components because of this kind of propaganda (for want of a better word).

The third point which I wish to make is that we have to live with the fact that we are going to pollute our environment, but we do not have to regard these pollutants as a nuisance or something to be cleared away. There is now developing an increasing awareness that pollution is a source of new materials for our food, and that this pollution can be harvested instead of released into the environment—by impounding, treating in appropriate ways, and eventually encouraging fish to develop in previously polluted water. Such a system is being [tried] in our palm-oil factories in Malaysia, because they are the sources of the most intense water-pollution in the country (the BOD of palm-oil effluent is over 25,000 ppm). Here, at the end of a series of ponds, can be reared various categories of fish—e.g. surface-feeders, plankton-feeders, etc., on aerobic or somewhat anaerobic waters. This is a system incorporating not altogether novel principles, and can be investigated in analogous farming systems, e.g. under Chinese market gardening where all kinds of wastes are recycled and made useful.

Sigurbjörnsson (Panellist)

I certainly agree with the conclusions or series of actions which Professor Borgstrom has suggested, except that I would put the control of world population growth first—ahead of food production [and everything else]—because I believe that the evil [lies] there. If you could stop population growth, I think our worries about the environment and food would [gradually] disappear. So this is No. 1. But even if we could stop [such growth], if we could bring it to zero today, we would still face [great increases in some countries, amounting to] almost a doubling of the present world population. This is because of the [structure] of the population—because of the number of young people, and especially the number of young girls who are going to be pregnant and have their two [or more] babies.

This doubling of the population will equal zero growth [the parents being generally still alive], so whether we like it or not we will have to expand our technology to meet the demands of at least a doubling of the population. Indeed I am afraid we must worry rather about the quadrupling of the population, if we are not successful in bringing population growth to zero. We are thus left to face the problem of producing more food, as implied in the title of this Session, and I do not see any real alternative outside of agriculture; for the total catch of fisheries is only about three times the world's laying of eggs by hens. Even if we exploit all the oceans for all the known fish, we might possibly double the present catch or [at most do] a little bit more.

Thus we are left with agriculture, and have to consider our options within agriculture. Some of these have been pointed out by Professor Borgstrom, and I think one which he presented is extremely important, namely the use of human food for animal feed. This may not be easy to change, but I think this is something we must strive for. The equivalent of the animal feeds, mainly grain, that is now being fed to the world's livestock, would suffice to feed all the Chinese and all the Indians—almost one-and-a-half thousand million people [or more than one-third of the total human 'family']. Most of the livestock are ruminants, and ruminants like roughage. This is an area that we should explore: to utilize our non-arable lands, turn vulnerable and erodable areas into grassland, and use the roughage so produced for fodder for animals, thereby reducing their need for grain. I can see a lot of alternatives in animal feeding, and I think this can include also poultry and pigs. This is an area, an alternative, that we have to look into very seriously, as was pointed out in the [keynote] paper.

Now turning to plants—and this is closer to my own speciality of plant breeding—I think this is where the greatest possibility lies, because ultimately we have to rely more and more on the direct utilization of the green plant and its products. I see a great potential here to meet the problem of a population growth to at least a doubling, or maybe more, of the present numbers. In addition, of course, we must adequately feed the 500 million people who are now said to be starving in the world, and potentially another 15 hundred million who may be on the brink. These are the people we also have to worry about, and a good example of what we can do is the 'green revolution', whether you like it or not. It has shown what can be done in expanding the production of cereals.

Of course, there is a great difference between areas of the world in the matter of how far you can expand production. I do not know how far we are towards the limit in Europe or North America, but in the latter we are up to about 14 tons to the hectare of winter wheat. If [we contrast that with] the average yield in Pakistan in 1975 of about 1 ton per hectare, we can see the tremendous potential we have towards multiplying grain production with known technologies. I have read that the Indians now have about 20 million tons of grain in reserve. We can thank that to the green revolution in the Punjab and some other areas, and to good monsoons, but without either we would not have these reserves.

The green revolution has its disadvantages. It does not adapt itself to all situations. Thus it fits certain situations but not others, and we have to develop other strategies for those other situations. But it certainly did multiply the yield per hectare, increasing the input of fertilizers and energy and often resulting in increased pollution! We have to face these problems, as we need fertilizers. Yet I do not think we should waste oil as a source of energy, but should use it for [making and distributing] fertilizers. The energy problem is something we have to face—as came out already on the first day of this Conference—and maybe the answer is nuclear fusion, which may be our cleanest source of energy. I hope the problem of nuclear fusion will be solved soon. It may take another 20 years but not likely very much more, and then we will look back on the energy problem with nostalgia because hopefully that problem will be solved.

What we fill our [automobile] gas-tanks with will be the question—*what*, not *if*. If we can solve this problem, and I am optimistic enough to think that we can, we shall need more energy input for an advanced agriculture, and this is a technological challenge which we have to face. But there are many other problems which we have to face, including that of how to stop ruining vast areas of land that are now being irrigated—for example on the Indian sub-continent [and in the Near and Middle East]. By irrigation you lose tens of thousands of acres [each one of 0.408 ha] of land every year, due to increase in salinity. This is a problem we have to tackle. Another is the use of pesticides: the growing resistance of pests and disease organisms to pesticides is [only one of the problems which their widespread uses invoke]. So we have to develop biological methods of control, and that I think will be the answer—rather than going for increased use of chemicals.

[Returning to my own specialist] field of plant breeding, there is a tremendous potential ahead of us. The badly named 'genetic engineering' I think will be continuingly useful in plant breeding; the creation of [new genera] such as *Triticale* has shown us that. [We also have] the adding of extra capabilities to cereal plants and other crops—for nitrogen fixation such as the legumes have—as well as the techniques of cell-culture and the induction of mutations to supplement natural variability.

All of these [and yet other] techniques will I think go a long way [in helping us] to shape our agricultural products. If the tomatoes do not taste good enough, as somebody was complaining, we have to spice them. We can breed them in [almost] any way you want—any shape—square, round, and undoubtedly with any taste. I may be labelled an optimist but I see great potential in agricultural production; however, I certainly want to agree with you people here that we must not increase agricultural production indiscriminately. We must appreciate that we have only this planet; we are not allowed to abuse it, and our agriculture has to be in harmony with our environment. I think we know this very well here in Iceland, which in many ways mirrors the situation on this globe. We are an island community. We have our limited resources and must not abuse them. We have to live within the particular ecosystems that we have, and I think that goes for the world as well. [Applause.]

Starr (Panellist)

I would like to go back and try to respond to the how and when questions which Miss Baroody put to us [in the preceding Session], particularly with respect to the meteorological and biological professions, by noting some of the information that they might be able to provide to the agricultural community for moderating the impacts of climatic fluctuations on agricultural production and perhaps even producing as well some modest improvements in total production.

As Professors Hare and Bryson pointed out in earlier remarks, the meteorological profession is notorious for making many forecasts—in fact, many poor-quality forecasts. I would like to add to their remark [by saying] that the *kind* of forecasts which this profession makes at present—even if they were perfect—is the wrong kind to be of any use to the agricultural sector. These forecasts are of very short range, [usually 24 hours or so]. They provide far too short a lead-time for the agricultural community to plan adaptively and act effectively in response to them. One possibility for action, then, which offers the promise of stabilizing and modestly improving agricultural production, is vigorously to pursue methods for long-range climatic forecasts. [These might have to be] crude in detail in comparison with the daily or weekly forecasts which are available now, and perhaps just characteristic of the [aggregate temperature and moisture conditions for the] entire growing-season, but making statements with lead-times sufficient to permit effective response.

As an example, I think of the growing-season for [maize] in Wisconsin. As one proceeds north through Wisconsin, [maize] is a crop that becomes more and more marginal, [because the growing-season becomes shorter and shorter]. There are many specific hybrid varieties of [maize] now available that have been selected for various lengths of time to maturity. In the northern part of Wisconsin, one must always plant 90-days maize varieties, [because the growing-season is rarely much longer]. [In the south,] close to the Illinois border, one can gamble and plant 105- and 120-days varieties, which are higher yielding than the 90-days varieties provided the growing-season is long enough for them to reach maturity, but which suffer the risk of being cut short by an early frost. [These long-season varieties were originally developed for Illinois, not Wisconsin, conditions, but farmers saw the higher yields and now often gamble by planting them.] I believe it was in 1974 that an unusually early killing frost occurred in Wisconsin and ruined much of the maize production, almost 75% being destroyed by this early frost in some areas. [Now, if the farmers throughout Wisconsin had planted 90-days maize, the disaster brought about by the early frost would have been minimized. But to achieve this, they needed an estimate of the length of the 1974 growing-season already when they were still deciding what varieties to plant—not on the day before the killing frost occurred.]

So I feel that the meteorological profession has an obligation to spend less energy on perfecting their short-term forecasts and more energy on long-range forecasts that [may be] crude in detail but provide the lead-times sufficient to permit the agricultural sector to plan adaptively and act on them. [The forecasts need not be perfect. Indeed, even if they are only slightly better than random chance, agricultural production would increase slightly as a result of taking them into account, while in the long run, production would also be less sensitive to climatic variation.]

I would put the same remarks to the biological profession with respect to reductionist attention to physiological detail. The kind of prediction of a single plant's response [or, even worse, a single leaf's response] that a physiologist likes to make under carefully-controlled laboratory conditions is of no utility to the agricultural community now, [and will be of little utility even when long-range climatic forecasts become available.] The problem is again one of too much attention to too much detail. Instead, we need estimates of the aggregate response of an agricultural region of some millions of hectares to correspondingly crude resolution forecasts of temperature and precipitation fluctuations. I would accordingly appeal that we all stand back, blur the focus of attention on detail, and try to come up with crude but usable models for the interaction of meteorological and biological factors in agricultural production.

Jóhannesson, Björn (Panellist)

I rather doubt that I can contribute very much to this discussion, simply because most of what I have to say has been expressed by [many] people before, but I must say that I found Professor Borgstrom's keynote paper timely—particularly his observation about 'ghost acres'. As a student of soils, I [could] call attention to the fact that this global natural resource deteriorates progressively through water and wind erosion, desertification, salinization due to deficient irrigation systems, and because of various construction activities. Yet there still remain sizeable areas of unused but potentially arable soils. Dr Charles E. Kellogg, of the US Department of Agriculture, who has studied this subject in detail, stated in 1975 that about 1.4 [thousand million] hectares were being used for farming in the world, and that this figure could be increased by about 1.6 [thousand million] new hectares of potentially arable soils. Yet many of these soils are inaccessible, and Dr Kellogg points out that the largest areas of potentially good but yet unused arable soils are not where the bulk of people needing food and jobs now live.

I agree with the observation made here yesterday to the effect that for some of the world's soils, particularly for those of wet lowland tropics, there have not been developed satisfactory agricultural production management systems. It makes little sense to impose on this tropical ecological environment, agronomic practices that were developed for soils in cold-temperate regions. Disastrous results in tropical lands may be cited from such blunders, some of which I have had an opportunity to observe. In this context it should be stressed that extensive research remains to be done on which to base sensible agronomic management practices in tropical and subtropical zones. This work calls for much expertise and, in order to succeed, such expertise must be based primarily on nationals. Unfortunately it takes considerable time and money [and also patience] to build up competent research expertise, and I cannot help but feel that the required or essential research efforts will be too little and too late, so that blunders leading to destruction of vulnerable ecosystems will continue on a critical scale.

Time does not permit me to touch on several important plant- or food-production factors other than the soil component. Yet I wish to express my agreement to the assessment in Professor Borgstrom's keynote paper of water as the most limiting factor in the world's agriculture. I could also touch on the fertilizer and pesticide industries, on which the world's agricultural production is increasingly dependent, whether we like it or not. To indicate the huge scale of the fertilizer industry, Professor Borgstrom observed in a book published in 1966 [but not cited in his References] that by the year 2000 the amount of industrial fertilizers entering into transport will amount to more than 500 [thousand million] tons, which is [many times] more than the total weight of the world's human population at that time [is expected to be]. The industries [involved] will exert increasingly heavy priority demands on energy and transportation resources.

If the world wishes to stretch its feeding or carrying capacity to the utmost, it seems to me that the following approaches will need to be adopted: (1) To re-use or recycle on an increasing scale both water and plant nutrients that are contained in sewage and other organic waste-products; (2) to take stronger measures to conserve fossil fuels and water, and in particular not to foul the environment; and (3) to develop and apply agricultural production practices that will sustain maximum yields and are at the same time commensurate with the above-mentioned approaches. Furthermore, efforts will need to be made to make the food-chain from the green plant to the human mouth as efficient as possible, as Dr Dasmann has brought out. Yet, if we are to implement effectively the above-mentioned approaches, I will have to agree with Dr Fridriksson that such development would not lead to a very pleasant world, but rather to a dreary and crowded one, with drastic limits on personal liberties. Yet probably there will exist few alternatives to such development as time passes on and the world's population multiplies.

Finally a few remarks prompted by my experience as an international civil servant with the UN Development Programme for some 13 years. During this period I visited several developing countries in subtropical or tropical regions in connection with UN efforts to help these countries to help themselves. Poverty, malnutrition, hunger and disease were often apparent in these regions, while the Western world is concerned because of over-eating. It is an established fact that rather feeble efforts by the human family—not only by rich countries but by the developing ones as well—have failed even to reduce, [let alone] eliminate, the disgrace of inequity between wealthy and poor nations. The UNDP is the principal international unit where the human family pools resources, mostly from the richer nations, and supports with these means technical assistance in the developing countries with the cooperation of UN specialized and executing agencies such as FAO, UNESCO, WHO, etc. This year the UNDP receives only about $500m to be distributed for technical assistance projects in about 100 countries. The UN General Assembly has urged the rich countries to contribute 1% of their gross national product to aid the developing world. Only one, Sweden, has attained that goal; the others, including the super-powers, are much or well behind.

It seems to me that the political leaders of both the developed and the developing countries, and indeed the vast majority of the citizens of the rich and favoured lands, lack the political will and cooperative spirit to bring this world's house in order—to try to diminish the untenable gap between rich and poor nations. And I might conclude these remarks by expressing the view that this Conference will not have been convened in vain if it can help awaken, even a bit, the conscience of the world's political leaders, and particularly those of the Western world, and thus contribute towards greater cooperative efforts to combat hunger, disease and ignorance on our globe.

Myers (Chairman)

Thank you all panellists [for so effectively attacking problems from different aspects of this vast and vital theme]. Before turning the discussion over to the floor, there is a rather splendid intervention requested by Professor Hare.

Hare

I apologise for interrupting, but this Conference would not have taken place if it had not been for the personal dedication and enthusiasm of our mutual friend Nick Polunin. [Now] I have arranged for him to be taken out of this room by his son, and if anyone sees him show up, kindly make sure that he does not get [back] in. His wife has already been squared, so she knows [what we are up to]. A small group of us, acting on your behalf, have commissioned a gift for Nick, to commemorate his work [for and] in this Conference. It is an Icelandic gift, [from this land which he loves—a special gavel of carved Icelandic birch (his favourite tree)] is being made at this moment. It is going to be [in a special wooden box with a gold plate] inscribed with an engraved notice that simply reads, 'To Nicholas Polunin with gratitude: The Reykjavik Conference on the Environmental Future, June 5 – 11 1977'. It is going to [be costly, to the] extent of about $4 per participant at this Conference. Now those of us who have commissioned this are quite prepared to pay the whole, if that is your wish, but [we thought many more] would like to make a contribution [and so be involved]. An appropriate sum would be $4 or 800 kronur, with an absolute maximum of $5 or 1000 kronur. If we are over-subscribed we will donate the balance to an Icelandic environmental cause. If we are under-subscribed we will pay the balance ourselves. So please, if you wish to, put something in an envelope and place it in my pigeon-hole for room 221 at the reception desk. Alternatively, Icelandic cheques can be made out to Dr Sturla Fridriksson. We hope to make the presentation either at the [Lord Mayor's] Reception or at the Prime Minister's luncheon.

Myers (Chairman)

What better way to spend part of this Session: glad to hear about the idea. Before I turn over the discussion to the floor, I would like to pass on to you a comment which I once heard at a gathering in Africa while I was with FAO. The meeting was concerned with the question of food and population, and the lifeboat ethic came up in discussion. You can imagine how African people view this topic. At any rate, one of the Ghanaian delegates said that, while he would no doubt find the water rather chilly if he were swimming around in it, he could understand, in a cold and objective fashion, some of the rationality of the lifeboat argument; but he would feel singularly [disgusted] if he swam over towards one of the lifeboats and was told to go away and practise altruism (good for your soul, etc., [and there's no room anyway]), whereupon he looks in the lifeboat and finds it full of cats and dogs.

Ladies and gentlemen, we now have 25 minutes left for discussion, though we could run over a little bit. The floor is open to you: ladies first.

Dodson Gray, Elizabeth

I found Dr Dasmann's summary of Professor Borgstrom's paper one of the most exciting things I have thought about in a long time— exciting because it is an example of genuine 'newthink' about our problems. So often when we get together, we say we have been thinking about the world in such-and-such-a-way; [then] we act because of the way we have been thinking about the world. [But] it does not work. We are heading for a precipice, and yet we seem incapable of going back and restructuring how we think about our problems, [which need to be tackled] in a different way.

Now what [Professor] Borgstrom has done is to help us think about human food-needs and agriculture as a connecting series of ecological processes, and he has come out with what I would call an agricultural economy based on symbiotic relationships. Earlier this morning we talked about two alternatives [which we have had] so far, and now Professor Borgstrom presents us with a third. We have talked about the industrial or economic system which we now have; Mr Goldsmith has suggested that whaf we need to do is deindustrialize and de-develop. That may be true; but there is a third alternative, which is entirely to re-think the way our present economic system functions, as [Professor] Borgstrom is suggesting.

It seems [that] the economic system we have now is constructed on a basically linear model. We pick up raw materials at one end, assuming there are always going to be these raw materials; then we feed them into what is basically a straight-line industrial process, and out comes a widget or gidget or a car and some waste. And we do not ever think what will happen to the widget, gidget, car or waste. Then we go back to the beginning of the linear process and we start again. As I have talked to economists about this, I have been fascinated. They do not really understand conceptually that what we have discovered about the Earth as it functions in terms of its life-systems, is that it is a *gigantic recycling machine*. Conceptually, the life-systems of the Earth are circular, and conceptually you cannot go on drawing and extending a straight line in the midst of what is basically a circle. It will not work. It seems to me, therefore, that we must go back to scratch—which nowhere have we started to do.

Our economists' quaint way of saying, 'We must internalize any environmental externalities' means to me that they have drawn their own kind of world around straight-line processes, saying: 'That's all of reality we take seriously. If you were to make a great scene over here about some ecological disaster, we would take that little bit of reality and internalize it into our system of reality by quantifying it (which is all we can do with it). But we are not really going to consider *all* your biosphere and ecosystem problems except one at a time.' Yet [I submit, Mr. Chairman] that just will not do.

I think that very shortly we shall have to go back and re-conceptualize the economic system itself in what I would call a circular or recycling manner, so that it does fit into the Earth's life-processes—so that we need not expect ever to start with 'virgin' raw materials, and so that, instead, we use the waste of one process as energy and raw materials for another process in the way Nature fits together in ecosystems. We have a tremendous lot of re-thinking to do, and I [believe Professor] Borgstrom has made for us a marvellous conceptual 'first leap.' I find it very encouraging that perhaps we can [now]start to do this.

Malone

I have found this particular Session to be one of the highlights of the Conference [—full of great sense]. Indeed Mr Chairman your panellists are to be congratulated, and particularly Dr Dasmann. I would, however, like to stamp out once and for all what appears to be a lingering impression that the discussion of fossil fuels and CO_2 on Monday morning led to the conclusion that nuclear sources of energy were the only alternative. In my summation I tried to point out that the alternative sources, including nuclear ones, should be subjected to the same kind of analysis that Professor Flohn gave so perceptively of the hazards of fossil energy. It is not either fossil fuels or nuclear energy: let us look at all sources of energy, objectively and scientifically, and then turn our findings over to the political decision-makers.

Myers (Chairman)

During the course of this Conference we have had many eloquent expositions of a vision down the road, maybe 10 years away, where we will live in ecological equilibrium and will have some environmental common sense. But the question I think we might direct more attention to is how do we get from here to there. What pragmatic steps do we need to take? Maybe we could try to focus during the remaining [20 minutes] or so on what institutional initiatives do we have in mind for, say, the next one year or by 1980. This would be of assistance to the Resolutions Committee.

Ducret

First, I wish to express my full agreement with the last speaker. Indeed, in energy questions, the alternative is not either fossil fuels or nuclear power: there are several alternative sources of energy. Our [session] tomorrow morning will be devoted to this subject.

With reference to the topic [now] under examination, I wish to clarify some aspects of the relationship between the energy and the food-production sectors. With limited and probably scarce energy resources, a first competition can be expected between energy utilization in agriculture and in other energy-consuming sectors. This competition has already [been in existence] since the so-called 'energy crisis', as utilization of artificial fertilizers has significantly decreased at the world level.

Another and maybe stronger competition will probably occur in the future between the utilization of agricultural products and wastes for food production, and their use in the production of fuels from biomass—particularly for making methanol and other alcohols for internal combustion engines. It can be forecast that, the more expensive energy becomes, the stronger will be these two kinds of competition.

Dodson Gray, David

Two points to follow up on the previous two speakers: the first is that we are learning, when we are thinking about nuclear energy, to talk in terms of examining the *entire* fuel-cycle. What we have just been saying about CO_2 build-up in the atmosphere suggests that we also have to examine fossil fuels in terms of their entire fuel-cycle. Indeed in relation to *any* energy source we need to look at its entire fuel-cycle.

A second thought is this: As we look at the energy sources that are now called 'the energy problem', it is very helpful if we recognize we are also talking about 'calories'—just as we are when we talk about food and about hunger. We must recognize that we human beings as a living species are now competing with technological 'species' for calories. That is really the issue. [Our hunger and theirs must stay in balance, if we are to sustain a happy symbiotic relationship.]

Royston

Just one point I would like to make to underline something which I found tremendously exciting about the paper and certainly about [Dr Dasmann's] interpretation and presentation of the paper. It really sums up something I have been getting increasingly worried about all this week. Here we are, the great bulk of us from the parasitic industrialized high-technology white races of the world. We have been sitting here coming up with resolutions and coming up with solutions for the rest of the world, and it seems to me that we have been really guilty—as has been suggested earlier—of the most rampant ethnocentricity. [It seems to me] that the greatest thing we can do in the western industrialized nations is to get off the back of the developing countries, not to [continue] coming up with a lot of pious sentiments and resolutions telling them how to solve the problems of the impending [ecodisaster].

We should start by cleaning up the situation in our own homes and we should, if we are thinking about any resolutions, start with some resolutions to deal with *our* problems—with the way that we are draining the resources of the world, and how we are responsible for the way in which the world runs. We should get that priority straight first. Then maybe, if we can sort our own problems out, this [will] resolve the basic problem, [namely that] the world has been brainwashed—absolutely brainwashed.

Self-help and self-reliance is an ideal which is springing up all around the world. Wherever one goes one sees from the village level—from the grassroots level—particularly in the developing countries, but also in the industrialized countries, in the city streets, in the ghettoes, in the Kalmar Car Assembly plants of Volvo, this idea of the community and of the self-help and the self-reliance of the group. It is an idea that is coming up and we need to apply it primarily in the industrialized countries, to realize that we have got to learn to live on our own resources and not act as a parasite living off the rest of the globe, and then coming up with these wonderful resolutions saying that it is that vast booming population out there which is the cause of all the trouble. It is not. It is the vast booming consumption right here, and with that I do not mean Iceland, because Iceland is a wonderful example of a country that in fact is living within the balance of its resources. But I was tremendously excited and thrilled by this paper just because it puts the problem right where it belongs, at [the door of] those nations that are responsible for the rape of the Earth.

Franz, Kristi

In the discussion of futures, [10 to] 50 years hence, I have yet to hear us address ourselves to the youth and children of today. Yet they are the ones who will inherit the futures. They are the ones who will be in decision-making positions of the future. I would like to call for proposals and suggestions for involving the youth of all cultures in the determination of global futures. They are entitled to experiential and educational systems that enable them to give input, gathering present information and developing their abilities to determine their futures. The youth and children of today will inherit the environmental futures of tomorrow. Let us address ourselves to them.

Fosberg

I agree in general with Dr Royston's suggestions that we put our house in order and reform our bad habits and so forth, but I do not like to see it coupled with the suggestion that we stop trying to [engender improvement] in the rest of the world. For a couple of hundred years, at least, we have been giving bad propaganda to the rest of the world and teaching our bad habits with incredible success. I do not think we should now just pay attention to ourselves, and when we have a few good ideas, keep [quiet] about them until we have got ourselves all straightened out. I think that is pretty bad practice all around; I think we should keep on having conferences like this and telling, not only the rest of the world, but also ourselves, what we should be doing and not doing.

Tan (Panellist)

I wish to add to Dr Royston's perspective. You are deliberating on how to formulate an environmental preservation code, and you are doing this on the premise that your part of the world has undergone the [cycle] of economic development. Then you are also attempting, perhaps without understanding our problems deeply enough, to convince us that we [in the Third World] should follow your ecological sense when we have not so far undergone the sequence of economic changes [that might put us in need of] ecological preservation of the same sort—to the same degree and at the same time that you are proposing. So in my opinion it is too idealistic an approach to propound to the developing world that they should emulate your ecological sense, when their experience and priorities are not of the same order. You cannot expect the developing world not to develop and industrialize, nor even to pollute and endanger their environmental stability, [still less] to stop such a process and follow you on the same mission, well-meaning though it is.

Björnsson

As was pointed out in Professor Borgstrom's paper, I think we have to be careful [about possibly] stopping the purchase of materials from the developing countries.

They would go bankrupt. They are producing cash-crops for the very purpose of getting currency to buy medicines and things that they need. I think this is the wrong thing to advocate here, or at least we should do it with due caution.

Rudolph

I want to point out something pertinent about the ecosystems of the world, namely that waste is going to increase, and I think we should be more concerned about the kind of waste we produce. In the natural world, microorganisms break things down; but our technology is producing things that microorganisms cannot work on. We ought to get the technologists busy thinking about biodegradable materials rather than non-biodegradable materials, and also we should get the microbiologists working on finding [and fostering] organisms that work faster in recycling, [to deal with this inevitable increase in many wastes].

Myers (Chairman)

[Thank you all for your valuable further comments. Now] let us turn to [Mr Goldsmith again].

Goldsmith

I agree totally with [most of the recent speakers, though] I am an optimist, by the way. I hope that we never develop fusion [energy] because to do so would not solve [the] problem. Energy consumption is possibly the best measure of the damage we are doing to our physical environment. The problem therefore is not to find new sources of energy but to learn to live with less. It is not realistic to suppose that we can use more energy and further expand the technosphere without further devastating our environment. We cannot have our cake and eat it. However, we can have our cake and eat someone else's. Unfortunately, as Professor Borgstrom has pointed out, we have already eaten everybody else's cake. The lady over there from Malaysia suggests that the Third World must be allowed to make the same mistakes as we did. I cannot agree with this. We got away for a while with making all the mistakes we did because there were, at the time, other cakes to eat, [whereas] today there are not. The mistakes we made are simply not repeatable.

A new energy source, in any case, cannot contribute very much to the solution of world problems. Thus it is a pure act of faith that modern agriculture can make any [major] contribution to solving the world food problem. There is absolutely no evidence that it can. According to the US Department of Agriculture, [during] 1940–50 pests ate on average 31% of the US harvest, while in the decade from 1951–61 they consumed more than 33%. According to the Environmental Protection Agency (EPA), American farmers 30 years ago lost 17% of their crops before harvest; today, though they use 12 times more pesticides, losses to pests have almost doubled.

The notion that fertilizers can make a permanent contribution to the world food problem is equally unproven. In any case, at the present price of fertilizers, less than 10% of the farmers in India can afford them. Even in the US, [fewer and fewer] farmers are likely to find it economic to use artificial fertilizers—especially when fertilizer manufacturers start paying a realistic price for natural gas, which is the main fuel they use today.

As for machines, how do they contribute to producing more food? Tractors do not produce more food, they simply save labour, and labour is not exactly scarce today, at least on the world [scale]. Perhaps one field in which machinery did produce more food was in fishing. Here, however, we have reached the point where the fishing industry is not only meeting with diminishing returns on technological equipment but [actually getting] negative returns. World fishing catches went up steadily after the war, largely because of the increasingly sophisticated machinery used by fishermen. [But then] in 1971 fishing catches started to fall, and have continued falling. The reason was that the seas had been over-fished. To introduce more technology today would simply enable us to deplete fish stocks further, thereby further reducing world fishing catches. We are now in a situation, in fact, in which the more energy we use and the more technology we introduce, the lower the fishing catches will fall [, and I fear this will soon be the case in other fields].

Olindo

I would like to compliment Dr Dasmann on his excellent [rendering;] however, I was astounded at the major exclusion, in the discussion this morning, of the negative impact of religion on feeding and living styles. Starvation does not mean there is no food. Starvation can mean that someone has been so disorientated in thinking that he or she is dying—quite adjacent to nutritious food—because of what religion has done to such individuals. I feel that one of the solutions to be considered is that, in view of the influence of religious thinking and outlook on life, we address ourselves to this issue. Mr Chairman, I seem to see a lot of merit for us to look again at the life-styles and feeding habits of basic human communities, which I am prepared to submit are free from waste in so far as their consumptive attitudes and practices go. Who knows? there may be an answer there!

Jensen

On the subject of ethnocentricity, I feel very strongly that much of what we are saying—and this is now dramatically substantiated by the paper we have just heard—derives from the curious notion that we can draw lessons from experience and expect others to draw similar conclusions without having had similar experiences. I lived for many years among the Dyaks of Borneo as the only non-Dyak, and I can assure you that to try to have told them that there was no point in having a bicycle—that it was easier to walk—would have taken an inordinate amount of persuasion. To have told people that there was no point in sitting in a car when you could ride a bicycle, would have been impossible. Now for us to sit here (one could, incidentally, ask how many of us came under sail or how many approached the hotel on foot or by bicycle) and tell people elsewhere in the world that they should not enjoy the easy life, displays a certain gall.

While I agree with many of Mr Goldsmith's splendid ideas about de-development, we speak from the experience of having had development. The fact that we give up certain things, and that we realize in so doing that we lose something, does not entitle us to deprive others of the same 'advantages'. We may [argue] that we are unhappy with our way of life. Is this not like the old story of the young girl who was about to marry a rich man? Her father advises her, saying: 'My dear, wealth doesn't make you happy'; to which, of course, the answer is 'No, but how much better to be unhappy in comfort'. I do not know that happiness is a very meaningful notion for us to discuss, but I do think you can talk about a certain ease of life. That we, with our washing-up machines, our television sets, our motorcars, our aeroplanes, and all the rest of our paraphernalia, tell other people not to enjoy these things, is honest only if we are prepared to go without them ourselves. I had a discussion yesterday with someone here about a TV programme and I said 'I don't have a TV set'. There was a note of surprise at my response, [but I wonder] how many here do not have TV sets? How many do not have cars? How many do not travel by aeroplane? I feel that we must ask these questions.

Kuenen

We have collectively been blamed during this discussion for not really applying ourselves to the real problem, which is [collectively] poverty, misery and hunger. We are here to discuss that; but what we have come here to try to do also is to make it clear that some techniques suggested by ecologists, technologists, agronomists and politicians are not the right way to get the hungry people fed. For instance, the cutting down of the tropical forests is not the way to improve agriculture, and there are quite a number of people from the Third World who know this, and who ask us to support them to convince their politicians that this is not the way to do it.

I know—in fact I have said it myself—that we are responsible for a great deal of this destruction; but this was because we know that so many people can only think in the short term, and I have heard repeatedly during this Conference the notion that 'let us watch long-term developments'. Any blame [that may be] put on this group here for not really considering serious matters is wrong. We are searching for a long-term solution for a problem which too many economists believe they can solve in the short term. As has been [indicated clearly] this morning, it is the incorporation of ecological thinking into the economic steering of this world that may save *us*, which is not important; but it may save the world.

I think many of us here are absolutely convinced of the fact that we have been raping this Earth—we have been exploiting poor peoples [and gobbling up their rightful home possessions]. If we were not convinced that something was wrong, we would not be here. And I sincerely hope that the rest of the discussion will not be in this style of 'Let's get rid of a washing machine, that will help the world'. It will not. [What will help to save the world] is a long-term global plan which has got to be developed, and we are here to do our [bit of] duty towards trying to help to solve that problem.

Sigurbjörnsson (Panellist)

I certainly agree with the last speaker on the way I think this discussion should continue or end. But I think our 'machine gun' was not listening to my rifle, in the way he responded, because I was not talking about expanding economic growth. I was talking about feeding starving people—the ones who are now starving—and the inevitable increase in population that *has* to be fed. This is a problem that the agronomists have to face.

We can think and dream as much as we want, but this is a real problem to be solved. In that context you may have [noticed] that I did *not* talk about pesticides as a solution. The thing I advocated was biological control of pests. I never said anything else. And concerning fertilizers, it was a very good example [we were given], because in the fisheries we do not use fertilizer, and the catch has not increased. In agriculture we *do* use fertilizers and the production has multiplied, and I will predict that if we stopped using fertlizer today, we would all starve next year.

Myers (Chairman, concluding Session)

It is now past 1 o'clock and we shall have to [stop] in a moment or two. Could I have a show of hands to indicate how many people would still like to make an intervention? Then we could consider not winding up this session but adjourning it to another time. A few? Well, let us see if we could not slot in an extension, if not late this evening, then at some other time. [But in any case] I would like to thank the people on the panel for their contributions and the people on the floor for what has been a very stimulating and worthwhile discussion. The Session is adjourned [amid shouts from the floor of 'why not now'; nor was it possible to resume later on, owing to the following excursion, dinner, and the Baer–Huxley Memorial Lecture, the discussion of which continued until midnight].

Session 11

ENVIRONMENTAL ASPECTS OF ENERGY ALTERNATIVES

Keynote Paper

Environmental Aspects of Energy Alternatives*

by

AMASA S. BISHOP

Director, Environment and Human Settlements Division, United Nations Economic Commission for Europe, Palais des Nations, 1211 Geneva 10, Switzerland

&

CLAUDE G. DUCRET

Environment and Human Settlements Division, United Nations Economic Commission for Europe, Palais des Nations, 1211 Geneva 10, Switzerland.

Introduction: The Context

Since the so-called 'energy crisis', practically all countries in the world have dedicated considerable attention to energy strategies. Some, such as France and Spain, have made straightforward commitments—notably towards nuclear power—whereas most have adopted series of measures aiming at further diversifying their primary and secondary energy sources.† All are, in addition, trying to develop ways and means to increase the *efficiency* of energy production, conversion and utilization. A great deal can be done in that field: indeed, a recent study carried out by the United Nations Economic Commission for Europe (UNECE, 1976*a*) indicates that the overall efficiency of the energy sector as a whole, currently

* *Presented by Dr Ducret in the absence of the senior author.* The views and ideas in this paper are personal expressions of the authors and should not be considered as necessarily representing those of the United Nations Organization.

† The distinction between these may be exemplified by coal, which is a primary source of energy, whereas electricity *derived from burning coal* is a secondary source.—Ed.

about 15%, could reach 20 or even 30% in the early 1990s by the application of a set of quite stringent but feasible energy-saving measures. Economic activity of the 1990s could thus be twice as great as it is today, without requiring any increase in our present energy consumption.

Although some countries have faced—and in certain cases are still facing—energy supply difficulties, the world *as a whole* is not now critically short of conventional energy reserves, as is indicated in Table I. Its currently known reserves could in principle permit continuation of the historic energy-growth for many decades before their exhaustion (Table II). With the exception of a few specific cases, the energy crisis was and is primarily a geographical, economic and political problem—particularly in terms of balance of payments and economic independence, further complicated by environmental difficulties—and not a supply-shortage problem. The real crisis is to find a transition to new types of energy sources within the next few decades. With respect to conventional-energy reserves, the situation in Europe, notably in the western part, is not particularly bright, as is shown in Table III.* Known reserves are very limited and consist mainly of coal, while the consumption patterns show a clear disbalance in favour of imported oil; in 1972, for example, Western Europe imported almost 62% of its energy requirements as liquid fuel while producing only 1.5% of its liquid-fuel consumption.

From the environmental point of view, this situation is not a favourable one: indeed, those primary sources of energy that create the least environmental problems are either (1) comparatively small (such as natural gas) or (2) do not appear among the currently recoverable reserves (such as solar

TABLE I

Conventional Energy Reserves (in 1972) in 10^9 Tonnes† of Coal Equivalent (United Nations Economic Commission for Europe, 1975, 1976*b*): figures have been rounded to the nearest five and are the latest available.

Region	*Solid fuels*	*Liquid fuels*	*Natural gas*	*Oil-shale*	*Uranium*[a]	*Hydro-power*[b]	*Total*
Western Europe	390	13	40	5	15	10	475
Eastern Europe	400	6	—	—	n.d.	1	410
North America	1,620	90	120	2,920	135	15	4,900
USSR	1,070	110	190	5	n.d.	25	1,405
Others	1,120	990	240	390	100	230	3,070
World reserves	4,600	1,210	590	3,320	250[c]	280	10,260

[a] Below $26/kg. Figures refer to use in *conventional* fission reactors (i.e. not breeders).
[b] In order to make hydropower (a renewable source of energy) commensurable with non-renewable forms of energy, its annual potential output has been arbitrarily multiplied by 100.
[c] Minimum values as no specific data are available for Eastern Europe and the USSR.

* On being asked whether they had not any later figures, the authors stated that such were difficult to obtain, especially from some countries, but that there had not been any marked changes in recent years.—Ed.
† All tons used in this paper are confirmed by the authors as being metric tons = tonnes.—Ed.

TABLE II

Fossil Fuels: Reserves, World Consumption Growth-rates, and Reserves Depletion (United Nations Economic Commission for Europe, 1975, 1976*b*; United Nations Statistical Yearbook 1975).

	Solid Fuels	*Liquid fuels and oil-shale*	*Natural gas*	*Total*
Present world reserves[a]	4,600	4,530	590	9,720
World consumption (1974)[b]	2.52	3.57	1.66	7.75
World consumption growth-rates (1964–74)	1.3%	7.1%	6.2%	4.4%
Depletion of present reserves in [indicated] years at 1964–74 growth-rates[c]	250	65	50	95

[a] From Table I; figures in 10^9 tonnes of coal equivalent.
[b] Figures in 10^9 tonnes of coal equivalent.
[c] Computed by projecting; figures rounded to the nearest five. Obviously, these figures have only a very approximate and indicative value.

power) because the technology of harnessing them on a large scale is still under development, uneconomical when competing with 'conventional' sources of energy, and often unreliable as they mainly result from cyclic natural phenomena. Thus about 90% of the world reserves that are considered as likely to be technically and economically recoverable by the year 2000 at the level of large-scale schemes, are constituted by those forms of energy which require relatively extensive treatment to reduce adverse

TABLE III

Primary Sources of Energy: Repartition of Imports and Production for Europe and the USA (figures as percentage of total consumption in the area). From UNECE (1976*a*).

Primary source of energy	*Western Europe (1972)*		*Eastern Europe*[1] *(1972)*		*USA (1970)*	
	Import	Production	Import	Production	Import	Production
Solid fuels	2.46	21.13	−4.74	73.00	−2.85	23.42
Gas	0.39	10.96	0.95	12.35	0.95	32.59
Liquid fuels	61.82	1.42	13.22	5.76	11.39	32.91
Hydro	—	2.01	[0.16][2]	0.22	—	1.26
Nuclear	—	0.58	—	0.01	—	0.32
Geothermal	—	0.01	—	—	—	0.01
Tidal	—	...	—	...	—	...
Total[3]	64.87	36.12	9.59	91.34	9.49	90.51

[1] Without USSR.
[2] Represents imports of electricity in general.
[3] Totals per area do not necessarily equal 100% because of stockpiling.

effects on the environment—for instance, emissions of sulphur dioxide and the related deposition of acid rains. Furthermore, the pattern of reserves in Western Europe seems to show that, at least until the end of this century, the incipient energy sources will be environmentally unfavourable (UNECE, 1975).

A second element that should be taken into account when considering future energy supplies and the possible environmental effects of their exploitation, is the potential occurrence of Man-induced climatic modifications at local and—perhaps only at a later stage—global levels, resulting from the large-scale release of several 'pollutants' (WMO, 1976*a*). Without attempting to review this very fundamental topic, Table IV gives a rough summary outlook of five main factors emitted by the production and use of conventional energy-sources that, if released in very large quantities, could potentially initiate unintended climatic changes or magnify naturally occurring climatic fluctuations (Flohn, 1973; Schneider & Dennett, 1975; Kellogg, in press). It is worth remarking that practically all these man-made causes of climatic change are leading towards a warming effect on surface temperatures.

Estimating how and when such causes could have a significant impact is an almost impossible exercise, as no satisfactory experiment can well be carried out. However, climate modelling *(see* Schneider (1974) for review) and calculations seem to converge on the estimation that a 1% increase in the heat that is normally available to the system (world average: 100 W per m^2 at ground-level) would result in an increase of about 2°C in the mean surface temperature, with a very much larger temperature change in the polar regions—perhaps three to five times as great. It now already seems that, from dissipated heat alone, several areas of the world are lying in that order of magnitude: indeed in 1970, man-made diffused energy released per square metre of land was 0.42 W in Eastern Europe, 0.32 W in Western Europe, and 0.24 W in the USA (Ducret, 1976). However, much higher values are found locally and may require rapid action (Flohn, 1973).

In summary, the present energy situation seems to be characterized by the following features:

(1) There is no energy shortage either in supplies or in reserves at the world level for the time being; however, the situation may change in a reasonably foreseeable future.

(2) If the world should continue to rely on conventional-energy reserves, energy consumption is bound to use environmentally unfavourable sources of energy.

(3) Man-induced climate modifications resulting from the production and use of conventional sources of energy* as well as from other sectors of human activity, seem likely to represent an ultimate barrier to the use of energy resources that are not currently accounted for in the natural energy-balance of the Earth. One of the most convincing arguments advocating the use of alternative

* Including nuclear fission but not fusion.—Ed.

TABLE IV

Major Factors Potentially Creating Unintentional Climatic Changes Resulting from Energy Production and Use.

Factor	*Origin*	*Mode of action*	*Influence on surface temperature*	*Apparent present level of importance*
Carbon dioxide	combustion of fossil fuels[1]	'greenhouse' effect[2]	warming	+
Particulates	combustion of fossil fuels[3]	albedo modification	varies with ground albedo but mainly warming[4]	+
Dissipated heat	all stages of energy production and use	alteration of regional and global energy balance	warming	±
Nitrogen oxides	combustion of fossil fuels	catalytic impoverishment of the ozone layer[5]	changes in ground solar spectrum	–
Water vapour	cooling of all types of power-plants, hydropower schemes[6]	increased cloudiness	probably more cooling than warming	–

1 But also respiration, volcanoes and combustion of all biological materials.
2 Caused also by chlorofluorocarbons (Ramanathan, 1975).
3 But also photochemical smog, wind erosion, volcanoes, and some agricultural practices.
4 Until recently it was felt that particulates would mainly lead to a cooling trend, but the inclusion of ground albedo in calculations has apparently changed the picture (Kellogg, in press).
5 Caused also by stratospheric flights and chlorofluorocarbons emissions.
6 But also plant transpiration, irrigation.

sources of energy—such as solar power—is the fact that they are already included in the natural planetary energy-flow and cannot therefore result in any large-scale climatic modification.

Consequently, although there are no really fundamental reasons for a hasty attitude that might lead to irrevocable choices in energy strategies, as humanity can reckon with enough known reserves to allow for reasonable policies, there is every justification for encouraging and enforcing energy-conscious programmes aimed *inter alia* at an increasingly efficient use of existing energy-resources. Moreover, there are no grounds for complacency with our present energy-saving measures, as virtually everything still remains to be done in that particular field in order to save limited resources (especially those that are least unfavourable from an environmental point of view). Finally, there are a wealth of arguments in favour of alternative sources of energy already included in the planetary energy-flow, as they appear to be the only means simultaneously of allowing an increase in, or the retention of, justified standards of living all over the world and to avoid having to face climatic modifications that would hardly be compatible with the growing inelasticity of many human institutions. This important question has been discussed by Lovins (1976).

Potential Role

Alternative sources of energy are those energy resources which would, at least on a theoretical basis, allow for energy strategies other than those that are currently enforced or adopted in the group of countries which nowadays consume the major part of world commercialized energy. In the context of this paper, the following energy sources are consequently considered as 'conventional': coal, gas, oil, hydropower, and nuclear *fission* energy. They will therefore not be examined, and incidentally no consideration is given in this paper to the possible development of breeder reactors. On the other hand solar, wind, tidal and geothermal energy, as well as ocean temperature gradients, biofuels and fusion power, are regarded as non-conventional (for the above-mentioned countries) and consequently could play a role in energy alternatives if appropriate policies should be adopted.

What could be the share of non-conventional energy sources in the total energy supply of, say, the year 2000? Strangely enough, the answer depends considerably upon the viewpoint adopted. Indeed, if alternative energy-sources are examined in the light of the present production standards, very likely the answer will be 'hardly significant'—at the best, 1% or 2%—as the appropriate technology for large-scale installations is neither available nor would it be competitive within the present energy-market. This technology requires considerable capital investments and in such a brief interval could only be carried out on a demonstration scale.

If, however, alternative sources of energy are examined from an end-use point of view, the picture looks quite different. Table V shows, for example, that about one-quarter of the total energy consumption results from the 'Households and others' sector. Solid fuels (mainly coal), gas, and liquid fuels (mainly oil), represent more than 80% of that consumption, and are principally used for space- and water-heating and cooking purposes—that is, for satisfying low-grade energy requirements which could particularly conveniently be met by such alternative energy sources as geothermal and solar energy. The technology for small-scale applications is simple and generally already developed—or requires only a limited research and development (R & D) effort in order to be applicable to most domestic needs. However, a too-optimistic approach should be regarded with some caution, as serious limitations exist at present: energy storage facilities nowadays severely hamper the use of most alternative energy sources and, for instance, only a mere fraction of existing buildings could accommodate

TABLE V

Consumption of Primary Sources of Energy by Various Sectors (figures as percentage of total consumption). From UNECE (1976*a*).

Sectors and primary sources of energy used	*Western Europe (1972)*		*Eastern Europe* (1972)*		*USA (1970)*	
Households and others:						
electricity	3.77		1.67		4.11	
solid fuels	3.88		15.80		0.63	
gas	5.27		2.38		11.08	
liquid fuels	15.97	28.89	4.27	24.12	7.91	23.75
Industry:						
electricity	4.61		4.48		3.80	
solid fuels	4.90		24.75		8.23	
gas	9.15		14.46		15.50	
liquid fuels	20.23	38.89	7.69	51.38	9.81	37.34
Transport:						
electricity	0.25		0.27		0.02	
solid fuels	0.16		3.19		...	
gas	0.10		0.03		0.95	
liquid fuels	17.69	18.20	4.71	8.20	23.42	24.39
Rejected heat from electric power-generation		16.30		14.02		14.54
Total		100.–		100.–		100.–

* Without USSR.

solar collectors. Nevertheless, there is no doubt that, if widely used at the small-scale level and, in most cases, with currently available technology, alternative sources of energy should be in the position of playing a significant role in the total energy supply at the end of this century—notably in the 'Households and others' sector.

Quantifying the potential role of small-scale applications of alternative sources of energy presents some difficulties. Indeed one should remember that energy statistics provide accurate data for commercialized energy but only rarely are designed in such a way that non-commercialized energy sources could be taken into account: dung in India, for example—where incidentally it is a 'conventional' energy-source but a 'non-conventional' one in the USA—does not usually appear in the energy balance although it seems to represent a very significant part of domestic energy-supply. Similarly solar energy, so far used for drying harvests and clothes, pumping water, recycling, or producing salt from sea water, is not included in the usual energy balances; indeed the amount of conventional energy that is saved by such practices is almost never even mentioned or assessed in statistics. Small-scale applications of non-conventional sources of energy are therefore practically bound to be left aside from energy statistics, even if they would in effect be playing a significant role in the *real* energy balance of a country or continent.

Finally it should be stressed that any significant development of alternative sources of energy—notably at its most likely level, i.e. by a profusion of small-scale installations—is bound to have deep and long-lasting social and political effects. These would result from an entire trend-reversion from a centralized and centralizing energy policy associated with regular and continuous consumption growth (which is more or less distinctive for the present situation), to a decentralizing and more and more decentralized one, characterized by a growing tendency towards a steady-state energy-flow in society, or even a decreased one as a result of conservation measures at least in already economically developed areas of the world.

Solar Energy

As such, solar radiations cannot be utilized and are always converted into another form of energy: heat in solar collectors, electricity in photovoltaic cells, chemicals during the photosynthesic process, or mechanical energy in wind power, hydroelectricity and sea currents. However, we will in this paper—on a recognized, purely arbitrary basis—restrict the use of the term 'solar energy' to only the immediate (as opposed to fossil, such as coal) and to the direct (as opposed to indirect, such as wind) forms of solar energy, namely to its thermal and photoelectric conversion.

Several types of domestic-size solar collector are marketed but all apply the same principle: solar radiations are absorbed by a black metal sheet enclosed in one or several shallow glass-enclosed boxes situated on the roof or the walls of buildings. Tubing attached to the black metal sheet carries heating fluid (usually water) which is stored in a tank that is connected to a heat exchanger. When needed, sanitary water and/or space-heating water is circulated through the heat exchanger as well as, in very advanced and still experimental devices, fluids for heat-operated cooling systems. Often, sophisticated systems are assisted by a heat-pump.

Various estimations have been made of the potential role of solar heating and cooling (Israel nowadays covers about 2% of its energy needs with simple solar collectors for domestic water heating). For instance, the Energy Research and Development Administration (ERDA) in the USA estimated that in 1985 the energy equivalent of about 5 million tonnes of coal could be provided by solar heating and cooling devices, reaching 72 millions in AD 2000 and exceeding 500 millions in AD 2020 (Hammond, 1975). In his energy programme, President Carter plans the utilization of solar energy in more than two-and-a-half million homes by 1985. For Denmark, it has been estimated by Sørensen (1975)—on the basis of a zero-energy-growth policy—that about 30% of the national energy needs could be covered by solar space- and water-heating. In addition, experimental houses allowing the economic assessment of solar space- and water-heating, improvement and testing of technology, and monitoring of possible environmental impacts, have been built in several countries (for example: United Kingdom, cf. Brachi, 1974: USA, cf. Böer, 1974; Federal Republic of Germany, cf. Sandscheper, 1975).

From an environmental point of view, the utilization of solar energy for space-heating and -cooling leads to very moderate or even insignificant problems. Indeed no air, water, waste, or noise, pollution need be generated.

Maintenance energy is limited to running a circulation pump and—if included—the heat-pump. However, the widespread use of solar collectors by the application of Equator-oriented roofs, would result in an urban area with a clear trend towards monotony in situation and appearance of buildings: main streets would have to be east–west oriented. This situation would certainly encourage, or even sometimes necessitate, chequer-board structures of 'solar cities'—probably giving them a similar appearance to those nowadays seen in many North American suburbs. From the climatic point of view, a widespread use of solar collectors over large areas could possibly lead to very minor but foreseeable local alterations. However, they certainly would not exceed those resulting nowadays from fossil-fuelled or electric heating and cooling devices of a comparable capacity.

For some time now, large-scale installations have been envisaged which would use solar heat for the production of electricity, but the only existing facility of this kind is a relatively sophisticated 'solar oven', in Odeillo in the French Pyrenees—erected mainly for investigating the resistance of materials at high temperatures rather than for studying electricity production. Several large-scale projects have recently been proposed for areas ranging from deserts to the Swiss Alps (cf. Meinel & Meinel, 1976). They consist of an enormous dish-reflector made up of hundreds or even thousands of synchronized mirrors tracking the sun, which concentrate solar radiations on a steam boiler. High-pressure steam is produced at temperatures of up to 600–1000°C and electricity is generated by conventional turbines. The first large-scale solar-powered electric-generating station, scheduled for completion in 1980–81, will be built near Barstow, California; it should have a 10 MW capacity and will cover more than 52 ha (130 acres) in the Mojave Desert. As, naturally, solar 'plants' can generate electricity only during daytime, they have to be linked to large-scale storage facilities or connected to a wide-range distribution network as was suggested for the USSR (Tarnizhevsky & Smirnova, 1974).

Some concern for the environment is obviously manifested with regard to such large installations. Land-use is certainly not the least worry; between 30 and 50 km^2 are required for 1,000 MW of peak electric output, and this could lead to local disturbance of temperature régime and probably a significant—but not necessarily negative—effect on plants through shading, temperature changes, etc. Aesthetic problems also result from the over one-hundred-metres-high 'sun tower' supporting the boiler, as well as from cooling towers, electric transmission lines, and other ancillary installations that usually accompany any type of conventional power-plant. However, from an environmental point of view, three factors basically distinguish solar-powered stations from fossil- or nuclear-fuelled ones: *(a)* when coal and uranium are mined and gas and oil are pumped, all require pipelines or tankers or railroads or enrichment facilities, etc., for eventually fuelling a conventional electric power-plant. No such preliminary step is needed with solar power, and the question could be raised as to whether, on a long-term perspective, the additional area required by a solar plant is not more or less compensated for by the land saved from surface mining, pipelines, harbour facilities, etc.; *(b)* no waste whatsoever is rejected; and *(c)* although local climatic modification can be expected from a large-scale facility, it will have no impact upon the global balance.

It could therefore be concluded that no really serious environmental problems are encountered with solar power-plants except with regard to land-use and from an aesthetic viewpoint. These impacts are apparently less serious than those raised by conventional power-stations, including gas-fired plants.

Photovoltaic conversion of electricity comprises limited, small-scale and extremely specific applications, such as photography or space exploration, which strictly have no common measure with large-scale cheap electrical production, and cannot supply normal applications on the domestic level (such as power for TV sets). The major objectives of the present R & D effort in this field are to develop low-cost, reliable systems capable of producing a significant amount of energy. The chief thrust is in the silicon-cell arrays, but other basic material is also being investigated—including gallium arsenide, selenium compounds, cadmium telluride, and some organic types of semi-conductor materials. Yields can vary greatly (Laguës, 1975), and costs are still astronomical but could rapidly decrease (Wolf, 1975). Very high capital investments, ranging nowadays in the order of magnitude of $2,000 per kW of peak output (Chalmers, 1976) and far above conventional means, represent the main problem of a wider use of photovoltaic conversion of solar energy—together with the elaborate and labour-intensive manufacturing and assembling technology and the possible scarcity of some basic materials. We also suggest that the real energy budget for photovoltaic conversion of solar energy be investigated, as the present manufacturing technology for solar cells involves complex and energy-intensive steps such as vacuum pumping, high-voltage acceleration, hydrogen annealing, thermal binding, etc., which might exceed the amount of energy produced during the lifetime of the cell. Furthermore, as for other solar-energy applications, photovoltaic power-stations are only able to produce electricity during sunshine. Therefore it has been calculated that a terrestrial solar installation providing the equivalent of a 1,000 MW fossil-fuelled or nuclear power-plant would need a peak output of 2–3,000 MW.

Some authors are visualizing solar energy as playing, on a very long-term basis, a major role in human economy with the help of huge geostationary satellites with a capacity of up to 10–20,000 MW—far enough from the Earth to be able to operate around the clock. The energy produced could either be transmitted to Earth after conversion into microwaves by ultra-high-frequency generators and picked up, regardless of weather conditions, by a very large receiving antenna probably located at sea (Glaser *et al.*, 1970), or it would be used in space by large-scale industrial satellites (O'Neill, 1976).*

From an environmental point of view, photovoltaic conversion of solar

* That is, future Earth satellites aimed at performing industrial operations such as those requiring a 'very high vacuum' or even for 'mining the Moon for minerals'.—Ed.

energy shows the same major advantages (no mining, no waste, etc.) as the thermal conversion of solar energy into electricity. It should, however, be warned that the making of some types of solar cell requires basic materials which are known for their toxicity to Man and to the environment—such as arsenic, cadmium, selenium, etc.—and consequently that the manufacture of these cells might be quite polluting. Space power-plants with terrestrial receiving antennas would, in addition, take up large areas for the antennas themselves and for maintenance facilities (space-shuttle bases, etc.), and would probably represent a risk to human beings who accidentally crossed the microwave beam. Furthermore, artificial transmission of large quantities of solar energy, together with those normally received on Earth, would eventually have the same potential climatic disadvantages—in terms of heat-pollution—as conventional methods of producing similar quantities of energy. These disadvantages would, however, be overcome by use in industrial satellites of the energy produced .

The need for further extensive R & D in large-scale applications of solar energy for the production of electricity, as well as the high capital investment required and other possible obstacles that might have to be overcome, lead us to believe that—if they ever become feasible—these applications will probably not occur before the end of the century.

Wind Energy

Wind-energy utilization has accompanied all civilizations since a human dared to sail for the first time. Well before the time of Christ, Persians used windmills for grinding wheat and, up until the 1920s, tens of thousands of windmills were pumping water in the American Midwest. They were put out of work by the advent of cheap electricity. Wind is mechanical energy and therefore a noble form of energy. In addition to sailing and pumping, the most promising utilization of wind energy is the highly efficient conversion of its mechanical energy into electricity by the help of a rotor. Theoretically, large machines should be built, as the output of a rotor increases according to the square of the blade-length (and the cube of the wind velocity); but the strength of materials, together with the intermittent character of the energy produced, constitute major limitations (Bruckner, 1974; Bockris, 1975).

There are, nowadays, two types of wind generator. *(a)* Those having a horizontal-axis rotor, with blades facing the wind and slow rotation-speed. Their theoretical efficiency is 59%, while practically only 40% of the wind's kinetic energy is converted into electricity (Black, 1976). Several large installations have been erected (McCall, 1973; Simmons, 1975), the largest operating one being a 100 kW machine built by the US National Aeronautics and Space Administration (NASA) as part of the ERDA wind-power programme (NASA, 1975), while a 2 MW windmill is under construction at Tvind, Denmark (Hinrichsen & Cawood, 1976). *(b)* Those having a vertical-axis turbine, with two or three rigid or flexible blades turning at high speed. They are derived from an idea first expressed by G. Darrieux in 1927. Wind direction is immaterial, and the conversion efficiency reaches 35%. This device is particularly suitable for small-scale and local applications (Brown, 1974).

As usual for alternative sources of energy, wind power requires capital investments ranging from $200 (Tvind) to over $2,000 per installed kW, but in a few years' time it could probably become competitive with conventional energy-sources (Wade, 1974). From an environmental point of view, wind power is obviously a clean source of energy: no air, water, or waste-energy, pollution, no waste produced, and limited land-use and noise problems. Its major drawback is the aesthetic concern with regard to large-scale installations which could considerably disfigure sites. In that respect we should wonder what would be the reaction of nature-lovers and environmentalists if, for instance, the British coasts, which are a suitable potential site for large-scale wind-power utilization (Musgrove, 1976), should bristle with dozens of 100-metres-high towers in order partially to cover national needs. A similar, large-scale project has recently been investigated in the USA (Widger & Derrickson, 1976).

There is no doubt that landscape protection will be the major issue raised on the large-scale use of wind power if significant amounts of electricity are to be produced by large wind-generators. Now ERDA, for example, has planned a 1,000 MW wind capacity for 1985 in the USA and a 60,000 MW wind capacity for the year 2020 (Hammond, 1975). In Denmark, Sørensen (1975) has calculated that wind power could apparently provide directly or indirectly (through hydrogen production) nearly 70% of the energy required for meeting domestic needs in the year 2050. It would not be surprising if, in the long term, nature conservation and landscape protection associations and other similar citizens' groups come to be among the harshest opponents to such large-scale development programmes.

Geothermal Energy

The amount of energy contained under the Earth's surface is simply enormous, being about 2,000 times all known reserves of fossil fuels (Marinelli, 1974). Its major manifestations are earthquakes (mechanical energy), which derive from continental drifts, and geothermal energy. The latter is probably originated by three factors: *(a)* mainly the decaying heat of radioactive materials contained in very large quantities in the inner part of the globe, which are slowly dissipated by conduction to the surface; *(b)* some fossil heat dating from the time of the Earth's formation, about five thousand million years ago, and, finally, *(c)* probably some heat generated by gravitational force. Geothermal power is therefore only partially a renewable source of energy and is only included in the natural energy-flow on the surface of the Earth to the extent of heat that is naturally dissipated by conduction. The spectacular manifestations of geothermal energy, such as volcanoes, geysers and hot springs have from ancient times aroused Man's interest in tapping them for his own benefit.

The Romans used geothermal energy for domestic heating purposes:

warm water from hot springs was circulated under the floor of houses through a network of pipes. Nowadays, geothermal heat is mainly used for space-heating purposes and, when steam is available, for electricity production. Space-heating is probably one of the easiest ways of using geothermal power: warm water is usually pumped from a hot or warm aquifer and either directly or indirectly (heat exchanger) circulated to the heating systems of buildings, greenhouses, swimming pools, etc. After use, the water is normally reinjected into the aquifer through a separate well. Geothermal district-heating systems have been developed in several countries following the pioneering experience of Iceland in this field since 1925 (Einarsson, 1974). Such heating grids nowadays exist in France, Hungary and USSR, and substantial savings in oil for heating have been recorded: in Iceland, the Reykjavik district is said to save annually about 200,000 tonnes of oil through geothermal heating (Spurgeon, 1973), and France expects to save by 1990 between one and one-and-a-half million tonnes of oil by an extensive utilization of warm aquifers (Clot, 1977).

From the environmental point of view, geothermal space-heating leads to effectively no problem if used water is reinjected into the original aquifer: land utilization is minimal, noise is very limited—a few pumps—and there is practically no aesthetic concern. The amount of diffused heat released from the heated facilities is probably of the same order of magnitude as, when not lower than, from buildings heated by conventional means. There is therefore a good deal of both economical and environmental justification for using geothermal heat for space-heating, even on a large-scale level. When available, it is therefore highly preferable to both fossil-fuelled central-heating systems and to the use of electricity.

When geothermal steam is available, it can be used for the production of electricity: several industrial geothermal power-plants exist in various locations around the world, notably in Iceland, Italy, Japan, Mexico, New Zealand, USA and USSR. Steam supplied at the wellheads must first be cleared of water and of minute fragments of rock that would otherwise damage the turbines. It is then used directly for electrical generation. The economic efficiency of power-plants using naturally-occurring geothermal steam is higher than fossil-fuelled or nuclear-fuelled ones as there is no mining, transportation of fuel, etc. The present very modest share of geothermal electricity in the world energy balance is due to the scarcity of naturally-occurring, high-grade, 'dry' steam sources and their location in remote places. Its importance, however, will certainly grow rapidly in the years to come (Marinelli, 1974), particularly with the help of remote-sensing techniques which nowadays allow extensive surveys at the planetary level (Barnea, 1972).

From the environmental point of view, the production of electricity from geothermal energy can generally be considered as less clean than space-heating (Battelle Institute, 1973; Arnorsson, 1975). The situation mainly results from the fact that the solubility of salts and gases in water regularly increases with rising temperatures and pressures. Now, as commercially exploited geothermal reservoirs have temperatures ranging from 150 to more than 300°C (Jaffé *et al.*, 1975), their steam contains sometimes high levels of various minerals and gases, some of them being harmful to the environment (in fact the first commercial exploitation of the geothermal field in Lardarello, Italy, was undertaken for the extraction of emanating boric acid, not steam). As the reservoir is under pressure, it is not always feasible to reinject the condensed water from the cooling process into the steam-producing reservoir; instead, it must be disposed of otherwise. Geothermal power-plants therefore lead to some air and water pollution, although the level and relative importance of this pollution are very dependent on the site involved. However, the most common air-pollutants emitted are mainly H_2S, with its characteristic rotten-egg smell, traces of nitrogen oxides and of radioactive elements (such as Radon-222), and various amounts of ammonia, boric acid, hydrogen, methane, carbon oxides, etc. With respect to water, the major pollutants are various salts (such as sodium, potassium, calcium and magnesium chlorides and sulphates), silicates, carbonates and waste heat, representing about 80–90% of the extracted energy. Pressurized steam emerging at the wellhead also makes a very loud noise, comparable with that of a jet aeroplane, and thus represents an important problem, particularly for workers.

Aesthetic concern has also been expressed, as geothermal energy is generally found in non-industrial areas where access roads, drilling rigs, pipes, cooling towers, high-tension lines, etc., may alter considerably the original landscape. Potential microseismic effects and local subsidence might also appear, although both seem unlikely when using natural steam sources.

As a general assessment, it can be argued that the production of electricity by geothermal energy conversion is by no means a pollution-free process. It should, however, be kept in mind that there is a definite 'natural' impact upon the environment in any case, due to the nature of emerging geothermal energy: flora, notably, can in some locations be seriously affected by sulphur emissions from fumaroles. Furthermore, when not used, 100% of the energy is released to the environment. As a consequence, the use of naturally occurring geothermal power should—in theory at least—lead to an improvement of local situations with respect to thermal and chemical pollution, if methods for reducing the latter are used, although now problems of noise and landscape disturbance are created.

Although not proven feasible, either technically or economically, the artificial attainment of deep high-temperature geothermal fields (500–600°C) has been seriously considered (Harlow & Pracht, 1972). Two parallel holes, several hundred metres apart, would be drilled and connected within the geothermal field by artificial geological fractures—a technique commonly used in oil exploitation. A fluid (generally water) would be injected into one of the holes, recovered in the form of superheated steam at the outlet of the other hole, and then carried to turbines. In addition to drilling techniques used in the oil industry, other methods have been envisaged, such as underground nuclear explosions in the American project 'Plowshare' (Marinelli, 1974; Hammond *et al.*, 1973), or the use of a 'subterrene'—a device consisting of a very small remote-controlled nuclear reactor, housed in a pointed cylinder approximately 2 m in diameter, which would melt the rock, passing through it by gravity and leaving a glass-lined pipe (Robinson *et al.*, 1971).

There is no particular reason for believing that the exploitation of deep geothermal fields would be giving rise to a lot of more serious environmental problems than those deriving from the utilization of naturally-occurring geothermal sources—particularly if closed-circuit systems should be made necessary in order, for example, to avoid contamination by radioactive by-products resulting from underground nuclear explosions. It should, however, be stressed that the energy which could be extracted from deep geothermal fields is definitely not included in the natural flow of energy at the surface of the Earth and, consequently, even if it could be made perfectly 'clean' with respect to chemical pollution, large-scale utilization of this source of energy would contribute to modifying the natural climate-balance evolution.

Ocean Temperature Gradients

For a long time, the enormous power-potential of oceans has fascinated the imagination of engineers: the principles of its utilization were originated by J. d'Arsonval in 1881 and, in 1929, G. Claude built a shore-based ocean gradient power-plant in Cuba which, although producing 22 kW of utilizable power, was an economic failure. Nowadays, projects envisage ocean power-plants as very large semi-submerged offshore structures. All use the same principle, which is the difference in temperatures between warm surface water—about 25°C all the year round within the Tropics—and the fact that, even in equatorial areas, sea-water temperatures drop always to 4–5°C at depths of around 1,000 m. Although this temperature difference is not large, in comparison with usual power-plants, it could be employed for generating electricity. The efficiency of conversion would therefore remain very low, probably in the range of 2–3%. The warm water from the ocean surface would flow through a heat-exchanger where it would cause another fluid—probably propane, 'freon' or ammonia—to boil. The vapour would then power a turbine and condense back to a liquid in a condenser cooled by deep-sea water (Metz, 1973; Haber, 1977). A similar device has also been designed for use in polar regions, using the temperature difference between sea water (about 0°C) and the colder air (Isaacs & Seymour, 1973). Other designs using foam have also been studied (Zener & Fetkovich, 1975).

Environmental aspects of such installations have only been studied superficially. Several potential impacts should be considered: the major one apparently results from the very large discharges of unusually cold water to the sea-surface, leading to an increase in absorption of heat by the ocean's surface layer. This might eventually induce alterations in ocean thermal circulation, which is of paramount importance from a climatic point of view: changes in the path of the Gulf Stream, for instance, would be dramatic for Western Europe. However, in view of the tremendous amount of water involved in ocean currents—the flow of the Gulf Stream is

2,200 km^3 per day—climatic alterations as a result of ocean temperature-gradient power-plants would not seem likely unless numerous and very large-scale installations were put into operation.

Nitrates, phosphates, and other nutrients, are widely dissolved in large quantities in deep-sea water. When brought into the light at the ocean surface, these elements constitute some of the raw materials for plant growth, thereby increasing the ocean's primary production. Extrapolation of results from small-scale experiments conducted in the Virgin Islands, USA, indicate that these 'artificial upwellings' may lead to the production of 20 times as much algal protein per hectare as the proteins produced by alfalfa, the highest protein-producer in land-based agriculture (Roels *et al.*, 1976). This impact would thus be beneficial if a mariculture complex were to be associated with the power-plant. On the other hand, accidental releases of ammonia, methane and, more important, 'freons', from the heat exchanger, could cause significant pollution problems, notably in view of the likely detrimental effect of freons on the ozone layer (Cicerone *et al.*, 1974; Wofsy *et al.*, 1975; WMO, 1976*b;* Maugh II, 1977).

Tidal Power

Tidal power is probably the only large-scale source of energy which is at least partially provided by the Moon. Its practical utilization for electricity production requires a minimum of 4 m of tidal difference in the sea-level. This considerably restricts the areas in the world where this source of energy could be used. The same water can, in principle, pass twice through the turbines, at the flood- and ebb-tides. Only two plants are in operation in the world: la Rance in Brittany, France, and Kislogub near Murmansk in the USSR.

The main economic drawback of tidal power is the high investment cost, amounting to over $1,000 per installed kW.* It is therefore in the same range of cost as wind or solar power, and the power produced is of a regular but cyclic although inelastic nature. As a consequence, it seems likely to be economical only when used with complex reservoir systems and in connection with hydroelectric storage lakes.

Nowadays, tidal power plays a very minor role and it will certainly continue to be only a minor source of energy despite major projects such as those in the Bay of Fundy, which is said to have the highest tides in the world. Other possibilities include Ungava Bay, Canada (Godin, 1974; Electrical World, 1977*a*); the Bristol Channel, England (Shaw, 1975, 1976); the Golfo Nuevo and Golfo San José, Argentina; Cook Inlet in Alaska, USA; and, in the USSR, the White Sea and the Sea of Okhotsk (Energia, 1972), which last is said to have a 25,000 MW potential.

* According to a recent report (Electrical World, 1977*a*), the Canadian Government is considering a project for the development in the Bay of Fundy of about 4,915 MW of tidal power, which would cost an estimated $6.5 'billions'. The investment costs would therefore be around $1,300 per installed kW.

There are only limited environmental problems associated with tidal power, but it should be recognised that a strong perturbation of the water régime might be very detrimental to wildlife: concern has, for instance, been expressed about the Bay of Fundy, which is considered to be one of the world's richest wildlife and nature areas—where beautiful marsh plains have been built over the ages and are continually washed by the incoming tides (Electrical World, 1977*a*). In urban areas, stagnation of sewer water could give rise to serious problems with respect to marine pollution. Finally, the aesthetic aspect may be regarded as more or less important, depending on personal opinion, beauty of site, etc. Tidal power, despite its attractive nature, is and will remain very limited in its practical application.

Biofuels

The utilization of various biological materials for the production of liquid or gaseous fuels has been extensively investigated in recent years, and several pilot and demonstration plants are now in operation; the most promising outcome seems to be in the production of alcohols from agricultural wastes such as sugar-cane molasses, the production of methane from sewage-water treatment, plant sludges, animal feedlots etc., and the pyrolysis of solid municipal wastes to a fuel gas or oil.

Both alcohol and methane production are based on a microbiological fermentation of organic matters, in the absence of air, followed by an enrichment step such as distillation for alcohols (mainly methanol and ethanol) or separation of gases (Saverny & Cruzan, 1972; Ghaswala, 1975; UNESCAP, 1975; Konstandt, 1976; Seguier, 1976). Both the alcohols and the gases are easily stored by conventional means and have numerous applications in various fields where they can be substituted for fossil fuels for various energy utilizations, such as transportation: the alcohol content of Brazilian gasoline has ranged from 2% to 8% in recent years*—a level at which no adjustments for or of automobile engines are required (Hammond, 1977). Alcohols have numerous applications in chemical industry, and methane can substitute for natural gas in practically all its utilizations. Experimental 'biofuel farms' have been suggested—and a few are even in operation—for growing specific plants uniquely for their ability to provide easily the basic materials for biofuel production (Chedd, 1975; Hammond, 1977). It seems likely that such biofuel farms will provide a significant part of the future market for liquid fuels—notably in the automotive transportation sector (principally cars and aeroplanes), and probably in competition with hydrogen in that sector.

The pyrolysis of solid municipal wastes—which involves heating refuse to a temperature of between 500 and 1,000°C in the absence of air—has reached the economic exploitation level: a solid-waste disposal plant situated in Chicago, USA, will convert refuse to low-sulphur fuel at a rate of 250,000 tonnes annually. The fuel will be purchased by an electrical power company and mixed with coal to produce electricity (Power Engineering,

* *See* page 384.—Ed.

1977). Similar but bigger facilities are also planned elsewhere (Wilcox, 1973; Electrical World, 1975, 1977*b*). The saving on conventional incineration expenses and the sale of fuel to power companies, as well as the sale of recovered metals, makes pyrolysis of solid municipal wastes attractive from an economic point of view, although the fuel produced has a lower heating value than conventional energy sources (Power Engineering, 1977).

Comparisons in performance and the usual pollutant emissions between cars using gasoline or methanol as fuel have shown that an engine burning the alcohol emits, for slightly diminished performances (acceleration) and for the same mileage, only about one-sixth to one-tenth of the amount of unburned hydrocarbons, carbon monoxide and nitrogen oxides, as compared with a gasoline one. Furthermore, methanol has a high-blending octane value, i.e. its addition to gasoline at a suitable percentage can obviate the use of tetraethyl lead as an anti-knock agent (Reed & Lerner, 1973). Utilizing methanol as an additive to, or as a substitute for, gasoline in internal combustion engines, could therefore be a potential way of alleviating both energy and pollution problems. As methanol constitutes a significant part of natural gas, its combustion properties are well known and its utilization would not require any new technological development (Hammond, 1977). It would certainly not lead to more serious problems than does natural gas, and could therefore be considered as 'clean'.

The pyrolysis of wastes is also attractive from the environmental point of view: the land required for conventional refuse disposal is dramatically reduced, or the sea is far less polluted, than when garbage is simply dumped. Sulphur emissions at the stack of the power-plant are similar to those from low-sulphur coal combustion, and the process permits the recovery of metals. Other air-pollution problems are probably similar to those which emerge from coal or oil combustion, although significant chlorine, and maybe fluorine, emissions should be expected as refuses generally contain higher levels of these elements than do usual fuels. Therefore, the pyrolysis of solid municipal wastes represents a better solution to refuse disposal problems than landfilling or dumping at sea, but it is in no way a perfectly 'clean' solution from the environmental viewpoint. The perfect solution would be not to produce any municipal waste!

Nuclear Fusion

Much hope has been placed in controlled thermonuclear fusion. However, despite considerable effort in many countries, fusion power has still not reached the demonstration stage, and its first commercial applications are expected to take place only at the beginning of the next century. Fusion power can be considered, on the human scale, as a practically inexhaustible source of energy: the fuel—consisting of deuterium and lithium, or (in more sophisticated reactors) of deuterium alone—is in abundant supply. In brief, the simplest fusion reactor would function as follows: a gaseous mixture of deuterium (a naturally-occurring isotope of hydrogen, found in water) and tritium (an artificially-produced and radioactive isotope of hydrogen) is ionized to form a plasma, heated to several hundred million degrees centigrade, and confined at that temperature—through the use of magnetic fields—for a sufficient time to permit fusion reactions to take place. The reaction products consist of helium (a useful and non-radioactive gas) and neutrons (which are used to produce tritium from a surrounding lithium blanket) (Bishop, 1958, 1969; Gough & Eastlund, 1971; Rose, 1976).

The only potential chemical pollutant from a fusion power-plant would be due to leakages of tritium fuel, which has a half-life of about 12 years. This leakage could be held to very low levels and, accordingly, should not pose a significant environmental hazard. Furthermore, the careful selection of construction materials would minimize radioactivity induced in the core by neutron activation and should avoid long-term storage problems of derelict reactors (Hirsch & Rice, 1974). It is important to note, however, that the energy produced by fusion power is not included in the natural energy-balance of the Earth, and its use would thus have to be maintained within the acceptance capacity of the climatic environment.

Conclusions

Energy plays a role of paramount importance in the environment: it sustains life and maintains a delicate climatic balance on which most human activities and institutions depend. The large-scale utilization of conventional sources of energy, which are not accounted for in the natural energy-balance of the globe, results in the emission of chemical and physical pollutants which are potentially altering life-patterns and modifying directly or indirectly the properties of the atmosphere—causing, consequently, inadvertent climatic modifications. The utilization of alternative sources of energy is a way of avoiding these problems. Their potential role until the end of this century has been discussed in this paper, particularly taking into account the feasibility and availability of large-scale *versus* small-scale applications.

Several alternative sources of energy, which could be in a position of potentially playing a significant role in future energy supplies, have been reviewed, their technical features being rapidly examined, and their potential effects upon the environment assessed. The following conclusions can be drawn:

(1) With the exception of geothermal, satellite-solar and fusion power, the reviewed sources of energy are included in the natural energy-flow, and could consequently be used more extensively than at present without creating large-scale climatic problems.
(2) None would create very important pollution problems.
(3) Small-scale applications are preferable to large-scale ones.
(4) Several applications can produce electricity with reasonable efficiency, but often with a significant environmental impact. Electricity, therefore, appears to be an environmentally expensive form of energy, and its utilization should be restricted to those cases where it is the only applicable form of energy.

(5) Most alternative energy-sources seem to have target applications for which they appear more suitable than others: geothermal for district heating, solar collectors for water-heating, methanol for internal combustion engines, etc.

In spite of the so-called 'energy crisis', humanity is not, at the present time, short of energy, but faces a completely new economic, social, and political, situation. The crisis is not a supply problem but rather a need to find a transition to new types of energy sources within the next few decades. Policy decisions have to be made, but there is no justification for them to be made too hastily and while options are still open in a long-term perspective. Most alternative energy technologies, provided with reasonable R & D financing, could be operational in the near future, thereby permitting a real possibility of policy choice for energy alternatives. It is our view, therefore, that there is little doubt that humanity has a lot to gain from retaining a maximum of flexibility in energy options by saving and diversifying its energy resources and, finally, by allocating the most suitable energy source to each energy utilization.

References

ARNORSSON, S. (1975). Geothermal energy in Iceland: utilization and environmental problems. *Naturopa,* **15,** pp. 23–6.

BARNEA, J. (1972). Geothermal power. *Scientific American,* June 1972, pp. 70–7.

BATTELLE INSTITUTE (1973). *Environmental Considerations in Future Energy Growth:* Study prepared for the US Environmental Protection Agency, Washington, 1973 [not available for checking].

BISHOP, A. S. (1958). *Project Sherwood: the U.S. Program in Controlled Fusion.* Addison-Wesley Publishing Company, Reading, Massachusetts: vii + 216 pp., illustr.

BISHOP, A. S. (1969). *The Status and Outlook of the World Programme in Controlled Fusion: Presentation Made Before the National Academy of Sciences.* National Academy of Sciences, Washington, DC: 25 pp., illustr. (mimeogr.)

BLACK, T. W. (1976). Megawatts from the wind. *Power Engineering,* March 1976, pp. 64–8.

BOCKRIS, J. O'M. (1975). Possible means of large-scale use of wind as a source of energy. *Environmental Conservation,* **2**(4), pp. 283–8, illustr.

BÖER, K. W. (1974). The solar house and its portent. *Ekistics,* **38**(225), pp. 96–9, illustr.

BRACHI, P. (1974). Sun on the roof. *New Scientist,* 19 September, pp. 712–14.

BROWN, C. A. (1974). New interest in an old power-source. *Cooperation* (Ottawa), August, pp. 12–19.

BRUCKNER, A. (1974). Taking power off the wind. *New Scientist,* 28 March, pp. 812–14.

CHALMERS, B. (1976). The photovoltaic generation of electricity. *Scientific American,* October, pp. 34–43.

CHEDD, G. (1975). Cellulose from sunlight. *New Scientist,* 6 March, pp. 572–5.

CICERONE, R. J., STOLARSKI, R. S. & WALTERS, S. (1974). Stratospheric ozone destruction by man-made chlorofluoromethanes. *Science,* **185,** pp. 1165–7.

CLOT, A. (1977). La géothermie 'basse énergie'. *La Recherche,* **8,** pp. 213–23.

DUCRET, C. G. (1976). Environmental aspects of energy conversion and use. Pp. 645–60 in *Aspects of Energy Conversion* (Eds J. Blair, B. Jones & A. Van Horn). Pergamon Press, Oxford & New York: xvii + 847 pp.

EINARSSON, S. S. (1974). Chauffage urbain grâce aux sources d'eau bouillante. *Courrier de l'Unesco,* February, pp. 24–5.

ELECTRICAL WORLD (1975). Utility will buy power from refuse plant. *Electrical World,* 1 August, pp. 78–80.

ELECTRICAL WORLD (1977*a*). Canada again studies tides of Bay of Fundy. *Electrical World,* 15 February, pp. 27–8.

ELECTRICAL WORLD (1977*b*). Pyrolysis units get bigger. *Electrical World,* 15 February, p. 28.

ENERGIA (1972). The Kislogub tidal power-station. *Energia* (Moscow)—*see Water Power,* January 1974, p. 35.

FLOHN, H. (1973). Globale Energibilanz und Klimaschwankungen. *Schriftenreihe der Rheinisch–Westfälische Akademie des Wissenschaften,* **234,** pp. 75–117.

GHASWALA, S. K. (1975). Cattle dung offers energy relief for rural India. *Energy International,* April, pp. 25–6.

GLASER, P. *et al.** (1970). Series of papers in *The Journal of Microwave Power* (special issue on satellite solar power-stations and microwave transmission to Earth), **5,** 296 pp. [* Several papers involving numerous authors.]

GODIN, G. (1974). The power potential of Ungava Bay and its hinterland. *Water Power,* May, pp. 167–71.

GOUGH, W. C. & EASTLUND, B. J. (1971). The prospects of fusion power. *Scientific American,* **224,** pp. 50–64.

HABER, G. (1977). Solar power from the ocean. *New Scientist,* 10 March, pp. 576–8.

HAMMOND, A. L. (1975). Solar energy reconsidered: ERDA sees bright future. *Science,* **189,** pp. 538–9.

HAMMOND, A. L. (1977). Alcohol: a Brazilian answer to the energy crisis. *Science,* **195,** pp. 564–6.

HAMMOND, A. [L.], METZ, W. [D.] & MAUGH II, T. (1973). Pp. 53–60 in *Energy and the Future.* American Association for the Advancement of Science, Washington, DC: xii + 184 pp., illustr.

HARLOW, F. H. & PRACHT, W. E. (1972). A theoretical study of geothermal energy extraction. *Journal of Geophysical Research,* **77,** pp. 7038–48.

HINRICHSEN, D. & CAWOOD, P. (1976). Fresh breeze for Denmark's windmills. *New Scientist,* 10 June, pp. 567–70.

HIRSCH, R. L. & RICE, W. L. (1974). Nuclear fusion power and the environment. *Environmental Conservation,* **1**(4), pp. 251–62, illustr.

ISAACS, J. D. & SEYMOUR, R. J. (1973). The ocean as a power resource. *International Journal of Environmental Studies,* **4,** pp. 201–5.

JAFFÉ, F., CUENOD, M. & VERCELLINI, R. (1975). Utilisation de l'énergie géothermique. *Bulletin Technique de la Suisse Romande,* **22,** pp. 1–9.

KELLOGG, W. W. (in press). Global influences of mankind on the climate. In *Climate Change* (Ed. J. Gribbin). Cambridge University Press, New York, NY. (in press).

KONSTANDT, H. G. (1976). Engineering, operation and economics of methane gas production. Pp. 379–98 in *Proceedings UNITAR Seminar on Microbial Energy Conversion,* Göttingen, FRG, 4–8 October 1976 (Eds H. G. Schlegel & J. Barnea). Erich Goltze, Göttingen, FRG: 643 pp.

LAGUËS, M. (1975). Les cellules solaires de demain. *La Recherche,* **6,** pp. 870–3.

LOVINS, A. B. (1976). Energy strategy: the road not taken? *Foreign Affairs,*

October, pp. 65–96: see also *Environmental Conservation*, **3**(1), pp. 3–14, Spring 1976.

McCall, J. (1973). Windmills. *Environment*, **15**, pp. 6–17.

Marinelli, G. (1974). L'énergie géothermique. *La Recherche*, **5**(49), pp. 827–36, illustr.

Maugh II, T. H. (1977). The ozone layer: the threat from aerosol cans is real. *Science*, **194**, pp. 170–72.

Meinel, A. B. & Meinel, M. P. (1976). Solar photothermal power generation. *Environmental Conservation*, **3**(1), pp. 15–21, illustr.

Metz, W. D. (1973). Ocean temperature gradients: solar power from the sea. *Science*, **180**, pp. 1266–7.

Musgrove, P. (1976). Windmills change direction. *New Scientist*, 9 December, pp. 596–7.

NASA (National Aeronautics and Space Administration) (1975). Towers of power. *New Scientist*, 27 November, p. 518.

O'Neill, G. K. (1976). Space colonies: the high frontier. *The Futurist*, **10**, pp. 25–33.

Power Engineering (1977). Solid waste-to-fuel conversion facility opened. *Power Engineering*, January, p. 74.

Ramanathan, V. (1975). Greenhouse effect due to chlorofluorocarbons: climatic implications. *Science*, **190**, pp. 50–52.

Reed, T. B. & Lerner, R. M. (1973). Methanol: a versatile fuel for immediate use. *Science*, pp. 1299–1304.

Robinson, E. S., Potter, R. M., McInteer, B. B., Rowley, J. C., Armstrong, D. E., Mills, R. L. & Smith, M. C. (1971). *A Preliminary Study of the Nuclear Subterrene*. Report LA-4547 from the Los Alamos Scientific Laboratory, Los Alamos, New Mexico: 12 pp., illustr.

Roels, O. A., Laurence, S., Farmer, M. W. & Emelryck, L. van (1976). Organic production potential of artificial upwelling marine culture. Pp. 69–81 in *Proceedings UNITAR Seminar on Microbial Energy Conversion*, Göttingen, FRG, 4–8 October 1976 (Eds H. G. Schlegel & J. Barnea). Erich Goltze, Göttingen, FRG: 643 pp.

Rose, D. J. (1976). The prospect for fusion. *Technology Review*, December, pp. 21–43.

Sandscheper, G. (1975). Sun shines on family of four. *New Scientist*, 14 August, pp. 382–3.

Saverny, C. W. & Cruzan, D. C. (1972). Methane recovery from chicken manure digestion. *Journal of the Water Pollution Control Federation*, **44**, pp. 2349–54.

Schneider, S. H. (1974). Climate modeling. *Review of Geophysics and Space Physics*, **12**, pp. 447–93.

Schneider, S. H. & Dennett, R. D. (1975). Climatic barriers to long-term energy growth. *Ambio*, **4**, pp. 65–74.

Seguier, F. (1976). Energie: on cherche des hommes de paille. *La Recherche*, **7**, pp. 672–4.

Shaw, T. L. (1975). Tidal power and the environment. *New Scientist*, 23 October, pp. 202–6.

Shaw, T. L. (1976). Tidal power: closing the gap. *Water Power and Dam Construction*, May, pp. 24–8.

Simmons, D. M. (1975). Windpower. In *Energy Technology Review* No. 6. Noyes Data Co., London, England: vi + 300 pp., illustr. [not available for checking].

Sørensen, B. (1975). Energy and resources. *Science*, **189**, pp. 255–60.

Spurgeon, D. (1973). Natural power for the Third World. *New Scientist*, 6 December, pp. 694–7.

TARNIZHEVSKY, B. V. & SMIRNOVA, A. N. (1974). [Mentioned in a paper entitled 'Synthesized Soviet Solar System'.] *New Scientist*, 22 May 1975, p. 446.

UNECE (UNITED NATIONS ECONOMIC COMMISSION FOR EUROPE) (1975). *Environmental Aspects of Energy Production and Use, With Particular Reference to New Technologies*. United Nations, Geneva, document ENV/R.43: 42 pp., 2 annexes.

UNECE (UNITED NATIONS ECONOMIC COMMISSION FOR EUROPE) (1976*a*). *Increased Energy Economy and Efficiency in the ECE Region*. United Nations, New York, document E.76.II.E: vi + 102 pp.

UNECE (UNITED NATIONS ECONOMIC COMMISSION FOR EUROPE) (1976*b*). *Environmental Aspects of Energy Production and Use, With Particular Reference to New Technologies*. United Nations, Geneva, document ENV/R.43/Add.l: 10 pp., 1 annex.

UNESCAP (UNITED NATIONS ECONOMIC AND SOCIAL COMMISSION FOR ASIA AND THE PACIFIC) (1975). *Report on the Workshop on Bio-gas Technology and Utilization*. United Nations, Bangkok, document E/CN.11/HT/L.18: 56 pp.

WADE, N. (1974). Windmills: the resurrection of an ancient technology. *Science*, **184**, pp. 1055–8.

WIDGER, W. K. & DERRICKSON, R. A. (1976). New England wind power... coastal or mountain. *Power Engineering*, December, pp. 43–7.

WILCOX, D. (1973). Fuel from city trash. *Environment*, **15**, pp. 36–42.

WMO (WORLD METEOROLOGICAL ORGANIZATION) (1976*a*). The World Meteorological Organization's statement on climatic change. *Environmental Conservation*, **3**(3), pp. 227–30.

WMO (WORLD METEOROLOGICAL ORGANIZATION) (1976*b*). WMO statement on modification of the ozone layer due to human activities. *Environmental Conservation*, **3**(1), pp. 68–70.

WOFSY, S. C., MCELROY, M. B. & DAKSZE, N. (1975). Freon consumption: implications for atmospheric ozone. *Science*, **187**, pp. 535–7.

WOLF, M. (1975). Cost goals for silicon solar arrays for large-scale terrestrial applications—update 1974. *Energy Conservation*, **14**, pp. 49–60.

ZENER, C. & FETKOVICH, J. (1975). Foam solar power-plant. *Science*, **189**, pp. 294–5.

Discussion (Session 11)

Fuller (Chairman, introducing)

Some 50 years ago [,when I was about the age I suppose of many of the younger ones among you], I began to feel that it would be wise to develop some definition of environment, and I finally invented a little poem:

Environment to each must be
All that is, excepting me.
Universe in turn must be
All that isn't me—and me.

Thus the only difference between universe and environment is, for each of us, 'me'—the observer, or doer. Yet the environment which most people think of is I believe quite local, being little concerned with the metaphysical and physical universe surrounding us and lying beyond.

Being born eight years before the Wright Brothers flew, I was brought up with the dictum of the times that it was inherently impossible for humans to fly. Yet when my daughter was in her baby carriage in Lincoln Park in Chicago, a little cloth-wing aeroplane flew by, and by the time her daughter was born in New York, the sky seemed to be getting ever-fuller of larger and faster ones. Consequently my granddaughter, in her parents' apartment on the third floor of an old wooden house on the flight-path of planes [outbound from] La Guardia airport, heard everybody speak of airplanes all the time and so it was not surprising that her first word, instead of being 'mama' or 'dada', was 'airp'. She probably saw thousands of airplanes before she ever saw a bird, let alone a farmyard animal.

Thus the younger world sees things very differently from the way the older world sees things, and possibly the greatest hope for us at present, because we are in a bad dilemma on our old planet, lies in the fact that our young world is being born free from the conditioned—or misconditioned—reflexes of the others.

When I began to try to [size up] the environment 50 years ago, I liked to be able to see things in a large way—to realize, just looking at our planet, that the air extended around the Earth, that the wind did not stay in one place, that the water did not stay in one place, and that the information did not stay in any one place but went right around the world. So I came to think in terms of the world itself [and as one whole]. Yet if we look at a globe, we may think we see a half of it but we do not read the tangential: we find if we make a study that we can only really read one-quarter of the Earth's surface [at any one time].

Finding that all the known methods of projection, such as Mercator's, were very much distorted, I nevertheless wanted to see the whole [world map] at once—if possible without breaks in the great continental contours. So I [discovered] a mathematical method of projecting the Earth's surface in 20 equilateral [and other] triangles, employing the [icosahedron which has] the largest number of identical triangles into which we can divide a sphere [with corner angles of 72°, and holding to what is called a uniform boundary scale. By symmetrical shrinkage or projection], I arrived at a method of unwrapping the world in a way in which there was no [visible] distortion of the relative shape or size of any of the parts. But it took me two years to find the 12 vortexes in the main oceans of the world, which very excitingly turned out to be close to the earthquake centres. Of the 20 triangles, one, in [Southeast] Asia, bears 52% of all humanity.

Human beings are born naked—helpless and ignorant but given curiosity and the drive to learn by trial and error. We have to learn that our muscle is nothing whereas our mind is everything—that our intellect has the capability to discover principles, to understand [them, and to put them to work]. But although, at present, muscle is in control of human affairs, we come to beautiful free conferences like this and [plan to put Man's collective mind properly into control].

This is a very extraordinary meeting we are having, in which I find myself in the privileged position [practically on arrival, owing to illness of the intended chairman], of introducing the keynote speaker, Dr Claude Ducret [, of Switzerland and the United Nations], whom I have not met before. By reading his biography I could seemingly be expert about him, but I prefer to manifest forthrightness in every way and merely admit that I have just met him but he is obviously a very charming man. [Before calling on him to speak, I would like to name two panellists who are already known to you: Dr Maria Buchinger, from Argentina, and Dr Pierre Laconte, from Belgium, and to introduce the two others on the platform—Dr Dieter G. Altenpohl, Technical Director of Alusuisse, and Dr Torgny Schütt, of the Energy Research and Development Commission of the Swedish Ministry of Industry.]

Ducret (Co-Keynoter, presenting paper)

First I would like to convey to you the regrets and disappointment of Dr Bishop, with whom I had the pleasure of preparing this paper, that he is unable to attend our Conference [: so I have to do my best alone].

Dealing with the huge subject of the 'Environmental Aspects of Energy Alternatives' in a limited time [is a daunting task that seems best attempted in the form of] a summary of alternative sources of energy—alternative, that is, to the conventional ones currently in use in those countries of the world which nowadays consume the major proportion of energy. I will further restrict my presentation to the alternative energy sources which, in view of the present level of technological development, [seem likely] to play a significant role in the future, provided that appropriate policies are adopted. As conventional sources of energy I consider coal, gas, oil (including oil-shale and tar-sands), hydropower, and nuclear fission-power, and as alternative sources of energy that I will review briefly there are solar, wind, geothermal, ocean temperature gradients, tidal power, biofuels and nuclear fusion.

I would like now to [comment] on a few issues. What was called the energy crisis was not an energy crisis: there was no real shortage either in supply or in reserves. The 'energy crisis' was a political crisis and shortly thereafter became an economic crisis—the end of the era of cheap energy. Reserves are not unlimited. In the reasonably foreseeable future human society will be confronted with the following elements: the pattern of conventional reserves available for the next century is constituted of environmentally unfavourable sources of energy: that is, coal, shale-oil and tar-sands. Their widespread use would lead [to further build-up of carbon dioxide in the atmosphere and hence] to problems of the 'greenhouse effect', and so the obvious trend is to say that fission power might be the good solution. Yet nuclear power is far, I think, from being without technical, economic and ethical problems. In addition it does not represent in any way the ultimate solution to the problem of the thermal situation of the Earth. In fact if nowadays we were suddenly to replace, by nuclear power, all [energy resulting from the] consumption of fossil fuels, we would dramatically increase the amount of heat—the heat load—to the environment because of the low efficiency of power plants. In addition, the resources available for nuclear power are not unlimited either, so it is not an ultimate solution—maybe it could be a long-range or short-range or mid-term solution, but not the ultimate solution.

The ultimate solution [to me would be] using ultimate sources of energy, although all of them are not necessarily suitable for that. I would say that, by definition, large-scale utilization of fusion power and geothermal power are not the ultimate solution, because the two sources of energy are not included in the present energy-balance of the world, and consequently they could lead, if very widely used, to climatic disturbances. Space solar power, and [transferring] this energy from space to the Earth, leads to exactly the same problem. Tidal power is attractive, but it is hopeless because its potential is so very limited. Consequently the energy sources which could be widely used in the future are: ground solar* applications (particularly small-scale ones), wind energy*, ocean temperature gradients and biofuels. This does not mean that geothermal energy or fusion power should not be used at all, but that their ultimate utilization has to be carefully maintained within the acceptance capacity of the climatic environment. Here I need only refer you to the presentation by Professor Flohn [on our first day].

The energy crisis, consequently, is ahead of us. It is not the accident which took place a few years ago; and by the way I think we should be quite grateful to OPEC countries for what they did at that time, because [it made us] very conscious of oil/energy utilization.

The real energy crisis will be [according to my understanding], to resist the temptation of choosing easy, short-term but irrevocable choices—energy choices—such as, in my view, fission power. In fact I believe that flexibility in energy choices should be the fundamental elements of any energy strategy. We may be wrong in our strategy, and we must keep the flexibility for the generations to come to change the orientation of this energy strategy. We should avoid giving future generations the burden of our own mistakes, and in fact we should avoid doing to future generations similar things to those which we did to the Third World—as was pointed out yesterday.

The real energy crisis [for me] will be to choose and develop new energy sources even if they would not be commercially competitive with conventional sources of energy. The energy crisis might also be a competition between raw materials for biofuels and production of food, and to change our angle of vision from the producer point of view to the consumer point of view. Indeed, considerable energy savings can be made by applying the proper energy sources to the proper utilizations. It could in fact be argued, although I do not want to go too far into this now, that the major source of energy in the so-called developed world is energy savings.

I would like to close my presentation with a few remarks about the economic system, energy and the environment. The laws of economy are based on mechanics. In mechanics, energy has a limited role of getting material circulating in the system. In economics there are therefore, according to the ease of their production, cheap and expensive energies. The environment, on the contrary, obeys the laws of thermodynamics. In thermodynamics, as in the environment, there is no cheap or expensive energy: there are high-grade and low-grade energy.

Forests have life-cycles on the scale of centuries, Man on the scale of decades, birds on that of years, Bacteria on the scale of hours, and biochemical processes on one of seconds. In reality in the environment, what was true in the past will not necessarily prove true in the future. Everything evolves and changes. In a thermodynamical economy, the tropical forests would not have been cut because their life-cycles would have a very high value. Whales would not have been nearly exterminated because their capacity of harvesting Krill for us efficiently would have been taken into account and computed. There would be no air–water pollution because waste would not be waste but raw material. Air and clean water,

* Here I would like to add two items that pressing duties elsewhere prevented me from being present to mention in this Session: salt-gradient solar ponds, as described by Professor Carl. E. Nielsen in *Environmental Conservation* (Vol. 2, No. 4, pp. 289–92, Winter 1975), and wave power (from wind), as it is being developed *inter alia* by Sir Christopher Cockerell, inventor of the hovercraft.—Ed.

because of entropy, would have a very high value, and so on. And we would not have spent a lot of effort and money in looking for energy sources which are apparently inappropriate solutions. I believe that we have made a great many mistakes. We are in our childhood in this field; but the major thing is to recognize the mistakes that have been made and to choose the right direction [henceforth]. In any case I believe that one of the major things we should keep in our minds is that the eyes of the future generations are upon us. [Applause.]

Fuller (Chairman)

Based on a scenario of Nature producing petroleum, an eminent geologist once computed that the amount of energy expended over the period of time which it took Nature to make that petroleum, should presuppose a cost of over $1m a gallon [3.78 litres to the US gallon]. But this is a cosmic accounting, [whereas on Earth all] we do is stick pipes in the ground and the oil is for free. Then a large part of it is used for transporting people to work, without producing any life-support. Getting to some of this energy economics is really extremely important, and I hope we shall keep [our sights] on that level.

Buchinger (Panellist)

Referring to your division of the world into western Europe, eastern Europe, North America, USSR, and 'others', I am speaking now for the 'others'—or, rather, for Argentina, as you mentioned a possible project for use of some of our tidal energy. This project was first [discussed] in 1919 and is based on the fact that the tides in the Golfo San José (780 km^2) and Golfo Nueve (3,400 km^2) are not simultaneous. Connecting the two bays through the narrow Valdés Peninsula would make an abundant energy-supply available, but at the same time it would produce an ecodisaster. Two—until now completely separated—aquatic ecosystems would be connected, with unforeseeable consequences to the rich biota. For this reason the project has been vetoed from the beginning. However, there are other projects under way to use the tidal energy of Patagonia. The Golfo San José could produce 8,900 million kW/h per year, the Golfo San Julian could produce 423 million kW/h per year—and so on. The interconnection of power-plants using this energy could supply permanent energy for the needs of Argentina and Chile, producing five times as much energy as is calculated to be needed [locally and over a wide area] for the year 2000.

Another energy source of Patagonia is wind power. Between the 45th and 48th parallels of latitude, the average wind-intensity is 46 km per hour. The First Latin American Solar Energy Congress, held recently in San Miguel, Province of Buenos Aires, should also be mentioned here, as 14 institutes presented their experiences and working plans. In the last few years the domestic use of solar energy as well as its application in industry has been constantly growing. [Recently] one single firm sold 3,000 solar tanks in a month. I would like to stress the point that the solar appliances are getting less expensive, as the mass-production of solar artefacts is correlated with the lowering of their prices.

Here again we are reminded that the words 'ecology' and 'economy' derive from the same base—the Greek word for a house or place to live. Ecology is the science or knowledge of 'okiós' or home, and therefore good economy has to be based on sound ecological practices.

Laconte (Panellist)

I [have little to add to the excellent survey given by Dr Ducret] but would like to contribute a remark to bring into perspective the overall view presented by Chairman Buckminster Fuller in the economic cosmic accounting which he suggests.

Three important reports have recently appeared on the supply and demand perspectives for energy in the years to come, shedding a little more light on the variables to be taken into account in deciding what is worth while. After reading the report by Professor [Carroll] Wilson of MIT, I foresee a short-term advantage for our countries, i.e. countries around the North Sea, in the next 10 years. It is the only part of the world which will have an energy surplus, thanks to North Sea oil (although this is a very short-term resource). I believe advantage should be taken of this to capitalize on ways of having alternative sources of energy for the coming years of distress, say after 1985 [or 1990]. This suggestion has already been made and it is, of course, of a political nature.

My last point is to emphasize again that problems encountered with regard to energy consumption are not only technical but also psychological and political. For example, when we were driving around this beautiful country of Iceland yesterday, the [question] was raised as to why there were so few trees—to which someone replied, 'Well, before planting trees in the ground, you must plant trees in people's minds'. People must be convinced that Nature is important. Educating people on the need to control energy consumption is equally important as, or perhaps more important than, the technical means of energy-saving, so well explained by Dr Ducret. On the day when people are proud to have a car that has run 100,000 miles rather than a new model, a further step will have been taken in the right direction. This involves the whole process of education to use the mind rather than the muscles, as Mr Chairman said, to save our planet from the energy crisis.

Fuller (Chairman)

In relation to what you say, in the United States—which of course has a very small percentage of the human population but is out in front in industrialization—the overall energy efficiency is only 5%, that is, out of every 100 units of energy we [practically] throw away 95. At all times in North America there are [a calculated] 2 million cars standing in front of red lights with their engine going, and over 200 million [animals] jumping up and down or going nowhere, while having the Arabs and Israelis fighting each other adds to the nonsense, [when already] we are up against the environmental hazard of all those who are making incredible amounts of money out of petroleum—not to mention the amount of propaganda they can handle, and the number of human beings whose jobs are related to that sphere. So getting human beings really to care about their environment—apart from many young people who are deeply concerned—is difficult. It is particularly hard to get the power structures, the bureaucracies, in any way educated so that they really care.

Altenpohl (Panellist)

I have been dealing with energy from the viewpoint of practitioners for a number of years and have come to realize that a reassessment in a triangle is necessary. The corners of the triangle are the public, government and industry, with industrial activities and related matters. In the centre of the triangle are some honest brokers, such as academia. I believe that this group here should mostly belong in the centre of the triangle, being not committed to any corner of this very dramatic assessment.

If we consider what energy does for us—the locomotives, the cars running around, the heating of our homes, heating of the schools children go to, and of hospitals—everything averages 2 kW per hour per person around the clock around the year. But the United States—the classical villain—uses 10 kW per hour per person. The industrial nations average 5 kW, and India only 0.2.

In conclusion I would like to remind you that we cannot tell the world's 4,000 million people to use [less] energy; they will [want to] use at least the same as their brothers and sisters did. They will ask for more, and it is to be hoped that more energy can be provided—but how?

Schütt (Panellist)

Time is running out and I have to restrict myself to some special points, even though I feel that a lot more should be said to clarify the energy question. The energy shortage is a typical example where each country at the moment is trying to solve its own problems in collaboration or in competition with others, so that it can serve as a fundamental example of how to tackle problems which will come in the future in food and other areas.

The Sun produces such an enormous amount of energy [at the Earth's surface] that it corresponds to [about 100] million normal nuclear reactors; unfortunately this energy is diluted all over our Earth's surface, and varies with time and geography. It is easy from this figure to understand that we produce a very small amount of energy compared with the sunshine—the radiation energy heating the Earth every year, and indeed every moment. About 50% of this energy reaches the Earth's surface, the rest being reflected in the atmosphere. At present, if you divide the energy produced and used in the world by the population, you will end up with the figure of around 1.5 kW for each person. This means that we have an energy amplification factor of about 15, because the human being has a rest-energy need of about 100 W, and, when working, [a need of] 200 W. In the near future we [expect to] use approximately 4 kW per person.

In Sweden, for instance—and this goes for many industrialized countries—we are using 40% of the energy to heat our homes and buildings, 45% for industry and 15% for transportation. For heating the buildings and to [provide energy for] industry, we use mainly oil at the moment, because it is such a cheap resource. In transportation, oil is used to the extent of almost 100%. By comparing the paper presented to us today, and the recent report from the 'Workshop on Alternative Energy Strategy (WAES)', you will see that there are quite different opinions about how long into the future we will have access to cheap oil. The WAES report predicts 10–15 years, after which there will be a lack of oil. At the moment we are not sure how to support ourselves with energy when we run out of oil. Of course coal is a possibility, but the environmental consequences of using coal are nothing to wish for. Even if every nation is mainly concerned with its own problem, it has to take its neighbours and future generations into consideration. But how to do this is quite a problem.

You could argue that the industrial world should use nuclear energy if possible, so as to spare the oil for the developing countries, because oil is much easier to handle on a small scale than nuclear energy, and in the developing countries it is perhaps cheaper. If we run out of oil in 10–15 years we will really be in trouble, because we cannot do much [to safeguard ourselves] in such a short period. To project and produce a new technical system takes 5–7 years, and to introduce it on the market takes another 10 years. If you would like to reduce your dependence on imported oil by building a new import harbour for coal, for instance, it will take at least 10 years, proceeding in a normal peace-time way.

I see only one possibility to solve these energy questions in the short period of time [available now] that we realize we are really in trouble. But I do not think people at the moment are feeling this way. Many of us think that technology and our future knowledge will save us from the problem, but this you can doubt.

At the moment 10,000 of the lakes, for instance, are spoiled in Sweden, because of sulphur dioxide emission from the use of [oil]. One of our future alternatives, when oil is lacking, is to produce and use biomass energy from highly productive short-rotation forestry. This means that we have to burn a fuel which probably will have severe environmental impacts, both in production and in use. We will be in a position where we have to decide if we shall accept these [impacts] because of the advantages we get from the energy produced, while at the same time suffering from the environmental consequences. How to evaluate and balance advantages against drawbacks will be our main concern.

Fuller (Chairman)

Our informative discussions relating to sources of energy have reflected the specialization of those taking part, but we have not heard much about the energy uses and how they may really be reduced. I think this is very important in relation to what was said about India and her low energy consumption. The [extremes are exemplified by] India, where you do not have heating, and North America, which is being inefficient in its uses of energy. Now [let us have] statements from the audience.

Savin

I found Dr Ducret's paper very stimulating in drawing our attention to the complexity of the problems confronting us, and the quantification provided by Dr Schütt and others was quite refreshing in reminding us of the very real [severity of these] problems.

To take first the [apparent] discrepancy between the alternative energy studies' report to which Dr Schütt drew our attention and the keynote paper presented by Dr Ducret, they are in fact saying the same thing. The confusion arose as, in the paper, the oil and oil-shale reserves are aggregated, [whereas] taking known liquid fuel reserves alone would make the figure of 15 years correct. This refers to discovered [and proved] reserves and does not take into account any hoped-for discovery of further reserves.

Problems then confronting us with these orthodox fuels will be the problems of setting them to work. We have to discover and develop more oil reserves. We have to extract coal. The problem, mentioned by our Swedish colleague, of lead-time to develop even existing known reserves of coal, and the problems of converting oil-shale into economic materials, have still got to be solved, and although the hydrocarbons are there, we still have not got the economically viable technology to enable us to do this. So I think my main point would be not to accept too readily Dr Ducret's suggestion to hasten slowly. I would tend to agree more with our Swedish colleague—it is later than we think, and in this connection the reduction in use of energy is perhaps the biggest immediate contribution that we can make.

In this connection the experience of our United States friends is not encouraging. The United States consumes something of the order of one-third of the world's currently produced oil. It imports 45% of its requirements at the present time. President Carter recognizes this, and puts it to the American people. From the radio this morning it appears that a committee of the Congress thought perhaps President Carter's [advisers] had been too optimistic. And yet when a modest proposal to reduce the amount of gasoline consumed in the United States is being put to the American legislature, it is advised that the proposal is unlikely to be accepted. Therefore the President's policy has on the one hand been suggested as needing strengthening, whilst on the other hand the first modest steps to try and reduce energy consumption apparently have been shelved at least for the time being.

Tan

Actually, mine is more a question than a comment. We were told by the Food and Agriculture Organization [of the United Nations] in the 1950s that 80% of the power used in the developing world came from draught animals, and I think that until recently in Malaysia half our power was derived from animals. Yet I notice [from the paper that its authors do] not consider animal power as an alternative source of energy, obviating the use of fossil fuels, despite the fact that many countries are still largely dependent on [animals for traction and motive energy]. We are now rapidly replacing a lot of our animal power with little tractors, and for these gadgets to work we have to import at least some of our fuel. So I would like to address this question to Dr Altenpohl, for another viewpoint: how would you consider that developing countries can solve their energy problems in terms of this particular resource which was and still is abundantly available, but which is not being encouraged to be developed by scientists from the developed world, who are invariably approached for advice on development in tropical countries?

Altenpohl (Panellist)

Perhaps I can answer this very quickly by referring to Dr Schumacher, who like Professor Roger Revelle, of the University of California at San Diego, has travelled extensively in India, Pakistan and other [least developed] countries, looking into the intermediate technology of how to provide energy (for example to pump up water, as it is still done mostly by cows). This is a country-by-country approach to determine whether a particular country can afford to replace animal power. The Egyptians, for instance, right now think about drawing current from the Aswan plant surplus electricity to relieve the cows from the water pumping—so pumping more water all day long and around the clock. This has to be looked into, country by country, also to assemble statistics of how much animal power is to be put into total energy accounting.

What we might call the grey energy market, of people burning wood and using animal power, nobody knows about in detail. We have only very rough estimates. But I do not think it is good advice to the developing countries to use more animal power, because these animals have other values.

Ducret (Co-Keynoter)

Just a short comment here: [perhaps in our paper we did not sufficiently point] out the inadequacy of the majority of energy statistics—they include only commercialized energy. You are referring to non-commercialized energy, which in some countries indeed plays a major role by providing the people with the major portion of the energy that they really 'consume'. Energy statistics are, however, interested only in commercialized energy. There is a revolution to carry on in this field; I fully agree with you.

Miller

I wish to join those who applaud [this] paper and appreciate especially the diversity which it gives to energy uses and possibilities of action for the future in exploiting the various sources of energy towards satisfying our needs.

The Chairman spoke of the supremacy or the satisfactions of the mind over muscle. We have concern in developing, for Man, various kinds of energy from various sources. We should look as well to the consequences. This brings us back to what we covered the other day in the meteorological discussions on the heat budget of the Earth, where we were considering carbon-releasing uses and their consequences to the heat budget of Earth and the build-up of CO_2. Can we not intensify our use of such energy-sources as wind, tides and gravity, which have no by-product or other chemical consequences? If only we had established the baselines years ago, so that we could measure against them what we are doing to the heat budget of the Earth, I think we would have some very valuable guidelines for today. Unfortunately we only have eyeball experience; hard data are lacking. We know only the little that is most evident—smog, inversion and heat domes. What the effect on human beings and other organisms may be in the environment of the future is the matter we ought to pay attention to.

What might this mean to Man? If we use wind energy for transportation, as was done traditionally, with Man acting as some sort of a sailor, his vessels might be slower than methods of today's shipping, and yet there would be [some recompense and especially] some satisfactions. Wind energy could be an accessory, just as turbine engines became an accessory to sail in the early days.

If we look back to the point of muscle *versus* mind, we should recognize that there are satisfactions to the mind which come from work of the muscle. There are satisfactions also which we could look at in regard to transport by the various users. The satisfaction of an engineer or a mechanic listening to the hum of an efficient engine driving a ship through the water can be tremendous. He has real feeling of harmony and satisfaction of power, with the sense of completeness in that experience. Likewise the man who is rowing a boat can have a similar feeling of satisfaction; and also we know that there is quite a culture throughout the world of people who are sailing vessels, and sailing boats of all kinds, in their time off—as a recreational measure bringing great satisfaction.

I am interested in what Dr Laconte said about planting the tree first in the mind of Man. This seems [to be carrying further] where we were last night, searching for a way to plant in the minds of the world's population the considerations that have to be faced by every citizen if we are to reach solutions. So as an educator I am getting back to the point that we need to reach the minds of men and women—to show what satisfactions there are in exploiting the alternative uses of energy, including the combination of muscle with mind. [And there is joyful satisfaction in Nature and in wilderness.]

Thorhaug

I am rather disturbed that we have just had a discussion on environmental aspects of energy applications and I have been given the feeling by speakers that there are many of these new alternative energy sources which have little or no environmental impact. That is not so. There is no energy resource which does not have environmental impact either at the source of obtaining the energy, as with coal mining, or at the place where you are producing [or using] the energy, as with thermal pollution. It may be that we have not yet developed solar, wind or geothermal energy on a large enough scale to see what type of environmental impact there will be, but there is certain to be some such impact.

The second thing I would like to say is that organisms in the environment can be thought of as topological surfaces constructed of many kinds of parameters. When the organism is in the middle of these surfaces it is in a fairly balanced kind of place, where much change can occur with small effects. But when the edge of the surface is approached, it is near the brink of disaster—whether we are concerned with an organism, a population, or an ecosystem—and it can easily go over the edge. What we need, which I have not heard very much about during these entire meetings, are alternative solutions, so that we may have the needed energy, but also stay within boundaries in the environment. We know from many years' study that there are environmental effects of energy utilization. There is thermal pollution; there are radiation effects on nuclear power-plants, and for instance in Florida we have leakages of radioactivity. On the other hand, we do require energy, and what we need now are environmentalists who will sit down with good statistical characterizations of parameters' effects on organisms and discuss their needs with the engineers before coming up with solutions. We need better alternatives advanced by people who have been trained to find rational solutions and make trade-offs in energy.

Dr Altenpohl said that if environmentalists protest we would be deadlocked, because we only have a limited amount of time. But environmentalists must protest, though in goodly numbers and with alternatives, so that we can give some help in this area. Technocrats can take a little here and give a little there. We have got to come to some solutions—some environmentally acceptable solutions—and we must have trained people to make these decisions. We all get up and stand on our own soapbox here at this [Conference] and then go back home having listened only to ourselves. But what we need are people who can bridge the gap between energy needs and environmental needs, with numbers to show us how and when we can alter Nature's topological surface without going over the brink of disaster. We especially need individuals who will take the responsibility of rationally balancing the environment that we need with the energy that we need.

Schütt (Panellist)

It is quite true that problems exist on different levels. We have the world-wide problem of carbon dioxide having effects on our future climate, but we also have regional problems today from sulphur dioxide. In Scandinavia we have the acid precipitation, which not only destroys a large number of lakes but also affects the growth of the spruce forests by acting on the soil's flora and fauna. We have different environmental legal systems in the Scandinavian countries [from those which obtain] in France, West Germany and the United Kingdom. For example, most of the sulphur dioxide which makes the rain acidic in Scandinavia is let out to the atmosphere outside Scandinavia, so that international agreements are required to control this environmental problem. Establishment of international legal agreements for sulphur dioxide emission, for dumping of oil or radioactive wastes at sea, etc., is fraught with political problems, which have to be solved if we are to solve the environmental ones.

Fuller (Chairman, concluding Session)

At the present time our limited planet Earth is in a very extraordinary condition—showing [exponential growth despite] its finite possibilities. Human beings were originally deployed remotely from one another and learned to manage under special environmental conditions until, quite unexpectedly, we have suddenly all been integrated [by modern developments]. Now we have the extraordinary condition when we have one ship with 150 sovereign admirals, and we have the port side trying to sink the starboard and the stern trying to break off; so we are in a real predicament. Yet nobody [seems to have any real authority over any one country, let alone over the total planet].

Our spaceship has resources and accumulated knowledge, and a prospect that, with these resources and our know-how, within 10 years we could have all humanity enjoying a higher energy income than was enjoyed exclusively by the profligate North Americans in 1972, while also bringing humanity to a higher standard of living. All this could be done within 10 years, while concurrently phasing out all use of fossil fuels and atomic energies. We could live entirely on our energy income. But the political incentives and the monetary incentives are not in that direction, though technically and incontrovertibly we now know how it could be done, and with a large factor of safety. We did not know before that we had this option. Yet sadly our great political systems, all the great ideologists, are still working on the assumption of a fundamental inadequacy of life-support.

We are not on this planet for the enlightenment of human beings, and as such could not be more inconsequential. But we seem destined to have a function in the universe, and the beautiful function which we have rests in our mind and not in our muscle. So I think we are in a final examination of what is really critical—and your concern here could not be more important—probably in a 10-years' final examination of whether or not humans are going to stay on this planet. We are in revolution; if it turns bloody it will be all over for everybody. But if we use our minds, we may win through. We have the option.

I thank you for a stimulating discussion [and declare this Session closed].

Session 12

INDUSTRIAL ALTERNATIVES: NON-WASTE TECHNOLOGY

Keynote Paper

Industrial Alternatives: Non-waste Technology

by

MICHAEL G. ROYSTON

Faculty Member, Environmental Management, Centre d'Etudes Industrielles, 4 Chemin de Conches, 1231 Conches, Geneva, Switzerland.

Introduction

Waste in production and in products themselves leads to depletion of resources on the one hand and pollution of the environment of Man on the other. Waste is not only non-aesthetic, it is also uneconomic. The wisdom of non-waste is seen in the old Icelandic concept of 'thryfa', from which, *via* Old Norse, come the English 'thrive'—to grow vigorously—and 'thrift', the saving or avoidance of waste.

Western society seems at last to be turning its back on Vance Packard's 'Wastemakers' and returning to the necessarily non-waste ways of its forebears and of the rest of the world, which is the only solid basis for sustainable development.

The International Conference on Non-waste Technology, organized in Paris in December 1976 by the United Nations Economic Commission for Europe, marked a turning-point in this respect, summed up perhaps by the 'Pollution Prevention Pays' policy of the 3M Company (cf. Ling, 1976), which enabled them to apply their theory that

$$\text{Pollution} + \text{Technology} = \text{Potential resources}$$

and thus not only to eliminate 70,000 tonnes of air-pollutants and some 2,000 million litres of waste waters, but also to save $10m as a result. Non-waste makes both economic and ecological sense!

Waste Characteristics of Technological Systems

Figure 1 shows different technological systems operating with different degrees of 'non-waste', based on the classification of Ananichev (1974).

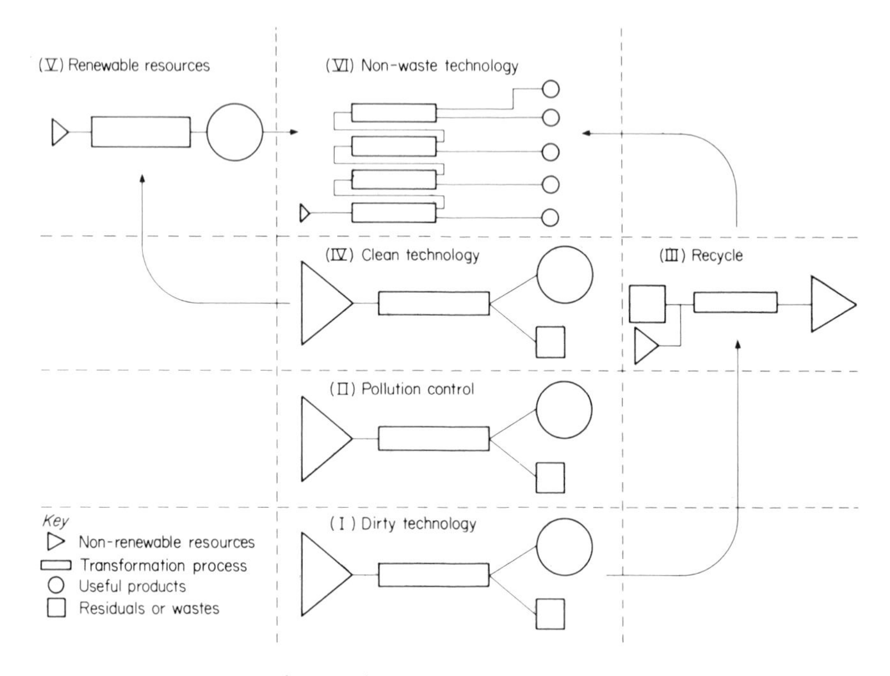

Fig. 1. The different technologies.

First (I), comes the '*Dirty Technology*' of the past, in which the residuals (such as emissions, degraded air, water, land, biota, disease, etc.) may be more significant than the useful products—particularly when the resource input is counted. In such a situation real 'progress' is difficult to justify in terms of desirable development, as costs outweigh benefits. It becomes a *negative-sum game*.

Secondly (II), there are the current technologies using '*Pollution Control*' as an added-on system. Pollution is reduced—and, incidentally, GNP, employment, and sales rise, owing to this new economic activity—but more resources are needed to construct and run the new pollution-control equipment. So, from a global viewpoint, some resources are saved from pollution while others are used up. Here we have a *zero-sum game*.

The obvious approach is to '*Recycle*' (III) the residuals rather than destroy them with expensive pollution-control equipment. Now, by looking at the total system of resource consumption, residuals handling, etc., we can see that it becomes worth while to recover a lot more than the classical scrap-iron, copper, lead or gold. Individuals feel this urge to recycle, as evidenced by the voluntary systems that are springing up very widely in Sweden, the Netherlands, Germany, France, Switzerland, the

UK, etc.—and all of them requiring considerable personal inconvenience to segregate glass, paper and metals at the source before they get mixed up in the garbage system or 'urban ore' (as the US Bureau of Mines calls it).

Nations are mounting programmes against waste—e.g. the Department of the Environment (1974) 'War on Waste' in the UK and the Ministère de la Qualité de la Vie (1974) 'La Lutte contre le Gaspillage' in France. At the industrial level, 'Waste Exchanges' have been set up in France, Belgium, Germany, the Nordic countries, etc., which consist of a computerized information system putting those with available industrial wastes in touch with those who could use such wastes as raw material—e.g. arsenical wastes from non-ferrous metallurgical processes that can be used for paints, agricultural chemicals, etc. As reported by Hammond (1975), the first such 'Abfallbörse', set up in Hamburg in 1973, successfully met 34% of all offers of waste and 60% of all demands for waste in its first year of operation.

The risk with recycling is that it can take more energy and equipment to recycle a waste than to win the corresponding material from virgin ore or raw materials. The challenge of recycling is to design the total system from the beginning—so that the residuals are easily converted into new resources, e.g. segregation of the copper wiring of automobiles, the two-piece one-metal can, or the returnable bottle. It is when recycling is designed with a total view of an integrated raw-material → product → residual → recycle system, that it becomes a true *non-waste technology*.

The alternative—and better solution—to recycling is not to produce the waste at all. Resource saving and pollution control are best dealt with on the drawing-board by making sure that the problem does not arise in the first place—through designing a '*Clean Technology*' (IV).

Improving Prospects

With a Clean Technology, the useful outputs outweigh the wastes and minimize on the inputs. It is truly a *positive-sum game*. The approach at the production level is to:

- use waste materials as inputs;
- reduce water-consumption by operating on a closed water-cycle, segregating drains, etc.;
- minimize energy consumption by selecting process conditions, using low-energy-consumption equipment, integrating heat and power (steam and electricity) systems, grouping plants and insulation so as to avoid energy losses, etc.;
- improve yields by correct selection of specifications and process conditions, by collecting spillage, by segregating and re-using wastes, by internal recycle, etc.;
- choose the right, clean, low-energy and long-life products and process routes so as to minimize waste.

This was how the 3M Company eliminated pollution and saved $10m.

The same company switched from a cotton herbicide whose production generated 12 kg of pollutants per kg of product to one which generated only 2 kg of pollutants and was cheaper to make! They also designed a unique inert gas-drying system which enabled them to recover hydrocarbon solvent, and they developed a mercury-free catalyst for resin manufacture as reported by Ling (1976). Dr Hans Gysin reports (pers. comm., 1977) that Ciba–Geigy took advantage of the recent recession to close down and overhaul all their chemical plants in Basel, cutting down on water-use and energy-use and so improving yields that, not only was pollution reduced considerably, but the overall economics of the operation was improved.

As Peterson (1977) has pointed out, Dow Corning now recover chlorine and hydrogen that were previously lost to the atmosphere from their silicone plant—a net saving of $900,000 per year. Hercules Powder have reduced their discharges to the Mississippi and have saved $250,000 in materials and water costs. A Goldkist poultry-plant cut water-use by 32%, wastes by 66%, and saves $2.33 for every $1 ultimately spent. Peterson reports the chairman of Hanes Dye and Finishing Company as saying 'cleaning up our stacks and neutralizing our liquids was expensive but, in balance, we have actually made money on our pollution control efforts—the Environmental Protection Agency has helped our bottom line'.

The recent ECE Conference cited earlier produced hundreds of examples from virtually all industrial sectors to show how Clean Technology can work and does pay. For example, from Canada, Rapson & Reeve (1976) report that a 'clean' closed-cycle bleached-kraft mill cost 5% more to build than a 'dirty' open-system mill, but costs 5% less than a mill fitted with

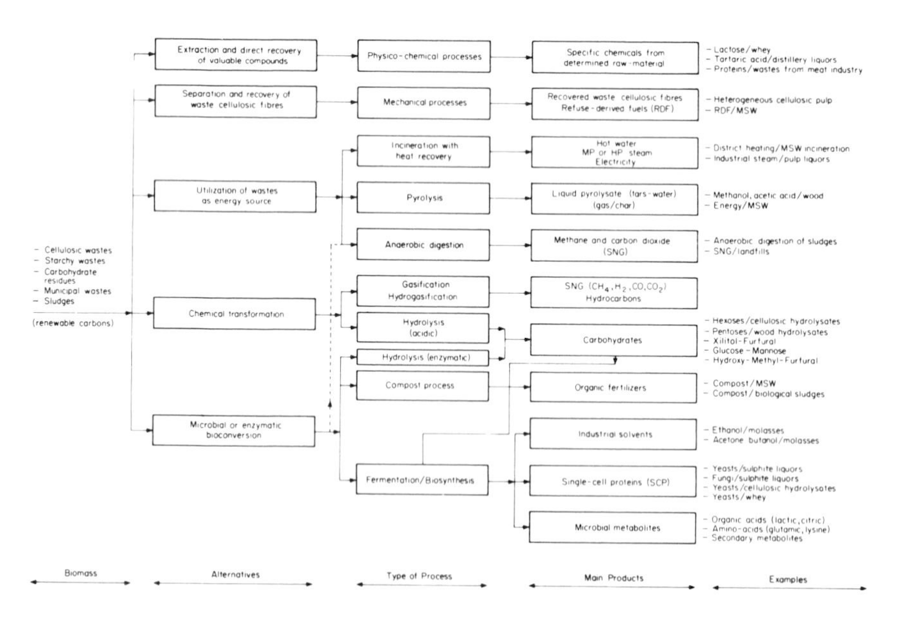

Fig. 2. Chart indicating technological alternatives for upgrading cellulosic wastes, carbohydrates and other residues from agricultural and food industries (from Baret, 1977).

add-on pollution control equipment. The operating cost of the clean mill is less than that of an existing 'dirty' mill. In the the Soviet Union, according to Torocheshinkov (1976), closed-system working, energy-saving, and water-saving, in a 1,360 tonnes/day ammonia plant, cut production costs to 40–45 roubles/tonne NH_3—one-third of those of the open system. The French Ministère de la Qualité de la Vie (1976) has more instances where, for example, recovery of hydrocarbons in an oil refinery, recovery of methionine in a food factory, recovery of protein from a yeast factory, and recovery of iron-dust from a steel works, each brought in extra revenue of millions of francs per year. One of the best examples is the Swiss cement industry which not only paid for its dust filters in less than three years (with the recovered dust) but increased productivity by 60% because of the better working conditions, as shown by Coskuner (1975).

However, even with the cleanest technology, you cannot get out more than you put in—unless you use '*Renewable Resources*' (V). Clearly, with some 100,000 million tonnes of dry-weight biomass being produced per year by photosynthesis, this natural process presents the world's main renewable resource for energy and raw materials—particularly when some plants, such as the Puerto Rican Napier Grass *(Pennisetum purpureum)*, also known as Elephant Grass, can yield up to 80 tonnes of dry-weight biomass per hectare per year (Shorrock, 1978). Figure 2, after Baret (1977), shows the general range of products which could be made from organic materials, or even from organic wastes. Figure 3, from Battelle Geneva Research Centre (1976), shows how, even in the specialized area of petrochemicals, most of the products could be derived from green plants—especially if they are like the giant Hawaiian Ipil-Ipil or Lepili (*Intsea bijuga*) which, according to Governor Leviste (1977), of the Philippines,

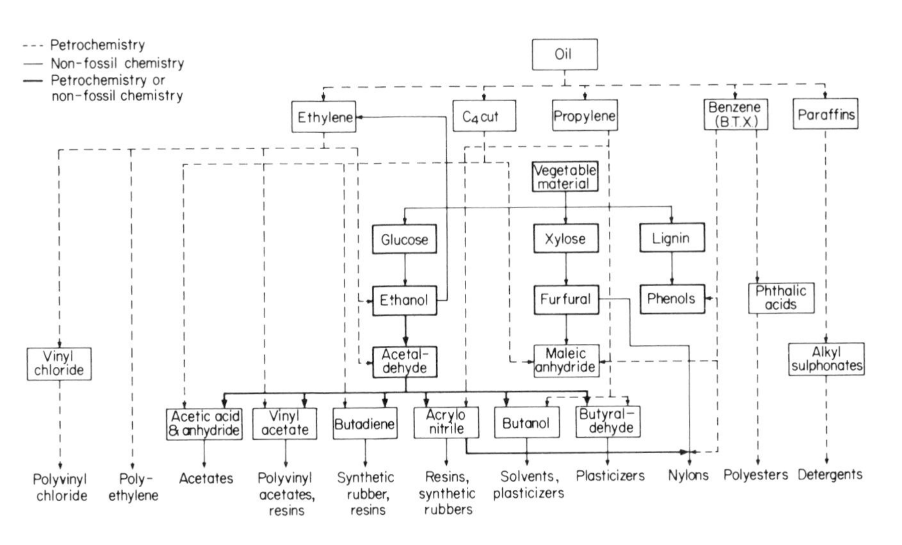

Fig. 3. Chart showing range of petrochemicals which could be replaced by wood by-products or other plant materials (from Battelle Geneva Research Centre, 1976).

grows to 15 m in height and a circumference of 60 cm in two years and is leguminous, adding combined nitrogen to the soil. Brazil is already planning to replace gasoline with ethanol derived from sugar-cane and manioc (cassava) within ten years, as Hawrylyshyn (1976) reports.

The prospect of producing from such renewable resources:

— fuels such as methanol, ethanol, methane, and hydrogen;
— fertilizers based on 'green manure', or composts;
— foods and feedstuffs based on fermentation biomass;
— detergents based on sugar;
— chemicals from the sugars, alcohols, aldehydes, ketones, organic acids, amino-acids, etc.;
— construction and packaging materials based on cellulose, lignin, and such secondary derivatives as nylon, resins, etc.

provides one of the major challenges for technology today. And it is a challenge which is finding a response, all over the world, in private and public laboratories in the industrialized as well as developing countries.

At another level, the challenge is not only to extract as many useful products as possible from the world's biomass; it is to manage deliberately that biomass so as not only to ensure a future supply of energy and raw materials, but also the vital water and food. These depend ultimately and totally on the way in which we manage the land and its covering vegetation, since they constitute a unified whole—as most countries have found when widespread deforestation has been followed by erosion, drying up of springs, loss of crop yield, silting of irrigation dams and channels, floods, desertification, and so on almost *ad infinitum*.

In other words, the exploitation of renewable resources should be based on the augmentation of those resources by the deliberate extension of vegetation cover—so as not only to increase the resource base, but also to regenerate fertility of the soil, retain moisture, ameliorate the climate, etc., by growing 'fuel forests' and 'raw-material plantations', especially of fast-growing and mixed leguminous trees, canes and grasses. Thus should we do everything in our power to conserve the integrity of the total ecosystem.

In fact, all industrial processes should be designed to be analogous to natural ecosystems, i.e. by ensuring that each unit contributes to the integrity of the whole system and by ensuring that its wastes can be used by the next unit in the chain—which implies once again a revision of technology in terms of the integrity of the total system, or '*Non-waste Technology*' (VI).

Encouraging Examples of Integration

The most striking examples of the elimination of waste all seem to involve the integration of two or more processes which are normally (and wastefully) operated separately.

Thus, in Finland, Härkki (1976) reports that flash-smelting of copper has been practised in conjunction with sulphuric acid manufacture. By combining these two processes, 97.8% of the sulphur is recovered and over 50% of the energy requirements are met by recovered heat and electric power generated from steam produced internally. As is now being practised in a number of countries, the addition of a fertilizer plant increases the useful production while using lower inputs of energy and raw materials and yielding lower outputs of pollutants.

Another example of how industrial systems get more efficient (less wasteful) as they become more complex is the case of garbage incineration. Kreiter (1975) has shown, in a study in the Netherlands, how a garbage incineration plant handling 140,000 tonnes p.a. could generate electric power and thus save 14,700,000 m^3 of natural gas p.a. If, in addition to burning garbage and generating electric power, the low-pressure steam were used in a waterworks, a further 11,900,000 m^3 of natural gas could be saved per annum.

Thus we can begin to see how we should be building industrial plants as local industrial complexes rather than remote 'monocultures'. Pyrolysis of garbage is also effective because it links fuel production with metals recovery. For example, a study has been made by Yevstratov & Kievsky (1976) of an integrated water-system for a town of 40,000 inhabitants in the Soviet Union, which includes a central-heating and power-plant, a chemical plant, a construction plant, and food factories as well as a non-ferrous metal plant and other service industries. The totally integrated system enables 95.66% of the water to be recycled, with the complex piping, drainage and treatment plants paying for themselves in less than a year because of the savings in water consumption and by the recovery and re-use of 'pollutants'.

A classical example of combining two entirely different processes so as to reduce waste is that of whisky distilling and animal-feed production. The North British Distilleries Co. Ltd, in Edinburgh, was faced with an ultimatum of stopping the pollution due to the spent wash from its still-bottoms or being closed down. The company decided to evaporate the wash from 2% to 40%, using mechanical vapour recompression followed by drying, as reported by Gray (1974). Processing costs in terms of steam, water and electric power were £312,000 p.a. Capital charges at 20% depreciation and 15% interest were £211,000 p.a. The revenue from the sale of the dried wash for £60 per tonne as high-quality animal feed was £1,200,000. The net profit of the operation of integrating an animal-feed plant with the whisky distilling is £687,000 p.a. for an investment of £602,000 *and* it eliminates the pollution.

As reported by Royston & Royston (1976), the Ahlstrom forest-product plant at Varkaus in Finland produces sawn timber, plywood and pulp, and says 'the pulp mill also makes use of the chips and sawdust from the sawmill and plywood mill. The loss of raw material is consequently small, and the economic and technical requirements of the production entity are better than if the product units of various types were acting independently'.

Another example of integration is that of a Swedish pulp-mill which is linked with a metallurgical plant—Airco Metals—to make common and integrated use of power-plants, steam-generating plants and water-systems.

In many cases, adoption of integrated energy-systems enables chemical plants to meet their own power and process-heat requirements at lower cost and with lower levels of pollution than through buying and running steam boilers.

In general, 'energy sense makes non-waste sense makes economic sense'. A good example is the divergent development of aluminium smelting technology, the pre-bake pot having been developed for its energy-saving and then being found suitable for fluoride recovery and cryolite reconstitution—compared with the Soderberg pot, as shown by Nestaas (1975).

Other examples of entering different manufacturing areas are the production of magnetic-tape-grade iron oxide powder from titanium dioxide 'acid–iron' wastes, and the production of a range of pigments from spent pickle-liquor from an alloy steel-tubing plant.

A number of plants have been built to integrate phosphoric acid manufacture with that of gypsum plasterboard and thus avoid the normally massive calcium sulphate pollution resulting from the former operation. The plasterboard plant, built by Rhone Progil at Rouen, paid for itself from its sales revenue of 73,500,000 French francs per year, whereas running a pollution abatement operation would have cost an extra 5,000,000 Ff p.a., as shown in the study by the Ministère de la Qualité de la Vie (1976).

New Industries from Wastes

The Soviet steel industry manufactures a wear-resistant glass-ceramic called Slagsitall—made from blast furnace slag—which can be used for wall panels, tiles, pipes, etc., and is expected to save the building industry 400m roubles p.a., as reported by Scholes (1973).

In China, as reported by Orleans (1976), the late Chairman Mao Tse-tung was always urging the 'struggle against waste' and the need 'to change wastes into treasures and turn harmful into beneficial'. Among the many Chinese examples is the effort in Shanghai which produced annually 2 million tonnes of building materials from waste materials. In a Chiaoutou power-plant, fly-ash was turned into 180,000 Yuan worth of insulating bricks.

Mineral wastes can become whole new business areas, predominantly in the manufacture of building materials. Thus English Clays pioneered the prefabricated Cornish Unit House as a way of disposing of china-clay wastes, and other uses include general aggregate, fill, road building, founding sand, calcium silicate bricks, floor tiles, and fibreboard made by combining china-clay waste with paper-mill waste. Colliery spoil can be used for roadstone, bricks and cement manufacture. Slate waste can be expanded for lightweight aggregate or made into bricks or cement. No less than 5.4 million tonnes of pulverized fuel-ash from British power-stations were turned into lightweight bricks in 1972. Waste calcium sulphate from phosphoric acid and hydrofluoric acid manufacture can be mixed with colliery waste to become self-combusting, while producing cement and sulphuric acid and thus winning two useful products from two waste materials. Slags can be made into bricks, roadstone or insulation. Red mud from alumina manufacture can be made into pigments, iron, bricks, roadstone or cement, as indicated by Gutt *et al.* (1976).

Non-waste mining operations are characterized by the conversion of mine sites into recreational areas, as has happened so successfully with the Wigan Alps in Lancashire and the tin mines around Kuala Lumpur in Malaysia, as well as the gravel pits around London—on one of which, at Thorpe, the 1975 World Water-ski Championships were held—and, of course, the famous molybdenum mine in Colorado that was jointly planned so successfully by American Metal Climax (Amax) and the Sierra Club.

Peterson (1977) gives examples from the chemical industry—including Union Camp, which now sells for $1 per lb flavours and fragrances based on wastes which it used to sell for 8 cents per lb. Westvaco has converted into chemicals, which it now sells for $22.5m per year, wastes that it used to dump.

Further Encouraging Uses

What is probably the clearest evidence of the virtues of an 'ecological' integration of different operations lies in waste-heat utilization (cf. Royston, 1975). Power-plants at Battersea, situated on the River Thames in London, England, now supply space-heating to low-cost apartment blocks across the River, while the nuclear plant at Ageste in Sweden supplies the equivalent of 55 MW of hot water to the Stockholm suburb of Farsta, 4 km away. At Hinckley and Hunterston in England, and at Cadarache in France, fish-farming in the hot water coming from nuclear plants has demonstrated rapid growth and high yields. In the United States, extensive work on building agro-industry complexes around nuclear plants has shown promising results and, in Germany, a system has been developed to warm the soil of agricultural land with nuclear waste-waters and, hence, extend the growing-season.

In developing countries, the corollary of this is biogas production, in which, as Srinivasan (1977) shows, cow-dung, that was traditionally dried and burnt as fuel, is now subjected to anaerobic digestion to yield methane, which can be used for cooking, and a residue that contains the fertilizing value. The non-waste technology here consists of visualizing the complete animal waste–fuel–manure system and exploiting it completely, as is attempted in all of these technologies (Table I).

The characteristic basis of all these non-waste technologies is the putting together of two or more processes which are traditionally different but which are symbiotic in that they can live off each other's wastes—a concept which Royston (1976) has termed 'co-productivity'. Examples which he cited include the Port of Ashkelon, Israel which went into the oil business to dipose profitably of hydrocarbons recovered from oily ballast-waters; paper mills in the protein and alcohol business to dispose usefully of sulphite liquors; flour, frozen food and sugar manufacturers going into the protein feed business to take up their fermentation biomass; steel mills going into the municipal waste-water business to get sewage to

TABLE I

Summary of Examples of Non-waste Technologies

Integrated systems	*Wastes avoided*
Copper smelting–sulphuric acid –fertilizer	SO_2, heat, resources
Garbage disposal–power generation–drinking-water production	Land, heat, resources
Garbage disposal–fuel production–metal recovery	Land, resources
Domestic water–industrial water–pollutant recovery	Water pollutants, resources
Whisky production–animal feed	Water pollutants, resources
Timber–plywood–pulp	Water pollution, resources
Heat–power	Air pollution, heat, resources
Metallurgy–paper	Air pollution, heat, resources
Aluminium–cryolite	Air pollution, energy, resources
Alloy steel tubes–pigments	Water pollution
Titanium dioxide–pigments–magnetic tape	Water pollution
Phosphates–plasterboard	Water pollution
Steel–ceramic	Solid wastes
Mining–building material (bricks, cement, aggregates	Solid wastes
Electricity–insulating bricks	Solid wastes
China clay–prefabricated houses	Solid wastes
Mining–recreation (fishing, boating, walking, etc.)	Land
Electricity–heating (homes, fish-ponds, glasshouses, fields)	Heat
Animal waste–gas	Water pollution
Food–fine chemicals	Air pollution, resources
Chemicals–fine chemicals	Water pollution
Harbours–oil	Water pollution
Paper–alcohol–protein	Water pollution
Foods–protein	Water pollution
Steel–municipal waste-water	Water pollution
Cheese–pigs	Water pollution
Chickens–pigs	Water pollution
Electric power–sulphuric acid	Air pollution

use in their plants; cheese manufacturers and chicken farmers going into the pig-raising business to dispose of their wastes; and the US electric power-plant which produces sulphuric acid, etc.

If non-waste technology, or symbiotic processes, or industrial ecosystems, are so effective, why are they not more widespread? At least in part the reasons appear to be as outlined in the next section.

Obstacles to the Implementation of Non-waste Technologies

There is one major problem in developing the type of non-waste technologies mentioned above. This is simply that most individuals and most companies, and indeed much legislation, tend to relate to single-industry or sectorial lines. Thus, a mining engineer knows little about recreational areas and does not want to think about them, a copper company does not want to be in the fertilizer business, and legislation often prevents a chemical company from supplying surplus electric power to the local community. And yet, as we have seen, it is by an 'ecological' grouping of different industries that we can maximize production with minimum damage to the total resource-base, in a somewhat similar manner to that in which the total protein yield of the natural ecosystem comprising savanna is much higher than if all wildlife were cleared and replaced by a herd of Hereford cattle.

Another obstacle is structural–strategic. As most of the integrated systems will be complex, and as the balance of production must be made to prevent wastes from accumulating while optimizing the effective use of resources rather than maximizing production alone, the system is best managed at the local level, i.e. from the bottom upwards. Unfortunately, it is a characteristic of both the free market system, in terms of centralized marketing policy, and the socialist system, in terms of centralized resource allocation, that planning and management tend to be the other way round, i.e. from the top downwards. Indeed, one might say that many of our environmental ills stem from just this common characteristic of the world's two main economic systems.

Finally, there is the fault of most environmental strategies in placing excessive emphasis on emission standards, and, by so doing, giving the impression that pollution abatement is the answer to industrial environmental problems—at any level of resource or financial cost.

A Strategy for Non-waste Technology

The first element of the strategy is to switch from an environmental protection policy based on emission standards to one based on effluent charges. Apart from anything else, it is well established that such a policy enables one to get the same degree of environmental protection in a diversified environment at one-half to one-third of the cost of a policy based on uniform emission standards, as has been shown by Kneese & Bower (1968).

A policy of effluent charges—which, of course, should set charges to correspond to the total social and economic cost of damage caused by pollution—has a major advantage in that it demonstrates both the economic impact of pollution and the economic way out. Thus, if pollution clearly has a price, pollutants clearly have a value (as potential resources) and, hence, industry should work to realize the true value of these 'wastes'. So, if you cannot sell your 'pollution', you might try to find a partner who could build a corresponding production unit to absorb your wastes. But who is to act as marriage broker to arrange for this fruitful and non-polluting union? Not the enterprises themselves, nor central government, but probably the local community, who have the most direct interest, in both the economic and the ecological impact of local industries.

Here we can see a specific action in the movement of co-determination by industry, government and the community, in the planning and execution of economic projects and the safeguarding of environmental quality. This bottom-up type of organization reflects also trends in devolution of political power in most countries and decentralization and democratization trends in many industrial enterprises, as well as a lessening of the grip in both centralized marketing and centralized planning.

The other aspects of the strategy require changes in government policy which controls industrial activity—both to open up hitherto closed areas of monopoly and, at the same time, to be more flexible in applying classical anti-trust legislation to the creation of local conglomerates. Enterprises in fact are, in many ways, ready to move in this direction, due to becoming progressively blocked in classical areas of monolithic expansion or vertical integration. Thus, the current industrial trend for horizontal diversification fits in very well with the ecological need for a conglomerate system of symbiotic industrial processes.

Moreover the trend against over-specialization among scientists and technologists, and a growing tendency for all specialists to receive some form of environmental education, will work towards opening conciousness to the possibility of designing and operating multiple-production systems. However, a much greater technical and educational effort is needed in this respect to show the feasibility of such systems, and particularly in designing backwards from the wastes which have to be eliminated to the products which can be made to use those wastes.

Finally, local communities should be given the right to plan their local economies and environments, so as to come up with plans for completely integrated agricultural, residential, and industrial, environmental systems which make maximum use of local resources to meet local human needs—without destroying either the sustainability of the local resource-base or the beauty of the surroundings which Man depends on for his well-being.

Conclusions

The most effective way of maintaining a productive technological system which conserves resources and minimizes pollution is by the integration of manufacturing operations so that the wastes of one operation become the resource-bases for another operation. The number and type of operations depend on the availability of local resources and the extent of local needs.

These industrial operations plus the manufactured products, raw material and energy supplies, agricultural, residential, and commercial, services and wastes, should be further integrated into a larger technological 'ecosystem' based on the total community and its total environment.

Development planning should, therefore, start at the community level, based on a 'holistic' approach to all the technologies, resources, products and wastes involved, and should be 'co-determined' by industry, government and the community itself.

Thus, development and the technology used in development, should go with Nature rather than against it, and should be characterized by diversity, coexistence, interdependence, hierarchical ordering of resources and wastes, and the survival value of each element based on what it contributes to the resource-base of the total system rather than what it takes away from that resource-base.

Developing and implementing such truly non-waste technologies will open up the chance of a whole new era of human well-being, with the opportunity at last of banishing disease, starvation, misery, despair and ugliness from the world—without destroying the environmental resource-base on which human survival itself depends.

References

ANANICHEV, K. V. (1974). *Non-waste Technology*. UN Economic Commission for Europe, 15 November 1974, ENV/AC. 4/R3, 11 pp., illustr. (mimeogr.).

BARET, J. L. (1977). *Considerations Concerning the Upgrading of Cellulosic Wastes and Carbohydrate Residues from Agriculture and Agro-industries*. Proceedings UNEP/FAO Joint Seminar on Waste Management: Utilization of Agricultural and Agro-industrial Wastes, Rome, 18–21 January 1977. FAO, Rome, Italy: 44 pp., illustr. (mimeogr.).

BATTELLE GENEVA RESEARCH CENTRE (1976). *Non-fossil Carbon Sources*. Battelle Research Centre, Geneva, Switzerland: 17 pp., illustr. (mimeogr.).

COSKUNER, U. (1975). *Treatment of Solid Wastes at Source—Industrial Wastes*. Seminar on the Collection, Disposal, Treatment, and Recycling, of Solid Wastes. Hamburg, 1–6 September 1975. UN Economic Commission for Europe, Geneva, Switzerland, ENV/Sem3/R3/Add 1: 11 pp. (mimeogr.).

DEPARTMENT OF THE ENVIRONMENT (1974). *War on Waste: A Policy for Reclamation*. Her Majesty's Stationery Office, London, England: 33 pp.

GRAY, R. M. (1974). Cut evaporation costs. *Chemical Processing*, September, pp. 294–8.

GUTT, W., NIXON, P. J., SMITH, M. A., HARRISON, W. H. & RUSSELL, A. D. (1976). *A Survey of the Locations, Disposal and Prospective Uses of the Major Industrial By-products and Waste Materials*. Seminar on the Principles and Creation of Non-waste Technology and Production, Paris, 29 November – 4 December 1976. UN Economic Commission for Europe, Geneva, Switzerland: 81 pp. [*See also* Pergamon Press (1978).]

GYSIN, H. (1977). *Environmental Management in Ciba–Geigy*. Proceedings of Third International Environmental Seminar. Centre d'Etudes Industrielles, Geneva, Switzerland: 13 pp. (mimeogr.).

HAMMOND, B. (1975). Recycling begins at home. *New Scientist*, **68**, pp. 152–3.

HÄRKKI, S. (1976). *Outokumpu Flash Smelting Method*. Seminar on the Principles and Creation of Non-waste Technology and Production, Paris, 29 November – 4 December 1976. UN Economic Commission for Europe, Geneva, Switzerland: 7 pp., illustr. [*See also* Pergamon Press (1978).]

HAWRYLYSHYN, G. (1976). Brazil spends $500 MM on use of alcohol for automotive fuel. *World Environment Report*, New York, **2**(25), p.l.

KNEESE, A. V. & BOWER, B. T. (1968). *Managing Water Quality: Economics, Technology, Institutions*. Resources for the Future Inc. and Johns Hopkins Press, Baltimore & London: x + 328 pp., illustr.

KREITER, B. G. (1975). *Energy Recovery from Municipal and Industrial Waste*. Seminar on the Collection, Disposal, Treatment, and Recycling, of Solid Wastes, Hamburg, 1–6 September 1975. UN Economic Commission for Europe, Geneva, Switzerland: 22 pp., illustr. (mimeogr.).

LEVISTE, GOVERNOR A. (1977). *Lepili, a Timely Versatile Tree which Brings Hope to Developing Nations*. Governor's Office, Batangas Province, Batangas City, Philippines: 11 pp.

LING, J. T. (1976). *Developing Conservation-oriented Technology for Industrial Pollution Control*. Proceedings of a Seminar on the Principles and Creation of Non-waste Technology and Production, Paris, 29 November – 4 December 1976. UN Economic Commission for Europe, Geneva, Switzerland: 6 pp. [*See also* Pergamon Press (1978).]

MINISTÈRE DE LA QUALITÉ DE LA VIE (1974). *La Lutte contre le Gaspillage: Une Nouvelle Politique Economique, Une Nouvelle Politique de l'Environnement*. La Documentation-Française, Paris, France: 100 pp.

MINISTÈRE DE LA QUALITÉ DE LA VIE (1976). *Usines Propres: La Technologie au Service de l'Environnement*. La Documentation-Française, Paris, France: 58 pp., illustr.

NESTAAS, I. (1975). *A Survey of Pollution Problems in the Aluminium Industry*. Industry Sector Seminars, Aluminium Meeting, Paris, 6–8 October 1975. UN Environment Programme, Paris, France: 38 pp.

ORLEANS, L. A. (1976). China's environomics: backing into environmental leadership. *Environmental Policy and Law*, **2**(1), pp. 28–31 and 98–101.

PERGAMON PRESS (1978). *Non-waste Technology and Production*. Pergamon Press, Oxford, England: 678 pp., illustr.

PETERSON, R. (1977). Money from wastes. *Development Forum*, United Nations, Geneva, Switzerland, January – February, **5**, p. 3.

RAPSON, W. H. & REEVE, D. W. (1976). *Non-waste Production of Bleached Kraft Pulp*. Proceedings of a Seminar on the Principles and Creation of Non-waste Technology and Production, Paris, 29 November – 4 December 1976. UN Economic Commission for Europe, Geneva, Switzerland: 6 pp. [*See also* Pergamon Press (1978).]

ROYSTON, M. G. (1975). *Some Examples to Illustrate the Concept of Non-waste Technology*. 20 October 1975, UN Economic Commission for Europe, Geneva, Switzerland, ENV/SEM 6/PM/R3: 14 pp. (mimeogr.).

ROYSTON, M. G. (1976). *Eco-productivity: A Positive Approach to Non-waste Technology*. Proceedings of a Seminar on the Principles and Creation of Non-waste Technology and Production, Paris, 29 November – 4 December. UN Economic Commission for Europe, Geneva, Switzerland: 23 pp. [*See also* Pergamon Press (1978).]

ROYSTON, M. G. & ROYSTON, E. (1976). *The Economic Benefits of Investment in Pollution Control*. International Programme in Environmental Management Education, Centre d'Etudes Industrielles, Geneva, Switzerland: 96 pp. (mimeogr.).

SCHOLES, S. (1973). Blast-furnace slag-wastes to build on. *New Scientist*, July 26, pp. 206–7.

SHORROCK, C. (1978). *Chemicals and Fuels from Renewable Resources*. Proceedings of the Fourth International Environmental Management Seminar, Centre d'Etudes Industrielles, Geneva, Switzerland: 30 pp., illustr. (mimeogr.).

SRINIVASAN, H. R. (1977). *Bio-gas (Gobar Gas) and Manure from the Waste of Farm Animals*. UNEP/FAO Seminar on Waste Management: Utilization of Agriculture and Agro-industrial Wastes, 18–21 January 1977, FAO, Rome, Italy: 15 pp.

TOROCHESHINKOV, M. F. (1976). *State of Non-waste Technology: Industrial Experience*. Proceedings of a Seminar on the Principles and Creation of Non-waste Technology and Production, Paris, 29 November – 4 December. UN Economic Commission for Europe, Geneva, Switzerland: 8 pp. [*See also* Pergamon Press (1978).]

YEVSTRATOV, V. N. & KIEVSKY, M. I. (1976). *Experience in Designing a Complex Scheme for Refining and Re-use of Waste Waters and the Creation of a Drainage-free Scheme of Water Supply and Sewerage in an Industrial Enterprise*. Seminar on the Principles and Creation of Non-waste Technology and Production, Paris, 29 November – 4 December. UN Economic Commission for Europe, Geneva, Switzerland: 13 pp., illustr. [*See also* Pergamon Press (1978).]

DISCUSSION (Session 12)

Gilliland (Chairman, introducing)

This Session is devoted to 'Industrial Alternatives: Non-waste Technology', and I am Jim Gilliland, Director of Environmental Control for the Amax Environmental Services Group, which is headquartered in Denver, Colorado. Mr Stanley Dempsey, who was shown on the printed programme as Chairman of this Session, is Vice-President and Director of Environmental Affairs for Amax, and has recently assumed additional responsibilities as Senior Vice-President of External Affairs (which is currently undergoing structuring) within the company. Consequently he has asked me to be his substitute here and to convey his regrets for not being able to participate directly at this Conference and in particular for missing the opportunity [which he had long looked forward to, of renewing past acquaintances]. My own background is engineering, and up to 10 years ago I dealt primarily with problems of occupational health, being with the University of Colorado Medical Center where I serviced this question of occupational health as a consultant to industry and as a research worker. Slightly more than 10 years ago the Amax management sensed the [imminency] of the environmental movement and I was subsequently hired to provide guidance and to expedite the environmental programme of the company.

Amax is a multinational diversified natural resources company involved in the production of metals and energy minerals [, and to show you that we are not all unmitigated baddies I will tell you a little about it]. We literally cover the spectrum from A to Z—from aluminium to zinc, with in-betweens of coal, iron ore, lead, nickel, copper, potash, oil, and gas—and are the world's largest supplier of important alloys of molybdenum, which we mine in Colorado. Amax had been publicly recognized for achievements in environmental improvement even before Earth Day 1971, while already before the enactment of the United States National Environmental Policy Act of 1970, the impact of modern industrial development on the environment was a concern which had become an integral part of our corporate planning philosophy. We view, and try to practise as best we can, sound environmental management as a responsibility and an opportunity.

We do not deny that natural resource development has an impact on the environment, and so pursue our programmes with open planning, public involvement, and whatever other skills are available, to minimize the adverse impacts and to optimize the positive impacts. Following the keynote paper and presentations of the panel, it is to be hoped that there will be time for us to consider briefly some of our specific environmental programmes which exemplify the thesis of non-waste technology.

Our main speaker is Dr Michael Royston, [Programme Coordinator in Environmental Management of the Centre d'Etudes Industrielles, Geneva, Switzerland,] and two of our panellists, Dr Tom Malone and Professor Philip Jones, you will recognize as former outstanding chairmen. The other two represent multinational corporations but are here in their own right by the Conference's invitation: they are Mr Charles T. Savin, of the Environmental Control Centre of British Petroleum, and Mr George H. Robbins, Corporate Environmentalist of the Luscar Group, based in Edmonton, Alberta, Canada.

Royston (Keynoter, presenting paper)

It is for me somewhat of an overwhelming experience to be here speaking to you today, as a mere engineer who started his career over a quarter of a century ago in the engineering workshops of GEC, London, and to be here in front of so many eminent and pre-eminent scientists. It is also for me an inspiration to be in Iceland. I think all those of us who have been here this week have seen how Iceland has coped with the issues of living, as has been said already so many times, on the margin of the habitable world, and it is interesting also for me to recollect that in the Icelandic language the concepts of development and conservation are so closely linked. The old Norse concept of *threfa*, which exists almost intact in Icelandic, means 'to grasp', but with derivations meaning on one hand 'to clean' and, on the other, to 'develop' or 'thrive', which we have in English in the two words 'thrift', meaning to save or conserve, and 'thrive', meaning to develop or grow vigorously.

It is for me particularly inspiring to be in this Nordic country which has presented this essential conservation ethic for the last eleven hundred years, and also to see the consequences of that in the Icelandic life around us. It is interesting to note that this concept of thrift and thrive—of non-waste—in fact has been present in most societies up until very recently, and indeed is still present in many societies around the world. [Dr Royston thereupon presented his paper more or less as printed on the foregoing pages].

Jones (Panellist)

I have been accused by many of my colleagues of being an incurable optimist, but I think I have met my match in Dr Royston. I admire the very positive approach of his paper, and the very positive note [he has struck throughout], but I am reminded when we talk about non-waste technology of a little poem in the front of a calculus book that I once studied, from about 100 years ago. [It was merely a couplet but complete in itself:]

All naturalists observe that fleas hath smaller fleas upon them prey;
And these have smaller fleas to bite'em, and so proceed *ad infinitum*.

How this poem relates in my view to non-waste technology is that you never really get rid of waste. Ultimately there is always a little bit left, and while I cannot in any way criticize the very positive and optimistic theme of the paper—and I will probably use it methodically with some of my students and some of my colleagues who are usually very sad people—I feel obliged on this occasion to speak a little bit about the other side of the coin. First of all I think there is an implication in the paper that the laws of thermodynamics can be repealed, whereas I am not quite sure that this is so. The laws of thermodynamics, put in very simple terms, say that you cannot get anything for nothing. This being the case, complete and perpetual motion of the recycling type [is impossible]—there is a cost somewhere, and perhaps we should look very carefully at this cost.

The second problem that I see is that most of the industrial engineers and process engineers that I know who are involved in the development of these industries, do not read the literature which supports this document. Perhaps we should try and encourage them to do that, but the literature which supports Dr Royston's paper [largely comprises] reports of international meetings, and a lot of my process engineering colleagues get their noses very deeply into process engineering news and digests and chemical engineering and things like that, so they do not often look beyond and into these optimistic areas.

One of the points mentioned in the paper, which I agree has worked very well in many cases, is the voluntary separation of solid wastes. Many of you may have witnessed that during times of national emergency people tend to work [hard and consistently together], but I found that in many of the larger countries, such as the United States and Canada, people tend to operate rather on a flash-in-the-pan basis. Everybody gets all turned on about this idea of separation of putrescibles and glass and metals, and of coloured glass from white glass, and you will probably end up with 1–2% of the population that are continuously committed to this project. But then [commonly] the whole thing dries up, because the receptacles [get lost] and the industries that were involved in receiving this 'process material' tend to become a little discouraged, and so they tend to go back to the old system again and dig holes like the old tomcat and bury [the whole lot].

The waste exchange process which was mentioned in the paper—and I thoroughly agree has a tremendous amount of promise—is usually characterized by limited and diverse industrial settings such as those of the UK, perhaps Switzerland, or perhaps even smaller countries. I am not entirely sure that it would work in areas where there are a lot of similar industries, all of which produce a similar kind of waste. Even where waste exchange has worked well—and probably the UK have one of the best records for this—I think the best proportion that they ever managed to place was something like 40% of the available waste, and this only occurred in one particular instance.

Non-waste technology is highly desirable but clearly requires motivation. Now the first motivation that industry recognizes in the capitalist countries, and indeed in the non-capitalist countries, is the profit motive. If somehow a profit could be shown, not in the general sense but in the specific sense—'it can be demonstrated that *this* industry will make *this* profit by following *this* course of action'—there would be a very good chance that the board of directors would adopt it. There is an alternative way of acquiring this motivation and that is a negative motivation by, if you like, taxing new materials and virgin raw materials. Recovering and internal recycling tends to be somewhat energy- and capital-intensive, and this of course is an area where we must also be very alert—to make sure we have not created a system which recycles our waste but which consumes energy at a greater rate than, in fact, it produces new material. But of course we must consider all the external energy uses when we examine this situation, such as bringing the new material from long distances—the sort of thing that Dr Royston mentioned as frequently forgotten.

Industries generally seem to be a bit leery about external recycling, because they are not entirely certain about the quality, quantity and reliability of these sources. Industries above all like to have their products produced, and their materials available, in the quality and quantity that *they* want when they want them, and they do not want to have to hold up a production because something is not available. This of course then implies storage to maintain consistent supplies.

One of the most important issues when we are talking about by-product recovery and by-product manufacture—which I suspect is probably an underlying theme of this whole paper—is the question of the market. I am immediately reminded of a small pulp-and-paper plant in Southern Ontario which has the capacity of producing vanillin as a by-product of its sulphite-paper waste process. The only problem is that that very small pulp plant could produce enough vanillin to supply the world's needs twice over. So one has to say: what else are we going to produce besides vanillin? Thus when we [consider] these very logical by-product processes, we must take a pretty careful look to see whether anybody really wants to buy those things, because that is what industry is certainly going to do.

Many of the processes that were described with such enthusiasm by Dr Royston can and should work, and somehow we have got to try and make sure this information gets to the right people, and that they are properly motivated to try them. Irrigation with municipal sewage treatment-plant effluents, for example, is obviously one of the very best ways to remove nutrients; and yet in North America, particularly right now, we are going to enormous capital expenses to remove nutrients. We have got operating costs and capital expenses involved in removing nutrients, so that we do not green our lakes any more than they already are. However, the cost of transportation is very high, and sometimes the climate does not lend itself to this type of recycling.

One of the problems that I suspect exists but is not really spoken of in the paper, though I think it should be mentioned, is the technological lag. There is indeed a significant technological lag between what is sometimes called 'laboratory curiosity' and the world reality. There are many processes and procedures which have been developed in laboratories and peoples' minds but have never been adequately tested, and of course the first thing that a major industry wants is to be sure that what they are about to indulge in in a capital sense is going to work. Indeed I think one of the major problems is the lack of available funds to demonstrate some of these innovative technologies.

In closing I would like to quote Dr Royston's own paper, which I think is an excellent paper, when he says: 'If non-waste technology, or symbiotic processes, or industrial ecosystems, are so effective, why are they not more widespread?' I think that is a very good question, and one which we should dwell upon here today, because there is a great deal of useful meat in this paper, and I hope somehow or other between us we can answer that very question. 'If we are all so smart, why ain't we rich', as Li'l Abner would put it. And I think that that is the key question which we should consider today.

Savin (Panellist)

I have to say first of all that I am here in a private capacity and not representing any institution [or particular interest], though I think [Secretary-General] Polunin makes it very clear to us all that we are here as invited individuals; he also adjures us to speak out freely as individual specialists, while giving us the opportunity to retract in any case of serious need.

Having said this I can assure you [that] I found Dr Royston's paper both stimulating and interesting. Certainly we must all agree that waste is uneconomic, and one of the uncovenanted benefits of the [so-called] energy crisis was that it made us all more aware of the value of materials which in the past were too lightly discarded, and in this way I think it partly reinforced the environmental concerns which are expressed with varying degrees of force. There has been over the past few years a considerable transformation of the industrial scene relating directly to these increasing costs of raw materials.

It seems to be a fact of life that those who have polluted most have the most to save and to gain, and here I share the slight concern expressed by Professor Jones that Dr Royston's paper was perhaps a little too positive. Wherever human beings engage in any activity there will be pollution, if by that we mean disturbing the state which existed before the human activity started. Process residual cycling has been known of course for many years, and is increasingly being both used and considered. The recycling of used products, however, requires care to ensure correct specifications both of the starting material and of the final products, and to avoid resultant harm from those products. To take just a few examples from my own sphere of knowledge, fermentation alcohols—both ethanol and methanol—can be used, well and effectively, in the internal combustion engine, but careful standards of quality are needed. The engines require some modification, and a very great deal of care has to be exercised in storing and moving the alcohols concerned. There are particular problems, for example, from water contamination, which do not exist with the use of gasoline.

Another problem—perhaps another illustration—is the recycling of used lubricating oils. Again it is very worthy and right that we should collect waste oils and use them in the most effective manner possible, but there are problems if you recycle and reuse. If refining cannot be done under very careful control [to keep down the contents of carcinogenic-type materials], I would prefer to see such oils burned under controlled conditions to provide heat and energy and at the same time to destroy perhaps potentially harmful products.

Another problem of recycling [is that of maintaining some] positive cost–benefit. In the United Kingdom, for example, we have had, on and off over many years, very enthusiastic and dedicated volunteer bands of people who are prepared to collect waste paper and sell it for recycling and beneficial use. This moreover benefits the many voluntary organizations on whose behalf the paper is collected, but there is a problem in that periodically the bottom 'drops out' of the market and nobody is prepared to take the waste paper that has been so assiduously collected. This probably happens in many cases of waste-cycling [potentially and similarly in other countries]. It is not, I think, that the enthusiasts and volunteers lose heart, but there are problems if there is no positive fiscal incentive for them to continue with the collection, or no physical outlet for the materials collected.

One must agree about the importance of emission standards but should beware that those imposed upon industry, or upon communities, are not in fact far higher than are necessary to preserve the environment [or even than are practicable at all]. Moreover they may inhibit the adoption of newer, better and more effective technology. If we are wedded to the 1960s, then the technology of the late 1970s will sometimes not be adopted [as being anathema to our ways].

Therefore I personally prefer the use of the so-called 'best practicable means', whereby each case is examined on its merits in the light of technology and taking into account particularly the cost–benefit considerations and local circumstances, to agree on the most effective means at the particular time for the required purpose. I am not altogether happy with the alternative proposal of effluent charges which, it seems to me, [may be] another way of providing a licence to pollute on payment of money, which is really not acceptable in this day and age. For I suspect on a more mundane level that if once governments found they had another tax-base, they might start to forget the ecology and remember the revenue-raising potentialities. I think there has been a recent case of this in Germany, where the Federal Government attempted to establish economically viable pollution-charge levels but in fact found that this had (or would have had) a very adverse effect upon the economies of the industries concerned. So revised and much lower charges have had to be proposed, which presumably defeats the object of the exercise.

One final point and, if I may, one example of wholehearted support of Dr Royston's [conviction] that these problems must at the end of the day be dealt with at the local community level: I believe that in the United Kingdom, with our system of land-use planning, we do have a better opportunity than in some countries to ensure participation of interested parties. However, it must not be forgotten that these local communities' concerns are not always for the environment and local ecological considerations but may be more concerned with the problems of bringing industry to their neighbourhood and so maximizing jobs.

In illustration of this I would like to touch on an interesting situation in the Shetland Islands. Many of you will know that, with the development of North Sea Oil, the Shetland Islands are scheduled to become one of the big focal points for collecting oil brought in from various parts of the North Sea, with the resultant establishment of a very large oil port in the Islands. Now clearly, in a closed community such as Shetland, which in very many ways is similar to Iceland, such developments could have a very real and potentially adverse impact upon the community and upon the environment. Shetland Islands Council themselves had the forethought to secure some unique legislative provisions which enabled them to have very good control over these developments. Perhaps arising out of that, and certainly arising out of the shared concern [of the developers], the Sullóm Voe

Environmental Advisory Group* was established, on which representatives of industry, the local authority, and leading conservation concerns such as the Nature Conservancy Council, the Countryside Commission, and the Royal Society for the Protection of Birds, also sit, together with distinguished academics. They attempt to reach a consensus view on the way that the oil-related development should proceed, so as to minimize the effect on the environment whilst at the same time allowing the development to go forward to the very real benefit of the inhabitants of the United Kingdom [of which the Shetlands form an integral part].

Gilliland (Chairman)

Your remarks are being recorded as being your own and not necessarily those of British Petroleum. Our next panellist is Dr Tom Malone, [the eminent meteorologist and general environmentalist] who is Director of the Holcomb Research Institute of Indianapolis, Indiana, but unfortunately has to return to Washington this afternoon [on urgent US Senate advisory business], so we certainly want to hear his observations and comments.

Malone (Panellist)

I thank you, Mr Chairman, and would first like to assure you that I am not leaving of my own volition but in the interests of implementing some of the recommendations of this Conference. Some rather important hearings are being held before our United States Senate tomorrow, and I have been asked to testify [in a manner that amounts virtually to a Presidential command]. They bear directly on the problem which we discussed on Monday morning—climate and the impact of fossil fuels on climate.

I would like to share three points with the group here. The first has been developed by my colleagues in the United States, under the auspices of the Rockefeller Brothers Fund [see *The Unfinished Agenda*, edited by G. O. Barney and published by Thomas Y. Crowell Company, New York, 1977], and contends that it is desirable to integrate the management of solid wastes with energy and materials recycling systems. This requires not only technical measures but, perhaps more importantly, institutional innovations with respect to utility practices, market regulations and building codes.

Point No. 2 is that it is deemed desirable to recycle organic nutrients as a supplement to, and partial substitute for, chemical fertilizers. This should have the following advantages: *(a)* of reducing the energy used in the production of fertilizers, *(b)* of reducing the adverse environmental effects of fertilizers and, not irrelevant to our discussion, *(c)* of helping the disposal of urban garbage and sewage sludge.

[Point No. 3 concerns] nuclear waste management, on which I have to comment that perpetual diligence is required. In this matter the experience in our country is *not* encouraging, because no nuclear waste management programme in the United States over the past 30 years has demonstrated the kind of diligence [that would be] required for a much longer period. Here I would cite the leakage at Hanford Works of [some 4 million litres] of high-level radioactive waste and the ground-water and rain-water movement off-site of low-level wastes which were planned to migrate not more than a few inches in millennia.

With regard to fusion energy and its sometimes attractive potential, we must remember that plans are not yet even on the drawing-board, but it is the considered view of many that it is likely to be expensive in capital, material, energy and human skills.

* *See* the account by its then Chairman, Peter G. Brackley, published in *Environmental Conservation*, Vol. 2, No. 3, p. 222, Autumn 1975.—Ed.

Finally, as a closing note, I think the question of effluent charges—and I am glad it was brought up—is not a closed book yet. The Ruhr Valley experience was probably oversold. There appear to be a lot of practical difficulties, but I would not be too quick to close out [effluent charges] as an option.

Robbins (Panellist)

While I concur entirely with what Dr Royston has brought forth, I would like to offer one additional viewpoint that hopefully will shed further light on this very important subject of non-wastage. My comment is simply that, when considering waste and the waste influences of our industrial society, we must remember how the problem is really two-sided. Dr Royston has adequately addressed one side, namely wastage at the effluent stage of our industrial processes. There is no doubt that it is wasteful to dispose of available energy and/or potentially useful resources which might be wrapped up in the form of by-products of any particular industrial operation. Such action to me [is analogous to, and as wasteful as, a] 'keep the change' concept. To conduct further this disposal without any consideration of alternative uses for that energy or by-product resource, should only be labelled as foolish.

The other side of the problem concerns the wastage so commonly occurring at the front end or pre-development stage of the industrial operation—namely the destruction of already-existing natural resources that are or could be useful [in other ways to] our society. An example that immediately springs to mind in this regard is the wholesale clearing and burning of forested land to prepare sites for farms and housing developments such as is so prevalent throughout [Canada and many other countries]. Another example is the problem of the porpoises being caught up in the tuna fishing nets, and yet another, which is of particular interest to myself as a reclamationist, is the loss of soil materials and soil profile characteristics during the overburden removal phase of strip-mines. Here I might interject for the record that I am not the individual responsible for a recent North American slogan that 'strip-mining prevents forest fires'!

I would like now to turn briefly to the topic of industrial alternatives. Dr Royston has identified several technologies and alternatives that are available to industry to avoid wastage at the effluent or by-product stage of various industrial processes. But are there any technologies or alternatives available to avoid wastage of the type that I have mentioned, namely for pre-development wastage? I feel that the answer for the most part is 'yes'. Of course there are certain situations—the porpoise problem, for example—where technology may still require further development; but in most cases the answer is 'yes'. The solution is relatively simple. All that is usually required is wise planning, a little foresight, and a sustained conscientiousness throughout the industrial venture.

To end on a cheery note, I reflect on our pertinent Canadian industry as well as on others that I am familiar with. I submit that [the above reclamation problem and its] solution have been readily accepted, and that significant advances in conservation are continuously being made on the industrial front [—especially if we compare the situation with not so many years ago when, admittedly, there were fewer people and consequently less demophora on our Earth]. Nevertheless, in reflecting on the industrialization of the world at large, I would say there are still some serious shortcomings. I cannot offer solutions to these deficiencies at this time, [but would] appeal to this eminent forum to consider whatever resolutions are necessary in that regard.

Gilliland (Chairman)

I would like to speak on just one other aspect of non-waste technology that really has not been addressed here. In the United States, for example, to secure the necessary authorizations for a new-facilities construction, involves some degree, and sometimes a substantial degree, of environmental [impact] assessment, including such considerations as resources monitoring to establish the background status of the area, and the projection of socio-economic effects of the development. When once the permits are secured, however, there frequently is inadequate follow-up to ascertain whether or not the short- or long-term impacts did indeed happen as anticipated. To promote, therefore, a non-wasteful utilization of the skills and technology which went into the planning of the development, means that a post-operational monitoring programme must also be intelligently planned and implemented. By this means we will learn how better to plan future projects, and be able to address more accurately the anticipated results which many times are arrived at just by pure hypothesis.

I am afraid our enthusiastic keynote speaker and panel [not to mention your Chairman] have used up most of the discussion period, but please proceed from the floor.

Franz, Eldon

I am developing an increasingly severe mental cramp and it has nothing at all to do with the fact that I have been sitting still! I am one who believes, as Dr Fuller pointed out this morning, that our greatest asset is our mind. I also believe that there is a definite complementarity between mind and body, but that is another topic altogether. What concerns me here is to reflect upon a few of the topics that we began to discuss when we were talking about life-support systems. In particular, using Dr Royston's terminology, we are engaged in a two-factions game—Man against what is not Man, by Dr Fuller's definition of environment. Only a part of [the environment] is capable of bioregeneration, and [that is made up of] the symbiotics with which we occupy this planet, and which are absolutely essential for our own life-support. Now I submit that the kind of recycling technology which is proposed here is simply another example of the ostrich putting its head in the sand, my reason being that it completely ignores the products of that technology and where they are subsequently distributed into the environment.

In the example which was given, one of the outputs was PVC [which can be] an extremely dangerous material, and the fact that the by-products of its production are controlled at the production site does not mean that the material does not eventually enter the food-chains of our life-support system. One of the most vital characteristics of the biosphere is that it has evolved with various species partitioning different elements from one another in various specific ways. What we are essentially doing as people in the world is modifying the nature of those biogeochemical cycles by bringing into them greater and greater quantities of materials to which the system is not adapted. That is extremely important.

It is essential that we do not lose track of this perspective [but realize] that the systems into which we are dumping these various commodities are not adapted to them and, as many people have stated, these systems are beginning to show the signs of degradation [more or less] universally. So the problem is not internal recycling in technology; the problem is in fact many of the outputs of technology itself, and one of the biggest doubts about this kind of internal recycling is that it simply makes it easier for us to live with the ultimate problems, because our minds are temporarily assuaged to believing that we are doing what is right. We simply must not stop trying to integrate technology, economics and ecology. It is very difficult to maintain this perspective but ultimately our lives all depend on it.

Sigurjónsson

I think Dr Royston and much of this discussion really gives us the clue of how to survive without ecodisaster—how we can go on living this life we have made, having all these things we want, and yet keeping the nation clean. Problems of some by-products will be resolved, as has been done in the past, and this paper opens up new aspects. And if they are thought to be too energy-expensive, I need only mention that there is now coming on a new process for aluminium production that is absolutely closed, without any contamination at all of the surroundings, and using only one-third of the energy that is used at the moment for aluminium processing.

So surely as Man has been able to go to the Moon [and come back safely] he will be able to recycle everything we are wasting now, and this will be to the benefit of all, while making this life more pleasant and comfortable.

Kuenen

You must all be aware what a terrible optimist I am, and I was therefore very pleased with Royston's paper. He showed the way in which we could go, and I was somewhat disappointed that some of the panellists seemed to raise objections which might indicate that they were looking too much to the past and not enough into the future. But that is for them to sort out.

The question I wanted to put is this: has this vegetable plant which I know as a zoologist, which was invented some 3,000 million years ago and has supported life for all that time, been sufficiently studied and compared with the technological plant that our speakers have been discussing, and is it possible that the efficiency of energy utilization by the [green] plant is perhaps slightly neglected by our technologists? [They should consider] not only quantities but also qualities.

Royston (Keynoter, responding)

All chemical engineers or chemical processors use the word 'plant' to describe the complex installation in which raw materials and energy come in and products and waste go out—somewhat similar to your plants. It is very interesting, if one looks at the evolution of technology, and as I was suggesting of non-waste technology, that we see there increasing convergencies of technology with the complexity, the interacting cycles, of natural plants. That last picture I showed of that Russian town: maybe if you stood far enough back you might think it was the Krebs cycle or something like that. Technologists are learning rapidly and learning extensively from the way [green] plants work.

Someone mentioned thermodynamics. One of the essential parts of course of the natural system is that you do not go in one step from your highest energy level to your lowest energy level; you go through a series of steps, and the thermodynamic efficiency of the overall process is very high. One of the reasons why our industrial plants have up to now been so low in efficiency is because we went from a high energy level to the low energy level in one step. Non-waste technology, or multiple re-use, is a process of staging those energy levels so that we in fact increase the overall thermodynamic efficiency of the total process.

In this way again we are learning from Nature. Technology has got to go with Nature if it is to be an intelligent technology that will help Man to survive and will help the [world] ecosystem to survive.

Buchinger

We here all agree that clean industry is possible—that in the initial stages it is expensive, but in the long run very economical. This means that in a country with guided economy, investment should be made by the country. In another country, where there is so-called free economy, there is of course competition, and with good legislation the dirty industry can be compelled either to go clean or go broke. Now I do not know, if there is free competition, that one should feel so sad if a group of industries, which is not capable of adjusting itself and lives at the expense of the entire community, should go broke. I am not very sorry for the people who lose their jobs, because I am quite sure that this industry will be replaced soon by a clean one and so they can go on working. It does not take too long to see such changes. Just recall the glass industry: when the Pilkington process was introduced in England, within a year everybody in the United States changed their method—even such factories as were brand-new and had a big capital investment.

Gilliland (Chairman)

That is a very good comment. Actually, the only problem is that by closing down some of the inefficient industries there is [need for a] translocation of people. Did you have your hand up back there? [Voice] No, I was only scratching my ear.

Olindo

I think that the conflict of present-day development with the environment may have resulted from the emphasis of what has been repeatedly [claimed] this morning, namely that the most important resource we have is our mind. My own thinking is quite different; the heart, the mind and the body *combine* to make the most important resource, but not any one separately. I would like [you] to share the view that perhaps the greatest asset we have is the heart in so far as it moderates the mind and the body.

Franz Eldon

My point is a very simple one, but I think it is so easy to overlook. With reference to the aluminium plant that was discussed*: the problem is not limited to what is [happening] on the plant site. The problem is that segments of the biosphere have simply not evolved to cope with increasing burdens of [unaccustomed] materials. [Even aluminium itself may become a problem as there is evidence of its association with Alzheimer's disease.] My mind and my heart and my body are integrated in my idea of the way in which we should approach Man. I agree with that 100%.

Tan

Professor Jones mentioned that the first motivation of industry is profit, but I am afraid this particular element is interfering, for example in Malaya, with the way in which we try to solve the problems of polluting our Earth. Another speaker asked: Why are these [recycling and other desirable] methods not more widespread? We go again back to industry for the answer: Because industry has not found any effective treatment system which makes them enough money! Recently I came across an example of how big industry is doing precisely the opposite of what ecologically-minded people are advocating. We have not [popularized] a way of treating our palm-oil waste in an economic manner, i.e. without fancy gadgetry, and recently a European company came to Malaya to advocate a treatment system that was more expensive than the entire original production factory, which is a ludicrous situation. So until matters are controlled by enforceable legislation, the rest of the biosystem, including the poor fishermen and farmers who depend on clean rivers, must put up with the continuous flow of poison to their environment in order that a small group of people can get 'economic returns' from their investments.

That is why I found Dr Royston's paper most relevant. It is something we can look at and say, 'Yes, this [offers several] practical means by which we can

* In a special workshop held on the Tuesday night in connection with, but not as an integral part of, the Conference.—Ed.

solve [some] problem; perhaps we won't take it *in toto*, but certainly there are some practical elements in it which we think would suit our system and we can tackle it in a similar way'. So I wish to applaud Dr Royston for his pragmatic approach which is I think more important than the philosophical arguments that I find very difficult to understand. I also applaud them for their idealism, but I am afraid I find it very difficult to appreciate how they can be effectively applied to the problems we have, of rural folks deprived of their livelihood, their food, their drinking water, etc.

Gilliland (Chairman)

Might I comment that there is sometimes another layer to a problem area? You have said that industry must operate at a profit, and that is true; but sometimes the reason why we do not do more of the things which Dr Royston [and others have suggested] is due to regulatory restrictions. I think we would all like to locate and have industrial complexes where we could cycle materials, and in fact we do bring waste materials in from far distances, having sited certain of our plants where we have a market for some of our by-product materials. Regulatory restrictions, however, can have a limiting role in this regard, at least in the United States system.

Dodson Gray, David

The question keeps coming up: Why do companies not do [more of the sort] of things which Dr Royston has talked about? Certainly one of the constraints often is financial, but there are some non-financial barriers too. Companies tend to view themselves as being in a certain business, and [for example] they are not likely to go into a dairy-farm business in order to deal with their waste. There are [often] managerial barriers which, though not insuperable, need to be recognized. Often it is not the financial constraint; for instance, one proposal has been made which would deal with these sorts of problems in industrial terms—not on Dr Franz's terms—[namely] that if each company were required to deal with all wastes and keep them on its own site, it would be under a constraint to turn those wastes into useful products.

That is very much the sort of simple principle that Maurice Strong was looking for last night*, and if we had to do this we would jolly well find a way to use our waste, and we would then have to leap that organizational–managerial barrier and find a profitable way, and also learn [if necessary] how to go into the dairy business in order to deal with it.

Goldsmith

[Watching my ear-scratching this time,] I would like to make two points. John Davoll, Director of the Conservation Society in the UK, once said that if anything was politically feasible and economically viable, then it was not worth doing, and if it was worth doing, then you could bet your bottom dollar that it was not politically feasible or economically viable!

One must realize that there are a lot of constraints on our activities, but the most important ones are not exerted by political or economic pressures but by the forces of Nature and these, it is worth noting, are as indifferent to the exhortations of our politicians and economists as they were to those of King Canute.

If social policies are to be governed by political or economic constraints only, the results, in the long term, cannot be anything but catastrophic. In the case of economic constraints, the reason is that economic theory has been developed in a veritable void. This means, among other things, that economic costs bear little relationship to real costs—by which I mean biological, social and ecological ones

* In the Baer–Huxley Memorial Lecture, printed on pages 613–25.—Ed.

and—when once these have been incurred, they must necessarily be translated into economic costs. In other words we must one day pay for the biospheric damage that we do. Biospheric costs are long-term economic costs, and the costs which are taken into account by our economists are simply short-term economic costs.

Secondly, we must redefine 'pollution', for the current use of the term is too narrow. Pollution really means introducing randomness or noise into a system. We pollute the biosphere not only by releasing toxic wastes into it but also by depriving land of its tree cover and causing it to lose its topsoil, or by covering it with cement to favour the urbanization process. Such activities lead to the contraction and deterioration of the biosphere, and thereby to a reduction of its overall stability. In other words, it is not just the exhausts from motor-cars that are pollutants but motor-cars themselves—and all the other parts of the technosphere which play no [useful] part in the strategy of Nature and are random to the biosphere. It would be a stupid parasite indeed that killed its host, so clearly we must limit this pollution.

Borden

As an environmental psychologist, something that has concerned me here in the last couple of days is the tendency to define environmental problems primarily in technological terms. There are a number of us in the United States who are beginning to view the so-called environmental crisis in much more psychological terms, and who therefore see many of the problems as instances of maladaptive human behaviour. For example, nearly 40% of our resources and energy are consumed by people in their homes. Consequently, a substantial proportion of recoverable resources and conservation of energy can be realized by redirecting our emphasis away from technology and towards people.

Understanding how people feel about environmental problems, finding out what they know and what personal sacrifices they would be willing to make, and determining what environmentally responsible behaviours they currently engage in, are essential before such programmes can be initiated. I do not want to develop this theme in detail now, except to point out that much of it falls in the area of psychology. One practice, at least in the United States, that I find particularly upsetting, is something one of my colleagues calls 'eco-pornography'. In particular there has been a recent explosion of television and newspaper advertisements in which various industries portray themselves as environmentally concerned, and in so doing attempt further to placate the public's worries regarding environmental and energy issues.

Recently I examined some statistics from a Harris Poll and it sounds as though in the United States people are becoming very environmentally aware. Yes, people are becoming more aware of these problems, but the new advertising techniques of many industries may serve to short-circuit the chances that individuals' heightened concern will lead to conservational activities within the home. For example, there is one oil company that shows off their impressive new technologies with a film of an enormous floating derrick which can be towed to a drilling site, flooded, and rapidly installed for production. To illustrate their concern for the natural environment, they also show fish swimming around one of their recent installations and conclude by saying that now the fish have a home—as if, for 30 million years, fish have been searching the waters of the globe for an oil-derrick to live under!

Another example of the use of ecological half-truths that capitalize on the public's environmental worries and portray industry in a seemingly favourable light, goes something like this: A large strip-mining company tells how they will care for our energy needs of the future while at the same time 'helping the environment'. They explain how the topsoil that they remove is set aside during mining, and then is carefully replaced following the operation. They do not, of course, mention that the topsoil also fostered a complex climax ecosystem of hardwood trees, mosses, ferns, insects, birds, [mammals], etc.—instead, they show their staff members planting fast-growing softwood conifers that will become, in only 20 years, 'a forest'. Particularly [appalling] is their inference that these fast-growing trees are 'happy trees' because they are growing fast: apparently an old oak tree lives a dolorous existence!

I think that these are very insidious activities for industries to engage in. At a time when the general public desperately needs honest ecological awareness, people see instead a collection of expensively produced presentations of ecological half-truths or falsehoods that are intended to justify further industrial exploitations. Finally, my own research has shown that, unless a person attends ecological or environmental studies programmes at a university, the chances of obtaining such facts through the media are very low. Consequently I think that there must be a thrust to present honest environmental educational information to people in their homes, utilizing all communications media. [Applause.]

Jones (Panellist)

I am a little disappointed that Professor Kuenen, as another perennial optimist, apparently does not wish to examine the other side of the coin, but I believe that part of our environmental problem today is that we tend to sit down and make these esoteric pronouncements of what ought to be, without looking at some of the realities. I am also disappointed with our charming lady from Malaysia who does not like the word 'profit' [despite her expressed disagreement]. I am talking about the machinery of industry; whether it be in eastern European or in capitalistic societies, the motivation of industry is still profit, and if, as it appeared, the paper was starting to demonstrate that one can show [how] certain of these non-waste technologies are indeed profitable, that is the method you should use to sell them—rather than altruism. If, however, you cannot show that profit is going to be generated, you must create a profit motive by applying tax to new materials, but you must not manipulate the economy to do this.

[Despite my earlier-expressed approval I now think that the paper in places paints too rosy a picture; indeed if I had] written it myself, being an incurable optimist, I would have expected some critic in the audience to have got up and carved my jugular into several pieces. For if really we say that we are going to increase the cost of things as a result of clean technology, it is going to increase inflation, which is probably the No. 1 criminal in the view of the public—the same public who elect the political leaders who in turn implement the kind of plans we talk about here.

If our machinations result in increased costs, increased inflation and increased unemployment, we might as well have all stayed home and [twiddled our thumbs] this week. For unless we set a tremendous stage before them, so that they fully understand the consequences of not accepting these evils, Society is not going to like [these proposals, which would] fall right on our heads like rocks.

Gilliland (Chairman, concluding Session)

That concludes the Session and I thank you all for your indulgence.

Session 13

SCARCITIES AND SOCIETAL OBJECTIVES

Keynote Paper

Scarcities and Societal Objectives*

by

CLAIRE RUSSELL

c/o David Higham Associates Ltd, 5–8 Lower John Street, Golden Square, London W1, England.

&

W. M. S. RUSSELL

Reader in Sociology, Department of Sociology, University of Reading, Whiteknights, Reading, England.

INTRODUCTION

Scarcity is essentially an excess of population over resources. 'One cannot insist too often', wrote William Vogt in his classic *Road to Survival* (1949), 'that . . . population increase is, fundamentally, as much a physical process as though one burned down storehouses containing food.' In short, we may sum up the concept of scarcity in the simple words spoken by Grettir in Louis MacNeice's poem 'Eclogue from Iceland' (MacNeice, 1938, pp. 29–39): 'too many people.'

Crises of overpopulation and scarcity have been frequent in human prehistory and history. Occasionally, these crises have arisen through a relatively sudden drop in the supply of resources without an immediate

**Presented by Peter B. Stone, Editor-in-Chief,* Development Forum, *Palais des Nations, 1211 Geneva 10, Switzerland, in the unavoidable absence of both Authors.* We acknowledge with warm thanks the valued part played by Mr Stone in this and other respects, and are asked to mention that the Russells' (and also Mr Glasser's) literary agents are David Higham Associates Ltd, 5–8 Lower John Street, Golden Square, London W1, England.

matching drop in the population consuming them: the crisis itself normally restores the balance by raising the death-rate. This has happened to at least two societies established in climatically marginal conditions—the Norse colonies of Iceland and Greenland. They came into being during the climatic optimum of AD 400–1200, which reached its peak in 800–1000 (Lamb, 1966), when the North Atlantic was exceptionally free from both ice and storms. Iceland was discovered by Celtic explorers in the eighth century, and settled by the Norse in the ninth century; Greenland was discovered in the early tenth century by Gunnbjorn Ulf-Krakason, and settled in 986 by Eirik the Red (cf. Jones, 1973).

In the early Middle Ages, Iceland was the site of a flourishing and complex civilization, and even southern Greenland was thickly settled, with some 300 farms, 16 churches and a cathedral. From the thirteenth century on, however, the climate in high-northern latitudes began to deteriorate, culminating in the 'Little Ice Age' of the sixteenth to early nineteenth centuries (Lamb, 1966; Bryson, 1975). 'The main landmark in Greenland towards which the Vikings sailed had been called Black Mountain, but became known instead as White Mountain, as the snow- and ice-cover increased' (Bryson, 1975, citing O. Pettersen). Icebergs and storms made communication with Greenland increasingly difficult, and some of the land where the colonists grew crops is now permanently frozen (Lamb, 1966; Jones, 1973). Towards the end of the fifteenth century, the Greenland colony disappeared, after suffering prolonged malnutrition to judge from the small stature of the later skeletons (Lamb, 1966; Deuel, 1967).

Meanwhile, conditions were deteriorating in Iceland. 'According to early Icelandic law', writes Hermann Palsson (1975), 'ownership was an absolute right. But this law was changed in 1281 to the effect that no one with a surplus of hay could refuse to sell it to a needful neighbour; in such circumstances it became a criminal offence to defend one's property.' The saga of *Hen-Thorir*, in which a feud is started by the refusal of a landowner to sell his hay surplus, 'may have been written in order to pave the way for this revolutionary innovation' (*Ibid.*). This shows not only the growing imbalance between population and resources as the climate worsened in Iceland, but also the social tension and decline in individual freedom that always accompany population crisis. From the sixteenth century, according to Lamb (1966), 'growing of cereals completely ended in Iceland, only to be resumed in the 1920s and after'. Parish records of the early eighteenth century document the advance of the glacier over farmland, and Iceland endured several centuries of misery and foreign oppression until the climate improved again (Lamb, 1966; Simpson, 1975).

The example of the Norse colonies is important in view of the concern that is currently being expressed about the world climatic situation. 'Combining the nature of the recent climatic change with the present narrow margin of world food-grain reserves,' writes Bryson (1975), 'an urgent need to consider and react to the possibility of continued climatic variation is indicated.' It may be added that fisheries are particularly sensitive to climatic change, which affects them in complex ways (Jensen, 1939; Dickson & Lamb, 1972; F. S. Russell, 1973). Clearly it is not enough for population to be in balance with present resources; ideally, population should be reduced well below this point, to provide an ample safety-margin.

Population Crisis and Population Cycles

Far more often than otherwise, crises of overpopulation and scarcity have arisen through population growth. The more favoured human societies have been able repeatedly to increase their supply of food and other resources through technological advance, and hence support an increased population. Whenever this happened, however, their populations tended to increase beyond even the new resource level, producing scarcity: in the words of H. G. Wells (in *Men Like Gods*), mankind 'spent the great gifts of science as rapidly as it got them in a mere insensate multiplication of the common life'.

The essentials of the situation were clearly worked out by Robert Malthus in publications between 1798 and 1830, and were outlined even earlier, in 1793, by Hung Liang-chi, educational commissioner of Kweichow (W. M. S. Russell, 1976). For all the complexities of demography, in the long run (migration apart) population change is a function of ('crude') birth-rate and death-rate, reckoned as numbers per hundred or thousand per year. If the birth-rate is higher than the death-rate, the population grows. The trouble is that the growth is by compound interest, as the more people there are, the more offspring they can breed. A calculation has been made which shows the fantastic implications of this (Cipolla, 1962). If mankind had sprung from a single couple, living about 12,000 years ago, shortly before the coming of agriculture, and if there had been one more birth than deaths per hundred per year (a modest 1% increase per year), then today the world population would form 'a sphere of living flesh, many thousand light-years in diameter, expanding with a radial velocity many times faster than the speed of light'. In real life, as opposed to the wonderland of mathematics, nothing of the kind can happen. So in real life, when a population increases even at this modest rate, sooner or later one of two things *must* happen—either the birth-rate comes down, or the death-rate goes up, and the increase is checked.

Until recent times, human civilized societies tended to have high normal death-rates (3–4%) but even higher birth-rates (3.5–5%); hence their populations grew (Cipolla, 1962). They obviously did not end up as expanding spheres of flesh, etc. Instead, every so often, they began seriously to outgrow their current supply of resources. They then entered a period of what we have called *population crisis*, with well-defined characteristics (C. Russell & Russell, 1968, 1969, 1970*a*, 1970*b*, 1971; W. M. S. Russell & Russell, 1976; C. Russell, 1970, 1975; W. M. S. Russell, 1971). The economic effects included price inflation and fall in real wages (which can be shown in many times and places to depend closely on population growth), unemployment, and often grandiose building projects (designed to absorb labour) which further depleted resources. The political effects included reduced freedom of the individual and a tendency to tyrannical government. The cultural effects included narrow specialization, distrust of new ideas (especially if simple and wide-ranging), superstition, intolerance and restrictive censorship. The social effects included greater inequality between classes, greater difficulty in moving from lower to higher classes, social tension and a tendency to redirect resentment to defenceless minorities. The behavioural effects included increases in all kinds of violence, involving violent crime, class conflict and persecution of minorities: the usual protective attitudes to women and children were impaired.

Such a society as a whole was apt to break down into anarchy, and prove particularly vulnerable to foreign invasion and conquest. Meanwhile the inadequate food-supply would result in famines and chronic malnutrition, which, combined with crowding and all kinds of social stress, would render the people liable to high mortality-rates from epidemics.

As a result of such a population crisis in a society, the numbers would be drastically cut down—which, for a time, improved the situation, because the population was much reduced relative to its current resource supply. Hence such *relief* periods were times of economic, political, social and cultural advance, often called renaissances (W. M. S. Russell, 1976). These relief periods were particularly fruitful in Europe in producing sustained cumulative technological and social progress—not because of any inherent superiority of Europeans (there is no evidence for, and plenty of evidence against, any such inherent differences between peoples), but because until recently Europe had a far lower population density than that of any other civilization (W. M. S. Russell, 1967, 1968). However, in Europe as elsewhere, these relief periods were always brought to an end by renewed population growth, which eventually ushered in another population crisis. In this way, history is everywhere broken into *population cycles* of crisis and relief.

We have described population cycles of this kind in the history of Europe, North Africa, the Near and Middle East and Central Asia, China, and the civilizations of the New World (W. M. S. Russell, 1967; C. Russell & Russell, 1968). In more recent studies (unpublished), we have found population cycles in the history of India, Sri Lanka, South-East Asia and Japan, and (from archaeological evidence) in prehistoric societies in (for instance) southern Britain, the southwestern United States and the Marquesas Islands. The saw-tooth curve of population change resulting from the cycles is well shown in a graph of the changing population of Egypt, published by Hollingsworth (1969).

In Table I, by way of illustration, we have outlined the population cycles of China, which include at least seven major population crises in historic times. In the second column, only the major dynasties are mentioned; numerous lesser dynasties commanded larger or smaller parts of China for longer or shorter periods during the crises. The longer relief periods between crises are marked by the great dynasties—Han, T'ang, Sung, Ming and early Ch'ing—whose names recall the glories of Chinese art, science and literature. The third column shows important stages in the growth of Chinese food production. The advance of the great engineering works of water control, on which Chinese food supply and civilization have long

TABLE I

The Population Cycles of China.

Dates	*Period*	*Food production*	*Water control works**	*Population* (millions)
BC 481–206	*Crisis 1:* Warring states, Ch'in, civil war	N.W. Loess area developed	1.6	
BC AD 206–221	Han (actually 2 dynasties)	N.E. Lower Yellow River developed	6.6	
AD 221–618	*Crisis 2:* 3 kingdoms, Tsin, S. and N. dynasties, Sui	S.E. Lower Yangtze developed; Grand Canal	7.6	
618–755	T'ang (at its height)			
754				50
618–907			43.9	
755–960	*Crisis 3:* Civil war, later T'ang, 5 dynasties			
839				30
907–960			12.3	
960–1280	Sung	Early-growing rice	174.4	
1100				100
1280–1368	*Crisis 4:* Yuan (Mongols)		175.6	
1290				60
1368–1644	Ming	Potatoes, sweet potatoes, ground-nuts, maize	411.2	
1393				65
1600				150
1644–1683	*Crisis 5:* Fall of Ming, Manchu conquest			
1661				100
1644–1912	Ch'ing (Manchu)		603.4	
1700				150
1779				275
1794				313
1850				430
1850–1880	*Crisis 6:* Taiping, Nien, Moslem revolts			
1872				330
1900				430
1912–1949	*Crisis 7:* Warlords, civil war, Japanese invasion			
1931				450
1950–	People's Republic			
1953				580

* Number of engineering works of water control per 50-years' period.

depended, is shown in the fourth column. Chi (1963) has listed, for the various periods shown, the number of items of water control works to be found in the Provincial Gazetteers of Imperial China. For easier comparison between periods, we have in each case computed the number of works per 50-years' span. These are minimum figures, but they give a fair picture of the differences between earlier and later periods, and between periods of relief and population crisis, when the increase in water control activity slowed down or even went into reverse.

The population estimates in the fifth column are chiefly taken from the study by Ho (1959) of census and other materials. The absolute figures cannot be accurate, but there is no doubt about the increase in population resulting from the advances in food production listed in the third column, or about the fluctuations accompanying the crises.

The population crash that ends each cycle reflects a prodigious quantity of human misery and waste. What population cycles mean for the individual may be seen by looking at his or her longevity (W. M. S. Russell & Russell, 1976). Between the mid-thirteenth and the mid-fifteenth centuries in England, for instance, the expectation of life at birth for a male dropped steadily from about 35 years to about 17 during the population crisis, and rose again to about 33 during the recovery period. On the world stage, population cycles are associated with such great discontinuities of history as the fall of empires.

TABLE II

The Population Cycles of China, Northern India and Europe.

Century AD	*China*	*N. India*	*Europe*
1st	Relief	Crisis	Crisis
2nd	Relief	Crisis	Relief *
3rd	Crisis	Crisis	Crisis
4th	Crisis	Relief	Crisis
5th	Crisis	Relief	Crisis
6th	Crisis	Crisis	Crisis
7th	Relief	Relief	Crisis
8th	Relief	Crisis	Relief
9th	Crisis	Crisis	Crisis
10th	Crisis	Crisis	Crisis
11th	Relief	Crisis	Crisis
12th	Relief	Crisis	Relief
13th	Crisis	Relief	Crisis
14th	Crisis	Relief	Crisis
15th	Relief	Crisis	Relief
16th	Relief	Relief	Crisis
17th	Crisis	Relief	Crisis
18th	Relief	Crisis	Relief
19th	Crisis	Relief	Relief†
20th	Crisis	Crisis	Crisis

* Due largely to importation of resources.
† Due to emigration of people and importation of resources.

In the past, the population cycles of different regions were not necessarily in phase with one another. In Table II, we have illustrated this by showing the population cycles of China, northern India and Europe, side by side, over the past two millennia. We have greatly oversimplified the sequences, chiefly for the convenience of listing the conditions of relief or crisis by centuries. For instance, the population crisis that began in Europe in the sixteenth century ended in fact half-way through the seventeenth century. Towards the end of the eighteenth century, another crisis set in; but the full development of this crisis was averted, from the mid-nineteenth century onwards, by massive imports of resources from outside Europe, and large-scale emigration to other continents.

In spite of omitting such fine details, Table II does show how cycles were staggered in different parts of the world. The great Classical period of north-Indian civilization under the Guptas, for instance, coincided (in the fourth–fifth centuries) with crisis in both Europe and China. Cycles could be out of phase even within continental regions. Thus the first and worst outbreak of the Black Death in Western Europe (1348–9) caused few casualties in Bohemia, where population was still in better balance with resources, while the peoples of the rest of the region, weakened by malnutrition and stress, died in their millions (W. M. S. Russell & Russell, 1976).

The effects of recurrent population crises were thus essentially regional. Within a region that was rendered vulnerable by its climate, these effects could include lasting and devastating damage to the natural environment and its resources. This has been the fate of a great extent of land stretching from Morocco through the Near and Middle East to Turkestan. This region, subject to irregular rainfall but being extremely fertile when irrigated, has been called by W. M. S. Russell (1967) the Dry Belt of the Old World. For millennia it was among the richest and most productive regions on earth.

It was within this region of abundance that civilization began, and its cities continued to be greater than any built in Europe before the 16th century AD—except for those, such as ancient Rome, which imported their food from the Dry Belt. Yet by the early twentieth century AD, the whole Dry Belt lay in ruins, thanks to a long succession of population crises. 'Destructive wars and invasions led to the neglect and collapse of irrigation works, and in these times of anarchy the shepherds swarmed over the land, cutting down trees and overgrazing the vegetation. For lack of protective plant cover, large areas dried up, and others were devastated by soil erosion' (C. Russell & Russell, 1968). By the early twentieth century—and this typifies the history of the whole Dry Belt—'nothing but wilderness covered the area where, early in the 2nd millennium BC, "Lot lifted up his eyes, and beheld all the plain of Jordan, that it was well watered everywhere, even as the garden of the Lord" ' (*Ibid*.). We have told this story in detail elsewhere (W. M. S. Russell, 1967; C. Russell & Russell, 1968). Every population crisis brings with it a variety of immediate scarcities; the example of the Dry Belt shows that a succession of population crises may create lasting scarcity in a region which was formerly of overflowing abundance.

The Modern World Population Crisis

The modern world-wide population crisis has a number of new features. First, the population growth this time is largely due to a fall in normal death-rates, brought about by modern medicine, so that to stop the growth and its unpleasant consequences we have to get birth-rates down much further than ever before. Second, populations all over the world are now so absolutely large that very small rates of growth add very large numbers of people. Third, some countries (such as Britain and Japan) have reached very high absolute population densities, and the supply of useful land has become a serious limiting factor all over the world. Fourth, modern industrial civilization provides amenities and facilities for its citizens, notably education and health services, on a scale without precedent in human history; these wonderful achievements are all threatened and impaired by population pressure (C. Russell & Russell, 1967, 1968; W. M. S. Russell, 1977). Fifth, modern industrial civilization needs unprecedented amounts of such resources as minerals, energy supply and means of waste disposal. Sixth, modern technology has provided methods of exploitation in peace and destruction in war that can seriously deplete or damage environmental resources in *any* part of the world—not only in arid lands. Seventh, whereas formerly crises in different regions were staggered (Table II), the cycles are now all synchronously in phase, and the present crisis *is* world-wide. Although the pattern of specific scarcities may differ in different places, no country in the world is free from threats of overpopulation and scarcity—expressed, for instance, in price inflation.

In the words of Lester R. Brown (1976): 'the global population increase of nearly 700 million during the 1960s was roughly the same as that occurring between 1800 and 1900. . . . For each of us, there was nearly one-fifth less fresh water, mineral reserves, arable land, fossil fuel reserves, living space, waste absorptive capacity, marine protein, and natural recreation areas in 1970 than in 1960.'

The present world population is plainly too high to assure everyone a decent standard of living. To quote Brown (1976) again: 'in countries containing one-third or more of the world's people, average food intake is today below the minimum required for normal growth and activity.' The one grain of comfort about world population growth is the fact that the *rate* of growth is a little lower in the 1970s than it was in the late 1960s, when it exceeded 2% p.a. Still, absolute annual increments of population continue to be enormous. According to United Nations estimates (1977), the world population was 3,632 million in 1970 (mid-year); by 1976 (mid-year), it had grown by 413 million to 4,045 million.

Such is the gross scale of the world population crisis; we can now consider in more detail the pattern of scarcities made up by the demand and supply of the world's various resources.

The Natural Resources of the World

The scarcity problems of the world have to be considered as a whole, not only because the modern population crisis is world-wide, and because problems of air and water pollution are often of international scope, but also because the world's resources are so unevenly, and so differently, distributed. Industrial civilization is so demanding of raw materials that industrial nations now have to import even so common and widely distributed a mineral as iron ore from the poorer countries (Couper, 1972). Other minerals were less evenly distributed to start with. Northwestern Europe must import most of its copper, phosphate, tin, nickel, manganese ore and chrome ore (Brown, 1976). Four countries (Chile, Peru, Zambia and Zaire) supply most of the world's exportable surplus of copper; three others (Bolivia, Malaysia and Thailand) account for 70% of all the tin in international trade (*Ibid.*). A jet engine cannot be built without small amounts of columbium and cobalt; most of the columbium comes from Brazil, and most of the cobalt from Zaire (Klare, 1972).

In 1969, at the then rates of use, it was reckoned that *world* supplies of a number of minerals would run out by about the end of the present century (Cloud, 1969; Goldsmith, 1971): these included known sources of gold, silver, platinum, lead, zinc, tin, mercury and tungsten. They may last somewhat longer with a reduced acceleration in industrial demand, but they are still a cause for concern.

Until the 1940s, the poorer countries outside the regions of industrial civilization exported food to these regions. But owing to their huge population growth, they eventually all became food importers: by 1964, they were importing an annual total of about 25 million tonnes of grain (Paddock & Paddock, 1968). The surplus populations that were driven from the countryside by lack of land and employment, swarmed into the coastal cities where the incoming grain shipments were unloaded (Cloud, 1969). In 1964, the Soviet Union joined the food importers (Paddock & Paddock, 1968; Goldsmith, 1973). Japan and most of Europe have been importing food ever since they became industrialized. Thus the world is very widely dependent for grain on Australia, Argentina, Canada, and above all the United States (Paddock & Paddock, 1968). The United States is not only the biggest exporter of wheat and rice; it provides 90% of the exportable supply of protein-rich soyabeans, which since 1973 have been imported even into China (Brown, 1976).

To round off the story of distribution of resources, the two main sources of energy in current use, coal and oil, are also heavily monopolized by particular countries. Of the world's probable coal reserves, more than half are in the Soviet Union, and 93% are in the Soviet Union, the United States or China; of the world's *proven* oil reserves in 1974, more than half were in the Middle East, and a further 12% in Russia (Foley, 1976). As Gerald Foley (1976) has pointed out, the countries with large coal or oil reserves are unlikely to risk depleting their own stocks by exporting more than they need to for earning imports, and this factor may well keep world oil production from rising far above its present level.

The problem of world resources, then, is complicated by the fact that different countries control different resources, so that no country can be considered wholly independently of others. Nor can any one resource be considered in isolation, for the different resources in turn depend on one another in a complicated way. We can see this if we consider the problem of world food-supply. Since about 1830, the world population has quadrupled (Brown, 1976). Now of course nobody can survive without eating, and this large population growth has only been possible because of a large increase in world food-supply. It is important to realize how this happened. Land food-supply (the main source of food energy) can only be increased in two ways: by farming more land, or by getting higher yields on the land that is farmed. In the last century-and-a-half, both things happened. First, the manufacture of steel ploughs made it possible, for the first time, to cultivate the world's great temperate grasslands—in Australia, Argentina, Canada, and above all the United States (W. M. S. Russell, 1967). As we have seen, these countries are still the world's main suppliers of grain: they constitute the main granary of the world; no such bonanza will ever happen again. For, as Mark Twain put it, 'they're not making land any more'.

The large areas of land that have not yet been permanently cleared in the tropics (for instance in Brazil) generally have soils that are poor and lacking in plant nutrients (W. M. S. Russell, 1967), or, worse still, soils that bake into a kind of brick after a few years' exposure to the tropical sun (McNeil, 1964; Gourou, 1966). As for the reserve cropland in the United States, most of it had to be brought under cultivation in the early 1970s (Bryson, 1975; Brown, 1976).

The other means of increasing food production on land has been to increase crop yields per unit of land area. Between 1900 and 1973, for instance, the United Kingdom more than doubled its wheat yield, Japan more than doubled its rice yield, and the United States more than tripled its maize yield (Brown & Eckholm, 1974). But the high land- and labour-productivity of modern agriculture is only attained by the use of massive supplies of energy that are without precedent in the history of agriculture. It was estimated in 1972 that American farm tractors consumed about 8 thousand million gallons* of fuel per year, whose energy value is equal to that of the food-crops consumed in the United States; in 1969, American farms consumed about 7 million tonnes of nitrogen fertilizer, requiring for its production the equivalent of about 1.5 thousand million gallons* of petroleum (Goldsmith, 1973). In the words of Howard T. Odum, cited by Foley (1976): 'a whole generation of citizens thought that . . . higher efficiencies in using the energy of the sun had arrived. This is a sad hoax, for industrial man no longer eats potatoes made from solar energy; now he eats potatoes partly made from oil.'

So far as American maize production is concerned, it has been calculated by Carol and John Steinhart that 'further applications of energy are likely to yield little or no increase in this level of productivity' (Brown & Eckholm, 1974). So the huge yield increase, like the huge land increase, is not going to happen again. The world has been using up its grain reserve since the early 1960s: in 1975, current reserves could already have been exhausted in about three weeks (Bryson, 1975). Simply to maintain present production levels, very large amounts of energy are required—not to mention other prerequisites.

Fisheries, although an important source of protein, do not add greatly to the world's supply of food energy, and use even more energy than agricul-

*1 US gallon = 3.785 litres.

ture. The output of food energy from the British fishing fleet in 1969 was only 5% more than the energy expended (Foley, 1976). Once again it is clear that, simply to maintain present food-supply levels, a continued high energy supply is essential, and this brings us to the energy problem.

Industrial civilization has been created by using the fossil fuels—chiefly coal, crude oil and natural gas. The Burmese were drilling for oil by the tenth century AD, the Persian kings had their food cooked by natural gas in the fourth century BC, and the mammoth-hunters of Czechoslovakia were burning outcrop coal in the Later Old Stone Age (W. M. S. Russell, 1977). However, substantial *world* production of fossil fuel began only at the end of the eighteenth century AD, when British coal mines reached an annual coal output of some 10 million tonnes. Coal production increased spectacularly in the next couple of centuries: one-half of all the coal ever produced up to 1969 was mined after 1938. Meanwhile, commercial oil production began in Romania in 1857, and from 1890 onwards, world oil production increased steadily at 6.9% per year until the early 1970s (Thomas, 1956; Cloud, 1969; Foley, 1976).

The geological formation of the fossil fuels from plant remains took millions of years, and they constitute in practice resources of finite amounts. On the basis of continued growth in production at existing rates, it was calculated in the 1960s that world oil production would peak in the 1990s, and thereafter decline rapidly as the reserves became depleted; coal production was expected to peak during the twenty-second century (Cloud, 1969). These calculations overlooked the distribution of resources between independent nations. In Foley's words (1976): 'The end of the era of steady growth in oil consumption was very close in 1973, even without the action taken by the oil-producing countries.' For combined ecological, economic and political reasons, world oil production and consumption is evidently nearing an upper limit. Coal is also very unevenly distributed, and no vast future increases in production are to be expected. The remaining plant fuel, wood, is in dangerously short supply, especially in poor countries where it is the main heating fuel (Eckholm, 1976).

Solar power, river power, tidal power, wave power, wind power and earth-heat (geothermal) power are all worth developing, and are all being developed or at least experimented with (Simon, 1975; Foley, 1976). Wind power is a particularly attractive form of energy production, being free from pollution problems and unsightly constructions, and requiring relatively small initial outlays of capital and energy (Bockris, 1975; Foley, 1976; Musgrove, 1976*a*, 1976*b*, 1977). Wind can also be used to save fuels for shipping: Greenhill (1972) has discussed the possible use of merchant sailing-ships that could be more efficient than even the clippers of the later nineteenth century, thanks to modern advances in materials, aerodynamics, meteorology and navigation. However, even wind power is not yet envisaged as supplying more than a fraction of world energy resources, and the other energy sources listed above are limited by geographical distribution and/or technical or ecological problems. According to Zuckerman

(1977), 'it is still doubtful whether any of these sources can be of more than marginal help to the basic problem'.*

Nuclear fission reactors have been on the scene since the first one was opened at Calder Hall, Cumberland, England, in 1956. In the 1960s it was feared that these reactors would use up the world's uranium reserves in a couple of decades, making it necessary to build more of the very dangerous fast-breeder reactors, which need far less uranium (Cloud, 1969). This has not happened simply because the growth of fission power has been drastically slowed by technical difficulties and safety requirements, and nuclear power in the middle 1970s still supplied only about 0.3% of the world's primary energy (Foley, 1976). The waste-disposal problems and accident risks are terrifying. In the United States, the fast-breeder reactor has now been 'relegated . . . to the category of long-term research', but 'the USA will continue to build safe light-water reactors' (Zuckerman, 1977). However, the safety of even these reactors is in dispute, and there remains the formidable problem of waste disposal (Edsall, 1974, 1975).

Nuclear *fusion* power, it is true, would probably supply hundreds of thousands of times as much energy as the fossil fuels, present few pollution problems, and produce the useful and now scarce gas helium as a by-product. But unfortunately nuclear fusion power is little more than a dream; not all the experts are sure it will ever be possible, and none expect it to be at all widely practicable during this century (Foley, 1976).

The Limits to Growth in Resource Production

In all these circumstances, no major increase in world energy production is likely in the foreseeable future, and this of course limits the further increase of food production. At least we may learn to stop recklessly wasting our resources at source, as with the flaring of natural gas emerging from oilfields to get rid of it! According to Foley (1976), 'the Middle East gas flares and the Great Wall of China were the first distinguishable man-made objects sighted by one of the returning moon missions'—monuments, respectively, to human waste and tyranny.

The virtual impossibility of substantially increasing energy production sets an obvious limit to the production of food. But this does not mean that the solution of the energy problem—even by nuclear fusion power—would enable food production to increase much further. We have already seen that both farmland area and crop yields are approaching their limits anyway. There are in fact a whole set of limiting factors for the increase of world resource production: some of them have been discussed by Meadows *et al*. (1972), Mesarovic & Pestel (1975) and Lovins (1976).

The connection between energy and food is only one link in a very complex network connecting all the different natural resources and human uses of them. Virtually all resources and uses depend in the last resort on land area, which sets a particularly obvious finite limit. Land area has to be

* *See*, however, the paper by Amasa S. Bishop & Claude G. Ducret and the discussion following it, in Session 11, 'Environmental Aspects of Energy Alternatives'.—Ed.

shared between food production, water storage, mining, processing and manufacture, housing, transport, recreation and waste disposal. According to a Government White Paper (Anon., 1975), for instance, the area of British 'agricultural land, currently about 47 million acres*, is steadily declining . . . over the five years to 1972/73 the average loss of agricultural land in the United Kingdom was about 144 thousand acres a year. Of this, 69 thousand acres a year was transferred to forest and woodland, whilst most of the remainder was permanently lost to urban, industrial and recreational uses'. The annual loss may now be even greater (Goldsmith, 1977). And all this does not take into account the limits set by the actual or potential effects of land, water and air pollution.

Scarcity problems may be eased to some extent by greater efficiency in the use of local resources, which could benefit both the society achieving it and also the rest of mankind, by reducing demand on the world resource pool. In recent years, for instance, there have been several studies of the feasibility of making Britain self-sufficient in food (W. M. S. Russell, 1977). If this could be achieved—it would clearly only be possible if encroachment on farmland ceases—it would not only benefit Britain in terms of balance of payments and fuel and mineral imports, but it would also help to relieve the pressure on the world's exportable food supplies.

Similar local and global benefits could be achieved by avoiding waste of energy and other resources, which (by the very definition of waste!) should be possible without reducing anyone's standard of living. It is important that economies should not be based on the kind of arbitrary paper cost-accounting which is too often another name for 'spoiling the ship for a ha'porth of tar'. Thus a small branch railway-line, making a paper loss on arbitrary narrow criteria, may make an indispensable contribution, as a service, to the economic and social life of the people of the region. Hence cost–benefit analyses should be made in terms of real resources, such as energy, and if possible in terms of real human benefits.

To be sure, there are difficulties in the way of measuring human well-being, as Lord Ashby has discussed recently (1976). 'Rousseau', he begins his article, 'records that he used to walk to Bercy, an hour away from his home, solely for the pleasure of hearing a nightingale.' We can cost this after a fashion—writing-time lost, shoe repairs, etc.—but how do we measure the benefits? However, Ashby himself cites a serious attempt, by the (British) Programmes Analysis Unit, to estimate the economic *and social* costs of living in an area that is exposed to high levels of sulphur dioxide. It should at least be relatively simple to classify proposed schemes as detrimental, indifferent or favourable to human well-being. It is probable, for instance, that reducing the manufacture of very large cars, which are highly demanding of energy (Foley, 1976), would do nobody any real harm.

Increased efficiency and reduced waste are highly desirable objectives for any society. But these are palliatives, which are not going to solve the world scarcity problem. It appears inevitable that world resource produc-

*1 acre = 0.405 hectare.

tion and consumption cannot now increase very much and may have to be substantially reduced.

Growth in Prosperity by Reducing Population

If resources cannot increase and may decline, how can we initiate or maintain growth in prosperity? Fortunately, there is evidence from history that growth in prosperity per head is possible without increased resource consumption, and even with *reduced* consumption, provided there is a proportionately greater drop in population. The great renaissances of history have all been based on growth in prosperity *per caput:* this has generally come about through increase in resources and/or drop in population (W. M. S. Russell, 1971, 1976). All that has to improve is the population–resource balance. It is therefore theoretically possible for prosperity to increase even if resources decline, and this is exactly what happened in the Florentine renaissance.

We have so far used the word *renaissance* for those flowerings of art, science and social life over whole regions, that have occurred during periods of relief from population pressure and crisis—for instance the European renaissances of the eighth, twelfth and fifteenth centuries AD. There have, however, been two outstandingly creative *local* renaissances, namely that of Athens in the fifth century BC and that of Florence in the fifteenth century AD. A number of factors can be shown to be common to these two cases, although here we are only concerned with the high level of individual and civic prosperity in both cities; the favourable population–resource balance that brings about prosperity is the *sine qua non* of a local as of a regional renaissance. The interesting thing is how this came about in Florence.

The chief Florentine industry was the manufacture for export of finished woollen cloth, dyed in brilliant colours (Cronin, 1967). Florence imported from England the wool she made up into cloth. In the late fourteenth century, the English began to keep more and more of their wool at home to make up into finished products. These proved more lucrative when exported in competition with Florence, which thus lost both raw material and markets. The sequence of events is shown in Table III. The English export figures are simplified from the detailed graph published by Carus-Wilson (1967). The Florentine figures are based on less exact and regular estimates from various sources (Luzzatto, 1961; Cipolla, 1970; Cochrane, 1970). By 1378, Florentine cloth output had fallen to less than half its peak amount, and the wool-carders rioted, anxious for their jobs (Carus-Wilson, 1967). As Table III shows, the new, much reduced level of output was maintained for most of the fifteenth century.

With such a drop in production, how could the city be so prosperous? The answer lies in a great reduction in population, and hence in the number of people sharing the profits of export. The city population was something like 100,000 in 1347—94,000 people were reported on the bread-line in that famine-year (Bowsky, 1964; Cronin, 1967). The Florentines of that period of bloody civil conflicts (in one of which the poet Dante was banished for

TABLE III

English Wool and Florentine Cloth, 1350–1550.

English wool export (thousand sacks/year)		*English cloth export* (thousand pieces/year)		*Florentine cloth output* (thousand pieces/year)	
1350–60	30	1350	3	1350	70
1375–90	20			1380–1485	30
1410–50	15	1410–70	30		
				1530	19
1550	3	1550	120	1550	15

life) were well aware of being crowded. In the thirteenth century, when threatened with a visit by the Dominican friar John of Vicenza, who had the reputation of raising the dead, the Florentines protested that they already had too many of the living in their city (Cronin, 1967). However, the succession of famines and epidemics, which continued until 1402, reduced the city's numbers to about 55,000 (Bowsky, 1964; Cronin, 1967). This drop in population gave rise subsequently to prosperity and civil peace, and to the wonderful Florentine renaissance under the early Medici.

Even within the life-time of Lorenzo dei Medici, the English woollen cloth manufacturing industry began to boom and English wool exports to dwindle—a trend that continued drastically in the following century (Table III). After Lorenzo's death (1492), with the population rising again to perhaps 70,000, there was serious unemployment (Cronin, 1967), and Florence fell to the rule of the demagogue friar Savonarola, who expelled the Jews who had been welcomed to Florence by the Medici for their learning (Roth, 1965), and made a bonfire of many works of art. Despite a brief recovery of cloth output in the 1560s, to about 30 thousand pieces, the Florentine economy continued to decline—less than 14,000 pieces by the 1590s, and only 6,000 pieces by the mid-17th century—as other northern nations followed the English example and developed their own textile manufacturing industries (Luzzatto, 1961; Cipolla, 1970).

For a lasting renaissance, evidently, a community may have to be self-sufficient in essentials; for trade, although a splendid stimulus, is an unreliable basis for lasting prosperity. But the population–resource balance is seen to be decisive. Without a favourable balance, prosperity and outstanding creative activity cannot arise; without a *sustained* balance—hence a stable low population—they cannot last. The importance in this connection of 'a stationary condition of population' was apparently first discussed in 1857 by John Stuart Mill, the great Victorian pioneer of feminism and birth control (Harte & Socolow, 1971). But the essential lesson to be learned from the Florentine renaissance is that growth in prosperity can come about through reduction in population *even with reduced production and consumption of resources*.

The Prime Objective: Voluntary Birth-control

All human population crises in history except one have ended in a reduction of population by a sharp rise in the death-rate. The exception was the crisis averted in nineteenth-century Europe by imports from, and emigration to, the new lands opened up in other continents. As we have seen, no such escape from population crisis is possible now or, probably, will ever be open again. It follows that, if world population growth continues much longer, there can only be sharp rises in death-rates in some or all regions. In the words of Ehrlich & Holdren (1975), 'believing that there will be 8 thousand million people in the year 2010 is somewhat akin to believing in Santa Claus'.

If population reduction by high death-rates occurs on anything like a world scale, the accompanying anarchy and environmental damage will probably preclude any benefit from the reduction itself. On the other hand, if we can stop population growth and start population reduction quickly by humane and civilized means, we can not only avert disaster but achieve growing prosperity. The prime objective for all societies must therefore be the reduction of their populations by voluntary birth-control—as the only alternative, and a very rewarding alternative, to their reduction by involuntary death-control.

At present, world expenditure on birth-control is estimated at about 3,000 million dollars annually, as compared with well over 200 thousand million dollars on military expenditure (Black, 1976). This is really very heartening, as it indicates how easily birth-control expenditure could be greatly increased. It also shows clearly that the responsibility for continued population growth rests with governments and not with peoples. Ordinary people are often accused of resisting birth-control, and are said to need extensive education on the subject, but many studies show this to be a misconception. In Britain there is evidence that reduction in population would begin if all unwanted births were eliminated (W. M. S. Russell, 1977). According to the Eugenics Society (1975), for instance, 'mean family size in England and Wales today would be below replacement level, perhaps by about 10 per cent, if births which are unwanted, in the sense that the mother regrets that the birth ever occurred, had been prevented'. The occurrence of so many unwanted births shows clearly that the present British contraception and abortion facilities are not reaching all the people who need *and want them*.

Similar evidence comes from all over the world, including the poorer countries. 'The people of these countries have often turned out to be much readier to accept birth-control than anyone expected' (C. Russell & Russell, 1968). 'The occurrence of traditional methods of contraception and the knowledge and use of induced abortion (usually illegal) in nearly all known rural communities is unequivocal human evidence of the desire of at least some groups to control their fertility' (Senanayake, 1977). Udry *et al.* (1973), studying two industrial and eight poor countries, estimated the effect which removal of unwanted births would have on birth-rates. They found that with such removal the birth-rates would be much lower in all cases, and would be reduced by well over half in some of the poor countries.

In family planning programmes in Taiwan, the eagerness of the people for birth-control facilities was shown by the fact that the new methods spread widely outside the target areas as people in other districts heard news of the programmes (Berelson *et al.*, 1966). Deys & Potts (1976) have described the case of a very poor district of Calcutta, with a population of 30,000. As soon as oral contraceptives were distributed in 1968, 2,552 women began to use them—a higher rate of use than in the United Kingdom. Even after the government stopped the free supply, it was found in 1973 that 643 women were *buying* contraceptives, spending on them 10% or more of their tiny disposable incomes. More than 90% 'eagerly agreed that they would still be using contraceptives if they had access to the original free supply'.

Still more impressive are the estimated 3.8 million induced abortions each year in India, with an estimated 180,000 maternal deaths. According to Lester Brown: 'survival odds for an American soldier going to Vietnam in 1970 at the height of the Vietnam war were much higher than for a pregnant Indian woman considering an induced abortion. Clearly the willingness to take such grave risks is an index of the desperation among women wishing to prevent unwanted births.' The provision of abortion on demand should be an objective of high priority for all countries. As Deys & Potts (1976) observe, it would be well worth delegating large-scale abortion work to 'properly trained and adequately supervised non-doctor personnel'. This system seems to be working well in China (Brown, 1976).

The benefits of a voluntary birth-control campaign begin at once, with less pressure on food, housing, and all kinds of natural and social resources. In Aberdeen, Scotland, where a phenomenally successful birth-control campaign was operated in the 1960s, the immediate results included a reduction in school class size and a spectacular drop in the percentage of crowded housing (Service, 1973). The saving in social welfare costs is also enormous and immediate, so that such campaigns pay for themselves almost at once, and very soon begin to increase the prosperity of the region (Berelson *et al.*, 1966). The ratio of immediate benefit to programme cost has been estimated as five to one in Barbados (Balakrishnan, 1972) and 10 to one in Britain (Laing, 1972). A birth-control programme started in Mecklenburg County, North Carolina, in 1960, is estimated to have saved $250,000 in welfare payments within three years—20 times the operating costs of the programme (W. M. S. Russell, 1971).

Even these large benefit–cost ratios may be underestimates. Fossman and Thuwe studied 120 Swedish children 'born after their mothers had applied for therapeutic abortion on psychiatric grounds and been refused, comparing them with an appropriate control series of the same size' (in Petersen, 1972). Compared with the controls, the unwanted children were more often registered for psychiatric services, and listed more often for criminal behaviour and drunken misconduct; more of them received psychiatric care, were educationally sub-normal, and got public assistance; and more of the unwanted females than the control females married early and had children early. It is clear that unwanted children are likely to *cost* a society more than wanted ones, and therefore avoiding unwanted births will save even more money than is suggested by the existing estimates—not to speak of the saving in burdens other than financial ones.

It is clear that voluntary birth-control programmes bring large and immediate benefits of many kinds; even in simple financial terms, they can rapidly save many times the small outlay required, which is surely within the means of the world community. If only we could eliminate all unwanted births, growth in prosperity could be achieved without depleting or destroying the resources on which it depends, and many scarcity problems would be progressively solved.

References

ANON. (1975). *Food from Our Own Resources.* (Cmnd. 6020) HMSO, London, England: 21 pp.

ASHBY, Lord (1976). Protection of the environment: the human dimension. *Proceedings of the Royal Society of Medicine*, **69**, pp. 721–30.

BALAKRISHNAN, T. R. (1973). A cost–benefit analysis of the Barbados family planning programme. *Population Studies*, **27**, pp. 353–64.

BERELSON, B., ANDERSON, R. K., HARKAVY, O., MAIER, J., MAULDIN, W. P. & SEGAL, S. J. (1966). *Family Planning and Population Programs.* Chicago University Press, Chicago, Illinois: xiv + 848 pp.

BLACK, T. (1976). More bombs or fewer babies? *Ecologist*, **6**, pp. 147–9.

BOCKRIS, J. O'M. (1975). Possible means of large-scale use of wind as a source of energy. *Environmental Conservation*, **2**(4), pp. 283–8, illustr.

BOWSKY, W. M. (1964). The impact of the Black Death upon Sienese government and society. *Speculum*, **39**, pp. 1–34.

BROWN, L. R. (1976). *In the Human Interest. A Strategy to Stabilize World Population*. Pergamon Press, Oxford, England: 190 pp.

BROWN, L. R. & ECKHOLM, E. P. (1974) *By Bread Alone.* Pergamon Press, Oxford, England: xiii + 272 pp.

BRYSON, R. A. (1975). The lessons of climatic history. *Environmental Conservation*, **2**, pp. 163–70, illustr. [Reprinted the following year in *Ecologist*, **6**, pp. 205–11.]

CARUS-WILSON, E. M. (1967). *Medieval Merchant Venturers* (University Paperback edn.) Methuen, London, England: xxxvi + 314 pp., illustr.

CHI, C.-T. (1963). *Key Economic Areas in Chinese History*. (Reprint.) Paragon Book Reprint, New York, N.Y.: xxiv + 168 pp., illustr.

CIPOLLA, C. M. (1962). *The Economic History of World Population.* Penguin, Harmondsworth, England: 126 pp.

CIPOLLA, C. M. (Ed.) (1970). *The Economic Decline of Empires.* Methuen, London, England: viii + 280 pp.

CLOUD, P. (Ed.) (1969). *Resources and Man* (A Study and Recommendations by the Committee on Resources and Man, National Academy of Sciences–National Research Council, USA). W. H. Freeman, San Francisco, California: xii + 259 pp.

COCHRANE, E. (Ed.) (1970). *The Late Italian Renaissance 1525–1630.* Macmillan, London, England: 462 pp.

COUPER, A. D. (1972). *The Geography of Sea Transport.* Hutchinson University Library, London, England: 208 pp., illustr.

CRONIN, V. (1967). *The Florentine Renaissance.* Collins, London, England: 353 pp., illustr.

DEUEL, L. (1967). *Conquistadors without Swords: Archaeologists in the Americas.* Macmillan, London, England: xxii + 647 pp., illustr.

DEYS, C. & POTTS, M. (1976). Attitudes to population control in developing countries. *Eugenics Society Bulletin*, **8**, pp. 8–13.

DICKSON, R. R. & LAMB, H. H. (1972). A review of recent hydrometeorological events in the North Atlantic sector. *International Commission for the North-West Atlantic Fisheries, Special Publications,* No. 8, pp. 35–62, illustr.

ECKHOLM, E. P. (1976). The other energy crisis: firewood. *Ecologist*, **6**, pp. 80–6.

EDSALL, J. T. (1974). Comments on the hazards of nuclear fission power and the

choice of alternatives. *Environmental Conservation*, **1**(1), pp. 15–24.
EDSALL, J. T. (1975). Further comments on hazards of nuclear power and the choice of alternatives. *Environmental Conservation*, **2**, pp. 205–12.
EHRLICH, P. R. & HOLDREN, J. P. (1975). Eight thousand million people by the year 2010? *Environmental Conservation*, **2**(3), pp. 241–2.
EUGENICS SOCIETY (1975). *Quantity and Quality of the British Population*. Eugenics Society, London, England: 11 pp.
FOLEY, G. (1976). *The Energy Question*. Penguin, Harmondsworth, England: 344 pp.
GOLDSMITH, E. (Ed.) (1971). *Can Britain Survive?* Stacey, London, England: 260 pp.
GOLDSMITH, E. (1973). World grain outlook. *Ecologist*, **3**, pp. 361–2.
GOLDSMITH, E. (1977). Planning for starvation. *Ecologist*, **7**, pp. 42–4.
GOUROU, P. (1966). *The Tropical World* (4th. edn., transl. S. H. Beaver & E. D. Laborde). Longmans, London, England: xii + 196 pp., illustr.
GREENHILL, B. (1972). The sailing ship in a fuel crisis. *Ecologist*, **2**(9), pp. 8–10.
HARTE, J. & SOCOLOW, R. H. (Ed.) (1971). *Patient Earth*. Holt, Rinehart & Winston, New York, N.Y.: xviii + 364 pp., illustr.
HO, P.-T. (1959). *Studies in the Population of China*, 1368–1953. Harvard University Press, Cambridge, Massachusetts: xviii + 341 + xxxii pp.
HOLLINGSWORTH, T. H. (1969). *Historical Demography*. Hodder & Stoughton, London, England: 448 pp.
JENSEN, A. S. (1939). Concerning a change of climate during recent decades in the Arctic and Subarctic regions. *Det Kgl. Danske Videnskabernes Selskab Biologiske Meddelelser*, **14**(8), pp. 1–75.
JONES, G. (1973). *A History of the Vikings*. (Paperback edn.). Oxford University Press, London, England: xvi + 504 pp., illustr.
KLARE, M. T. (1972). *War Without End*. (Vintage edn.) Vintage Books, New York: xx + 464 + xix pp.
LAING, W. A. (1972). *The Costs and Benefits of Family Planning*. Political and Economic Planning, London, England: vi + 23 pp.
LAMB, H. H. (1966). *The Changing Climate*. Methuen, London, England: xii + 236 pp., illustr.
LOVINS, A. B. (1976). Long-term constraints on human activity. *Environmental Conservation*, **3**(1), pp. 3–14.
LUZZATTO, G. (1961). *An Economic History of Italy* (transl. P. Jones). Routledge & Kegan Paul, London, England: viii + 180 pp.
MACNEICE, L. (1938). *The Earth Compels: Poems*. Faber & Faber, London, England: 64 pp.
MCNEIL, M. (1964). Lateritic soils. *Scientific American*, **211**, pp. 96–102.
MEADOWS, D. H., MEADOWS, D. L., RANDERS, J. & BEHRENS, W. W. (1972). *The Limits to Growth*. (Potomac Associates edn.) Earth Island, London, England: 205 pp.
MESAROVIC, M. & PESTEL, E. (1975). *Mankind at the Turning Point*. Hutchinson, London, England: xiv + 210 pp.
MUSGROVE, P. J. (1976*a*). Energy analysis of wave-power and wind-power systems. *Nature* (London), **262**, pp. 206–7.
MUSGROVE, P. J. (1976*b*). *Windmill Research at Reading University*. University of Reading, Reading, England: 5 pp. (mimeogr).
MUSGROVE, P. J. (1977). *The Variable Geometry Vertical Axis Windmill*. University of Reading, Reading, England: 4 pp. (mimeogr.).
PADDOCK, W. & PADDOCK, P. (1968). *Famine—1975!* Weidenfeld & Nicolson, London, England: x + 276 pp.
PALSSON, H. (1975). *The Confederates and Hen-Thorir: Two Icelandic Sagas* (transl.). Southside, Edinburgh, Scotland: 139 pp.

PETERSEN, W. (Ed.) (1972). *Readings in Population.* Collier-Macmillan, London, England: xii + 483 pp.

ROTH, C. (1965). *The Jews in the Renaissance.* (Torchbook edn.) Harper & Row, New York, N.Y.: xii + 378 pp.

RUSSELL, C. (1970). The concept of pseudosex. *Guy's Hospital Gazette,* **84**, pp. 241–5.

RUSSELL, C. (1975). The chimpanzee carnival: food, space and social behaviour. *Biology and Human Affairs,* **40**, pp. 77–100.

RUSSELL, C. & RUSSELL, W. M. S. (1967). Population and behaviour in animals and Man. *Memoirs and Proceedings of the Literary and Philosophical Society of Manchester,* **109**(5), pp. 1–16.

RUSSELL, C. & RUSSELL, W. M. S. (1968). *Violence, Monkeys and Man.* Macmillan, London, England: xii + 340, illustr.

RUSSELL, C. & RUSSELL, W. M. S. (1969). Sociological factors in fertility control. *Journal of Biosocial Science,* **1**, pp. 289–96.

RUSSELL, C. & RUSSELL, W. M. S. (1970*a*). The sardine syndrome. *Ecologist,* **1**(2), pp. 4–9.

RUSSELL, C. & RUSSELL, W. M. S. (1970*b*). Overcrowding: a human problem. *Australian Natural History,* **16**, pp. 429–32.

RUSSELL, C. & RUSSELL, W. M. S. (1971). Primate male behaviour and its human analogues. *Impact of Science on Society,* UNESCO, Paris, France, **21**, pp. 63–74.

RUSSELL, F. S. (1973). A summary of the observations on the occurrence of planktonic stages of fish off Plymouth 1924–72. *Journal of the Marine Biological Association of the United Kingdom,* **53**, pp. 374–55.

RUSSELL, W. M. S. (1967). *Man, Nature and History.* Aldus, London, England: 252 pp., illustr.

RUSSELL, W. M. S. (1968). To seek a fortune. *Listener,* **80**, pp. 365–7.

RUSSELL, W. M. S. (1971). Population and inflation. *Ecologist,* **1**(8), pp. 4–8.

RUSSELL, W. M. S. (1976). The origins of social biology. *Biology and Human Affairs,* **41**, pp. 109–37.

RUSSELL, W. M. S. (1977). Britain and the ecological crisis. Pp. 28–38 in *Britain's Crisis in Sociological Perspective* (Eds M. B. Hamilton & K. G. Robertson). University of Reading, Reading, England: iv + 75 pp.

RUSSELL, W. M. S. & RUSSELL, C. (1976). The history of the human life-span. *Update: Journal of Postgraduate General Practice,* **12**, pp. 571–88.

SENANAYAKE, P. (1977). Providing family planning services for rural populations. *Eugenics Society Bulletin,* **9**, pp. 15–18.

SERVICE, A. (1973). *The Benefits of Birth Control: Aberdeen's Experience 1946–1970.* Birth Control Campaign, London, England: i + 25 pp.

SIMON, A. L. (1975). *Energy Resources.* Pergamon, Oxford, England: x + 165 pp., illustr.

SIMPSON, J. (1975). *Legends of Icelandic Magicians* (transl.). Brewer, Cambridge, England (for the Folklore Society): viii + 120 pp.

THOMAS, W. L. (Ed.) (1956). *Man's Role in Changing the Face of the Earth.* University Press, Chicago, Illinois: xxxviii + 1193 pp., illustr.

UDRY, J. R., BAUMAN, K. E. & CHASE, C. L. (1973). Population growth-rates in perfect contraceptive populations. *Population Studies,* **27**, pp. 365–71.

UNITED NATIONS (1977). *Monthly Bulletin of Statistics,* **31**(3), pp. xxvi + 258 pp.

VOGT, W. (1949). *Road to Survival.* Gollancz, London, England: xvi + 335 pp., illustr.

ZUCKERMAN, Lord (1977). Nuclear energy: the burning question. *Radio Times* (London), **215**(2792), pp. 13–15.

DISCUSSION (Session 13)

Goldsmith (Chairman, introducing)

Dr Ducret this morning said the best form of energy was energy-saving, for whatever our resources may be, they are eventually going to run out. At present the accent is on finding further resources; yet whatever these may be, they too are going to run out eventually. The ultimate problem is really not to find more resources but to learn to live with less of them, and this is really the subject we should be concentrating on. We need a new set of solutions based on a totally different interpretation of our problems and hence require the use of a totally different set of resources whose very existence we have almost forgotten. I refer to biospheric resources, [rather than our usual] technospheric resources.

Of biospheric resources I will deal with [two] only: the family and the small community. If we are going to learn to deal with less resources it will mean reducing our need for them, and clearly [the way to do this is through employing the] astonishing biospheric systems which previously satisfied all human and other basic needs. For some two million years Man used practically no technospheric resources at all; our most sophisticated technology was the use of digging sticks, bows and arrows, and things of this sort. To say that Man *needs* much more for existence is simply untrue. It is only in the last century or so that we have 'needed' the amounts of fuels and minerals we are using today. It is *industrial* man that needs all these things, though previously Man's needs were perfectly well satisfied by the normal functioning of the natural systems in which he lived.

We may illustrate this by looking at the problems of the family, which, as a resource, is infinitely more valuable than all the oil wells that have ever been found or are yet to be found. If one were to restore the family to its proper place it would solve a whole lot of problems at once, whereas every time we try to build up the technosphere in new, clever ways we solve one problem but create others. That is a problem multiplier.

If we restore the family, we solve a very large number of different problems, thereby reducing the need for technospheric solutions for which we do not have the resources. Why do we have a housing problem in Britain? Because we have 2.3 people per house. We used to have 8 or 10 people. The number of houses required is multiplied by 4. Why? Because the family unit has disintegrated. If the family unit were to be re-integrated, we would cut down the need for houses [to one-quarter—and that means also the cost of maintenance and wasteful use of agricultural space].

If you restore the family unit you will also, to a large extent, solve the problem of crime and delinquency, which is one of the major problems in America, [where from] a recent Gallup Poll I see that the problem of crime is considered to be far more important than the problem of inflation or unemployment or any other problem. And if you look at it carefully, you will see that a very, very large proportion of delinquents and criminals come from broken or otherwise unsatisfactory family units. If you solve the family problem, you will solve a whole set of other problems—such as the educational problem, which is becoming increasingly difficult because, in industrial countries, an increasing number of children are emotionally disturbed and simply cannot be educated. The teacher becomes a policeman to keep order, because these children cannot concentrate. Meanwhile the cost of education goes up and up. The solution lies in restoring the family unit and exploiting a biospheric resource whose very existence we now largely ignore. If we want to talk about a form of growth—a new growth—there is to my mind no other possible form than biospheric growth, instead of the technospheric growth to which we have been committed for the last 150 years.

[The authors of our Keynote Paper on 'Scarcities and Societal Objectives', Dr and Mrs William Russell, of Reading, England, are unfortunately unable to be with us today, and their paper, which you already have, will be presented by Mr Peter Stone, an Oxford-trained geologist who is Editor-in-Chief of the UN *Development Forum*. He was largely responsible for the successful publicity of the famed Stockholm Conference under Maurice Strong. Our panellists are Dr Gordon Butler, Director of the Division of Biological Sciences of the National Research Council of Canada, Mr David Dodson Gray, Co-Director of the Bolton Institute, of Wellesley, Massachusetts, and Dr Richard Borden, of the Department of Psychological Sciences, Purdue University, West Lafayette, Indiana.]

Stone (Presenting paper for Keynoters)

[Mr Stone commenced by warmly recommending the Russells' paper (of which copies had been circulated earlier) and explaining some salient points before speaking] about methods by which societal objectives [might] be achieved.

Although the Russells deal with the need for birth-control programmes, they do not deal with the problem of how to persuade people to adopt them—the problem of communication and motivation. I personally believe there are several other desirable societal objectives in addition to voluntary birth-control, although I am quite prepared to concede that in the long term birth-control is the only way to survival. Other societal objectives [that are widely] recommended include motor cars which last for 30 years, recycling limited resources, reduced consumption of energy, better behaviour with regard to pollution, and so on. It is very important to define the objectives, but they [remain] academic unless you mobilize the population at large to action. But how do you do this? Can it be done at all?

Ecodisasters discussed here all have one thing in common—a very rapid change in the conditions of life as experienced by a mass of the people. These rapid changes may be, for example, something as simple as a sudden rise in prices. The government may begin a communications effort and call for voluntary wage-restraint or a tightening of belts. We have already had the Government of India pleading the cause of family planning, for people to 'stop at two'. These efforts may be successful, and result in passive adjustments by the broad mass of people. Instances are familiar enough. Thus in the last war the British voluntarily decided to eat less and to re-apportion the country's resources in a voluntary fashion to meet war-time needs that everybody agreed were pressing.

The recent world-wide inflation tells us that something, whatever it is, has not been voluntarily accepted by the broad mass of the people. The oil price-rise which accentuated the already growing inflation was regarded by almost everybody as a sort of caprice—something that was not the result of a 'natural' or inevitable scarcity. The [concomitant] inflation seems to be an example of a rapid change in the condition of life which has damaged and is damaging societies very grievously indeed. It is reasonable to argue that this could have been avoided by a proper communications effort.

One can quote the standard comments dwelling on the difficulties: who would do it, how are you going to plan it? It is terribly difficult with a commercially-orientated 'free' press that is [chiefly] interested in football, skiing or tennis; people want to watch their favourite character on television! The energy problem is not like a war; not like the situation when the British, with the big identifiable bogey Hitler, did eat less and did redirect their efforts in other directions. In communication terms the difference is largely one of degree. People need to have an environmental issue communicated to them so subtly but deeply that it feels like a war—that the matter is as urgent as a state of war.

Quite apart from the business of communicating environmental concern to everyone, is that of communicating across income gradients so that the rich North understands the concerns of the poor South, and the two are not for ever locked into the present adversary relationship. [Much the same applies to ethical or tribal barriers, and ones of class, religion, or mere custom and habit.] I think we all need to watch once a week, for the foreseeable future, a television show called 'How is our world getting on—will it last?' We [now] have the technology to do it in the shape of direct-broadcasting satellites.

Let us take another aspect of the media which is rather more abstract. For many years we have had bodies such as the British Broadcasting Corporation and Italian television, which were major public broadcasting corporations with public responsibilities. They used to [give us] an hour's religion a week, and they felt they had to do something about the environment because they had public responsibilities to discharge. Unfortunately, the trend is in favour of the dissolution of these public-service bodies. People want to break up the BBC and disband it. Italian TV has almost broken up as a result of a court order which said that its monopoly of the air time was in fact illegal. We all know what American television is like—mostly a padding between the advertisements. We know, too, that the public broadcasting service in America behaves like a poor parish church trying to repair the roof. It hardly has the money to get on the air.

[Still with the media], we have a debate about the free flow of information. Third World countries do not like the way that wire services ignore the problems of development, and they do not like the Western domination of the world news. But suddenly this valid criticism from the Third World has turned into something quite different. It has threatened to [cause] governments to express a desire to censor information that leaves their country for distribution elsewhere. Of this there are already many time-honoured examples. During the Stockholm Conference on the Human Environment, for instance, a number of countries said, 'Oh, we don't have any oil pollution on our beaches'. Everyone knew that they did, one could read about it in their local papers, but no government could be expected to kill its own tourist industry. This kind of thing happens regularly with the World Health Organization: countries deny that they have an epidemic of some disease, even when it is very evident. The free flow of all sorts of information has long been at serious risk, and the situation is getting worse in some important sectors. One example is the way in which the press copes with the responsibility of world-wide reporting. There is not a publication distributed world-wide of a non-specialist kind that is not losing money, or is not very close to it. Newspapers everywhere are reducing their overseas staff in favour of agency service—the substitution of personal reportage by impersonal mass-produced news.

If our society is going to hang together, as it swings through the periods of stress already upon us, a radical change in our attitude to information and communication is essential. Communication is the nerve-system which allows societies to have the agility and speed of response to deal with rapid rates of change in the environment—with what we call an ecodisaster. Now you do not have to mobilize people to act voluntarily and quickly, because you can *force* them to do things. Thus you could force people to have lower wages or fewer children, and so on; but, if you do so, you will not mobilize them to pull together—in fact, you may well cause a crash in morale and a bigger disaster instead [of an improvement]. One may recall the Russells' paper and its description of the rise and fall of Iceland's population. In each of those falls there was a situation in which the population was forced to respond because it had not done so voluntarily [beforehand].

We hear a good deal about China and how the Chinese have met their [potential or actual] ecodisasters, [being] forced by circumstances into doing something about them. But I do not think it was quite so simple. The Communist revolution was the [latest] attempt to tackle ecodisasters; force of circumstances had already deposed the emperors and gotten rid of the war-lords. The Chinese people had already witnessed the efforts of Sun Yat-sen and then those of Chiang Kai-shek. One of the most striking features of [present] China is the effectiveness of its information system. The imperatives of living with, and extricating themselves from, what is essentially environmental disaster, are discussed every day in the cells—the lowest communal level possible, just larger than a family unit. Chinese people are told every day that bicycles and railways are right and that big roads and big cars are wrong. They know why late marriage is encouraged, and they know why they need to work for the improvement of their environment. As a result, they do so voluntarily and are mobilized, not coerced.

If the world community is to steer its way out of the coming storms, I think it has got to do so by a cooperative and voluntary effort. It will not get through by *élites* forcing changes at gunpoint. Ships work best with volunteer crews who know what they are doing and why they are doing it, with officers who tell them what it is all about and also, my last and perhaps most important point, with officers who listen to the crew.

I had the honour to be with Maurice Strong at the Stockholm Conference, with responsibility for public information. I never really knew what Maurice himself thought public information was about, but I very soon found out what the others on the Conference secretariat thought it was—the dishing out of press releases after some boring intergovernmental working-group. Press releases were never to mention disagreements; there could never be anything juicy or funny in them. And as for communications working in reverse—for the UN or the assembled governments, listening to what *they*, the recipients of the press releases, had to say to *them*—not a chance! My subsequent experience in international communications has confirmed me in the view that most people involved in governmental and intergovernmental work have a disastrously limited and mechanical understanding of the subject. We should indeed hope for 'new communications' to go with 'new growth', but I fear we are not likely to get it. Perhaps we need an expert meeting to review the state of the art. It is quite possible that the new UNESCO 'International Commission for the Study of Communication Problems' will address itself to the need to make some global preparations for the problems to come. It is [now] up to environmental scientists to make sure that communications get proper recognition as a vital element in our defences against ecodisaster.

Goldsmith (Chairman)

Inflation may be, from the point of view of environmentalists, a very good thing—in a sense a necessary step—because, [to the extent that] the world has become industrialized, it has largely been because it has been cheaper to use machinery than people. Industry has undoubtedly been, above all, the substitution of machinery and other technospheric inputs for labour. When once these inputs became too expensive, then we might see suddenly a transition to the opposite situation. Thus it might become economic to put people back instead of going on using machines, and then we might get what must [seem to many] the ideal thing—a gradual transition to a much more sustainable and more sensible world rather than [have such changes thrust on us] as a result of disasters of one sort or another. Inflation in fact may be a necessary step in this direction, towards a more satisfactory society.

Butler (Panellist)

I have read the erudite and interesting paper of the Russells and I commend it to you. Some of you will have perceived that this paper has focused on one of the central issues of this Conference, when it points out that, in recorded history, in various regions of the world, Man's population and welfare have had a cyclical course with time. There have been periods of population growth, followed by increasing problems and [then] crises—until the population curve turns over and descends as the difficulties intensify. Finally, when the population had declined sufficiently, food becomes more plentiful (sometimes also due to an improvement in climate), the population increases again, and the cycle is repeated.

One of the things we have been talking about in this Conference are the various signs, varying with the perceptions of each of us, that the world is approaching the end of a cyclical increase in human population and activities, to be followed by a levelling-off and decline. Mr Stone has just described some of the societal ills which accompany this levelling-off, and which may presage an ecodisaster. This is what we fear to contemplate.

The Russells' paper shows that whereas, in the past, this phenomenon was restricted to countries and regions, it has now become a greater possibility to occur throughout the world, because of increasing communication, trade and migration.

The question before this Conference, already discussed [to some extent], is, 'How, for the first time, can we alter this historical pattern?' We would like to stabilize the population somewhere between the minimum and the maximum of this cycle. The desirable level for stabilization depends very much on individual perceptions. Some persons would like to go back to the conditions of a century ago, others would choose today's level, and still others would choose the [projected] higher level of 50 years hence.

Whatever level we choose, to achieve it will require a tremendous effort, personal hardships and acrimonious argument. Only thus can we reverse a pattern [that has been] established through thousands of years of human civilization.

Goldsmith (Chairman)

Our next panellist is an engineer turned theologian, who has involved himself very considerably in the whole limits-to-growth debate, and is co-author of a two-volumes book called *Growth and Its Implications for the Future*.

Dodson Gray, David (Panellist)

I thought Mr Stone's presentation of the summary was very interesting, [as was surely] his own outlining of what seems like a very significant direction of social invention.

[It seems clear to me that we have a] need to find new social inventions. One of the key problems which we face in a cyclical set-up like this is that there are long time-delays; that, for example, the population can have kept on growing past the point of disaster and then, after the disaster-point becomes apparent, the population still keeps on growing, and it takes a very long time before mid-course corrections actually take effect. Among the things we can look for, in addition to more deaths or drastically fewer births, are social inventions that give us some time in between. [It is clear that improved communications would help to accelerate] the rapidity with which the word gets out and people realize the problem and take their own lines of action.

I would like to suggest several other social inventions which have been developed in times past for dealing with scarcity crises. It is quite clear, as the Russells suggest, that scarcity *equals* too many people. Yet the United Kingdom dealt with this with a great social invention; it invented the [British Empire and then the] Commonwealth, and it began to import food which it could not grow itself from New Zealand and from other distant parts of the [Empire and subsequently] Commonwealth. When Britain abandoned colonialism, we all got into world trade. When it was working fairly, the idea was that we were exchanging surpluses. [There was] a great deal, for instance in the UK, of the goodies of technology. And other people had a great deal in the way of goodies of food and raw materials. We would give up some of our goodies in exchange for some of their goodies. It was an invention and it resulted (at least in theory) in regional specialization of function in order to get least-cost production. The result we have lived with. This is a classic example, I think, of a social invention which expands the natural carrying capacity of a given region. I think we are going to have to look for social inventions which will either expand the natural carrying capacity or expand the adjustment time, or else speed up the adjustment process.

Sitting here these last few days I had an idea for a social invention and I am going to share it with you. You are the first ones to hear about it. So bear with me, and I hope you will find it provocative if not yet completely worked out. It is an example of the sort of thing we are going to have to do and, as I understand it, Mr Strong last night was suggesting that we get on with.

We in the United States late in the nineteenth century had several disastrous financial panics which were financial equivalents of what today in the biological realm we would call ecodisasters. There were sharp breaks in the webs of relationships that were sustaining us. Finally, one of those financial disasters was prevented, though in a way which I do not think could happen in an ecological disaster. This was done by one man: a financial panic was stopped by J. P. Morgan. When we in the US realized that one man had been the social invention that had [saved the situation], we had several years of Congressional hearings, from which we were all appalled [to find] that one man had such power. We were extremely grateful that he had [saved the situation], and we decided we had to institutionalize this social invention with a semi-public, semi-private, trustee agency to overlook the well-being of the financial fabric to prevent future financial disasters. Now what I am going to suggest is that we do a similar sort of thing in relation to our ecological disasters.

We in the US call this invention the Federal Reserve System. The rest you know, I hope. The Federal Reserve System is subdivided by regions, so that in itself it is a kind of financial version of local ecosystems that work reasonably well. It is interesting that one of the major functions of the Federal Reserve System is to monitor and maintain the well-being of the financial network of relationships by controlling the rates of growth in the money supply. I think that if we were to have something comparable in terms of the well-being of ecosystems in a region, it would have to control the rates of growth in demand upon the ecosystems. [Ideally the entire world should ultimately be so covered.]

It is also interesting that one of the things which the Federal Reserve System does is to require local banks, which are key factors in local financial systems, to have reserves. I think that a similar sort of thing might be made of a Trustee Agency in terms of local regional ecosystems—that indeed they would have to have ecological reserves, or ecosystem reserves, so that we would have ways of safeguarding and allowing time to mend potential breaks in the fabric which supports us in life.*

Now that is just one idea. I think there are going to be others. But from the idea of a Trustee Agency looking after the financial fabric, we can see that we also have similar [needs] in terms of looking after the general economic fabric, [and yet others in terms of looking after the environmental fabric]. Ever since [the time of] Lord Keynes we have [become more and] more aware of the financial need, and what I am now suggesting is that we need to safeguard environmental productivity and environmental capacity to meet demands upon the environment. I think that something like a Federal Reserve System for the environment begins to loom as

*Would it be too much to dream of a Global Environmental Trustee Agency, perhaps linked with a system of Biosphere Reserves?—Ed.

necessary if we are going to damp the heights and depths of the swings, so that we are not simply left with an ecosystem with fixed boundaries. Ecosystem limits can be moved; we are not stuck with a given structure, for we can create new structures [as well as more of the same].

One other thing I want to add and then I will conclude. The picture that the Russells anticipate, of a society pressing against its resource limits, is one of social disintegration. An alternative strategy in such a situation is to take measures to increase social cohesion. You can do this by a variety of means, and the World War II experience is the one I keep coming back to. Certainly war-time helped us to deal with scarcity by increasing social cohesion, and an essential prerequisite for that is to have an increased sense of fairness. As I understand it, in England it was because everyone felt that it was being done fairly that people were willing to get along on much less than formerly and do their part. In President Carter's energy proposals (April 1977), approximately 20% of his whole message was devoted to the need to do what we did *with fairness*, and much of the complexity of his proposal was a consequence of his trying to deal with that fairness issue.

So I think we need to recognize that social disintegration such as the Russells envisage in a scarcity society is *not* inevitable. It is only inevitable if we continue present social structures with their inequalities, their resentments, and their perceived injustices. One of the ways in which society buys time is by increasing societal cohesion. People did this in the Battle of Britain; we can do it also in this moral equivalent of war, and I think we need to recognize in a variety of ways that we are going to have to come up with societal strategies and inventions which will make this an all-win effort—so that the environment will win, and everyone in society will win.

It will no longer suffice to say it is going to be a win-or-lose situation in which the environment loses or we win, or that one part is going to win and somebody else is going to lose. What idiocy [, when we are all in the same boat]! We need to recognize this and put our wits together, which means in part putting our consciences where they ought to have been all along.

Goldsmith (Chairman)

I now call upon our last panellist, Dr Borden, who is an environmental psychologist. I think this may be a field which will have something to teach us. The problem we are faced with is basically a psychological and a social one—one of values—at least if we take a culture to be a controlled mechanism, which I think can be shown. A society is governed by its culture and this you can see in a tribal society's stability, with its specific environment clearly maintained by its particular cultural pattern, in which are embodied its values. If we are going somehow or other to re-establish some sort of stability, it clearly will be also by somehow modifying our values and developing that sort of cultural framework which will enable us to do so.

The world of environmental psychology appears to be a very important one.

Borden (Panellist)

I want to take a couple of minutes to outline, and give you some flavour for, what environmental psychology is. But before I do that, there are two comments I would like to make on Mr Stone's presentation. First, I think it is futile to look for some 'general equation' to try to solve social problems. Instead it would probably be much more practical to look towards what might be called 'small situational solutions' in whatever area you are involved. Oftentimes the solutions to psychological problems are quite simple and obvious: all you really need to do is to ask people what is wrong. For example, in the instance of population control, many population problems are really a function of birth-control, and birth-control is the use of contraceptives. In America, unwanted pregnancies among teenagers create a most serious population problem. If you ask a teenager 'Why don't you use contraceptives?', which is what one of my colleagues has recently done, they say they do not use them because it is embarrassing to purchase them. It is embarrassing because in America they are dispensed by pharmacists who keep them behind the counter and are well known for leering at youngsters who are bold enough to ask for them. What we find is that if you put them out in the open where they are readily available, and where interaction with the pharmacist is unnecessary, the sale and use of contraceptives increases dramatically.

In other instances, psychological processes and solutions are much more complicated. For example, Mr Stone's reference to President Carter's so-called moral equivalent of war on environmental problems makes a nice analogy. However, in the case of a real war, it is easy to mobilize the public psychologically by developing enmity towards the enemy. Further research in group processes has repeatedly demonstrated that such a building up of enmity also produces strong feelings of amity and cohesiveness within the [home] country. Unfortunately, the analogy breaks down when applied to environmental problems because the enemy is not somewhere else or even someone else—*the enemy is us*. This is further complicated by the powerful ways in which social forces work to inhibit interactions that could help alleviate environmental problems. In particular, it is easy to walk along the street and see someone throw a beer can from their car or drop litter as they walk. It is very hard to walk up to that person and say 'Pick it up!' None the less, that is what we must do if we are to enforce environmental beliefs interpersonally.

Now let me return to what I really came here to say, namely what environmental psychology is and what it promises to become. Actually, environmental psychology is broken into two areas; the first is somewhat behaviouristic and attempts to understand the ways in which environmental influences act upon individual and/or group feelings and behaviour. Here we find people examining such issues as stress effects associated with noise, crowding, air quality, and so on. Other people are also [investigating] the immediate and long-term psychological effects of massive natural events such as floods, tornadoes, hurricanes, and so on. Finally there has been considerable research in recent years concerned with the influences of the build-up environment on psychological factors. These architectural–psychological investigations are concerned with quality of life in the home environment and also with group processes and group productivity in work environments.

The other breed of environmental psychologists take a more proactive or humanistic approach, based more on the assumption that the human being is an anticipative constructor of outcomes. Lately, this field has been referred to as the 'Psychology of Environmental Concern'. Briefly, this field of study focuses on such things as the measurement of environmental attitudes, the structuring and evaluation of environmental education programmes, the determinants of environmental voting behaviour, and environmentally motivated purchase and/or boycotting behaviour. My own interest has been directed towards the way in which people perceive (or perhaps do not perceive) environmental problems, and if they do see environmental problems, what kinds of feelings would they have. Further, I am much interested in the ways in which such feelings lead to various kinds of individually responsible environmental behaviours. After a couple of years of work we have been able to narrow down to a few important phenomena.

I do not want to sound mystical, but I do want to introduce a concept that is called 'Primitive Beliefs'. By primitive belief I do not mean some ghost-like archetypical image lingering in the recesses of our monkey brains, [but rather in the manner that] anthropologist Clyde Kluckhohn develops the concept in his theory of value structure. In this way a primitive belief is used to mean a belief that is very central to a culture—so much so that it is not even conscious, i.e. most people are unaware of it. None the less it is felt by the holder to be true and not in need of defence. Such beliefs are not usually acquired by tuition but are so pervasive in the culture that they are 'absorbed' by its members.

In terms of environmental concern and environmentally responsible behaviour, two such primitive beliefs are especially important. The first of these has been elaborated in a well-known paper by Lynn White and may be described by the phrase 'Man is apart from Nature and should utilize Nature to his advantage'—or, more briefly, *Man over Nature*. According to White and others, this belief derives primarily from the Judaeo–Christian concept of mankind and Nature which sought to remove animistic spirits from the non-human environment and stressed Man's superiority to Nature. It was under the influence of this belief system that science developed, and which Descartes promised would make us 'Lords and masters of Nature'. More recently we find a corollary belief in technology. A serious consequence of these beliefs is that they obscure the fact that we *do live in ecosystems*.

I shall return to discuss other influences of this belief but first I would like to outline one other important primitive belief that exerts a countervailing influence on environmentally responsible behaviour and/or conservation. This second belief is that 'growth and progress' are natural and right. The most important consequences of this belief are: first, a vague sense of optimism that accompanies the idea of growth, or conversely, the uncomfortableness about the idea of no growth. Another consequence of this belief is that growth justifies growth. In other words, if someone, say a 'developer', attempts to persuade local residents to permit the building of a new airport, factory, suburban sprawl or whatever, his most persuasive argument is that it is 'necessary for growth'.

I have recently been examining transcripts of re-zoning meetings and in virtually every instance the arguments made by developers rest on the need for 'growth and/or progress'. Psychologically, by reason of the centrality of these beliefs, they are highly persuasive because they short-circuit other, less central counter-arguments. Clearly, people are easily upset by the notion of no growth. Apparently if something is not growing, [it is somehow] wrong. One instance, at least in America, where institutions [commonly are] not growing, is in academia. Surprisingly, academics seem to have adapted quite well. Instead of building more dormitories and bigger programmes, American academics seem to be focusing on the quality of their programmes—for a change, and quite happily as it appears.

I would like now to return to the Man over Nature belief and some of its consequences. There was some talk at this Conference yesterday of poll results indicating that Americans are becoming very concerned with the environment. I feel that there may be some false optimism derived from those statistics. True, there are some people in America who are very concerned about the environment, and have altered their lives in the direction of self-imposed conservation. Unfortunately there is a substantial proportion of the population that is reacting with the opposite response. Here we find that the sale of large automobiles, in the face of less but more expensive petroleum, has increased dramatically. What they are saying is 'damn it, I have always wanted a big car—I am going to have it *now* and I don't care'. So what we have is a polarization of environmental attitudes. Politically this could become a serious confrontation, as we have recently seen in the public demonstrations at nuclear power-sites and in regard to supersonic transportation. Given the financial power that these technologies can exert, we may in fact have some very bad environmental decisions taking place at the political level.

Let us take a moment to look at environmental attitudes. In some of my own recent research I have interviewed several thousand people and found many who do not care at all about environmental problems, many who care but do not do anything, and some who have changed their lives to help in whatever ways they can. Some of these differences are a function of personality. Most of them are linked to belief in technology, and the way in which this belief works in environmental commitments goes something like this: It appears to be akin to the well-known *diffusion of responsibility* process that occurs in [potentially] altruistic situations. For example, if you are ice-skating and fall through the ice, the more people who witness your dilemma, the more likely it is that you will drown. This paradox first came under investigation following a brutal beating and knifing of a young woman in New York City who was disrobed, stabbed some 30 times, and cried for help in the street for half an hour before dying. Many people witnessed her situation, but none helped. Since then, numerous controlled experiments have determined that if only one person sees another person in need, that one person feels highly responsible and usually renders assistance. But as increasing numbers of people see someone in need, their feeling of responsibility is diffused, because they feel someone else will help. In fact, the overall probability of the victim getting help from increasingly large groups of bystanders rapidly approaches zero.

Belief in technology 'short circuits' environmental concern in a very similar fashion. As long as people feel that technology can solve environmental problems, they lose their motivation to make environmentally altruistic self-sacrifices. The dramatic nature of this phenomenon is revealed in the results of a recent poll which showed that 88% of the people questioned agreed with the statement: 'Technology got us into environmental problems, but technology will get us out.' As long as people retain such [blind faith in technology], there is no reason to expect them to take energy-conservative actions in their home, to use their automobile prudently, to recycle their waste products, or otherwise personally to commit themselves to the solution of environmental problems.

Goldsmith (Chairman, interrupting)

Though your time is up, you are getting down to the nitty-gritty. We are talking about societal goals, and I think you can define a society in terms of its value system—in terms of the things it believes in, including its taboos—[but please conclude briefly].

Borden (Panellist)

I am currently testing some notions regarding the use of the media for environmental education. I believe very strongly that the only way we are going to be successful in this regard is to use some of the available technology, and obviously the most important one is the communications media. I found that there are very large differences between 'ecophiles' (people who are concerned and motivated in environmentally responsible ways) and 'ecopaths' (people who are indifferent to environmental problems). The most important difference is that the former are willing to doubt the supremacy of technology. One asks then, where does doubt in technology come from? The answer seems to be that doubt [in the supremacy] of technology comes from learning ecology, and we find that people who go to universities to take ecological and environmental studies courses become less infatuated with technology and more environmentally responsible.

Now, how are you going to educate an entire population environmentally? The only way that you can do it, as they cannot or will not all come to universities, is to get into their homes. The best way to get into their homes is through the media, and in Western society the most powerful medium is television. I am currently constructing a pilot project to be used in a university town and nearby mid-western city with about 60,000 residents. Using public-service television we are constructing short educational messages that focus on two issues: (*a*) recycling of refuse, and (*b*) the use of public transportation. In a 30- or 60-seconds 'spot' we first develop concern by dramatizing the compound effect of irresponsible behaviour and then provide directive information which emphasizes (*a*) that there is no technological solution, and (*b*) what people personally can do. In this last part the people are educated about specific activities concerning how to recycle, how to find their local recycling facility, who to call or write to for more information, bus routes, bus fares and so on. This I believe is the key. We must, in a systematic, repetitive and well-produced fashion, (*a*) teach the basic lessons of ecology, (*b*) generate concern, (*c*) personalize rather than technologize the solutions, and (*d*) provide specific instructions regarding activities that people can engage in.

As I mentioned at the beginning of this presentation, environmental concern is increasing. Our job, as I see it, is to help the public to translate this concern into responsible action.

Tan

I have been pleased to hear mentioned in an enlightened way the prime objective of voluntary birth-control, which is not difficult to implement in [many] developing countries where people seem [readier] to accept this form of behaviour—as I have encountered among normally uneducated people who nevertheless seem quite keen to limit their families. Many of the women had had 6–8 children before they were 30 years old; they wanted to limit their families but were put off birth-control—I believe more by [some forms of] Christianity than by Women's Lib. Also, Asian women can be very sensitive to western medicine and many had very unfortunate experiences with the [anti-natal] Pill, which was originally formulated for women of a different constitution, well nourished and generally of larger stature. The lesson has at least been learnt, that what may be beneficial and applicable in the developed societies can turn out to be a farce—even a near-calamity—in the more ancient societies of the rest of the world. In Kuala Lumpur we were encouraged to limit our families by taking advantage of a fairly informal practice now available in hospitals: after four children a woman can voluntarily have a tubal ligature and I know that many illiterate women have taken advantage of this although not all such operations have been successful.

Myers

I thoroughly enjoyed the substance of the paper and the comments of the panel, and feel that this has been very much of a vintage day in the Conference. But whereas the paper dwelt at some length on the idea that the principal form of scarcity is precipitated by excesses of human numbers, and I do agree that this is *a* principal if not *the* principal factor, let us not go overboard on that one. Let us also bear in mind that another very relevant form of scarcity is lack of intelligence in allocation of resources which is often associated with a lot of people. For instance, in 1949 China had 500 million people who were hungry, illiterate, malnourished and unhealthy, and the country's natural environments were a wreck. Today the situation has been restored in significant measure in several directions, and China has 800 million human inhabitants. This in my view is partly because they have been making excellent use of one resource which is still a very long way off being totally exploited, let alone over-exploited, and that is the resource of the human mind, which was mentioned this morning. I am referring to the inventive capacity of the human mind—not the technological capacity of the human mind which the panel has so well warned us against.

My other point is that there has been a good deal of talk about war—the spirit of war and the social cohesion that a war-time situation engenders. I think this is rather interesting as it looks as though, given our styles of living and over-consumption of resources, we may be heading for something that will approximate to a siege economy. In fact in some respects we are well into that state already.

I sometimes wonder whether it might be an interesting exercise to look at the present military expenditures, which I believe are about 400 thousand million dollars per year world-wide, and which are generally justified in terms of security of one kind or another; I also wonder whether, supposing 1% of those military expenditures were allocated to other forms of security, such as safeguarding our environment or safeguarding supplies of exhaustible resources, might not this add at least as much security at national level and at global level as would have been perhaps lost through such a 1% saving on military expenditures?

By the same token, there have been in history at least three occasions when society has managed to mobilize a vast array of its resources—human, capital, financial skills, and other forms of resources. The three that come to mind are China's building of its Great Wall (which took many decades), the pyramids of Egypt (the construction of which I believe extended over several generations), and, more recently, the American effort to send a man to the Moon. I wonder if, were it possible to look into the minds of people who were engaged in those [great] efforts, we could find out what levers of their psyche were being pulled in order to encourage them to pool their efforts into a common purpose of that scale. That is why I am very glad to have seen on the panel this afternoon (and I was very interested to hear what he had to say) an environmental psychologist—though what I am also looking for is a social psychologist who can tell us something about how people should cooperate as a community, especially in a situation where we are into non-zero sum games as opposed to zero sum games (as was so well pointed out this morning).

Jones

I am encouraged to hear noises being made about getting some kind of scientific education into the living-room and into the family, because I do believe that a lot of the environmental troubles which we tend to discuss here, and in fact the problems themselves, are so exceedingly complex that society generally needs a great deal of assistance to understand them. However, getting involved in 'educational programmes' on mass media is something which tends to be entirely foreign to many of our scientific colleagues, and in particular to academics—they hate to get involved in the 'Children's Hour' approach to science. Yet this is really where [most]of this information is transferred.

A good example of public misunderstanding was demonstrated when a colleague of mine was involved in a survey to evaluate the perception that people had of air pollution. In a number of Canadian cities we have air pollution indexes, and he was trying to determine whether or not people perceived air pollution [when] they walked out of the house in the morning and their eyes started to water, or whether their eyes started to water when they read the newspaper which cited a high air-pollution index. In this questionnaire was a very clear explanation of how the air pollution index was constituted—what components were involved in this magical number—and it was explained that it consisted of particulates and sulphur dioxide. It was a very simple, primitive index. Then the questionnaire went on to say: in the event that you were told or warned that we were going to have an inversion, that the air pollution index would rise above the critical number, that certain industries would be asked to switch to low-sulphur fuel, and that a rather serious problem was about to arise, what would you as an individual feel that you could do to help?

The most popular answer in the returned questionnaires was, 'I would leave my car behind'—which makes sense, even though the questionnaire did explain that the air pollution index had absolutely nothing to do with motor-cars which do not normally produce sulphur dioxide, though they do produce some particulates. But the second most popular answer, which was the one which had everybody really in despair, was that they would use a low-phosphate detergent. And this, Mr Chairman, suggests to me that somehow or other we should try and get some level of scientific expertise into the living-room *via* the family, *via* the television programme, and I think that perhaps our eminent scientific colleagues should not look down their noses at those other colleagues who are brave enough to stand in front of the 'Children's Hour' type of programme and provide this kind of information in a clear and lucid fashion—so that the children can, in fact, explain these matters to their parents.

Dodson Gray, David (Panellist)

Other interesting and creative areas of social invention right now [in America] lie in the kinds of curricula which are being developed for use originally in Junior High School. But these curricula are also now being picked up in the broader community, where people are trying to solve problems. I think this is an important example of public communication—not by the television or the radio but in relation to particular issues. I have with me a copy of a curriculum which was developed in the United States, and I am going to put it out [as an exhibit and for anyone who wants a copy to leave his or her name]. It has to do with *Energy, People, and the Environment*. It is a coal-mining issue, dealing with the history of coal-mining, and it is designed to help young people. But it is also being picked by other groups and communities that are concerned about strip-mining.

Goldsmith (Chairman)

As Chairman I think it may be my prerogative to guide the discussion in a particular direction, and I would like to do so in that of societal goals, which I think is the principal subject of this discussion, though communication and scarcities enter too. I think the discussion was started in the right direction by our environmental psychologist here. Does anyone have any further comments to make in the same direction?

Thorhaug

In Florida, where I come from, there is an extremely strong environmental movement among the lay people, and this involves individuals ranging from extremely old, retired people, of whom we have a great many, to very small children. These people really have changed their life-styles and have worked endless hours for the environment: for example, picking up litter, or going to hearings. They have developed many of the kinds of traits that we would like to see most of society emulate. These people come from the upper classes and from the lower class; they range from about 5 years old to about 90—literally. How they became so environmentally conscious might well be an interesting area of environmental psychology, [especially as they] are really the keystone of how we have gotten our environmental measures passed. The legislators are very conscious of the numbers in these groups: for instance, they have 6,000 in the Audubon Society in Florida and about 1,000 in the Sierra Club. They have pushed our environmental goals further than any [other section of society].

There must be something unique about these people; a good many others just could not care less, and persist in driving their air-conditioned cars to their air-conditioned offices, burning energy. It occurred to me that since this rapid development of 'environmental consciousness' in the last 10 years in Florida, California, Oregon, and other [areas] has occurred, it may well be worth looking at the reasons and taking the findings [elsewhere with a view to promoting] the same kind of consciousness for the whole of society.

Bryson

We are supposed to be talking about societal goals, and presumably these will be the goals that we, in our wisdom as ecologists, perceive. This we will have to communicate somehow, getting these goals across effectively, and perhaps we can do it: I can suggest what seems to have been, for us, a successful method of doing [what is really our duty if we have the opportunity—and if we have not, we should make it].

Working on the assumption that people like things free, and that people like to have work done for them, the Institute for Environmental Studies at the University of Wisconsin put together the equivalent of one person full-time to produce a series of seven two-minutes' radio messages on the environment each week. These were [given daily, and were] written out so that they could be used in newspapers as well—free. All we had to provide was the equivalent of one person full-time to do it and a whole bunch of little tape cassettes that we could mail to radio stations. We gave it to them free [too]. As a consequence, 179 radio stations and over 50 newspapers carry these little two-minutes-a-day 'earth-watch messages' and enormous numbers of people get them. Thus we are reaching a very large audience with the environmental message.

It has been suggested that we have to reach the children. I agree. We have now started a little comic strip about the environment, that is free to the schools. The cost to us is very small. Many of the people assembled here are associated with organizations that could provide the equivalent of a full-time person, whatever it might cost, in a particular country, and because the radio stations get it free they love it. The messages are short enough to put in a place where they did not happen to sell an advertisement, the newspapers can use it as a filler, the message gets across, and the price is low. Try it; it works!

Borden (Panellist)

We have done several years of research on the determinants of pro-environmental beliefs and have [found that] very little of this belief system is a function of childhood experiences. This is undoubtedly embarrassing to our more behaviouristically inclined theorists. Instead, it seems to indicate that the human mind operates in a dialectical rather than a linear fashion. I will give you an example. If you are a general in the army and you want to raise your son to be a general in the army, you may attempt to shape your son's beliefs along those lines—but very quickly, almost overnight, many generals find their sons have turned into hippies. There is little in behaviouristic theory that can explain such an oppositional shift. From a dialectic point of view, however, whatever lessons you thought you taught [your son may actually] have taught him the opposite.

I am currently working with a topological model that is popularly known as catastrophe theory. According to such a model, people who have highly polarized beliefs are most susceptible to dramatic or catastrophic changes. In the case of environmental beliefs we would expect very dedicated technocrats, upon gaining relatively few ecological insights, to have the most dramatic life-style changes. Support for this notion comes from the surprising number of 'back to earth' or 'self-sufficiency' enthusiasts, and even environmental movement leaders, who used to be high priests in some area of technology.

Another promising feature of this phenomenon is that a dialectical shift from 'big or new is beautiful' to Schumacher's theme of 'small is beautiful', may be much more likely and easier to effect than many people would think. This I feel is one of the most promising goals of the psychological approach to environmental issues.

Buchinger

I am sorry to be the Devil's Advocate but it may also be that we have a different sort of society [in Latin America]. It seems to me that people do not like things which are free, but often prefer ones which have a price-tag. For instance, we were making a booklet which was handed out free and landing in the wastepaper basket. But when we put a price-tag on the same booklet and gave it as a gift, everybody noticed it and kept it in their libraries [: it had an indicated value!]. On Buenos Aires television one of the most expensive children's shows has an Uncle Ecology, while another 'personality' is a crocodile that is having a lot of problems with the river, with snakes, and with monkeys. It is so good aesthetically that it has stimulated other TV stations to follow suit.

Dodson Gray, Elizabeth

I am distressed when we try to formulate social and environmental objectives without having a dialogue between the *élites* of society and the people. I am repeatedly reminded in our US setting of the great divorce between the *élites* and where ordinary people's heads are. This is nowhere more evident than in regard to the so-called 'popular culture' which television manifests. And popular preferences in turn shape the television offerings. [Most members of our managerial *élite* are too busy to watch much television.] I happen to like television, and I find from my own watching that I come to conclusions about what is happening to the popular consciousness which are quite different conclusions from those based upon research surveys and questionnaires, most of which are based upon very narrow questions [that are in turn] based upon the researchers' prior interests and hypotheses.

We are all living in an era in which our publics are changing quite rapidly in their views. What happens on television and in pop culture is an indication of this. But I submit that you cannot formulate societal and environmental objectives successfully when there is this kind of division between what is in the heads of the leaders, supposedly, and what is in the heads of the people who are to be led.

[The Chairman thereupon closed the Session.]

Session 14

LIFE-STYLE ALTERNATIVES

Keynote Paper Part I

A Culture of Poverty

by

E. F. SCHUMACHER*

Chairman, Intermediate Technology Development Group Ltd, 9 King Street, London WC2, England.

INTRODUCTION: THE DOCTRINE OF DEVELOPMENT

'Only the rich can have a good life'—this is the daunting message that has been drummed into the ears of all mankind during the last half-century or so. It is the implicit doctrine of 'development', wherein the growth of income serves as the very criterion of progress. Everyone, it is held, has not only the right but the duty to become rich, and this applies to societies even more stringently than to individuals. The most succinct and most relevant indicator of a country's status in the world is thought to be *average income per head,* while the prime object of admiration is not the level already attained but the current *rate of growth*.

It follows logically—or so it seems—that the greatest obstacle to progress is a growth of population: it frustrates, diminishes, and offsets, whatever the growth of gross national product (GNP) would otherwise achieve. What is the point of, let us say, doubling the GNP over a period, if population is also allowed to double during the same time? It would mean running faster merely to stand still; *average income per head* would remain stationary, and there would be no advance at all towards the cherished goal of universal affluence.

In the light of this doctrine, the well-nigh unanimous prediction of the demographers—that world population, barring unforeseen catastrophes, will double during the next 30 years—is taken as an intolerable threat.

* Died 4 September, 1977. The general theme of this paper being already elaborated as Dr Schumacher's contribution to *Voices for Life*, edited by Dom Moraes (Praeger, New York, 1975), and/or in his earlier *Roots of Economic Growth* and *Small is Beautiful* (Blond & Briggs, London, 1973), no further references are offered.—Ed.

What other prospect is this than one of limitless frustration?

This modern assumption that 'only the rich can have a good life' springs from a crudely materialistic philosophy that contradicts the universal tradition of mankind. The material *needs* of Man are limited and in fact quite modest, even though his material *wants* may know no bounds. Man does not live by bread alone, and no increase in his *wants* above his needs can give him the 'good life'. Christianity teaches that Man must 'seek *first* the kingdom of God, and His righteousness', and that all the other things—the material things to cover his needs—will then be 'added unto' him. The experience of the modern world suggests that this teaching carried not only a promise but also a threat, namely that 'unless he seeks first the kingdom of God, those material things, which he unquestionably also needs, will cease to be available to him'.

Our task, however, is to bring such insights—supported, as I said, by the universal tradition of mankind—down to the level of everyday economic reality. To do so, we must study, both theoretically and in practice, the possibilities of 'a culture of poverty'.

To make our meaning clear, let us state right away that there are degrees of poverty which may be totally inimical to any kind of culture in the ordinarily accepted sense. They are essentially different from poverty, and deserve a separate name; the term that offers itself is 'misery'. Some 13 years ago, when I began seriously to grope for answers to these perplexing questions, I wrote the following in *Roots of Economic Growth:*

> All peoples—with exceptions that merely prove the rule—have always known how to help themselves; they have always *discovered a pattern of living which fitted their peculiar natural surroundings*. Societies and cultures have collapsed when they deserted their own pattern and fell into decadence, but even then, unless devastated by war, the people normally continued to provide for themselves, with something to spare for higher things. Why not now, in so many parts of the world? I am not speaking of ordinary poverty, but of actual and acute misery. . . . Poverty may have been the rule in the past, but misery was not. Poor peasants and artisans have existed from time immemorial; but miserable and destitute villagers in their thousands, and urban pavement-dwellers in their hundreds of thousands—not in wartime or as an aftermath of war, but in the midst of peace and as a seemingly permanent feature—that is a monstrous and scandalous thing which is altogether abnormal in the history of mankind. We cannot be satisfied with the snap answer that this is due to population pressure. Since every mouth that comes into the world is also endowed with a pair of hands, population pressure could serve as an explanation only if it meant an absolute shortage of land—and although that situation may arise in the future, it decidely has not arrived today (a few islands excepted).
>
> It cannot be argued that population increase as such must produce increasing poverty because the additional pairs of hands could not be endowed with the capital they needed to help themselves. Millions of people have started without capital, and have shown that a pair of hands can provide not only the income but also the durable goods, i.e. capital, for civilized existence. So the question stands and demands an answer. What has gone wrong? Why cannot these people help themselves?

'EPHEMERAL' AND 'ETERNAL' CATEGORIES

A culture of poverty, such as mankind knew in innumerable variants before the industrial age, is based on one fundamental distinction—which may have been made consciously or instinctively, it does not matter—the distinction between the 'ephemeral' and the 'eternal'. All religions, of course, deal with this distinction, suggesting that the ephemeral is relatively unreal and only the eternal is real. On the material plane we deal with goods and services, and the same distinction applies: all goods and services can be arranged, as it were, on a scale that extends from the ephemeral to the eternal.

The extremes are easily recognized. An article of consumption, such as a loaf of bread, is *intended to be used up*, while a work of art, such as the Mona Lisa, is *intended to be there for ever*. Transport services to take a tourist on holiday are intended to be used up and therefore ephemeral, while a bridge across the river is intended to be a permanent facility. Entertainment is intended to be ephemeral; education (in the fullest sense) is intended to be eternal.

Ephemeral goods are subject to the economic calculus; their only value lies in being used up, and it is necessary to ensure that their *cost* of production does not exceed the *benefit* derived from destroying them. But eternal goods are not intended for destruction; there is no occasion for an economic calculus, because the benefit—the product of annual value and time—is infinite and therefore incalculable.

When once we recognize the validity of the distinction between the ephemeral and the eternal, we are able to distinguish, in principle, between two different types of 'standards of living'. Two societies may have the same volume of production and the same *income per head of population*, but the *quality of life* or life-style may show fundamental and incomparable differences: the one placing its main emphasis on ephemeral satisfactions, and the other devoting itself primarily to the creation of eternal values. In the ephemeral life-style there may be opulent living in terms of use of ephemeral goods but starvation in terms of eternal goods—eating, drinking and wallowing in entertainment, in sordid, ugly, mean and unhealthy surroundings. In the eternal life-style, on the other hand, there may be frugal living in terms of ephemeral goods and opulence in terms of eternal goods—modest, simple and healthy consumption in a noble setting. In terms of conventional economic accounting they are both equally rich, equally developed—which merely goes to show that the purely quantitative approach misses the point.

FAILURE OF AFFLUENCE

The study of these two models can surely teach us a great deal. No one, I suppose, would doubt or wish to deny that the life-style of a modern industrial society is one that places primary emphasis on ephemeral satisfactions and is characterized by a gross neglect of eternal goods. Under certain imminent compulsions, moreover, modern industrial society is engaged in a process of what might be called ever-increasing ephemeralization; that is to say, goods and services which, by their very nature, belong to the eternal side, are being produced and used as if their purpose were ephemeral. The economic calculus is applied everywhere, even at the cost of skimping and paring on goods that should last for ever. At the same time, purely ephemeral goods are produced to standards of refinement, elaboration and luxury, as if they were meant to serve eternal purposes and to last for all time.

Nor, I suppose, would anyone wish to deny that many pre-industrial societies have been able to create superlative cultures by placing their emphasis in the exactly opposite way. The greatest part of the modern world's cultural heritage stems from these societies. The affluent societies of today make such exorbitant demands on the world's resources, create ecological dangers of such intensity, and produce such a high level of neurosis among their populations, that they cannot possibly be imitated by those two-thirds or three-quarters of mankind who are conventionally considered underdeveloped or 'developing'. The *failure of modern affluence*—which seems obvious enough, although it is by no means freely admitted by people of a purely materialistic outlook—cannot be attributed to affluence as such but is directly due to mistaken priorities.

Culture of Poverty

In the light of these considerations, it is not difficult to understand the meaning and feasibility of a culture of poverty. It would be based on the insight that the real needs of Man are limited and must be met, but that his wants tend to be unlimited, cannot be met, and must be resisted with the utmost determination. Only by a reduction of wants to needs can resources for genuine progress be freed. The required resources cannot be found from foreign aid; they cannot be mobilized *via* the technology of the affluent society that is immensely capital-intensive and labour-saving, and is dependent on an elaborate infrastructure that is itself enormously expensive. Uncritical technology transfer from the rich societies to the poor ones cannot but transfer to poor societies a life-style that, placing primary emphasis on ephemeral satisfactions, may suit the taste of small, rich minorities but condemns the great, poor majority to increasing misery.

The resources for genuine progress can be found only by a life-style that emphasizes frugal living in terms of ephemeral goods. Only such a life-style can create (or maintain and develop) an ever-increasing supply of eternal goods.

Luxury and refinement have their proper place and function, but only with eternal, not with ephemeral, goods. This is the essence of a culture of poverty.

One further point has to be added: the ultimate resource of any society is its labour power, which is infinitely creative. When the primary emphasis is on ephemeral goods, there is an automatic preference for mass-production, and there can be no doubt that mass-production is more congenial to machines than it is to men and women. The result is the progressive elimination of the human factor from the productive process. For a poor society, this means that its ultimate resource cannot be properly used; its creativity remains largely untapped. This is why Gandhi, with unerring instinct, insisted that 'It is not mass-production, but only production by the masses, that can do the trick'. A society that places its primary emphasis on eternal goods will automatically prefer production by the masses to mass production.

This brings the whole human being back into the productive process, and it then emerges that even ephemeral goods (without which human existence is obviously impossible) are far more efficient and economical when a proper 'fit' has been ensured by the human factor.

Keynote Paper Part II

Cultural, Social and Political Goals

by

NICHOLAS GUPPY

Botanist and author, 21a Shawfield Street, London SW3, England.

INTRODUCTION

In thinking ahead to a world where ever-increasing rewards in money and goods can no longer be postulated, and where shortages of raw materials and of energy are more than likely, we would do well to look at the rewards available in other, non-industrial societies—including our own in the past. We commonly find that social and cultural activities and rewards are far more important in them than they are in present-day industrial communities.

In this paper I do not propose that we should attempt to redress the balance by creating new Departments of Culture, or new Festival Theatres or Halls, but rather that we should reorientate our attitudes to life and work. This is because, for health, a large part of the life of every human being must be concerned necessarily with types of thinking and activity that we now label as creative or cultural, and may even class as élitist; moreover we must change our social goals, because some of our most precious current ideals lead in directions that are ecologically, economically and humanly grotesque.

From such changes would come once more a world in which Man would be able to live on income resources, as part of a stable ecosystem.

WHOLE-THINKING

When we consider 'culture', we have become accustomed to think of music, opera, ballet, theatre, paintings, cinema, or whatever—as something that a small group of people is presenting to us, the spectators, for our entertainment. Even our sports, such as football, are largely spectator events. Only a few individuals are able to feel that these are a living expression of their everyday existences.

This was not always so. Even today there are a few living cultures in which dance, music, theatrical performances and sport are an integral part of life. In Bali (Covarrubias, 1937) the little girl who sells shirts by day becomes a skilled ballerina at night, dancing a part in some epic of Balinese history or religion; the man who mends bicycles is also a musician or a sculptor; the religious festivals are times for games and dressing up. And it is the same in varied degrees with the remaining tribal and pre-industrial cultures throughout the world, almost all of which societies will disappear within the next 5–15 years as a result of aid programmes, tourism and industrial development. Tragically to relate, very few of their peoples even possess rights to their own lands, and from the point of view of the developer or the government official who is out to make money or get promotion, they are just 'in the way' (Guppy, 1958).

I speak with particular feeling about these so-called backward societies, because from personal experience I have found them to be culturally often much richer than our own. It is wrong for us to think they represent previous, earlier stages in the evolution of our own sort of society. Indeed they often represent something quite other—the end-points of parallel lines of cultural evolution. As such they have had and still have a great deal to teach us, particularly nowadays, about ourselves and our life-styles—among other things because, mostly, they are cultures which are in balance with their ecosystems. Indeed I suggest that the most important philosopher of the last hundred years, in creating what we call modern thought, was not Marx, or Engels, or Bertrand Russell, or Sartre—he was so-called primitive-man.

From looking at African sculptures, Picasso began a revolution in modern art. From studying child-rearing in New Guinea, Margaret Mead triggered off a revolution in education. What we consider the most advanced modern thinking about sex, dance, childbirth, almost anything one can mention, again and again one finds has derived from the alternatives shown to us by studies of other, non-industrial cultures (Bateson, 1972). The peoples of these cultures, almost everywhere doomed to extinction, can still show us the way towards a better style of life. If for this reason alone, we should try to ensure that they survive.

It is above all in learning how to *think,* that such non-industrial cultures have most to teach us. For we have largely forgotten what 'thinking' in the whole sense—perhaps one should call it 'whole-thinking'—really means. Scientific and mathematical thinking have yielded staggering results during a mere 150 years, have revolutionized the world, and have carried everyone along in a runaway flood of discovery and development. Yet the world so produced is largely incomprehensible to its citizens. In 1877 nearly everyone could understand the industrial processes of the day—steel production, gasification of coal, etc. This was the age of great popular excitement at discovery. Today, in industries such as the Chemical Industry, it is reckoned that at most 5% of the general population is capable, with education, of such understanding.

This is only one out of many reasons why, in large part, the modern industrial world is positively distasteful to so many of its inhabitants. And, as if to exacerbate matters, many processes have been made unnecessarily complex in the interest of eliminating human labour (Schumacher, 1973). All this represents an unbalanced attention to, and concentration on, only one type of problem-solving, one type of thinking—to the total neglect of those 'human' aspects of Man that science has not yet taken upon itself to measure (Marcuse, 1964).

The unbelievable hideousness of our cities and the meaninglessness of so many lives indicate the abysmal waste amid apparent productivity that characterizes so much of our activity. They also indicate how narrow a part of the brain has been involved, dealing with only one range of problems, in creating our present industrial world. I believe that the increasing concern among so many people—including scientists—with the arts, with religious search, with the irrational, and with more natural life-styles, all indicate the need we have to think in these other, neglected ways, and eventually to remedy this imbalance.

Articles and books on visual thinking, body language, and non-verbal thinking (Arnheim, 1970; Hinde 1972), are beginning to explore the edges of this subject, but I believe that we must go much further than any of them has done. For many people, words or signs—with which alone we customarily associate the word 'thinking'—do indeed provide the most important imageries. But sounds, shapes, patterns, colours, tastes, smells, the sense of touch, physical movements, and even the movements of the bowels or other organic functions, also provide highly important imageries. All of these imageries can be experimented with, and so we must accept that each—even muscular movement—may be a mode of thought. The relative importance and degree of development of these different imageries account for various forms of creative activity and expression—dancing, cooking, acting, chemistry, music, mathematics, literature, painting, etc. They may also account for many of the differences between the characters of individuals—including their ambitions, motives, relations, ways of life, and thought—and between the cultures that different peoples have evolved (Atkinson, 1958).

Animal behaviourists acknowledge that dogs must think largely in terms of smell, and that the sense of smell of ants must give an imagery no less precise to them than that of sight to us. We are only now beginning to acknowledge that humans also may communicate by smell, and that hair, for example, is a scent transmitter. In Bali this is part of the culture—a woman who loves a man goes near him 'so that he can smell the scent of her hair', while in Germany a lover may say to his or her beloved: 'Ich kann dich gir riechen'—'I like how you smell' (personal communication from E. F. Schumacher). Bats 'see' with their ears, bees use dance imagery to convey specific directions to other members of the tribe, and in India temple dances convey entire systems of philosophy.

Creative Thinking and Health

Thinking, then, is experimenting with images—the mental reproductions of the perceptions of *any* of our senses—and often we find that an individual may not be able to think at all adequately in any but one special imagery. Thus a painter or a mathematician may be almost incoherent in words, and a writer may be insensitive to music. Yet I believe that everyone has to solve problems in *all* his imageries—not only his prime imagery—if he is to maintain his health.

This brings us to the whole knotty problem of the origins of creativeness—the uniquely human quality, the quality that separates Man most from all other creatures—which is certainly among the least explored of psychological topics (Taylor & Barron, 1967). It is important for us to look at this for a moment, in our exploration of potential life-styles and their rewards.

Creativity arises, I believe, because images cannot be held still. In the brain they are constantly changing and being blended with other images, influenced by stimuli—such as noises, stomach aches, insect bites, or fears and wishes—for that is how the brain works. When conscious control of the mind is removed, all these shufflings and transformations (which represent the brain's efforts to fit imagery together into a coherent whole) are apt to manifest themselves as dreams. But dream solutions, even if satisfactory, cannot always be remembered (McClelland, 1964). So the mind has an urgent need to control its images—to fix them in forms in which they can be examined, criticized and altered if necessary. The only way in which it can do this is by trying to reproduce them—and then perhaps improve them—and this I believe is the origin of creativeness, and indeed of all human achievement in whatever direction (Guppy, 1958). For unless a creator can record his thoughts with his hands or body, at each stage, he can make no progress in thinking along to the next.

I believe indeed that until *any* disturbing phenomenon in the environment has been established satisfactorily in the mind's model by means of 'whole-thinking', the mind remains in a state of disequilibrium. Normally, interest is aroused, thought is fixed on the disturbing image, and it is compared, consciously or unconsciously, with what is already known—until it can be fitted in to the individual's reality model. But if for any reason the individual cannot do this, he will remain to some degree out of balance. Thus it is that a Nijinsky who is denied dancing will lapse into a state of rigidity—because, for him, dancing is a very precise and above all necessary form of thinking. Even though he may be able to think in other imageries, unless he can think also by dancing he is unable to fit his experiences together into a coherent whole, and feels incarcerated in a hostile world (Lowenfeld & Brittain, 1947). Games, gymnastics, shouting, fighting, dancing, love-making, are all means of such non-verbal thinking, which can be so important that mental health may be impaired if they are denied (Lowen, 1958, 1967). These are the aspects of thinking in which non-industrial cultures are so rich, and which have had so much influence on our own greatest artists and thinkers.

Viewed this way one can see why western civilization produces so many neurotics, depressives, delinquents, drug and sleeping-pill addicts, alcoholics, lunatics, psychotics—because in our society most people are denied opportunity to think in the varied imageries which are *necessary* to them. This is no mere fancy. It provides a general framework within which we can better understand many problems—such as the fact that, in the United States, apartment buildings have had to be demolished because it is now realized that their very design led to crime. The great hopes of urban renewers and city planners have led to vast, wasteful, human-disaster areas being built all around the globe through neglect of the sensitivities that comprise non-verbal thinking—areas where people only go to live when they cannot afford to go anywhere else.

Social Goals

One of the phenomena of the past decade has been the appearance, particularly in wealthy societies, of numbers of people loosely called 'hippies' or 'drop-outs'. They it was who first used the term 'alternative life-style' about themselves. They seem less conspicuous now, perhaps because they have influenced vast numbers of others who, while not strictly hippies, have moved towards the hippy position in their search for individual life-styles.

Clearly there are many different kinds of drop-outs, so it is hard to generalize about their motives—some may be noble, others ignoble. But they have been very widely misunderstood (Guppy, 1970). Desmond Morris, for example, has compared them with caged beasts in the human zoo seeking new stimuli to escape boredom; and elsewhere (in the same book!) as seeking to dampen down their perceptions to avoid over-stimulation. He has even proposed building a Palace of Pleasures to cure them (Morris, 1969). I think both these diametrically opposed views, and the various other explanations that have been offered, miss the target and confuse the issue. The one point in common which is shared by all these individuals, whether they exist alone or in communes and collectives, is that they aim at self-sufficiency 'outside the system', but in fact are living on the surpluses of our society—either of money, or of waste materials which they collect and use, or of land or houses which they occupy. Ecologically they are entrepreneurs and they represent a direct assault on the idea that there is a universal minimal living standard, and upon the social goal of egalitarianism.

This particular social goal, egalitarianism, so close to many people's hearts, I believe to be the actual, though at first glance unlikely-seeming, cause of many of the ills of industrial societies. From the beginning of time people have dreamed of a golden age in which all would be equal; where the lion would lie down with the lamb; where a man would give according to his abilities and receive according to his needs. The great religions see men as equal before God, and envisage a paradise in which this is the case—'in the next world'. But some humans have thought such a paradise could be created on Earth: St Augustine's 'City of God' (fifth century) set an ideal for the whole Middle Ages; Sir Thomas More's *Utopia*', in 1516, gave the name to this whole field of thought. And in the past 100 years the ideals of Marx have reshaped the lives of whole nations, and led to class struggles in others—all in the interests of an ultimate egalitarianism.

Equality before the law seems obviously desirable. But unfortunately equality, particularly in material terms, has also been much confused with fairness, when in fact they are different ideas, often leading in different directions. It is obviously unfair, for example, to impose equality upon individuals differing in ability or ambitions, energy or taste.

When there is a tremendous disparity between the poor and the rich, greater material equality may seem a very desirable state of affairs. Some or all the wealth of the rich should be taken away and distributed to the poor, or invested in industry, whose thereby increased productivity should produce goods for distribution to everyone. This has been the benevolent ideal in both socialist and capitalist countries. But it seems to produce a positive feedback—the gap between poor and rich remains, and productivity accelerates out of control.

It is at first hard to see which came first in our troubled world—the frenzied over-productivity, or the pursuit (but non-achievement) of equality. But when once we try to analyse what is happening from an ecologist's point of view, I believe we can show that this positive feedback loop arises as a direct result of social motivations and resultant well-intentioned legislation designed to produce or foster material equality, accentuated in its effects by our present-day means of production.

First let us examine the dynamics of human societies. Sociologists, anthropologists and economists have, on the whole, failed to draw general conclusions as to how societies change and evolve (Parsons, 1966). On one hand there is the 'progress of understanding' view, which sees societies changing as ethics, manners and knowledge change, so that the eighteenth century outlook gave place to the nineteenth, and that in turn to ours of the twentieth century. From this point of view, pre-industrial societies necessarily represent stages in development towards societies such as our own, which represents an advance on all those which came before. And then, on the other hand, there is the view that class conflict, centred on the means of production, has shaped societies throughout history. These are perhaps the two most widely-held views, but neither has been of value in predicting change, and this reduces confidence in them. Their principles may be important, but are not fundamental. In the present paper I wish to propose a more ecological approach—to view societies as the dynamic product of the relationships between Man, belief and environment, in which the fundamental force is the release of human energy (I use 'beliefs' as including knowledge, technology, ethics, religion, social codes, etc., and 'environment' as including resources), measurable as inventiveness:

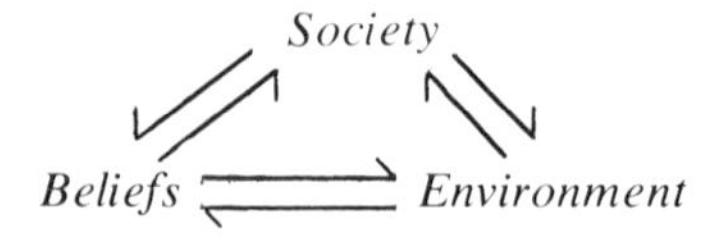

From such a schema it will be seen that 'progress' should only be applied to any change which, over the long-term, produces increased release of energy—and this may well include destructive events, such as wars or revolutions. It may also be noted that, while the *natural* course of history can always be predicted as moving in the direction of such increased energy-release, other human activities may seek to shape or alter this in accordance with pre-held beliefs or interests. Obviously, if this schema *does* describe the events, it will be important to locate where energies are not being released freely, so that means can be found to allow development to take place constructively.

A century ago, in 1877, the growth-point of energy release was in the finding and exploitation of new resources. In the intervening period it was in the area of beliefs, as scientific knowledge grew; and in the next 30 years I believe it will be in the reshaping of society—particularly if resources dwindle as anticipated.

The most important single point to be made is that the dynamics of societies depend on the dynamics of individual humans, even when these are working through institutions. Left alone, all humans will develop to their highest capacity (ecologically speaking, by occupying the largest possible niche), and all societies (Margalef, 1968) will develop to their maximum information content—i.e. to the maximum complexity and diversification of occupation, given the limitations imposed by resources, available knowledge, and their systems of beliefs. Stratifications within such an ideal society would be based upon individual qualities and abilities, and life-style would depend on individual choices among what is available.

Such a society would be diverse and inegalitarian, but there is no intrinsic reason why it should not be contented. My own observations suggest that discontent comes less from envy than from frustration—from limitation of the options that are open to individuals, whereby they can express their own abilities. And such limitations may be as severe as, or more severe than, when the goal of the society is equality, as compared with when position and wealth are largely inherited.

Would the inequalities of such a society necessarily lead to the exploitation of the weak by the strong? Man has been described as a 'pioneer species', and in pioneer societies equality is indeed important. But examples are known of societies where individual behaviour might be viewed as more mature. In Nature, egalitarian societies tend to be either pioneer communities which rapidly give place to other, more mature communities, or else totalitarian systems, where the more varied exploitation of available resources (which might lead to richer and more diverse societies) is suppressed. (Examples: a reed-bed is a pioneering community of almost identical plants which colonizes shallow water, builds up debris, and is then invaded by more diverse species; a cabbage field is a totalitarian system where all are equal, and where the dictator, Man, removes diversifying elements, namely weeds).

In a pioneer community which wishes to remain egalitarian, the differences between individuals' abilities will indeed be considered undesirable—because these differences will lead to the overthrow of that society. But in a mature society, where diverse elements are closely interwoven, the differences between individuals contribute to stability and are mutually supportive. What is important is that this must be recognized, so that *all* non-criminal occupations and life-styles carry dignity, status and respect. By all means let Man be a pioneer species in the colonization of other worlds, but on Earth, with a growing population and limited resources, he must learn to behave as a mature species, and so to utilize most fully, economically, and prudently, the available resources of the environment.

Man has already done this outside of the industrial West. In many pre-industrial societies (some, as in South-East Asia, at high levels of civilization), representing the mature stages of different belief-systems, Man has achieved equilibrium with his environment. The greatest task of Man today is to achieve this while using, and not abusing, the knowledge and power given by science.

The Social Mosaic

It is commonly assumed that the people in a modern country are all part of one great society, with common goals and dynamics. However, I believe this contention obscures certain important relationships, which become clearer if we view a country such as Britain, France, or the USA, as a mosaic of communities of different sizes, with different energy-relations, both internal and external, and with different ranges of life-styles. By borrowing some terms used in plant ecology (Shimwell, 1971), we can place such communities in a 'sere', showing their stage in succession towards a mature or 'climax' society (Table I).

Successions proceed because each earlier stage produces surpluses which allow growth by invasion (other terms could include immigration, colonization) or by population multiplication (Kershaw, 1964), and at each higher stage there is an increased specialization of occupations or roles. Production of surpluses and rate of growth are typically greatest per unit in the earlier stages of a succession, and progressively less in the later stages until, when the climax is reached, inputs and outputs are minimal, there is recycling of nearly all waste, and a high state of stability and predictability is reached. The removal of surpluses or suppression of role-proliferation will artificially keep any society at a more juvenile level than it would otherwise attain, with an associated greater energy and higher productivity. And this, of course, more or less summarizes the goals of most modern nations: to produce and remove surpluses ('increase trade'), and to reduce role-proliferation ('increase equality').

In vegetational successions at all stages the substratum exploited is that upon which the community stands, and with human society the situation at first is not dissimilar—until we reach the level of cities and metropolises, at which point we find that their substratum for exploitation is often the entire world, and that there is no question of their reaching a state of balance with minimal inputs and outputs. Indeed their very aim is growth, and they are

TABLE I

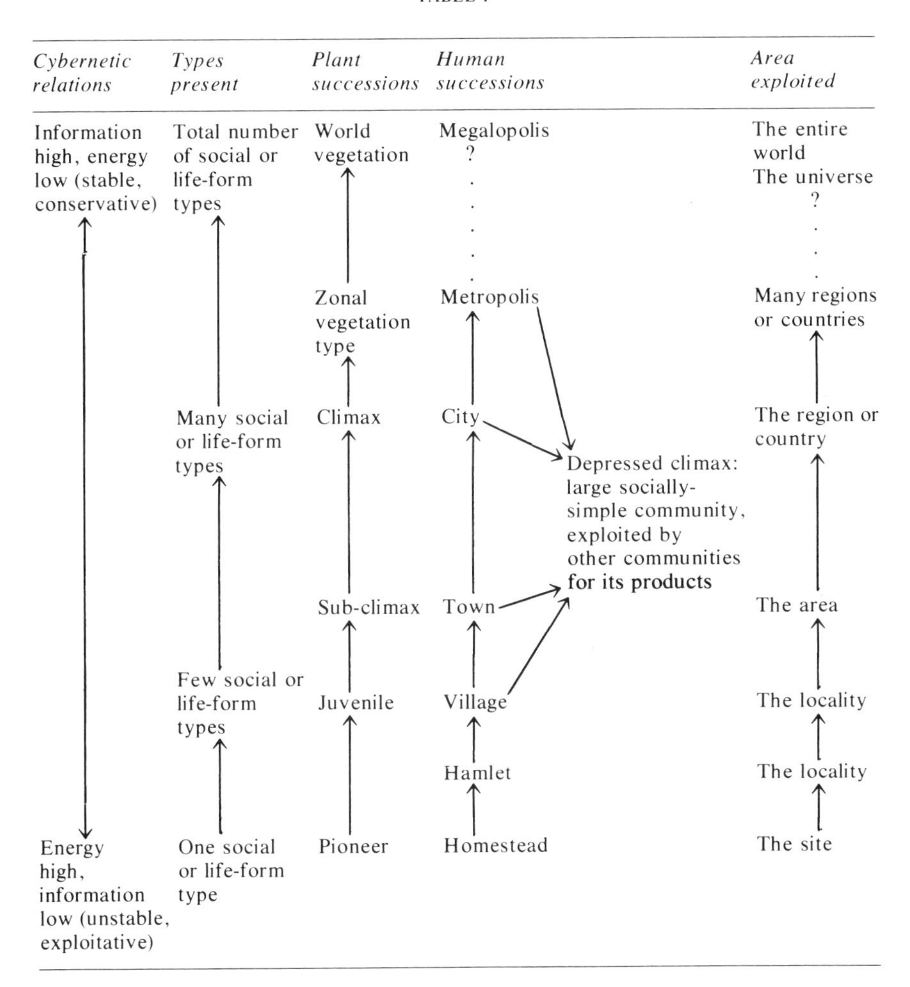

Cybernetic relations	*Types present*	*Plant successions*	*Human successions*		*Area exploited*
Information high, energy low (stable, conservative)	Total number of social or life-form types	World vegetation	Megalopolis ?		The entire world The universe ?
		Zonal vegetation type	Metropolis		Many regions or countries
	Many social or life-form types	Climax	City	Depressed climax: large socially-simple community, exploited by other communities for its products	The region or country
		Sub-climax	Town		The area
	Few social or life-form types	Juvenile	Village		The locality
			Hamlet		The locality
Energy high, information low (unstable, exploitative)	One social or life-form type	Pioneer	Homestead		The site

able to achieve this by importing the surpluses of other societies, whether in their own or other countries—whose own growth is thereby stunted.

In the face of a likely shortage of resources, whether through the using up of all available supplies or because the supplying countries will see the light and decide to use their own raw materials to develop themselves (Hutchinson, 1970; Goldsmith *et al.*, 1972; Meadows *et al.*, 1972; Marshall, 1974), can the existence of even those metropolises which already exist be sustained? Or will population growth or other factors inevitably (as has been predicted) lead to the appearance of yet vaster megalopolises of possibly hundreds of millions of inhabitants, with all the concomitant problems of supply, transport, communications and distribution—which will depend for all that they need, even more than our present metropolises do, upon distant sources, and upon the continued suppression of community development in those sources?

Looked at simply in such a way, this seems an absurdity—particularly when we consider the anonymity, amorphousness, ugliness and sheer inconvenience of so many of the existing vast urban areas, which typically have grown up around manufacturing towns, and which despite their size lack the richer social structure and culture of older, often far smaller centres.

The most extraordinary examples of this type of growth are the 'shanty towns' around the great cities of the Third World, built by the destitute of the countryside, who flock to the cities in the hope of work, and finding none, stay on. Why? Because the industries of these cities are highly overproductive of surpluses and wastes of materials and money, some of which such people can acquire and use as, in more opulent terms, was the case with the hippies we looked at earlier. In fact, when we look at the structure of the largest cities of today, we are seeing societies which not only exploit the whole world, but give so little back that the world itself has to rush to live on the scraps which they let fall.

If this process was making everyone in the world well fed, clothed, housed and happy, and if the world possessed inexhaustible supplies of energy and raw materials, it might just be tolerable. But neither of these conditions prevails. Furthermore there is another anomaly: the overproductivity of these cities does not arise because their entire population is engaged upon productive work. E. F. Schumacher (1973), in *Small is Beautiful*, has estimated that in Britain roughly 3½% of 'total social time' is so spent; and Britain is not an extreme case. In countries where production is highly automated, or where there is a vast, uneducated, poor, out-of-work population, the percentage of 'total social time' spent on productive work is much less.

Thus high industrial productivity at present depends upon a tiny part of the population—a mere subculture upon whom the rest depend, and who do not even reap the rewards of their work. This is so in all industrial societies, whether socialist or capitalist. Indeed, no people on Earth are so poor, spiritually, as are the unfortunate workers of industrial societies everywhere. They are the deprived of the Earth, depending almost entirely upon the state or some large enterprise to find them a job, and for many other things in life; from the psychological point of view, large numbers of them seem to be permanently in the child (and not the adult) ego state (Berne, 1975), finding it hard to make decisions for themselves and being easily taken over by political or other organizers, advertisers, etc.*

Let us consider an industrial town in which the great mass of the inhabitants are workers, drawing somewhat similar wages, living in somewhat similar homes, and having similar possessions. They are importing raw materials, and exporting finished products, far beyond their own personal needs. They draw wages. But do they develop a richer society, with more

* This seems to us to be true also of some employees of big international organizations which look after them practically throughout their working lives.—Ed.

roles and occupations, with cultural institutions and outlets for self-expression in all the fields of their thinking and creativity? No—they remain workers. Other people may do these things, using the surpluses which the industrial workers produce (including leisure); but they will probably do so away from the industrial area, in some pleasanter environment. Thus we see the development of what I have shown in Table I as a 'depressed climax': cities, metropolises, or parts of them, which, despite their size, are not mature societies, but simpler more egalitarian communities. These, like their equivalent stages in the plant world, are highly overproductive; but instead of developing socially, they export their surpluses and so remain socially static.

My proposition is that, at present, most Western industrial nations are mosaics of varied social structure, all kept alive by the high productivity of the areas of depressed juvenile social growth scattered among them; and that we should think of such productive, socially juvenile regions as being regions in which humans are being exploited for the benefit of the more complex societies of other areas. I also contend that this whole system, based upon the deadly combination of egalitarian social goals and mass-production methods, is draining the rest of the world in ways that cannot endure. Ought we not therefore to rethink the idea that egalitarianism is a desirable goal, when it seems clearly to equate with 'exploitation of the masses', and with the reduction of human energy and potential? Should we not aim at allowing more complex social growth to develop out of the use of individual energies, and concentrate rather on the idea of individual self-development instead of egalitarianism? And should we not change our exploitative production methods?

Of course it may be argued that developed or overdeveloped countries need juvenile, highly productive, depressed areas, in order to compete in the modern world. But to compete for what? In order to keep such ghettoes going or to wage war against other societies? Is it indeed *necessary* for our methods of production to be so environmentally destructive, and so degrading in human terms that even monetary or material incentives are beginning to lose their appeal? With the knowledge that we now possess, surely we can find better alternatives? I believe that we must do so, whatever the position regarding resources, because if we do not, the division of the world between privileged technocrats and dependent aborigines—which term will eventually include most of the populations of developed countries—will deepen in all industrial societies.

In *Small is Beautiful,* Schumacher (1973) has pointed the way to a solution. He sees our modern productive processes as being too big, too complex, too costly, too violent, and too neglectful of human beings. He calls for the return to a gentler life-style, pursuing once again quality, dignity, freedom and fulfilment. Today it is no longer any use having the sort of small iron-mine that can supply a single steel-foundry. You have to have an entire mountain, and upon it build a whole city, which will disintegrate as soon as the last of the iron has been shipped to the USA or Japan, Britain or Germany. Then you have to find another mountain and repeat the process—as long as you can!

Such vast enterprises are highly efficient, and produce vast profits—so long as supplies and markets hold up. Ecologically they are a disaster. If they fail, it is a titanic disaster. But even if they succeed, they are still a disaster. In the first place, ideally they employ very few people indeed, because in them automation replaces human labour, while furthermore they replace all the smaller enterprises which preceded them, and throw their employees out of work. High productivity in Birmingham, Düsseldorf and Pittsburgh is throwing millions of craftsmen out of work in India and Peru, who then have to be fed—as if it were a favour—out of our surpluses, or those of their own urbanizations.

Schumacher calls for us to use our resources to reverse these processes—to develop a cheap, universally accessible small-scale technology, which will allow work-places to be created that will cost no more than a man's yearly earnings; which will be relevant to local ecology and energy supplies; which will favour human scale and involvement; and which will allow work to be desirable and creative in itself. Economists, bankers, accountants, and others, must all learn to structure and service such small-scale units. If at first such a prospect seems like a dramatic reversal of everything we know, it is only because modern management 'ethics' have led us to believe that success means only two things—getting bigger and maximizing profits. Even from this viewpoint there may be surprises—for growth is greatest earlier in the succession. In fact, what we have been witnessing is failure of the most devastating sort—caused by continual investment at the top end of the scale—namely the growth of a vast network of greedy cancer-cell industries consuming the body of the Earth, reducing the quality of human life below that of our own favoured horizons (I speak as a middle-income Westerner); and as we now begin to realize, producing in us that hectic flush that looks like health but in fact precedes collapse. We have failed to see the pattern—which is more to do with the dynamics of human energy than with industrial production. It is time to change course.

Political Problems

In a world which is divided into competing nation-states, the changes in the direction of the humanization of industrial societies that I have described as desirable would at one time have been bound to be resisted in the interests of national strength and unity. But in a world where unemployment is high and seems bound to rise further, and where there is no clear sense of direction, I believe that the sort of model which I have presented is at least optimistic, and that movement in the directions which it predicates will release human energy and in fact create the greatest possible strength—happy people who know where they are going and believe in it. I do not believe that these changes will produce a decline of living standards. On the contrary, I believe they will raise them, in the *real* sense of the word.

In the past we have lived on the myth that eventually the rest of the world would catch up with the material standards of the West, and that what was good for the West would eventually be good for the rest of the world also. With the above model—allowing societies to mature, emphasizing individual self-expression, and keeping to smaller-scale units of production—we might actually be able to achieve this, and eventually a balanced world ecosystem. Given more rational social goals, the world's major problems could be seen then to be problems largely of boundaries (Margalef, 1968), and their solution to lie in the dissolving of boundaries. Where there is no boundary between people of different colours, there is no reason for racial conflict. We can see this in Brazil and its obverse in South Africa. Where there is no boundary between peoples occupying the same territories, conflicts may be acute but they are seldom long-lasting or insoluble: Jews and Arabs can live peacefully in the same London hotel. In the plant societies from which I draw so many of my analogies, there are boundaries between different societies, yet these boundaries are always fluid and dynamic—unless artificially constrained. Each grouping is changing, serving its function for the individuals that comprise it, and then is replaced when these individuals or needs change.

In human societies the frontiers tend to be made rigid, and the problem of how to change them easily has never been solved. Barriers such as those of colour, class, race, sex, national frontiers, languages, and above all political or religious beliefs, all cause trouble the more clearly they are defined. Even if we cannot tackle the actual differences that exist at the levels of ideology, culture, philosophy, nation or group, so long as we know that the point of conflict will always be at a boundary then at least we can work on trying to reduce the clarity of boundaries. Only by doing this can we hope that mature societies will be free to develop throughout the world, which then will offer the greatest possible number of differing roles for individuals.

It is at the much lower level of the individual that we need boundaries. Here are the barriers we must honour, re-erect and maintain, for these are the barriers that allow energy release to take place safely and rejuvenate our society.

Democratic institutions exist in some form in almost all societies, but those who operate them often suffer from a lack of perspective as to their own roles—whether to advance individual freedom as far as possible, to adhere to traditions, or to promulgate some idealogical solution. Much of this confusion arises from existing theories of the structure and dynamics of societies, and of the various extant models—structural, functional, class conflict, etc. I believe that a simple examination along ecological lines shows that many of our *idées fixes* have actually caused our problems, and have made it difficult for us to direct our enormous resources and energies in constructive ways.

In past times churches, mosques, temples, and the like, were community centres from which general principles about Man and his relations to the universe were given out, that served as guides in the general conduct of people's lives. Without in any way wishing to trespass upon the more spiritual domain of such establishments anywhere, I believe that this role should now be fulfilled on a simple, humble level by the ideas and ideals of ecology, which relate Man to his role in Nature and show men, women and children which actions are correct and life-enhancing; which give them a view of the cosmos which is still subject to change as knowledge advances, but yet salutary and inspiring. I believe that this is a role that ecologists should seek to promote, and that everywhere ecology should be taught as a new guide to enable people to think about themselves and about their world.

At one time human beings were considered to be the wealth of a nation, and we must come to feel this yet again. It is right that we should seek to control our populations, but we should do so because we want to better human life, and not just to maximize profits and discard those who are temporarily useless for making them.

The ecological view, which will seek to advance the maturing of societies everywhere, will promote the maximum recycling of waste, the maximum self-sufficiency of regions, and the maximum use of individual talents; it will also allow the maximum number of alternative life-styles to emerge, because it will maximize human energy and creativeness.

Conclusion: Ten Practical Recommendations

(1) Large-scale enterprise may frequently be more suitable in the future than small-scale enterprises. But the point of largest growth must lie not with heavy industry; it must lie with technology for the individual and small group throughout the world—in developed as well as undeveloped countries.

(2) The growing-point for investment should increasingly be not the corporation but the individual.

(3) The growing-point should not be the city but the countryside, the village, and the town of limited size.

(4) Alternative fuels and sources of energy—alcohol, biofuel, solar, geothermal, wind, tidal power, and yet others—will be needed on a vast scale, but also in units for the individual and small group.

(5) Cities must be kept as small and compact as possible. Problem areas of the future are the big, depressed, deflected-climax cities. They will dwindle when growth begins in the countryside.

(6) Tribal peoples, agricultural land and remaining natural areas must be treasured and enriched. They constitute much of the present and future chief wealth of the world.

(7) Individualism, not egalitarianism, should be the watchword.

(8) The wealth of the individual must again be regarded as the wealth of the nation, and must be encouraged so far as it can be exercised personally; for individualism in a mature society is quite different from individualism in a pioneer community. *Consumption* must be taxed, not income. This will reduce excess production, give an incentive for saving for individual purposes, and kill inflation at one blow.

(9) Ideological, cultural and social boundaries should be weakened to diminish sources of conflict; commercial and economic boundaries should be strengthened and multiplied to favour individual and local enterprise instead of big business.

(10) The most important form of energy is the energy of the individual. It only needs setting free.

References

ARNHEIM, R. (1970). *Visual Thinking*. Faber & Faber, London, England: 345 pp., illustr.

ATKINSON, J. W. (Ed.) (1958). *Motives in Fantasy Action and Society–A Method of Assessment*. Van Nostrand, Princeton, New Jersey: xv + 873 pp., illustr.

BATESON, G. (1972). *Steps to an Ecology of Mind*. Palladin Books, St Albans, England: 510 pp.

BERNE, E. (1975). *Transactional Analysis in Psychotherapy*. Souvenir Press (Educational and Academic), London, England: 270 pp.

COVARRUBIAS, M. (1937). *Island of Bali*. Cassell, London, England: xxv + 417 + x pp., illustr.

GOLDSMITH, E., ALLEN, R., ALLABY, M., DAVOLL, J. & LAWRENCE, S. (1972). A blueprint for survival. *The Ecologist*, **2**(1), 44 pp., illustr.

GUPPY, N. (1958). *Wai-Wai*. John Murray, London, England: xii + 375 pp. illustr.

GUPPY, N. (1970). View to a death. *Sunday Times*, Times Newspapers, London, 3 January, p. 1.

HINDE, R. A. (Ed.) (1972). *Non-verbal Communication*. Cambridge University Press, Cambridge, England: xiii + 443 pp., illustr.

HUTCHINSON, G. E. [and others] (1970). *The Biosphere*. Scientific American, San Francisco, California: viii + 134 pp., illustr.

KERSHAW, K. A. (1964). *Quantitative and Dynamic Ecology*. Edward Arnold, London, England: viii + 183 pp., illustr.

LOWEN, A. (1958). *The Language of the Body*. Macmillan, New York, NY: xiii + 400 pp.

LOWEN, A. (1967). *The Betrayal of the Body*. Macmillan, New York, NY: viii + 275 pp.

LOWENFELD, V. & BRITTAIN, W. L. (1947). *Creative and Mental Growth*. Macmillan, New York, NY: xv + 412 pp., illustr.

MCCLELLAND, D. C. (1964). *The Roots of Consciousness*. Van Nostrand, Princeton, New Jersey: v + 219 pp.

MARCUSE, H. (1964). *One Dimensional Man*. Routledge & Kegan Paul, London, England: 201 pp.

MARGALEF, R. (1968). *Perspectives in Ecological Theory*. University of Chicago Press, Chicago, Ill.: viii + 111 pp., illustr.

MARSHALL, P. T. (1974). *This Finite Earth*. George Harrap, London, England: x + 172 pp., illustr.

MEADOWS, D. H., MEADOWS, D. L., RANDERS, J. & BEHRENS III, W. W. (1972). *The Limits to Growth*. New American Library, New York, NY: 207 pp., illustr.

MORRIS, D. (1969). *The Human Zoo*. Jonathan Cape, London, England: 256 pp.

PARSONS, T. (1966). *Societies: Evolutionary and Comparative Perspectives*. Prentice-Hall, Englewood Cliffs, New Jersey: viii + 120 pp., illustr.

SCHUMACHER, E. F. (1973). *Small is Beautiful*. Blond & Briggs, London, England: 255 pp.

SHIMWELL, D. W. (1971). *The Description and Classification of Vegetation*. Sidgwick & Jackson, London, England: xiv + 322 pp., illustr.

TAYLOR, C. W. & BARRON, F. (Ed.) (1967). *Scientific Creativity, its Recognition and Development*. John Wiley, New York, NY: xxiv + 419 pp.

Discussion (Session 14)

[The fate of the discussion of this Session was even worse than that of Session 1, for through some undetermined mishap of non-recording, or possibly owing to the general strike, the tapes of Session 14 remained blank for considerable periods after their beginning and unclear before their end. Thus the entire presentation of the key-note paper (by Mr Guppy, speaking also for Dr Schumacher) was lost or unclear, and so was the Chairman's introductory speech. A similar fate befell Professor Hare's contribution and the later stages of the discussion, before it was truncated in favour of the Lord Mayor's reception, although fortunately the Chairman was able, from his notes, to reconstitute his own introductory speech, which we print below in square brackets as he warned that it may not be verbatim. To our great sorrow and the loss of the world, Dr Schumacher died rather soon afterwards.]

Jensen (Chairman, introducing)

[I shall forgo telling introductory pleasantries for two reasons: first, the shortage of time, since we have to finish at 5.45 in order to arrive punctually at the Lord Mayor's reception; secondly, because it was said yesterday that so serious a subject should be tackled seriously. With this I agree fully, although I believe that to concentrate seriously on a question does not imply dispensing with our sense of humour.

First I suppose I should introduce myself: I am Erik Jensen, of Anglo-Danish background but with a Malaysian passport. An anthropologist by training, I am now by calling an international civil servant with the United Nations, having been for some years Chef de Cabinet in the Palais des Nations, Geneva, Switzerland. The others on the platform are, on my right, our Keynoter Mr Nicholas Guppy (speaking also for Dr Schumacher, who unfortunately is unable to attend the Conference), whom we have the pleasure of knowing already, as we do also two of our panellists, Mr Glasser and Professor Hare. The others are most welcome new-comers: Mrs Franz, representing the younger generation, and Mr de Candolle, scion of the great botanical family and himself a noted Swiss banker and man-of-letters.

This Session's theme is 'Life-style Alternatives'. When we hear this, many of us think instinctively of communes or Teddy Goldsmith. But the subject is broader and, with respect, even more important. One could argue that it is at the heart of all our discussions, which have been largely focused—as Maurice Strong made evident—on two faces of the same coin: what will happen in the developing countries, and what is to be done in the industrialized world?

Let us assume for one happy moment that the growth in population will soon be contained; there has to be a solution to it, which will be catastrophic if ordered transition fails. Among the Dyaks of Borneo, when the population expanded beyond the carrying capacity of the land, the traditional answer lay in cholera epidemics or head-hunting. However, even with a stabilized world population, we still face the fact of vast numbers who aspire to live 'better', and through modern communications and the media are remarkably conscious of how 'the other half' lives; they crave a larger share of global resources. By all means let us allow them the benefit of our advice, but in the last analysis it is they who will and must decide. We cannot impose the new-found values of some in the West on the peoples of India or Africa, or on the tribes of the Amazon, as a solution to their problems as they perceive them. They have their visions of 'the good life' we have known, and they would consider it a piece of neo-colonialist conceit if we were to tell them not to develop economically.

What is a 'good life', one may ask? (Let us leave 'happiness' to fairy tales, where 'they' live happily ever after—although the happiness myth must, through unrealized promises, be responsible for untold unhappiness.) This 'good life' has been mentioned repeatedly, and we should try to define it. Whose lives are good? My namesake Erik the Red killed a man in a brawl, was outlawed, sailed from Iceland, discovered Greenland, and later established a successful settlement; he remains a hero to this day. Van Gogh cut off his ear and went mad but painted brilliantly. Beethoven, deaf and distrustful, may have contributed as much as anyone to others' happiness. The Dyak who in his youth took many heads and in consequence had many women, who became a respected member of the community living and farming in harmony with the spirits, married, had children and grandchildren, and after death passed his name to his great-grandson, was thought to have achieved a good life. Or have we?

The question is not frivolous. It is the conception of 'the good life' which largely determines values, dictates priorities, and steers society. Each person's appreciation of 'the good life' has roots deep in his culture, his religion, his philosophy of living—irrespective of whether he is able to articulate his convictions. People have a sense of themselves and of what they are capable of becoming which transcends material things; even for the very poor, the purpose of life reaches beyond food, drink and shelter. But as in the development of higher cultures, the 'higher values' presuppose material sufficiency if not surplus. All men and women have basic needs and most covet a standard of physical well-being—of comparative ease and security—which is not always found in the developing countries. To achieve this, the developing peoples will use all means at their disposal—and who among us should consider that unjustified.

Hence in a world of finite resources, changes must occur among the conspicuous consumers of the West, with their extensive ghost territories abroad. Will they in general be prepared to curb their appetites, to change a life-style which expresses their understanding of 'the good life', to recognize the rights of others to a larger share of what remains—in short, to contain their lives within shrinking limits?

In place and also in time, one man's freedom ends where the other's begins; the freedom of our generation to exploit resources has to end where the freedom of other peoples and of succeeding generations should begin. This implies immediate adjustments and a changed time-scale in our thinking, that should no longer be based on next month's new car or next year's new house—but instead, on the next generation. A new time-scale presupposes an important shift in the current values of the West, and only when these new values permeate society sufficiently will political action follow. Few politicians in a modern democratic society lead—except where the people want to be led. So it is the people, influenced by the media, books, films, current ideas and, of course, the educational system, who will determine the direction in which Western societies are to move in relation to other parts of the world and to the future.

If we in the West prove willing to restrict consumption—our 'burning' of the common capital of mankind—Mr Strong's 'new growth' should be possible. A fairer sharing of existing and potential resources would provide the Third World countries with the means to satisfy the basic needs of their populations and to develop in accordance with their own ambitions—in short, to achieve respect for their legitimate aspirations. 'New growth' would presuppose basic changes in the use as well as the distribution of global resources. And it predicates the right of peoples throughout the world to live decently—with respect for themselves and the respect of others for their way of life.

I now give the floor to Mr Guppy, who will be speaking also for Dr Schumacher.] [He will be followed by our four panellists.*]

* *See* introductory paragraph of Discussion on previous page.—Ed.

Glasser (Panellist)

I am grateful to you, Mr Chairman, and to my panellist colleagues here, for permitting me to make an extended unprogrammed contribution. I am concerned—disturbed, in fact—about the prevalence here, and at [other] international gatherings, of an attitude that presumes to prescribe, from the point of view solely of the industrialized countries, ways of life for other cultures and other peoples. Western Man, after all, cannot claim success in running his life, [such as might justify] him in trying to persuade others to follow his example: rather does he seem to be trying to infect the rest of humanity with his own malaise!

In my current comparative studies, I have taken the opposite direction to the hitherto received opinion about 'development' and 'progress', and have set out to discover, in a traditional culture, people's emotional perceptions of their inherited aims in life—the values and modes of behaviour by which they have preferred to pursue them. It is worth remembering that 200 years ago Edmund Burke courageously expounded the view that society was organic, and pleaded that one should not try to treat it in terms of mechanics as an engineer does; that one must accept the antiquity of tradition as proof of its 'necessity' and hence of its claim to continuance.

Man's social evolution is something like 40,000 years old. [I believe] that he is essentially a selfish animal, conditioned and constrained to apparently altruistic behaviour only because he realizes it to be in his ultimate best interests. It can have been no accident, surely, that the model identity that has evolved in various societies over the world—by that I mean the attributes of the ideal man—bear a remarkable mutual similarity. It can be no accident, too, that the great religions, in their respective visions of the road towards that ideal, whether it be the eight-fold path of the Buddhists or the 10 commandments of the Christians or Jews, are virtually unanimous. Western Man has no monopoly of wisdom in that respect. We should therefore devote our energies, insights and intelligence, to making it possible for other peoples to pursue their own perspectives of fulfilment in their own way.

With that preamble, let me set the scene by showing you some figures:

*Country**	*Growth-rates* (%) Population growth 1960–73	Population growth 1965–73	GNP per head 1960–73	GNP per head 1965–73
Bangladesh	2.6	2.4	−0.2	−1.6
Burma	2.2	2.2	0.7	0.7
Chad	1.8	2.0	−2.1	−3.3
El Salvador	3.5	3.4	1.9	0.8
Guinea	2.8	2.8	0.1	0.1
Mali	2.1	2.0	1.0	0.5
Niger	2.7	2.7	−1.9	−4.6
Peru	2.9	2.9	2.1	1.8
Senegal	2.1	2.2	−1.8	−2.8
Upper Volta	2.1	2.1	−0.4	−1.1
Venezuela	3.4	3.3	2.0	1.3
Zambia	2.9	2.9	1.7	−0.2
United States	1.0	1.0	3.1	2.5
United Kingdom	0.5	0.4	2.4	2.3

* *World Bank Atlas: Population, Per Caput Product and Growth-rates.* The World Bank, Washington, 1975.

You will see that the word 'growth' must here be interpreted in an odd fashion; thus Niger has a negative figure for GNP 'growth' of −4.6%. There are other minuses too—for instance Chad, with minus 3.3%. These are not growing at all but actually falling behind. You have here what economists charmingly call negative growth. Compare it with two advanced countries—I chose two only, for reasons of time—the United States and the United Kingdom. The United States have more or less reached equilibrium on population; for GNP their growth-rate is positive. It has gone down somewhat but it is still substantial. The UK is perhaps in a better state in regard to population—with the growth-rate down to 0.4% but the GNP growth-rate a little below that of the United States. Whatever else you may conclude from these figures, there is an enormous difference in terms of access to resources between the advanced countries and The Rest; and when I say 'The Rest' I am referring, of course, to the overwhelming majority of mankind.

So they have a long way to make up, if indeed they ever can. This brings us to the sombre reality that, on resource grounds, it is increasingly clear that they never can catch up. Indeed, we—the people of the advanced countries—will have to lessen *our* demands on the world's renewable resources. When discussing these issues, I must admit to a certain sensitivity [— especially when confronting audiences from the underdeveloped world]. I therefore preface further observations by quoting from a very experienced African psychiatrist, Professor T. A. Lambo, [Deputy Director-General of the World Health Organization, who became Chairman, following the untimely demise of Jean Baer, of the International Steering Committee of this Conference]. Lambo believes the central problem to be stress, frustration, and the destruction of social institutions facing the traditional cultures of Africa. He states his beliefs as follows—and personally I regard his views as valid in essence for most peoples in the Third World and, in retrospect, for the rest of the World:

> Within these societies there are: collision (and fusion in some cases) of two or more cultures; the disruptive effects of industrialization; the emotional and social insecurity and isolation of the individual who is transplanted from the rural to an urban environment; the assumption of new roles, for example by young politicians, with consequent erosion of the authority and power of the traditional elders. All this is coupled with the switch of moral and social values in the process of shedding of tribal life, thereby creating what may be termed existential frustration and an existential vacuum.*

Of course the punch-line is really at the end: '. . . thereby creating what may be termed existential frustration and an existential vacuum'. Here we face the unpleasant reality of acculturation. You have a breakdown of values, a collapse of certainty about life's goals; and the only way it seems that societies can attempt to tranquillize such terrifying stress is by seeking higher consumption, and by other forms of 'tranquillizer'—including violence, crime and war.

Let us look at another Lambo quotation:

> But disillusionment has come and we begin to doubt whether material prosperity is necessarily synonymous with successful living. . . . There is no doubt that this preoccupation with affluence tends to destroy traditional cultures. . . . While some cultures appear more adaptable than others to the needs of industrial society, many others show selective resistance and susceptibility. Yet in all cases the modified cultures weaken dependencies of individuals on family, clan, shrine and community, and it is this break in the affective bond which is at the root of the contemporary conflict agitating the mind of the African.*

* The African mind in contemporary conflict, *WHO Chronicle*, Vol. 25, No. 8, pp. 343–53, August 1971.

Conflict breaks up family and social discipline, undermines identity, and [destroys many of] the loyalties that made it profitable and worth pursuing; the individual ends up by being entirely alone. He loses his bearings. If I may go a little further, here is my final quotation from Lambo:

> There are clinical grounds for assuming that the process of constrained acculturation can impinge harmfully upon individuals, causing more susceptibility to psychological ills than if they had remained members of a stable, highly stratified community with a traditional culture.*

We should observe that the word 'culture' is used here in a wider sense than the popular [— as including not merely] arts and crafts and so forth but also the total pattern of the conventions of behaviour, evolved through the generations, whereby the ideal identity is pursued.

This function of culture is crucial to understanding the inherited fulfilment goals of a traditional community. Bearing that in mind, let me quote a key example from the research I have been doing in Calabria, [in southernmost Italy,] which illustrates how it came to pass that one of the most important institutions of the village where I lived, collapsed before the inflow of higher technology. The village, San Giorgio Albanese, stands on a mountain shoulder [about 460 metres] up. At its side is a ravine containing a stream called the Fikthe. Up to six years ago, all the women used to do their laundry down there on the banks of the stream. Of course the girls went there too, to look after the younger children and to help their mothers with the work; and because the girls went there the boys went too. Each day saw a dynamic, familial gathering down at the Fikthe—so developing, out of the daily needs of life, a prerequisite for true community. The washing had to be done. While it was going on, the older generation, in the course of casual speech and while exchanging anecdotes and gossip, conveyed down the generations the values of the community. (In that period, I should add, there was plenty of water for the village.) Then, about six years ago, following slightly greater affluence and the influence of television, the washing machine came. A lot of the women got washing machines, and no longer went down to the stream. Other women followed suit, and did *their* washing at home. So now nobody goes down to the Fikthe, and the water reservoir up on the mountain is no longer adequate. The new technologies are thirsty, and nowadays water flows from the taps for only four hours a day.

One result is that the women now work isolated in their houses, cut off from the life-enhancing interchanges with their peers. Another is that this once-effective means of passing down traditions to the younger people and to the children in the village is withering away. There is a general sense in the village that something important has gone, but they do not know what it is. It is very difficult to explain the process to them, as indeed, *mutatis mutandis,* it is to any other group of people. How could one go to them and say: 'Do away with those washing machines, or other like aids.'

This is only one of many examples I could give you of how the impact of technology—for the best of reasons one might think, to save effort and improve physical comfort—can have disastrous effects. I hope it will remain in your minds [and help to persuade you that] we must change our perspectives in this whole development debate, no longer accepting unquestioningly the dogma that technological extension and greater resource-usage open the way to fulfilment. [Technology should be fitted to people's needs, rather than the reverse.]

* The African mind in contemporary conflict, *WHO Chronicle,* Vol. 25, No. 8, pp. 343–53, August 1971.

Franz, Kristi (Panellist)

This Conference seems to be calling for greater awareness of, and sensitivity towards, natural processes and natural habitat. We have talked of the need for humans to recouple with their natural habitats and natural biological interactions to avert ecodisaster. Yet it seems that we humans are de-programming the very qualities of sensitivity, awareness, reverence and the respect that is needed to recouple with the environment. As a defence-mechanism needed to cope with the monstrous man-made and manipulated environment, we choose not to hold on to or develop these qualitites. Humans proceed to become immune and indifferent to the products and consequences of their behaviour. As our awareness of and respect for the delicate balances and diversity of the natural world become difficult to maintain, and [consequently] diminish, the human species moves closer to ecodisaster.

I maintain that as part of our approach to avert ecodisaster we need to be supportive of those elements in human culture that heighten [in each person] awareness of [his or her] surroundings. Therefore, in discussing alternative life-styles, let us give close consideration to a goal of developing and maintaining people's *direct* involvement with the arts of their cultures.

Music is one art-form to examine in achieving this goal. Charles Darwin perceived that music serves a purpose in our lives. This perception can be supported by the historical observation that music is a constant cultural attribute of human civilizations. Its very survival and continued presence suggests that it fulfils a basic human need. Accepting this assumption, we can then presume that its absence or distortion may make it more difficult for us to maintain a healthy existence and meet the challenges of life. Assuming that music is an important part of human culture, direct involvement with it may [even] be essential for survival.

Looking at life-styles in the United States, we observe that most people's [musical—if they can be so termed—] experiences take the form of passive listening to electronic reproductions of sound, which [may be] devoid of subtle harmonies, nuances of tone, colour and texture, that can express, communicate, and emote, human feelings. Furthermore, the increased level [or volume of this 'music'] is so great that it is no longer [really] music but noise that bombards the listener to the extent that the auditory system suffers injury or is even permanently destroyed. People continue to tolerate gradual increases in sound-level in music as well as sound-bombardment produced by the technology of their society. At some point, sensory overload takes place, the result being a conditioned reflex to ignore and tune out one's surroundings. As a defence-mechanism one becomes numb to the stimuli and experiences a sensory uncoupling from sound experiences in the environment.

Musical experiences, through electronics, become a distortion of an important element in human culture: a direct involvement with the production, as well as direct personal control of musical sound, is lost. Statistics presented in 1974 by the Music Educators' National Conference in the United States showed that, for the first time [in several decades], the percentage of people becoming directly involved in the study of music and musical instruments had decreased. If music indeed serves a purpose in our lives [— and I think we would all agree that it does—it seems] important to reverse a trend from indirect passive involvement in present life-styles to more direct participation with music in our cultures.

In developed countries we can see the trend of indirect involvement not only in music but in many essential elements of human life-styles. The assumption that indirect involvement with an essential element of one's habitat leads to an uncoupling from important functions in human culture can perhaps be observed in looking at musical experiences today. Alternative life-styles should surely include a recoupling of people and their music, such that they experience it directly.

Direct experiences with music should re-establish its diverse functions in human culture—such as for expression of emotions, communication with others, relief from drudgery (as was done with work-songs), development of mental and physical concentration, and development of listening skills to heighten sensitivity and awareness. Music serves a necessary function in the human habitat. [I submit that] there is a need to re-establish direct relationships between people and their music as part of a holistic approach to creating alternative life-styles that will avert ecodisaster. Let us reincorporate human arts as one small way [of changing our behaviour towards re-establishment of a sane and viable life-support system.]

Candolle (Panellist)

After listening for several days, there are some things I would very much like to say. But first I want to express sympathy with what has been [well said] today by Mr Guppy, and to some extent by Mr Glasser. Until they spoke, I had felt that not enough insistence had been given to the fact that most of the troubles we are facing [stem from] our Western civilization, and particularly from the belief which it has developed in the last 200 years in what we call 'progress'. The idea of progress [as being a veritable ideal] is applied to all fields [of human endeavour]. I myself have been described as a banker, which is no longer true, as for the last few years the centre of my life has been music, which is where I am very much in sympathy with the last speaker.

Permit me to describe to you what has happened in that field. Music, by its very essence, can only express a few things, and always the same ones: these are human emotions such as pity, love, joy, fury and grief—so that composers [and musicians] have always found themselves expressing these same things. But now for some generations a composer has invariably felt that he could no longer express those things in the way that his father did. There must be progress: he must use a new language—an improved language. [With music thus apt to be] severely limited by physical laws, contemporary composers and musicians have found themselves forced into positions where they make music in their own 'language', which has ceased to be a means of communication with the general public. They produce noises with no real meaning for a rational, ordered mind [and are widely incomprehensible to all but themselves].

It has been said of creative artists that they are 50 years ahead of their time. If that is so—and I think it is [in some cases]—and considering that composers have been [spawning] these noises now for 20 years, we can only be some 30 years away from something much worse than any ecodisaster—something more like real Armageddon, and much the same may be true of other arts. [Laughter and applause.]

Rather than depend on a 'language' of expression for which there is no public, and no prospects of one, musicians have been looking back at baroque and renaissance music, using instruments of the times after ridding them of any 'improvements' that have been made in the search for 'progress'. They have been trying to play in the manner of the past—even using the pitch which they imagine to have been that of a given period.

May I suggest that this is an indication—a very clear one—that the idea of 'progress' is on the way out: the consequences of this are indeed worth considering, [and in other fields than the auditory and visual arts]. If the golden age of 'progress' is past and we no longer believe in it, can there logically be such a thing as being 'backward'? How will our failure to believe in our own 'progress' affect the attitude all around the world of those who have traditionally looked up to us? Is there not a likelihood that, like musicians, others may start searching for things that have been linked with human happiness in the past? I think of such occupations as hiking, handicrafts, community singing, gardening, [and others which are similarly unharmful] to the biosphere. Indeed they are sometimes even beneficial. If only some of the energy which goes now into skiing on snow or water, to racing around in motor-cars or on motor-bicycles, or to swimming in pools, were spent on planting trees or building compost heaps!

These may be the dreams of an optimist but I believe in expressing them after so much [despondent] talk. I was very glad to hear Mr Guppy's figures about American opinion and Mr Goldsmith's views about the feeling in France [: is it really too much to hope that there may be on the way] some real turning of public sentiment, and should we not influence politicians and [policy-directing] officials who reach the public and could get results? Or would it be better to go to the public directly? [Applause.]

[As is indicated in the introductory paragraph of this Discussion, the remainder of the discussion, including the comments of the other panellist, were lost. The same fate befell the Chairman's concluding remarks, which were, however, remembered as being warmly favourable, and laudatory of the participants. A majority decision to resume the Session later was not implemented, largely because it was found at the first convenient opportunity that all participants had by then said what they wanted to say.]

Session 15

ENVIRONMENTAL MANAGEMENT TOWARDS BIOSPHERAL EQUILIBRIUM

Keynote Paper

Environmental Management and Biospheral Equilibrium*

by

MOSTAFA K. TOLBA

Executive Director, United Nations Environment Programme, PO Box 30552, Nairobi, Kenya.

INTRODUCTION

The subjects discussed in the earlier part of this Conference are far-ranging indeed. There has undoubtedly been much thought given to potential disasters for the present and future generations, which may be brought about by activities of Man impinging upon the health of the biosphere, and interfering with the dynamic processes of the planet's ecosystems.

The title of the Conference, particularly the word 'ecodisasters', suggests a solemn warning. The whole range of threats to the continuation of harmonious relationships between Man and the natural environment is daunting, to say the least.

Nevertheless it will be my contention that 'Growth Without Ecodisasters' is possible, at least without the ecological havoc that has sometimes stemmed from human activities. 'Development without destruction'—for that is the expression which we prefer to use—has become our goal in the United Nations Environment Programme (UNEP). We believe that goal will be reached through proper environmental management. To put it simply, we believe that the fundamental question confronting mankind is how to meet the basic needs of humanity without simultaneously destroying the resource-base—that is, the environment from which those needs must be met.

Let me assure you at the outset that my basic optimism is not lightly held, although it is our business in UNEP to keep a close watch on all potential threats to the environment—and there are many.

* *Presented by Dr Letitia E. Obeng, Chairman of Soil and Water Task Force Division of Environmental Management, United Nations Environment Programme, in the unavoidable absence of the author.*

Each year I present to the Governing Council of UNEP* a 'state of the environment' report in which I select certain topics for special attention. These topics must meet one or more of the following criteria—they must:

(1) Be of international importance;
(2) Have attracted public, scientific or governmental, attention;
(3) Be urgent;
(4) Have received insufficient attention from governments and the United Nations system; and
(5) Represent problems on which UNEP itself has done or intends to do some work.

The first topic this year was the *ozone shield*—that layer which lies in the high stratosphere 10–15 km above the surface of the Earth, and which forms a vital protective barrier against harmful ultraviolet radiation from the Sun. Certain activities and products may adversely affect the ozone layer. Examples are supersonic air transport; some aerosol propellants; nitrogenous fertilizers; and atmospheric nuclear explosions. These things are meant to meet perceived human needs such as, respectively, rapid travel, household insecticides, food and defence. Yet the satisfaction of these needs could inhibit the satisfaction of other, more fundamental needs—such as human health, health of livestock, and maintenance of the productive capacity of the Earth and the seas.

One likely effect of damage to the ozone layer could be an increase in skin cancer. But this is only one of many forms of cancer. In many developed countries, cancer today is responsible for around 17% of all deaths. Well over half of all cancers are now thought to be of environmental origin, initiated by chemical compounds released by modern technologies or life-styles. The second topic, therefore, is *environmental cancers,* resulting from the attempt to satisfy perceived needs but again at the expense of human health.

The third topic of this year's report deals with an even more basic threat to human well-being—*land loss and soil degradation.* Some forms of agricultural practices, such as the cultivation of steep slopes, slash-and-burn and deforestation, may damage the soil cover irreparably. In addition, activities designed to meet other human needs, such as housing, roads and airports, may remove valuable land from agricultural production.

The fourth topic is *firewood.* To a poor household, in many developing countries, firewood is the only source of energy for cooking and heating. Yet the extensive and indiscriminate use of trees for firewood is resulting in rapid and alarming depletion of a key renewable resource, as well as bringing in its train erosion and soil damage. The ill-advised felling of trees in certain areas for commercial purposes, or for increasing the area available for agriculture, adds to the problem. Flooding often occurs when tracts that should retain a good forest cover are denuded of trees.

But there are many other reasons for concern in addition to those

* Consisting of 58 member states elected by the General Assembly of the United Nations for periods of three years.—Ed.

featured in my latest report to the Governing Council. We have only to think of what is happening to the supplies that are available to Man of fresh water—a prime necessity of life. Deprive living things of the water which they need, and death results—often in a matter of days. Every country wants, and needs in some form, continued development; yet water consumption rises drastically with the spread of human settlements, with industrial development, and with the increase of modern agricultural practices. Along with that there is the rise in pollution, and other alterations of water-bodies—for example through eutrophication. Degraded water can contribute to ill-health and disease. The seriousness and the complexity of the water-related problems facing the world came under intensive review at the United Nations Water Conference in Argentina in March 1977.

Extreme scarcity of water is one of the major factors in the processes of desertification. But desertification is a global problem which cuts across many areas of environmental concern. When we speak of desertification we must speak of soil, water, forests, other plant and animal life, and so on. We are now in the final stages of our preparations for the United Nations Conference on Desertification,* authorized by the General Assembly in December 1974. As we all remember, it was the great drought which affected the countries of the Sahel some six years prior to 1973 that aroused world concern—first to help relieve the mass human suffering which it had caused, and secondly to try to understand what may lie behind such drastic weather fluctuations and concomitant problems. The conference aroused widespread interest in developed and developing countries alike.

We realize today, perhaps as in no previous generation, our utter dependence on pure, fresh air. We have been alerted to what we are doing to the planet's air by our industrial emissions, yet there is still very much that we do not understand about the ultimate effects of human actions on the atmosphere. We have but lately begun to think about the dangers associated with the depletion of the genetic resources of our world's plant and animal life-forms. Cultivated plants have less resistance to disease and drought than wild plants. We have been neglecting the latter so that there are now fewer extant species. There has also been a disappearance of some animal species. Wild beasts, like wild plants, have greater resistance to disease than do domestic ones. A wealth of attractive large mammals forms also one of the bases of the African heritage, yet some people think that, because of the pressures which Man is putting upon them, such forms of wildlife, although still found in such great abundance in eastern Africa, may be practically eliminated in a few decades.

Then there is the highly intricate question of reactions in microbiology. A growing weakness is apparent among many kinds of microbiological resources, although we do not understand very well the problems in this field. A great deal of public and scientific attention has also been given in recent years to assessing the possible risks and benefits of the controversial biological technology known as recombinant DNA research, which allows molecular biologists to transplant genes from one species of living organ-

* Which took place in Nairobi, 29 August to 9 September, 1977—*see* the account by Ralph Townley in *Environmental Conservation*, Vol. 5, No. 1, pp. 69–70, Spring 1978.—Ed.

ism to another. The issue is fundamentally one of unknown risks *versus* unknown benefits, neither of which can be specified for many years. Scientists have already, however, agreed to accept a degree of public control over such research.

But you are perhaps well aware of these and many other potential sources of ecodisaster. We live and work in a closed system in which 'everything is connected to everything else', as Dr Barry Commoner put it—and, I might add, in which 'everybody is connected to everybody else'. There are outer limits to what the planet can sustainably provide, and there are inner limits representing basic human needs which must be satisfied. These outer and inner limits, together with the rising aspirations of the peoples of the world, are the main components of the environmental issue. The limits in the ability of the biosphere to absorb the impact of Man's activities must be respected for Man's own long-term well-being—even for his survival. The 'sociosphere'—the sphere of Man's needs and aspirations—is in some ways more complicated even than the biosphere, but is perhaps in some ways more flexible.

Human societies are inherently teleological, that is, they work towards self-chosen ends. It is my belief that, as we see ever-more-clearly the end results of the choices we make, we will come to choose the roads that lead to a more abundant life for all, on a continuing basis, rather than those that lead to disaster in any form.

Why Development—and How?

Many of the negative aspects of development have so impressed themselves on the minds of those concerned about the environment that the question is often asked, 'Why development?'. Some, indeed, advocate arresting economic growth in the interests of environmental protection.

It is true that, in the past, industrial and agricultural development have created many environmental problems, with consequent costs to human health and well-being. But environmental problems are caused also by a lack of adequate development. Today there are hundreds of millions of people without the basic human necessities such as adequate food, shelter, clothing and health; hundreds of millions more lack access to even a rudimentary education or regular employment. This is not only an intolerable situation in human terms, but it also has serious environmental consequences. The relentless pressures that arise where, through poverty, basic human needs are not met, can be as significant in erasing the resource-base from which Man must inevitably gain his sustenance as the pollution created by industry, technology and over-consumption by the affluent.

I therefore conclude that, for the developing countries in the world—in which more than two-thirds of mankind live—there is no alternative to pursuing technical, economic and social change so as to meet basic human needs and secure better prospects for their citizens. I also believe—and this is a basic point in the philosophy of UNEP—that far from being in conflict, *environmental and developmental objectives are complementary*.

It has taken time to clarify what is the best approach to the wide variety of environmental threats that first began to capture widespread public attention, particularly in the West, some 10 years ago. It was already apparent, at the time when the UN Stockholm Conference on the Human Environment ended on 16 June, 1972, that there was a new climate for international cooperation to ensure environmentally sound development. The link between development and environment was further clarified in the years following the Stockholm Conference. Developing countries themselves came to see the need to incorporate environmental considerations into their development efforts. We have also seen the emergence of such terms as 'the new development' or 'alternative styles of development', which suggest a more rational way of looking at development—namely one in which environmental considerations come to play their rightful role.

There are, however, certain important differences between the form which a new kind of development might take in industrialized countries and that which would emerge in developing countries. In the industrialized countries, it might take the form of a reorientation of society's aims, so that the entire population has more opportunity for self-expression in the fields of culture, education, the arts and humanities—those non-physical areas of development which represent the highest levels of human achievement. This is a logical development. The psychologist Abraham Maslow has argued persuasively that there is a hierarchy of human needs. When needs such as hunger, clothing and shelter have been met, other needs—social, aesthetic and spiritual—assume increased importance, and life is truly impoverished if this level of human needs is neglected.

The new orientation brought about by such an expanding concept of human needs has the additional advantage of being less demanding, than its predecessor, on the environment—in particular on natural resources and energy. Present patterns of production and consumption involving widespread waste of resources, extravagance and planned obsolescence would be replaced by conservation and re-use. I am encouraged by signs from several countries that such a reorientation of life-styles and societal aims can now be discussed seriously, although clearly the change implied by this approach is an immense one which could be achieved only after many years.

The developing countries still lack the infrastructure required to meet the growing needs and aspirations of their peoples. But each country should be able to select the path to development which is best suited to its own human skills and natural resources, and which responds to its own needs and accords with its own culture and value-systems. Developing countries should have access to the technologies which they require, and should be enabled to adapt these technologies to their own needs—rather than to run the risk of distortion through imported technologies.

But the forms of development adopted anywhere in the world should be based on environmentally-sound practices, and should avoid the destruction of the resource-base. Everywhere, technology must be developed which does not pollute the environment or waste resources. The conservation of species is a common problem throughout the world. Human settlements problems everywhere have much in common, though solutions must take into account local circumstances.

Developing countries are in a unique position to avoid the situation from which many developed countries still suffer—namely, the consequences of growth patterns based mainly on cost-effectiveness, particularly concerning the siting, design and operation of industrial activities. We only need to look at the pollution prevailing in some industrialized cities, or read about the effects of some recent industrial accidents on the populations living in their vicinity, to be convinced of the need for new and better approaches to development.

These new approaches to development also have important international implications. The debate around what is called a 'new international economic order', with its emphasis on a more equitable use of the world's natural resources and on the need for development in a form that would be appropriate to each country's requirements, shows that there is already some recognition of these implications. I hope the international community, and especially the developing nations, in advocating and designing this much-needed new economic order, will not forget that there will be no sustained development or meaningful growth without a clear commitment at the same time to preserve the environment and promote the rational use of natural resources—that is, their use in ways best calculated to further the aims which society has set for itself, and which take account of all effects that are known or may be expected to follow from such action. It is hardly rational in today's world to use more resources than are necessary for a given purpose.

Careful assessment and wise management are the two essential concepts for the pursuit of development without destruction. It may be worth while to give a brief account of what assessment and management mean in the UNEP context, and what we are doing to assist the nations of the world to apply these concepts to their own interrelated problems of development and environment, and also to those which they share with other states.

Environmental Assessment

'Earthwatch' has become the term for global environmental assessment. It is a dynamic process of integrated environmental assessment by which relevant environmental issues are identified, and in which necessary data are gathered and evaluated to provide a basis of information and understanding for effective environmental management.

The components of Earthwatch are monitoring, research, information exchange, review and evaluation. *Monitoring* systems generate a continuous flow of information about the condition of the environment. *Research* enhances our understanding of environmental processes and facilitates interpretation of environmental data. *Information exchange* is an obvious requirement if the benefits of monitoring and research are to be enjoyed by all. *Review* is an analytical process by which problems can be defined and gaps in knowledge and understanding can be identified. *Evaluation* is a synthetic process involving the collation, correlation and interpretation of the results of monitoring and research. Thus evaluation is the culmination of the assessment process, and provides one of the main inputs to the policy formulation and planning functions of environmental management.

From its inception, UNEP began to develop and apply these concepts in practical terms. The most marked progress was made in the field of monitoring and information exchange. A Global Environmental Monitoring System (GEMS), an International Referral System (IRS), and an International Register of Potentially Toxic Chemicals (IRPTC), have been established.

The purpose of GEMS, which is a global network of monitoring systems that is coordinated through the efforts of UNEP but executed by a number of UN agencies and other bodies, is to establish the means to identify the trends and changes that are brought about in the global environment by human actions. In the agricultural and land-use areas, for example, GEMS is active in the monitoring of soil and vegetation cover. Collaboration has been established with FAO and UNESCO on work leading to a global assessment of soil degradation and degradation hazards.

Vegetation-cover monitoring has started, in close collaboration with FAO, with a Tropical Forest Monitoring Pilot Project in West Africa. This will soon be joined by a related project on tropical rangelands in the same general area. The overall aim of these vegetation monitoring activities is to develop a realistic methodology which may, in the next few years, be used in all areas of the world. Thus GEMS is attempting to address itself to some of the fundamental problems of today's world.

The IRS, through a directory and associated search procedures (whether computerized or manual), provides a 'switchboard mechanism' to put the seeker of environmental information in touch with those who can provide it. The IRS does not itself handle substantive information. The system is completely decentralized, depending for its functioning on the establishment of national systems for environmental information which the IRS links together. Now operational, and supported by a network of some 70 focal points, the IRS is a valuable tool in helping governments, decision-makers, professionals, research workers, educators, and others, to gain ready access to sources of environmental information. Well over 3,000 such sources from 25 countries and several UN agencies are now registered in the system.

The IRPTC provides the means to handle data and answer queries on potentially toxic chemicals. In addition to the many benefits which they have conferred on modern society, chemicals are also among the chief pollutants of the human environment. Some half-a-million chemicals are currently used in industry, agriculture, public health, food preservatives, additives and processing aids, drugs, and cosmetics. Chemicals also enter the environment as waste products of industrial processes, and some of these may find their way into Man through the food he eats, the water he drinks, and/or the air he breathes. Hazards induced by chemicals in the environment must be scientifically and objectively evaluated. For this, all available information on their physical and chemical properties must be collected, in order to identify them and to study their hazards for living beings—including Man. The ultimate purpose of the IRPTC is to supply information to those who are responsible for environmental protection, and to provide the basic data for evaluating and predicting hazards associated with particular chemicals. To achieve these objectives, the IRPTC relies on centrally-stored data, and also, through a network of cooperating data-bases and institutions, on data that are available elsewhere. The IRPTC operating centre is already established in its new headquarters in Geneva, Switzerland.

Assessment is the weighing of costs and benefits, and the assessment programme provides the tools that are continually needed for proper environmental management. The programme must also provide early warnings of any global threats to human welfare, or even survival, that may be brought about through Man's activities which transgress the outer limits to which we referred earlier. Thus as part of Earthwatch, UNEP has developed a programme on the outer limits which, among other things, deals with threats to the ozone layer and of climatic change. In these and several other areas, we are designing programmes of research and monitoring, coupled with an evaluation mechanism.

Our approach to the ozone question illustrates well the way in which each major component of Earthwatch—evaluation, monitoring, research and information exchange—is being used by UNEP to construct a comprehensive assessment programme. Thus *evaluation* reveals that the ozone layer, and hence its effectiveness as a filter of harmful rays from the Sun, may be reduced by the release of fluorocarbons, and, indirectly, by the manufacture and application of nitrogenous fertilizers. A coherent programme of *monitoring* of total ozone and its vertical distribution is under design. UNEP intends to support *research* into the effect of fluorocarbons and fertilizers on the ozone layer, and also into the consequences for life on Earth of an ozone reduction and a consequent increase in ultraviolet radiation. Aware of the important issues at stake, the Governing Council of UNEP requested me to convene a meeting to review all aspects of the question. This meeting, which was held in Washington on 1–9 March, 1977, brought together experts designated by governments and intergovernmental and non-governmental organizations—including representatives of the International Council of Scientific Unions. The meeting provided a forum for the exchange of *information* on all aspects of ozone assessment.*

The ozone programme, like other similar studies which UNEP supports, is meant to provide data for establishing priorities for environmental management, and for indicating the form which such management should take.

Environmental Management

The concept of environmental management, as advocated by UNEP, is a broad one. The most pressing objective is of course to meet basic human

* *See also* the account, by its Convenor Professor R. E. Munn, of the first meeting of the UNEP Coordinating Committee on the Stratospheric Ozone Layer, held in Geneva, Switzerland, 1–3 November, 1977, and published in *Environmental Conservation,* Vol. 5, No. 2, p. 155, Summer 1978, with details of its 14 Recommendations.—Ed.

needs within the potentials and constraints of environmental systems and natural resources. Environmental management brings two new dimensions to the development process: it calls for environmental quality to be taken fully into account, and it extends the time-frame—for development must be suitable over the long-term. At present environmental management is too often neglected because governments, and others whose activities can affect the environment, lack the necessary information, political will, or resources, or find it hard to respond to the long-term and transectoral nature of environmental problems; or because existing administrative institutions—whether professional, legal, financial, or other—are poorly adapted to the concept of environmental management.

In collaboration with a host of United Nations and other bodies, UNEP now has a great many activities under way which could help further the application of environmental management. To take a few examples: we are demonstrating new environmental approaches to the improvement of marginal settlements in Indonesia and the Philippines, and are advocating an ecological approach to the management of arid and semi-arid rangelands at the United Nations Conference on Desertification; we have helped to prepare a world plan of action for an environmentally-sound approach to schistosomiasis control; we are establishing pilot schemes to illustrate the ecodevelopment approach, which is the application at the local level of environmental management principles within the development context; we are establishing microbiological resource centres as storehouses of microbial strains and are seeking to apply the knowledge available in this area to everyday problems of rural development; and we are establishing demonstration centres to show how renewable sources of energy can be used to provide the basic needs of small rural communities in developing countries. These examples show that environmental management can be introduced into a broad range of human activities.

Let me now turn to the way in which we applied the concepts of development-without-destruction to the Mediterranean. There has been a serious, sharp decline in recent years in the quality of the marine environment of the Mediterranean Sea. If this is not soon halted and reversed, it could result in a veritable ecodisaster for the many communities that are dependent upon fishing and tourism; it would greatly detract from the quality of life of the people living in the region; and it would demonstrate appalling contempt on the part of modern Man for a region which has played such a central role in his cultural evolution. To let the Mediterranean die would be an outstanding example of the pointlessness of heedless development.

In January 1975, after several years of preparation by various UN organs and specialized agencies, UNEP convened a meeting of Mediterranean governments in Barcelona, which agreed upon an action plan for the protection of the Mediterranean.* This plan contains a major assessment programme involving monitoring of, and research into the health of, the

* *See* the account by Stanley P. Johnson in *Environmental Conservation*, Vol. 2, No. 3, pp. 235–6, Autumn 1975.—Ed.

marine environment, through the joint and coordinated efforts of many regional and national institutions. On the management side, the action plan contains two elements: integrated planning of the region up to the end of the present century, and legal measures. At a second conference in Barcelona in February 1976, the 16 nations present adopted unanimously a final act embodying a convention and two protocols to protect the marine environment; 15 nations signed the act and 12 of these also signed the convention and protocols. The proportion of states signing this convention immediately after its finalization was unprecedented. The meeting also agreed to establish a centre in Malta to enable states to combat major oil-spills in the Mediterranean.*

Such conspicuous achievements in a region as politically sensitive as the Mediterranean is striking proof that, where the political will exists, the goal of development without destruction is attainable. But we should not delude ourselves: that act of political will does not come easily, because it implies very large changes in public attitudes and expectations.

Development Without Destruction: Food

Nowhere is this kind of development more needed than in the provision of food. There has been mounting concern in recent years about food—one of the world's primary problems. We in UNEP feel deeply about the necessity of maximizing food production while at the same time preserving the ecological basis for sustaining such production.

Any rational strategy for increasing food production on a sustained basis must take explicit account of the complementarity of environment and development. Indeed, there is probably no area where the objectives of environment and development are more intricately interrelated. This view was reflected at the 17th session of the FAO conference, which stated: 'The major environmental problems facing agriculture, forestry, and fisheries, were not only the avoidance of environmental pollution but the ensuring, in the development process, of the maintenance of the productive capacity of the basic natural resource-base for food and agriculture through rational management and conservation measures.'

The urgency and magnitude of the task of attempting to double food production by the end of the century, and at the same time assuring the supply of basic food requirements to all, should not be underestimated. It is vitally important that any short-term measures to increase food production be effectively integrated with long-term policies. To achieve this the following ten considerations need to be borne in mind:

(1) Pressure to expand areas under agriculture—frequently aggravated by the loss of good agricultural land for industrial and human settlements purposes—has often resulted in serious environmental disruption. To cite but a few specific examples, the expansion of agriculture to steep hillsides leads to serious erosion in Indonesia; increasing pressure of slash-and-burn

* *See* the account by Peter S. Thacher, now Deputy Executive Director of UNEP, published in *Environmental Conservation*, Vol. 3, No. 2, pp. 152–3, Summer 1976.—Ed.

agriculture adversely affects tropical forests in the Philippines; deforestation in the Himalayas contributes to the increasing frequency and severity of flooding in Pakistan, India and Bangladesh; and overgrazing and deforestation assists the southward march of the Sahara in the Sahelian Zone of Africa. Experts estimate that the land which is being lost to agriculture by such processes as these may now exceed the acreage of new land that is being brought into production. Thus rational management of both arable and pastoral agricultural land is becoming increasingly urgent, and its limits must always be remembered.

(2) Technology transfers in the field of agriculture have often taken root with bad results because proper account was not taken of local social, cultural, economic and, above all ecological, conditions. Simple technologies, known over the centuries, such as the use of terracing to prevent soil erosion, were often ignored. Existing technical and scientific knowledge must be better mobilized and more effectively applied to ensure sustained rather than short-term benefits.

(3) There is an intimate relation between inputs of energy and the output of food. Scientifically planned inputs of energy to the land can yield extraordinarily favourable results, although a saturation point can be reached after which any extra inputs are wasted and may cause serious environmental degradation.

(4) Fertilizers are needed for increased food production but their excessive use poses environmental threats, the dimensions of which are not yet fully understood. While they should be used with maximum efficiency, it is never efficient to create dangers for Man or his environment.

(5) Pests cause significant world-wide crop losses, despite the ever-increasing use of pesticides. Through the processes of natural selection and evolution, new strains of pests have appeared, which are often more vicious and less susceptible to control by chemicals than were earlier strains. Neither larger doses nor alternative pesticides provide a permanent solution. Another major problem is that existing modes of applying pesticides have extremely low efficiency rates. A third element of concern is that the distribution of pesticides through ecosystems takes place most commonly by selective concentration as they pass through successive levels of food-chains and food-webs; thus high pesticide levels accumulate in the higher animals and in Man. It is therefore essential to develop alternative methods of integrated pest-management. Where chemical pesticides *must* be used, they should be used with the greatest possible efficiency.

(6) Vast food-losses occur each year in storage, processing and handling. New and better techniques for preventing loss and waste could do much to increase available world food-supplies.

(7) Weather and climate have always been important to crop production, often outweighing the factors that are subject to human management. Today there are increasing signs of possible changes in climate and weather patterns. Man's ability to predict and anticipate these changes has greatly improved, and he may soon develop a new capacity to influence them as well. Improved, long-term forecasting is needed to make crop planning more efficient.

(8) Irrigation schemes are undoubtedly needed in the developing countries, but, after several years, salinization of the soil may render it once again unproductive. In some cases, moreover, irrigation schemes spread water-borne diseases. When ecological and environmental principles are applied from the planning stage, these hazards can be averted, and the health, well-being and productive capacity of the human population can be improved.

(9) We know that the world can produce surpluses of food and that much more can be produced where the skills and capital are available. There is an urgent need to develop mechanisms to distribute the surpluses from the favoured to the poorer regions.

(10) Strategies to solve the world food problem must be developed in full knowledge of the web of interdependence that exists between this and the other major problems facing mankind—those of population, energy and other raw-material shortages, underdevelopment, and environmental degradation.

Conclusion

In the years ahead, we face the task of meeting the minimum needs of mankind and of avoiding environmental catastrophes. I have spoken encouragingly about the prospects because I am convinced that disaster is not inevitable. But the urgency is extreme; we have very little time in which to set right our approach to the environment and to meet the legitimate demands of the world's poor. We shall need to act far more thoroughly and speedily than hitherto if we are to redress human and environmental grievances, and we shall need to harness the energies of all sectors of society in this effort.

An extremely important role must be played in this by the scientific community and by non-governmental organizations such as your own. It is the duty of all concerned individuals and organizations to bring the problems and solutions to light, and to display them with appropriate objectivity. In particular, through the diligence and thoroughness which is the mark of all sound scientific endeavour, scientists can help Man to see the dangers which confront him, and to understand that it is essential for him to adopt ever-wiser and safer approaches to managing his planet.

DISCUSSION (Session 15)

Bryson (Chairman, introducing)

At the start let me warn you that I am a tough Chairman: I shall be most impolite and most unpleasant if we carry over too long. Now I do not know whether our Secretary-General and President knew how nasty I could be when they asked me to chair this Session, but let us try to keep to the schedule.

This [first] Session today being on 'Environmental Management Towards Biospheral Equilibrium'—management for the needs of Man and Nature—I would like to introduce the topic by suggesting several considerations that, to me, seem important when we talk about management. I have deliberately been wearing this silver 'Bolo-tie' during the course of this Conference, because it was made by Hopi Indians. Now Hopi Indians have a particularly interesting view of Man and Nature—to which I subscribe most heartily as it is one of the various world views which [seems in this case to be] as realistic as [any] you will find in any culture. The Hopi, unlike many of the North American Indians, does not regard Nature as separate from himself. The Hopi says: We are part of Nature, and if we observe the laws of Nature (of course expressed in terms of the Nature gods) and collaborate with Nature, living as part of Nature, all will go well.

For 8,000 years the Hopi have survived in the same place—unlike many, many tribes around the world who have at one time or another had to move or modify their culture. Their philosophy of Man and Nature seems to have worked. In the course of this Conference we have already talked about Man and Nature—sometimes as though Nature has more rights than Man, and sometimes as though Man has more rights than Nature. [But] we are also natural, in my philosophy, and our habitat is as natural as the habitat of an Alga or the habitat of a pine-tree. It is the imbalance brought about by our power that really concerns us.

The second point that I would like to make in introduction is a general principle which deals with management. We often talk about optimizing food production and, at the same time—in the Conference and elsewhere—we talk about optimizing diversity, optimizing stability, optimizing the viability of certain things, and optimizing development for the standard of living of mankind. But, ladies and gentlemen, there is a universal principle that we get from mathematical analysis, logical analysis and experience with trying to [arrive at such multiple maximizations]: it is not possible, in general, unless one is very very lucky and is dealing with simple systems, to optimize for more than one variable at a time. You cannot optimize food production *and* biospheric stability *and* development all at once. It is not possible in theory or in practice; there must be adjustments from the optimum.

A third principle is that there is variability, and in talking about management we must remember that we are dealing with a variable system—with variable climate, variable numbers of people, etc. A variable system with feedbacks can produce oscillations and further variabilities. A variable system cannot be effectively treated as though it were a steady system. We must always have a margin, so that the oscillations or variations do not amplify in the manner of populations going to great peaks and great crashes. If we are going to push the population of the Earth to its outer limits, and if we are going to push food production close to the outer limits, then we must take into account the stability characteristics of those outer limits. With a variable food supply, for example, we cannot maintain the population at the highest level that the food supply can produce. With adequate storage we can maintain it at the average level, and with inadequate management of those resources we can only maintain it near the lower limit. So we have a very serious question here, philosophically, when we talk about management and optimization in a variable system.

The last point I would like to make before I turn the Session over to others on the platform—and I hope we shall take this into account in our discussion—is that management may be antipathetic to freedom. Now the question that we must face is how much freedom we can have, and still maintain viability. But let us remember [all the time] that management, if total—and especially if necessarily total—means a minimum of freedom.

This morning's keynote address is by Dr Mostafa Tolba who, as you will all be aware, is the Executive Director of the United Nations Environment Programme. On one hand it is unfortunate that he is not here, for, as most of you know, he is a dynamic, forceful, charming, intelligent man [—and a distinguished biologist in his own right]. On the other hand we are fortunate because, [partly] as a consequence, we are blessed with the presence of Dr Letitia Obeng, who will present a summary of Dr Tolba's address. [Dr Obeng is originally from Ghana, where she made her name by establishing and directing a pioneering hydrobiological institute, and she will be followed by three outstanding panellists whom you already know: Perez Olindo, Barton Worthington and Donald Kuenen. Thereafter Dr Obeng will reply before the discussion is opened to the floor.]

Obeng (Presenting paper for Keynoter)

Dr Tolba requested me to present to you his views on the subject of 'Environmental Management Towards Biospheral Equilibrium'; but now I understand that his paper has already been circulated, and so distinguished participants will have had an opportunity to read it in detail; [consequently it does not seem] necessary for me to present it in full. However, I have here the responsibility of presenting Dr Tolba's views, and even if I held different views, I would prefer to put [his] across as I was asked to do. I shall therefore emphasize the salient points expressed in Dr Tolba's paper, in the hope that they will continue to underline the message that I was asked to [convey].

If I may be permitted to do so at this point, I should like to express my pleasure at seeing Mr Strong here. For me he symbolizes the essence of the world's effort to preserve the human environment, and his presence here reassures me that we are doing the right thing. [Dr Obeng thereupon emphasized the main points of Dr Tolba's paper, which is printed in full above.]

Olindo (Panellist)

The paper that has been so well presented by Dr Obeng is one that expresses views which I agree with almost completely. I also agree, Mr Chairman, with the approach, and there is very little really to say after [that. But] I have looked [again] at the paper [and considered your own comments sufficiently to agree] that if the idea of global management for biospheral equilibrium is pressed, this [may necessitate considerable] restriction of certain freedoms. As I read the paper, I noticed a reference to the continuing momentum for a redefinition of the economic concepts that have been held to date, and for a redistribution of the world's wealth.

The prospects of a new international economic order [look] good. In it we, of the developing countries, are asking for a bigger share of the world's wealth [, which we feel can hardly be denied to us]. At the same time we resist considering the concept of viable sovereign states. There seems to be an implicit agreement that certain states are not economically viable within the current economic institutional set-up. Are we hearing voices almost saying that they cannot exist as entities? If the idea of the new international economic order is to be discussed [properly and openly], the question of what constitutes a sovereign state may have to be redefined, and I am encouraged to suggest that the system which may emerge and prove workable [may be along such lines of equitable redefinition].

We have the United Nations, we have the regional Economic Commissions for Africa, Europe, Asia and Latin America and, Mr Chairman, the ecological and environmental systems do not respect these political set-ups. They transcend them. [The effective management will be undertaken at the national level, but international cooperation should be propagated first and strengthened on a regional] basis. Environmental management will have to transcend political boundaries. It must follow ecological boundaries through cooperative effort among nations. A majority of the people in the world will need to accept a lot of adjustments before the new international order takes effect, and one hopes that this point will be made when the developed and the developing countries agree to share the wealth of the world. There is much in this paper that we need to support. However, if my remarks are provocative to my political friends in this room or anywhere else, I would like to invite them to comment but also to ponder the issues raised.

Worthington (Panellist)

I fully agree with my friend Olindo that this [paper constitutes] an admirable statement, wide-ranging in topics relating to the environment and to its use for human needs. [But] I would like to develop one aspect covered in the latter part of the paper where it is concerned with environmental management and the development of resources without destruction. Dr Tolba takes the particular example of food; he might have taken fibres, or a number of other human needs which are provided as a result of biological productivity.

The ordinary sigmoid curve of growth helps us to understand problems in the development of resources and [also the incidence of] ecodisasters. I was first introduced to the sigmoid curve half-a-century ago in studying the then unexploited resources of Lake Victoria—by a great conservationist, Michael Graham, who developed it in relation to fishery research. But it relates equally to the subjects mentioned in Tolba's paper. The horizontal time-scale can be days, years or centuries. The vertical scale may be weight of a particular organism, the numbers of a population or an ecosystem. If you take the case of a tropical rain-forest, we are up where the curve flattens off. The ecosystem has developed and become stabilized with its great diversity of species. The human population of the world is where the upswing of the curve is approaching its steepest. The population of Blue Whales is very near the bottom, where the growth-rate is least, because they have been driven down there by over-exploitation. If maximum sustained yield is the objective, management should keep the resource at the steepest part of the curve, where the growth is most rapid. A sudden break in the smoothness of the curve usually spells ecodisaster.

Now look at this from the point of view of developing food-resources, and I would like to take fishery resources [as an example] because it is in that subject that I learnt my lesson. In an African lake you start with a native fishery, adapted over centuries and in nice balance with the resource, which maintains itself in the upper part of the sigmoid curve. You then introduce modern technology and, in the early years, you get terrific catches. You think the fishery is going to be marvellous, and it grows like wildfire. But unwittingly you drive the resource down the curve to near its bottom. You have started an ecodisaster, from which, owing to drastic changes in the ecological balance, the resource may never recover.

Such experiences illustrate what is a very simple and well-worn concept, but a concept which continues to hold a useful place in relation to the kind of management which Dr Tolba is advocating in his paper.

Bryson (Chairman)

Perhaps I can use the prerogative of the Chairman for taking one minute to add a consideration to illustrate the point I was trying to make in my introduction, namely that if the carrying capacity is indicated by a sigmoid curve and has levelled out, and the population has reached the carrying capacity, what happens if the carrying capacity drops? The population will exceed the carrying capacity and will drop too. A typical drop in a biological population is two-thirds mortality. Tropical rain-forests can be stable because the variability is small, but remember that the marginal lands which we are often talking about have large variabilities, and so management must level out their use before reaching too close to the maximum short-term capacity.

Kuenen (Panellist)

I want to come back to this [question of] variability which the Chairman introduced in his introductory talk, because I believe it is one of the essential things in which people do not really see what is wrong with our system. We want to exploit to a maximum the possibilities that there are, which is a reasonable thing to do, but we forget this fundamental principle of variability which means that you must always have something in reserve. This is because the set of values at a certain moment may develop a number of negative elements which together bring down the maximum possibility, and I believe that is very much more fundamental than even the introduction of Bryson brought to our attention. I feel strongly that, in considering such matters, we should be [alertly] aware of this point in our calculations and our projections for the future.

About Dr Tolba's paper, of course I agree with the other panellists that it is an excellent one. There are, however, one or two points to which I would like to put a question mark. He implied that we will certainly make the right choice. But it is quite obvious that this has not been the case up to now. Take the example of the tropical rain-forest—excuse me if I keep coming back to this point but it is so essential and fundamental for us all that I cannot stress it too much—it is being exploited for short-term profit without regard for the long-term destruction and without regard for the fact that it will not yield good agricultural land. It is a typical example of a misuse of a resource which is going on in spite of the fact that all those who know something about it are [painfully] aware that it is the wrong thing to do. And I believe there are more examples of this kind.

In my own country, Holland, the fishermen refuse to accept the limitations put by the government and the European Community on the catch of fish. They know quite well that if they continue to go on catching large amounts it will deplete the stocks even further. What they say is, 'Oh we'll see after that—the Government will help us out'. It is a typical example of misuse of a resource; while people know that what they are doing is wrong, still they continue because they cannot bring themselves to restrict their short-term profit. Therefore I am not as optimistic as Dr Tolba is, that we will actually be doing the right thing.

Disaster is not inevitable: no, it is not inevitable, but how slim is the chance that we will correct our activity in time. As Dr Obeng said in presenting the paper, 'We have little time: We shall need to act far more thoroughly and speedily . . .'. Indeed we shall, but how can we make our system, which Dr Tolba himself said is not adapted to this kind of decision, function in such a way that it really will do that?

I very much appreciate the optimism of Dr Tolba. He could not possibly have taken on the job he has, if he was not an optimist. We all believe that the chances are there; but are there really the factual data which adaptation to the situation requires, as is very clearly set out in the paper by Dr Tolba? [And can we ensure] that this adaptation [will go on] growing at a rate which is in the right proportion to the rate at which we are destroying our environment? Although I personally am particularly concerned with Nature conservation aspects, and these ones of food production are on a more direct level, I believe that the destruction of the soil, and of the productivity of the soil, by [widespread] mismanagement, is going on so fast that the adaptation of our way of thinking, the way in which we put our minds to this problem, is insufficient [for the world's needs]. There is great danger that the change in our way of thinking [and concomitant action] will come too late.

Bryson (Chairman)

One of the [members of the International Steering Committee of this Conference] who was, hopefully, going to be here as a participant, Professor Viktor Kovda, head of the Institute of Agrochemistry and Soil Science of the Academy of Sciences of the USSR, whom most of you probably know, would have enjoyed very much hearing those last comments. Because he is not here, I will quote one brief statement that he made as a result of his own [very extensive] studies. He says that, of all the arable soils that existed 5,000 years ago, half are now gone, world-wide.

Obeng (Responding for Keynoter)

Of course Professor Kuenen is right in pointing out that there are disasters, but I think there is a fine difference between what he says and what Dr Tolba's paper says. It seems fairly evident that many of the things we have done since we started using our resources have not taken the interest of the environment into account, and this has contributed to the problems which we now face, and about which we have come here to talk. The point that is being made here by Dr Tolba is that we do not have to continue doing the things which we have for a long time been doing wrongly. He also makes the point that, although it may not be easy, we shall have to change and do things the right way if we are to have a sustained and viable environment.

Regarding the point which Professor Kuenen makes about choices, let me [go back to that part of Dr Tolba's] statement, as I do not think he was so adamant. He said it is his belief that, as we come ever more clearly to see the end-result of the choices we have made, we will choose the road that leads to a more abundant life for all. He goes on to say that this is not easy. Nevertheless he believes it can be done, because he is encouraged by the advances—perhaps small, but undoubtedly important—that have been made since attention was focused on the preservation and protection of the environment [— particularly and most widely] in Stockholm in 1972. He quoted the Mediterranean where a number of countries, which [usually] do not even talk to each other, have come together in the interest of protecting the marine environment in the Mediterranean. They are prepared to stop the bad choices which have been in practice for generations and which have been 'killing' the Mediterranean, to pursue better environmental management. We already know enough on which to base this kind of environmental management.

[It is clear] that still a lot more information has come out of the deliberations and exchange of ideas that have taken place here about the environment. Knowledge is not the problem. What we need is the [corporate] will to apply our knowledge to do the right thing. If fishermen know they should not over-fish and yet continue over-fishing, it does not make over-fishing right. They do it because it is the easier thing to do to satisfy immediate and pressing needs. I believe that, similarly, some of the choices which we continue to make are not in the best interest of the environment and of the satisfactory use of our resources for development. We do know in some cases what we do wrong. We know in other instances what we should do right. But we also know that, sometimes, it is easier and more convenient to do the wrong thing; we choose the easier path. Again it often happens that we put the blame on politicians who take decisions for the wrong things to be done.

I believe that we as scientists and professionals individually and jointly should feel responsible for promoting good decisions. Like most distinguished participants, I have been to several meetings and conferences and enjoyed the brilliant papers that have been read. I think we should not let things stop just there. I think we should become better involved in what happens—because of what we are and what we know—and assist in making the right choices and doing the right things. Dr Tolba's paper implies that if we are able to apply our will to doing the right things, we can achieve development without destruction.

Polunin (Jr)

I would like to credit fishermen with more intelligence than perhaps has been afforded them by Professor Kuenen. In my experience, fishermen are not unaware of ecological principles, and a lot of traditional oceanic communities are perhaps some of the best places in which to look for such an ecological appreciation in traditional peoples. [This applies] particularly to those who are dependent on the sea for their livelihood. Thus we find in existence among some Pacific atoll communities, fairly elaborate taboos and other forms of regulation which help to protect biological resources from the depredations of people. I am not implying here that these taboos evolved from an ecological appreciation, but I do know from personal experience that fishermen, and particularly so-called traditional or primitive fishermen, often in fact have quite sophisticated ecological perspectives of the resources which they rely upon.

The difficulty [it seems to me] is not a lack of ecological understanding [so much as the stark] reality of competition. [Indeed these people] often have extensive ecological knowledge and they can frequently understand the need for conservation; but they see no point in restraining themselves when somebody else is going to come [from afar with incomparably better means to over-]exploit the same resource.

A case in point is provided by the western Indian Ocean Seychelles [archipelago where I had a job as their first Marine Parks Warden after graduating from Oxford]. In the Seychelles there has long been talk of protecting sea-turtles, and the fishermen know this resource very well and can see that within their lifetime there have come to be [drastically] fewer sea-turtles; but at the same time [they are faced with] the reality of fishing boats from other nations which come into their territorial waters or lie just outside and effectively exploit the same resource.

Bryson (Chairman)

There are a lot of hands. We have 24 minutes. So unless you desire to be overly competitive and exclude the others, please hold your comments to the shortest possible time.

Goldsmith

[To those of you who know me at all well I need hardly say that] I disagree with most of the points made by Dr Tolba. I think it is a typical example of making exhortations—of exhorting us to do things which we will not do. Many of [these things seem to me to be] ecologically impossible, others to be sociologically impossible, while some [risk being] physically impossible. I think we have got to realize that we cannot have our cake and eat it.

One of the points that has been made is that we have got to redistribute surpluses from the [more affluent] to the poor regions. I would be in favour of this, being for de-industrializing the favoured regions—if you call them favoured. On the other hand it is impossible for a reason of theoretical principle: as an ecosystem develops, more and more of the resources and energy used in its development go into maintenance, and less and less into growth. On the other hand it does not need any more growth, because it has reached a position of stability—i.e. a position in which discontinuities are reduced to a minimum. Unfortunately, as you develop the technosphere, i.e. cities, the opposite occurs. Your capacity for growth increases, because the availability of resources and energy falls, whereas the need for them increases because instability, rather than decreasing as in the case of the developing ecosystem, [tends to increase]. That is to say, the need for resources and energy increases as you develop. In any case, who wants resources for development more—the inhabitants of New York or the Eskimos of Greenland? Have you heard any people clamouring for money in Greenland? Not at all, but every large city in the world is in real trouble and needs more and more resources and energy.

So in fact you find the process continuing in the opposite direction. More and more resources will leave the so-called developing world to go towards New York and other places of this kind where the resources are really needed. So it is unrealistic to suggest that rich countries are going to give resources to the poor; the opposite will be continually recurring. To suggest, too, that we can have large-scale irrigation systems and yet avert ecodisaster is an act of faith. Are there any examples of large-scale dams where you do not have evaporation problems; where you do not have an increase of water-borne diseases; where you do not have earthquakes or problems with [aquatic] life, waterlogging, or salinity? The onus of proof is upon the people who say [that these large-scale works are practicable,] because there is no evidence that [they are] even possible. And if you think [they are] possible physically, you are going to run across economic constraints because lining your irrigation canals (which is the only means of avoiding waterlogging and salinity) is so extremely expensive that it is simply not done.

Altenpohl

I have only one big question to Mr Chairman. He made a very interesting statement and I wish for an example [in illustration. He] said it is not possible to optimize on more than one variable at the same time. Maybe he can outline an example? I would like to suggest the example of energy and pollution, or energy and damage to the environment. If I wish to give more energy to a certain region or country, to follow your golden rule I must concentrate on this and not put in too many variables at the same time.

Bryson (Chairman)

What I said was that you cannot generally optimize on more than one variable at a time—*generally*. Of course you can think of a multivariate situation where you have, let us say, population, pollution, economic standards, etc., so that you have an n-dimensional surface. That surface has some irregular form. If you find one peak on that surface, for one value—say, pollution or population—the peak for some other variable is generally somewhere else. In other words, generally speaking, unless you happen to have a particular interlinked system (and energy and pollution are interlinked), then you cannot optimize on more than one variable at a time.

Buchinger

I think everybody here believes that we know what ecodisasters are, but I would like to know what UNEP is doing to help to clarify the situation, because actually in the paper they are speaking about aerosols and so on? At the same time in Argentina we are, in spite of the protests of ecologists, setting up a factory of aerosols, while in Bolivia a DDT factory is going to be built. Now the aerosol factory is getting funds from the World Bank and the DDT one is being built with the help of FAO. [Laughter.] I would suggest that UNEP send out to each government, wherever somebody goes to give ecological or anti-ecological advice, [counter-advice] to tell the government that they should be careful, that this might be wrong. [Personally I was concerned and] also distressed [by this] paper, but this I think we can talk about in the coffee break, and I do not want to take anybody else's time.

Bryson (Chairman)

This seems to be a prime example of adaptive behaviour in a competitive system. [Laughter.] Would you care to respond to that?

Obeng (Responding for Keynoter)

Let me say that it would be wrong to assert that because we want to protect the environment, there should be no development; and in this context, the subject of the use of DDT and perhaps other insecticides against mosquitoes and malaria is always interesting. DDT emerged almost as a saviour in a number of different circumstances. For controlling mosquitoes, it has been one of the most effective weapons, although in the end the mosquitoes triumph. But if it is not used where it can kill the mosquitoes, what will the people do? There are other methods, like destroying the breeding places of the mosquitoes, killing the larvae by various means, protecting people from being bitten, and so on. But where mosquito density is so high that these measures alone cannot solve the problem, should people be deliberately exposed to bites and malaria? At the present level of our knowledge, as we still employ insecticides, we have to ensure that they are used efficiently and with the least possible hazard and damage.

You ask what UNEP is doing about this. UNEP has a definite policy about the control of pests, namely to promote an integrated approach which draws on all known and effective control measures. In the case of malaria, insecticide use is at present the most effective measure that we know. It would be unrealistic and indeed foolish to forbid its use and let people suffer. We seek and support improvement of measures, like biological control of the larvae [by predacious] fishes, sound ecological modifications to destroy larval habitats, and the development of efficiency in the use of insecticides in order to prevent unnecessary contamination and waste. Towards this effect we support the promotion of suitable application methods, training, education, and so on. For instance, we have been told about aerosol cans and the effects of their use in the environment. So we would encourage other means of application, with technicians duly assisted to undertake their work efficiently, and the public being well informed and encouraged to take part in the control programmes. Malaria is difficult to control—WHO admits it.

With some other diseases, such as schistosomiasis, the UNEP approach to pest management is more easily adaptable, using measures which can be effectively employed with or without molluscicides. We know that if water-supply and waste-disposal systems are improved and well instituted, like the destruction of the habitat of the snail vector, they may help to break the transmission-cycle. But there has to be a well-coordinated programme supported by health education, and so on.

In the course of development, however, where choices have to be made, invariably decisions are greatly influenced by financial constraints. Referring to dams, again the situation is similar. Because dams, especially in the tropics, bring so many headaches, we cannot say that they may not be built. They are useful and necessary. We have to make sure that, when built, they are environmentally safe. Invariably we know what to do to minimize adverse effects, but financial and other considerations rob us of the will to take the precautions and promote sound management.

Towards bringing down the level of these diseases, it is not so much that we do not know what to do as that we lack the necessary finance. We have to prevent the people from getting into contact with infected water, which can be done if you provide conveniently-placed sources of good water and educate the people about the consequences of going to [the bad]. It has been done in China. It can be done [elsewhere]. Now that is one way in which UNEP is approaching the subject. Instead of using chemicals to kill snails or DDT to kill mosquitoes . . .

Buchinger [interrupting]: They don't use DDT because [they] could not sell their meat because it had been . . .

Obeng: Exactly.

Buchinger: And it doesn't have any economic back-throw.

Obeng: Exactly.

Buchinger: Because we have a very good . . .

Obeng [interrupting]: How much malaria have you?

Bryson (Chairman): Ladies!

Obeng: How much malaria have you?

Buchinger [continuing regardless]: . . . against malaria; we had it before.

Bryson (Chairman): Since the recorder in the recording of all this is hooked to *this*, if there is going to be an argument let us please have it so that it gets recorded for posterity.

Obeng: Yes, you see . . .

Bryson (Chairman): One minute.

Obeng: Yes, don't worry, well I came all the way from Nairobi because of this . . .

Bryson (Chairman): So did everybody else.

Obeng: Yes, well let me at least just make a point, which needs to be made. This is important. You see, it is a matter like the lady said. They don't have malaria.

Buchinger [interrupting]: We had it before.

Obeng: But you don't have it *now*. You don't have it now. This is the point.

Buchinger: Because we take . . .

Obeng [interrupting]: And so do many other countries, which also take quinine and maladrine and all the other things, but they still have it. It depends on the conditions, and that is what is meant by making the choices—and, having made the choice, doing it in such a way that it does not cause more harm than it [ought to . . .].

In irrigation, for the producing of food, we know the things that will go wrong. We know also the things which may be done to improve them; but it's difficult, because sometimes it is financially difficult, sometimes it is physically difficult. It does not mean that it can therefore not be done.

Bryson (Chairman)

All of this proves my point that you cannot optimize on more than one variable at a time. [Laughter and applause.]

Fosberg

I put my hand up first with one comment to make; now unfortunately I have three. [Laughter.] But they are short. One is that I think subsequent interventions have refuted much of what has gone before. The second is that I think [young] Nicholas Polunin has given us perhaps the perfect example of Garrett Hardin's 'Tragedy of the Commons' in his comment. Third, I have a question for the Chairman, in regard to this business of not being able to optimize [a multiplicity of] components. Granted that the optimum in all components of a complex system cannot be reached simultaneously, I would ask whether, if one adopts a holistic view, there may not be an optimization of the resultant of all these components—in other words, an optimization of the functioning of the system as a whole?

Bryson (Chairman)

That, I suppose, depends on how you define the optimum—optimum stability or optimum [various] other things? If you define a composite variable that takes all these things into account, then you can maximize it. I shall not continue, though, because the Chairman should not use up time that other people want.

Flohn

When dealing with variability, most people think only of time-variability on one particular spot, and this is especially true with regard to climatic anomalies. It has frequently been said that if, for example, we have some climatic anomalies which are producing a loss in the productivity of wheat and other cereals, this may smooth because other areas have an opposite effect. But here I must say one [word] of warning.

We know now, and I think all my fellow climatologists could give examples of this, that the climatic anomalies over the globe as a whole are closely interrelated, and that it is not only possible, but we have several examples of this kind in recent times, of severe anomalies in several parts of the world occurring simultaneously. Now it is only very recently that we have come to observe these matters on a global scale. The last example of this kind was in the year 1972, when we had the lowest temperatures in the Canadian Arctic and the Greenland area, and, simultaneously, the major drought and failure of the Indian monsoon, the catastrophe of the fish-catch off the Peruvian coast, and the peak of the Sahel drought. All these events and some more are closely interrelated—not by random [chance] but by law. It is the interrelationship of these anomalies which we are calling teleconnections, and which are mainly produced by the distribution of land, sea and mountains on Earth. This may create, if we take the Earth as a whole, even larger anomalies, such as we have experienced in recent years.

Royston

[Even as your] eternal optimist [I must admit that,] if we look into the past, of course we will be pessimistic. But if we look into the future there are grounds for optimism. With UNEP [support at CEI] we have recently been working with 30 different developing countries. We have found that, in those 30 developing countries, over 12,000 top-level decision-makers and policy-makers have expressed a definite need for attending educational programmes in environmental management. Those, for me, are grounds for optimism.

Starr

Just a comment with respect to the sigmoid population curve for a K-strategy species in an ecosystem: the K-strategy is a workable one in a stable environment which has a well-defined carrying capacity that does not vary much for long periods of time. [In such an environment, the carrying capacity can be accurately perceived by the species in question, and it can thus make the fine-tuning adustments to its growth-rate which are necessary for a smooth and steady approach to this well-defined equilibrium population level.] But there is another strategy, the R-strategy, that is characterized by explosive growth inevitably followed by collapse. This is a workable strategy when environment is changing rapidly, i.e. when there is a great deal of variability—something that we are [experiencing now. In such a situation, the carrying capacity is not well-defined because it varies dramatically over short periods of time. In such a situation, one naturally 'makes hay while the sun shines'.]

Is not the human species today responding in precisely this way to its perception of a rapidly changing, technologically amplified (in the short run) carrying capacity of our planet? [Is not the only solution to our population, food, and resource, problems to make clear the true, long-term carrying capacity of this planet, which I believe to be well below our current population level? If we do not make it clear, and act quickly to accommodate our species numbers to it, in the long run it will manifest itself without our assistance, and the inevitable collapse of the R-strategists will be upon us.]

Misra

I would like to add a few words in support of [the Chairman's] observation that management requires self-discipline, and re-emphasize that without self-discipline it is not possible to manage the natural resources in the way that we would like to do. Management at the moment will require rearrangement of many of the world's resources: I do not believe in their distribution as of other kinds of wealth. It is important to manage our own resources *in situ,* which requires at the moment a lot of educational communication to be developed among the governments, our politicians, the electorates and the scientists. In each country they have to work together. Governments can be thrown out—can be replaced if necessary—but not the resources. The scientists and the public are aware of this fact. So I would like to say that we must share knowledge much more than any [material] wealth at the moment.

You will know that in India we have a very high respect for life—including trees and other plants, which we try to conserve very well. Conservation mostly means to the people merely the conservation of wild animals, and especially the larger ones; but of course they do say sometimes that the habitat should also be conserved. Without actually going into details of the habitat, I would plead here for a better conservation of all the plants; many of those which were becoming scarce in India from time to time were declared to be sacred plants, and even up to this day people worship these plants and do not cut them and consequently their numbers have increased.

It is absolutely true that [preserving] only large-scale areas for conservation alone will not do: conservation has to be brought to every home. Our behaviour has to be changed; indeed everywhere, all around, I would like to see little patches of forest, because in India, for instance, although we have 23% of the land in wilderness—which may be a high figure as compared with many other countries—yet it is ill-distributed. We have large tracts of lands where desert conditions are prevailing today. So in order to mitigate these difficulties we must try to redistribute and rearrange things properly, and there must be sufficient rapport and communication developed amongst the politicians and the scientists and the electorate.

Bryson (Chairman, concluding Session)

Thank you, Professor Misra, for pointing out that there are other strategies for management than simply redistribution.

Now I am not going to call on any more speakers because the time is up. But is it the desire of the Session that there be an adjourned meeting to continue the discussion this evening? Since I hear no 'nays', I declare the morning meeting closed and am going to suggest to our Secretary-General that we have an adjourned session this evening. [This, however, was not possible, owing to shortage of time.]

Session 16

ETHICS OF BIOSPHERAL SURVIVAL

Keynote Paper

Ethics of Biospheral Survival: A Dialogue

by

BEATRICE E. WILLARD

Professor of Environmental Science, Colorado School of Mines, Golden, Colorado 80401, USA;
formerly *Member of the President's Council on Environmental Quality, Executive Office of the President of the United States,*

EMMANUEL O. A. ASIBEY

Chief Game and Wildlife Officer, Department of Game and Wildlife, Post Office Box M.239, Accra, Ghana,

MARTIN W. HOLDGATE

Director, Institute of Terrestrial Ecology, Hills Road, Cambridge, England; currently *Director-General of Research, Departments of the Environment and Transport, 2 Marsham Street, London SW1, England,*

YOICHI FUKUSHIMA

Chairman, National Committee on Nature Conservation, Science Council of Japan, 7-22-34 Rappongi, Minato-ku, Tokyo 106, Japan

ELIZABETH and DAVID DODSON GRAY

Co-Directors, Bolton Institute, 4 Linden Square, Wellesley, Massachusetts 02181, USA.

Part I by
BEATRICE E. WILLARD

WITHOUT AN ETHIC?

Where will human beings end up if they have no ethics, no strong rules of conduct, to guide human activities in the ecosphere? What happens when we do not have agreed-upon standards for treatment of all living things? The answer is abundantly clear to us as we look around Earth in all directions—up and down as well as across land and water and back over the centuries of time. Western Man, especially, has predicated action upon an exploitative ethic towards the ecosphere, as I prefer to call the biosphere: 'the world is my plum to pluck, eat, enjoy, consume—as I see fit.' A few enclaves can be found where people have exercised a different attitude and ethic, and their way of life has usually been equated with primitiveness and regarded as a problem.

The results are clear if only we will read the landscape—deserts where once grasslands flourished, the desolation of brick-hard soil where once-vital tropical forests had produced their diversity, impoverished grazing areas where forests formerly stood; permanent streams have become seasonal from silting, and rivers flow for only part of the year. Many lakes are dead or dying, with declining fisheries where 'unending bounty' has offered itself through the ages. Major seas are horribly polluted, and even the oceans are widely threatened.

Yet where the ethic of reverence and restraint, of understanding biological needs and ecosystem interactions, is used it can allow and even help the inherent resilience of such populations and systems to heal, replenish and begin the long road to restoration. In this paper we shall look first at the scientific bases of such an ethic for biospheral viability, and then consider why the human species does not yet take these seriously.

THE RULES OF LIFE'S GAME

We learn from ecology that there are basic principles of the biosphere which are 'the rules of life's game', and for which human beings need to have a healthy, active respect. When we move in rhythm with them, they work for us. The great goal and challenge of human life is not to overcome or thwart the Earth's forces but to learn how best to harmonize our human activities with the dynamics and processes of Earth's ecosystems. We need to make the least alteration possible in these systems, so that we foster the well-being of Earth's ecosystems as well as the well-being of all mankind for generations to come.

It is my conviction, as someone who has taught a great many conservation education seminars, that the ethics of biospheral viability arise out of an understanding of ecosystems. As the principles, concepts and processes of ecosystems are demonstrated to people in a field setting, the meaning of these principles for human management of ecosystems begins to germinate and unfold in people's minds. Ecosystem principles begin to form the criteria for engineering designs. They begin to become the basis for agency and corporate decisions. With this integration of global ecological principles into human activities, these activities are brought into greater harmony with their environment.

Main Ecological Principles

What are the principles of ecology which underlie the operation of Earth's ecosystems? We learned long ago that there are natural principles—natural laws—operating in our environment and that we have to live and work with them. Gravity is perhaps the most compelling of these. Expansion of water as it freezes into ice is another. The various solvencies of materials introduce a host of others. We have learned, too, that to quarrel with these principles is useless, even damaging. So we have figured out how to adapt our operations to these principles and to their functioning.

There are also natural principles that apply to the operation of living systems which are just as compelling in their application but which are less immediate and acute in their operation than, for example, the law of gravity. Because the result of the operation of these ecological principles may take years, decades, even centuries, to be observed adequately, human beings with their short time-scale may not recognize them as principles. Therefore we may violate their operations by our actions—usually inadvertently, and seldom maliciously.

What are these natural ecological principles operating around each and every one of us all the time? Let us attempt to answer that question by treating eight of the most important that are not necessarily the most obvious.

1. *Everything Affects Everything Else*

First and of primary importance is the principle of holocoenotic environment, with everything affecting everything else either directly or indirectly. Nothing operates in isolation, but rather is everything more or less in concert.

One of the best examples of this principle of interrelatedness is to be found in the fact that DDT, PCBs and radionuclides have been found in the tissues of virtually all organisms tested to date. These substances have never been introduced by human activities on to the Antarctic Continent, yet the flightless penguins of the Antarctic have these substances in their flesh. The penguins must obtain them from outside the Antarctic Continent through a series of food-chains involving migrant organisms.

Is it possible that an intuitive sense of this interrelatedness is a part of all the Earth's major religions? It is stated perhaps most powerfully in Albert Schweitzer's phrase about 'reverence for all life'.

This principle of holocoenotic environment, or 'wholism', leads us to certain criteria for guiding human activities. It leads us to the practice of 'looking before we leap', and inculcates the need for each individual and group to engage consciously or otherwise in 'ecological reconnaissance'. This involves analysing ahead of time the ramifications of potential activities upon our immediate habitat or our ecosystem, upon present and successive generations of the human race, and upon living resources of all kinds—ranging from fisheries to forests and wildlife, and indeed to all forms of life no matter how insignificant they may be economically. Look before we leap, indeed: determine ahead of time our human impact upon potentially fragile ecosystems.

These analyses, and the data-gathering which must precede them, can be time-consuming and expensive. This is the case primarily because our awareness of the need has preceded our institution of adequate research into Earth's ecosystems. Therefore we must investigate each specific action and its site individually. But in time our knowledge and understanding of ecosystems will be adequate for developing generalized manuals of information and performance standards. Reaching such a threshold will make it far easier and more effective than hitherto to build environmental considerations into human society—more or less as other aspects of human behaviour have been incorporated into human policy and action.

When these two milestones—adequate ecological information and infusion of its significance into human policy and action—have been reached, we shall have reduced greatly the amount of time, stress and resources needed to harmonize human actions with Earth's ecosphere. Human beings will then begin to enjoy the dividends from present investments by occupying our proper niche in suitable ecosystems and bringing our activities into balance with various ecological processes quite consciously and intelligently—rather than by the seemingly unavoidable discipline of the usual balancing forces of the ecosphere, namely starvation, war or extinction.

2. *Ecosystems and Niches*

The second great principle of ecology is that the Earth is covered with a vast array of ecosystems, large and small, which interact with one another. Within these systems, each type of organism has a role to play—a 'niche' to occupy. To us these roles of other organisms sometimes seem detrimental, negligible or obscure. But research demonstrates that, despite our lack of immediate appreciation of their role, they are nevertheless significant in the total functioning of their systems.

A prime example of this is a cypress swamp bordering the Savannah River between an atomic energy plant and the river. Until recently the cypress swamp was considered of little use to Man. However, research of the material-flow pattern conducted in the swamp disclosed that, under the specific conditions currently prevailing—namely, a long-undisturbed ecosystem with water at pH 7—the kaolinite of the soil in the swamp was absorbing all the radioactive ^{123}Cs that was being released by the nuclear power-plant. Therefore the radioactive caesium would not reach the river unless the pH balance of the swamp was changed, for example by adding small quantities of near-by well-water. In that event, radioactive caesium would be released from the kaolinite into the river. This change would occur if the trees were cut. The cypress trees of the swamp occupy a vital but obscure niche in the total functioning of that ecosystem.

The criteria for human behaviour which can be drawn from the principle of ecosystem existence and composition serve to reinforce, refine and extend the criterion implied by the earlier principle that everything is interrelated. From an understanding of the qualitative and quantitative interrelations of the driving forces and manifold variables of given ecosystems, we should be able to derive criteria for wise management of living resources in virtual perpetuity.

A close analogy can be drawn between understanding specific ecosystems and how to manage them at optimum levels, and understanding the operation of the human body. We learn about our bodies' individual differences, their rhythms, their responses to changed conditions, their resilience, their 'outer limits', their recovery-times. Knowing these characteristics and many more, we can wisely design, plan and manage our activities in relation to others.

The state of ecological/environmental science is unequal across Earth's ecosystems. But in recent years several highly significant programmes have greatly advanced our understanding: they include the International Biological Programme (IBP), which first provided basic quantitative information on the productivity of major ecosystems; the Man and the Biosphere Programme, which will utilize IBP data to develop information about human influences on ecosystems; the ecology programme at the University of British Columbia and others elsewhere that are being extended by the International Institute for Applied Systems Analysis; and the research programmes of ICSU, SCOPE, UNEP and IUCN.

It is all-to-easy to see how sparse this network of scientific research and data-gathering really is, though it is already a vast improvement on the situation of only a few years ago. But it is highly important that we all recognize the potential of these programmes and many others, and that we foster the realization of their potential for the benefit of humanity and the viability of the ecosphere in a world where everything affects everything else.

3. *Material Cycling and Energy Flow*

The third great principle of ecology is that chemical substances in ecosystems cycle through and among components of those systems at varying rates. For this reason they are available for re-use again after time-periods which depend upon their paths of cycling and means of disposal. Conversely, energy follows a one-way downhill path, sometimes circuitous but always being dissipated eventually as heat, and never returning to its former usefulness.

What criteria for human behaviour follow from these principles? We recognize immediately that few if any things are waste to the ecosystem; materials are used, discarded and picked up by other ecosystem components for their use—on and on and on, in endless cycles. We should consciously analyse our disposition of materials to make sure they are being located in the place which will be most useful to future generations, while also allowing present generations a reasonable possibility of disposition provided they maintain a high quality of any effect on ecosystems. We need to examine the 'out of sight, out of mind' syndrome which tends to operate extensively in relation to the disposition of materials that are not currently in use. We need to recognize that, under certain circumstances, materials may be deposited in system-sinks for varying periods of time that can range up to hundreds of millions of years. Much more innovative long-range thinking and action is needed to bring human activity into harmony with ecosystem processes, so that *material flows* are fostered, and not material sinks.

For example, I am highly concerned that at present we are mining nitrate and phosphate deposits that took many, many millennia to form. We are distributing them to agricultural lands, increasing run-offs of nitrogen and phosphates into rivers, reaping crops and distributing them to people who use them for human foods, etc., the waste of which in many Western countries goes into lakes, rivers and eventually the oceans. Thus it may be removed from ecosystem benefit for millennia, as much of it will not be recycled until new phosphate and nitrate deposits are formed on the ocean floors and ultimately elevated in continent rebuilding millions of years from now. We can assist the operation of ecosystems by facilitating recycling and avoiding those semi-dead-end pathways that keep materials cooped up for long periods.

Energy on the other hand does not cycle. Each use leads to a loss of capacity to do useful work—until no such capacity remains. Therefore we need to make sure that each use of energy is essential, and that it is implemented in the most conserving manner. By analysing our personal life-styles, we can discover many ways of using our own body energy to our own bodily benefit, while conserving fossil fuel for higher uses. Walking around the corner instead of driving, brushing our teeth by hand, walking upstairs instead of using the lift or elevator, ventilating rooms by natural processes in summer to avoid use of air-conditioning, drawing curtains to retain heat in winter and block it out in summer, and so on. Our societies can do the same, matching much more carefully than we have done in the past the thermodynamic quality of the energy source to the particular characteristics of the end-use. Using electricity to heat our hot water, for example, wastes a very high-quality form of energy on an end-use that can be met by waste heat from many other sources.*

In making sure that each use is essential and implemented in the most conserving manner, we take cognizance of our descendants and their needs for energy for essential uses, including transport and special manufactures.

4. *Limiting Factors*

Within all ecosystems, specific features of the environment interact with the genetically controlled capabilities of living organisms in such a manner as to restrict or limit the functioning of those organisms. These interactions define the operating parameters or behavioural characteristics of that ecosystem and of the living organisms within it. Frequently it is a whole

* Moreover one shudders to think how much of this high-quality energy is totally wasted forthwith to cool again that heated water when, for example, it comes out of the warm-water tap far too hot for the human body to stand!—Ed.

constellation of physical and chemical factors in the environment which are interacting with not just one but a group of species to describe the limiting factors of the system.

The criteria implicit in the principle of limiting factors are more subtle and more difficult than most others to deduce from the operation of the principle, and we would welcome assistance in thinking about this. It is not that such criteria are any less important or demanding than others: rather the contrary. But the end-results of the operation of limiting factors may not occur within the life-span of individual people, so that the action of this principle in ecosystems can appear quite remote from human experience and interests. Yet one generation can set in motion ecological processes which will cascade through ecosystems in masked fashion over a period of hundreds of years. The end-result may appear several generations hence, and at that point the result will often not be understood because the originating processes may have been lost to history.

A good example of this subtle and paradoxical characteristic of limiting factors in relation to developing criteria for human behaviour is the human experience of decreased mortality-rates and resulting expansion of our human numbers. We do not know yet how this change in human mortality-rates over the past 50–100 years is going to work out; what we do know is that the effect of there being more and more of our species on Earth is ruining many of the ecosystems that are most affected by human beings, while few if any of the others escape entirely.

5. *Prolific Nature of Biological Reproduction*

We know that the vast majority of living organisms have a capacity to produce far more of their kind than can possibly be supported by the ecosystems in which they normally live. The probable evolutionary reasons for this stem from the fact that overpopulation ensures that some individuals survive to reproduce the species. These individuals survive the numerous exigencies of early life. They survive the various stresses of factors in the physical environment. They survive the competition with peers and with other species for food, water, light (energy). They survive predators, diseases and accidents.

If even one of these multiple interacting constraints lessens appreciably, then more individuals will survive. And they will grow, eat, live, reproduce and live longer; and then more of their offspring will survive, grow, eat, live, reproduce and live longer. The population of such a favoured species tends to grow by geometric proportions. This geometric rate of growth can readily be seen in the world's human population, which took all the time from 3 million years ago to approximately 1850 to reach 1 thousand million people. In the next 80 years (approximately 1850 to 1930) it doubled to 2 thousand millions. The population doubled again to 4 thousand millions, between 1930 and 1976 (approximately 45 years). Clearly, doubling times have been decreasing, which means that rates of increase have been accelerating.

Criteria which can be derived from this prolific nature of biological reproduction are numerous, and have special urgency, pungency and clarity of applicability. All wild species live with natural controls which bring reproductive processes into balance with ecosystem abilities to sustain the species. Some mechanisms are intrinsic to the species, and some, including disease and predation, are extrinsic. The human species has removed most controls. Human alteration of ecosystem components and driving forces has not been coupled with a deep respect for the prolific nature of biological reproduction in every species, including our own. The criteria drawn from the principle that everything is interrelated, principles about ecosystem niches, and the function of limiting factors, all have direct application here. But most compelling are criteria which warn against inadvertent, unthinking, unpremeditated, actions which significantly alter controlling factors affecting reproduction and population numbers.

Frequently, humans do not see the interconnections of one action that triggers another, often distant or subsequent result. We are apt to be unaware when we have altered significantly the driving forces or components of some ecosystem. For example, in the western United States we have sought to protect grazing sheep by poisoning or shooting Coyotes [*Canis latrans*, also called Prairie Wolves], but we have not seen these actions to be immediately traceable as the cause of a decline in local grain-crops. Yet the causal loop is closed when this situation is coupled with quantitative observations of a sharp rise in rodent and bird populations. Protecting grazing sheep is important, but respect for the interconnectedness of life and the driving force of biological reproduction would suggest other means of protecting sheep from Coyotes which will not have these other side-effects. Similarly, we must recognize that, one way or another, the human species must have its own population controls in order to bring our numbers into balance with Earth's capabilities to produce.

6. *Carrying Capacity*

All systems have a definite capacity or capability to sustain a given amount of life without collapse, or to uphold an optimum level of production. Ecologists call this the ecosystem's 'carrying capacity'.

This principle operates also in engineering systems, where transmission lines are designed to carry a given electrical load; in computers, which are designed to carry a given information load; and in mine-shaft cables, which are designed to withstand certain forces. Engineers are familiar with what happens to these systems if these loads are exceeded: they cease to function properly or productively. So it is also with ecosystems.

Because ecosystems are living, they have more built-in dynamic resilience, stemming from their vastly greater diversity and complexity, than do engineering systems. Because of this greater complexity and concomitant resilience, however, living systems may not give signs that the carrying capacity has been exceeded until the actual point is well past in time, numbers and balance. But a 'domino' effect may have been set in motion, with a stealthy cascading of effects through successive parts of the ecosystem.

That there can be criteria for human behaviour in ecosystems that will be indicated by carrying capacity is a fact which has been ignored, denied, disobeyed or misunderstood by humans for a long time. This is a strange phenomenon, because most people understand the advantages of budgeting time, money or even resources. Few people, on finding themselves in a lifeboat for an indefinite period, would deny the wisdom and necessity of rationing available resources among those present and of restraining themselves in the use of those resources. But only a few people see Earth as a lifeboat or spaceship with limited resources—some of which are renewable under optimum circumstances (e.g. forests, fisheries, soil and water), whereas others can never be restored when once they have been used or eliminated. Examples of these 'once-and-for-all', finite resources are fossil fuels, rare and endangered species, uniquely beautiful places on Earth, and treasures of past cultures and civilizations. Yet another category of resources is comprised of those which become tied up and therefore not available for use for long periods (e.g. gold and other precious metals, and gems, and copper). Few people seem to believe (and act on the belief) that their present generation should ration and restrain itself for the benefit of future generations.

These characteristics are strongly indicated by the nature of carrying capacity as it functions in all ecosystems—urban, rural, agricultural, and non-human-dominated ecosystems alike. Engineers quickly recognize carrying capacity as it operates in various physical systems—bridges, sewage treatment plants, power-plants and motors of all sorts. Seldom, except in emergencies, will engineers even consider overloading such facilities. Quite apart from engineering, a body of custom has grown up, often supported by law, about carrying-capacity limits. This assures us that people who have purchased a theatre or stadium ticket will have sole use of a given seat—its capacity being one person—without overloading it with two or more people. Further, if one purchases a hotel room that is capable of sleeping more than oneself, one still has the sole use of it. And if one is using a given table in a restaurant, one is usually assured that others will not be seated with oneself and certainly never seated on one's lap!

And yet, with numerous facets of urban areas, grazing lands, camping-grounds, national parks and wilderness areas, open spaces, recreation areas, and some other types of situations, we are only beginning to grope towards realization that each has an upper unit to its capacity for providing services and desired resources. Used beyond that undefined but very real zone of capacity, the resource suffers stress, deterioration, eventual alteration, and loss of capacity—even decline and demolishment. Being facetious in analogy, who among us would attempt to seat a highly trained elephant in a dining chair? Yet civilization does it all the time by not determining the upper zone of optimum functioning for all types of ecosystems, and by not managing human activities to keep them below this zone—unless a clear emergency arises.

We hear it said that 'technology can change the limits; therefore carrying capacity is not stable and certain'. This is cited as an excuse for not recognizing and obeying the criteria for behaviour prescribed by carrying capacity. Yet this same thinking is not accepted for steel with a given capacity to withstand stress—a capacity that is increased by the addition of molybdenum. The altered capacity is calculated and utilized—but recognized as to its limits. Similarly, ethics for biospheral viability mandate our determining the capacities of all ecosystems for the stresses to which they will be subjected—financial, social, maintenance, service, air and water quality, etc.—and then, for the good of the system and its components, all human activities would be designed, planned and operated below these limits.

The participants in the Club of Rome analyses of the implications of this principle for mankind are to be commended for their efforts, as are the many other scientists undertaking similar and related systems analyses to grapple with the huge and complicated problems of budgeting the world in these pioneer areas where no such attempts have ever before been made on a global scale. The United Nations is also to be commended for undertaking global conferences on several issues that are crucial to bringing human activities into balance with the sustaining powers of the ecosphere: these have concerned the Human Environment (Stockholm, 1972), Human Population (Bucharest, 1974), Food (Rome, 1974), Human Settlements (Vancouver, 1975), Water (Mar del Plata, Argentina, 1977), Desertification (Nairobi, 1977), and Environmental Education (Tblisi, 1977), with others to come.

7. *Ecosystem Development*

Over geologic time, ecosystems gradually develop from especially simple systems on biologically naked surfaces of rock, sand or water, through progressively more complex systems which in each case change the nature of the systems, until finally an ecosystem is reached which is highly complex, relatively stable, and more or less permanent. Natural processes exist for healing various naturally-occurring ecosystem disruptions such as fire, landslide or insect infestations. These processes are generally slow, but sure. They can heal perturbations due to humans, especially when the perturbations are introduced in a manner which complements or is integrated with existing processes of the ecosystem. But if processes are introduced which are diametrically opposed to existing ecosystem processes and 'cut across the grain', then these newly introduced processes can be as toxic to ecosystem operation as cyanide is to the human system.

For example, any serious surface damage to arctic tundra systems will take centuries or even millennia to heal. Would not building the Alaskan pipeline to pump heated crude petroleum destroy the tundra? But tundra is very easily transplanted, with little or no visible reduction in the vitality of the plants. Furthermore, Styrofoam sheet acts as an excellent insulation for maintaining permafrost. Therefore it is the most economic, efficient and effective thing to do to move tundra turf physically from where it will be disturbed to where it is needed to cover a barren surface. But one needs to be sure to move wet tundra to wet sites, snowbed tundra to snowbed sites, and dry fellfield tundra to windswept fellfield sites.

Criteria evolving from the principle of ecosystem succession or development are not obvious at first thought. But the ethic begins to emerge as we come to comprehend the amount of time which it has taken to develop soil on bare rock, water or sand—hundreds to tens of thousands of years, depending on the substrate and the region of the globe. We are guided then to value the time which ecosystem succession takes and what all that development time has resulted in. The emerging ethic guides us not to lay bare vegetated surfaces any more than is absolutely essential. In such cases as strip-mining, this ethic mandates that a portion of the profits shall be required to be spent to restore to viable ecosystems the land and water bodies.

The evolution and operation of this ethic, together with the others mentioned, was abundantly evident in the 'Experiment in Ecology' launched in 1967 by American Metal Climax, Inc. (AMAX, Inc.) to develop and demonstrate a new system for bringing on-stream, with a maximum of environmental consideration, a new molybdenum mine. A new large, valuable lode of molybdenum had been located 3,000 feet (914 m) below the ground surface near the Continental Divide west of Denver. The local company executives wanted this new mine to be better planned and designed than their Climax Mine which had been operating for 55 years, and had made a conspicuous large blotch on the beautiful high mountainous Colorado landscape which they all loved.

In order to gain a better understanding of how to accomplish their goal, these men consulted the Colorado Open Space Council. A small group of its members agreed to work with the Climax officials on developing an understanding of what needed to be done to fit this new mine into the high mountains west of Denver with a minimum of disruption to the water, air, vegetation, wildlife, aesthetics, recreation, and other resources, of the area. Care was taken to develop personal trust as mining executives and engineers talked with environmentalists about all sides of each problem associated with this massive undertaking. This process allowed environmentalists to examine the constraints associated with mining, and to learn something of its techniques and products. It allowed the miners to learn the fundamental objections to the old ways of the industry.

In a few months, the engineers began to evolve new ideas: they found they *could* build mine buildings of materials that would be *visually compatible* with the mountains; they *could* move boulders and fallen trees and 'walk' vehicles through forests instead of carving out roads with bulldozers. They *even* found they *could* build a tunnel from the lode nearly 1000 m below the land-surface on the *east side* of the Continental Divide all the way to a site for the mill on the surface on the *west side* of the Continental Divide. This required tunnelling uphill through Pre-Cambrian metamorphic granites for 9.6 miles (16 km); but it offered a much superior site to build a mill, to contain mill-tailings, to recycle process-water, and so retain the beauty of Upper Clear Creek Canyon near Denver.

The Experiment in Ecology had ripple effects. Two high-voltage power-lines were to be brought for miles across the mountain landscape to service the mine, mill and tunnel. The Climax men told the Public Service Company officials: 'We want you to construct these lines with minimum impact on the ecosystems of the State. We do not want any more of these "tin soldiers" across the landscape, like your new Cabin Creek Line.'

Challenged by this, Public Service began to examine and analyse anew their whole transmission-line installation process. They hired ecologists who told them where ecological changes would be major and where minor. The result was two new lines that are barely discernible, ecologically or visually—and design engineers who are proud of their new ways of doing things.

8. *Specialization, Diversity and Stability in Ecosystems*

Ecosystem development processes result in a system becoming inhabited by an increased number of species, which expand to fill the available ecological roles (niches). Competition for resources (air, water, food/energy, space, light/energy) brings increase in specialization, creating additional niches and augmenting diversity. As in economic and social systems, diversity creates stability by buffering the influence of any single perturbation in the system. Therefore the simpler a system is, the more responsive will it be to any far-reaching perturbation, while the more diverse it is, the less responsive will it be to perturbation.

One can see this in the experience with ice storms in the south-eastern United States in the winter of 1972–73. Natural forests sustained breakage that was proportional to their inclusion of trees which had brittle wood. On the other hand, plantations that had been planted exclusively with Longleaf Pine *(Pinus australis)* sustained the greatest damage, because of their brittle wood *and* long needles which held a considerable weight of ice.

The principle of ecosystem specialization and diversity leads to some obscure but highly important criteria for ethics of ecospheric viability. These criteria fall into certain categories.

(*a*) *Avoid contributing to the extinction of individual species.* With the extinction of a species, the total array of genetic materials is depleted, thus limiting the available resource materials for evolutionary processes to work upon. Again, as we come to comprehend the amount of time it has taken to develop the variety which we have of species and genetic material, we are guided to value not only the resulting diversity but also the development time involved.

From the strictly human standpoint, improvement of breeds and strains of domesticated species may require a drawing upon genetic materials that exist only in wild populations. For example, after hybrid corn crops in 1972 were severely reduced by an infection, the hybrid corn stock was further improved by crossing with wild maize that was more resistant to infection.

(*b*) *Protect and treasure a wide variety of all Earth's ecosystems as the 'basic documents' of ecological and environmental understanding.* Mature ecosystems are benchmarks for operation of ecosystem processes and for investigation of these processes pursuant to developing reliable predictive capabilities for determining impacts of human activities on a wide range of ecosystems, as well as for improving policy and practice in living-resource management.

Western civilization protects benchmark cultural and technological entities such as the Dead Sea Scrolls, the Gutenberg Bible, the Magna

Carta and the Rosetta Stone. Examples of all the Earth's ecosystems are similar benchmark reference points. Without these reference points, environmental scientists will be groping for the missing pieces of ecological puzzles. They are groping enough as it is, suffering as they must from the deletion of key segments of the puzzle because of natural elimination processes and the events of history.

Criteria for behaviour must therefore include careful planning of all ecosystem alteration, so as to avoid disturbing mature ecosystems any more than is absolutely necessary. A corollary of this is that if it is necessary, then technological processes must be invented to restore ecosystems as nearly as possible to their original mature state—either directly by transplant or indirectly by imitation of ecosystem processes which will culminate within a reasonable period (100–200 years) in reinstating the mature ecosystem. These criteria are not as far-fetched and out of reach as they may appear to be at first glance. In the Rheinbraun coal-mining area of West Germany, they are already in operation in fairly advanced form. The technology for some aspects already exists but must be redirected to this use, as witness the tundra transplant experiments in Arctic America.

Diversity of organisms in age-classes, life-forms, shapes and ecosystem integration has long been recognized aesthetically as more attractive and beautiful than ecosystems with greater uniformity in all aspects. Hence the ethic of maintaining the attractive and beautiful is highly important as a mental, emotional and even spiritual component for nourishing human life.

(*c*) *Understand the processes and thresholds operating among ecosystem components*. Recently-acquired knowledge of ecosystems is indicating that there are distinct thresholds for ecosystem processes, and that these processes do not function below a critical mass of species, or a critical mass of individuals within one species.

The precise criterion here for ethical behaviour will need to await further understanding of ecosystem operation. But such evidences as are now available indicate the need for caution when taking actions which decrease the size of populations of rare, endangered, or even merely threatened, species. And the evidence leads us to caution against any alteration of their habitat, until such time as the fabric of that habitat is understood adequately to guarantee viability of the species with any given change.

Ecological Principles Define the Choice Before Us

The above eight basic natural principles define one set of options for humans—options for the full benefit of humans for very long periods of time. Overlooking or ignoring the operation of these natural principles defines another set of options for humans—options which are narrow and which benefit only a few for a short time. What are some of these choices which we can make?

Are Our Actions Quite Separate from Other Living Things?

We can choose either to think that our actions are quite separate from those of other living things and from the landscape in which they live, *or* we can recognize the primary principle of ecology: that *everything affects everything else*. It is easy to believe that human actions have limited effect—especially in the wide open spaces of the American West. For many, it may be hard to see the effects of removing plants from desert hills. We may not actually *see* any effects of displacing deer and other creatures, but that does not necessarily mean that there are none.

If we observe more intensively, as we in the United States are having to do in preparing environmental impact statements in compliance with the National Environmental Policy Act, we find that with fewer plants, less oxygen is being produced, less carbon dioxide is being consumed by the plants—and so there is less air-purification. We find less soil-holding capacity, and more salt, silt and clay washing into the basins, the Colorado River and the Gulf of California; therefore we are getting less soil in which future plants can grow, and more salinity load upon river and gulf waters. So a downward-spiralling cycle of decreasing productivity is set in motion, causing short-range and long-range effects that can generate ripples of largely unknown magnitude and ramification.

The requirements of the National Environmental Policy Act are motivating some interdisciplinary team-research on ecosystems, thus putting into practice the first principle of ecology. One such effort was recently completed in Colorado on a new ski-area site. The team was composed of a geologist, a hydrologist, a soil scientist, a plant ecologist, an aquatic biologist, a wildlife biologist, a specialist in forest recreation and scenic analyses, an economist, a planner, an architect, an engineer and a business analyst. This team functioned from the outset in an interdisciplinary manner, its members interacting substantively with one another, with the property owners and developers, with the Forest Service representatives, and with the county planners and others. The operation was not Utopian, but it went a long way towards integrating various interest groups into the study, planning and design of a major project. It certainly recognized at every juncture the first principle of ecology, and that therefore a range of professionals must be involved in such a project and must interact and work together as a team in the study, planning, design and implementation of the project.

Are Only Certain Living Things Useful?

We can choose to believe that only such living things as cows, deer, sheep, trees, fish, wheat and Man are really useful, *or* we can choose to realize that *all living things* play a role in the life-support systems of which humans are merely a part. If we take the former choice, we might well exterminate such useful but unattractive plants as the fungus producing penicillin or the fruit that once was thought to be poisonous but is now considered a luxury food—the tomato.

If, on the other hand, we follow our first ecological principle of universal interdependence, we will surely *recognize the need to maintain all types of habitats* for living things, at the same time that we are developing some areas for human use. We will carefully avoid damaging more of each habitat than is absolutely necessary for our purposes. We will plot actions in advance, so as to detour and preserve any small and fragile or rare and endangered ecosystems. We will use land-disposal of sewage as much as possible. Where there is a choice of sites, we will stay with the drier ones, keeping out of the highly productive estuarine areas and grassy flats. This was, in fact, the choice of a company that was planning to mine phosphate in California when it moved its millsite from a productive river-bottom in order to protect the soil-holding ability, productivity and wildlife habitat of that ecosystem.

As Long as There is Space, Is There No Limit to Human Settlements?

We can choose to believe that, as long as there is space, there is no limit to the numbers of houses, highways, industries and reservoirs that can be developed on any given segment of land, *or* we can recognize, as livestock and wildlife men have long had to do, that any given piece of land or air or water has a definite optimum ability to 'carry' a given use and still remain productive.

We will see that, in each environment, there are limiting factors for growth and development. For plants, this capacity to carry the activity may be determined by soil nutrients, water, exposure to the sun, or the dynamics of air-masses. On the other hand, for housing developments and highways the capacity to carry may be established by the optimum psychological, social and environmental densities of people, by the jobs available, by the air conditions in relation to transportation patterns and industrial plants, and by the existence (or non-existence) of practices which will permit rapid restoration of the habitat by remaining segments of that habitat.

Another interdisciplinary study, done in the oil-shale region of Colorado, illustrates application of this principle. A team of similar composition to the ski-area study team, investigated the ecosystems of a segment of the Piceance Basin north of the Colorado River and east of Grand Junction. As the field data were analysed in the light of the project proposal, the scientists involved found that the air-pollution levels would exceed the State air-quality standards during a considerable portion of the year—unless the plant could be located out of the valley.

The engineers re-thought the whole matter, and undertook investigations into other, more acceptable sites for the plant. In this case, relocation of the plant would relocate spent-shale disposal also—thus moving another major environmental problem into a site where the environmental impacts would be far less than in the original site.

This particular study was instigated by an engineer who had found that project design had to employ *a three-pronged integrated approach* incorporating engineering, economics and ecology, on an equal and interactive basis—which he termed 'the 3Es tripod'.

Why Not 'Use Up' and 'Throw Away'?

We can choose either a linear path for use of all materials, and a highly-consumptive pattern of energy-use, *or* we can model our activities after the life-support systems of which we are a part, by cycling materials and goods and by using energy frugally and in the most efficient manner possible.

The latter choice would mean building homes with a great deal of insulation, so as to expend a minimum of energy to heat or cool them. It would mean stopping the dumping of useful organic nutrients into our waterways in such concentrations that plant growth increases to a point where navigation is hampered. The phosphates and nitrates involved can be cycled onto the land to fertilize crops, improve pasture and increase tree-growth, while at the same time reducing the need for so much mining of phosphates and nitrates. Cycling materials would mean transporting cans, oil and bottles, to recycling centres rather than to dumps. It would mean better mass-transit systems.

At this point, someone always says, 'But those measures are only a drop in the bucket to the energy or phosphate, or whatever, demands of the world today'. True, if such measures are only carried out in an unorganized and scattered manner. But they need not be unorganized and scattered—indeed, they must not be! For many of our resources are finite, and therefore our use of them should be systematically careful, organized and frugal.

Why Be Concerned about 'Mature Ecosystems'?

We can choose to foster a low diversity of organisms with low stability of community, *or* we can choose to maintain those segements of landscape that now support mature, diverse, stable and specialized habitats. We can also choose to assist in the regenerating processes that can heal and restore mature, complex ecosystems which have been damaged or disturbed. It is interesting to note that the value of diversity in the business community and in the financial and investments community is widely recognized and practised. The same principle should apply perforce to life-support systems.

We can choose to identify those life-systems which are permanent, mature and diverse, no matter what their size, so as to avoid damaging or destroying them. These mature systems provide a wide variety of niches for animals and plants—frequently including rare and endangered species. Recently, another ecologist and I had the opportunity to preserve diverse, stable ecosystems for a power company. I spent a day of ecological reconnaissance of the alpine tundra with the design engineer, locating for him within the route of a proposed high-tension line both the transient, unstable life-systems and the permanent, mature ones. With this distinction in mind, the engineer was able to locate transmission-line towers so as to avoid the mature stands—some of which are several thousand years old, judging from the depth of soil accumulated. Without this knowledge, the engineer would have eradicated thousands of years of tundra development with a few hours of work, for he had proposed a tower for the centre of one of these old stands!

So it is with all development. If ecological reconnaissance precedes intensive exploration, development design, construction of roads, installation of utilities, etc., it is possible to avoid damage and destruction of areas of high ecological value. With prior ecological investigations, design and development can be planned so as to maintain game migration-routes, fisheries, winter grazing-grounds, grass-beds, striped bass spawning areas, zones of seismic activity, mature old stands of vegetation, marshes, woods, prairies, and other habitats for rare plants and animals.

Topsoil can be removed from development sites to where it can cover subsoil and rock, often enhancing productivity. And this can be done at great saving, at least compared with the cost of rebuilding soil on raw substrate. Such restorative actions can be done best and cheapest when a comprehensive plan is developed in the earliest stages. It is much more difficult, for example, to accomplish habitat rebuilding or restoration of a species' population after construction on the project has begun. But it may be relatively simple to work around a segment of habitat which should be saved. In the United States, the National Environmental Policy Act has established in law the insistence that everyone is to 'stop, look, examine, analyze and redirect' actions so that those actions can harmonize with ecosystems, as a skilled hand slips adroitly, gently and smoothly into a glove—with little or no disruption of either the hand or the glove.

Are Not Our Environmental and Economic Objectives Incompatible?

We can choose to believe that our environmental and our economic objectives are incompatible, *or* we can recognize that they are compatible when viewed in long-range perspective. In fact, we can see that recognition of environmental objectives is essential to continued realization of even our economic objectives in the longer range.

Examples from the past, where profits have been reduced or forgone and sizeable capital investments have been made in regenerating renewable resources to produce greater subsequent profit, provide evidence that our environmental and economic objectives can indeed be compatible. One example is the rebuilding of the Pribilof Islands fur-seal herds from around 200,000 animals in 1912 to around 1.4 million today. This effort netted the US Treasury $25m in the first 50 years of operation.

Another example involves the Pacific halibut fishermen. They recognized that the halibut stock was very low, as they were catching little. They agreed to taking no fish for a period of years, during which investments were made in research into halibut life-history, populations and habitat. The Pacific Halibut Fishery Commission has calculated that the cumulative gain paid to the fishermen from these investments, coupled with management practices and regulations based on research findings, has amounted to 50 times the original dollars invested.

Another choice for compatibility of our environmental and economic objectives is investments made in restoring the Sockeye Salmon runs on the Fraser River in Washington and British Columbia. This had once involved some 30 million fish in peak years. By the mid-1920s the population had declined to 1 million fish in peak years because of over-fishing, pollution and modifications of habitat that blocked the salmon from going into their ancestral spawning-areas. In 1940, a long-range management programme was undertaken. It included restoring the habitat by removing obstructions such as the Hell's Gate rockslide, which kept salmon from moving to their spawning-grounds. It included also cleaning up the waters and regulating the take of fish at a level well below the annual production. An investment averaging $2m per year for a 15-years' period (total $30m) restored the Sockeye stock of the Fraser River to a point where it had increased four-fold, and the economic value of $30–$50m (boatside) is now harvested annually. So an *annual* boatside gross income resulted which was equal to or greater than this initial $30m investment in restoration.

It is clear that had we, in 1940, chosen to believe that our environmental and economic objectives were incompatible (and thus chosen not to make a financial commitment to this long-range Sockeye Salmon management improvement programme), we could not have realized a significant economic *and* environmental gain.

Our Choices Either Affirm or Deny Our Citizenship in the Community of Life

We have been exploring our choices and options. We face choices made and choices neglected, and have all seen the trouble that can stem from human failure to take into account the full consequences of human actions.

With a fuller understanding and conscious practice of recognizing what should be chosen and when humans can operate as positive, productive members of the community of all living things. We humans are not an organism apart; despite our unique mental gifts we are members of this living community, whether we recognize and accept this fact or not. As participants in this living community, we can and do make choices which either *deny* or *affirm* our citizenship in the community of life. This continuing opportunity of choice presents us with an ongoing opportunity to develop and enhance an ever-deepening sense of our 'kinship with' and 'reverence for' all life.

Yet ours is not just an ethic of choice but an ethic of accountability, because the choices must take into account the full consequences of human actions within complex ecosystems. It is high time we gave an affirmative answer to Aldo Leopold's question asked more than 30 years ago (Leopold, 1945):

> When will Man learn to treat the land as a community in which he is a member, rather than as a commodity he has a right to exploit?

Part II by
EMMANUEL O. A. ASIBEY

First I would like to speak of the religious basis for a biospheral ethic which can be found in Africa. The ethic of primitive Man, so arrogantly labelled and abused by modern Man, was the 'silken tie' that kept him in balance with his environment. Common law, religiously adhered to, runs through all races, being deeply rooted in ethics. Countries which have statute laws respect and recognize common law.

The few enclaves (to which Dr Willard refers) which do not practise the exploitative ethic are in those parts of the world where the exploitative ethic has not undermined the stewardship ethic. Ironically, the stewardship ethic is usually equated with primitiveness—something that hinders modern development. Modern development is the misunderstood and misapplied concept on which the exploitative ethic thrives.

Very different is the African sense of attributing souls to both animate and inanimate objects—such as the Earth, streams, rivers, rock outcrops, mountains, big trees, and so on—all of which must be treated with respect or else they die. There is also the African philosophy of Trinity enshrined in the ethical attitude of land stewardship. Land, the limited resource of mankind, belongs to the Trinity of the Dead, the Living and Generations Unborn. The Living has use-right, and the responsibility to leave it unimpaired for the benefit and enjoyment of posterity. The Living has no right to trade it off or abuse it to the point where Mother Earth will be killed. Failure in effective stewardship is punished by the Departed Souls which have the duty to avenge generations unborn. Thus the generation which abuses its stewardship in resource management should expect to suffer untold agony.

I would like also to comment on the socio-political basis for a biospheral ethic. Governments hold themselves responsible to protect the interests of their nationals—their trade interests and their desire to survive—even when they violate the fundamental rights and ethics of other nations and governments. The single law of the survival of the fittest has been translated into socio-politico-economic exploitative ethics. Hard trade agreements, hardening of peace treaties, war and the threat of war, all thrive on political power. It is common historical and modern knowledge that the suppression of weaker minorities, and the lack of stewardship ethics in socio-political relationships between nations, strengthen the exploitative ethics which have spelt the doom of modern Man and necessitated ethics of biospheric viability. The awareness and revival of the philosophy and commitment of the big man to the small man, and ultimately to posterity, is vital for modern Man.

In the few enclaves where people have exercised stewardship ethics, big men are keepers of their brothers. Where they have turned out to be the exploiters of their brothers, there has been woe and doom. Throughout the ages, Man's inhumanity to Man has made countless thousands mourn. Now Man's lack of stewardship ethic has undermined his very survival. It is essential to have political commitment to sustain ethics for ecospheric viability.

Part III by
MARTIN W. HOLDGATE

Why Is the Biosphere Viable?

The biosphere has been viable because: (*a*) solar radiation, the Earth's orbit and the Earth's rotation have stayed within certain limits for about 4,000 million years, and hence provided acceptable outer limits for life; (*b*) the biogeochemical cycles of oxygen, carbon, nitrogen, phosphorus, and sulphur, the hydrological cycle, and atmospheric circulation, have likewise remained within certain limits over most of this period; and (*c*) organic evolution has been able to proceed in a world that has been variable enough for natural selection to be faced with new challenges, yet constant enough for stress and physical disturbance rarely to cause extinction.

As a result a wide range of ecosystems has evolved. Some have very high diversity of co-adapted species existing in moderately stable situations; others evolved with a lower diversity of species that were especially adapted to the stress of disturbed situations such as deserts. Mobility and dispersal-capacity allow shifts in climatic and soil zones to be followed by biological responses, and the world's physical and biological diversity is such that there is almost always some species that is (and usually many that are) able to take advantage of environmental change.

What Now Threatens the Biosphere?

We cannot be sure of where the life we know is going, but we can at least glimpse potential threats. In addition to those which are commonly discussed, Man is now emitting carbon dioxide on a scale that *could* double its atmospheric levels within 50 years and concomitantly raise the mean world temperature by about 1.5°C. We are emitting some pollutants (e.g. suphur oxides) on a scale comparable with Nature. Nitrogen oxides and chlorofluorocarbons may affect the stratospheric ozone screen against ultraviolet radiation. We have drained wetlands, thereby reducing methane emissions (important as a curb on excess oxygen), and have injected into the environment some persistent and potentially toxic materials that were not present during evolutionary history.

The widening extent of ecological change by Man is also expressed in massive clearance of forests and conversion of wetlands, replacing both these natural desiderata with herbaceous communities that are either harvested by Man as crops or are consumed by his livestock. We know that these changes have fundamentally altered the aspects of many landscapes, but we do not know their more subtle effects.

No particular ethic has a monopoly of this man-engendered situation. Western European Christian, Eastern European Communist, Moslem, Jewish, Hindu, Chinese, Amerindian and African groups have all participated in forest clearance and wetland drainage, on varying scales reflecting varying technologies. Many have also become industrialized and have generated pollution.

The approach of traditional culture to environment may vary, but in today's United Nations there appears a universal dedication to development, tempered only by the recognition that it must be environmentally wise. This may make it possible to evolve certain common guidelines, if not a common ethic of biospheral viability, before threats to the biosphere become acute hazards.

The Approach to a Biospheric Ethic

There is a distinction between human *needs* (of shelter, food, companionship, a secure place in the community, and some creative outlets), and human *desires* or *aspirations* (e.g. for modern travel, consumer goods, rich diets or television). The boundary is blurred, but there is a common tendency to demand and actively pursue more than is required to satisfy the true needs of the individual even at a time when many individuals have their basic needs quite unsatisfied.

Similarly, there is a distinction between what Man *can* do to the biosphere (and get from it) and what he *should* do. In the past, the ethic that predominated in many developed and 'successful' cultures (especially the Western European one) was of Man's lordship of Nature—and hence his right to harvest or mine the resources of the Earth as he wished. Such an approach was containable when Men were few and their powers to alter ecological systems remained limited—and when the main environmental changes were directed towards replacing natural ecosystems by altered ones which were nevertheless biologically productive, as in intensive agriculture and forestry. Things changed with the widening of human impacts to activities that proved inimical to life, and resulted in widespread pollution or actual destruction of ecological systems in lakes, rivers, estuaries, in the ocean and on land.

GUIDELINES FOR A COMMUNITY OF NATIONS

The situation is complicated by the heterogeneity of the Earth's environment, of its peoples, and of their economic development. The frontiers of the Earth have not been drawn to give every community a 'fair share' of fertile soil, mineral wealth, water or ocean. It is impossible for people everywhere to experience the same environment. A major question that any environmental ethic must face up to is how far one community should exploit its national environment when this leads to a high standard of living existing alongside deprivation for no other reason than the historic accidents that led to a particular frontier pattern.

Are there any general rules or codes of conduct which might at least guide communities (through UNEP or some similar agency) to use the resources of the world more wisely than hitherto? For example, it would seem desirable to accept (following the Stockholm Declaration) that:

(1) Communities should maintain certain basic standards for the quality of air and water, set on the basis of the best scientific knowledge, so that damage to living systems of land, fresh water and ocean is kept below prescribed and acceptable limits. All nations should regulate their emissions to air and water in such a manner that these limits are not exceeded.

(2) Communities should accept an obligation not to damage, through pollution, the environment of other states or of waters beyond the limits of their national jurisdiction.

(3) The ecological systems of the seas should always be cropped in a fashion and under a management régime which recognizes the interlinkage of the world's seas, and in particular is based on the concept of optimum (not necessarily maximum) sustainable yield.

(4) The management of freshwater systems should always reflect the unity of catchments and the need for international agreements where water basins are shared. For example, forests may need conservation in upland areas not only to prevent soil erosion locally, but also to prevent flash-floods lower down. Abstractions of water for irrigation or industry should not so deplete river-flow that downstream communities suffer.

(5) The genetic wealth of the world, especially in forms of plants and animals that are important for human use, should be conserved as a reservoir for future forest, crop and livestock breeding; countries within whose territories such genetic riches are concentrated should accept a responsibility for stewardship of these genetic resources, with support by international agencies and aid where appropriate.

(6) The richness of the world's landscapes and wildlife constitutes an important part of Man's inspiration, and merits conservation for its role in human creative thought as well as in its own right as part of the stream of evolution. Modern science continually stresses the ecological importance of inconspicuous organisms, such as the decomposers involved in the great cycles of carbon, nitrogen, phosphorus and sulphur. The pharmaceutical value of little-known plants is increasingly appreciated. In our present state of ignorance, it is prudent not to allow the needless extinction of species—for utilitarian as well as moral or scientific reasons.

(7) The maintenance of the great biochemical cycles is fundamental to the viability of the Earth. Quality objectives and measures should be evolved to safeguard those cycles, both globally and regionally. International agencies, through 'Earthwatch' and similar monitoring programmes, should provide early warning of any changes which may threaten those cycles, so that co-ordinated action may be set in hand while there is yet time to arrest adverse trends.

(8) Because of the diversity of the world's environment, nations are bound to depend on the productivity of one another's land in order to have adequate diet (both in terms of quantity and quality). If people everywhere are to have at least basic minimum standards of health and well-being, then agricultural policies will have to be developed, nationally and regionally, with the needs of other communities also in mind, and common strategies will have to be evolved for major or particularly valued crops.

(9) Global Environmental Surveys, which define the ecological patterns, trends and processes of the world and the options for management, need to be evolved as a foundation for the wise use of global environmental resources. There needs increasingly to be international discussion of the way in which national strategies for environmental management may be harmonized.

(10) Time-scales for environmental management and the cropping of ecological systems should likewise take due account of the

response-time of the populations and ecosystems concerned. While the concept of maximum sustained yield has its weaknesses, it enshrines one valuable principle—that the use of an environmental system should not, without prior thought, involve extraction from that system of more than natural replenishment will make good on an acceptable time-scale.

Part IV by YOICHI FUKUSHIMA

The Japanese Way of Thinking About Nature

Prior to the introduction to Japan of the occidental idea of Nature, the Japanese way of thinking about Nature had a very unusual character. Japan had stepped on to the so-called capitalistic stage in 1868 with the Meiji Restoration. Before 1868, Japan had been ruled by the Shogun (the Lord of the Tokugawa family), who presided over hundreds of feudalistic lords all over the country. This Tokugawa rule had lasted for 300 years, prior to which there had been more than 400 years of feudalism, with several successive ruling warrior families. Still earlier, a succession of emperors had ruled Japan. But political power had been acquired gradually, first by the aristocratic classes and later by warrior classes—until finally a firmly feudalistic system had been established in which the emperors were sanctified as titular rulers over and above the actual rulers.

From earliest times the Japanese have had a strong tendency towards animism. In this way of thinking, there exists no boundary between the human, the natural and the spiritual worlds. For example, there was until very recently no word to express what in English is meant by 'Nature'. There are many folk-tales and legends about marriages between humans and other creatures—animals, birds, plants or even stones. Stemming from this traditional animistic thinking, the Japanese have enshrined as gods all sorts of creatures as well as non-living things and even natural phenomena, so that there are today a large number of divine trees, divine stones, divine mountains and divine rivers.

Responding to a Less Severe Environment

Europeans have had to struggle against the severity of Nature and have tried to conquer Nature. But in Japan the environment is generally mild, so that it has not been necessary or seemed appropriate to have an antagonistic feeling about Nature. On the contrary, in so far as people have not violated Nature or damaged their surroundings, Nature has always been expected by the Japanese to be favourable to human beings. Natural disasters were thought to be punishments by Nature (or gods) for their being violated by Man. There is a Japanese word 'Ten-batsu', which literally means the 'Heaven punishment', and the Japanese are accustomed to see the heavens as similar to the gods.

People worshipped Nature and took great care not to offend the will of

Nature or gods. In cases where it became necessary to cut trees or cut open mountain areas in order to construct roads or buildings—or even to kill wild beasts—people would hold ceremonies to calm the anger of Nature or gods. Such customs remain a part of Japanese civil life to this day.

Shrines were also built for famous feudalistic lords during the feudal era, and until recently it was the custom to enshrine heroes who fell in battle. For example, the generals of the war between Japan and Russia (1903–4) were enshrined and are worshipped to this day. Needless to say, major figures in Japanese history—such as the Emperor of Meiji—were enshrined, and each year millions still visit the Meiji Shrine.

We also have a large number of shrines to foxes and other creatures. People visit these shrines to seek divine favours. For example, you will find in most taxicabs some sort of amulet which the driver has bought at a shrine for protection against accidents. The drivers go to the shrines both to be purified and to acquire such amulets in return for an offering of money.

Let me recount one of the old legends which I believe will convey to foreigners the concept of the Japanese about the relation between Man and Nature. Once there lived a happy couple who had a young son. One day the wife confided to her husband that she was the spirit of a very large willow tree which her husband had saved from being cut down some years earlier. She was concerned, she said, because the people had decided to construct a building in which to pray for the Emperor's recovery, and she knew that the willow tree was destined to be cut for the ridgepole of the building. But by whatever means the husband tried to save the willow tree, it was in vain, and the wife disappeared.

After the willow had been cut down, the people tried to carry this huge timber to the capital. But despite the strenuous efforts of many strong men pulling, they could not budge it. Then the young son of the willow tree's spirit joined the group pulling, and lo! the great timber began to move very easily. The boy was, of course, greatly rewarded by the Emperor afterwards. This building was constructed in the middle of the thirteenth century and is still standing in Kyoto, the former capital city of Japan. It is a building 70 metres in width. In this legend you can see that, in the traditional Japanese way of thinking, there is almost no distinction between what is human, what is natural and what is spiritual.

The Recent Change in the Japanese Way of Thinking About Nature

This traditional Japanese way of thinking about Nature has changed drastically in recent years, particularly following the defeat of Japan in World War II.

Prior to World War II most villages had a grove of great trees in which were their own shrines, which were called 'Chinju' and which protected the village and villagers. After World War II this cult of Shintoism was strictly prohibited as a national religion by order of the general headquarters of the Allied Forces. This was done because of the deep connection between the Emperor system in Japan and the Emperor himself as the living god. As a result of this post-war policy, the management of the shrines came to be neglected and at the same time the forests were destroyed.

Moreover, in the years after 1960, under the policies of the so-called high-growth economy, large-scale development accelerated very seriously the destruction of Nature. The natural fabric of rural areas was severely damaged by the unlimited construction of residential areas, recreation areas and highways. This was, of course, the period in which the Japanese seashore was fatally injured by the construction of vast oil-refining facilities and other factories.

In short, what has happened is that the Japanese have lost their traditional respect for Nature and fear of the 'Heaven punishment'. However, many traditional ways of thinking remain, and in their personal lives most Japanese continue the custom of visiting shrines very frequently to pray for divine favour. But in their public behaviour the Japanese have lost their way of thinking that human beings should live in coexistence with Nature.

Traditional thinking has been lost, but the Japanese have not yet established a scientific basis for a new concept of protecting or conserving Nature. And it is a great pity that in these circumstances the destruction of Nature goes on more and more. We need to establish as soon as possible new norms or mores regarding our relation to Nature. I believe this is one of the most urgent tasks for us as scientists and humanists to tackle.

Part V by
ELIZABETH & DAVID DODSON GRAY

Dr Willard has stated the ecological principles and the criteria for human action which flow from those principles, and she has done this so convincingly that we feel compelled to ask: *Why then do we not live according to that obvious ecological ethic?* To be sure, people do not yet know these ecological principles sufficiently, and grasp their full import for our human activities. But our cultural problem of finding an adequate environmental ethic is deeper than that. We have confused ourselves about our fundamental relationship to the Earth by using world-views, pictures of reality, which are fiction. Let us now consider what are those pictures about reality which our Western Civilization has projected upon the walls of our universe.

THE BIBLE AND THE JUDAEO–CHRISTIAN TRADITION

One picture which has become deeply embedded in the 'lining of the mind' is that of the Spirit God of the universe, who is 'above all and beyond all', and who is said to have created Man (Adam) to have dominion over all the animals and the rest of 'creation'.

Many theologians have reminded us that 'dominion' can mean responsible stewardship. But it is clear that—however 'dominion' is interpreted—it always means 'above' and implies a right to exercise power over others. In that centuries-old and still mythically powerful story from the Book of

Genesis in *The Bible*, Man is conceived as definitely 'above' the rest of Nature. He partakes of the 'spirit' character of the divine, is created in the divine image, gives all other animals their names, and so on.

This same consciousness of Man's superior place in the cosmic scheme of things is recapitulated in Psalm 8:3–8 and is characteristic of one view of our relationship to the Earth—not only in biblical times but also in a manner that is still widely held today. The assumption is that reality is indeed 'up and down', and that God—imagined as 'Pure Spirit'—is at the apex of a cosmological pyramid. Just below come men—and we do mean males—created in the image of God by a specific act of God's creation. Then come women, and then children—so derivative that they are not even in the Genesis story. Then come animals, which do not have the unique human spirit at all—and thus, although they live and move, they do not have 'being' as humans do. Thus animals are below humans. Further down still are plants, which do not even move around. Below them is the ground of Nature itself—the hills and mountains, valleys and streams—which are the bottom of everything, just as the heavens, the Moon, and the stars, are close to God at the top of everything.

What is clearly articulated here is a hierarchical order of *being* in which the lower orders—whether female or child, or animal or plant—can be treated, mistreated, violated, sold, sacrificed or killed, at the convenience of the higher states of spiritual being that are found in males and in God. Nature, being not only at the bottom of this pyramid but being the most full of dirt, blood and such nasty natural surprises as earthquakes, floods and bad storms (not even to mention untold possible ecodisasters), is obviously a prize candidate for the most ruthless 'mastering' of all.

Evolutionary Ascent and the Scientific World View

Still another picture of the cosmos has been projected upon the cultural mind in the 100 years since Charles Darwin. This is a secular picture of our place in the world, and lacks any divine Being. It is a picture in which Reality began at the bottom with a very hot mass which, over millennia and far longer periods of time, cooled and developed into the 'primeval soup' from which came the compounds out of which all life slowly evolved through the ages, progressing slowly *up* the evolutionary ascent until we once again find Man at the top as 'the most highly-developed species'.

The similarity between these two pictures of cosmic reality is curious! Both assume quite unconsciously that Reality is hierarchical, that profoundly and at its essence the grain of Reality is 'up and down'. Whether life is visualized as secular and evolving up from primeval compounds to the highly developed brain of Man, or whether life is visualized as religious and beginning with a top-down act of the divine who puts everyone and everything in its place, both views are equally and curiously clear that Man stands in the topmost position on Earth.

It is interesting in this context to contemplate the root meaning of the word 'hierarchy', for it is derived from Greek words meaning 'holy order'. And certainly each of these cosmic pictures has been viewed as a sanctified order—one of them being legitimated by religious authority and the other by a newer priesthood whose authority is scientific and academic but perceived today as qualified to tell us about 'what is'.

These two leading cosmic visions of the order of things underlie the Western Tradition and its great scientific and technological achievements. All of these achievements—from our splitting of the atom to our going to the Moon, from our healing of diseases to the prolonging of our lives, from the multiplicity of new drugs, pesticides and laboratory-created compounds to the vastness of our industrial growth and productivity—are predicated upon our confidence that Man is truly 'above', and that his wants and needs and desires are the most important thing upon this Earth. Underneath it all there is a confident assurance that what-is-above 'calls the tune'—and that what-is-below will constantly be compliant and adapt.

Hear that confident assurance in this comment by a professor at a great technological institution, speaking at a public meeting:

> The ancient Greeks trembled in awe before the piteous might of their tribal gods, but the thunderbolts Zeus had to throw around are as nothing compared with the power which science has put in the hands of modern Man.

Hear that self-assurance again in this comment of a graduate student at the same institution:

> I came here to build computers. The question is not how to appreciate the environment, but how to master it.

Little Lower than the Angels

That power of science which is wielded with such confidence is born of the sense of himself which modern Man has been given by the religious and evolutionary visions, namely that he is indeed the most highly developed species with the best brain—one whose exploits must of necessity be 'onward and upward'. Man the thinker or Man the scientific researcher is not deterred even by the possibility of nuclear destruction or by the awesome hazards of DNA viral research. Man's 'right' to do these things is spoken of in terms reminiscent of earlier kings' defence of the divine right of kings.

Surely this Man whose head is 'crowned in flame' has in this view the right, if not always the wisdom, to blaze new trails—never to rest upon his laurels, but always to think new thoughts, produce new miracles which have never been seen before, to prolong life, and, yes, to create in the test-tube new life itself. Whether blessed by God or by the evolutionary ascent, such a man is only a little lower than the angels—and is justified in working to diminish even further that distance! Meanwhile, all that which is below Man will stand in awe of his vast and varied accomplishments.

But wait! It does not seem to be working out that way. The ozone layer may be thinning, the pollution level thickening, the soil nutrients getting less, and the water pollutants getting more. Species that we have done away with, never thinking they would be missed, it now appears 'do' some things which we did not know they did and which life needs to get done.

How can this be? All these things—ozone, water, soil, lower species—all are 'lower' than we. Their task is to provide us with what we want, to absorb our wastes, to get out of our way and not cause trouble. That is always what those 'below' have been expected to do, including so-called 'primitive' peoples. They have always been overpowered by superior strength and 'firepower'. They have had to adjust and accommodate to that which was 'superior'. How is it that now Nature will not do the same?

The answer comes back to us with the relentlessness of the tide: 'It is because you are not right about who you are. Your city has been built upon the sands of an anthropocentric illusion. Castles built upon hierarchies which do not exist simply cannot last.'

The Myth of Man-above

It is painfully clear that we have created a myth—a myth filled with anthropocentric illusion. It is the myth of Man-above-Nature. It is a myth very similar to the pre-Copernican, deeply-felt assumption that the whole universe revolves around us—around Earth. Only with great difficulty and much trauma for the religious establishment did we change our world-view after Copernicus. We *wanted* to believe that we were the centre of the Universe. Similarly, we want now to believe that Man, with his wonderful technological achievements, is truly *above* Nature—a position suggested by biblical faith and confirmed by biological evolution, not to mention technological successes.

Should such a Man have to 'fit himself' into ecological principles and processes like a hand into a glove? Nonsense! It is unthinkable! The Man at the pinnacle does not have to *fit* himself into anything.

I hope that you can see from this brief development of the myth of Man-above that: it speaks to the needs of many men; it seems to be confirmed by every technological success; it does not encourage anyone to believe that humans are merely one species among many living with natural processes, all of which must respect ecosystem processes and fit themselves into them. The myth of Man-above is in basic conflict with the ecological paradigm.

The Myth of Reality as Hierarchical

But even beyond the myth of anthropocentric illusion there is the further conceptual problem that we have 'mythed' reality in hierarchies—ladders of above and below, 'a great chain of being' in which that which is declared 'above' is somehow 'better' and fit to control that which is below. Thus we have an ingrained belief in men being over women, so-called 'civilized' over 'primitive', 'developed' over 'underdeveloped', human over Nature. Always, those thought to be 'above' have controlled and exploited those thought to be 'below'. And the Earth—pictured as below the human, and consciously or sub-consciously mythed as female, as in 'Mother Earth' and 'virgin resources'—has been controlled and raped.

Our conceptual problem, we are at last beginning to see, is that we are now discovering that reality is composed not of hierarchies but of systems. And in systems everything affects everything else. But that is not to be expected if the picture in our heads is of hierarchies. Our analogies have betrayed us.

Need to Re-myth the Earth in Systemic Realities

The truth is that we must re-myth the Earth into systemic realities in which all factions are interrelated and none is 'Lord'. Equally, we must re-myth the human family so that all are related and none is 'Lord'. The two processes are integrally related to each other, as Dr Asibey has suggested in his comments about the socio-political basis for the ecological ethic. As long as we cling to hierarchies in human social systems, we shall hold them to our hearts when we relate ourselves to the Earth.

For example, we are deeply suspicious of the interpretation of the environmental ethic as 'responsible stewardship'. That concept to us seems paternalistic, still clothed in hierarchical categories, and subtly related to old ideas such as 'enlightened slave-owners' and 'the white man's burden'. 'To be graciously responsible for that which is below us' is not *really* our situation on spaceship Earth. Yet it is widely touted as an acceptable version of the environmental ethic. Just as women did not wish to be owned by even 'responsible' husbands, and just as slaves did not wish to be owned by even the most enlightened slaveowner, so the Earth will not flourish as the 'property' of even 'responsible' stewardship wielded from 'above'. Power corrupts, and self-interest always motivates the 'above' to the detriment of the 'below'. And great is the power of human rationalization to justify the self-interest and make it appear to be 'responsible'!

The Myth of the Earth as Female

Old myths die hard. One of the last to die will be the imaging of the Earth as female according to the male psyche. This imaging of Earth as feminine has powerfully, yet subtly, influenced the way in which male patriarchal cultures all over the world have referred to 'Mother Earth'.

Repeated references to 'carrying capacity' make us wonder again about our possible overloading of the carrying capacity of natural processes. Perhaps it is because of the vastness of the oceans and the skies that we have not worried about them nearly enough. But this is a little like the way a small boy views his mother, who looms so large in his small universe. Does it ever occur to him that she may run out of physical strength to nurture or feed him? Does he ever imagine that she has a psychic stress-level beyond which she cannot go? Mothers can assure you that this never crosses young minds. The assumption is always and unconsciously made that Mother will eternally provide food, recycle waste, and endlessly clean up dirt, soiled dishes and littered houses.

Once again, our analogy has betrayed us. We have laid a myth of the eternally-feeding and endlessly-recycling Mother upon biosphere processes which are resilient but have threshold-limits of tolerance. The myth does not fit the reality. But we may destroy the Earth and ourselves before we discard *that* myth.

A Time for New Myths?

The sad fact that today we do not have an environmental ethic which is intuitively obvious and compelling to us all is a problem rooted in our myths and may well prove to be our undoing. Our old exploitative myths (such as Man-above, Reality as Hierarchical, and the Earth as Female) are still cluttering up our collective unconsciousness. Like old soldiers, our old myths fade away slowly. Yet they are fictitious illusions which do not match the reality of our ecological situation. The old myths tell us that we are 'Lord'—a little lower than the angels, with the animals and the Earth and the all-important plants below our feet.

We need a new myth to capture our psyche and give us a new and realistic human identity within the systems of the biosphere. That myth has not yet been born, and so we must wait for it. Meanwhile we are in a transition time similar to that after the calculations of Copernicus rendered the Ptolomaic myth useless as being untrue. Aldo Leopold asked: 'When will Man learn to treat the land as a community in which he is a member, rather than a commodity he has a right to exploit?' Our answer would be: 'When the old myths die, and new ones are born.'

Meanwhile in such a harassed period of transition we must not only educate ourselves and the world about ecological principles but also acquire a really sound environmental ethic, if we are to preserve the biosphere from ecodisasters.

Background Reading

Barbour, I. G. (Ed.) (1973). *Western Man and Environmental Ethics: Attitudes Towards Nature and Technology.* Addison-Wesley, Reading, Massachusetts: 276 pp.

Barney, G. O. (Ed.) (1977). *The Unfinished Agenda: The Citizen's Policy Guide to Environmental Issues.* Thomas Y. Crowell, New York, NY: 184 pp.

Cahn, R. (1976). The environmental ethic. Pp. 1–47 in *Values of Growth* (Critical Choices series Vol. 6). D. C. Heath, Lexington Books, Lexington, Massachusetts: xxiv + 161 pp.

Cobb, J. B., Jr. (1972). *Is It Too Late?: A Theology of Ecology.* Bruce, Beverly Hills, California: vi + 147 pp.

Dodson Gray, E. (1977). *Why the Green Nigger?* Bolton Institute, Wellesley, Massachusetts: 307 pp., Xeroxed.

Eckholm, E. P. (1976). *Losing Ground: Environmental Stress and World Food Prospects.* W. W. Norton, New York, NY: 223 pp.

Faramelli, N. (1971). Ecological responsibility and economic justice. Pp. 31–44 in Sherrell (*q.v.*)

Goulet, D. (1977). *The Uncertain Promise: Value Conflicts in Technology Transfer.* Overseas Development Council, Washington, DC: xiv + 320 pp.

Kormondy, E. J. (1976). *Concepts of Ecology* (2nd edn.). Prentice Hall, Englewood Cliffs, New Jersey: xiv + 238 pp., illustr.

LEOPOLD, A. (1966). *A Sand County Almanac: With Other Essays on Conservation from Round River*. Oxford University Press, New York, NY: xv + 269 pp., illustr. (1945 edn. London, xiii + 226 pp., illustr.); also Sierra Club–Ballantine Books, xix + 295 pp. (1970).

ODUM, E. P. (1971). *Fundamentals of Ecology* (3rd edn.). W. B. Sanders, Philadelphia, Pennsylvania: xiv + 574 pp., illustr.

PASSMORE, J. (1974). *Man's Responsibility for Nature: Ecological Problems and Western Traditions*. Charles Scribner's Sons, New York, NY: x + 213 pp.

POLUNIN, N. (Ed.) (1972). *The Environmental Future*. Macmillan, London & Basingstoke, England, and Barnes & Noble, New York, NY: xiv + 660 pp., illustr.

SANTMIRE, H. P. (1970). *Brother Earth: Nature, God and Ecology in Time of Crisis*. Thomas Nelson, New York, NY: 236 pp.

SHERRELL, R. E. (Ed.) (1971). *Ecology: Crisis and New Vision*. John Knox Press, Richmond, Virginia: 159 pp.

STEFFENSON, D., HERRSCHER, W. J. & COOK, R. S. (Ed.). (1973). *Ethics for Environment: Three Religious Strategies*. UWGB Ecumenical Center, Green Bay, Wisconsin: viii + 132 pp.

STONE, G. C. (Ed.) (1971). *A New Ethic for a New Earth*. Friendship Press, New York, NY: 172 pp.

WEISBERG, B. (1971). *Beyond Repair: The Ecology of Capitalism*. Beacon Press, Boston, Massachusetts: xii + 201 pp.

Discussion (Session 16)

Juel-Jensen (Chairman, introducing)

This Session is about ethics, [but] I shall be brief. Yesterday, the session on 'Life-style Alternatives' got somewhere near the rub: the riches or the potential riches of the permanent *vis-à-vis* the poverty of the ephemeral, the temporary, were touched upon. The increasing spiritual poverty of the West makes it so much more urgent that we revert to the permanent and that we do not export our poverty to other parts of the world.

Professor Bryson referred to his [Hopi Indian emblem], to how it reminds him of a society that has remained stable, and to how much he has learned from it. I, too, carry an object. Normally my pectoral cross is under my shirt. It was given to me by an Ethiopian Archbishop three years ago. Ethiopia, one of the oldest Christian communities on Earth, is a country where, until recently, when the worst and poorest ideas and material things were imported from the West, the Christian religion was not just something you did sometimes on Sundays but was part and parcel of your life. The community may have been medieval but it made a coherent rural whole that was rich in spiritual [ways]. The fighting in Northern Ethiopia is not just between guerillas and the forces of a military junta, but has deeper roots.

I have friends who were chased out of Tibet when the Chinese destroyed their culture. I am deeply impressed by the spiritual richness they have, which has sustained them both in northern India and elsewhere that they have settled. They may have material poverty but we have much to learn from the intangible resources which they [retain]. I have recently worked, not only with Ethiopians, but also with certain Arabs in the Gulf States, and I have been impressed by the sincerity of purpose of some Arabs who are frightened that the present oil riches—material, temporary riches—may destroy their country for ever. Indeed they are seriously inspired by their religion, Islam, to thinking long-term of how they can conserve their environment. So the opening note which I want to strike is that: unless the philosophy is right (and I do not necessarily hold that the Christian one is the only one, when indeed I am sure it is not—even though, for me, it is the right one), all efforts at conserving the world in which we live [may prove] in vain.

[For this paper on the 'Ethics of Biospheral Survival', which they have subtitled 'A Dialogue', we have a veritable galaxy of authors—from North America, Africa, Europe and eastern Asia. Consequently we have a very wide range of viewpoints expressed in their respective sections of the paper. Of the six co-authors it has only been possible to assemble two on the platform—Mr and Mrs Dodson Gray, Co-Directors of the Bolton Institute, of Wellesley, Massachusetts. By training they are an engineer and an historian, respectively, but they are both now theologians, having graduated together from Yale in their adopted discipline. Very proper key-note speakers for this Session, Mr Dodson Gray will speak first, followed immediately by Mrs Dodson Gray. The other contributors of parts of the paper are Dr Beatrice E. Willard (who headed up its drafting on her return home to Colorado after spending several years in Washington as one of the three full Members of the US President's Council on Environment Quality), Dr Emmanuel O. A. Asibey (Chief Game and Wildlife Officer of Ghana, based on Accra), Dr Martin W. Holdgate (Director-General of Research in the United Kingdom Department of the Environment, based on London), and Professor Yoichi Fukushima (Chairman of the Japanese National Committee on Nature Conservation, based on Tokyo). Also on the platform we have three panellists: Dr F. Raymond Fosberg, of the Smithsonian Institution, whom you already know, Dr Anitra Thorhaug, likewise, and, just arrived from Pakistan, Dr Khalid Hamid Sheikh, Chairman of the Department of Botany in the University of the Punjab.]

Dodson Gray, David (Co-Keynoter, presenting paper)

There is currently a vigorous ongoing dialogue about environmental ethics. In this dialogue there are a number of different positions [already represented in our paper which you would by now all have received if it had not been for the strike and some mechanical breakdowns]. What you will find there is a selection of views which is interesting, in part because it comes from several continents and traditions—Japan, Africa, UK and USA. We are thus confronted with a wide spectrum of views. There are those who say that an environmental ethic already exists, the problem being that we do not understand it [or even] see it. This is essentially Dr Willard's position in the first part of the paper which seems to hold that what we need is better spectacles and better understanding or education. Aldo Leopold had a similar view.

Then there is another view that, yes, there already is an environmental ethic: 'It is not that it has been tried and found wanting; it has been tried and found difficult.' That is a very different position, and John Muir, Robert Cahn, and perhaps Gerald Barney, are representatives of this part of the spectrum.

Closely related to this view is the one which sees that an environmental ethic exists, but is the wrong one [—at least for most of us]! For example, the parts of this paper by Dr Asibey (Ghana) and Professor Fukushima (Japan) sharply contrast traditional environmental ethics with the currently ascendant environmental ethic which has its roots in both modern technology and the capitalistic ordering of society and its profit motive. This view would say that there are a number of environmental ethics but there is a conflict among them. Then, of course, there is always a socialist critique which places the blame for our environmental problems on the capitalistic structuring of society. Barry Weisberg's *Beyond Repair: The Ecology of Capitalism* presents such a view.

There is also the view that there is not yet an environmental ethic—or at least not one which, in an integrated view, takes seriously the inclusion of human activities and human culture within the environmental setting. This view sees that an adequate environmental ethic must take account not only of interactions between the human species and other species but also of the interactions *within* the human species, which result in benefits and effects being *displaced to* other species and even non-living parts of ecosystems. Erik Eckholm, Norman Faramelli and Denis Goulet are representative of this view which sees injustice and social oppression as a frequent contributor to environmental degradation, so that an adequate environmental ethic must include a concern for justice.

Eckholm cites El Salvador, Chad, Ecuador and the American 'corn belt' as regions where the present land-tenure system fosters short-sighted environmental practices, in which marginal lands are brought into production (either to increase profits or to provide survival sustenance to marginal farmers). Where marginal farmers are displaced, by *latifundia* [large estates], from access to fertile lowlands, they do terrible things to the ecology of the hillsides and highlands from which they try to eke a precarious living. So, injustice in the human sphere often means that the effects are displaced to the environmental sphere.

Finally, there is another view, which comes closest to our own. It asserts that it is not enough that we have an environmental ethic on paper: it must exist in a compelling image. It must be something which, when we see it, grabs us in our hearts and we say, 'Yes, that is the way we must live'. This is what the Old Testament and the Ten Commandments had, and it is the truth-quality which characterizes all the major religions. It goes with a metamorphosis in perspective—the religious term for this is 'conversion'—and I think that anything less than a conversion of the heart is not yet a complete environmental ethic.

As people trained in theology and ethics, it has been for us a most exciting and rewarding experience to be here with you in these days. Dr Willard [was very sad not to be able to come in the end, for urgent family and other reasons. As expressed in her paper,] her concern is that we learn from ecology that there are basic principles in the biosphere which are the rules of the life-game. When we move in rhythm with them they work for us, but when we do not move in rhythm with them they do not work for us [but sometimes violently against our interests]. We need to learn how best to harmonize our human activities with the dynamics and processes of the Earth's ecosystems, remembering always that virtually *everything affects everything else*.

Dodson Gray, Elizabeth (Co-Keynoter)

Dr Willard presents us with an ecological paradigm of humans as a part of ecosystems, fitting themselves into natural systems without major disturbance [thereof]. The question that has been raised often at these sessions is: Why do we not [develop a binding] ecological ethic? It is not rational for the [usually] rational Western society to rush towards the brink of ecological disaster: Why then, do we do it?

Dr Borden yesterday gave us an interesting clue when he talked about what he called the primitive beliefs of the average non-ecologically trained person. I would call this 'the mythic nature of social reality'—the pictures that are in the average person's mind, head and consciousness about the relationship between humans and the environment. As Dr Borden [indicated], there is often a separation, because Nature is perceived as being on a different level from humans, and it is this perception of ourselves in relation to Nature that I as an historian and theologian want to look at.

There are three basic pictures, or myths, which have been reproduced by the social reality that we call Western Civilization. They happen to be in exact opposition to the ecological paradigm. They explain a great deal about *why* we do not fit within the ecological system and *why* it is difficult for many people even to conceive of our so fitting, let alone fitting in like an adroit hand into a matching glove.

The background for these myths is given to us in the West by the biblical view of the universe, in which God, as Spirit, is eternal and up above; Man is created in the image of God, but beneath, as a combination of Spirit and body. Then come women and children, and then, below them, [other] animals. Plants come beneath [again], and land at the bottom. The traditional Judaeo–Christian mythic vision of how things are, presents us with a pyramid of being in which Spirit and what is eternal is 'above', and what is natural and mortal and constantly changing is morally inferior and 'below'.

So in our own time this heritage [of three basic myths] is still among us as 'pictures in the lining of our minds.' The first of these is a myth filled with anthropocentric illusion. It is *the myth of Man-above-Nature*. It is a myth very similar to the pre-Copernican, deeply-felt assumption that the whole universe revolves around us, around the Earth. Only with great difficulty and much trauma for the religious establishment did we change our world view after Copernicus. We *wanted* to believe that we were the centre of the universe. Similarly, we want now to believe that Man is truly *above* Nature—a position suggested by biblical faith and confirmed by biological evolution, not to mention technological successes.

Should such a Man have to 'fit himself' into ecological principles and processes like a hand into a glove? It is unthinkable! The Man at the pinnacle does not have to *fit* himself into anything.

The second 'picture in the lining of our minds' is *the myth that reality is hierarchical*, and made up of ladders of above and below—'a great chain of being' in which that which is declared 'above' is somehow 'better' and fit to control that which is below. Men over women, whites over blacks and other colours, so-called 'civilized' over 'primitive', 'developed' over 'underdeveloped', human over Nature. Always, those thought to be 'above' have controlled and exploited those 'below'. And the Earth—pictured as below the human, consciously or subconsciously mythed as female, as in 'Mother Earth' and 'virgin resources'—has been controlled and 'raped'.

Now, our conceptual problem is that you cannot have a hierarchical view of reality and also think of reality as composed of systems. What happens when you move into ecology is that you perceive that you can never do just one thing, and that reality is, in effect, a system. In a hierarchy, that which is above impacts on that which is below—and that which is below 'takes it' and does not strike back. The problem is that Nature really is not that way. Natural systems have boomerang-effects in which impacts come back upon that which is 'above'. But this is inconceivable to people whose heads are still structured by hierarchical thinking.

The third myth is *the myth of the Earth as female*. I was interested to find, when I was looking at some of your marvellous Icelandic sculpture, that there was a striking [piece] of a woman bending over her child. He is sitting, nursing at one breast, and she is bending over kissing his forehead. It is entitled 'Mother Earth'. Now, Dr Willard in her paper raised the interesting question: Why do we never think that natural systems have carrying capacities? We know that everything else does—restaurants, automobiles, theatres. So we do not try to seat two or three people in one theatre seat; we abide by the carrying capacities of our built systems.

But still the ordinary person, the non-scientifically trained person, never thinks about the carrying capacity of *natural* systems. The reason, I would submit, is that we never think about the carrying capacity of Mother, who is always there to feed, sustain and love you, no matter what you do. Her energies and love are perceived as being as endless as the oceans and the skies. The assumption is always, and unconsciously, that Mother will eternally provide [for our needs] and endlessly recycle [everything possible].

This feminizing of Earth as our Mother is a myth, and it is a myth which has betrayed us because the Earth is *not* feminine. It is only mythed that way. The Earth's systems are natural systems, and they do not 'know' us or 'love' us as our mothers do. The Earth's systems do not 'feel' as mothers do. It is interesting that I do not find ecologists referring to 'Mother' Earth. Yet I find average people doing it all the time. For example, headlines about the weather during the extremely cold winter of 1976–77 in the northeastern United States frequently [referred] to Nature as either animal-like or a ferocious mother that has turned upon us.

These myths are barely beneath the surface consciousness of the average person. My feeling is that we will never do different things to the environment, unless these myths are killed off by the ecological paradigm. But we do not yet have a myth that speaks to the average person about the ecological ethic. What we have learned is now couched in scientific language. It is not yet expressed as a powerful image or metaphor or myth. Until we have a potent ecological image and myth, my opinion is that we will not overwhelm these other more primitive myths that are inadequate and incorrect, and we will not grab people in their gut where they need to be convinced about the environmental ethic.

Fosberg (Panellist)

I am not going to say very much about a paper that I have not seen [because of the breakdown of two printing machines and non-return of two taxis from the town but will] indulge in a generality that should be reasonably sound. I think that cultural features, including ethics, are products of evolution [and hence to some extent] of natural selection. In other words, they have been of survival value to the culture concerned. We of the Western European, including North American, Australian and so forth, culture have been consciously or unconsciously influencing, at least partially unfortunately or disastrously, but successfully, the older cultures of the rest of the world. Along with this, inevitably we have exported the ethics that have developed with and supported the materialistic culture in which we live. In my opinion, this set of ethics in our present position on the exponential curve that I referred to earlier in this Conference, has become obsolete. It [seems] un-ecological or even anti-ecological, and I would strongly advise the rest of the world to reject it; I would also urge and hope that we can ourselves reject and escape this set of ethics. Hopefully we can select from the ethical systems of older and perhaps wiser cultures, and synthesize an appropriate ethical system for our modern situation. Given time it would evolve, but we do not have this time. We must, at least as an interim measure, use the best features of existing systems.

Thorhaug (Panellist)

I am a lay person in the field of ethics. However, much of my own set of ethical beliefs and values has been derived from my study of biology, and one might say that I was quite Darwinian in my view of ethics. I believed that ethics have evolved for the good or for the survival of groups of primitive peoples—to effect social cohesion for and of the group, and against the individual's behaviour. Ethics are not fixed but rather evolving (as Dr Fosberg mentioned). An example of this is monogamy and adultery, which are so different from society to society. In Arab countries marriage can include several wives; the Afghanistan women may have several husbands. In our culture marriage must be monogamous, and anything outside monogamy is adultery; however, divorce allows a series of different spouses, which in other cultures might be termed polygamy. Some of the Polynesian cultures consider adultery to be intercourse outside of the family unit, not outside of a monogamous spouse-with-spouse relationship.

So this kind of ethical concept differs between societies and evolves in time and space. The methods of enforcing ethics are group-disapproval, ostracism, outlawing, legal sanctions and fear of the supernatural. Now many of these, of course, we no longer cling to, and we have given up our enforcing capacities in many ways and left it to courts and other bodies. The individual and the family unit have given up their enforcing capacities of ethics to the policeman or the legal system or the government—or to people who may not respond rapidly enough or be unable to enforce the ethics of society.

Ethics were accumulated group-wisdom that was gained by trial and error and transmitted on both a conscious and a subconscious level from generation to generation—which was fine as we passed slowly from the Stone Age to the agricultural and then the industrial ages. But now that technology—the atom bomb, flight to the Moon—is intervening in our society, things are happening so rapidly as to [blur] generation-gaps [and even some shorter gaps].

Three decades ago we did not have the awareness about the Third World that we have now. Two decades ago we did not have a consciousness about minority groups. One decade ago we did not have a consciousness about the population problem. We now have all those new technologies and facts—which have changed our realities and thus our ethics even within our own short lifetimes—to which we have to adjust, so that the whole process of the accumulated folk-wisdom that is being passed down from parent to child, and the information that kept the tribal units functioning, just simply is not really a viable situation any more. The father did not learn the awareness that might be of critical value for the survival of the son's world.

Coupled with rapidly changing realities and various value systems in various countries, we have people living now at many levels of ethical behaviour that were useful in the past. There are still Stone Age societies who have such Stone Age ethics as cannibalism. There are people living with the agricultural sets of ethics, such as valuing large families and many children, which are believed helpful to run the farm. There are people living in the industrial sets of ethics. While I was living in Israel and dabbled in archaeology, it was explained that when you go into a ruin to sort out at what level the situation was, simultaneously there would be people living with different types of pots and different kinds of fire-making. If you bombarded Jerusalem today and then, a thousand years from now, tried to reconstruct the civilization, you would find television sets and all modern devices in sophistication located in one part of Jerusalem and, elsewhere, people eating out of crude stone pots in stone houses with little fires and smoke coming out of a hole in the room—all living simultaneously in Jerusalem.

In a similar way, we have to cope with this great diversity of ethics, all existing simultaneously and originally for useful purposes in the long history of Man. So we are having to cope with a great deal of ethical change from technology at the same time as we are introducing these new weapons or technologies which are making us change now. But how do the changed ethics get transmitted, accepted, acted upon and enforced, in time to effect the survival of what we now perceive is the real group—that is, the human species and the other biota on Earth? Can we afford to let the old mechanisms go, and instead transmit ethics of violence on television? We must educate the children to be more sensitive to the environment [and perhaps await] an environmental prophet to come and give us, ready-made, a whole set of environmental ethics and be of such strong character and so fervent that we would all follow.

Sheikh (Panellist)

I arrived only last night due to circumstances over which I had no control and am sorry to have thus missed the earlier sessions of this very interesting Conference. Attendance this morning has convinced me of my tremendous loss due to late arrival. Now the inclusion of botanists on this panel suggests that they are particularly qualified to speak on the ethics of biospheral viability, for they are concerned not only with the structure and function of plants but also with their role as the basic producers in the biosphere, whose viability should be the concern of all human beings.

Religion teaches that God has created everything and one has to understand and appreciate the beauty of God's creations and to realize that there must be a purpose behind them all. Often we may not understand the functions of the components of the various ecosystem's of the biosphere, but although we may not have the capacity or the kind of knowledge needed to understand their functions, we do realize the golden principle of ecology—that everything is connected with everything else—and yet modern Man has apparently forgotten this rule of Nature and has unwittingly tried to interfere with Nature. This approach has put Man today on a collision course with disaster: we just cannot afford to forget the maxim that 'you cannot command Nature except by obeying her.'

Religion is a tremendous support to the people of the developing world, particularly those who are living on the bare subsistence level or even below that. Faith stems from religion. The faith of the poor people is like a shield to guard them against the odds they have to face in life. If properly guided, these people in the developing countries can surely be made to see the role they have to play for their own betterment. We keep talking about the inflow of materials from the developed world to the developing world, but we must realize that the people in the developing world have to make their own efforts also to improve their lot. I come from a developing country and I think that unless and until we try to help ourselves we cannot improve our situation.

The ethical approach to the problems of mankind today emphasizes that every individual is to be respected, and [should] be given equal opportunities. The lack of equality of opportunities in the different parts of a developing country has led to the problem of uncontrolled urbanization. This is what we are facing in Pakistan today. According to the 1972 statistics, 26% of Pakistan's population are living in urban areas, and in the light of the past four years' experiences it has been projected that by the year 2000, 66% of our population will be living in urban areas. We are all very familiar with the problems attendant upon uncontrolled urbanization.

Ethics of biospheral viability to improve the future of mankind also emphasize the need for the curtailment of the scourge of illiteracy and the encouragement of mass education about environmental care.

Another point that I would like to lay stress upon here is that the economic considerations do not always determine the pattern of development. The traditions of people have also to be kept in mind while talking about development in a certain region. Very often the development projects in the developing countries have come to grief because of a number of human obstacles. These obstacles may lie, besides at the economic, national or international levels, at the socio-cultural level.

Nowadays the tendency is to talk about large numbers rather than about the individual. I think that proper encouragement [should] be provided to the individual right where he is working, and he [should] be given due respect as well. With the removal of ignorance, with concerted efforts to teach people not only about the environment but also to make them realize that they have to help themselves, there could be tremendous improvement in future—otherwise the future, as has been said before, seems to be dark unless we take a very optimistic view such as we are trying to [maintain] here.

Buchinger

This time I would like to speak not only for Argentina but for Latin America in general, and share with you a song of an Indian tribe in Amazonia. Latin Americans express great concern for the wise use of their natural resources and preservation of their scenic wonders. This concern may take the form of an agreement among several countries to protect the Vicuña (*Vicugna vicugna*), or it may be expressed in a simple song such as that of the Huitoto Indians, of which the words run something like this:

> Fruits of grasses and trees,
> Animals that sustain us,
> Fish, birds which fly low and high;
> May our Gods keep them,
> So that our children can survive.

Goldsmith

The ethical systems of hunter-gatherer societies truly reflected basic evolutionary principles. For this reason it is difficult to avoid regarding hunter-gatherer societies as climax social systems—perhaps the only ones possible involving Man, as they represent the only human social systems that are capable of living within a climax ecosystem without destroying it.

For an ethical system to reflect evolutionary principles, I think one must identify God with the evolutionary process which after all is a very reasonable thing to do. God is said to have created the world, [so] is *the* evolutionary process. Hence the two can be identified. In a sense Gandhi implicitly made this identification. If we do so of course a religious man can then no longer countenance the present systematic destruction of the biosphere, which it has taken God 3,000 million years to develop. In a way it is surprising that the church has not seen things in this way and joined the ecological movement.

Miller

We see in modern times and maybe throughout all history that one of the things which takes the place of a personal ethic or a personal religion is a sense of nationalism. If you look at the school systems of any country, benign and intellectual as they may be, or [yielding] as they do an academic product, they mostly are very nationalistic in what they are providing, and as a result we have nations instilling a strong sense of patriotism in their citizenry. This gives us, on the world level, a complex of nations. When they get together in the United Nations they appear to be rather a club of rival nations, each trying to express its own opinions, get its own share, and do [all it can in its national interest rather than acting as a committee for the common global good. It must take an act of great diplomacy and leadership to entice committee results out of that kind of a forum—which is one reason why the whole effort seems to be largely wasted nowadays, though there are of course others].

Recently I had the opportunity to visit a number of schools in the People's Republic of China. I was hoping to see there examples of non-competitive education, meaning education towards attaining goals without personal competition against other persons, corporations or systems. I think they are attaining this in a way but I did not find what I was looking for. Dealing in a nationalistic sense in our education programmes, we are getting too much competitiveness which then builds into our sense of what a corporation must be and how we must act in harvesting or exploiting the global wealth.

In the discussion of ethics, as we see it here, we must go back a long way to find where we ought to be. I think China is trying to do this and perhaps it is an experiment in going back to determine the carrying capacity of the Earth. In that way they are extending the carrying capacity to the utmost to take care of a dense population. Eventually, population must come into balance with productivity. But it will take [extensive] experimentation for us to find the balance—to determine how far we are willing to go back from where we have been, in personal comforts and exploitation and consumptiveness, to levels where the land economy can sustain us properly.

Juel-Jensen (Chairman)

There was a question, which we will take last, and there was a statement, which I am sure people have strong views on. The question, if I understood rightly, summed up in one sentence, was: 'Is it possible to de-industrialize the world?' The statement concerned the propriety of our ethics. Has anybody on the panel got any views on this?

Dodson Gray, Elizabeth (Co-Keynoter)

Whether or not it is possible to de-industrialize the world, in an interesting way Darwin's theory of evolution reinforced the Judaeo–Christian myth of Man Above. Originally, evolution was seen as a great insult to the Christian community: it was very much resisted because part of the whole Christian myth is that Man is a little below the angels, and the rest of the Earth is set below his feet. But when [the theory of] evolution originally came along, it was a great threat to that view of Man Above. That we might be descended from monkeys was a horrible thing for the

Christian faith to think about, and so Darwin's teaching was strongly resisted by the Church.

But in a strange way the basic idea of evolution was domesticated. Instead of viewing reality as coming *down* from God by a creative act setting up the hierarchy, God was taken out, an evolutionary process inserted in God's place, and the hierarchy was seen as starting from *below*—with the primordial soup, and coming up through the evolutionary schema and ending with Man in the same position, namely, Man Above—the most complexly developed species and, we are convinced, the most brilliant.

In the average citizen's mind, this is the basis of the confidence in technologies that Dr Borden referred to. We are sure we are the smartest species. The ethic embedded in this is that, as we are the smartest species, what we do to the Earth will work out all right. The religious picture of Man's status coming down from above, and the evolutionary account of Man's status emerging from below, were now harmonized and served to maintain the picture that Man is at the top of creation and is the culmination of the evolutionary process.

Juel-Jensen (Chairman, interrupting, there being some unrest among the audience)

I would like to use my prerogative as Chairman and say that the view is perfectly tenable, that Man is the most highly developed creature on Earth. This is all right, so long as we remember that the freedom we have been given also carries with it an enormous responsibility, because freedom without responsibility leads to chaos and anarchy. I do not see that there is anything wrong with the idea of considering Man the most highly developed creature on Earth, but he has the freedom to make a mess, not only of his own life, but [also] of that of other creatures on this Earth, and this, I believe, is the dilemma. Incidentally, I happen to believe there are angels.

Dodson Gray, David (Co-Keynoter)

It is not at all clear that Charles Darwin, out of his writing and thought and study, saw that evolution led to a hierarchy. What is very clear is that, in the social working of Darwinism and by many of his followers, evolution was viewed as validating human supremacy over Nature; they made a jump from being 'most developed' to, therefore, 'Man necessarily above'.

Royston

While we are talking about evolution in theory and ethics, I am very struck by recalling the profound works of Teilhard de Chardin, the palæontologist and philosopher, who in a number of his books put forward very cogently the view that the path of evolution—certainly the path of evolution as far as Man is concerned—has moved away from Man, as an individual, being the essential unit of evolution, to the idea that the social group or community of Man is the unit of evolution. The implications of this of course are profound. It is particularly profound for the personal individual ethic, because essentially what Chardin is saying—and I think that we have heard this echoed many times this week, from many points of view—is that the fundamental ethic which we are looking for is a community ethic to replace the individual ethic. The individual finds his fulfilment within the community. His ethic is then based on what he *gives* to the community, rather than what he takes from the community, and the community ethic is based on what the community gives to its life-support system rather than what it *takes* from that life-support system.

In this respect Garrett Hardin's *Tragedy of the Commons* has been mentioned several times. It is interesting when one looks at the Alpine communities, which in fact have had the commons as their basic resource. They traditionally avoided the tragedy of the commons because the management of the commons is a matter for the total community—it is not a matter for the individual. If those commons were exploited by individuals, of course that would lead to tragedy. But the Alpine community, which led to the basis for the whole Swiss political and democratic system [which is so widely admired], avoided [such exploitation by individuals] because they had in fact developed a community ethic to group survival in a very difficult environment which was based on what they contributed to that environment—what they gave, and not what they took.

Ducret

Being a biologist by training, I am not very familiar with [the subject of] ethics. However, I will risk a comment on [the so-called] pyramidal structure which, according to her, Mrs Dodson Gray, had its origin in the Judaeo–Christian religion. But does it not have a much earlier origin?

Early human societies were hunter-gatherer groupings with apparently limited abilities to survive in an adverse environment. They probably had a strong hierarchical structure based on experience and leadership in various activities. The leaders of these societies were probably 'circumstantial', i.e. they were head man or woman for one given social activity and only as long as he or she was the best in the group. This circumstantial leadership created a flexible pyramidal structure which became, much later on, crystallized through the [development] of agriculture and its consequences, sedentarization and creation of heritage: people started to inherit land, plants, resources, and, progressively, leadership.

I am therefore not convinced by Mrs Dodson Gray's [contention that the] pyramidal structure resulted from the Judaeo–Christian religion, [but feel rather that it may have] evolved originally from much earlier circumstances in mankind's history which are still recalled in our behaviour towards the environment: we are indeed hunting the environment [and its products] rather than managing it. This means also that this behavioural pattern is a very [fundamental] part of our being, and that modifying it [would be] an extremely difficult—if not an impossible—task.

Dodson Gray, Elizabeth (Co-Keynoter)

[Dr Ducret] may [well] be right: it may have deeper roots. The crucial item, however, is the understanding of the soul. The Judaeo–Christian tradition was concerned to say that humans had souls, but that plants and the animals did not. This is very different from other religious traditions. Our colleagues from Ghana and Japan are careful to point out that, in their religious traditions, 'soul' was *in* Nature, and that therefore there was not this division or layering or even any separation between the 'soul' which humans have and the soul—or quality of being alive, if you will—which the natural environment has.

It seems to me very crucial for Western society that, in keeping with the Judaeo-Christian tradition, 'soul' was seen as related to the ethereal spirit of God above (imaged as 'in Heaven'), and the Earth which had no soul was related to a Hell which was imaged as 'underneath' and [so part of] the grubbiness of matter.

My position is that the Greek split or dualism between mind and matter was able to come in and make such inroads in our Western civilization because we already had this hierarchical structure of thought in place in our religious heritage.

Crabb

I want to address myself to the notion that somehow or other, because of our technological explosion or revolution, there is a concomitant explosion or revolution on the ethical plane. I think these are two completely different planes that in a conceptual sense are unrelated. But of course we have our differing—evolving, if you will—social malaise which changes with the technological and other aspects of the human environment. Yet that is not, I think, a question of ethics. [Rather do] I think that its is based upon the essentials of human nature, and that it is effectively demonstrable that human nature—from all the information that we can gather about it during our 5,000 recorded years of history *plus* the anthropological and geological regressions into the past—does not itself change. I think maybe a simple example would be in literature. The plays of Shakespeare, for example—some 350–400 years old now—are still packing the theatres where they are played. Also you can go back to the ancient Greek plays which are still, some of them, extant and still successful in the theatre.

It seems to me that this [continuity of human nature] is the kind of thing with which ethical concepts are concerned. We have the same human reactions of hate, fear, love and revenge, which are the grist of most of our theatre, and which we can relate to just as well as [did] the Greek societies of two to three thousand years ago—because the constant element has not changed. We are trying to relate that somewhat to the more specific, pragmatic or concrete concerns of a congress such as this. It does not at all quarrel with the various suggestions that have been made by way of how we should handle the environment and so forth, but I think it is detracting a great deal from that effort if the ethical concept [is brought] into it. It is a question of applying these ethics, or answering these ancient but always current philosophical questions, to the new facts which have evolved out of technology.

Now granted that technology has been evolving new problems of application much faster [in recent] than in previous periods, [it still] does not mean that the fundamental ethical equation has varied one iota. Another way of putting it, which is quite well known, I guess, is to say that the Greeks posed all of the philosophical questions which we know of today—that the philosophical problems are merely the application to the ever-changing facts of human society of these philosophical postulates of the Greeks—[whereas] the philosophical postulates as such have not changed. [As for Chardin who] was mentioned a moment ago, I do not know to what extent he is actually influential in the current philosophical establishment, but I suspect that it is just an updating or contemporary application of the Greek philosophy.

Juel-Jensen (Chairman)

One thing I would hope, in a gathering such as this, is that at least we would rid ourselves of a good deal of cant. One of the speakers talked about Stone-Age Man—I do not know why it should be Stone-Age Man—eating somebody else. We complain because bits of the Amazon Forest are removed, and yet we do not complain when a madman eats the liver of his Minister who has been dismissed. This is another blot on this Earth which ought to be removed. It is often said that you must keep politics out of the management of the ecosystem. I think this is very difficult to do and, if you are honest and if you have a set of ethics, inevitably politics sometimes will enter into it. But I think probably [our panellist] who should respond to what has just been said from the audience, is Dr Fosberg.

Fosberg (Panellist)

I am really not at all competent to discuss ethics in some of the terms that have been brought out here. I think perhaps much of what we have been saying is an elaboration of rather simple ethical ideas and beliefs, and I agree that the Greeks did express all or most of them. If I were required or asked to express my own ethical feelings or beliefs [at all] succinctly, I would have to throw something into this discussion that is—or would probably be regarded as—quite extraneous.

Juel-Jensen (Chairman, interrupting)
Please throw it in! [Amusement.]

Fosberg (Panellist)
I would say that we could be well guided, ethically, to avoid as far as is at all practicable, the diminishing or destruction of beauty [Applause] in any of its various forms. We must think about why we feel that things, scenes, personalities, and other aspects of life, are beautiful. This is certainly a product of evolution, too.

Polunin (Jr)
The Dodson Grays have listed social injustices as one of the major issue areas for ethics in environmental concern, and there has been a lot of talk also of the importance of developing some kind of social ethic. I would like to present the case, however, against the simplicity of such a question. We have heard that, before we can make progress in combating environmental problems, we must put our own 'social house in order', but it is clear that such a correction is vastly more complex than some would have us believe. On one hand we have heard from Mr Glasser and others that one of the sources of poverty in modern society is the breakdown of traditional communal ties between individuals. On the other hand there are groups who are fighting for what is essentially a complete breakdown of ties and interdependence of individuals within [our existing] society. Such are, for example, the extremist proponents of female 'liberation', who believe that to develop our individuality we must essentially break the ties between individuals—because such ties mean dependence, and dependence means exploitation of one individual by another [they claim].

I would like to submit that this latter ethic is yet another case where the developed world is attempting to thrust its ideas on the underdeveloped world. There are [too] many cases where people have gone to developing countries and have seen women, for example, performing certain functions which are only performed by men [in the overdeveloped countries whence they come], and the assumption automatically is that the women are being exploited by the men in the society. Concordant changes are made to try and adjust the situation, but the outcome is not always satisfactory. Women are often quite happy performing the roles which other members of society—particularly the males—are unable to perform, and such dependence I think is a very important part of the cohesion of the community as a whole. [To me this, and anything else which derogates from the stability of existing ecosystems—especially climax ones—is patently unethical.]

Franz, Eldon
A fundamental concept of human ecology is that of the adaptive strategy. Briefly defined, an adaptive strategy is a system of cognition and behaviour that maximizes well-being and minimizes the hazards to survival of human populations. In approaching an adaptive strategy, there are three essential elements, of which the first is our understanding of habitat, the second is our understanding of the nature of Man, and the third lies in our behavioural codes. The specific role of behavioural codes in an adaptive strategy is to translate Man's perception of himself in relation to his habitat into rules for resource management.

Behavioural codes have been of many kinds, and among them are characteristics of cultures which we ordinarily would consider to be within the scope of religion and ethics. But modern societies also have legal systems, and I would like to link where we are right now to the what will come [this afternoon] concerning legal systems. Within the context of the same general ethic we have seen, over the last 200 years, the incorporation into our legal system of the so-called external costs of many of the kinds of changes that have been made by technological man. Among the most important of these are the cost of labour, which was internalized when it was made basically impossible for people to hold slaves or to exploit labour as was done early in the industrial and agricultural revolutions. A second [such advance] is internalization of the costs of hazards to industrial workers—the actual health-hazards of exposure to toxins in the industrial environment, and so on. We are still [only] getting around to doing this in many areas—for example, we still have difficulties with black-lung disease in mining situations. A third level is coming round just now, namely the internalization of the costs which have been incurred through environmental degradation. So I think that this all is really occurring within the context of the same basic ethical and moral environment, and that these are basically legal charges.

Juel-Jensen (Chairman)

Professor Bryson, I am sure you must have some succinct and [pungent] comments to make.

Bryson

Of course I do, but I was trying to avoid making them. I am always a little bemused by listening to scholars getting themselves into a semantic morass when talking in finite terms about abstractions, and I was delighted with the summary that my colleague made. In fact, I was amused at how Man could *descend* from the apes to the *top* of a pyramid. [Laughter.] How you get up by going down gets into the realm of relativity. Personally I am a top carnivore and I enjoy it.

Rudolph

Someone has made a point biologically about thoughts, and I suppose ethics—that bad ones take as much energy as good ones to think out and imagine, and that therefore in a sense all these are equally selected by evolutionary processes. A thing about Man that [makes him] a little different from most other evolutionary organisms is that he accumulates things through tradition, and also he can predict the future a little better than other organisms. Therefore a bad thought [or tradition] may take a long time, as may a bad ethic, to be eliminated by selection. So I think we have got to make the point that a bad ethic is not going to be eliminated right away, unless we have an immediate ecodisaster [to take care of it].

Fosberg (Panellist)

That is [practically] what I meant by our not having the time to develop a new set of ethics.

Savin

This is a little outside my sphere, Mr Chairman, but when I hear ethics and morality spoken about, I take it the ordinary man in the street has a contribution to make. I have not had the advantage of reading the paper but when I heard the exposition it did seem to me that basic principles were becoming fudged and relativism was creeping in, and from this point of view I was very encouraged to hear Dr Sheikh's commentary on the theme. My worry is that evolution throughout the ages has taken place from basic principles. These have been elaborated, developed and expanded, and in this way it seems to me order of a sort has emerged. What we have been hearing of, or sometimes have [even] heard [advocated] during this Conference, has been revolution, and in that way I firmly believe lies chaos.

If anything was learnt from the eighteenth century, it seems to me, it has been the arrogance of human beings and the aridity at the end of the day of relying on human reason to bring ourselves to truth. And I hear something of this today. I am perhaps old-fashioned and naïve but speak from a committed Christian standpoint, and I find it difficult to relate this to the use of the term 'myth'. I think that we should to some extent take cognizance of our brothers in developing countries and perhaps revert from our all-human wisdom to the divine and seek a source of revelation, which seems to me preaches the uniqueness of the individual in the pattern of creation, and from this develop, as Professor Misra told us this morning, a respect for all forms of life. [Indeed] it seems to me that, unless and until we revert to an acceptance of the divine in our discussion, Man will have no constraints in dealing with the environment and this must be to his harm and that of the environment in which he lives.

Juel-Jensen (Chairman)

Shall we let Mr Goldsmith have [another] go? [Audience assents with amusement.]

Goldsmith

I am open to physical constraints like everybody else, but have one or two small points. First, in view of the fact that Man is above all [other creatures] as an ultimate carnivore, he is infinitely less important than the primary synthesizers on which he depends [, and so] you can remove the carnivores. If you want to establish the relative importance of the different forms of life in an ecosystem without any question whatsoever, it is the top carnivore that is the least important. If you remove the top carnivores, you just get increasing, slightly [bigger] oscillations, if you remove the second range of carnivores, you get [still] bigger oscillations; but if you remove the primary sythesizers, the plankton and the grass, the whole thing disappears. [Similarly with the world if you remove the life-support system, everything will wither and starve to death.] So Man is, in fact, considerably less important than the grass which feeds the herbivores which he [in turn] feeds on.

Secondly, we were talking about the Judaeo–Christian tradition and the idea of God. As far as I understand, the principal difference between our type of religions—the so-called upper religions of Christianity, Judaeism, Islam, etc.—is that our gods are largely asocial and they have also left the ecological systems in which they lived. In tribal societies, whether they be African or Asian, etc., we find that gods are in fact part of the community. A god is usually so; in fact the Supreme God plays only a very small part, and there are practically no cults of supreme gods in tribal societies, where they are regarded as the moulders, the creators, but nobody takes much notice of them.

The real gods in tribal societies are the Nature Gods or the Ancestral Gods, who maintain their position within their tribe; their total identity is as members of the families, as members of communities, and they take a tremendous interest in normal communal matters, their main function being to maintain the traditional normal culture pattern, which they do very well. What is important I think is that, as normal tribal societies have collapsed, so have the stable relationships which they succeeded in maintaining with the environment. So, you see, the accent has shifted to these Supreme Gods, and we have desanctified both the social structures and the ecological structures which they previously sanctified. What we have actually done is to desanctify Nature, to desanctify social structures, and to sanctify Man in their place. This is where we really have gone wrong.

One last thing I have to say is that this whole notion of sanctification is a very important one, which one can again examine in cybernetic terms. What is sanctification? It means, in cybernetic terms, that information which has been sanctified [becomes] non-plastic—you cannot modify it. And if the goal of all behaviour within the biosphere is increasing stability, it is quite clear that no stable behaviour can be based on unstable information. If information is organized hierarchically, the generalities must be more stable—hence [less plastic]—than the particularities. You [can] change the particularities—not the generalities. If you were to change the generalities, you would be adapting to freak situations the whole time. In the same way, it is extremely important that genetic information should be non-plastic, and just as important that cultural information or the generalities of a cultural information pattern should be non-plastic too. Therefore, when scientists suggest that behaviour should not be based on non-plastic generalities, or ethics, they are in fact introducing instability into human behavioural patterns.

Dodson Gray, David (Co-Keynoter)

The [section of our] paper by Professor Fukushima of Japan was a very interesting documentation of some of the things that Mr Goldsmith has been talking about. He recounts the experience of rapid industrialization in the fifties and sixties, and he concludes with a rather wistful concern for exactly the sort of things we have been talking about. I commend his very interesting but brief paper to you.

Juel-Jensen (Chairman)

Our time is running out but is anyone else on the panel bursting?

Thorhaug (Panellist)

I would like to make a very brief remark about the Darwinian view of evolution and ethics. I believe we can revive Darwinism, eliminating its hierarchical nature (with Man on the top), and looking at Nature simply as an evolving system—as a whole system, complete from the microbes to the higher carnivores. We might take his basic principle of 'survival of the fittest' for the whole ecosystem and eliminate the 'pyramidal' concept of the original Darwinism, thus having a useful concept for long-term survival which I think biologists and concerned citizens [could] use.

Fosberg (Panellist)

I do not think I have anything of much importance to say beyond what I said in my last comment, except perhaps to point out that what we are dealing with here is the world ecosystem, and I think that anything which tends in any way to degrade or impair this ecosystem [should], from the broadest possible view, be regarded as unethical. [Prolonged applause.]

Dodson Gray, Elizabeth (Co-Keynoter)

I would like to clarify that I do not think we have to throw out all of religion or even our Judaeo–Christian tradition—[rather] that creation was erroneously imaged as hierarchical in the Genesis myth in the Old Testament. This hierarchical picture of the nature of the universe was carried on [traditionally] in such a way that many theologians throughout Christian history viewed Nature as a mere backdrop—a stage upon which the drama of salvation was being played out. And in that too-narrow view, salvation was seen as basically a transaction between God and Man involving the soul—rather than salvation involving God and the world and encompassing all of life.

I do not think Nature has to be seen as a mere backdrop for human history. We can have a religious sense of the Creation in which, yes, the human species is unique, but also, every [other] species is unique. [As our Secretary-General has said somewhere, other biota on Earth have an equal right to live.] I would submit that the human species exists *within* the natural systems, not above them or apart from them, and that the human species *is* limited and must fit *within* those systems, as Dr Willard suggests.

Dodson Gray, David (Co-Keynoter)
—and within our heritage of Judaeo–Christian religious groups there is a great sense, not only that there was individual salvation waiting for fulfilment but also that all creation is groaning and travailing and waiting for this fulfilment.

In our deliberations here it has been abundantly obvious that a very important ethical resource which needs to be conserved is time. With sufficient time, ecosystems can recover; if consequences come quickly after action, there is time to mend our ways. It is when we waste time, not by being frivolous but by doing things that have long-delayed effects, or when we do things to systems that have a very long recovery-time, then we are asking for massive trouble in the biosphere.

Juel-Jensen (Chairman, concluding Session)
I think you will all want to join me in thanking the keynote speakers and our panellists [, as well as other participants and an exemplary audience, for what has been a provocative and interesting session which is hereby closed]. [Applause, followed by announcements about films concerning Iceland and an evening tour of Reykjavik.]

[Earlier, the Secretary-General had stressed that the papers of the Conference were the copyright of the sponsoring Foundation for Environmental Conservation and, following consultation with the President and others, that, the Conference being of necessity apolitical, participants were recommended to ignore some political-partisan papers which outside bodies were attempting to introduce—apparently on the assumption that the Conference was something to do with NATO and/or the United Nations.]

Session 17

COMMON LAWS FOR EARTH AND MANKIND

Keynote Paper

Common Laws for Earth and Mankind: A Glorious Hope?

by

GARY L. WIDMAN

Professor of Law and Director, Natural Resources and Environment Law Program, University of California, Hastings College of the Law, 198 McAllister Street, San Francisco, California 94102, USA; formerly *General Counsel, Council on Environmental Quality, Executive Office of the President of the United States*

&

GUNNAR G. SCHRAM

Professor of International and Constitutional Law, Faculty of Law, University of Reykjavik, Reykjavik, Iceland; Chairman of the Icelandic National Committee of the 2nd ICEF; Chairman of the Icelandic Environment Protection Committee and *National Adviser to the United Nations Conference on the Law of the Sea.*

NOTE

This Conference was originally scheduled to be held in 1976. At that time the keynote paper of Session 17, 'Common Laws for Earth and Mankind', was to have been given by H. E. Ambassador Edvard I. Hambro, of Norway, whose premature death saddened the world community of those concerned with Nature and with mankind's role on the Earth.

Ambassador Hambro had been a member of the Norwegian Storting, Chairman of the Board of Appeals of the Council of Europe, Member of the Permanent Court of Arbitration at The Hague, Ambassador and Permanent Representative of Norway at the United Nations, and President of the United Nations General Assembly. At the time of his death he was Norwegian Ambassador to France. His deep interest and effective work in conservation of the Earth's resources were widely known.

The authors are honoured by the invitation to prepare and present this paper, but greatly saddened by the circumstances of the invitation. We shall feel that we will have succeeded if we can bring some fractional part of Ambassador Hambro's insight to this important issue of harmonizing Man's laws with Nature.

GARY L. WIDMAN

GUNNAR G. SCHRAM

INTRODUCTION

In one sense mankind is like rain, silt, redwoods or whales—it was shaped by Nature's processes and at the same time is part of those processes. Man was not the engineer of those forces that produced him nor, in any intentional way, was he an architect of his own design. Man must look back, if he is to understand the birth of his species, to the soft green fabric on the lands of the Earth. He must look back still further, to Cambrian and Pre-Cambrian tides, if he is to understand the genesis of his animal ancestry. And he must also look back if he is to understand the forces and limits in his dynamic future on a finite planet.

Man may understand that his body reproduces on the rhythms of the moon and tides, that it makes and carries the saline water of those tides in its veins, that it is supported by a framework of minerals, and that it lives by combining oxygen from air with fuels torn from the natural fabric around him. He may see that he was precisely engineered for this Earth, but that his own body is an artefact. More precisely, his species was 'engineered' to use the renewable life and resources of the Earth at the time when the species evolved. For his body still needs the air, water, minerals and fuels that were prescribed by the natural processes of past eras. But Man learned *some* of his natural lessons well, and was able to stave off Nature's usual checks on population. Now, of course, there are some thousands of millions of people who need those same substances for their lives, and mere hundreds of those millions (in the developed world), use a larger share of the Earth's finite resources for their individual comforts and collective enterprises than do others.

In the past, Nature's limits forced Man into retreats. Natural boundaries of advancing ice and surrounding ocean at times restricted populations to land areas and food-stocks that were inadequate. Droughts, floods and overpopulation have in our own time caused suffering and death for many. Nature's ways of confining populations to fit limits often involved starvation and bitter conflict over lands, food and resources. Survival sometimes came at the cost of lives of the weak and unlucky. At other times it came through the sacrifices of some members of the threatened populations, or through the sharing of surpluses from other areas.

Man's study of his own physical history, and of the Earth around him, now shows him that he has approached, and may even have surpassed, some of the Earth's limits in supply of food, resources and climate for growing populations, with their present life-styles and short-sighted technologies. Now, projections of world food supply and demand show us that each year there is less surplus to share. The only real option for survival will be one of coordinated public and private decision and action, accommodating the limits of land and life-supporting resources. The power of the law to alter these decision equations may make a substantial difference. Law permits Man to design decision structures that may offer him a *limited* opportunity to be the architect of *part* of his own future. It may permit him a choice other than the classic, often cruel options of Nature.

The scenario of an animal population expanding beyond the carrying capacity of its territory has been frequently described in ecological literature and exemplifies one type of law contemplated by the assigned topic, that is, the 'Law of the Earth' (*see* Brown, 1978). The harsh consequences predicted by some of the Earth's laws are rooted in governmental, corporate and individual decisions that can be influenced by Man's laws. Consequently this paper suggests some ways of avoiding those effects, by guiding Man's law (and the decisions made under it) into a working partnership with the Earth and its laws.

The authors believe that the topic prescribes a critically important goal for Man's law; but without an understanding of the differences between the 'laws' of Man and Nature, its prescription could sound deceptively simple.

The Word 'Law'

The word 'law' has different meanings in the two contexts described in this paper's title. If different-sized metal balls are observed to fall many times from the Tower of Pisa at the same rate of acceleration, a scientific 'law' predicts that other metal balls will fall at that same rate in the future. 'Laws of the Earth', then—such as the 'law' of gravity, or principles describing consequences of populations exceeding the carrying capacities of their habitats—are shorthand statements of deductive logic predicting future events, based on repeated past observations of similar events and on assumptions that the forces driving these events will not change. The 'jurisdiction' for such scientific laws is the measurable *physical* universe. Any repeated event or process that can be seen or measured might become the subject of a 'law of the Earth'.

In one sense, the 'laws of Man' differ in that their scope is more confined. They do not comprehend the full universe of measured physical events. They can affect the future only through their impact on decisions by individual human minds. These laws work by altering the community and government responses to individual human actions and decisions. They may impose disincentives, such as taxes, or may use more direct enforcement by fines, prison terms, civil liabilities, or judicial orders halting or mandating certain actions. Alternatively they may work through incentives, such as specific financial incentives, or through social rewards for compliance.

Man's laws may not reflect rigorous physical observations of the past, and may not assume that the future will follow past patterns. A law may reflect its author's random or detailed observations of physical events, or his appraisal of the intangible values of culture and community in the future that he is trying to create. Instead of predicting futures that are unchanged from the observed past, Man's law describes futures which are *intended* to be different from the futures that would follow in the absence of that law. It is this very need to *change* the probable course of future events that makes human laws necessary. Therefore, a law's effectiveness is determined by its author's understanding of past and probable future events, and by his ability to use the limited arsenal of legal powers to change that future.

In a second sense, Man's laws are less confined. They need not focus on the physical world but may operate in the worlds of will, creativity or spiritual, artistic or reasoned thought, where they may encourage or dampen the spread of any ideas or values that move men to decision and action.

Man's laws, then, may focus on the universe of intangibles, and that is their strength. But it is also their limitation. Man's laws structure decisions for the individuals who live under them, but the laws do not make human decisions. Man's laws can be ignored or intentionally broken. They can be invalidated by courts and amended or repealed by legislatures. Such laws then become paper artefacts memorializing temporary and unacceptable goals. They failed to change futures as their authors had intended. So it is not enough to enact a law, and assume that one can build upon it in the next day's work, as one sometimes can in science. Man's laws cannot be effective without common understandings of problems, of the needs for change and of solutions that are acceptable to both the governed and the governing. The effects of Man's laws, then, must be constantly monitored to see if in fact conduct is changing to construct the futures that were contemplated by their authors.

In still a third sense, both types of law are similar: they are adjectival. Thus they describe and articulate structures and limits for Nature's processes and for Man's decisions, and thereby make futures understandable. They provide the framework for new insight, and they permit predictions of the consequences of disregarding both types of law.

Both Man's and Nature's laws are composed of elements of description and prescription, although for scientific laws the goal is often assumed to be purity of description. But when one reads the scientific treatises of a century past, especially in biology, one can perceive just how far Man's images of himself, and of his gods, coloured what he thought to be objective scientific description. Man's laws, on the other hand, are sometimes thought to be purely prescriptive, as when they tell the citizen not to do something he had contemplated. But when one reads the history of laws that were dramatically prescriptive—that required abrupt and unexplained changes in long-established patterns of conduct—one realizes that part of Man's law must also be descriptive if it is to be effective and survive. This means that it must realistically describe or assume the community's values and its understanding of the future.

For these reasons, any implication that the two types of law could be entirely 'common' cannot be literally accurate. Nevertheless one goal of a human legal system could well be a human future in harmony with the Earth and with its laws.

Man's law alone cannot create the human energies, character or knowledge which it seeks to channel, any more than a law of the Earth creates the natural processes which it describes. But those natural initiatives and energies that began in the green fabric of the Earth, will eventually create our futures in the frameworks of both types of law.

The Challenge to Man's Law

With that understanding, we turn to our task. We see the challenge of our topic expressed in these excerpts from Mesarovic & Pestel (1974), even though they did not use the word 'law':

> [The] rapid succession of crises which are currently engulfing the entire globe are the clearest indication that humanity is at a turning point in its historical evolution. And the way to make doomsday prophecies self-fulfilling is to ignore the obvious signs of perils that lie ahead—which indeed are already felt—and rely solely on 'faith'. *Our scientifically conducted analysis of the long-term world development based on all available data points out quite clearly that such a passive course leads to disaster.* It is *most urgent* that we do not avert our eyes from the dangers ahead, but face the challenge squarely and assess alternative paths of development in a positive and hopeful spirit. . . .
>
> In Nature organic growth proceeds according to a 'master plan', a 'blueprint'. According to this master plan diversification among cells is determined by the requirements of the various organs; the size and shape of the organs and, therefore, their growth processes are determined by their function, which in turn depends on the needs of the whole organism. . . .
>
> *[The] options facing humanity contain the genesis of an organic growth.* And it is in this sense that mankind is at a turning point in its history; to continue along the paths of cancerous undifferentiated growth or to start on the path of organic growth. . . .
>
> Will mankind have the wisdom and will-power to evolve a sound strategy to achieve that transition (from unbalanced world growth to organic growth)? In view of historical precedents, one might, legitimately, have serious doubts—*unless the transition evolves out of necessity. And this is where the current and future crises—in energy, food, materials, and the rest—can become error-detectors, catalysts for change, and as such blessings in disguise. The solutions of these crises will determine on which of the two paths mankind has chosen to travel.* [Emphasis in original.] *

* In the pages from which these statements were taken, those authors saw a need for a 'master plan' for organic world growth. We think it is not useful to employ the term 'master plan' in legal writing, as it connotes to some a plan imposed from the top of the government hierarchy on to a public that is not yet ready to accept its burdens. If such a step were in fact taken, it would be an equally objectionable counterpart of the 'undifferentiated growth' which those authors condemn: that is, it could become an 'undifferentiated' restraint on growth which, because of the circumstances of its imposition, would be unlikely to be effective. There is nevertheless a need for a *strategy* to inform and move the citizenry and legal and political institutions in ways that will lead unbiased decision-makers in many settings to choose far-sighted options such as those of organic growth and organic restraints, as described in the book cited. In that way, the law *could* hold common benefits for both Earth and Man.

To us, the 'turning point' of those authors necessarily requires that mankind for the first time find a way to get out of the position of a passive observer of its own growth. Man must recognize the collective consequences of his many decisions that ignore the Earth's limits. These are individual, corporate or governmental decisions that now appear to be unrelated, but that, one by one, inevitably create demands which the Earth cannot satisfy. The law of Man must play an important part, as it is one of the few institutions that can affect the millions of private and public decisions which lead us down the roads of excessive demand and inadequate life-support. But it would be a mistake to assume that law alone will be enough.

The Law of Limits and of Life-support

The law works in many ways, some less direct than others. But various elements of the legal system, as described below, could help move both governments and citizens to accommodate Nature's limits.

A. *International Proposals*

Because natural systems (especially those carrying pollutants) range across international borders, international legal tools are a necessary part of the 'law of limits'. To some extent these tools have been used. Treaties regulating the 'take' of whales (Convention for the Regulation of Whaling, 1931; International Whaling Convention, 1947), and others limiting dumping of industrial wastes into the world's oceans—such as the International Convention on the Prevention of Marine Pollution by the Dumping of Wastes and Other Matter (United Nations, 1972)—are examples.

But Man's relationship with Nature is not covered either by comprehensive treaties or by the rules of customary international law developed through the historic past. Certain general tenets of international law are, however, relevant to the question before us. The maxim of territorial sovereignty is one, and in this context it has a two-fold aspect. First, it guarantees the state full powers within its boundaries, but at the same time imposes a duty on the state not to allow its territory to be used to create a nuisance to other states (cf. *The Trail Smelter Arbitration* 3 UN Intl Arb. Awards 1905 (1949)).

Secondly, the tenet of state responsibility prescribes payment of damages for injurious acts performed by the state or its agents. Thus a state may be liable if it fails to control a source of harm to other states. This rule was reflected in Principles 21 and 22 of the Declaration of the 1972 UN Conference on the Human Environment, held at Stockholm, where it was said that states have the responsibility to ensure that activities within their jurisdiction or control do not cause damage to the environment of other states or of areas beyond the limits of national jurisdiction. States are also charged to cooperate in developing international laws of liability and compensation for victims of pollution and other environmental damage caused by activities within the jurisdiction or control of such states to others in areas beyond their jurisdiction (cf. Brownlie, 1974).

This suggests another approach which could be considered for concerted action at national, regional and global levels, as encouraged by the Stockholm Conference. Some of the Principles of the Declaration adopted at that Conference (especially Nos. 21 and 22) could be incorporated into international treaties.

Given the complexity of environmental problems around the world, it is evident that some conflicts will have to be resolved between states on a regional or bilateral basis. Some such advances have already been made. The Nordic Environmental Agreement of 1974 and the regional Oslo Anti-dumping Convention of 1972 are good examples.

In the past few years, regional organizations have shown increased interest in solving problems through treaties. The Council of Europe has drawn up three such instruments, the latest being the European Convention on Protection of International Fresh Waters against Pollution. The European Economic Community has issued a number of directives for the protection of the environment, on the basis of Art. 235 of the Treaty of Rome, which are binding upon all Member States. NATO has approached the issue through the creation in 1969 of its Committee on the Challenges of Modern Society, which has been active on such issues as marine pollution, air pollution, clean engines and road safety (cf. Huntley, 1971).

Such bilateral and multilateral resolutions of environmental problems through treaties should be encouraged. However, it is evident that regional instruments alone will not solve environmental problems on a global scale. They do, though, assist in the development of conventional international law on the subject, and thus strengthen the jurisdictional basis for other world-wide agreements, as well as for parallel national legislation on environmental matters. If Man is to be a far-sighted custodian of the Earth, he must take measures to guarantee for future generations a clean environment with adequate resources.

Structuring of the right kind of political action through international cooperation is, of course, of paramount importance. But no less important is the creation of the right jurisdictional basis for environmental action. What should be the rights and duties of states in the field of the environment? How are they to behave in their international relations, and where lie the limits of their responsibilities towards other states and the international community as a whole?

These questions cannot be satisfactorily answered unless we have created an institutional framework, which can spell out in considerable detail the rights and obligations of individual states. This is one of the greatest challenges of modern times, and much scientific progress will have been wasted if the fruits of research cannot be directed through the channels that only an adequate legal framework can provide.

But who should be responsible for writing a treaty for the environment? Obviously this is not a task that can be undertaken by any one government but must be approached by the international community as a whole. The United Nations would be a logical choice, since such a treaty could be a continuation of the work initiated by the 1972 Stockholm Conference. As it is possible that the United Nations Environment Programme may not at present have the resources necessary for preparing a draft convention on the environment, the task might be assigned to the International Law Commission, which meets in Geneva. This body has, in the past, drafted a number of international conventions that have later been adopted at UN international conferences. A similar process might be possible here.

An alternative, less comprehensive, approach could focus on particularly critical parts of the environment. One might begin with the 'small end' of the spectrum of life, by considering how a treaty could effectively protect the plankton and the oxygen exchange systems in the oceans. A further possibility might be a treaty designed to protect the stability of the upper atmosphere—not only from fluorocarbons but from other substances that may be found to have the potential for causing damage. Both might operate by establishing an international structure for regulation, empowered to act on the basis of any newly-developed knowledge, without waiting for negotiation and ratification after each new technical or ecological discovery—a concept roughly following the prototype of the International Convention on the Prevention of Marine Pollution by the Dumping of Wastes and Other Matter (IMCO, 1972).

B. National Proposals

National statutes might also limit demands on the Earth's life-support systems. Some of the principles of the Declaration of the 1972 UN Conference on the Human Environment (*supra*) would be appropriate for national statutes. Energy conservation laws proposed in various countries are also examples. Such proposals include, among other measures, increased prices of energy to discourage demand, and tax incentives for the installation of solar heating and improved insulation. Similar incentives or restraints on demand for other resources might be fashioned as well, but a complete catalogue of such proposals, and of other possible life-support legislation, would be well beyond the scope of this paper.

However, we have picked two issues as examples of opportunities for statutory change to protect the long-range interests of the public. In the absence of effective legislative effort by other professional groups, we believe the legal profession has special responsibilities for these two issues. Both involve obstacles that are peculiarly legal, and both require legal creativity for their resolution.

First, building codes, in this age of scarcity, could be revised to bring our constructed environment and the activities that use it more in line with natural limits. In the United States, the legal framework is found in local building codes governing private construction and in federal and state regulations of governmental construction. These laws often limit the options of the builder, in some cases through explicit prohibitions, in others through cost factors which require or at least encourage resource-consumptive building practices. Model building-code provisions could be drafted to encourage the use of renewable energy systems (including solar and wind) and, generally, to encourage the use of resource-conserving equipment (such as heat pumps and alternative toilet systems). One commentator has suggested that improved design of new and remodelled structures could save one-third of current US energy-use by 1990 (Lovins, 1977).

The second example is protection of prime argicultural land for the long-term future, in countries such as the United States where it is not so protected at present. A recent Rockefeller Brothers Fund report (Barney, 1977) has called for protective zoning and tax assessments for such land. But zoning, at least as it is typically practised in the United States, is probably not up to such a rigorous institutional task. Land may be taken for public projects or re-zoned for other uses with little or no regard for the incremental effect of its loss from the critical inventory of land resources or for its effect on food supplies. And new highways and other developments often carry with them burdens that must be borne by agricultural land in the area, land which can be quickly re-zoned. The history of zoning in the United States shows that re-zoning comes easily when the land market determines that some new use will produce more short-range return than the use zoned. The need is for a stronger, but at least as equitable, institution which can assure that the farmlands, which will be needed more with each passing year, are not lost today.*

Proposals for laws protecting life-support could benefit from new perspectives in their presentation. Many life-support problems involve the cumulative effects of thousands of individual, corporate and government decisions, where the perceived requirements of defence, profit or comfort (for example) are clearly focused and balanced against the unfocused priorities of life-support. When weighed in such balances, long-term life-support needs have been consistent losers. The problems of altering decision practices, where the short-term needs of the decision-maker are balanced against the longer-term needs of the community, have been well described by Hardin (1968). In addition to that problem, there is little information available to decision-makers, or their staffs, that puts the effects of their decisions into a far-sighted perspective. They need appraisals of the limits of specific life-support resources, and of the incremental effects of their decisions on those resources. Only when that information is used, will we change the patterns of incremental effects of these decisions.

Finally, there is an opportunity for some judicial role through statutes prescribing general principles of judicial decision-making which courts may, or must, use when deciding cases—see the general discussion of judicial canons etc. in the United States in Llewellyn (1960). The principles guiding judicial action may be identified in various legal systems as 'canons of construction', 'principles of equity' or 'policies' of the judiciary—all drawn from constitutions, statutory systems or judicial experience. These principles usually do not compel specific results but instead set priorities and provide authoritative guidelines for judges who find, in the reasons and values underlying the principles, some guidance for the cases then before them. They may need to weigh in their 'judicial balance' unmeasured

* Some innovations in taxing practice have been tested, as for example California's 'Williamson Act' (Cal. Gov. C.Sec. 51,200 *et seq.*), but none can yet claim adequate success—*see* Overview Corporation (1973).

values or long-range interests that are not articulated by competing parties in litigation.

In a related setting, several US states have taken tentative first steps in developing constitutional principles of 'life-support law'. For example, the Constitution of the State of Montana includes the following charge to the legislature (though it is not a self-executing principle of judicial construction):

> Art. IX, Sec. 1.(3). The legislature shall provide adequate remedies for the protection of the environmental life-support system from degradation and provide adequate remedies to prevent unreasonable depletion and degradation of natural resources.*

But no matter how they are stated, these judicial principles will be useful only to the extent that the judiciary understands them, and chooses to apply them in the thousands of decisions that may incrementally erode life-support resources.

The following are a few suggestions that might guide judicial decisions, although they are not presented here as a complete list, nor as finished products ready for immediate legislative action. In each case, it would be assumed that more specific legislative action or constitutional definition of property and other rights suggesting contrary results, would be controlling. The principles of construction would not change such explicit law, but would be used to 'fill gaps' in legal authority, where the law leaves several choices open to a judge.

The authors believe that, in general, it is wise to use law sparingly—that is, to permit commercial transactions and individual initiative a wide range of opportunity. But where the probable outcome of such private decision-making is likely to endanger the long-term interest of the general community, then it is important to understand just how critical the long-term interests can be. These principles have been drafted in that spirit of permitting the widest feasible range of transactions, limited by the sparing use of law to protect essential long-term public interests—cf. 'The Tragedy of the Commons' (Hardin, 1968):

A. Because the life-supporting resources of the jurisdiction and of the planet are finite, and because future demands on those resources can be presumed to be greater than present demands, judicial action should avoid destruction of such resources, and should favour their protection or their use at those later times when needs will probably be greater than they are today.

The principle is designed to 'keep options open' and avoid premature loss of critical lands and resources. It might be appropriately used when courts consider injunctions pending the resolution of other controlling issues.

B. Judicial action should favour protection of the diversity of the species of plant and animal life, and of the habitat and life-support systems for those species in the absence of explicit legal requirement to the contrary.

* For discussions of other proposals of this type, *see* Platt (1970) and Howard (1972).

C. Where the law requires any protective management of a wildlife species, it shall be assumed to require management for the maximum sustainable yield of that species in the absence of authority requiring otherwise.

Both principles B and C are designed to enhance the prospects for the survival of wildlife species—see, for example, Myers (1977). These principles reflect the assumption that in most cases where a species is endangered, it is endangered because of Man's taking for his food, sport or ornament. In such cases it is in the interest of the greatest number of people (as well as in the interest of the wildlife species) to see legal ambiguities resolved by these principles.

D. Measures that permit or promote the reuse of manufactured items or their components or materials are to be favoured.

E. Where courts may be called upon to decide among competing uses of land, they shall tend to favour that use which is most conducive to keeping it in a natural state, or, if it is not now in such a state, shall favour that use which promotes renewable life and life-support purposes—in preference to uses which do not, in the absence of explicit authority to the contrary.

Again, the purpose of such judicial policy is to prevent the avoidable, unreasoned loss of a finite resource-base. Agricultural lands are the best example of life-supporting terrain which should not be lost unless the law clearly requires it.

These proposed 'canons' of judicial policy would not stand in the way of government decisions to build projects, resettle populations, or to do anything where the government acted deliberately. Nor would they alter rights already defined in constitutions or statutes. But where statutes were ambiguous, where resource problems arose in courts before they arose in legislatures, and where private disputes risked environmental damage, then proposals such as these could lead the courts towards decisions in harmony with the laws of Nature.

Adequate Information

Since we are concerned with the harmful cumulative effect of many thousands of public and private decisions, any change of course would require, among other measures, the development and use of information about both the incremental and total effects of these decisions. The information that is usually available on such issues is typically weighted in two ways. Because experiences of the past and values of the present usually motivate proposed actions, information about those experiences and values is typically more comprehensive and more easily understood than predictions about the future consequences—especially when those projections raise the risk of opposition to a project. And as noted earlier, decision-makers—whether presidents, legislators, corporate executives or judges—are seldom given projections of the incremental risks created by a project; nor are they advised of the larger picture of the extent to which that particular resource is in jeopardy.

While the development of useful, adequately qualified information about future effects may involve expense, and also the use of new methodologies (such as forecasting or assessments of probabilities, for example), it may nevertheless be essential for far-sighted decisions. But as the usual ways of doing business—especially government business—have failed either to encourage or to force decision-makers to invest time, status or money to acquire such information, decisions have usually been made without the benefit of easily developable information about the probable future.

It was partly an effort to correct this imbalance that led to the environmental impact statement process in the United States and some other nations—cf. US law in National Environmental Policy Act of 1969, 'NEPA', Pub. L. 91–190, 42 U.S.C. 4321–4347 (1970), as amended.

Decision information can also be biased by differing states of the art of measurement of different types of values. The monetary side of a decision equation can be adequately described through conventional economic analyses, using widely-accepted methodologies. But, typically, environmental effects cannot be assessed by equally reliable methodologies, even though such effects are equally 'real' and often of greater magnitude than the more easily assessed economic impacts. This problem, too, was recognized in the US 'NEPA' legislation (*Ibid.*, Secs. 102(2)(a) and (b)), though no one would yet claim that these clauses have fully solved the problem. The state of the art of mathematically-structured decision analysis has advanced substantially in recent years, but these new decision methodologies have not yet gained wide acceptance.*

The reliability of information used in decisions is improved when it can be tested in the public—that is, by those with different experiences and values. In the United States perhaps more than in other nations, private citizens and groups have used the courts to force agencies to develop facts and consider long-term effects that would have gone unmentioned in earlier times. This has resulted partly from judicial review of agency decisions pursuant to the National Environmental Policy Act (*supra*), and partly from the Administrative Procedure Act (5 U.S.C. 551–576, 701–706). Recent appellate decisions in the United States have required agencies to document reasoned decisions before they proceed (*see* Pitts *v.* Camp, 411 U.S. 138, (1973). The role of the judiciary in this setting was summed up by Judge Bazelon:

> I cannot believe that Congress intended this court to delve into the substance of the mechanical, statistical, and technological, disputes in this case. . . . Thus the court's proper role is to see to it that the agency provides 'a framework for

* For an example of one such analysis on the question of whether to invest major US government funding in the development of synthetic fuels, *see* the Report of the Synfuels Interagency Task Force (1975). This report projected costs and benefits of investments in synthetic fuels research under different sets of conditions, and gave what seems to us fair, though rough, assessments of intangible costs. In practice, this report focused the Congressional debate on the known facts and reasonably estimated probabilities to the benefit of all concerned.

principled decision-making'. Such a framework necessarily includes the right of interested parties to confront the agency's decision and the requirement that the agency set forth with clarity the grounds for its rejection of opposing views. But in cases of great technological complexity, the best way for courts to guard against unreasonable or erroneous administrative decisions is not for the judges themselves to scrutinize the technical merits of each decision. Rather, it is to establish a decision-making process which assures a reasoned decision that can be held up to the scrutiny of the scientific community and the public. [International Harvester Co. *v*. Ruckelshaus, 478 F.2d. 615, 651, 652 (D.C. Cir. 1973).]

There are clear benefits that flow from public review of articulated, reasoned analyses of proposed government actions, and of the probable effects of those actions on finite resources and life-support systems, and these benefits have made the impact statement process a popular one. Environmental impact statement proposals have been made in Japan, France, West Germany and the UK, and have been adopted in whole or in part by statute or regulation in New Zealand, Australia, Canada, Israel and East Germany.

This critical need for decision information leads to a final suggestion for harmonizing the systems of human and natural law, namely that the judiciary might be given powers, in nations where no one now possesses such power, to compel the development of new information. This could be especially important in nations that do not have impact statement requirements.* It could help in articulating the uncertainties facing the courts, and in analysing both the magnitudes and the probabilities of uncertain consequences. Courts, or others looking to future effects, may then need to weigh the cost and time required to acquire more facts, against expected impact of that information to determine if it could be decisive. In short, the decision-makers could conduct rough 'value of information' analyses to determine whether the probabilities of improving their knowledge of future effects justified delay while parties acquired the necessary facts.

The judicial as well as the administrative branches of the legal system, then, should explicitly recognize the extent to which their decisions may be affected by existing information and by 'developable' information. They might be empowered to acquire information in various ways from direct sources, or from the parties, depending on the procedural safeguards involved—at least where the value of information about the future appeared to be worth its cost in time and money.

Conclusion

This paper has focused on some of the opportunities and limits in using Man's law to lead us to a future where Man plays his role as an organic part of the Earth's systems. The ideas roughly sketched here could begin to bring both systems of 'law' into harmony.

* For a related proposal for providing technical information to courts, *see* Leventhal (1974).

There are opportunities for treaties—and for statutory, constitutional, and judicial, steps—to bring Nature's limits into the decisions made under Man's law.

One answer may lie in the judicial and governmental decisions that affect Man's future on the Earth. When those decisions are made in a legal structure, which is in turn shaped by Nature's laws, Mankind has its best chance for a long tenure on this fragile planet. The law of Man can be made to serve the above ends and work concertedly 'for Earth and Mankind'—for Man and Nature and their joint life-support.

References

Barney, G. O. (Ed.) (1977). *The Unfinished Agenda.* (Task Force Report Sponsored by Rockefeller Brothers Fund.) Thomas Y. Crowell, New York, NY: 185 pp.

Brown, L. [R] (1978). *The Twenty-ninth Day.* W. W. Norton & Co., New York, NY: x + 363 pp.

Brownlie, I. (1974). *International Environment and Law.* [Not available for checking.]

Convention for the Regulation of Whaling (1931). Geneva, Sept. 24.

Hardin, G. (1968). The tragedy of the commons. *Science*, **162**, pp. 1234–8.

Howard, A. (1972). State constitutions and the environment. *Virginia Law Review*, **58**, pp. 193–229.

Huntley, J. R. (1971). *Man's Environment and the Atlantic Alliance.* NATO Information Service, Brussels, Belgium: 47 pp.

IMCO (1972). International Convention on the Prevention of Marine Pollution by the Dumping of Wastes and Other Matter. [Not available for checking.]

International Convention for the Regulation of Whaling, Washington [cited as International Whaling Convention] (1947). Dec. 2nd, 1946, and as amended 1972, 1973 and 1974.

Leventhal, H. (1974). Environmental decisionmaking and the role of the courts. *Univ. of Pennsylvania Law Review*, **122**, pp. 509–55.

Llewellyn, K. (1960). *Common Law Tradition.* Little Brown, Boston, Massachusetts: 505 pp.

Lovins, A. (1977). *Soft Energy Paths (Toward a Durable Peace).* Friends of the Earth International and Ballinger Pub. Co., Cambridge, Mass.: xv + 231 pp.

Mesarovic, M. & Pestel, E. (1974). *Mankind at the Turning Point: The Second Report to the Club of Rome.* E. P. Dutton, and Reader's Digest Press, New York, NY: xiii + 208 pp.

Myers, N. (1977). Garden of Eden to Weed-patch. *Natural Resources Defense Council Newsletter*, Vol. 6, Issue 1, pp. 1–15, Jan.–Feb. 1977.

Overview Corporation (1973). *How to Implement Open Space Plans in the San Francisco Bay Area.* Association of Bay Area Governments, San Francisco, California: Vol. 3, 146 pp.

Platt, R. (1970). Toward constitutional recognition of the environment. *Am. Bar Association Journal*, **56**, pp. 1061–4.

Synfuels Interagency Task Force (1975). *Recommendations for a Synthetic Fuels Commercialization Program.* US Government Printing Office, Washington, DC, Stock H 041–001–00111–3., 3 Vols. (*See* especially Vol. 2.)

Discussion (Session 17)

Bárdarson (Chairman, introducing)

First, in conformity with the request of our Secretary-General, I would like to introduce myself. My name is Hjálmar R. Bárdarson, of the Icelandic State Directorate of Shipping, which is the governmental department responsible for various maritime matters, including marine pollution. As an engineer in ship design and machinery by education, [I am privileged to be the Director; but] my excuse for having accepted this honourable chairmanship is that, since 1954, I have been participating as delegate of Iceland in international conferences on marine pollution, which is always closely related to legal problems.* In my introductory remarks I would, therefore, like to make a few comments, including additional remarks with reference to an item that is also mentioned in the paper.

It is sometimes criticized that it takes far too long a time for an international convention to come into force after a [competent conference or other body has adopted it], as the ratification by a sufficient number of countries often delays matters for an unreasonably long time, and, furthermore, amendments to existing international conventions are also very slow to become effective. Thus international technical conventions are apt to lag behind technical development, and this can be said of several existing conventions. [So I thought it might be] of interest to note how several recent international conventions during the last 10 years or so have been gradually developed into a form different from older conventions. As considerable amounts of my work in the international field have been connected with international instruments under the auspices of the UN maritime organization, IMCO, with headquarters in London, I will refer to IMCO conventions on marine pollution as examples.

The problem of oil pollution was the first of the marine environment to attract attention internationally, as immediately after the end of the First World War the rapid increase in the number of ships using oil for propulsion and carrying oil as cargo revealed the seriousness of the problem. The first international conference on oil pollution was held in Washington, DC, in 1926, but no agreement was reached then to secure an international convention on avoiding oil pollution.

The second international conference on oil pollution was held in London, England, in 1954. The outcome of this conference was the basically still existing International Convention for the Prevention of Pollution of the Sea by Oil, which came into force on 26 July, 1958, thus about *4 years* after the conference. The 1954 convention and practical experience of its application was the basis for the IMCO Conference held in 1962, when the 1954 convention was extended and amended. On the 18 May, 1966 an acceptance of the 1962 amendments by the Government of Iceland was deposited with IMCO as the twenty-first acceptance, and, with the Icelandic acceptance, the required number of acceptances by two-thirds of the contracting governments to the 1954 convention had been reached. The 1962 amendments came into force just one year later, on 18 May, 1967—*5 years* after the 1962 conference.

Then in 1967 came the *Torrey Canyon* disaster. So far the convention on preven-

* According to *Who's Who in the World*, Director Bárdarson designed the first steel ship to be built in Iceland, where he has been State Director of Shipping since 1954. As his country's Delegate to the United Nations Inter-governmental Maritime Consultative Organization [IMCO] since 1956, he has been Chairman of their Maritime Safety Committee's Subcommittee on Safety of Fishing Vessels (1963–76) and President of the Assembly (1969–71). He is also a leading member of the Icelandic Nature Conservation Council, whom he represented at our Conference.—Ed.

tion of oil pollution had only covered oil pollution from the normal running of ships. Maritime accidents and disasters were not covered.

At a special IMCO Assembly in 1968 it was decided to convene The International Legal Conference on Marine Pollution Damage in Brussels in 1969. The outcome of this conference was the adoption of two international conventions. The first one: The International Convention Relating to Intervention on the High Seas in Cases of Oil Pollution Casualties, deals with the right of a coastal state to intervene and take measures to protect its coastal waters and other related interests where a casualty occurs on the high seas. The second convention adopted was the International Convention on Civil Liability for Oil Pollution Damage. This convention aims at ensuring that adequate compensation is available to persons who suffer oil pollution damage resulting from maritime casualties involving oil-carrying ships, and places the liability for any such damage on the owner of the ship from which the polluting oil escaped or was discharged.

Contracting states are required not to permit ships under their flags to trade without insurance. As the owner according to this convention may, however, limit his liability in any one incident to an aggregate sum based on the size of the ship, another IMCO conference was held in Brussels in 1971 on The Establishment of an International Fund for Compensation for Oil Pollution Damage. This 1971 convention is supplementary to the 1969 conventions. The purpose of the compensation fund is to provide a further compensation when the protection by the liability convention of 1969 is inadequate.

The 1969 IMCO Brussels Conventions came into force in 1975, namely *6 years* after the conference was held, but the 1971 Convention on the International Fund for Compensation is still not in force (as of June 1977).

The last IMCO conference on Marine Pollution was then convened in 1973. This convention for the prevention of pollutions from ships contains provisions aimed at eliminating pollution of the sea by oil *and* other noxious substances which may be discharged operationally, and at minimizing the amount of oil which would be accidentally released in such mishaps as collisions or strandings. It covers all aspects of intentional and accidental pollution from ships by oil or noxious substances carried in bulk or in packages, as well as by sewage and garbage, but does not deal with dumping, which is covered globally by the London Dumping Convention of 1972, that is also under the auspices of IMCO in London.

The 1973 convention has not yet come into force, but much work is now being undertaken to ensure its acceptance by a sufficient number of countries, and special meetings are being held in IMCO to try to speed up the coming into force of this important new international convention, which will then supersede the 1954 International Convention for the Prevention of the Pollution of the Sea by Oil.

[To sum up regarding the time taken after a convention has been adopted by a competent body until it actually comes into force, experience has shown that this time is usually 4–6 years. This time is used by the different countries, first to examine the implications under a convention and translate the text into the language of the country concerned. Then appropriate legal action has to be taken, as in all democratic countries the permission for the government to ratify a convention has to be passed through its legislative assembly.

In some countries, also, later amendments to an existing convention have had to be passed as a separate bill through parliament. This is not the case in Iceland, as generally the permission by law for the Icelandic Government to ratify a convention also includes permission to ratify later amendments to the same convention—which makes ratifications much easier to handle.

Every international convention of the kind here discussed requires a certain number of countries of a minimum individual total size of fleet to ratify it, before a convention can come into force. To speed up the coming into force of a convention, the number of ratifications and total gross tonnage could be reduced, but this would not in my opinion solve the problem, as a convention only accepted by few countries, possibly with a rather limited total tonnage, would not have the effect to be expected from an international treaty.

Recent oil-tanker accidents on the United States' and some other coasts have caused considerable concern about the safety of tankers and their causing of oil pollution. Action in IMCO has now been proposed by the USA Government, based on President Carter's Message to Congress. This proposal for further action was discussed at the last Maritime Safety Committee meeting in IMCO, and will be discussed later this month at the next meeting of the Marine Environment Protection Committee. Then in the autumn there will be a special joint meeting of those two IMCO Committees, followed by the IMCO Assembly in November 1977, where most likely it will be decided to call for a special international conference to be held early in 1978 to discuss stricter standards for oil tankers' construction, equipment, inspection, and crew standards and training.*

The problem has rather often been connected with accidents of tankers registered in so-called 'flag of convenience' countries, [of their] not having sufficiently effective national departments for safety of ships and for control of maintenance and training of crews. One possibility is a proposal, by the British Government, to start international control of ships under the IMCO leadership. Then it is proposed that the conference on Standards of Training and Watchkeeping, currently scheduled for the fall of 1978, be advanced to the spring of 1978.

All these proposals are now under consideration to speed up further the action on prevention of marine pollution for the protection of the environment.]

The 1973 convention is of the most recent type of international technical IMCO conventions. The convention itself is actually only the legal framework, containing articles within the general terms of the convention. Then follows the technical content in separate annexes and appendixes. A diplomatic conference is needed to amend the articles, but a particularly important feature of this new convention is that it incorporates provisions which will enable *amendments to be made* to the technical requirements contained in the annexes and appendixes much more expeditiously than is the case in the existing convention. For many such amendments, it will now be possible to employ what may be called a '*tacit acceptance' procedure* which will enable them to be brought into force after a predetermined period. This will considerably facilitate the task of keeping the 1973 convention up to date and in line with modern technological progress. Proposed amendments to the convention will be submitted to an 'appropriate body designated by the IMCO Assembly' for consideration and adoption. The eighth IMCO Assembly in November 1973 designated as the appropriate body to exercise this function a new committee called The Marine Environment Protection Committee.

It took about 10 years of discussions between the technical and legal sides to find the solution now adopted as a principle both in the 1973 Marine Pollution Convention and in the 1972 London Dumping Convention—to make amendments possible more expeditiously than in older conventions, by considering that 'no answer' means an acceptance of an amendment. In my opinion this is a very great step forward in ensuring that technical development can be taken care of in international conventions in a reasonable time—without having to wait for a new diplomatic conference, and then wait for a sufficient number of acceptances before amendments become effective. This problem is discussed in general terms also in the

* This conference was held in February 1978 and appeared destined to have some useful outcome.—Ed.

paper now before us, but I thought it might be of interest to mention, [from my fortunate vantage-point,] the specific case of international conventions related to the marine environment.

It is now my pleasure to invite Professor Gary L. Widman, the senior author of the keynote paper 'Common Laws for Earth and Mankind', with the cautionary sub-title of 'A Glorious Hope?', to give a summary of the paper. Professor Widman is Professor of Law and Director of the Natural Resources and Environment Law Program at the University of California's Hastings College of the Law, San Francisco, and Professor of Law at Boalt Hall College of Law, University of California at Berkeley. He recently completed a three years' term of office in Washington as General Counsel of the Council on Environmental Quality in the Office of the President of the United States. He is also, in the leadership of the American Bar Association, a member of their Council and Natural Resources Section, and a member of the President's special committee on Energy and Law. It is a great privilege to pass the floor on to Professor Widman for his introduction of the keynote paper, following which the co-author, my countryman Professor Gunnar G. Schram, [of the Faculty of Law of the University of Reykjavik,] will have the opportunity to make further comments. [Also on the platform we have three panellists: Professor John H. Crabb, an international lawyer with wide environmental interests, Harvard-trained but now based in Geneva, Switzerland, Mr Charles T. Savin, of British Petroleum, whom you already know, and Mr Jack P. Attias, another bright legal light with special concern for Nature as well as Man.]

Widman (Co-Keynoter, presenting paper)

I have been interested to see the progress of things through this [Conference, and] think in a way we have something we might call an 'ecosystem of information'—information that is used in one discipline after another for different purposes but with the same information generating new research, new activity, and leading ultimately to different actions by the public. The law has taken part in this ecosystem of information, and the law itself is changing a good deal—particularly in the area of environmental law.

I will not attempt, in the short time available, to undertake any sort of cataloguing of environmental laws or any sort of cataloguing of things that we might do, because the list would be so long as to be limited only by our imagination. Instead I thought it would be more useful for me to talk about ways in which we could harmonize the energies and directions of the legal system with the energies and directions of your [environmental] work. The title that was assigned, of 'Common Laws for Earth and Mankind', I felt was broad enough to allow me that sort of scope. [My colleague Professor Gunnar Schram, from his wide practical experience *inter alia* as legal adviser of his country on the United Nations Conference on the Law of the Sea and Chairman of the Icelandic Environment Protection Committee, will fill in some further details following my summary of our joint paper.]

Let me start, then, with a little bit of discussion of terms, a little bit of discussion of what I think the human law system is all about, and point out the ways in which Man's laws are the same as, and the ways in which they are different from, the laws of Nature that you work with in your scientific professions. [Summary of paper (printed above) followed.]

Schram (Co-Keynoter)

My co-author Professor Gary Widman has given you an excellent résumé of our paper, so I will only take up about five minutes of your time. There are two main points which I would like to make at the beginning of my remarks. The first one relates to the following question: which is the best way of utilizing the enormous amount of [environmental] knowledge existing in the scientific community today in a great number of countries? This is a pool of knowledge which increases every day, every month, every year. The question before us is, therefore, how can we best get this pool of knowledge to the attention of the decision-makers in every country and in the international community as a whole? There are two main approaches to this [problem]. One is through direct political action, by establishing direct lines of communication to the administrators and to the decision-makers. The second line of approach is possibly a firmer ground for action, although not so immediately effective. This is through the making of new law. I am here underlining a most important aspect which I do not think has been given sufficient attention in many countries, namely that a significant amount of knowledge—of scientific knowledge—has not been made use of because the lines of communications have not been available or open enough.

The second point which I wanted to make in these few remarks concerns a matter which is obvious to lawyers, but I am not sure whether it is as obvious to scientists and others on the whole. This is that we have practically no law on the environment—we have, rather, a lacuna in international law as regards this important field of human relations. All we have here are a few conventions on specialized aspects of the environment, such as the Oil Pollution Conventions of 1954 and 1962, the 1972 Oslo Anti-dumping Convention, and the 1972 London Convention, but these are only limited instruments. What we therefore need to do through international action is to formulate new laws on the environment, which are in most fields non-existent. This of course is a very large task and it will take us many years. But the sooner we embark on this venture, the better [the results will be].

General International Law, the traditional law as we know it today, contains practically no customary rules or regulations relating to the protection of the environment as such. This is a fact which is not often recognized. We are here dealing with a new but vital element of Man's relationship with Nature—which is not reflected in the rules of customary international law as it has developed through the ages.

Certain maxims of general international law have, however, some relevance to the questions before us at this Conference which are of concern to environmentalists in every country. The maxim of territorial sovereignty is one, and in this context it has a two-fold aspect. It guarantees the State full powers within its boundaries while imposing at the same time a duty on the State not to allow its territory to be used for creating nuisance for other states. Secondly, we have the tenet of State-responsibility, which prescribes payment of damages for injurious acts performed by the State or its agents.

These rules were reflected in Principles 21 and 22 of the Declaration of the UN Conference on the Human Environment, held in Stockholm in 1972, which said that States have the responsibility to ensure that activities within their jurisdiction or control do not cause damage to the environment of other States or of areas beyond the limits of their national jurisdiction.

This leads us to a question of methodology. What is obviously here called for is concerted action on national, regional and global levels, as was emphasized in 1972 by the Stockholm Conference. And what are the most effective methods to be used in this respect which will promise us at least fair prospects of success? Basically, I would like to suggest two lines of approach: firstly, much wider use can be made of the traditional treaty form between individual States, and secondly, a global environmental code of law or convention should be written, containing *inter alia* the non-binding principles of the Stockholm Declaration of 1972. Such a code would be a quasi-conventional type of legislation intended for adoption by the international community through the United Nations.

As to the first point—the treaty law—it is evident that, given the great diversity of environmental problems around the world, complex situations will have to be resolved on a regional basis or even bilaterally between those States which are chiefly affected by these problems. We have already seen quite a few examples of such regional-approach solutions. One is the Nordic Environmental Agreement of 1974, which is the main Nordic instrument in this field. We also have the 1972 Oslo Anti-dumping Convention, which is concerned with the North-East Atlantic, and similarly a number of conventions drawn up by the Council of Europe. The European Economic Community has also been active in this field, and we have the NATO approach with its Committee on the Challenges of Modern Society.

Much less advanced than the treaty-law evolvement is the second approach: the creation of a new system of law—a general convention of law for the environment. While the case-law and the treaty-law is in constant development, this other task requires innovative action by the international community as a whole. Man's obligations as the custodian of the world demand that he take measures which will guarantee for future generations a clean environment and which will, at the same time, prevent the depletion of the Earth's fast-vanishing resources. The structuring of the right kind of political action through international cooperation is here, of course, of paramount importance. But no less needed is the creation of the right legislative basis for environmental action—the basis which is so noticeably lacking today.

What are, or what should be, the rights and duties of States in the field of the environment? This is the basic, most fundamental question. How are they to behave? What guidelines should States follow in their international relations, and where lie the limits of their responsibilities *vis-à-vis* other States and the international community as a whole?

These questions cannot be satisfactorily answered unless we have created the right institutional framework, spelling out in detail the rights and the obligations incumbent upon individual States—much as every citizen in society is affected by the rules of his national or local law.

My last point to you, ladies and gentlemen, is therefore that we should take action now—which is easy to say but may be difficult to implement. Who should be responsible for that action, for the code of law or the new Convention for the Environment? Obviously this is not a task that can be undertaken by any one country, by any one government, but must be approached by the international community as a whole. The United Nations therefore are here in focus, and it would only be a logical continuation of the work in the environmental field initiated by their Conference on the Human Environment five years ago if they were entrusted with this task, which could therefore be allotted to the United Nations Environment Programme in Nairobi or, perhaps, to the International Law Commission which meets in Geneva. The latter body has in the past written a number of international conventions that have later been enacted by conferences attended by members of the UN, and a similar process can be envisaged here.

My final word is that governments should be strongly urged by this Conference to take up this matter within the United Nations, with a view to initiating the law-making process in the environmental field which in my opinion—and I hope you all share this with me—is so urgently needed. The Reykjavik Conference could well serve as a catalyst on this issue, and in one of our resolutions we should draw the attention of governments to this important task of law-making for the environment.

Crabb (Panellist)

Professor Widman threw out a lifeline in terms of what I call the general field of legal philosophy or jurisprudence. He mentioned for example our distinct and different kinds of law; we have the law in the very broad general sense, which includes the laws of Nature, and then we have what we can call the human law, which deals with the affairs of Man and their future societal rapport. So it is worth emphasizing that here in the field of environmental law, which we wish to develop, we are dealing only with human beings. The objects of our legal concern or enactments, or indeed of any other kind of activity in the environmental field, do not themselves have any rights whatsoever. Thus for example we talk about an endangered species that we desire to preserve in the environment and of course the law intervenes in that respect. But this does not confer upon, for example, the Trumpeter Swan [*Cygnus buccinator*], any rights—it merely confers rights on other human beings who have an interest in preserving the Trumpeter Swan. Perhaps that is a very banal and obvious point but nevertheless I find in the course of discussion that it is sometimes lost to view.

As mentioned in the paper, environmental law is a new field: it is new as a general or independent concept. To a limited [extent] the problems which we now classify as environmental were handled in other ways by the law, and they were treated under different special rubrics. Thus we can cite for example the field of law known as nuisance. Another would be the field of law known as trespass. Those are still going to continue to exist, but those *plus* other fields that might be mentioned have in recent years been grouped together in a burgeoning field—[hitherto a largely] nascent field—that we can call environmental law. Now the development of a new legal field has a great many significances that I cannot take the time to go into right now. But I might just summarily say that when a new legal field develops in the professional consciousness, it immediately plugs in the lawyers' thinking, orienting it into a certain type of analysis. With the existence of something known as environmental law, a particular kind of a problem of which a lawyer might have traditionally thought: 'well, let's see, here's that problem of nuisance', [he will now regard and categorize very differently. For] he has got a new, convenient classification known as environmental law which can make a significant difference to the way he is going to perceive and analyse the problem.

Any field of law, in order to become solidified and mature, has to have a period of time for development. This has only barely begun in the field of environmental law. For example, the judicial decisions which Professor Widman referred to, take a long time to develop. We have perhaps for a few years now been getting cases going through the courts that are talked of expressly in terms of environmental law, and as time goes on, and assuming the trend continues in a burgeoning and developing way with respect to environmental law, we can look forward to becoming a great deal richer in terms of experience and, to that extent, a great deal more effective.

Savin (Panellist)

I think we all owe a debt of gratitude to the authors of what seems to be a most comprehensive paper. [Applause. But to deal with a few points rapidly, as we are warned that time is very short], I am not an expert on the UK land planning law, nor would I claim that it is perfect; but I think that, under it, people would find it difficult to change land-zoning by votes taken 'late at night', and I would have thought it should not be beyond the wit of lawyers to build in procedural regulations which would enable all those with interests in a particular zoning problem to be able to make their contribution to the debate before any changing in zoning is made.

As for laws applicable to the world environment, clearly these must be attainable and generally acceptable, and this must arise by securing general agreement on principles and objectives, leaving the details for local implementation. My particular worry is, how do you get states to adhere to these conventions and, when once they adhere to them, ensure that they are carried out. In this latter regard I think we heard from Professor Kuenen and Polunin junior this morning about some of the problems concerning competition between nations [that could complicate such] carrying out. We also heard in the discussion of the new international order that this means different things to different people. I would have thought that UNEP might well devote itself more in the future to trying to ensure that there is more universal observance of the obligations as well as the benefits.

Clearly a lot of world conventions and regulations relating to the environment are best carried out on a regional level, and in this we have the examples of some of the Nordic initiatives, the Barcelona Convention, and the Bonn Convention on the exchange of information. Clearly these conventions in the end must be applied at a local level, and I end finally with a plea that when the States involved come to implement the detailed environmental legislation [of the future], they continue to go for the objectives and do not try to lay down from their bureaucratic armchairs how things will be carried out—because, by the very nature of things, they will not know. These objectives must, at the end of the day, be achieved locally—by reconciling the interests of the community, the environmentalists, and industry, acting all together.

Attias (Panellist)

I, also, would like to thank the [keynote speakers]: I think they were excellent. I will make my comments very short. They are merely observations.

First of all my fantasy is somewhat similar to Professor Schram's. I think he really laid it out well as to the ways he conceptualized what should be done in setting up at least a major body whereby there can be some coordination between all the nations of the world in regard to a problem that is clearly world-wide. My only problem with that fantasy is that, had we reached the level of maturity where we could undertake something like it, if we did undertake it, could we see it through? I am really sunk with respect to the vested economic interests that would come into play in fighting such an attempt in many and various ways. I look very bleakly at that—very negatively. I do not have much faith in getting extensive cooperation from vested economic interests that would be losing money by [adherence to] strict environmental laws.

In regard to law and science, which broadly is our topic. I think the law likes to think of itself as a scientist. Lawyers like to feel that they work on facts, that they have some basis for their decisions, that they are smart, and that they need scientists; they [certainly] need facts—they need something to rely on. Yet my feeling is that, before the law becomes very effective in the area of environmental control, and before you can persuade a judge, at least on the intellectual level, you will need many more facts, many more proofs, and unfortunately these are going to cost a lot in the form of research and in the form of efforts by many, many people—including scientists, of course.

A significant point also in regard to environmental cases is I think a very subtle one dealing with environmental psychology. When you approach a judge, you are very sensitive to the fact that he has a wide range of discretion. In other words he is operating from a body of information that you really have no basis of determining unless you are intimately involved with him and acquainted with his background, whereby you might know the law, you might know [other] cases, you might know the facts; yet in your case, you do not know what is running him, and I think the best approach in regard to these judges is based upon [recognition of their having] discretion. So a significant approach to them would be [through] their sympathy for the environment—their appreciation of the sun in the morning, of a breeze in the evening, of the trees—something which I think is very valuable.

Bárdarson (Chairman)

[As we have already at least five more would-be speakers and the Secretary-General can give us only 10 minutes more—after 8 minutes for tea for which we will now adjourn—please limit your interventions to really brief comments of 2 minutes each at the very most.]

Misra

I am not a student of law, but I do understand a few things. Prompted by the IUCN and UNEP, we in India established a National Committee of Environmental Planning and Coordination. This began giving advice to the government, on the basis of which our constitution was amended. The amendment provided certain obligations and duties towards our environment: love and respect for, as well as protection of, the environment, including forests, wildlife, and so on, were [written into] our constitution. But the enforcement of the laws has been very weak indeed. Even after [making the constitutional amendment] we find that many of the States have not yet adopted it, though some have done so.

Another matter I would like to raise is that of Indonesia, where the contractors from Japan and the United States [and other] developed countries have clear-felled and removed the forests, for coffee plantation, in a very short time. If Indonesia herself had had to do the same job, it would have taken at least 10 years to clear the forests, but modern machines have played havoc there. Now can there be [established some] mechanism by which Japan and USA and other countries could prevent their contractors from undertaking such work in another country? With the provision of certain legal constraints, this would perhaps not have happened, at least on such a large scale.

On the point of the relationship between UNEP and the Geneva[-based] International Law Commission, I think that, although UNEP has no direct responsibility for doing such things in particular countries, still its recommendations or a plan of action could well be fostered through the International Law Commission.

Widman (Co-Keynoter)

On the question of restricting the activities of US corporations abroad, I think you should realize that this is primarily a political thing in the sense that the United States will do whatever the host government encourages it to do. If the host government imposes no restrictions, it is not easy to get a restriction order in the United States. The way it would be discouraged now—and by 'now' I mean within the last 6 months, and from this time on, hopefully—has to do with the way AID funds are channelled, and now AID is required to prepare impact statements on activities which it [finances] abroad.

This at least would mean that information would be developed on activities which have significant effects; the information would be available at least in the United States and presumably to some members of the public of the host country, and to the extent that the broad daylight of publicity [on such] happenings could defer action, this would happen. But certainly no firm prohibitions or guarantees [could be handed out] under the existing law.

Sheikh

The need for international legislation for environmental control has been very well emphasized here. I would like to add that there are certain problems at the national level, too, in this regard—particularly in those developing countries which have been under such colonial powers as Britain, Belgium and France, which are temperate countries. Some of their former colonies in Asia and Africa have large arid and semi-arid areas, and to cite one example, the water management rules framed by those former masters may need modification in order to accord with the requirements of the environmental conditions of these [developing] countries.

Obeng

After this morning's rather depressing comments, I have been happy to listen to the lawyers, as they make me feel that there is still hope for the environment. I think it is a pity that the paper by Dr Tolba was probably not read fully by participants because it explains what UNEP is attempting to do as a follow-up of the 1972 Conference on the Human Environment. Little bits and pieces have been picked out of context from the statement which was made. Particularly there have been adverse comments on the subject of sharing resources which was not part of the statement.

UNEP believes in achieving 'development without destruction', and maintaining a satisfactory environment through development which does not destroy. [It is] very much a personal thing—I mean, before it becomes national or regional or global. [Unless] we, as individuals [also], make personal commitments, we shall finish up in such a mess with our environment that we would not know what hit us.

Tan

I was not able to read through this whole paper, but I am struck by certain statements made by the panel and the main speakers. I would like to translate it into the manner in which we would react, or wish it to be applied, although I am not speaking for my country, Malaysia, but as an ordinary layman who would like such sentiments transmitted to the grass-roots level. The panel members mentioned about taking action now—that solutions should be on a bilateral and regional basis, and that there should be built-in regulations and organized information. I do not want to appear to be frivolous when I introduce the subject of Mohammed Ali, but I think [such a boxing champion] can serve a useful purpose too.

Television is one of the most powerful ways of influencing the average person, especially the uneducated (which can include politicians), and visual means of communications know no discrimination and can have a lasting impact. I think this is an effective way in which you could translate your ecology lessons or your idealism to the peoples of the developing regions. We receive all manner of television programmes, from the United States in particular; we import them as package-deals, because it is the cheapest way of acquiring foreign films. Is it possible for you in the developed countries to produce films on ecological destruction and preservation in your part of the world, and let us have a chance to see what you have done and can undo? When, say, a Mohammed Ali fight is shown live, during the half-hour or so when we are waiting for this [hero-star] to come on, viewing of one of these films could have the greatest impact, because he attracts such large audiences and also because the subject is such a contrast to the usual cigarette advertisements, for example, and can therefore [secure] attention. If such environmental films could be available in community halls, even rural people (including children) would [understand] what you are trying to tell them and so should learn to articulate their objections—to stand up for their well-being when their environment is exploited without regard for their rights.

Royston

I would just like to underline the importance of national action, or national law, in terms of this question of global environmental law. In Japan now, a Malaysian, for example, can sue a Japanese company that has caused damage to the Malaysian environment by operating in Malaysia. In Holland, a Dutch court has been given jurisdiction to [recompence] Dutch farmers against damage caused by irrigating their land from Rhine water which has been contaminated in France by the Alsatian potash mines. In Sweden it is illegal to export equipment abroad which does not conform to the Swedish environmental standards.

So there is a great deal to be done in national law to protect other countries and in fact the global environment. [Applause.]

Olindo

I just wanted to point out that, at the last UNEP Governing Council meeting, an important resolution was passed [to the effect] that if a chemical, an industry or a technology is condemned in the country of origin—and this morning we were talking about such chemicals—then that country of origin should be held responsible, and should indeed hold itself responsible to all humanity, to stop its nationals from proliferation of such polluting industry, chemical or technology.

This resolution will be put to the forthcoming General Assembly of the United Nations, and the point I am making is that we should take upon ourselves the collective responsibility of alerting this active group to ensure that [this resolution] is translated into world action. The lawyers among us may [move on] towards developing an international legal code which may help to ensure development without stress to the environment.

Bárdarson (Chairman)

This concludes the formal discussion, but in view of the late distribution of this interesting paper I would like to ask the Secretary-General whether it would be possible for further comments to be entertained in writing?

Polunin (Snr)

Perfectly, sir. They should be written clearly, signed, and passed to the Secretariat. If [they are suitable and] there is room, they will be included in the Proceedings in square brackets. This is the device we have used previously without objection [, and, wearing my editor's hat, I feel is particularly desirable with such a pioneering paper. Please note, however, that there ought to be a further opportunity to get any pertinent matters off your shoulders in the discussion of the concluding paper].

Bárdarson (Chairman, concluding Session)

Thank you Mr Secretary-General. As the panellists and authors do not wish to comment further, it only remains for me to thank you all for participating in this Session [and to declare it] closed.

Session 18

CONCLUSIONS FOR THE FUTURE

Keynote Paper

The Environmental Future: Have We Reached Consensus?

by

F. KENNETH HARE

University Professor and Director, Institute for Environmental Studies, University of Toronto, Toronto, Ontario, Canada; formerly *President, Royal Meteorological Society* and *Director-General of Research Coordination, Department of the Environment, Canada;* sometime *Master of Birkbeck College, University of London.*

Have we reached consensus in Reykjavik? If so, what are the details? Do we foresee a sounder, safer and happier outcome for mankind?

Perhaps we have not quite reached consensus, but there does seem to be a surprising level of agreement. I am far more used to hard words than to the gentle cadences of our week in Iceland. Perhaps we were lulled by our own gift of the gab, and should have muttered to ourselves an apocryphal beatitude: 'Blessed are the highly articulate; for they shall convince themselves.' Certainly we had many gifted and elegant speakers, as well as polite listeners. But this very fact should have made us suspicious.

Consensus is not always admired these days. For something like a decade, in the wake of Johnsonian politics in the USA, it has been fashionable to argue that confrontation between adversaries is a better way of doing things than agreeing in advance. The environmental movement itself has many members who believe that an overstated case will in the long run get us further than dispassionate reasonableness. In the academic community there are so many shades of opinion that one mathematical colleague of mine used gloomily to say, 'an academic consensus is identically zero!'. The ayes, the nays, the buts, and the ifs all combine to silence one another.

Yet this has not happened to us in Reykjavik, in spite of the preponderance of academics in our group. We have shown considerable support for what one of us called the ecological paradigm—the view that mankind can survive in the long run only if it lives by Nature's rules, within Nature's constraints; and if we learn how to give up the marauding life of a pioneer species. As many of us are ecologists, such a view is not surprising. But we have others present, notably from the business community, from the world of practical politics, and from the media. None of these has challenged the basic truth of the ecologists' claim—at least not while I was listening. One gets the impression, reinforced by remarks from some of our least ecological members, that economic growth has been widely challenged and rejected as a panacea. Two of our members, both from overdeveloped countries, claim that their own democratic governments have silently agreed that this is so. If growth is to endure, it must be a new, qualitative growth, as Maurice Strong so eloquently told us.

Of course we are a hand-picked group. It would not have been hard to pick others who would have scoffed at such opinions. Economists, developers, industrialists and financiers are still, by a majority, adherents to the growth principle, arguing that only by growth can higher standards of living, and a fair distribution of employment and income, be achieved. Economists, in particular, find it hard to accept the reality of a finite Earth or a stable society. Marxist philosophers argue that environmental degradation and resource depletion are the results of capitalism and imperialism—that they would vanish in a socialist millennium. And finally the environmental activists would put the case for conservation far more stridently. There are few such people here in our midst: and so we get a near-consensus expressed in the draft Reykjavik Declaration.*

I cannot hope to distil the whole contents of this achievement. The Reykjavik Imperative, as the Declaration came to be called, practically does that, at least as far as principles go. But certain broad conclusions do need ramming home, and I shall try to do this here.

Political Action Engendered by Crises

First and foremost we seem to agree that we must seek *political* solutions to environmental and resource issues. However much we may still enjoy talking about them, we have to learn how to go far beyond mere talk, so as to find remedies for evils that are threatening to kill us all off—burgeoning population, depleting resources, weakening institutions, loss of biospheral equilibrium, and several others. We have come to see, since Jyväskylä,† that environmental degradation is only one aspect of a mortal sickness which is suffered by humanity. In other species—and in our own in other epochs—such things solved themselves by pestilence, war and famine. Can we, as rational beings, devise a political escape from these regulators?

If we can, it is because we have arrived at a time of crisis. When the facts are stark enough, even democracies budge. I once asked the chairman of a major energy utility how he expected to solve his problems, in the face of

* Amended and passed at the Final Plenary Session of the Conference and, after sanctioned editing, published speedily in *Environmental Conservation* (Vol. 4, No. 3, pp. 161–3, Autumn 1977). Entitled 'The Reykjavik Imperative on the Environment and Future of Mankind', copies have been in wide demand. A slightly revised version is printed later in this book.—Ed.

† The town in central Finland in which the first ICEF mainly took place.

capital shortages, environmental objections and endless red tape. He replied: 'I'm waiting for disaster; we have excellent contingency plans.' We all know how long it takes for such moments of truth to arrive. But we seem to think that crisis is indeed at hand.

Four aspects of this crisis have emerged clearly during our discussions. These were the population–food–resource sufficiency question; energy supply; ecosystem stress and degradation (notably in the tropical rain-forests and desert margins); and climatic change—probably man-accelerated, like the rest. The details are spelled out in the 22 formally-prepared papers, and there is no need to repeat them. Although these questions are seen quite differently in different countries, at Reykjavik we seem to feel that in each case the impending crisis is real, is near at hand, and is now visible to many politicians. If this is true, the moment is ripe for the scientific and scholarly communities to step forward with suggested solutions—if we know how. But is our knowledge of politics up to the test? The Maharajah of Baroda warned us that his fellow politicians are bored by pious resolutions. We have to be able to answer Lyndon Johnson's query to a group of academic experts: 'therefore, what?'

The Pessimist's View of Politics

If we agree that the future lies with politics, law and socio-economic management, as we have repeatedly said, we have to be aware of politics' limitations, as well as our own. It is useless to expect more work from a horse than the horse knows how to give. And we have as often exaggerated the capabilities of politicians as we have thrown mud at them. Politicians are, on the whole, good conscientious souls; but they are not supermen. Nor is politics the art of the sublime.

We seem to be agreed, for example, that systems solutions are to be preferred to one-dimensional interventions. 'Everything connects', speaker after speaker has said—calling this interconnectedness the first rule of ecology. Often we have appeared to say that good environmental strategies *must* be systems-based.

But systems solutions may just not be feasible in a democracy, or even in an authoritarian state. The elaborate complexities of systems models baffle most voters, and hence most politicians. The layman will concede the interconnectedness, but still reject the idea of trying to manage things as a whole. Things achieve political visibility one at a time. The media contribute to this, by their determination to simplify complex matters so that unskilled readers, hearers and viewers can get the message. The United Nations, for example, ran the Habitat Conference at Vancouver in 1976 on the principle that human settlements ought to be studied in all their complexity, and due weight can be given to the various interconnections between mankind and the built, natural and institutional environments. Canadian reporters and commentators simply scoffed at such high-flown stuff. It was a conference about housing, they said, almost to a man—because that question, and not the others, had visibility in Canada at the time.

The stuff of day-to-day politics has this happenstance, almost stochastic, look about it. Legislators rarely find themselves confronted with tidy, orderly, connected material. Much of their time is devoted to sudden crises (or, more properly, to bad situations suddenly perceived). There is no time to devise comprehensive responses. Instead small, incremental measures are taken in the hope that they will suffice. Very occasionally a larger proposal surfaces in time for action—such as the introduction of the United Kingdom's comprehensive social service system in the wake of Lord Beveridge's report, or the revision of a complex legal code. But this rarely or never happens in the complex environmental–resources area.

The bigger the governmental unit, the harder it is to achieve the systems outlook. In my own country, Canada, one finds the ecosystem concept universally accepted at only one level of government—the tribal or village councils of Inuit, Indian or Métis communities. The native peoples are still close enough to ecosystems to *know* that they function as complex wholes. And their lives are simple enough for them to find comprehensive solutions occasionally desirable, credible and enforceable. At the federal government level, in dire contrast, the tangle of regional, occupational, institutional, and constitutional, vested interests, plus the clamour that permanently surrounds Ottawa, the capital, virtually preclude such luxuries.

Economists, of course, still cling to the hope that they can tell politicians how to manage the economic system—which they strive to see as a whole *via* their econometric models. Most developed countries use such models in their economic planning. But actual political interventions in the economy still depend on crude, one-dimensional regulators, like variable money supply, interest rates, taxes, and income redistribution *via* transfer payments. The failure of such interventions to curb inflation, to lower unemployment, or to correct sectoral imbalances, suggests that economists are scarcely ahead of environmentalists in getting results. In any case economic policy is incomplete unless it includes sound environmental and resource measures; yet it rarely does.

So the pessimist is apt to say: political institutions cannot and will not cope with issues as large as those we have been discussing in Reykjavik. My favourite pessimist, with whom I largely agree, is the American social commentator, Michael Harrington, who sees the West in a state of advanced decadence. Its institutions always achieve something quite different from what they intend. It is the accidental century, he says, in which something enormous (but unspecified) has died, and something enormous (but also unspecified) is being born. Things happen without our intention And if we intervene, we do not get what we seek.

In a far less gloomy mood one can see how the bureaucratic structure of government is at variance with the systems approach to the environment. Environment is a characteristically broad, ill-defined, horizontally-organized complex—like all real political problems. Academics say that it is interdisciplinary, and set up institutes like my own to try to grapple with it. But government, like a university, is vertically organized. The Cabinet portfolios correspond to specific, narrow interests, such as agriculture (which works for farmers, not consumers), or labour, or commerce. It is admitted that all real issues are broad, and hence orthogonal to this vertical-tube structure. But the theory calls for the ministers to balance the competing interests in the Cabinet room. National policy on real questions is supposed to emerge from such horse-trading. From personal experience I know that environmental policy does *not* so emerge. It is overridden by the claims of those who believe that their special interests will be prejudiced by such measures.

Nevertheless, in country after country, the realities of pollution, energy shortages, loss of forests, of grasslands and of wildlife, and the perceived fall in the quality of life, have compelled reluctant and incompetent governments to grope towards a better situation. And now, as Gunnar Schram so forcefully told us, it is going to be necessary to create an effective body of international law to keep up with these national efforts. So many environmental problems transcend national boundaries—such as the state of the oceans, or the fall of acid rain—that national efforts are badly weakened unless they are supported by international measures.

It may well be that the UN's International Law Commission is the right body to take the lead in this direction. But it cannot do so alone. Law is a technical, professional arena. We need craftsmanship and experience in drafting new legislation, or in adjudicating disputes. Lawyers cannot, however, know what *ought* to be drafted. The scientific principles that should underlie good environmental management must influence or even dominate the hard content of such law. Hence other UN agencies, and the ICSU spectrum, should make sure that they are listened to—as should conferences like our own.

For all countries, democratic or totalitarian, rich or poor, such new law poses major problems of personal liberty and national sovereignty. As Reid Bryson picturesquely said, management may be inimical to freedom. The ownership of property does not confer unrestricted freedom in its use, even when the owner is the State. What does sovereignty mean where no one is sovereign, as in the oceans? Or when there are conflicting claims to sovereignty, as in many federal countries? When once we concede that the situation demands an extension of management *for the long-term future*, we put in doubt many of the present assumptions on which good law is based.

Pessimist or optimist, I must inject one plea. It is that scientists stop quarrelling with politics and politicians. To sneer that some action 'is only a political solution' is to raise the question: 'what other kinds of solution are there?' If we do not like the tone and content of politics, we have to change them ourselves. And further I plead for better use of the UN and its agencies. In one of my professional fields the World Meteorological Organization achieves an extraordinarily high level of international cooperation to monitor a particular bit of the environment. Other agencies do less well, perhaps: but again the fault is more often ours than theirs. We have not demanded clearly enough that they do their vital work in a way which keeps pace with our problems.

POSTSCRIPTUM

Other thoughts kept passing through my mind as I listened to our talk. Flippancy kept surfacing, for example—and was sternly admonished by one member who felt that we should not joke when the fate of mankind is at stake. But it kept coming to the surface, not least in my own remarks. We heard some excellent academic put-downs. Perhaps speaking English tends to make one act in such a way, since the language makes verbal play easy and enjoyable.

I conclude, however, that some of us are light-hearted because we do not really believe in our bones that these dire things will happen in time to worry us personally. If we have achieved consensus on crisis, that consensus is intellectual, not emotional. Others among us are more consistently grave: perhaps, for them, belief has gone deeper. Still others are true optimists. Like Edward Goldsmith, they can visualize a new life-style for the world that can get us out of the box. Or like Buckminster Fuller, they still believe in high technology. Some of us can foresee escape. Others cannot.

Among our optimists are the young members, who have confronted us with an appeal that should be irresistible. But even here we cannot close our eyes. In my own country, and I suspect in others, many students are now impatient with some of the best things in liberalism—for example, racial equality. Racist incidents have been numerous in many western cities—and they are often the work of the young. Much of the litter strewn along our highways comes from young hands. Misled by their elders? Perhaps; but I do not think that this is always so. I would conclude: thank God for the young, for they have the energy and the time. But I do not believe they are either better or worse than their elders.

'Thank God' is, on the whole, what we did not. We have heard some spirited pleas for an environmental ethic and for a religious approach to these issues—although we know the difficulties involved, not least the decline of western belief in transcendental things, and their official proscription in some countries. Yet an insistent religious theme has kept surfacing—that our behaviour towards Nature depends on our belief about the *purpose* of life. Mechanistically it has done; but there are few of us who still take a mechanistic view of life.

Personally I believe that the issues are root-and-branch religious, although I do not have a clear-cut religious position myself. In a world without serious purpose I doubt if any system of politics can solve our problems. But how do we achieve belief? To that question I have heard no answers in Reykjavik.

Our many arguments about pessimism and optimism are in any case futile. Faced with what seems impossible, one must still seek to overcome. Kuenen was good enough to quote William of Tell, in the wake of my mistranslation that 'in the absence of hope it is still necessary to strive'. What William actually said (in 1568) was

> On n'a point besoin d'espérer pour
> entreprendre, ni de réussir pour persévérer—

which ought to settle the arguments once and for all.

We are meeting here in this beautiful, vulnerable and hospitable northern country, where poetry lives on as part of polite and apt behaviour. Otherwise I would not dare to end with verse. In 1919, the year I was born, all Europe lay devastated, and there were many who thought that the lights would not come on in their time. The poet and hymn-writer Clifford Bax surveyed the ruin, and wrote (to a great sixteenth-century Genevan tune, the old 124th) these words:

> Now, even now, once more from Earth to sky,
> Peals forth in joy Man's old undaunted cry,—
> 'Earth shall be fair, and all her folk be one.'

We have not quite reached this level of exaltation in Reykjavik. But neither have we despaired.

DISCUSSION (Session 18)

Polunin, Snr (Chairman, introducing)

Having been placed in this position through force of circumstances, I nevertheless find it a triple pleasure to call this, our eighteenth and last normal, Session to order—firstly because I feel we have been having on the whole a useful Conference which has been well worth convening, secondly because our speaker on this occasion is one of my dearest friends and most respected associates, and thirdly because I can now wield again this splendid gavel of carved Icelandic birch which you have been kind enough to give me and which is fast becoming the emblem of our Foundation for Environmental Conservation while remaining in my proud possession to the end of my time.

On this occasion we have no paper with which to be late in circulation—not, for once, because of strikes or breakdowns of printing machines or transport, but because the assembling of material for its drafting has only just been completed with our latest Session. As indicated on the printed programme, the paper is by Professor Kenneth Hare, now of the University of Toronto but needing no introduction to any of you. [Suffice it to say that, with his eminence as a meteorologist and climatologist, goes a fine biological sense which is such a vital part of the equipment of most leading environmentalists.] Superb synthesist that he is, even Professor Hare could scarcely be expected to concoct his conclusions for the environmental future before we have given him our suggested inputs, and so you may have noted him scribbling away in detached seclusion more or less throughout our Conference. As we know of nobody anywhere who is better qualified professionally and equipped in his abilities to do this most difficult job, we can feel confident that we shall get another fine [and perceptive] paper.

Hare (Keynoter)

We have had several days of listening to siren voices—our own tempting voices—and as I myself listened, a new beatitude occurred to me: Blessed are the articulate, for they shall convince themselves; and I think that they probably have [done so. It is only to be hoped that we have not convinced ourselves of unwise or otherwise undesirable things!] For some 35 years I have taught a course to first-year undergraduates in various countries, and so feel I have kept in contact with contemporary youth. But in the last few years [at discussions with students and] in their papers that I mark, I have noted rising racism and [reactionary antagonism in place of understanding liberalism that used to be tempered with idealism. That seems to be fast becoming *their* consensus and] therefore I have to conclude reluctantly that all is not well with the youth that I meet today. [Nor am I much more sanguine about the future, for] I believe we could repeat, over and over again, the experiment that we have just done, of bringing together leading [environmentalists] to discuss major problems quite freely and openly, before arriving at a friendly concensus. But unless we find a way, as today's sessions have suggested, of leading into effective action, such conferences as this are not very satisfactory as weapons with which to attack the [mounting] problems of humanity.

Yet I feel we have achieved a substantial measure of consensus from our enlightened and often highly articulate discussions here, and so I will try to pull out what I think was the consensus from my still unfinished written synthesis [, the outcome of which will I understand be published as the paper heading this Session. But I should warn that if we had brought together a group that was not based predominantly on concern for the future of Man and Nature, but had assembled] for example a group of economists, it would not have been like this at all. If we had brought together a group of industrialists, or developers, they would have come to a [very] different conclusion—because while we are here in Reykjavik, they are back in the capital cities lobbying their governments to get on with the serious business, which [for them] is development! If we had put together a group of theologians, no consensus would have been achieved at all. If we had put together a group of Marxists, a consensus would have been achieved as it is part of their 'logic' that a situation of the kind we are describing is really not possible, and so 'why should we bother about it?'. And if we had brought together a convention of activists (which we are not), we would have had placards or be trying to convince one another by slightly different techniques. But instead we chose the ecological paradigm [, which seems clearly to be the decision that most of us accept as following what we firmly believe to be the correct pattern].

Now for my choices of where I think we have reached consensus: (1) we need political solutions to our problems, (2) this is a time of crisis, which has been brought home to us since our earlier Conference in the form of an impending energy crisis that means some sort of an industrial crisis, (3) there is a crisis in the relations between the advanced countries and the rest, (4) there is a crisis in the protection of Nature, (5) there is a crisis, which is obscured by market surpluses in international trade, in the world food system, (6) there is a crisis approaching us in climate, which means probably throughout the Earth's ecosystems, or (7) the interconnectedness of things should be borne always in mind and somehow written effectively into the political systems of the world.

So I conclude that the future lies considerably with political action and with [due processes of] law—as well as, of course, with continued research and vigilance on the part of the scientific community. But ending these sessions with excellent statements of the legal and political problems that confront us, is I feel particularly appropriate, because I believe that the next Conference which we hold—and I devoutly hope that we shall hold one—should be devoted to these two subjects (of course without exclusion of ecological aspects), as they are absolutely essential to where I think the future lies. But the scientific community must also be involved, and somehow we must make sure that their input is broad enough for the politicians and lawyers to have a proper basis for their leadership. Let us conclude by saying that it matters not whether we are optimistic or pessimistic for, to quote a great leader of the past, 'In the absence of hope it is still necessary to strive'. [Applause.]

Polunin, Snr (Chairman)

Thank you, Professor Hare, for your characteristically penetrating analysis which has given us much to think about and discuss [—not least being put in our places in the matter of choice of participants for these Conferences]. As you can all see, we decided to dispense with special panellists on this occasion and so I hope will manage to hear all who signify by raising their hands that they wish to speak—but as briefly as possible, please, if we are to be in time for the State Luncheon and are to find space for a goodly proportion of the best comments in the book of Proceedings.

Buchinger

After such a marvellous résumé there is I feel nothing left for us to say—except in a rather egoistic way, that when we go home we want to take something with us which we can show to our politicians, and so it would be very useful if some of the recommendations which are made from here would give us certain indications of how we can avoid ecodisasters.

Whilst I seem to have been very critical about certain international agencies, I would still like to maintain that they [exist] because we need them and that their job is helping [the world by transforming the kind of things] we are learning and discussing here in Reykjavik into action programmes. That need will surely always be present. It is very difficult for a citizen of a country to get to his or her government and say, 'You are doing wrong', though, naturally, non-governmental conservation groups try to make this point. But with the international agencies presumably present in each country, even if they do not want to interfere with internal politics, if something is going wrong they should at least bring in and provide all the [pertinent] data—so that when the government takes a decision, it should have all sides of the problem [under consideration] and be able to avoid some ecodisasters caused by partial information.

Polunin, Snr (Chairman)

Thank you very much Dr Buchinger, *inter alia* for reminding us that we are here, as our title implies and our programme demands, to look into the future—the Environmental Future—as best we possibly can. We have been asked from various quarters to do just that. The second and third points, about specialized agencies and getting at politicians, I will not comment on here but hope others will.

Hare (Keynoter)

I deliberately spoke in very general terms so as not in any way to scoop the Resolutions Committee [whose recommendations we will be considering here] tomorrow. I am very well aware that they have something specific to [propose], and was simply trying to provide you with a digestive pill with which to wash down the feast that you have had.

Polunin, Snr (Chairman)

Mr President, I do not know whether you would like to add anything now, as Chairman of the Resolutions Committee or otherwise—probably not, as we know that your committee has not finished sitting; but if you wish, sir?

Pauling

Well, I shall not try to scoop the Resolutions Committee [either], but while I have been attending this Conference and listening to the summary discussion today, I began to ask myself what the environmentalists have done in studying the environmental impact of militarism. I know that when there are wars, tremendous damage is done to the environment—for example through use of defoliants in south-east Asia; but this is not what I am talking about now. I remember having read about what a large part of the United States and the rest of the world is under the control of the military forces—the army, the navy and the air force. I know that sometimes operations are carried out which have not been planned very well.

During the Second World War we [lived in a] house which was on the edge of a canyon. While I was at work one day and my wife was at home, she was startled by several tremendous explosions, which knocked the flowerpots off the wall. When I came home I saw that the great oak trees on the floor of the canyon below our house had been blown up. This canyon had been taken over temporarily during the war by the military—several thousand acres there—and we learned that the plan was for the floor of the canyon to be used as a poison-gas training field. Then one day when we were having lunch with our children and Sir Macfarlane Burnet, an immunologist from Australia, we smelled mustard gas. A white-faced young soldier came rushing up and said that we would have to be evacuated, so we got into our car and drove a couple of miles away, having put our lunch into some baskets, and we finished our lunch. After some hours we were able to return. This was the only time that this area was used as a gas training field, yet the trees had been blown up just for that purpose and because of the lack of an environmental impact statement.

I raise two points: to what extent do we know what is the impact on the environment of the military in peacetime, and [secondly], could we not ask for environmental impact statements on all military operations?

Polunin, Snr (Chairman)

Thank you very much, Mr President. That brings to mind another somewhat comparable situation, with regard to the ozone shield, which there is very widespread concern about. I think we helped a little bit in our first Conference, because we had in it Stewart Udall, [formerly Secretary of the Interior], and Reid Bryson, who is with us this time again. They were fresh from the deliberations in Washington which resulted in the canning of the great American SST project; but, they said—and probably this is still true—they do not know what goes on in military connections in the way of high-flying jets and everything else. In other words, the [authorities] try to suppress it but [some scientists] are too bright and persistent.

Flohn

Four years ago I was asked to speak before the Swedish Research Council about this question of the future evolution of climate under the impact of Man's activities. The meeting took place in the Cabinet Room in Stockholm under the chairmanship of the [then] Prime Minister, Olof Palme, and was followed by a rather lively discussion at the end of which Prime Minister Palme asked me a question, saying: 'We are to some extent here mere politicians, whereas you are a scientist. If you should give first priority to one or another action, what would you prefer?' This being an unexpected question for me I said, after a while: 'Well, it seems to me that at least in western countries society is not prepared to accept the general idea that economic growth [cannot go on] for ever. So the first issue should be to convince the public and direct public opinion to the effect that growth cannot be for ever, but if it goes on must necessarily overshoot.' It seems to me that, at least in the western societies, we must do this job because politicians cannot act without some consent of public opinion.

Polunin, Snr (Chairman)

Thank you very much Professor Flohn for that quite vital point. I think without question we all agree, but we will just have to go on repeating it until it is heard. Forgive me if I missed you before but there was such a sea of hands.

Goldsmith

I was going to say very much what Professor Flohn has just said, but in a slightly different way. The subject of this Conference really is: Can there be development without ecodisasters? But the term development has never been adequately defined. If development means the continued substitution of machinery and other inputs for human labour, so as to increase *per capita* production and hence consumption and thereby foster economic growth—then surely the answer is that there is no possibility of development for very much further without ecodisasters. I think we have reached a stage where it is extremely difficult for serious people to continue to assume that somehow, with the aid of science and technology, we can go on indefinitely enjoying economic growth without changing the climate in a disastrous way, without large-scale famines, annihilation of wildlife, etc. In fact we are heading towards a whole series of crises which can only be exacerbated by further economic growth.

So surely the answer to the question, 'Can we have development without ecodisasters?' must be, 'No, we can't—unless we redefine development'; but then we have got to be able to do so very precisely, which we have not yet managed to do. I think this Conference must be moving very strongly in such a direction, and that the whole notion of economic growth is beginning to be doubted in political circles. In fact, a Minister of the French Government told me himself that the government is beginning to realize that it is now no longer possible to increase the consumption level of the French people. He told me in so many words that the government knows it cannot be done, but the question is, how do you admit it to the people? It is an extremely difficult thing to admit; but if we do not face this one central question, then forget about the rest. It is not a matter of details any more—of management, as we call it—because every time we manage these problems in one way, we create problems in another way. It is a question of facing this one fundamental fact: that the continued substitution of the technosphere for the biosphere is no longer possible without catastrophic ecodisasters on a global scale.

Polunin, Snr (Chairman)

Thank you very much, Mr Goldsmith. For once, in the overall, I agree with you entirely.

Fosberg

I merely wanted to comment in response to Professor Pauling's question on what the military are doing in environmental affairs. I have had a bit of experience with the military, and have a friend who was more or less within the military. He was very concerned about this, and tells me they are desperately trying to suppress all the information on what they are doing, no matter whether it has any military significance or not, because they are scared to death of what the public is going to say [and do] if they find out.

Polunin, Snr (Chairman)

Mr President, as Chairman of the Resolutions Committee, I hope you noted that. It is possibly one of the things we ought to make a recommendation about, though of course it is up to you, Sir[, at least in the first instance].

Royston

On this question of consumption, I would like to point out that there is I think strong evidence that it is not so much the fact that individual people want to over-consume, and so not a question of dissuading them from over-consumption, but rather that there is a whole system established to persuade them—even to force them—to over-consume. It is not so much a question of changing their values as of restraining those forces which are causing them to over-consume.

Following from that, we have talked a lot in this Conference of developing countries and developed countries. I wish we had talked about developing countries and overdeveloped countries, because the problems are in overdevelopment and over-consumption in certain parts of the world, while in much larger parts of the world even the basic needs of safe water, sanitary housing and adequate nutrition have not been achieved.

This brings me to a more general point which I would like to make, and a question I would like to put to Professor Hare in relation to his admirable address. Going around the world I see, in parallel to this great surge of concern over environment, an equal concern for people to have a direct say in those decisions which affect their lives.

In the developing countries, whether it is in Papua New Guinea or the Philippines, or in Malaysia or South Korea or Sri Lanka, as also in the Jura of Switzerland, in Sweden and in France, there is a call and rapid movement for a more decentralized political system. Thus there is an almost universal call for a 'bottom-up' development process, which starts at a village or community level. As Professor Hare has indicated so clearly, when once people are at that level, they see clearly what is their relationship with the environment—what is the life-style that corresponds to the carrying capacity of that local environment. There is [local] wisdom with the people at that level, and it is beginning to be realized that the global crisis has to be resolved from the village level upwards, by building a political system which starts with the community and moves up to the region and then progresses up further to the national and international levels—rather than being imposed from the central government or central capital downwards, so leading to alienation and frustration. Then people [are easily] persuaded to take to excessive material consumption as an anodyne to their frustration and psychological distress.

So the question I would like to ask Professor Hare is: to what extent would he see in the future the feasibility of discouraging less this process which seems to be taking place universally for greater autonomy at the community level and so move into a bottom-up development process, compared with the complementary process that he has shown, of a re-structuring of the top-level of over-specialized government? As he indicates quite correctly, the problems are horizontal but our infrastructure is vertical.

Hare (Keynoter)

These are exceedingly difficult questions and ones that I ask myself a great deal. In my own [adopted] country, which is a federal country, I have tended to take the view that there was too much power in the hands of the provincial governments and too little in the hands of the central government, because through much of my life I have been obsessed by things that could obviously only be tackled nationally or internationally. I have been worried about the fact that the Canadian Federal Government has progressively lost authority during my lifetime. But now I am not so sure: big government is so bureaucratic, so institutionalized, and so rigid, with so many vested interests in it. If we want to change course in my country, one of the major problems is the industrial agreements which the government has with the trade unions [as part of] its bureaucracy.

There is no doubt in my mind that wisdom is [relatively] easily deployed at the level of the tribal council, the village council, even the family—most family lives are very well conducted—whereas wisdom is least easily applied in big corporations and governments. So I have gone all the way with Schumacher, personally, and [even with Mr] Goldsmith in much of what he says. But I have to ask this question: You know the ancient chestnut of the yokel who was asked the way to some place and said you can't get there from here—you know that is a profound truth! How do you unwind the mechanisms of a complex bureaucratic state? How do you do it [without precipitating the] actual failure of the institutions, so that everybody can see that it is not working? In Ottawa, if we were to try to disband a major federal department, we would be taken apart by the professional institution of civil servants and the other bodies that protect the vested interests.

One last point, I was speaking to a Chinese official a little while ago. He came to visit Ottawa as part of the first scientific delegation from the Chinese Government to Canada after diplomatic recognition was resumed between the two countries. And he said he was the president of a local body which, roughly speaking, is translated as 'Institute for Science and the People'. He said his function was simple: the people had great wisdom and they understood what life is about. On the other hand, Science had great wisdom too. It had comprehension of the actual functioning of things. He said, 'My job is to bring them together'. And I said, 'Do you mean you have some colossal institute in Tientsin?' 'Oh no', he said, 'we do this in the bush. We do this at little decentralized institutes around the country, to which we can send farmers for three months and scientists for three months, and have them exchange wisdom.' And I said to him: 'How many of these institutes are there?' And he said, 'One hundred and one thousand at the most recent count.'

Polunin, Snr (Chairman)

Wise people, the Chinese, though I wondered just before Professor Hare came on to that last point whether we do not need, and whether even the Canadian Government does not need, a kick from Paul Ehrlich's mule.

Vilhjálmsson

I want to comment here on the question of globality, or rather what I may call the non-globality or lack of globality, of the Conference. I think we must basically all agree that the problems we are dealing with are global in character, but it seems to me that we do not quite agree on how far-reaching are the conclusions which we have to draw from this globality. It seems to me that this discussion is very much under influence of the western way of thinking—of western culture or of European and American cultures, if you like. This limits the scope of the discussion. One example of this is the widespread tendency to put a sign of identity between progress and growth, which I think people do too often here. Of course there is no such identity between progress and growth; on the contrary, they are quite different phenomena. And development and growth are not the same thing either.

We are also dealing very much with the situation of the developed countries, as can be seen in the emphasis laid on such things as the Conservation of Nature rather than on population questions and just the simple one of hunger. I think this is relevant also when we discuss the possibilities of solving the problems of the future. I can imagine in this context the possibility of an unsolvable conflict between the north and the south. We may be able to solve the problems of the developed countries by themselves, and maybe we can also find a model for solving the problems of the developing countries. These two models might work quite well separately, but there might be a conflict between them so that they could not operate simultaneously on the same planet.

I would like to thank the President for bringing up the questions of nuclear arms and militarism, which of course are quite relevant here in Iceland, being matters of great concern to the Icelandic people as I hope you know. There are rather many people in this country who would, for example, consider the American military base in Keflavík to be the most important item of pollution in Iceland. As regards nuclear weapons, we must at least touch on that question as we must be able to agree, all of us, that the most important potential pollutant on the Earth is nuclear weaponry.

Cunningham

I just want to make a brief statement—possibly in the form of a question. I came to this important Conference primarily as an observer, being from a Third World country—Jamaica. I have been extremely impressed by the way most of our eminent scientists handled themselves under the barrage of information and questions that have evolved throughout the Conference so far.

[One observation I have, which has the support of some friends here with whom I have had discussions, concerns the fact that [much of the topics dealt with relates directly or indirectly to the] problems affecting developing countries. [But I notice that] representatives from these countries are somewhat limited. A better showing from developing countries would, I believe, set the stage for a more reasonable and meaningful treatment of the topics discussed in so far as they apply to and affect those countries.]

The majority of participants represented here are North Americans or Western Europeans. Representation from Latin America, from the Caribbean and even from Africa is somewhat marginal.

Polunin, Snr (Chairman)
I would like to have a moment on this problem and say that the reasons are numerous but particularly two: first of all the availability or [geographical] nearness, and secondly the sheer economic and financial constraints. About two weeks ago, after a year-and-a-half of struggling for support, I was suddenly informed by the United Nations Environment Programme that they wanted to send, or asked me about how they might send, people from about a dozen of the so-called less developed or developing countries. They very thoughtfully asked me for a list of people whom we had chosen to invite but who had indicated in replying to our invitations that, although they really wanted to come, they could not do so unless their basic expenses were paid. Alas, it was too late to do much; but in the end several did manage to come—the latest arriving only last night—and I am confident you will agree that they have been among our very ablest and best participants, our most keen and active colleagues. We would have had many, many more if we had not been [so sadly disappointed] in the matter of financing, and I personally will not undertake this kind of exercise again, much as I enjoy it as I believe it to be really worth while, without adequate financial backing which would involve, as my budget did originally, the wherewithal to bring in a fair proportion of people from many of those not-overdeveloped countries.

Altenpohl
Did we not miss one point that needs looking after, [namely that of] self-regulating mechanisms? I am still convinced that more than three-quarters of all existing or potential environmental problems are local or regional—in the manner of the pollution of Mexico City—or else are concerned with a certain industry, and that locally a self-regulating mechanism can be realized in many cases. Thus when the oil crisis comes, the Swiss bring out their old wood-burning stoves so as not to need too much oil. Or again, if the wealth of a nation decreases, so does the spendable income per person and with it the life-expectancy—people die earlier, and that is another self-regulating mechanism.

All manner of self-regulating mechanisms take care of matters in the manner of the free market-place. Thus those industries that are polluting are in all kinds of trouble and have to close down, whereas non-polluting ones flourish on. I put a lot of confidence in this kind of self-regulatory mechanism and I did not want to go from here without giving out this bit of hope which I trust others will share with me.

Polunin, Snr (Chairman)
Splendid: we have recruited another optimist/activist! [For I understand your house in the Valais is run largely on solar energy.] Thank you, Dr Altenpohl, for joining the sun-worshipping ranks and strengthening us with rays of hope: how often are the seemingly small and long-unobserved things among the really important—and yet obvious when they are pointed out!

Sigurjónsson
I would like to express my gratitude to this Conference [for accepting the invitation of the Icelandic Government and] coming here, and I would like to [emphasize] that this place might be very well chosen. As I think you will have noticed, we are living on the edge of the possibility of life. We are here really

fighting for our existence, and the only possible way to live in Iceland is to use the technique to which I am now calling your attention.

We have been talking about overfed people and undernourished people, and about overdeveloped and underdeveloped countries. But if you look carefully around here, the possibility for us to exist has been because we have got a technique to use the [offerings of] Nature, and our greatest danger now is that we might over-use this Nature—that we do not exceed the capabilities of our Nature around us. [We must remember always] that we are a part of a biological system and that the human race cannot exist if we do not take care of this biological system. We cannot over-exploit it without incurring disaster. And this is what I think will be the conclusion of this Conference: that we have to take care to our reserves and do not over-exploit them but instead use science for a better way of living. [Applause.]

Polunin, Snr (Chairman)

Absolutely, and that goes in my humble opinion for everything from [the geothermic resources of] Hekla, who has been kind enough not to erupt us though it is almost the only thing we have not suffered in getting this Conference into being, to the appalling erosion, Sir, which you have in places and which we hope you will soon succeed in controlling. You have good people, good minds working on it, and we hope good hands [as in the days of your heritage the Vikings].

Fosberg

This may be a little bit late to express what has been bothering me, but we have heard a good deal about village wisdom and tribal wisdom and so forth, and I certainly agree that it is a very real thing—that it exists and should be utilized as much as possible. However, in a good many countries the vast majority of people live in cities, not villages. There may be 10% of people who do live in villages and have all this wisdom, but how are they going to transmit it and use it to influence the tremendous masses of unwise people, or people who are unwise enough to live in cities?

Schütt

It is hard to determine what knowledge has been gained during this Conference, the course from the observation of a problem [to effective combative action being so long and complicated in Western government]. We use the resources of Nature and scatter them all over the world, so that they cannot be recovered again, or we destroy some part of the productive system of Nature, so that we cannot regain the productivity (for example of a forest) in the future. In describing what is really going on, the ecologists and other scientists play a very important part, because they are trained to see the relationships between causes and overall effects and can, from their scientific base, analyse and understand the situation before suggesting scientifically sound actions towards solutions.

You can ask yourself what are the main causes of the environmental effects which we see around us today. We already know that there are two main effects—one is from overpopulation, mainly in the developing countries. They are increasing very much in population, for which new resources are needed. Secondly, the industrial world is using too much [in the way of] resources, not because of overpopulation or increasing population but because of overconsumption. This has been obvious for a long time and the problems have been penetrated scientifically as we have seen today [and earlier in this Conference]. But when we come to implementation of the solutions, and determining what we should do to prevent further development in these undesirable directions, there seem to be no practical possibilities [and clear courses of action].

I know from Sweden, a small country where we are quite well off in many respects, that it is very hard to stop people voluntarily from consuming more [and more, though we are supposed to be 'modern' and hence enlightened]. Everyone agrees when talking in private that present consumption is enough, [or often far more than enough,] but society behaves in quite another way—demanding more production and higher income. And how is the increasing population in many places going to be stopped? [Surely not by telling people to have babies on religious or other grounds and actually paying them to do so!] Well, these are political questions and that is why I [say again that] the scientific penetration of the problems has come far enough to be handed over to the politicians [, doing everything in our power to ensure that they are good ones].

Hare (Keynoter)

Let me crystallize my anxieties about the point which Professor Schütt has raised—how do we get there from here? An economist will say: 'My colleagues at Resources for the Future in Washington contend all the time that if you wish to vary consumption, there are many ways you can do it.' If you let the market do it—whether it is a free market or a monopolized market—prices will do it. If prices become high enough, consumption diminishes—so there is one, the classical, way of letting prices determine. The fact that almost every price is an administered price, not a free price, makes nonsense of the arguments which flow from that fact; but still, that is one way of doing it. You can also put a gun at people's heads: in other words you can be authoritarian and say, 'You will not'. You can devise rigid controls. People cannot buy marijuana in my country: they do, as a matter of fact, but they are not supposed to. You can ration; this is the bureaucratic solution, which is widely used in the socialist countries, and in wartime is used in the democracies. You can alter public taste; this is what the propagandists do. It is also what the advertising agencies do, but for what most of us think are the wrong reasons.

This leaves us as I see it with only one way to [limit consumption and yet retain freedom of] choice, namely to make people *spontaneously* want to give up cigarettes, give up big cars, and give up other things that you think are bad. Here I might add that we have not had an adequate input from the point of view of environmental health, and I wish we had.

Allow me to make a last point: you *can* do the kind of things I have just been saying you cannot do—in a disaster situation! There was such a disaster in the Tennessee Valley area in the depression years, [when things practically] fell apart. It was not just that there was massive soil erosion, uncontrolled flooding, silting up of a floodplain and destruction of forests; there was also no income. The people were starving. Their entire society—the social fabric—had fallen apart, and the US Federal Government had no choice but to step in. What it did was to create a body—the Tennessee Valley Authority—with the mission of generalized regional development, starting with the water-power in the Valley, but controlling soil erosion and doing a number of things that would have been done under a full free economy if matters had been otherwise; but they were not, so this was a necessary device [that has become a famous model]. This action did involve the suspension of a considerable number of freedoms, and it was bureaucratic; it was not democratic, as we should consider it, except in that it was originally established by the Federal Government in agreement with the states [concerned].

It took a disaster to do this, and the question that has been raised, and that I raise again [as] I do not know the answer to it, is: How do you take a free society that is not bankrupt, that is not desperate, that is not starving, and alter its life-style in such a way that it will go in the direction that we think it ought to go?

Polunin, Snr (Chairman)

Thank you, Professor Hare: now there are so many hands going up, [of would-be speakers, that we will have to take our *topics* seriatim. Starting with that of environmental health, on which we would like to hear first Dr Juel-Jensen, followed by] Mr Eyjólfur Saemundsson, who was so [fearlessly objective] we thought the other night in our workshop on local environment and health matters—he is Adviser to the National Institute of Environmental Health, and [as he is evidently not in the audience] we have put out a call for him to be brought in if possible. Following that topic we will consider the *use* of disasters, for it is indeed 'an ill wind that blows nobody any good'. [That happens to be one of my abiding beliefs, another being in the law of supply and demand.]

Juel-Jensen

I think it is perfectly true that you should have had an environmental medical man here, and I am [scarcely that]; moreover, I have kept relatively quiet about medical matters because the medical profession has a lot to answer for. It has been quite irresponsible over the years—particularly those of its members who have gone into scientific medicine, for like most other scientists they have been pursuing their own goal because of the fascination of it. It is fascinating that you can keep people going on a tin kidney for years; as an intellectual exercise it is tremendously stimulating. It is fascinating that you can cure virus diseases, first in pots, then in rabbits, and sometimes in Man. But the law of diminishing returns has long since started to operate, and too rarely do we ask what sort of quality of life we are giving to people on the tin kidney, for instance, or what sort of horrors are we subjecting people to who have a serious virus disease, or cancer, when we treat them. I am not advocating that [pertinent] medical research should stop, but I do think that we, as a profession, should really pause and look at the quality of life that we give to people.

Somebody mentioned before that, with rising income and rising prosperity, so the age at death rose, and this I think is one of the curses of the overdeveloped world, because Man [has not evolved], or so I believe, to live much beyond three-score years and ten. The machinery starts to wear out and I think that you, as environmentalists, should consider seriously whether it is right to spend as much as we do on keeping people alive. I am not advocating for a moment that, if something should go wrong, we should take active steps to remove people, but perhaps sometimes we could spend less on keeping them alive without looking at the quality of their life.

Certainly medicine has a lot to answer for when we come to the population problem. That has already been touched upon, and I do not believe we have any right to [interfere with Nature by] eradicating mass scourges unless, at the same time, we also limit the population growth.

Finally, I think it is the duty of the medical profession to point out the dangers [of such activities as smoking], but it is not within the terms of reference of the medical profession to put a gun to people's heads and say: 'You must not smoke cigarettes because every time you smoke one that's five minutes less of your life.' If somebody chooses to smoke cigarettes, well, that is his funeral and they are his coffin-nails.

Polunin, Snr (Chairman)

Thank you very much for putting us in the picture in these respects. Mr Saemundsson not having been found, Mr Goldsmith now has the floor and, after him, our patient colleague Páll Bergthorsson, of the Vedurstofa Islands—the Icelandic Meteorological Office that I suppose is the source of the chronic reports of 'depression over Iceland'.

Goldsmith

I am interested in answering the question, 'How do we get there from here?' It has been my main preoccupation since, with [colleagues], I wrote *A Blueprint for Survival*, a document which [already some years ago] tried to answer this question [interjection from Professor Hare:] 'But you didn't ask me to sign it! I would have done so if you had asked me' [though of course some others would not].

I feel that there has been a lot of convergent thinking on this subject, and that a new and fairly coherent approach is fast developing—one that we might refer to as the ecological approach. Needless to say, it is not original. Tolstoy, for instance, and also Kropotkin, saw things very much the way we do, and so did Gandhi. The trouble is that Gandhi lived at the beginning of what we might call the Age of Technological Euphoria. People really believed in progress in those days. The next decade saw the development of antibiotics, the 'miracle' drugs, and also of pesticides. Then we flew to the Moon. All these astonishing achievements seemed to confirm the main tenets of the 'Religion of Industrialism' with which probably all of us here were once imbued. Indeed in the days of Gandhi and even later, to launch an ecological movement would have been to swim against a powerful tide. It would [probably] have been impossible. Today the situation has changed. The Religion of Industrialism is disintegrating. We are living in what has come to be an ideological vacuum. Also, physical conditions now favour the spread of ecological ideas. I think we have already reached that point at which it is more difficult to maintain our society on its present course towards ever more economic growth than to satisfy ecological requirements and reverse the process.

This is reflected in the inability of our politicians and economists to solve the problems that society faces today. Technological, material and institutional solutions—the only ones reconcilable with the world-view of industrialism—no longer seem to work. Take the case of health. One of [our speaker's] colleagues on the Canadian Environmental Council told me, if I remember rightly, that 80% of the health budget of Canada is spent on people who are in any case going to die within the next 10 months. It is a terrifying thought.

In France [the average] family now spends over £1,000 a year on health care, and this sum is expected to double within the next 15 years. Clearly it cannot do so because the money will not be available. The French Government realizes this, as do other governments, and most of them are now actively looking for different solutions to the health problem. The only satisfactory ones, needless to say, must be based on a very different view of health—namely that which the ecological approach to the problem would provide. Such an approach would mean setting out systematically to reduce the need for the high-technology medicine which we can no longer afford. People would seek to become healthier by leading a healthier life—by taking more exercise, eating better food [in more appropriate quantities than at present], and breathing less-polluted air. This of course would mean changing a lot of things—in fact the very way our society is organized.

Bergthorsson

You all know the story of the boy who shouted 'The wolf is coming!'. As there was no wolf visible at that time, people were not aware when the wolf actually came. The story says the boy was lying, but maybe he was not. Maybe he knew that the wolf would be coming but he did not choose the right moment to shout. I think it is necessary for the environment-minded people to shout, but it is important always to shout the truth at the right moment. [Shouts of 'Here, here!']

We must never be accused of frightening people without reason, or otherwise our action may be not only useless but even harmful. This may be another way of formulating Professor Hare's statement that very little can be done before the state of disaster is reached, and I think he is right. Let us therefore in the meantime prepare people to listen to us when it becomes necessary. [Applause.]

Polunin, Snr (Chairman)
Thank you very much: that has the wisdom of your gallant country behind it.

Sheikh
I would like to go back a little to the self-regulating mechanism *vis-à-vis* growth without ecodisasters. A self-regulating mechanism in Nature perhaps will not grow the way we seem to understand it. It grows at its own pace while we try to synonymize growth with development.

For developing countries we keep talking about their development, but, looking at the example of the so-called overdeveloped world, we see that there are ecodisasters which have been faced and which could multiply with the passage of time. In view of this, what should be the advice to the developing world? Is it better to develop or to remain static and let Nature take its own course?

Hare (Keynoter)
That makes me think furiously about what I am going to put in the synthesis I am to write.

Polunin, Snr (Chairman)
In calling on the next speaker I would like to explain that there is, in my old trade of botany, a pair of great names [towering above others in a manner that was good for our discipline]: one is Linnaeus and the other is de Candolle, and you now have the direct descendent of that line, Monsieur Roger de Candolle, s'il vous plaît.

Candolle
I would like to point out that my great-grandfather had [long] debates with Sir Francis Galton, who believed in intellectual heredity whereas my great-grandfather did not. [Objection from Chairman: 'Well I do, knowing you and many other shining examples'.]

As a mere onlooker I have hesitated for some time to say this: I feel we are often failing to face essential facts. As an instance I can mention the insistent use of the words 'developing countries', even though Mr Glasser has shown us with figures that the fastest 'developing country' is the USA. We also make no mention of the fact that our Western civilization is in rapid decline. I would just like to quote Mr Soustelle, the well-known ethnographer; in a lecture which I heard him give some time ago he outlined what he considers to be the signs that a civilization is on the decline, based on his study of global history, and they were:

(1) A shrinking of its 'Lebensraum'.
(2) A tendency for populations to gather in large centres.
(3) An ever-increasing control by the central authority.
(4) Most important of all: a loss of faith in basic tenets accompanied by great attention paid to details (people no longer go to church but must have a washing-machine).

Personally I feel that this decline is likely, in the fairly near future, to face us with events in the social field which, though I cannot foresee what they will be, will occur before, and perhaps greatly influence, the biospheric developments we are discussing.

Polunin, Snr (Chairman)
Thank you very much my very dear friend for opening up further [vital issues. Now] I am going to suggest that we decide on whether to adjourn or continue if we want to for another 20 minutes or so, if that suffices. Continue? So be it. Could we have more on that last theme, please?

Candolle
One instance which the ethnographer gave us was of a very civilized Roman who for some reason went up among the Huns and there he found a Roman citizen who was living among the Huns; he was wearing a sheepskin like everyone else and he had married a Hun woman. [So the first Roman] said to him: 'What are you doing here? You are cives Romanus?' And the [second] Roman said: 'Well, here I am not bothered with any laws.' [Shouts of 'Bravo, bravo!']

Hare (Keynoter)
I must say, [Chairman], having met Mr de Candolle in your apartment many times in the past, I had not realized he was so competent a practising anarchist, but now I am delighted to hear it [in this sophisticated way. But could I emphasize] how profoundly I agree that we, in the West, are in a state of decadence, in the true sense of [the word]. Now if you want to know what I mean by this word, and to quarrel with me, go and read Michael Harrington's book *The Accidental Century*. Harrington is a socialist, Catholic intellectual, who had a profound influence on American thinking—I believe more than Galbraith has in many ways, though people do not know him as well as Galbraith. He was the man who shamed the White House into the Poverty Programme years ago, and then helped to run it. He withdrew because he did not think it was going the way it should. His book starts with a sentence that I think Mr de Candolle would love and I hope the rest of you will. I cannot quote Harrington precisely, but this is roughly what he said: In the 20th century something enormous *but unspecified* died and something enormous *but unspecified* was born. His concept of decadence springs from this moment [of realization] that the Western societies—and please note that I am not talking about any other societies than those in which I myself live—these so-called advanced countries, are in a state of political disintegration because of their loss of faith, their loss of comprehension in their own mission and motive.

In a sense I am almost sorry that I am an environmental scientist, because it commits me to too narrow a part of the spectrum in which to be a social critic. I think of the environmental conundrum as being a part of a very much larger problem—the problem essentially of the purpose of life. What we have done in the West is to become affluent, but [at the expense of destroying] ourselves as compassionate human beings. If any one reads that as a [plea] to the peoples of the Third World that they should avoid economic growth, I did not say that and I do not mean it, as the developing world is not developing in that sense. This is also one of the tragedies of the age, because there is a clear need for the kind of growth that we have been discussing here; but for us in the West, the fundamental question is to find some idea large enough to capture our own imaginations, and to make us believe in ourselves again.

Polunin, Snr (Chairman)
Thank you very much. If we might add some dimensions to that, one would be discipline—in the nicer sense. We have lost discipline very largely [and come to dismiss it as an undesirable concept and even a dirty word]. Now Dr Miller has been raising his hand for so long that we might presume him to be a stern disciplinarian [—at least with himself—for not stomping out. May I express the hope that he will bring us back to considering the Chinese, among whom he recently sojourned].

Miller

Mr Chairman and friends, I have to read a little of this to get us off the ground. Our far-seeing Secretary-General and his sustaining menage convened this Conference and the one before it six years ago, and closed that Jyväskylä, Finland, First International Conference on Environmental Future in much the same spirit and theme that we are in right now. It is hard to believe that six years have passed—that we have had Stockholm and many things since—and one shudders an editorial shudder to think that it may be another six years before a like forum resumes. But resume [it must, or so I feel as a veteran of both occasions]. For although there are many conferences—some say far too many—this I believe is the only really free high-level series of its kind in the entire world].

May I ask this Conference if we can consider some appropriate action, or have we done so with that Coffee Group we talked about, to authorize at least an interim action body, such as the Foundation for Environmental Conservation, to be an Ombudsman—a prod and a conscience in the world scene during the coming months and years? This proposal, it seems to me, does not have to have an answer here and now, but it would be good if we could think about it and maybe say something before we adjourn tomorrow. I have not yet consulted with anybody to determine whether this is considered a practicable suggestion; maybe you think it is not and it can be dismissed. I am sorry to intrude on philosophical themes which are much more interesting; but the clock is going around, tomorrow we will be gone, and some of us will then regret that we did not do more when we [were together and] had the chance.

I am not interested in this in a personal way; nor do I wish to be a decision-maker. But my experience as a leader of Foresta Institute indicates to me that much could be done. In concert with other leaders or directors of nationwide groups, we signed letters of protest or recommendation to various bodies in government and in the communities where we felt that it was wise to let those in charge know that a conscience was speaking, and that there were people watching and caring. I have confidence that it had some notable effect. It had some good results both with appointments by US Presidents and over the actions of bureaucrats and ministries, and I think we would not be amiss to consider that means here. But I do not know whether we can assume the arrogance to do so, or whether we can leave it to the conscience of the Council of the Foundation for Environmental Conservation. But I do think we ought to give some kind of thought to this suggestion before we adjourn.

Polunin, Snr (Chairman)

Apart from responding to the gratifying reference to the Foundation for Environmental Conservation, of which I happen to be President, I do not know what to say to this proposal except that we are a very small and young body (which I like to refer to as a 'non-organization') with at present barely sufficient resources to foster our two main chronic activities, namely these Conferences and the quarterly journal *Environmental Conservation*. I suppose if somebody were to finance us adequately we could [start this] job [largely by correspondence] from our present headquarters in my home just outside Geneva, Switzerland.

Meanwhile considerable portions of the activity I can envisage are being carried on already by other bodies, including two of our own three co-founders, IUCN and WWF (I myself was the third), and by two of our collaborators, INTECOL and WERC, as indicated on the cover of our journal. And then, of course, there is always UNEP. Actually, I was approached earlier today by two of our Conference participants with a somewhat similar-sounding proposal which seemed to constitute a hiving-off from the Coffee Group (already!), so there appear to be many ideas and incipient actions milling around in other heads than Dr Miller's much-respected one. [As the other two members asked me to treat their overture in confidence, I cannot divulge their names or any further details without their permission. But altogether I wonder about the wisdom of setting up any more bodies for the time being—at least without the dual force of prestige and funds whose lack is causing the death of others, and especially as there are plans afoot to approach ICSU about the Coffee Group initiative, which I personally think would be a most appropriate adoption if it could be arranged, as ICSU is surely in some ways the world's environmental summit and has its own very active Scientific Committee on Problems of the Environment (SCOPE). But it certainly seems important to watch the forthcoming United Nations Conference on Science and Technology for Development—especially in view of the last two words in its title, concern over which was a leading tenet of the 'Coffee Club' (as the Group soon came to be called).]

Buchinger

I am a little sad that many of us seem to understand this whole meeting as an 'either' or 'or' affair. I personally go away very happy because it is not a question of not to develop or to develop in the wrong way. From quite a few of the papers presented we saw that we do not have to repeat the mistakes which the so-called developed world has made, and to suffer similar consequences. Now we know that there can be technologies which [practically speaking] do not pollute. We also know that there is something wrong with aerosols, because the technology of this century has made it possible to know these things, and, as Stewart Udall said, 'Learn from your mistakes, don't repeat them'. And now we are learning how to repair damaged ecosystems [as described for example in several recent papers in *Environmental Conservation*], and so altogether I am going away more optimistic than I was when I came.

Fosberg

I merely want to ask one question that has been recurring to me since hearing some of the comments about self-regulating mechanisms. I was just wondering if the ecodisasters that we are discussing at this Conference may not be the self-regulating mechanisms of our [human-dominated] ecosystem? I have, of course, to continue to be one of the pessimists of the Conference.

Obeng

As an individual attending this Conference from a developing country, I feel that, before there is any prodding of the international community in the manner which has been suggested, it is important that we should know what we are going to prod, as well as why and how. For on one side we have been discussing the developed world, which I would rather think of as the *industrialized* world—as opposed to the other [world], where the problems we face [are different]. There it is the basic needs [of life] that we are talking about, such as good drinking water, sanitation, and having food to eat from cultivating [the] land. They are two different 'worlds'. We view environmental concerns quite differently. What I think of when I talk about the environment [is] what should be done to make it possible for all the poor to continue to exist—whether or not they multiply a million times more than they should—with each getting the basic things which they are entitled to as human beings [is also what I feel we should be considering—how to feed them as it is a solemn duty to do].

As for the question of transfer of technologies, [you may say] it is easy to think we can see where the industrialized world has gone wrong and somehow avoid getting into [a similar] mess, but it really does not work out that way because conditions are so different and, when it comes to the crunch, whoever takes the decision invariably asks for the same things that caused the problem in the [industrialized] countries [because of the development they have achieved. Only we need to ensure, this time, that they are not destructive].

We may talk of village wisdom or tribal wisdom as Mr Polunin Junior and a few others have done—of what people whom you refer to as primitive have been doing for a very long time—and this enables me to retain some possibilities for optimism. But we have first to retain such wisdom—or, where necessary, salvage it and put it together again in order to [use them to] preserve and protect our resources.

Polunin, Snr (Chairman)

You are absolutely right that these resources, including wildlife, must be preserved, but mounting population pressures are going to make it increasingly difficult, and I really do not see how you can expect us to support you as you seemed to imply if you multiply a million times.

Goldsmith

I would like to try and answer this. I think we are all concerned with the problems of the Third World, especially [to the extent that they may be] caused by us. But the question is whether development, as I understand the term, can solve them: I am quite sure that it cannot! In order to develop, Third World countries must earn foreign currency. How are they doing this? Unless they have oil or copper or some other valuable mineral resources, they can only sell off their forests for lumber and grow cash-crops for export on valuable land that could be producing food for their often starving inhabitants. In the Philippines, for instance, 55% of the arable land is apparently used for producing cash-crops—largely for export. In Santo Domingo the figure is about 25%. And what do they do with the currency earned in this way? It is largely used for putting up air-conditioned office blocks, for building power-stations and motorways, [for importing huge prestigious cars and electric tooth-brushes], and for buying the rest of the useless paraphernalia of the profligate industrial age.

The fact is that trade with the Third World is negative aid—it involves selling the indispensable in exchange for the totally superfluous. If I were running a Third World country, the first thing I would do would be to cut myself off from the industrial world and foster self-sufficiency at every level down to that of the village. In fact, one should not be developing the Third World but de-developing it.

Polunin, Snr (Chairman)

The hour is getting late and we have several more would-be speakers. We will take the most insistent one who was first—briefly please—and then Dr Obeng again before adjourning until 9.30 tomorrow morning in the Crystal Room where the final plenary session will commence at 10 o'clock.

Juel-Jensen

Mr Chairman, I shall be brief, [but feel compelled to say that] I think it is a great pity Mr Goldsmith spoke again, because I can only disagree most profoundly with him. I happen to have responsibility, though intermittently, for between 80,000 and 100,000 people in one of the poorest countries of Africa, and, if he thinks it is funny to have amoebic dysentery because you have not got decent water, and if he is going to withhold appropriate technology which could be supplied very cheaply to help these people, then I think he is a cruel and heartless person. What one needs, in a poor country, is very little [beyond] use of intelligence and simple measures, whether they be technical or medical. But do not withhold the best things that the West can give to other countries, as was implied should be done.

Obeng

I shall be very brief because I think the gentleman here has said exactly what I was going to say. It is a pity that when we talk about conditions in the developing world on such [an occasion as this] we can even refer to such useless things as electric toothbrushes—particularly after my previous intervention on the problems that face developing countries. Such a comment shows that we prefer to see the problem from different angles. It is a pity [but it seems that] when it comes to interpreting such things, we of the developing countries will see our problems differently from people like Mr Goldsmith.

Polunin, Snr (Chairman)

Dr Obeng, we must have another such Conference [and hope] to hear you [further; meanwhile we will resume this Session at] 9.30 tomorrow morning in the Crystal Room.

[Adjournment.]

Renzoni

Before going to sleep last night I thought a great deal about our last hour of discussion, and after a while I finished with the question of the late Lyndon Johnson mentioned by Professor Hare: 'And now what?'

On one hand, during the various sessions of this week, in talking about the developing countries it has been repeated many times that we of the developed countries should respect their culture, their habits, their civilization, and so on. . . . But on the other hand, Dr Obeng yesterday said that they need tremendously the fundamental things, such as clean water, better housing, sanitation and other essentials—not electric toothbrushes. Now who knows at what point the introduction of this, let us say, civilization should be stopped. I cannot really figure out how the clean water could reach them without making reservoirs and pipes and roads; I cannot figure out how they can have better sanitation and housing if they do not have schools and universities to prepare physicians and train technicians to perform the respective works. And we can go on and on.

At what point, I repeat, should this form of exportation of goods from developed to developing countries be stopped? And who should decide [if and when] to stop it—the people of the developing countries? I doubt whether they will, because it seems to me that after clean water, they will want to try beer or Coke, and after better houses they will want to have radio, TV, telephones, [Mercedes or Cadillacs], and so on. And to 'improve' their life in this direction they have to buy all these things—unless they are to live on charity, and I doubt that they want that!

Hare (Keynoter)

I simply [wish] to say how much I agree with Professor Renzoni. There is a paradox here. I do not think that anybody with any sensibility in the advanced countries wants to export the flippancies of our civilization. To be sure there are the people who manufacture them, but for most of us I think clearly we would like to see the advanced countries choose something else. You heard what Mr Strong said the other night.* There is no question in my mind that the Western world is going to have to think out what are the true benefits of growth, and how they may be achieved [in reasonable degree] without stress and pressure upon resources and environment; how, for example, can you maximize the service input. [Mr Strong's and] my country is a backward one when it comes to personal service. The services that enrich life have atrophied at the same time as we have become a two-sets coloured television society. I have two television sets in my house—not because I

* In his Baer–Huxley Memorial Lecture, printed on pages 611–25.—Ed.

want them, but because my son insists [on 'keeping up with the Joneses']. Yet the difficulty I find, Professor Renzoni, is that the last thing I want to do is to tell people in the developing world what they [should] want, and therefore I am mute on this. If I go to a developing country I simply do not say: 'You shouldn't want electric toothbrushes, you shouldn't want coloured television, you should want clean water.' I want people from these countries to come to that conclusion themselves.

I think what [we] are talking about here is one of the most profound difficulties. In spite of the obvious warts on the face of affluence, in spite of the fact that affluence has not brought either happiness or prosperity to the West, nevertheless people choose it. They choose it in the West, and they will [surely long continue to choose it so far as possible] in the East.

Glasser

Returning to my earlier account of the village in which I lived some years ago in southern Italy, where the social event knitting the community together centred on the daily washing down in the ravine*, owing to television and the impact of the media inculcating other choices of what was supposed to be a better life, people's attitudes altered. They no longer went down to the river every day to do their washing, but instead went back to their homes, [using washing-machines and soon finding that they had not] enough water.

So you see it is not only a technological matter of the sort of processes of life [that we are involved with, but rather one] of altering the attitudes of how you achieve the objectives of living. It is only when these things happen that you discover that you have landed yourself in a position in the cultural sense from which it is very difficult to find a way out.

Polunin, Snr (Chairman)

Thank you Mr Glasser. This reminds me of a rather close parallel in our erstwhile village in westernmost Switzerland, where the Laiterie was the great meeting-point and social melting-pot. People gathered there to get their milk each early morning and late afternoon, and when our children were small they would say, if we were out, 'Hurry up, Mummy and Daddy, we must get back or we shall be late for the Laiterie'. This ceased to exist a very few years ago and we had difficulty in getting milk because it was not always available in the only shop that was within a reasonable distance. [Moreover, having gone through a series of pooling, packaging and other operations, it was never really fresh any more.]

Ducret

In the course of the week, several speakers have qualified themselves as optimists or pessimists. At first glance this looks like a trait of character: one has an optimistic or a pessimistic character. But when one tries to understand this sort of attitude, the impression emerges that [the basis is] something quite different from simply character. Indeed, as none of us can reasonably claim to know the future, being an optimist or a pessimist in the context of possible ecodisasters seems to be an expression of confidence in the socio-political system: I am an optimist if I believe that it will act in time and in such a way that ecodisasters will be avoided, or a pessimist if I do not so believe. In the context of this Conference, therefore, being an optimist or a pessimist seems to be more a political attitude than simply a trait of character.

[But we have to ask,] is this sufficient? Let us imagine a situation where ecodisasters have occurred. The only consolation of a pessimist would be that he was right in

* *See* Discussion of Session 14: 'Life-style Alternatives'.

being a pessimist. An optimist would be wrong but would have done what was feasible at his level. In any case, there will not be much difference [when once major holocaust or other] disasters have occurred. Let us now turn to the opposite situation, where ecodisasters have been avoided. The pessimist did not believe this was feasible, remained inactive, and is therefore not only wrong but also faulty in his inaction while knowing the situation. The optimist on the other hand was right, and helped in avoiding the occurrence of unfavourable events.

Whatever the future may be, there is no real ground for pessimism. One *must* be an optimist actively involved in the protection of our living world.

Polunin, Snr (Chairman)

Thank you, Dr Ducret. Let us all hope we can be optimists as [you have so persuasively indicated we ought to be]. Now I wonder if Professor Bryson has one of his pungent comments—or perhaps a poem, as our Foundation's Poet Laureate, with which to inspire us further?

Bryson

I cannot concoct a poem without a few minutes at least to think—so here is a comment. To me the question of optimism or pessimism is not quite in the same framework as politics; rather is it a matter of scientific certainty and urgency. I myself reject the idea that I am a pessimist or a doomsayer. I say I am a *realist*, and that there is urgency. The optimist does not see urgency—at least not those of my friends who say they are ecological or environmental optimists. They say that the trouble is not upon us *now*. The question to me is how certain we are of the time-scale of an ecodisaster. On one hand is the technical scientific assessment of where we are with regard to a point of no return, and on the other is how certain are we of that which sets the time-scale? And I think those of us who take the viewpoint that has been described as one of pessimism are saying that time is short, or that we may have already passed a [potential ecodisaster's] point of no return.

This is scarcely a political matter but rather like buying life-insurance. If you are an optimist, then why buy insurance and pay that extra now—wait until you are closer to death. If you are a pessimist you say, as we are using this term: 'I don't know when I am going to die so I'd better take out the insurance now.' It is as simple as that, I think—not one of 'Is there a solution or is there not a solution?', or 'We will survive or we won't survive'. But if we have concern for our fellow men, then we have to be concerned about whether we can do something in time to avoid trauma.

It is this sense of urgency that most of us express when we take the viewpoint that things are not good, and that the chances of easy solutions are slim. We are simply saying that there is urgency, not that there is no hope. If there is no hope, then eat, drink and be merry. To those of us who try very hard to find solutions because of this urgency, I say there *may* be solutions. In a sense we are being optimists, but we are urgent optimists only in the sense of 'Let's get with it now because, if we do not, then we must all be pessimists'.

Polunin, Snr (Chairman)

I thank you as ever, Professor Bryson, and, as I see you there, I recall vividly the occasion of our first ICEF in 1971 in Finland. [When you were on the platform, the aforementioned Stewart Udall nudged me as we sat together and said: 'Where did you find these remarkable speakers?' Since then you have become, whether you like it or not, a major television and other platform personality in the best of causes—truth.]

Now at last I see no further raised hands and so close the general discussion in order to give Professor Hare what I trust is still adequate time to respond and wind up the Session, meanwhile thanking him and you all most heartily for fine support and lively attention.

Hare (Keynoter, concluding)

Let me say first of all that I do not intend to take more than four minutes, but I was disturbed at the end of yesterday afternoon's prolonged discussion to have a good friend of mine say to me: 'I was amazed how pessimistic you were.' This really bothered me—not because he felt I was pessimistic, or rather not because he put that judgement upon me, but because I was disturbed that the rest of you might have done so too. May I extend what Dr Ducret said so beautifully, and [Professor Bryson] has just said [in his inimitable way]: I spent a considerable part of my life convinced that [our] species will probably eclipse itself—destroy itself—I grew up as an undergraduate under the shadow of Hitler, and was quite certain that Hitler was going to overrun Europe. My elders were not. It was the Chamberlain and Baldwin era in Britain, which is where I grew up; and I thought this was an appalling time. People were willing to tolerate atrocities in the interests of accommodating themselves comfortably to the conditions of the moment. They were afraid, and so they compromised. Subsequently, of course, one has lived under the shadow of nuclear warfare—Professor Pauling's point, with which I am profoundly in sympathy.

It seems to me that one is a fool if one enters into action without an adequate appreciation of what one is up against; and so yesterday I made it clear that I think we are politically in a desperate condition because I believe that democratic governments have proved themselves incapable of facing the realities that we have to face. That sounds pessimistic. It is not [so much pessimism as] realism! What one has to say is that the times are out of joint. Then I did not go ahead and say 'O cursèd spite', but went out of my way to say that a time of crisis is a good time—it is the [best time to] budge things. So I should like to say that I did not intend to be pessimistic, I did not intend to be optimistic: I intended to be *activist*, because it seems to me that the thing which has been lacking from our discussion is any kind of expression of determination of the sort that I hear all the time from our young student activist colleagues.

I happen to be a Trustee of a body called Energy Probe, which is a student organization, and those students are absolutely determined to deflect the path of Canada from what they regard as nuclear suicide. Now I personally do not agree with them that nuclear power is incapable of being made safe. I think it is capable of being made safe as a technology. Whether it can be made safe in the presence of militarists who want to use it as a military weapon is another matter. The real issue is that of nuclear warfare. But the point is that these [young people], though I do not agree with the view they take, are extraordinarily dedicated to action, and I would like to declare myself very much on their side.

BAER–HUXLEY MEMORIAL LECTURE

The International Community and the Environment*

by

MAURICE F. STRONG

Chairman of the Bureau, IUCN, 1110 Morges, Switzerland, and *Chairman of Petro-Canada, PO Box 2844, Calgary, Alberta, Canada;* formerly *Secretary-General of the United Nations Conference on the Human Environment* and subsequently *Executive Director of the United Nations Environment Programme;* current address: *Chairman, International Development Research Centre, Box 8500, Ottawa, Ontario, Canada.*

INTRODUCTION

It is a signal honour to have been asked to deliver this lecture to commemorate the outstanding contributions of Jean Baer and Julian Huxley to the cause of conservation. Both of these great men devoted their genius as scientists to helping us to broaden our perceptions of the human condition and their talents as leaders to establishing the institutions needed to safeguard and to improve the conditions of life on this planet. No two individuals did more to establish the foundations on which environmental conservation can now be seen to be firmly based and dealt with as a critical global issue.

It is therefore perhaps fitting that I use this occasion to assess the progress which has been made in the evolution of conservation as a major global issue and point up what I see as the major challenges that now confront us—particularly those of us who are committed to carrying on the work which Jean Baer and Julian Huxley so notably made their own.

I speak as a layman and an internationalist, but one who has been privileged in recent years to be directly involved in the processes of dealing with environmental conservation at the world level. I am neither scientist nor politician. But my work has been on the interface between science and

* Slightly updated from the version published in *Environmental Conservation* (Volume 4, No. 3, pp. 165–72, Autumn 1977). The lecture was delivered in the Theatre of the Hotel Loftleidir, Reykjavik, Iceland, on the evening of 8 June, 1977, following the mid-Conference excursion. It was applauded with great enthusiasm which was followed by lively discussion that extended to midnight. Including the parts played by the Lecturer and Chairman, there were in all in the discussion 32 interventions, etc., by some 20 different participants: *see also* pages 627–8—Ed.

public affairs, and in it I have been involved closely with both—with the result that I have formed a deep appreciation of the importance of both. To me it is clear that any solution to the present human dilemma requires the closest and most sensitive as well as positive interaction between the practitioners of science and those of public affairs.

First let me say a few words about my own use of the terms 'conservation' and 'environment'. I will use the former in its traditional sense as denoting the preservation and care of Nature and natural resources and also their economical use. There is still no generally accepted definition of environment. In the preparatory committee for the Stockholm Conference, long discussions took place in an effort to agree on a satisfactory definition, and the effort finally had to be abandoned. When I use the term 'environment', I use it as the umbrella term embracing the whole spectrum of issues which derive from Man's interactions with the natural environment and the effect of these interactions on both Man and Nature.

Although 'conservation' and 'environment' are far from being synonymous, it was the conservationists who pioneered the modern environmental movement and laid its foundations. Conservationists continue to provide the core of the leadership of concern for the environment; indeed their voices are heard and respected as never before. Thanks to the pioneering work of men like Sir Julian Huxley and Jean Baer, the protection and improvement of the environment have become global concerns.

Stockholm: Before and After

The United Nations Conference on the Human Environment, held in Stockholm in June 1972—just 5 years earlier—moved the environmental issue into the centre of the world political arena, and established it as an important item on the agenda of governments throughout the world. The concern for environment which led to the convening of the Stockholm Conference came almost exclusively from the industrialized countries. It was rooted in the growing evidences that the processes of urbanization and industrialization, which had produced such benefits for their societies, were also producing unexpected costs—pollution of air and water, mounting threats to the survival of wildlife, desecration of coastal, forest and wilderness areas, and deterioration of the quality of urban life.

These were all issues of small interest to the developing countries. They insisted that, if they were going to take 'Stockholm' seriously, it had somehow to be made relevant to their primary preoccupation with development. They made it clear that they thought underdevelopment and poverty constituted the most acute and immediate threat to the environment of their peoples. Accordingly, the agenda of the Conference, and the very concept of environment, were broadened—quite rightly—to embrace such issues as loss of productive soil and the march of the deserts, the management of tropical ecosystems, provision of pure water for human use, and the problems of human settlements. Thus the less-developed countries forced a clear recognition of the relationship between environment and development—of the truth that it is through the process of development that the environment is affected, either positively or negatively, and that it is through the planning and management of the development process that we must deal with the care and protection of the environment.

This broadened approach to the environment became the basis for participation of developing countries in the Stockholm Conference and for the rapid evolution of their interest in environment. It also had a decisive effect on the shaping of the recommendations of the Stockholm Conference and on the content and direction of the subsequent United Nations Environment Programme. The use, by the developing countries, of their majority vote in the United Nations General Assembly, to have the headquarters of the United Nations Environment Programme established in Nairobi, Kenya, provided dramatic evidence of the influence of the developing countries in shaping this new international organization devoted to an issue in which they initially had little interest.

The strong role of the developing countries at Stockholm and in the United Nations Environment Programme has been accompanied by a degree of interest and activity at the national level which is nothing short of remarkable in the light of the attitudes of these countries prior to Stockholm. Virtually all of them today have established at least the beginnings of national policies, legislation and governmental organizations to deal with environmental issues. At the international level, in addition to the United Nations Environment Programme, which grew directly out of the Stockholm Conference, UNESCO's Man and the Biosphere Programme, the International Union for Conservation of Nature and National Resources, the World Wildlife Fund, the Foundation of Environmental Conservation, and the International Institute for Environment and Development, all of which (or their prototypes) preceded Stockholm, have assumed increasingly important roles.

Environment-related activities of most concerned international organizations, including such major UN agencies as the World Bank and FAO, WHO and UNDP, UNESCO and WMO, have grown significantly. UN conferences on Food, Population, Habitat, Water, Desertification and the Law of the Sea have focused (and in their outcomes continue to focus) particular attention on a series of special issues related to the environment. The scientific community has become increasingly active through such organizations as the International Council of Scientific Unions and its SCOPE Committee, the International Institute for Applied Systems Analysis, and the International Federation of Institutes of Advanced Study. A number of new awards have been created to recognize achievement in the field of environment. Here I would like to pay special tribute to Professor Polunin and those who have worked closely with him to enable these International Conferences on Environmental Future to make an especially significant and useful contribution to the international environmental movement.

The growth of institutions, on both national and international levels, and the proliferation of activities since Stockholm, testify to the fact that environment has arrived as an issue. In the last 10 years the amount of progress that has been made in creating awareness and concern can be a source of very considerable gratification—particularly for those whose pioneering work provided the foundations on which this progress has been made. Here we should recall such influential journals as *Biological Conservation* and particularly *Environmental Conservation*, both established by Nicholas Polunin. But we must now ask ourselves hard questions about where we currently are and where we are going.

Environmental Awareness and Protection

Granted that environmental awareness has been extended to virtually every corner of the globe, we must ask how deeply it has penetrated and the extent to which it is actually influencing the behaviour of people and the actions of governments and institutions. We must examine soberly what we are actually doing to change the habits and the practices that have produced the environmental deterioration and risks of which we have become so aware. We must try to see where we are headed and what we must now do to influence the shape and direction of our future.

Against this set of questions I have to say that all the progress we have made represents only a beginning: the actions we have taken so far have been the easy ones!

It is true that there have been some impressive examples of how specific environmental problems can be dealt with by enlightened and concerted action. For example, there have been some very considerable advances in some places in the forward planning of resource use. In Malaysia, where much land still remains under tropical rain-forest, careful previous planning of land capability for mining, agricultural development, forestry, and other uses, has ensured that forest is not opened up before its use is finally decided on. Moreover, careful attention is being given to maintaining chosen areas in a natural state as national parks and reservoirs of the rich genetic material of the tropical rain-forest.

In the United Kingdom, after a head-on collision between conservationists and proposals for a new reservoir ('the Teesdale affair'), regular advance consultation took place between those planning for the development of water resources and those concerned with the conservation of Nature. As a result of this consultation long before plans become public, there has been no further direct conflict since 1965. Full account has also been taken, in planning alternative approaches to large-scale water developments (such as the Wash and Morecambe Bay barrages), of the ecological effects and the implications for conservation.

Enormous progress has been made in Iran to demonstrate how rapidly and effectively natural ecosystems can recover with protection. Results have included great increases in the numbers of wild sheep, ibex and gazelles, and there has been widespread recovery of vegetation even in very arid regions. The next stage is the difficult one of controlling use of these ranges, which should be capable of cropping at a much higher level than before they were protected. If this can be accomplished, it will provide a clear demonstration of the speed and economy of natural recovery. The success of conservation programmes, mounted to reverse the drastic decline in the populations of Vicuña and Polar Bears, shows clearly that timely and concerted action can ensure the survival of other endangered species of animals.

These examples are encouraging in that they demonstrate that we do have the capacity to deal successfully with many of the specific environmental problems which are now facing us. *The technological and managerial capabilities of industrial Man, and the recuperative powers of Nature, can be powerful allies in remedying environmental ills.* Indeed, I would go further: Man has the capacity to deal with all the major environmental problems which now confront him. Ecodisaster is not inevitable; it is within our power to prevent it. Yet, I cannot help but feel that technological Man is proceeding along a course that can only lead to disaster. Consider the evidence.

Ominous Tendencies

While we have shown the ability to deal with individual environmental ills, curing them in many ways, this is like treating a sore on the body which is caused by an infection that has spread throughout the body. It may bring some temporary relief; but unless the underlying ailment is dealt with, the body cannot be restored to good health.

Consider some of the Earth's oldest and richest ecosystems, the tropical rain-forests. About 80% of the world's tropical rain-forests has been effectively destroyed, and the remainder is being felled and burned at the rate of 11 million ha a year—more than 30,000 ha a day, or 20 ha a minute. In Africa more than half, and in India, Sri Lanka and Burma almost two-thirds, of the original rain-forest cover is now gone. The richest lowland areas are the most vulnerable: the plant communities of south-east Asia are among the most species-rich in the world, but those of the Philippines and Malaysia are expected to have vanished completely within the coming decade.

This tragic waste is proceeding unimpeded by all the efforts of conservationists. Even in areas where they have used their most eminent names to persuade and cajole the local decision-makers and wielders of power, conservationists have as yet made no decisive difference. In the Philippines, for example, the late Charles Lindbergh and Tom Harrisson worked long, hard—and in the end successfully—to persuade the government to establish the Mt Apo National Park. Mt Apo Park protects both one of the finest Dipterocarp forests in the Philippines (and hence in the world) *and* the only home of the endangered Monkey-eating Eagle. Yet today four-fifths of the park has been designated for logging and settlement.

Destruction of the world's tropical forests would constitute a loss more critical to the human future than the depletion of its oil and gas reserves. The IUCN and the World Wildlife Fund have mounted a large-scale programme aimed at reversing this dangerous trend before it is too late.

What may be the final session of the Law of the Sea Conference is now in progress.* Its results will determine how some 70% of the Earth's

* Again as we went to press in April 1978, though we understand there are likely to be yet others.—Ed.

surface—the sea-bed, the sea and its living resources—will be cared for, exploited and managed. It may well have more profound effects on the future security and well-being of the human species than any other issue now before the world community. Yet most people seem scarcely interested or aware of the issues at stake. Few realize the danger that the narrow, short-term interest of those primarily concerned with exploiting ocean resources will override the larger interest of all mankind in the careful management of the oceans and conservation of their resources. But this is precisely what now seems likely to happen—unless there is a dramatic last-minute change in the attitudes of the governments negotiating these issues.

The oceans and salt seas constitute the world's biggest wilderness, where hunting is a multi-million dollar activity involving some of the most sophisticated technology, where the long-established terrestrial practice of domesticating animals and plants has scarcely a toehold, where conservation and environmental management have most to offer—and have failed dismally. A growing number of fish stocks are being depleted: for example, north-east Atlantic hake and herring, north-west Atlantic haddock, north-east Pacific salmon, Mauritanian–Senegalese hake, eastern-central Atlantic squids, and south-east Atlantic pilchards. In addition, some 12 species of whales, dolphins and porpoises, 9 species of seals, all Dugongs and Manatees, all coastal crocodiles, all sea-turtles, and 30 species of sea- and shore-birds, are more or less threatened with extinction.

No attempt is being made to relate the effects of harvesting different species on one another, or the impact of harvesting with other impacts on the sea (such as pollution and incidental 'take'). So the assumed potential harvest of a given stock may often be put too high. Similarly, decisions on other uses of the sea (for example, on whether to mine or dump or drill for oil) are made in ignorance of the likely impact on fish, marine mammals or the populations of other organisms on which they feed.

Until the seas are managed for what they are—dynamic systems of biological and physical processes extending widely over time and space—they will continue to be managed badly. Management authorities with jurisdiction and interests confined to a few species or to a single nation are no longer enough: regional management authorities, capable of looking after whole ecological areas, are now essential. The recent convention on the Mediterranean is a promising step in this direction.

Our technological societies have also been shamefully irresponsible and insensitive in their response to the fate of the many indigenous peoples whose societies are being relentlessly decimated in the name of growth and progress. In both industrialized and developing countries, these peoples, who desire to live their traditional lives as nomads or hunters and gatherers, today constitute endangered 'species'. Yet they are the repositories of most of the evolutionary experience of the human family, much of which we must now re-discover and learn to apply to the shaping of a viable future. The destruction of these societies and the demise of these peoples would constitute a tragic and irreparable loss to the entire human community.

Need to Control Growth

All of these things are occurring as a result of the increasing demands which the human population is placing on the resources of the biosphere. And Man's impacts on the environment result primarily from growth—growth in his numbers and growth in his appetite for material goods. Both must be brought under control.

Population growth in the industrialized societies has already levelled off; growth-rates in the less-developed parts of the world continue to be high, but are also showing signs of subsiding. Nethertheless it seems possible that, by the early years of the next century, planet Earth will have to provide a home for double the present population—about four-fifths of them in the developing world—and this will create enormously increased pressures on natural resources and on the biosphere. Add to this the pressure that will result from the drive of the poor—who make up the major portion of the world's population—to improve their conditions of life, and the tremendous pressures already being placed on the system by the insatiable appetites of the rich minority, and you will see that something has clearly got to give way. In my view, that 'something' must be the appetites of the rich.

Population growth in the developing world must be stabilized, and this will happen either through voluntary methods of population control or the traumatic and tragic consequences of famine and disease or conflict.

Even the most hopeful assumptions about limitation of population growth in the developing countries lead to the conclusion that their claims on world resources will escalate sharply within the next three decades. For quite apart from their increasing numbers of mouths to be fed, they cannot be denied their right to achieve better conditions of life, and surely natural justice dictates that their claims be given top priority—firstly, to assure the meeting of their basic needs for a life that is compatible with human dignity and well-being, and secondly, for equal opportunity to share more fully in the benefits which our technological civilization now makes possible. This can only happen if the industrialized societies reduce the pressures which they are exerting on the biosphere and on natural resources.

There is little sign that the developed countries will reduce their demands voluntarily, for their existing commitment to continuous growth in gross national product is built right into the economic system by which modern industrialized societies function. It is based on the assumption that more is better—that the well-being of the societies can only be assured by continuous growth in the material sense. The expectations of consumers, the creation of employment, the incentives which motivate investors and managers, are all geared to this system in which material growth is the prime mover. And while there has been some serious questioning of this concept of growth during recent times, it remains the dominant theme on which modern societies operate.

To the people of the wealthy countries of the industrialized world, the suggestion that they should reduce their demands on global resources is immediately equated with calls for 'no-growth'—a state which none of them could conceive of as tolerable. Indeed, no-growth in the sense of the cessation of economic growth is not perceived as a viable alternative by any nation: it is equated with stagnation and resulting fears of unemployment, social unrest and reduced standards of living.

It is this attitude towards growth—of human population and its production and therefore demands on the life-support system—that is at the heart of our present dilemma. This is the disease which has spread through the body of modern technological societies. It is this growth disease which has within it the potential for self-destruction of our society; and it is to the curing of this disease that we must direct ourselves if we are to deal fundamentally and effectively with the need to create a sustainable balance between Man and the natural environment.

NEED FOR A 'NEW-GROWTH' SOCIETY

The challenge we now face is nothing less than that of creating a whole new approach to the growth of our society—to the goals of growth, to the processes of growth, and to the systems of incentives and penalties which determine our patterns of growth. Environment cannot be seen as an issue separate and distinct from growth—as an issue which can be dealt with simply by adding another element to our present growth practices. It is, after all, through the process of growth that we affect the environment, either positively or negatively, and environmental resources are both a contributor to and a product of the growth process. Thus preservation and improvement of the environment can only be ultimately effective if it is integrated into the complex of forces by which the growth of our societies is determined. Environmentalists must now take the lead in effecting the transition to a *'new-growth' society*.

It is understandable that the early emphasis of environmentalists was on cleaning up some of the more obvious cases of environmental damage, on stopping development which threatened further damage, and on subjecting existing sources of pollution to controls or 'add on' technologies designed to minimize their environmental impacts. Now, some environmentalists have called for a halt to growth in our highly industrialized societies. At the other extreme, there has been a tendency to regard the costs of environmental protection or improvement as simply an extra or added cost—something that cuts into economic return, often to the point where we cannot afford it. We are told that protection of the environment is nice if you can afford it, but that when there is a conflict between economic and environmental factors, economic reality must prevail.

Yet what kind of economic reality is it that leaves out of the cost–benefit calculations (on which economic decisions are based), the entire cost to society as a whole of maintaining the natural environmental capital of air, water, soil, and plant and animal life, on which human life and well-being depend? By what economic logic can the amounts required to preserve environmental values in a given economic situation be considered a 'cost', while the impairment of environmental capital that results from failure to make these expenditures is not seen as a cost? Surely, the reality lies in recognizing that in any activity which damages the environment, environmental costs cannot be avoided—that it is simply a question of how and when and by whom they are to be borne. *Environmental realities cannot be avoided by ignoring them in the name of economic needs.*

Economic growth has made possible some unprecedented benefits to society. But it has also enabled us to avoid facing up to certain basic truths which we must now confront. We now know that economic growth which depends on the running down and impairment of our natural environmental capital is neither sound nor sustainable. The bills must come in eventually. The growing damage, to our natural environmental capital, that is resulting from the same processes which produce the wealth of our societies, should make it clear to us that we cannot continue to grow as we have been doing in the past. A society which bases its growth on the degradation and using up of its natural environmental capital, will be no more viable than a business enterprise which does not provide an adequate depreciation and amortization account to maintain its capital and productive capacity. Yet, this is exactly what is happening today in varying degrees in almost all countries in both the industrialized and the developing world!

Individual examples of this wantonness are legion in the fouling of rivers and lakes, contamination of air, desecration of coastal areas, destruction of bird and other animal life, and deterioration of urban areas in the United States and other industrialized countries, while in the developing world we have massive destruction of forests, large-scale loss of productive soil through erosion, march of the deserts and other consequences of ecologically unsound land-use practices, and the appalling conditions of life in 'exploding' urban areas through lack of adequate water supplies, waste disposal and sanitation facilities. There is also the growing threat to rich and poor alike from such 'outer limit' risks as possible climatic change, damage to the ozone shield, contamination of human food-chains, pollution of the oceans and overexploitation of their living resources.

Surely it must be clear that present growth-patterns and practices are self-destructive and cannot be sustained! Is no-growth, then, the only answer? Let me say with all the force I can muster that no-growth is *NOT* the answer. The real alternative to no-growth is new-growth—a new approach to growth, in both the more industrialized and the less-developed societies.

The new-growth approach must be based upon removal of the artificial and self-defeating conflict between ecology and economics, which is now built into our system of economic decision-making. *We must make environmental concern and economics the allies that they can and should be in evolving an approach to growth, the prime goal of which will be to produce a better quality of life for people.* We must build into the system by which growth is generated and managed, wise measures that assure preservation of the resource- and environmental capital-base on which continued development depends.

In the more wealthy industrialized societies, this will mean a major transition to a less physical kind of growth that is relatively less demanding of energy and raw materials—one that is based to an increasing degree on the satisfaction of Man's intellectual, moral and spiritual needs and aspirations in such fields as culture, music, art, literature and other forms of individual self-development and fulfilment. These, after all, are the areas in which Man achieves his highest levels of growth in human terms.

Conservation and Development

Conservation must become a way of life, and incentives for it must be built into our economic system. In practical terms, this will mean a redesign of industrial systems to introduce super-systems in which the residues of one process become the raw materials of another. Technologies for recycling and re-use of materials and abatement of pollution must be integrated into such systems, not merely added on to them.

The conservation–development gap must be bridged. Conservationists must acknowledge that conservation is as much about people as it is about animals and plants, if not more so. It's not the presence of animals and plants that makes conservation necessary, but the presence of people. Therefore 'people experts' are practically as essential to conservation as are experts on Nature.

The proponents of development should recognize that conservation is a prerequisite for development, being the means whereby people can make the best use of the living resources on which they depend. But they will not be persuaded of this while so many conservationists appear to devote themselves to issues for their emotional appeal rather than attempting to concentrate on clearly-defined priorities that have been carefully arrived at through objective appraisal. Conservation is past the stage where being enthusiastic is enough, and should now seek to be as highly professional as are the promoters of reckless expansion.

On the global level, the 'new-growth' approach will require new dimensions of cooperation between industrialized and developing societies. It will require a re-vamping of the present international system of arrangements and institutions in such a manner as to enable them to support better and serve more effectively the interests and aspirations of the developing world. Interdependence, which is now a physical and environmental reality, must become a working reality in economic, social, and political, terms. This, in essence, is what the drive of the developing countries for a 'New International Economic Order' is all about. The healthy functioning of our interdependent technological society requires the full participation and active cooperation of the two-thirds of the Earth's people who live in the developing world, and this dictates that we heed their demands for a more just and equitable share of the benefits which this technological civilization makes possible.

Most new industrial capacity—particularly that which is resource- or labour-intensive—must be built in the less-developed parts of the world, and under conditions which enable developing countries to avoid many of the environmental and social costs that we have paid for our industrial development. The marriage of ecology and economics which I call 'ecodevelopment', would be designed to assure that the precious natural resources of soil, forests, water, and plant and animal life, in the less-developed countries are exploited in ways which make the best possible use of their own skills and labour, and harmonize with their own culture and value systems to produce maximum benefits for their people without destroying the resource-base on which sustained development depends. It means, too, assuring that they have full access to the latest technologies and support for the development of their own scientific and technological capabilities, so that technology will serve rather than determine their own growth-patterns.

The transition to a new-growth society which I am proposing has, perhaps, its best analogy in the human body. From the birth of a child to the time when it achieves physical maturity at the age of 18 or 19, the principal emphasis is on physical growth. Indeed, healthy physical growth and continued physical health are essential pre-conditions to the growth of human personality in its social, cultural, intellectual, moral and spiritual dimensions: yet growth in these non-physical aspects of human development has only nicely begun at the time that physical maturity is reached. The real growth is still to come. Our industrialized societies are very much like the physically mature human being. For us to continue to pursue purely physical kinds of growth would be as unhealthy and self-destructive to our societies as it would be for an adult person to pursue ways which simply added to his physical dimensions. And it would be just as wrong to say that societies must stop growing when they reach the stage of physical maturity as it would be to say that people stop growing when they stop growing physically.

The real growth of our societies in human terms is still ahead of us. But it demands that we change our ways and adapt to a more mature kind of growth that is less physically oriented and less demanding of resources and of the environment. On the other hand, developing countries are at a much earlier stage of growth, in which they must continue to grow in physical terms if they are to meet the needs and aspirations of their people. But they, too, must emphasize the kinds of physical growth which are healthy and sustainable, and which provide expanding opportunities for self-expression and fulfilment in human terms for their people.

Of course, acceptance of the need for the new-growth era will not automatically make it happen; indeed one is apt to wonder whether, in fact, it is a practical possibility? I believe that it is. But to practise it is a question both of the will and of the way, and I am convinced that the way must be easier to find than the will.

Needed: Changed Incentives and Penalties

The public-policy levers which governments can deploy today are capable of altering the system of incentives and penalties to which our economic life responds. We have clearly demonstrated this in fighting, or preparing to fight, wars. It is not the operation of the free market economy which produces the massive market for war materials. The market is created by an act of public policy by government, responding to the belief of their people that their security is at stake.

Today, the threat to our security through the physical and social imbalances generated by present growth-patterns, is to my mind as great as the threat of nuclear warfare. Indeed, it is even more difficult to deal with because it seems less immediate and less traumatic. The threat of a nuclear war may be averted right up to the moment when the button is pushed. But the threat of an ecodisaster can only be averted by foreseeing it far in advance and acting to prevent it. By the time it is upon us, it will be too late.

So, if we accept that the risk of ecodisaster is as great a threat to our security as that of nuclear warfare, we must use all the levers we can commend to prevent it. And I am persuaded that, if we consider it important enough, we *can* prevent it. For the means are available to re-gear the system of incentives and penalties which motivate our economic life, in such a manner as to make it profitable to carry out those activities which are environmentally sound and socially desirable, and unprofitable to do those things which impair environmental quality, destroy resources and detract from desirable social goals. Far from being negative to the economy, a commitment to the 'new-growth' society would unleash new and dynamic economic forces and would stimulate creativity, innovation and economic activity across a broad front. If expenditures on war materials—which are inherently wasteful, whether or not they are used—can be a major stimulus to the economy, surely expenditures on building better and more liveable cities, improved cultural and educational facilities, and recreational areas and opportunities for meaningful leisure, can be just as stimulating to the economy, while at the same time adding positively to the real capital stock of our human societies.

It must be made economically attractive to conserve energy and materials, with corresponding penalties for activities which are destructive or wasteful of resouces. Conservation must no longer be considered a fringe activity which is nice if you can afford it, but cannot be allowed to interfere with economic growth. It must become an essential element in all economic activity, and the whole process of economic decision-making must be designed around the need for a total commitment to conservation.

Let me say a special word about nature conservation as it is in danger of becoming the orphan of the environmental movement; for environmentalists concentrate on issues with more political clout, such as energy and nuclear power. It is worth remembering that environmental concern sprang from the efforts of the conservation movement, and that nature conservation remains the best organized and most cohesive part of a highly disparate collection of interests. The potential of nature conservation to wield political influence has been masked by the reluctance of many conservationists to confront directly those relevant issues with the greatest impact on ordinary people—fisheries being the outstanding example.

Governmental action must always be undergirded by a change in the attitudes, values and expectations of people. We should applaud and look up to those who adopt life-styles that are modest in terms of the amount of space which they monopolize or the amount of materials and energy which they consume; ostentatiously wasteful and indulgent living should become socially reprehensible. There should be an acute sensitivity to all activities which create risks of damage to our natural heritage, or impair the quality of life for others. People of industrialized societies in particular must again nourish their communal values and downgrade their competitive drives.

For clearly the transition to the 'new-growth' society implies some very profound changes in our present attitudes, expectations and behaviour, and in some of our traditional notions of distinctions between private and public rights and responsibilities. Up to now, the human species has changed its ways only after having been chastised by bitter experience. Man's history has been based on repeated cycles of advance, tragedy inflicted by Nature or by war, collapse, and rebuilding on a higher level. Now that, for the first time in our history, we possess the means of total self-destruction, can we risk repeating these cycles? Even if we could, it is doubtful whether the wholly unprecedented scale and nature of the risks which we now face, would enable us to have another chance if we were to wait until environmental tragedy was imminent.

So while there is strong evidence that we can make the transition to a 'new-growth' society, there is a very real question as to whether we will do so. Our future really turns on whether we will have the enlightened moral and political will, and the sense of our own ultimate self-interest, to change our ways before it is too late.

It is in these terms that I see the challenge to our generation of environmentalists. Those who pioneered this movement have succeeded in placing the environmental issue firmly in the agenda of the world community. An increasingly concerned public, press, and set of media, have helped to keep it so, and there can be few major countries left in the world without a ministry or department of environment or whatever. The challenge which we now confront is to move from the fringes of debate and action to the very centre of the search for new approaches to growth. For it is in the need to reshape our attitudes towards growth, and to redirect the growth process itself, that the battle for the environment will be won or lost.

Wolf-Wind*

A wolf-wind wails through the wilds of time,
a red dawn colours the sky.
The grass turns brown, the beasts grow lean,
the blossoms wither and die.
We earth-men know what the omens foretell,
but power-men wrangle and vie.

Gather in, earth-men,
trim down your flocks;
Hark to the counsel
of trees and of rocks.

Gather your blood-friends
(Let chieftains glower!),
For ours is the harvest
of Terra's deep power.

* Based on the manuscript given by Professor Reid A. Bryson to Mrs Elín Pálmadóttir at the Prime Minister's (our Patron's) superb luncheon following the Final Plenary Session which terminated the Conference on Saturday 11 June, 1977.—Ed.

FIRST BAER–HUXLEY MEMORIAL LECTURE (Introduction and Termination)

[As founder of this series and chosen representative of the fostering Foundation, the Secretary-General made arrangements to circulate to the assembled audience reprints of the obituary notices of Jean Baer and Julian Huxley which had been published in the second volume of *Environmental Conservation* (1975), before taking the Chair to commence the proceedings and introduce the speaker approximately as follows:

Polunin (Snr) (Chairman)

President Pauling, Your Excellencies, Ladies and Gentlemen: The other members of the still-embryonic standing committee for the Baer–Huxley Memorial Lectures have asked me to preside on this initial occasion and I consider it a very special honour to do so. The idea of memorial lectures named after departed notables is not new. It has long been practised with distinction by my own old University of Oxford where, for example, we have such recurrent events as the Romanes Lecture, which I recall hearing given on one occasion by Sir Julian Huxley himself.* So when two of our pioneering leaders in the environmental movement died almost simultaneously, to our great regret, early in 1975, methought it would be appropriate to establish a prize or lecture or something in their memory, which I may say was encouraged enthusiastically by both Mme Baer and Lady Huxley. This ought to be endowed for continuation in perpetuity; and perhaps, who knows, some of you may have some ideas, or even something better than ideas, about this? Meanwhile it is comforting to note that the remuneration for the Romanes Lecture was—and probably still is—only £25, and yet it is looked upon as a notable event which is an honour to give. On the present, pioneering occasion of our own series we are fortunate in having arranged to manage practically without financing, as indeed has latterly become our enforced custom in most other respects.

If time had allowed, both for thought and presentation (and unfortunately we have neither: we just walked in from the excursion a few moments ago, since when I have had to deal with cables and overseas calls and other things), I would have liked to have described for you some of the outstanding work and influence of these two towering figures of our environmentalists' world: but time does not so allow—especially as we have to cope later with two adjournments from this morning. So we are circulating at this moment, for those of you who may be interested in having them as I hope you will be, copies of obituary notices on both those whose memories we wish to honour, and also one of another colleague and Advisory Editor of our Journal, *Environmental Conservation*, namely the outstanding Greek ecologically-minded planner, Constantinos Doxiadis. In fact we did think of trying to link him in some way with this lecture; but we did not feel we should string on too many names, and he was not quite so

* Actually in this case the name derives from a member of Christ Church, George John Romanes, who died in 1894, and gave 'an annual sum of £25 for a lecture to be delivered once a year on some subject, approved by the Vice-Chancellor, relating to science, art or literature'.

much one of us and special in my own life for many years as, particularly, Julian Huxley or, latterly, Jean Baer. I might say that Jean was one of the most outstanding members of our 1971 Conference in Finland, and was Chairman of its winding-up, subsequently becoming the unanimous choice as Chairman of the International Steering Committee that started soon afterwards to plan our present Conference. Though a reclusive chap who shunned publicity, so that you could not find details about him in the usual works of reference, he was to my mind among the very greatest in all the environmental movement, and certainly one of its leading pioneers.

But you came here to listen to a far better speaker than I; and so, with deep reverence for our departed colleagues, whose indomitable spirit and feeling for Nature and Man—the order is deliberate on my part—will ever remain with those of us who were privileged to work with them, I call upon Mr Maurice Strong, Chairman of the Bureau of IUCN and also Chairman of Petro-Canada, to give the first Baer–Huxley Memorial Lecture. Mr Strong needs no introduction to any among you or to anybody anywhere in the environmental field—or shall we now say the wider environmental conservational world—or indeed to anyone else who is really interested in the future of Man and Nature. So Mr President, Your Excellencies, Ladies and Gentlemen, our friend and environmental protagionist *par excellence,* Maurice Strong.

[Mr Strong thereupon delivered the lecture printed above and it was followed by long-continuing and highly enthusiastic applause. Subsequently there was a very lively discussion which extended to midnight and involved, besides the Lecturer and the Chairman, at least the following (in the order of their first speaking), making in all over 30 interventions, etc.: Mr Ralph Glasser, Dr Bent Juel-Jensen, Dr Pierre Laconte, Professor Arthur D. Hasler, Professor Edward D. Goldberg, Mr Peter B. Stone, Dr E. Barton Worthington, Dr Thomas F. Malone, Mr Nicholas G. L. Guppy, Professor Gary L. Widman, Mr Edward Goldsmith, Mrs Elizabeth Dodson Gray, H. H. Fatesinghrao P. Gaekwad, MP (Maharaja of Baroda), Mr Perez M. Olindo, Dr Michael G. Royston, Dr Richard G. Miller, Mr Errol A. Cunningham, and Professor Torgny Schütt.

As the typescript for our book of Proceedings is already considerably longer than either the British or the American publishers desire, and as many of the main points of the discussion appear in the book, it is not felt necessary to publish this discussion here—despite its general erudition and, in places, delectable wit.

The Reykjavik Imperative on the Environment and Future of Mankind

On the fifth anniversary of the United Nations Conference on the Human Environment, held in Stockholm, we, the 130 environmental scientists and other* scholars from 20 countries participating in the Second International Conference on Environmental Future, Reykjavik, Iceland, 5–11 June, 1977, have prepared† and approved‡ the following statement:

Thousands of generations of human beings have lived on Earth. Now, for the first time, Man has become so numerous, and has achieved so much power over Nature, that the world may be destroyed, and meanwhile is being changed inexorably.

Governments, as well as people, tend to concentrate on immediate problems, and to consider only short-term goals. It is unusual for a country to develop a ten-years plan, and unheard of to have a 100-years plan. Yet there is now a high probability, almost a certainty [if the world continues as at present], that within 50 years there will occur, exceeding past ecodisasters, the greatest catastrophe in the history of the world. This might be a world-wide famine or war, which could destroy civilization and Nature, or it might be a combination of effects caused by unwise actions. But there is also the possibility that such a catastrophe can be averted if the right decisions about the environmental future are taken. Clearly decisions made now and during the next 30 years will determine the future of mankind and of the world. These decisions may well be the most important of all time.

* With approved editorial adjustments such as this insertion of 'and other'; afterthoughts are placed in square brackets.—Ed.

† Through a Resolutions Committee elected at the opening session of the Reykjavik Conference and consisting of Donald J. Kuenen, Thomas F. Malone, Letitia E. Obeng, Linus Pauling (Chairman), Gunnar G. Schram and E. Barton Worthington, with powers to co-opt and consult.—Ed.

‡ At the final plenary session attended by *ca* 100 remaining participants on 11 June, 1977, with Maurice Strong in the Chair, flanked by the Chairman of the Icelandic National Committee and the Secretary-General of the Conference.—Ed.

Goals

The principle on which decisions can be based is that of achieving the well-being of all people—those now living and those of all later generations—by working in partnership with Nature rather than against Nature. The ultimate goal must be a world in which every person, in every generation, century after century, has the opportunity to lead a life with as little suffering and as much fulfilment as possible. This life should include adequate food, clothing, shelter, education to the extent that he or she can benefit, suitable employment, leisure, and the opportunity to develop to the fullest extent. Thus will it be possible for individuals to safeguard and develop their own inherited cultures free from external pressures.

Freedom of choice in personal actions is essential; so also is preservation of the world's natural wonders, of millions of species of plants and animals, of the wildernesses, of the rain-forests and other woodlands, of the mountains, of the unpolluted lakes, streams and seas, and of all the other valuable and unique parts of our environment which should not be sacrificed to short-sighted exploitation and inappropriate technology.

Resources

We must rely on Man's labour and intelligence and on the world's natural resources for our future. Some of these resources are renewable, even though limited in amount: water, which is purified through evaporation and then falls again as rain or snow; energy from sunlight; wood from growing trees; and food from plants and animals. There are also resources that are non-renewable, being available only in limited supply: petroleum, coal, uranium, phosphate rock for fertilizer, and certain metals. It is unjust to future generations for the supply of a non-renewable resource to be used up by people living now. We must plan to base our survival almost entirely on renewable resources.

Population

While most of the damage and risks to the environment result from the actions and policies of the industrialized, rather than the less-developed, countries, a large proportion of the world's problems are exacerbated by the existence of more people than can be properly supported and by the present high rate of their increase as well as by their ever-increasing demands on the world's finite and other resources. There are now more than 4,000 million people living on Earth, and this number may be doubled by the year 2,000 AD. We consider it to be essential that this increase be prevented, and that the world's population be ultimately stabilized at a number which will permit everyone to lead a good life. The social, economic, political, religious and psychological aspects of the population problem are so complex that effective international action is impossible. We call upon each nation and local group to undertake vigorous action to educate its people about population problems.

To limit population-growth will not in itself save the world from impending disaster, but it is an essential part of the solution to the problem and its importance cannot be overemphasized.

Notwithstanding such considerations, the projected doubling of the population might fail to occur because of processes that would involve great human suffering. Hundreds of millions of people are now close to starvation. Methods of increasing food production, such as the 'green revolution', are partially effective, but are negated in many countries by adverse consequences and by the increase in numbers of those to be fed. The death of 1,000 million people from lack of food, some by outright starvation, but mostly by disease caused by severe malnutrition, may well be the tragedy of the closing decades of the twentieth century.

Energy

The seriousness of the problem of providing energy to meet the needs of the world is now known to everyone. The most important conventional sources of energy—oil, coal, natural gas and nuclear fuels—constitute non-renewable natural resources. There is no doubt that they will continue to be used for some time, but we urge that a vigorous effort be made through appropriate research and development to replace them as rapidly as possible.

The transition to other sources of energy should be achieved within 30 years—not only for reasons of conservation, but also to protect the environment from damage that might prove to be catastrophic. For example, a serious change in the climate might result from the warming of the surface of the Earth through several mechanisms. These include not only the dissipation of heat, which alters the regional and global energy-balance, but also marked increase in carbon dioxide and other substances in the atmosphere. The influences of various factors on the climate are not completely understood, so it is prudent to reduce those activities that could have serious consequences. One important way is to decrease the required amount of energy by eliminating wasteful and unnecessary uses. The misuse of energy and other resources must be made socially reprehensible; so must attacks on the biosphere and wanton desecration of Man's habitat or Nature's ecosystems. [To all this, education is the key.]

We urge that great efforts be made to develop all of the renewable sources of energy—especially those which depend either directly or indirectly on the greater use of the Sun's rays that fall on the Earth. The problem of energy can be solved by attacking it vigorously in all of its aspects.

Research and Development

Among human activities, militarism is one of the most wasteful; it is also damaging to our environment, threatening to our future, and *unworthy of our intelligence*. Approaching 10% of the world's income is now being spent on militarism. We recommend that a significant part of the money currently spent on armaments be transferred to appropriate research on, and active solution of, environmental problems. We further recommend that environmental impact statements should be required for all peace-time military activities.

We welcome the suggestion of the World Bank that an environmental impact statement should be required before grant aid is made available to any development schemes, and we consider it essential that a true environmental impact assessment be required in all major development schemes, whether or not they are aided by grants.

New Growth

We call for a fundamentally new approach to economic growth—a New Growth which emphasizes quality rather than quantity, reduces the demands on the Earth's resources, reduces the risks to its life-sustaining systems, and is compatible with human survival and well-being. We recognize that this effort will require major changes in the incentives and penalties which motivate the economic life of our societies. Diversity of Nature must also be sustained to the utmost extent possible. We encourage and support the preservation and continued development of a multiplicity of life-styles, thus enlarging the possibilities for diversity and variety in human self-expression and fulfilment.

We call upon all governments, all communities, all people, to take immediate action to avert the disaster that looms ahead.

- Following this general statement we wish to express our conviction that a fundamentally new approach must be taken to economic growth—one which will reduce the demands on the Earth's resources and the risks to its life-sustaining systems to levels that are fully compatible with human survival and well-being.
- Politicians must [listen to scientists and] initiate an integrated approach which is essential to solving so many environmental problems.
- We must learn far more about the interference which ecosystems can tolerate without suffering irreversible depredation, and about the exploitation [that they can support without losing their continuing function of producing] the required resources. To these ends considerable segments of all ecosystems should be set aside for subsequent ecological observation, monitoring, and study. The biotic life-support system must be sustained in every possible way throughout the world.
- Further action should include the establishment of trees or other appropriate vegetation to cover soils—either to regenerate or to recreate damaged ecosystems.
- In any further development of technologies, 'no release' policies for substances on the International Register of Potentially Toxic Substances should be strictly followed. We must be very careful not to do anything just because we do not know what the effect will be. An example from chemical technology is possible destruction of the stratospheric ozone shield, without which life on Earth could scarcely be sustained.
- The damaging effects of persistent insecticides can be reduced by alternative ways of controlling pests—such as diversified agriculture, use of less-susceptible varieties, biological control, use of short-lived chemicals, or manipulation of the environment (including changes in

the water régime). Financial help should be forthcoming to establish improved systems where applicable.

- In all cases of major technological projects, not only [should] we require environmental impact statements beforehand, but also surveys afterwards of the actual effects that such projects have had for the environment and the people affected.
- More strict legislation, nationally and internationally, on pollution is essential for further reducing the dangers of environmental degradation.
- The exploitation of the remaining tropical forests for timber for the rich countries should be restrained, and other ways found to help the 'developing' countries to balance their financial budgets. As there is little to be gained from agriculture on soil that was formerly occupied by rain-forest, the destruction of forests for agricultural extension by current technologies in these areas should be abandoned as counter-productive.
- The oceans are not merely the Earth's last untapped repository of natural resources and a source of food and recreation but also a vital component in the buffering of atmospheric processes without which life on Earth would be insupportable. The Laws of the Sea which are currently being defined must ensure that the oceans continue to fulfil their indispensable role and not become the ultimate cesspool of the world.
- The promotion of family planning is essential, and women should take the lead in explaining to other women what they can gain from its practice and what they stand to lose by having many children.
- It is necessary for public opinion to be informed continually of the dangers of wrongly applying technical possibilities. Governmental and non-governmental organizations in the field of education and public awareness must persist in their efforts to show political leaders, teachers and youth in what ways they can help to save the world from disaster.
- In addition, universities should engage more widely in training scientists, economists, sociologists and lawyers in the multidisciplinary approach of how specialized knowledge on details of our way of life can be combined to form a basis for continuing development within the limits set by the environment.
- Having heard of examples of wanton destruction of the environment by military exercises in peace-time, we ask that more information be gathered about such activities and published widely.
- The fundamental idea that one of the most important sources of well-being now, which is mineral oil, should be used to support the conservation of the environment for the future by setting aside one cent per barrel of oil, is an initiative so vital for the future of mankind that it should be taken up by all those involved in the exploitation of this non-renewable resource, with the objective of transforming at least part of this short-term profit into long-term benefit.
- Pursuant to the policy outlined in the Reykjavik Imperative, an activist

group of this Conference, drawn from the disciplines of law, administration, engineering, physics, and biology, has initiated arrangements for joint action by several international organizations. The initial objective is to prepare a statement of views on science and technology in relation to the environment, and their transfer between countries in different stages of technological development. This statement will be submitted to the United Nations for use in the preparations for the UN Conference on Science and Technology for Development, to be held in 1979. In addition, suggestions have been advanced for convening a high-level meeting of scientists and technologists concerned with the environment before the UN Conference, in order to express the viewpoint of the scientific community. [A separate 'catalyst' organization, for continuous follow-up of this and other productive environmental conferences, has been proposed as another outcome of this 2nd ICEF.]

- The participants of the Reykjavik Conference agree that they, individually and collectively, will work for the achievement of these purposes through their personal example and by seeking the involvement of their organizations as well as by active cooperation with other people and institutions.

The Foundation for Environmental Conservation

THE REYKJAVIK IMPERATIVE ON THE ENVIRONMENT AND FUTURE OF MANKIND

The Reykjavik Imperative constitutes the final statement of the 2nd ICEF. It was drafted during the Conference by the special Resolutions Committee which was set up at the opening session (*see* pages xxv–i), received suggestions for consideration throughout the near-week of the Conference, and was privileged to have as its Chairman the Conference President, Professor Linus Pauling. Their statement was considered, amended, and ultimately passed *nemine contradicente,* as the main business of the Conference's Final Plenary Session, under the chairmanship of Mr Maurice Strong, who was supported on the platform by the Conference's Secretary-General and by the Chairman of its Icelandic National Committee, Professor Gunnar G. Schram.

The Session opened on time as the adjourned discussion of Session 18, 'Conclusions for the Future', terminated, and the President having expressed the desire to remain in the audience to make his report as Chairman of the Resolutions Committee, the Secretary-General invited Mr Maurice Strong to occupy the Chair as prearranged for the Final Plenary Session.

Strong (Chairman)

Well, thank you. The only function of a Chairman, I believe, is to do as the Secretary-General tells him!

Polunin (Snr)

I am not telling you *how*, sir.

Strong (Chairman)

But that is the way we used to choose our chairmen in the United Nations, and they did not always do as they should, though I would like to. As I have been given no other instructions, I believe my only role this morning is to ensure that the proceedings proceed, and on time. But I will exercise one prerogative, and that is to make one contribution to the last debate. I cannot resist, on this definition of pessimist and optimist, [recalling that] in commercial terms a pessimist is someone who has loaned money to an optimist. [Laughter.]

While the group is assembling I will tell you one other story that I think is a good reminder for us; when we are worried about being accused by our friends of being doomsdayers or pessimists, I am reminded of the time when the concert hall was completely full of people and the concert was just about to begin, when someone yelled 'Fire!'. Someone else, with great presence of mind, said 'Don't believe it! Don't panic. Don't worry. Everything is going to be all right'. And he led the chorus in the singing of a beautiful piece of music, and everybody relaxed back in their seats and all were burned to death. [Laughter.] So it is not always the optimist who leads us in the right direction!

Mr Secretary-General, I believe it is my duty at this stage to call on our President, Professor Linus Pauling, for his report as Chairman of the Resolutions Committee.

Pauling

The Resolutions Committee decided to prepare a Statement which would be accompanied by a set of Resolutions [which we know are awaited quite widely in the world]. The Statement was drafted early, before many contributions from Conference participants had come in, but there was some opportunity to revise it a little, and the specific resolutions that accompany the statement were written after the [suggestions] from the participants had come in—or at least after most of them had come in—and those that were legible and grammatical were taken into consideration in preparing the [amended] Statement and additional resolutions which have been circulated and I hope you all have. I think the Statement reflects the sentiment of the whole Conference. I can read it out, paragraph by paragraph, if you would like and then deal with the Resolutions one by one.

Strong (Chairman)
Has everyone got copies of it? Good: I would suggest that it might be most useful [, now that everybody has them], for us really to take them paragraph by paragraph, Mr President, or [the six pages one at a time], and get the comments that way, perhaps giving a chance first of all to anyone who wants to make a comment that would apply to any major omission in the paper as a whole. I think, Professor Pauling, it would be best if you just conducted this part of the exercise as you see fit.

Pauling
Very well, we have time I believe to read it?

Polunin (Snr)
We have time, sir. This is the main item until we leave for the Prime Minister's, except that if I could have a few minutes before the end, about the termination or continuation of these Conferences, I would be very grateful.

[Thereafter there was detailed debate of many points (in more than 100 interventions, etc., of nearly 50 speakers*) until the following memorable and propitious intervention by one of our youngest participants, speaking on behalf of a group of our younger members (mostly in their middle-twenties) who had submitted a timely 'Environmental Resolution for the World's Youth and Children':]

Franz, Kristi
As a member of a generation that plans to be here 50 years [hence], I recommend that we strongly support this statement. We can spend many more days, as we have in the past [week], discussing specifics, [and outlining] basic philosophies, or basic directions. [Now it is up to each of us to take the specifics of this statement and act upon them.] I urge you to get this out to the people [of the world] and to accept this statement as a beginning-place of perspective and commitment to responsibility. Please do this for our young people, for those of us who are going to be here in 50 years and do not want to experience an [all-ending] ecodisaster. [Lively applause.]

[It was thereupon decided *nemine contradicente* to call the amended statement 'The Reykjavik Imperative on the Environment and Future of Mankind'. As such it was further amended during editing, suggestions of even minor changes being sent to all six members of the Resolutions Committee and only being adopted if a majority approved (the President having given the Editor the power of free choice in the event of any equal division of those responding). The Reykjavik Imperative was published—as a matter of urgency, following numerous requests—in *Environmental Conservation* (Vol. 4, No. 3, Autumn 1977), where it was, as now below, followed by a statement on 'The Foundation for Environmental Conservation: Auspices, Objectives and Needs'.

* There being no space to spare in these Proceedings, even for the cream of the discussion at the Final Plenary Session, it is hoped to summarize it elsewhere—Ed.

The only other event of the Final Plenary Session which may be worth recording is that, the Chairman having asked me as Secretary-General to wind matters up as there were a few minutes remaining before we had to leave for the farewell State Luncheon, when I began a little speech to thank people for coming and participating so actively in what I felt had been a useful conference which should be the last of the micro-series, they started to shout 'No' and applaud—whereupon the Chairman, Maurice Strong, checked that it was in unanimous approval of a request for continuation. So I thanked them all and thereupon undertook to convene more ICEFs in future if I should be spared and properly financed another time—otherwise, No! There seemed to be general support and even enthusiasm for my emerging idea of possibly holding together, separated by a full day or preferably a week-end break, two conferences of three or four days each on the themes already indicated on the notepaper of the 2nd ICEF, namely 'Industry and Environmental Concerns' and 'After the Oil is Gone'. Later on, another urgent theme emerged—'Global Resource Strategy'.—Ed.]

The Foundation for Environmental Conservation: Auspices, Objectives and Needs

The Foundation, which had started operations unofficially some years earlier, was finally established legally in 1975 in Geneva, Switzerland, as non-profit and tax-exempt by authority of the Council of State of the Republic and Canton and perpetually under Swiss Federal Government surveillance by the Department of the Interior, Berne. Its headquarters are at 15 Chemin François Lehmann, 1218 Grand-Saconnex, Geneva, Switzerland.

Founders and Governing Board

The Founders were the International Union for Conservation of Nature and Natural Resources (IUCN), the World Wildlife Fund (WWF), and Professor Dr Nicholas Polunin. The Governing Board of the Foundation consists of the last-named for life and a representative each of IUCN and WWF, with powers to co-opt other members for periods of three years at a time.

Objectives

(*a*) To undertake, in cooperation with appropriate individuals, organizations and other groups, all possible activities to further the ends indicated in its title, and specifically:

(*b*) To own and promote pertinent publications, in particular the Journal *Environmental Conservation*, and to derive therefrom revenues to be used towards coverage of publication costs—in mind are certain other journals, supplements to the present one, further Plant Science Monographs, a series of environmental monographs, a Multilingual Systematic (running) Glossary of Environmental Terms, and ultimately a major environmental encyclopaedia;

(*c*) To foster pertinent conferences, in particular the International Conferences on Environmental Future (ICEFs);

(*d*) To organize specialist 'workshops' to deliberate and pronounce freely on urgent aspects of environmental change or other causes for concern;

(*e*) To promote studies on environmental change and ecosystem future; and

(*f*) To accept and administer (under Swiss Federal surveillance by the Department of the Interior, Berne, and the authority of the Council of State of the Republic and Canton of Geneva) funds for the above purposes, including creation and bestowal of suitable awards.

Adopted Projects, etc., and Agreed Financial Needs in US dollars

1. Environmental Conservation

To bridge the gap between printers' bills and receipts from the publishers, about $7,000 p.a. will be needed during the present and next two years*; in addition there should be available $10,000 p.a. for the long-needed extension of each issue from 80 pages of text to 100 pages, and when possible about $10,000 p.a. to offer to the publishers to make available a substantial number of 'individual' subscriptions at half-price or less.

2. Office and Allied Expenses

About $11,000 p.a. during the present and next two years to repay essential office expenses of part-time secretary, postage, stationery and supplies, telephone and a modicum of necessary travel, etc.; also about $10,000 p.a. during the present and next few years for payment of part-time editorial assistance, attendance at conferences and meetings, rental and services of secretariat, etc.—all of which I donate at present but may not be able to continue; also about $8,000 p.a. for a part-time general assistant and his or her expenses, and about $2,000 p.a. to continue (and more if needed to expand) the International Environmental Consultancy.

3. Conferences Roster and Reports

To maintain (through the above International Environmental Consultancy) world-wide vigilance and compilation of all environmental and related congresses, conferences, symposia, colloquia, high-level discussion groups, meetings, etc., and extension of the Conferences & Meetings section of *Environmental Conservation* to be fully comprehensive and include regular publications of lists of all pertinent future events and reports on past ones, an estimated total of $14,880 p.a. will be needed at early 1977 prices.

4. International Conferences on Environmental Future

To cover advances and balance the budget of the 2nd ICEF (which was held in Iceland in June 1977 and resulted *inter alia* in the above 'Reykjavik Imperative') a further $20,000 is sought. Further substantial sums will be needed to underwrite the planned 3rd and 4th ICEFs, though it is expected that the bulk of essential financing will be contributed by or through the host country or countries. The first ICEF in 1971 cost approximately $44,000 (apart from relieving gifts) and the 2nd ICEF in all *ca* $65,000

* This need is now being covered for 5 years from 1 January, 1979 through the generosity of the publishers, Elsevier Sequoia S.A., of Lausanne, Switzerland.

(including preparation, and publication of the Proceedings in adequate detail).

5. *Multilingual Glossary on Environmental Terms*

Preparation, checking by specialists, editing and publication of a Multilingual Systematic ['running'] Glossary of Environmental Terms, probably more than 3,000 in number in English, French, Spanish and German, with a separate edition in Russian: $30,000 for completion of compilation, help with publication and possibly also translation into further languages.

6. *Other Publication Projects, etc.*

(*a*) A subvention of preferably not less than $40,000 to acquire other journals and in time start publishing single-volume Supplements to the present one, of which several attractive possibilities have already emerged.

(*b*) Preparation and subvention towards publication (with coloured fold-out map) of completely revised edition of *Introduction to Plant Geography* [*and Some Related Sciences*] to take another year and include new sections on pollution effects and ecology, plant conservation, introductions and their effects, competition, continental drift, and the role of plants in environmental conservation: $5,000.

(*c*) 'Conservation of the Plant World' (two volumes currently planned), including preparation, editing (sse) and subvention to help keep the price down, $25,000.

(*d*) Preparation and publication, in collaboration with suitable specialist colleagues, of books on 'Conceivable Ecodisasters', 'The Ozone Shield' and 'Desertification—fore and aft': $3,000 each. Others are contemplated for next item.

(*e*) Environmental Monographs and Symposia: $10,000 to start series with leading international publisher.

(*f*) Subventions for further *plant science monographs* of an environmental nature: $15,000. Under consideration or more are—besides item (*c*) above—volumes on 'The Biology of Old-World Desert Plants', 'The Biology of New-World Desert Plants', 'Atmospheric Pollution and Plant Life', 'Tree Planting Throughout the World', and others.

(*g*) Subventions, etc., for further *world crops books*—especially to concern new crops to help feed an increasingly hungry and demanding world: $25,000.

7. *Research Projects*

(*a*) Study of ecology of Dugongs and Green Turtles in south-east Asia, towards possibilities of sustained-yield cropping and natural utilization of sea-grass pastures: $30,000.

(*b*) Research project on Nature in a Concrete Jungle: $5,000.

8. *Other Items*

(*a*) Feasibility study for a proposed International Association (or World Council or Academy) of Environmentalists: $3,000.

(*b*) Endowment of the Baer–Huxley Memorial Lectures on topics of contemporary environmental concern, of which the first, on 'The International Community and the Environment', was given by Maurice F. Strong in Reykjavik, Iceland, during the Second International Conference on Environmental Future, and was first published in *Environmental Conservation*, Vol. 4, No. 3, pp. 165–72, Autumn 1977: $50,000.

(*c*) Endowment of prize for the best paper published each year in *Environmental Conservation:* $50,000.

(*d*) Funds to convene working groups of leading specialists to pronounce quite freely on major environmental threats. An endowment of $5 millions or S.frs. 10 millions would suffice for this and meanwhile cover several of the above needs from income.

Further projects are under consideration, including urgent research proposals, a major World Heritage Series of volumes, establishment of a biennial award for demonstrated environmental concern and concomitant action by a multinational corporation, and establishment of associated foundations, etc., in other parts of the world.

Operation and Needs

With its running costs already contributed or negligible, and in the absence of any institutional overheads or taxes, the Foundation constitutes what has been called a uniquely economical vehicle for support of the environmental/conservational movement—whether generally or through specific projects which can be put in train as soon as funds become available. Contributions are warmly welcomed and individually acknowledged, and can best be received by cheque in any negotiable currency made out to The Foundation for Environmental Conservation, c/o Banque Populaire Suisse, 1 Quai des Bergues, 1211 Geneva, Switzerland. Plans are afoot to recognize special services or substantial donations to the Foundation by the award of an engraved certificate designed by a world-renowned draughtsman.

The Foundation for Environmental Conservation
(Nicholas Polunin, President)

SECOND INTERNATIONAL CONFERENCE ON ENVIRONMENTAL FUTURE

List of Participants with their Preferred Addresses

ADALSTEINSSON, Dr Stefán, Agricultural Research Institute, Keldnaholti v/Vesturlandsveg, Reykjavik, Iceland.

ALTENPOHL, Dr Dieter G., Technical Director, Alusuisse Schweizerische Aluminium AG, Feldeggstrasse 4, 8034 Zürich, Switzerland.

ARADÓTTIR, Svava, Reynimelur 58, Reykjavik, Iceland.

ARNASON, Dr Ingvar, University of Iceland, Reykjavik, Iceland.

ASGEIRSSON, Eggert, Secretary-General, Icelandic Red Cross, Nóatúni 21, Reykjavik, Iceland.

ASGEIRSSON, Thórdur, Ministry of Fisheries, Reykjavik, Iceland.

ASIBEY, Dr Emmanuel O. A.,* Chief Game and Wildlife Officer, Department of Game and Wildlife, PO Box M.239, Accra, Ghana.

ATTIAS, Jack P., 600 Grapetree Drive, Suite 4EN, Key Biscayne, Florida 33149, USA.

BÁRDARSON, Hjálmar R., State Director of Shipping, PO Box 484, Reykjavik, Iceland. (Also representing Icelandic Nature Conservation Council.)

BAROODY, Leila J., c/o Saudi Arabian Permanent Mission to the United Nations, 6 East 43rd Street, New York, NY 10017, USA.

BERGMANN, Stefán, Association of Icelandic Nature Conservation Societies, Reykjavik, Iceland.

BERGTHORSSON, Páll, Vedurstofa Islands (The Icelandic Meteorological Office), Reykjavik, Iceland.

BISHOP, Dr Amasa S.,* Director, Environment and Human Settlements Division, United Nations Economic Commission for Europe, Palais des Nations, 1211 Geneva 10, Switzerland.

* Contributor or co-author of paper but unable to attend personally.

BJARNASON, Dr Jónas, (Associate Professor, University of Iceland), Icelandic Fisheries Laboratories, Skúlagötu 4, Reykjavik, Iceland.

BJÖRNSSON, Sveinn (Representing International Organization for Standardization), Industrial Development Institute of Iceland, Skipholti 37, Reykjavik, Iceland.

BORDEN, Dr Richard J., Department of Psychological Sciences, Purdue University, Stanley Coulter Annex, West Lafayette, Indiana 47907, USA.

BORGSTROM, Professor Georg,* Department of Food Science and Human Nutrition, Michigan State University, East Lansing, Michigan 48824, USA.

BRYSON, Mrs Frances E., c/o Professor Reid A. Bryson (below).

BRYSON, Professor Reid A., Director, Institute for Environmental Studies, University of Wisconsin at Madison, 1225 West Dayton Street, Madison, Wisconsin 53706, USA.

BUCHINGER DE ALITISZ, Dra Maria, Uriarte 2456–8°–'33', 1425 Buenos Aires, Argentina.

BUTLER, Dr Gordon C., Director, Division of Biological Sciences, National Research Council, Ottawa, Ontario K1A 0R6, Canada.

CANDOLLE, Roger de, 41 Chemin du Vallon, Chêne-Bougeries, 1224 Geneva, Switzerland.

COLLIER, R. V., Chief Warden, East Midlands Nature Conservancy Council, George House, George Street, Huntingdon PE18 6BG, England.

CRABB, Professor John H., Facylteit Rechtsgeleerdheid, Katholieke Universiteit Leuven, Tiense Straat 41, 3000 Leuven (Louvain), Belgium.

CUNNINGHAM, Errol A., Santa Maria Ltd, PO Box 38, Ocho Rios, Jamaica, West Indies.

DASMANN, Mrs Elizabeth, c/o Professor Raymond F. Dasmann (below).

DASMANN, Professor Raymond F., Environmental Studies Office, University of California at Santa Cruz, Santa Cruz, California 95064, USA.

DODSON GRAY, David, 4 Linden Avenue, Wellesley, Massachusetts 02181, USA.

DODSON GRAY, Mrs Elizabeth (address as above).

DUCRET, Dr Claude, Environment and Human Settlements Division, Economic Commission for Europe, Palais des Nations, 1211 Geneva 10, Switzerland.

DYNE, Professor George van, College of Forestry and Natural Resources, Colorado State University, Fort Collins, Colorado 80521, USA.

EAST, His Excellency Mr Kenneth, Ambassador, British Embassy, Reykjavik, Iceland.

EINARSSON, Eythór, Vice-Chairman, Icelandic Nature Conservation Council, Laugaveg 105, Reykjavik, Iceland.

EIRÍKSSON, Karl, Vice-President, Landvernd, Society for Land-Reclamation and Nature Conservation, Skólavörodustigur 25, Reykjavik, Iceland.

ELÍSSON, Gunnlaugur, Industrial Research and Development Institute, Keldnaholti v/Vesturlandsveg, Reykjavik, Iceland.

* Contributor or co-author of paper but unable to attend personally.

ELKINGTON, John B., Transport and Environment Studies, 24 Floral Street, London WC2 E9DS, England.

FLOHN, Professor Dr Hermann, Meteorologisches Institut der Universität Bonn, Auf dem Hügel 20, 53 Bonn 1, Federal Republic of Germany.

FOSBERG, Dr F. Raymond, Senior Botanist, National Museum of Natural History, Smithsonian Institution, Washington, DC 20560, USA.

FRANZ, Dr Eldon H., Environmental Research Center, Washington State University, Pullman, Washington 99163, USA.

FRANZ, Mrs Kristi R., S.E. 920 Sunny Meade Way, Pullman, Washington, D.C. 99163, USA.

FRIDRIKSSON, Dr Sturla, Head, Agronomy Department, Agricultural Research Institute, Keldnaholti, Reykjavik 110, and Chairman of the Genetical Committee of the University of Iceland, Ingólfsstraeti 5, Reykjavik, Iceland.

FUKUSHIMA, Professor Yoichi,* Chairman, National Committee on Nature Conservation, Science Council of Japan, 7–22–34 Roppongi, Minato-ku, Tokyo 106, Japan.

FULLER, Dr R. Buckminster, (University Professor Emeritus, University of Pennsylvania), 3500 Market Street, Philadelphia, Pennsylvania 19104, USA.

GAEKWAD, H. H. Fatesinghrao, M.P. (Maharaja of Baroda), 7 Dupleix Lane, New Delhi 110011, and Laxmi Vilas Palace, Baroda, India.

GARDARSSON, Professor Arnthór, Icelandic Nature Conservation Council, Reykjavik, Iceland.

GIBBS, Angela (Honorary Conference Secretary), 9 rue de la Cité, 1204 Geneva, Switzerland.

GILL, Professor Don, Department of Geography, University of Alberta, Edmonton, Alberta T6G 2HQ, Canada.

GILLILAND, James C., Technical Director, Environmental Services Group, Amax Inc., 4704 Harlan Street, Denver, Colorado 80212, USA.

GLASSER, Ralph, 96/100 New Cavendish Street (Apt.16), London W1M 7FA, England.

GOLDBERG, Professor Edward D., Geological Research Division, Scripps Institute of Oceanography, University of California at San Diego, La Jolla, California 92093, USA.

GOLDSMITH, Edward, Editor, 'The Ecologist', 73 Molesworth Street, Wadebridge, Cornwall PL27 7DS, England.

GOSSEN, Dr Randall G., Manager, Environmental Monitoring and Research, Canadian Arctic Gas Study Ltd, 1270 Calgary House, 550 Sixth Avenue, S.W., Calgary, Alberta T2P 0S2, Canada.

GRAY, David Dodson—*see* DODSON GRAY, David.

GRAY, Elizabeth Dodson—*see* DODSON GRAY, Elizabeth.

GUDBJARTSSON, Professor Sigmundur, National Research Council, Laugavegi 13, Reykjavik, Iceland.

GUPPY, Nicholas G. L., 21a Shawfield Street, London SW3 4BD, England.

GUTTORMSSON, Hjörleifur, Icelandic Nature Conservation Council, Reykjavik, Iceland.

* Contributor or co-author of paper but unable to attend personally.

HALLDÓRSSON, Ragnar, Director-General, ISAL, Straumsvik, Iceland.

HALLGRÍMSSON, Geir (Patron of the Conference), Prime Minister of Iceland, Forsaetisrádherra, Reykjavik, Iceland.

HALLGRÍMSSON, Helgi, Association of Icelandic Nature Conservation Societies, Akureyri, Iceland.

HARE, Professor F. Kenneth, Director, Institute for Environmental Studies, Haultain Building, University of Toronto, Toronto, Ontario M5S 1A4, Canada.

HASLER, Professor Arthur D., Laboratory of Limnology, University of Wisconsin, Madison, Wisconsin 53706, USA.

HERMANNSSON, Steingrímur, National Research Council, Laugavegi 13, Reykjavik, Iceland.

HINDS, Dr W. Ted, Ecosystems Department, Battelle Pacific Northwest Laboratories, Battelle Boulevard, Richland, Washington 99352, USA.

HOLDGATE, Dr Martin W.,* Director-General of Research, Departments of Environment and Transport, 2 Marsham Street, London SW1, England.

HOLT, Dr Sidney J.,* Fisheries and Environment Adviser, Department of Fisheries, Food and Agriculture Organization, Via delle Terme di Caracalla, 00100 Rome, Italy.

HUGASON, Reynir, National Research Council, Laugavegi 13, Reykjavik, Iceland.

INGÓLFSSON, Professor Agnar, Icelandic Nature Conservation Council, Reykjavik, Iceland.

JAKOBSSON, Jakob, Assistant Director, Marine Research Institute, Skúlagötu 4, Reykjavik, Iceland.

JENSEN, Dr Erik, Chef de Cabinet, Office of the Director-General, United Nations, Palais des Nations, 1211 Geneva 10, Switzerland.

JÓHANNESSON, Dr Björn, Soil Scientist, Laugarnesvegi 110, Reykjavik, Iceland.

JÓHANNESSON, Magnús, State Directorate of Shipping, PO Box 484, Reykjavik, Iceland.

JOHNSON, Stanley P., Head, Prevention of Pollution and Nuisances Division, and Adviser, Environment and Consumer Protection Service, Commission of the European Communities, Rue de la Loi 200, B-1049 Bruxelles, Belgium.

JONES, Professor Philip H., Institute for Environmental Studies, University of Toronto, Toronto, Ontario M5S 1A4, Canada.

JÓNSSON, Hördur (Representing International Organization for Standardization (ISO)), Reykjavik, Iceland.

JUEL-JENSEN, Dr Bent, Medical Officer to the University, Radcliffe Infirmary, Oxford, England.

KNOX, Professor James C., Department of Geography, Science Hall, University of Wisconsin, Madison, Wisconsin 53706, USA.

KUENEN, Professor Donald J., Rector Magnificus, Rijksuniversiteit te Leiden, Stationsweg 46, Leiden, The Netherlands.

LACONTE, Dr Pierre, Director for University Expansion, University of Louvain, 13 avenue G. Lemaître, B-1348 Louvain-la-Neuve, Belgium.

* Contributor or co-author of paper but unable to attend personally.

LINDAL, Páll, Legal Counsellor, City of Reykjavik, Reykjavik, Iceland.
LÖVE, Leó E., Breiavangi 34, Hafnarf, Iceland.
LÚDVIKSSON, Vilhjálmur, Icelandic Nature Conservation Council, Reykjavik, Iceland.
MACPHERSON, Dr Andrew H., Director-General, Western and Northern Region, Environment Canada, 1242, 10025 Jasper Avenue, Edmonton, Alberta T5J 3A7, Canada.
MAGNÚSSON, Dr Gudmunder, Professor of the University of Iceland, Reykjavik, Iceland.
MAGNÚSSON, Dr Jakob, Marine Research Institute, Skúlagötu 4, Reykjavik, Iceland.
MALONE, Dr Thomas F., Director, Holcomb Research Institute, Butler University, Indianapolis, Indiana 46208, USA.
MARSTRAND, Mrs Hanne, c/o Petro-Canada, PO Box 2844, Calgary, Alberta T2P 2M7, Canada.
MILLER, Dr Richard G., Foresta Institute for Ocean and Mountain Studies, 6205 Franktown Road, Carson City, Nevada 89701, USA.
MISRA, Professor Ramdeo, Department of Botany, Banaras Hindu University, Varanasi 221005, India.
MÖLLER, Dr Alda, Icelandic Fisheries Laboratories, Skúlagötu 4, Reykjavik, Iceland.
MYERS, Dr Norman, PO Box 48197, Nairobi, Kenya.
NICOLA, Erico, Les Bois Chamblard, 1164 Buchillon, Vaud, Switzerland.
OBENG, Dr Letitia E., Chairman, Soil and Water Task Force, Division of Environmental Management, United Nations Environment Programme, PO Box 30552, Nairobi, Kenya.
ODUM, Professor Eugene P.,* Director, Institute of Ecology, University of Georgia, Athens, Georgia 30602, USA.
ÓLAFSSON, Dr Sigurjón, Science Institute, University of Iceland, Reykjavik, Iceland.
OLINDO, Perez M., Science Secretary, National Council for Science and Technology, PO Box 30007, Nairobi, Kenya.
PÁLMADÓTTIR, Elín, President, Board of Environment of Reykjavik, Morgunbladid, PO Box 200, Reykjavik, Iceland.
PÁLSSON, Professor Einar B., Faculty of Engineering & Science, University of Iceland, Reykjavik, Iceland.
PÁLSSON, Páll S., Lawyer (Representing World Peace Through Law Center), Bergstadastraeti 14, Reykjavik, Iceland.
PAULING, Mrs Ava Helen, Linus Pauling Institute of Science and Medicine, 2700 Sand Hill Road, Menlo Park, California 94025, USA.
PAULING, Professor Linus (President of the Conference), Linus Pauling Institute of Science and Medicine, 2700 Sand Hill Road, Menlo Park, California 94025, USA.
POLUNIN, Mrs Helen E. (address as below)
POLUNIN, Professor Nicholas (Secretary-General of the Conference), 15 Chemin F.-Lehmann, 1218 Grand-Saconnex, Geneva, Switzerland.

* Contributor or co-author of paper but unable to attend personally.

POLUNIN, N. V. C. (Trinity College), Zoological Laboratory, University of Cambridge, Downing Street, Cambridge CB2 3EJ, England.

RENZONI, Professor Aristeo, Istituto di Anatomia Comparata, Via delle Cerchia 3, 53100 Siena, Italy.

REYNISSON, Arni, Icelandic Nature Conservation Council, Laugaveg, 105, Reykjavik, Iceland.

ROBBINS, George H., Environmentalist, Luscar Ltd, 800 Royal Trust Tower, Edmonton Centre, Edmonton, Alberta T5J 2Z2, Canada.

ROYSTON, Dr Michael G., Faculty Member, Environmental Management, Centre d'Etudes Industrielles, 4 Chemin de Conches, 1231 Conches, Geneva, Switzerland.

RUDOLPH, Mrs Ann W., Environmental Library, Battelle Columbus Laboratories, 505 King Avenue, Columbus, Ohio 43201, USA.

RUDOLPH, Professor Emanuel D., Director, Environmental Biology Program, College of Biological Sciences, Ohio State University, 1735 Neil Avenue, Columbus, Ohio 43210, USA.

RUSSELL, Mrs Claire,* 12 Downshire Square, Reading, Berkshire RG1 6NH, England.

RUSSELL, Dr William M. S.,* Reader in Sociology, Department of Economics, University of Reading, Whiteknights Park, Reading, Berkshire, England.

SAEMUNDSSON, Eyjólfur, Ministry of Health, Arnarhvoli, Reykjavik, Iceland.

SAVIN, Charles T., Environmental Control Centre, British Petroleum Limited, Britannic House, Moor Lane, London EC2Y 9BU, England.

SCHRAM, Ellert B., M.P., Tryggvagata 8, Reykjavik, Iceland.

SCHRAM, Professor Gunnar G. (Chairman, Icelandic National Committee), Faculty of Law, Háskóli Islands, Reykjavik, Iceland.

SCHUMACHER,† Dr E. F.,* Chairman, Intermediate Technology Development Group Ltd, 9 King Street, London WC2E 8HN, England.

SCHÜTT, Professor Torgny, Energy Research and Development Commission, Ministry of Industry, Sveavägen 13-15, S-111 57 Stockholm, Sweden.

SHEIKH, Dr Khalid Hamid, Chairman, Department of Botany, University of the Punjab, New Campus, Lahore, Pakistan.

SIGTRYGGSSON, Hlynur, Director, Vedurstofa Islands, (The Icelandic Meteorological Office), Reykjavik, Iceland.

SIGURBJÖRNSSON, Dr Björn, Director, Agricultural Research Institute, Keldnaholti v/Vesturlandsveg, Reykjavik, Iceland.

SIGURDSSON, Ingimar, Ministry of Health, Arnarhvoli, Reykjavik, Iceland.

SIGURDSSON, Páll, Secretary-General, Ministry of Health, Arnarhvoli, Reykjavik, Iceland.

SIGURJÓNSSON, Pétur, Director, Industrial Research and Development Institute, Reykjavik, Iceland.

* Contributor or co-author of paper but unable to attend personally.

† Died 4 September, 1977.

SKÚLADÓTTIR, Unnur, Association of Icelandic Nature Conservation Societies, Reykjavik, Iceland.
STARR, Dr Thomas B., Center for Climatic Research, University of Wisconsin, 1225 West Dayton Street, Madison, Wisconsin 53706, USA.
STEFÁNSSON, Dr Unnstein, University of Iceland, Reykjavik, Iceland.
STEINDORSSON, Steindór, Munkathverárstraeti 40, Akureyri, Iceland.
STONE, Jennifer, Crozet, Gex 1170, France.
STONE, Peter B., Crozet, Gex 1170, France.
STRONG, Maurice F., Chairman of the Bureau, International Union for Conservation of Nature and Natural Resources (IUCN), 1110 Morges, Switzerland.
TAN, Dr Koonlin, 346 Lrg. 10c, United Garden, Jln. Klang 5 Batu, Kuala Lumpur 21–10, Malaysia.
THORHAUG, Professor Anitra, Department of Biological Sciences, Florida, International University, Tamiami Campus, Miami, Florida 33199, USA.
THORMAR, Hördur, Industrial Research Institute, Keldnaholti, Reykjavik, Iceland.
THORS, Dr Kjartan, Marine Research Institute, Skúlagötu 4, Reykjavik, Iceland.
THORSTEINSSON, Geirhardur, Veltusund 3, Reykjavik, Iceland.
THORSTEINSSON, Ingvi, Landvernd, Skólavördustíg 25, Reykjavik, Iceland.
THORVALDSSON, Gudlaugur, Rector, University of Iceland, Reykjavik, Iceland.
TINBERGEN, Professor Jan,* Haviklaan 31, The Hague, The Netherlands.
TOLBA, Dr Mostafa K.,* Executive Director, United Nations Environment Programme (UNEP), PO Box 30552, Nairobi, Kenya.
VILHELMSDÓTTIR, Vilhelmína, Marine Research Institute, Skúlagötu 4, Reykjavik, Iceland.
VILHJÁLMSSON, Thorsteinn, Science Institute, University of Iceland, Reykjavik, Iceland.
WIDMAN, Professor Gary L., Professor of Law and Director, Natural Resources and Environment Law Program, University of California, Hastings College of the Law, 198 McAllister Street, San Francisco, California 94102, USA.
WILLARD, Hon. Beatrice E.,* Professor of Environmental Science, Colorado School of Mines, Golden, Colorado 80401, USA.
WORTHINGTON, Dr E. Barton, Scientific Director, IBP Publications Committee, c/o The Linnean Society, Burlington House, Piccadilly, London W1V 0LQ, England.

* Contributor or co-author of paper but unable to attend personally.

Editor's Postscript

Looking back on the Conference through these Proceedings, one retains the abiding impression that Man now has the knowledge and means to save his world but still shows inadequate signs of acting in time.

Index of Personal Names

Subject Index